Contents

Acknowledgements

I would like to thank colleagues in Trinity College, Dublin for their advice in the preparation of this book. Dr G. R. Brown undertook the task of reading the entire typescript. His comprehensive knowledge of chemistry, and timely advice saved me from many errors. Dr D. A. Morton-Blake read the draft of the unit on entropy and his advice resulted in at least a local diminution of chaos. Dr A. P. Davis gave invaluable help with photographs of molecular graphics as did Mr B. Dempsey with other photographs.

Of the many texts and articles I have consulted in the process of writing this book, I particularly acknowledge my debt to the following: (i) the Nuffield *Advanced Chemistry* books whose enquiry-based approaches used in the Nuffield courses have greatly influenced the approach I have adopted, and (ii) the text *Chemistry: An Integrated Approach* by R. S. Lowrie and H. J. C. Ferguson (Pergamon Press, 1975), now sadly out of print, which was the origin of my treatment of the 'CORN' rule in section 123.6.

The consistency of style and presentation of the text of both volumes is almost entirely due to the timely advice and clinical eye for editorial detail of Geoff Amor. At the Cambridge University Press, Lucy Purkis has been consistently supportive, even when faced with a sometimes tardy author. Similarly, Callie Kendall has been invariably helpful in researching the photographs.

My thanks go to all these people, but especially to my wife Margaret and two boys, Alastair and Euan, who have all too often taken second place to a computer and printer.

Text

535, 609, Science Museum Library; 558 Popperfoto; 562*t*, Jeremy Hartley/Oxfam; 562*b*, Kemira Fertilizers, Ince, Chester CH2 4LB; 563*l*, Wildlife Matters; 563*r*, Ecoscene/Ian Harwood; 571, © MacQuitty International Collection; 577, 583, 586, 590, 658*b*, Andrew Lambert; 578*t*, Courtesy of British Cement Association; 578*b*, Reproduced with the kind permission of Blue Circle Cement; 579, Courtesy of Smith and Nephew Medical Limited; 592, Redferns/photo by Andrew Putler; 593, Photo: Building Research Establishment/Crown copyright reproduced by permission of the Controller of Her Majesty's Stationery Office; 602, Courtesy of Pilkington plc; 628*tl*, by courtesy of Ohmeda, a BOC healthcare company; 628*tr*, NASA/Science Photo Library, 628*b*, Courtesy of British Oxygen Co Ltd; 649, 800, 853, Biophoto Associates; 658*t*, Nigel Luckhurst; 667, Alex Bartel/Science Photo Library; 784, Courtesy of The Scotch Whisky Association; 888, Courtesy of British Plastics Federation; 890, Sally & Richard Greenhill.

Figures

91.3, by permission of Paul Hamlyn Publishing, part of Reed International Ltd; 91.7, by permission of Oxford University Press; 94.5, by permission of John Wiley & Sons Inc., New York; 97.1, by permission of Professors Gordon and Zoller; 99.5, by permission of Chapman & Hall; 123.7, reprinted with permission of Macmillan Publishing Company from *Central Concepts of Biology* by Adela S. Baer, William E. Hazen, David L. Jameson and William L. Sloan. Copyright © 1971 by Macmillan Publishing Company; 123.8, 123.9, reprinted with the permission of Macmillan Publishing Company from *Biochemistry*, 2nd ed. by Geoffrey Zubay. Copyright © 1988 by Macmillan Publishing Company; 128.4, 128.6, 128.7, Harry R. Allcock/Frederick W. Lampe, *Contemporary Polymer Chemistry*, 2nd ed. © 1990, pp. 9, 505, 515, 516. Reprinted by permission of Prentice Hall, Englewood Cliffs, New Jersey.

Colour section

tropical rainforest: Ecoscene/Sally Morgan; crystals/minerals: Courtesy of the Natural History Museum; smog: Wildlife Matters; fire: Courtesy of Chubb Fire Ltd; algae: Ecoscene/Sally Morgan; Piccadilly Circus: Philip Craven/Robert Harding Picture Library; cyclist, climbers, mayonnaise, paints: Andrew Lambert; detergent in river: Ecoscene/Gryniewicz.

How to use this book

About units

Advanced Chemistry is divided into two books, which are each subdivided into two parts. Book 1 covers physical and industrial chemistry; book 2 covers inorganic and organic chemistry. In turn, the four parts are split into many fairly short units, rather than into a smaller number of long chapters. Each unit is designed to cover a compact area of chemistry, which you should be able to study over a period of one or two hours. Take some comfort from the fact that you are unlikely to have to know the content of every unit. All A level and AS level syllabuses cover a basic core, and then they emphasise different aspects of chemistry. If you pay attention to the syllabus that you are using, you should be able to avoid unnecessary work. Some units contain features that may not be compulsory on any of the syllabuses; mainly this is because I find those parts of chemistry especially interesting.

I have written the units with the aim of helping you understand the work, rather than presenting you with a large number of isolated facts. Of course, like any subject, chemistry contains a great deal of information that you will have to learn; but you will find learning much easier if you can understand how the information fits together.

Each unit is split into sections, and near the end of almost every section is a set of study questions. No doubt you will be tempted to pass these by; but avoid temptation! The questions will allow you to test your understanding of the work as you go along. They are designed to make you think, and to discover if you have understood what you have read. It would be best to regard them as puzzles, rather than as 'trick' questions designed to catch you out. Answers to all of these questions are given near the end of each unit so you can check your progress. (There really is little point in cheating by looking up the answers until *you* have tried to work out the answers.)

As well as the shorter end-of-section questions there are questions from past AS, A and S level examinations.

These are arranged at the end of the books. Only answers to numerical parts are provided. For help you will have to consult another book, or seek the advice of your teacher or lecturer.

Another feature is that each unit ends with a summary. This will provide you with a guide to what the unit covers, and it should serve as a useful aid to revision. However, do not expect to find explanations in the summaries.

How to find information

One of the key things that determines how well you learn chemistry is your motivation: do you really want to find things out and understand the work you are doing? If you are not in the right frame of mind, it would be better to leave study to another occasion. However, when you decide to study, study hard and for short periods.

A second point to bear in mind is that you will only make best use of your time if you know what you are trying to achieve in your study sessions. For example, if you decide that you need to 'learn about molecules' you are likely to waste a lot of time. This objective is too vague. It would be far better to aim at a clearer target. For example, you might wish to learn about 'covalent bonding in molecules' or 'the reactions of alcohol molecules'. One of the best places to find the right targets is the syllabus for your chemistry course. This will give you a detailed list of the things that you need to know about.

Once you have identified your target you should move to the index at the back of the book, or the table of contents near the front. The table of contents will, for instance, lead you to units on covalent bonding and the reactions of alcohols. If, as will often be the case, you need to look up specific pieces of information, the place to look is the main index.

INORGANIC CHEMISTRY

87

The Periodic Table

87.1 The origin of the Periodic Table

Chemists have long tried to find patterns in the properties of the elements. Some were discovered fairly readily; for example, elements were classified as metals or non-metals, and many of their compounds as acids, alkalis, or salts. However, it was widely believed that there had to be an underlying reason for the patterns. One of the first suggestions was due to Prout. Prout's hypothesis was that all elements were made from a whole number of hydrogen atoms. (Be careful here: an 'atom' in Prout's time was a very different thing to our understanding of the word.) According to him the atomic masses of the elements should be a whole number of times that of hydrogen, i.e. they should be integers. Unfortunately, from Prout's point of view, the results of experiments showed that the atomic masses of many elements were not integers.

However, as more elements were discovered, and their atomic masses determined, the search for patterns among the masses continued. Döbereiner drew attention to the fact that there were groups of three elements that not only had chemical properties in common, but also showed trends in the way their atomic masses changed. Some examples are shown in Table 87.1.

Observations like this explained little, but were intriguing, and began a search for other connections between the chemical properties of elements and their

Table 87.1. Some of Döbereiner's triads

Element	Atomic mass	Comment
Lithium	7.0	The difference
Sodium	23.0	between the values
Potassium	39.2	is nearly 16
Calcium	40.1	The difference
Strontium	87.6	between the values
Barium	137.4	is around 48 and 50
Chlorine	35.5	The difference
Bromine	80.0	between the values
Iodine	126.9	is around 45 and 47

atomic masses. A major attempt at making a link was made by Newlands in 1864 (Table 87.2). He grouped elements into sets of eight and claimed that every eighth element in the pattern was chemically similar. Newlands' *law of octaves* was largely ignored, or at best treated with mild amusement.

The first thorough attempt at relating chemical properties to atomic masses was made by the Russian

Table 87.2. Examples of Newlands' octaves*

No.		No.		No.		No.		No.
H 1	F	8	Cl	15	Co & Ni	22	Br	29
Li 2	Na	9	K	16	Cu	23	Rb	30
G 3	Mg	10	Ca	17	Zn	24	Sr	31
Bo 4	Al	11	Cr	18	Y	25	Ce & Le	32
C 5	Si	12	Ti	19	In	26	Zr	33
N 6	P	13	Mn	20	As	27	Di & Mo	33
O 7	S	14	Fe	21	Se	28	Ro & Ru	35

*This is part of a table that John Newlands presented in a talk he gave to the Chemical Society on 9 March 1866. The talk was entitled 'The Law of Octaves, and the Causes of Numerical Relations among the Atomic Weights'. (You might like to work out which elements G, Bo, etc., stand for.) Here is part of an account of the talk:

> The author claims the discovery of a law according to which the elements analogous in their properties exhibit peculiar relationships, similar to those subsisting in music between a note and its octave . . . . Professor G. F. Foster humorously enquired of Mr. Newlands whether he had ever examined the elements according to the order of their initial letters?

Newlands was not at all happy about the credit that went to Mendeléeff over the discovery of the periodic law. In 1884 Newlands wrote:

> Having been the first to publish the existence of the periodic law more than nineteen years ago, I feel, under existing circumstances, compelled to assert my priority in this matter . . . . As a matter of simple justice, and in the interest of all true workers in science, both theoretical and practical, it is right that the originator of any proposal or discovery should have the credit of his labour.

PERIODIC SYSTEM OF THE ELEMENTS IN GROUPS AND SERIES.

SERIES	0	I	II	III	IV	V	VI	VII	VIII
1	—	Hydrogen H 1·008	—	—	—	—	—	—	
2	Helium He 4·0	Lithium Li 7·03	Beryllium Be 9·1	Boron B 11·0	Carbon C 12·0	Nitrogen N 14·04	Oxygen O 16·00	Fluorine F 19·0	
3	Neon Ne 19·9	Sodium Na 23·05	Magnesium Mg 24·3	Aluminium Al 27·0	Silicon Si 28·4	Phosphorus P 31·0	Sulphur S 32·06	Chlorine Cl 35·45	
4	Argon Ar 38	Potassium K 39·1	Calcium Ca 40·1	Scandium Sc 44·1	Titanium Ti 48·1	Vanadium V 51·4	Chromium Cr 52·1	Manganese Mn 55·0	Iron Fe 55·9 Cobalt Co 59 Nickel Ni 59 (Cu)
5		Copper Cu 63·6	Zinc Zn 65·4	Gallium Ga 70·0	Germanium Ge 72·3	Arsenic As 75	Selenium Se 79	Bromine Br 79·95	
6	Krypton Kr 81·8	Rubidium Rb 85·4	Strontium Sr 87·6	Yttrium Y 89·0	Zirconium Zr 90·6	Niobium Nb 94·0	Molybdenum Mo 96·0	—	Ruthenium Ru 101·7 Rhodium Rh 103·0 Palladium Pd 106·5 (Ag)
7		Silver Ag 107·9	Cadmium Cd 112·4	Indium In 114·0	Tin Sn 119·0	Antimony Sb 120·0	Tellurium Te 127	Iodine I 127	
8	Xenon Xe 128	Caesium Cs 132·9	Barium Ba 137·4	Lanthanum La 139	Cerium Ce 140	—	—	—	— — —
9		—			—	—	—	—	
10	—	—	—	Ytterbium Yb 173	—	Tantalum Ta 183	Tungsten W 184	—	Osmium Os 191 Iridium Ir 193 Platinum Pt 194·9 (Au)
11		Gold Au 197·2	Mercury Hg 200·0	Thallium Tl 204·1	Lead Pb 206·9	Bismuth Bi 208	—	—	
12	—	—	Radium Rd 224	—	Thorium Th 232	—	Uranium U 239		

	HIGHER SALINE OXIDES							
R	R_2O	RO	R_2O_3	RO_2	R_2O_5	RO_3	R_2O_7	RO_4

		HIGHER GASEOUS HYDROGEN COMPOUNDS		
		RH_4	RH_3	RH_2	RH

Figure 87.1 *Mendeléeff's Periodic Table of 1905*

chemist Dimitri Mendeléeff. In 1869 he published a table that formed the basis of the Periodic Table that we now use. A version that he published in 1905 is shown in Figure 87.1. The claim was made that:

> the properties of the elements vary in relation to their atomic masses.

This statement became known as the periodic law. However, Mendeléeff was well aware that there was at least one anomaly in the table. You will see that he quotes the atomic mass of argon as 38. However, the experimental evidence was that the true value was about 39.6. Mendeléeff was so convinced that the periodic law was correct that he changed the figure to one which would fit the law. The reason he gave was:

> . . . argon represents a slight discrepancy . . . .. This leads one to think that argon still includes some other gas of high density in admixture with it.

In other words, he thought that the measurements of the atomic mass of argon were wrong because there was an impurity present. The anomaly would not, however, go away. It was joined by other exceptions to the periodic law; for example, cobalt (58.9) and nickel (58.7), and tellurium (127.6) and iodine (126.9) were in the reverse order of their relative atomic masses.

We now know that the reason for the anomalies was that the periodic law as Mendeléeff set it out is not correct. It was Moseley who showed that the position of an element in the Periodic Table depended on its *atomic number* not on its atomic mass. The correct *periodic law* is:

> **The properties of the elements vary in relation to their atomic numbers.**

(You might need to look at Unit 3 for information about the work of Moseley and atomic number.)

> **87.1** What do you think about Mendeléeff's attitude to the atomic mass of argon?

87.2 The modern Periodic Table

The modern Periodic Table (Figure 87.2) differs in several respects to Mendeléeff's table. The most obvious difference is that it is stretched out. No longer are there two columns of elements in the Groups, nor are there so many rows. The two columns are separated into two sets of Groups recognised by the letters A and B after them. The rows, which Mendeléeff called series, are now called Periods. There are three short and four long

Dimitri Mendeléeff, looking suitably Russian.

in the Period below the lanthanides. These elements have, respectively, the 4f and 5f orbitals filling. Their chemistry is complicated owing to the large number of oxidation states that they can adopt. Some of them have been isolated from nuclear reactions in remarkably small quantities. (In some cases literally no more than would fit on the head of a pin.) The American chemist G. T. Seaborg and his coworkers were responsible for developing extremely accurate techniques for dealing with such small quantities. The actinides coming after uranium are called the transuranic elements. Their tendency to be radioactive makes dealing with them even more of a problem. We shall make life a little easier by ignoring them henceforth.

87.2 In the modern Periodic Table the elements in Groups IB and IIB are shown in the d block. This block is the place where the transition elements are found. Why do you think some chemists argue that the Group IB and IIB elements should not be considered as d-block elements?

87.3 The Periodic Table and electron structures

In Unit 12 we discovered that the quantum theory gives us a neat account of the electron structures of the elements. Here we can relate the filling of electron shells to the position of the elements in the Periodic Table. Figure 87.3 illustrates the orbitals that are being used in the various parts of the table. The table splits into four main regions depending on the type of orbitals that are being filled. The regions are: s block, p block, d block and f block.

87.4 The Periodic Table, metals and non-metals

The majority of the elements in the table are metals, with about 20 being non-metals. The metals are to be found to the left of the zig-zag line in Figure 87.4, and the non-metals to the right. However, it is not always helpful to try to classify an element as either a metal or a non-metal. Many elements, particularly those close to the zig-zag line, show the properties of both categories. These are the *metalloids*: boron, silicon, germanium, arsenic and tellurium. A particularly important property of silicon and germanium is that they are semiconductors.

In Table 87.3 are gathered the typical properties of metals and non-metals. Metals that show these properties to the fullest extent are to be found in Groups I and II. These are the *s-block* metals. Especially, the metals to the bottom of Group I are the most powerful reducing agents. This is a result of the outermost s electron being extremely well shielded from the nuclear charge. The

Periods. The long Periods show the elements that we now call the transition metals. In Periods 6 and 7 are to be found the lanthanides and actinides, many of which were not known in Mendeléeff's time. Although it can be useful to distinguish the A Group from the B Group elements, we shall only do so when we discusss the B metals. These are the metals that lie by the side of the transition elements – especially Group IIB – or at the bottom of the other B Groups. For the most part, as earlier in the book, we shall leave out the A and B labels. For example, we shall normally refer to the elements lithium to francium as the Group I metals rather than Group IA metals and the elements carbon to lead will be Group IV, not Group IVB.

Sandwiched between Group II and III are the transition, or d-block, elements. These are all metals, some of which you will be familiar with owing to their widespread uses, e.g. chromium, iron, nickel, platinum. The transition metals have many properties in common, e.g. they often give coloured compounds, and make complexes rather easily. You will find out more about them in Unit 105.

Writing the Periodic Table in the way it is normally done, there are two sets of elements that are not shown in their rightful positions. These are the lanthanides and actinides. The lanthanides start at lanthanum (atomic number 57) and fit in the gap up to hafnium (atomic number 72). The actinides appear in a similar position

GROUPS

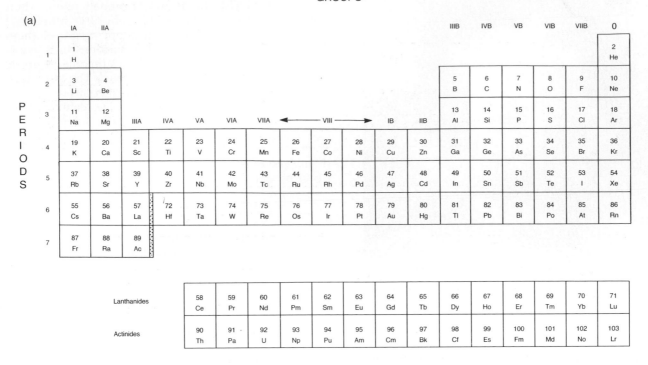

(a)

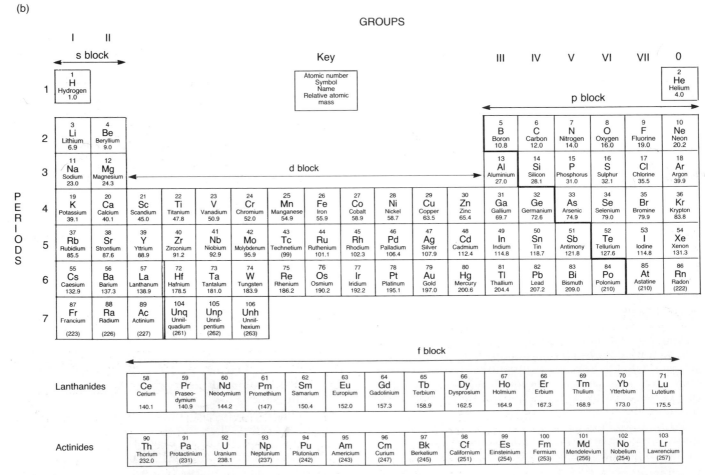

(b)

GROUPS

Figure 87.2 (a) The modern version of the Periodic Table showing the A and B groups. The atomic number is shown above each element's symbol. (b) The Periodic Table using a simpler notation for the groups. This is the table we shall normally use

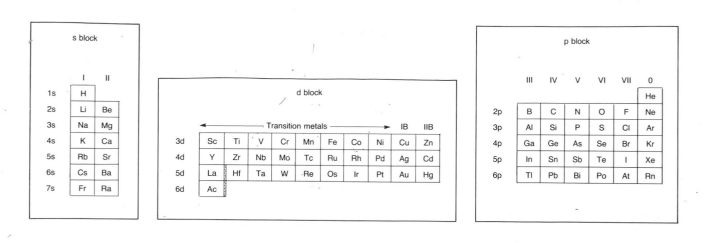

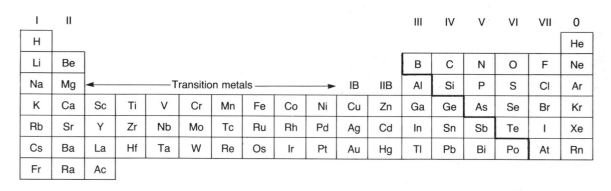

Figure 87.3 *The four main blocks in the Periodic Table showing the orbitals that are being filled*

I	II				Transition metals					IB	IIB	III	IV	V	VI	VII	0
H																	He
Li	Be											B	C	N	O	F	Ne
Na	Mg									IB	IIB	Al	Si	P	S	Cl	Ar
K	Ca	Sc	Ti	V	Cr	Mn	Fe	Co	Ni	Cu	Zn	Ga	Ge	As	Se	Br	Kr
Rb	Sr	Y	Zr	Nb	Mo	Tc	Ru	Rh	Pd	Ag	Cd	In	Sn	Sb	Te	I	Xe
Cs	Ba	La	Hf	Ta	W	Re	Os	Ir	Pt	Au	Hg	Tl	Pb	Bi	Po	At	Rn
Fr	Ra	Ac															

Figure 87.4 *A slightly shorter version of the Periodic Table. The heavy black line marks the division between those elements normally regarded as metals (on the left) and those thought of as non-metals (on the right)*

electron is only weakly held to the atom, so it is easily lost. The Group I metals react with water to give strong alkalis; they are known as the *alkali metals*. The metals in Group II tend to make weaker alkalis, and are known as the *alkaline earth metals*.

The *halogens*, in Group VII, show the most complete set of properties of non-metals. Fluorine, at the top of the Group, is the most powerful oxidising agent. It has a nuclear charge that is not very well shielded by the seven electrons it possesses. Fluorine will readily take an electron from a metal and make a fluoride ion, F^-.

Going across a Period there is a point at which the properties of the elements change from being primarily metallic to primarily non-metallic. You can see that this change-over point lies further to the right the lower down the table you go. Indeed it is true to say that:

> **Metallic nature increases going down any Group.**

This is summarised in Figure 87.5. The reason for this is to do with *shielding*. The further down a Group an element finds itself, the more protons it has in its nucleus, but the more electrons there are surrounding the nucleus. We saw in Unit 13 that the evidence of ionisation potentials is that shielding wins over the effect of increasing nuclear charge. Thus, going down a Group,

Table 87.3. Typical properties of metals and non-metals

	Metals	Non-metals
Appearance and properties	Solids, some with high melting points; lustrous, malleable and ductile	Gases, or solids with low melting points
Conduction of heat and electricity	Very good	Poor
Compounds	Ionic compounds with non-metals; alloys with other metals	Ionic compounds with metals; covalent compounds with other non-metals
Charge on ions	Positive	Negative
Chemical nature	Reducing agents	Oxidising agents
Electro-negativity	Low	High

the outermost electrons become progressively easier to remove. In other words, the tendency to make positive ions, and to behave as reducing agents, increases. We say that metals which show these properties to a marked extent are the most *electropositive*.

87.3 In Table 87.3, what do the words 'lustrous', 'malleable' and 'ductile' mean?

87.4 There are exceptions to the properties listed in Table 87.3.

(i) Which metal is a liquid at room temperature?

(ii) Which non-metal is a solid with an extremely high melting point?

(iii) Which non-metal is a liquid at room temperature?

87.5 Which two elements in the Periodic Table would you expect to combine in the most violent fashion?

87.5 What are the differences between the A and B Groups?

First, you can see from Figure 87.6 that the majority of the A Group elements are metals, and the majority of the B Group elements are non-metals. The most interesting differences between the A and B Groups come with the elements known as the B metals. These are shown highlighted in Figure 87.6.

In the later units we shall discuss the individual properties of these metals. For the moment you might like to look at Table 87.4, which summarises the key differences between the B metals and the other metals in the Periodic Table.

You may find that the last point in the table, about the inert pair effect, needs some explanation. Some of the B metals that show more than one oxidation state have a preference for the lower oxidation state. For example, lead(IV) compounds are often oxidising agents and convert into lead(II) compounds. In the $+4$ oxidation state, lead makes use of its two 6s electrons as well as its outermost two 6p electrons. In the $+2$ state, only the 6p electrons are involved, with the 6s electrons unaffected. The observation that the $+2$ oxidation state is more favoured by lead has tempted people into saying that the pair of 6s electrons is 'inert' (inert, that is, compared to the s electrons of the s-block metals);

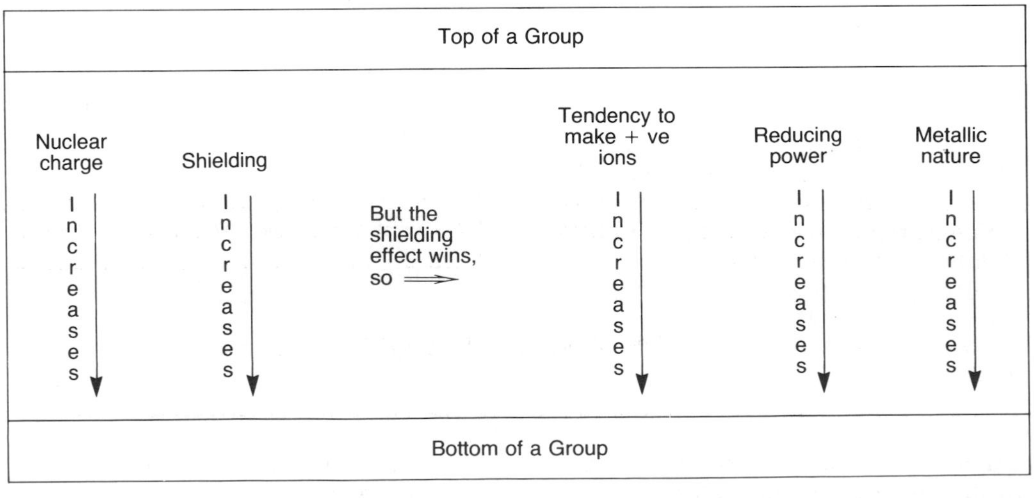

Figure 87.5 *Metallic nature increases down any Group in the Periodic Table*

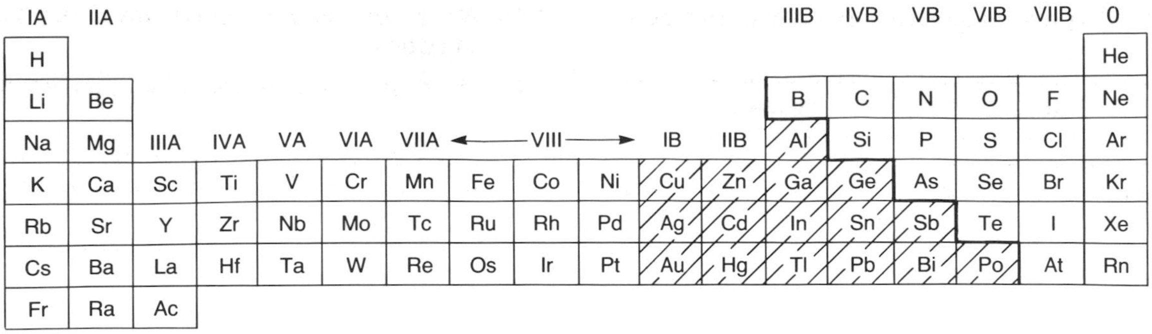
IA IIA IIIB IVB VB VIB VIIB 0
H He
Li Be B C N O F Ne
Na Mg IIIA IVA VA VIA VIIA VIII IB IIB Al Si P S Cl Ar
K Ca Sc Ti V Cr Mn Fe Co Ni Cu Zn Ga Ge As Se Br Kr
Rb Sr Y Zr Nb Mo Tc Ru Rh Pd Ag Cd In Sn Sb Te I Xe
Cs Ba La Hf Ta W Re Os Ir Pt Au Hg Tl Pb Bi Po At Rn
Fr Ra Ac

Figure 87.6 *The B metals in the Periodic Table are shown shaded*

Table 87.4. Comparison of properties of s-block and B metals

Property	*Example*
B metals are less reactive than the s-block metals	Potassium violent with water; silver and gold totally unreactive
B metals are weaker reducing agents, e.g. their $E^{\ominus}$ values are less negative than the s-block metals	$E^{\ominus}_{Na^+/Na} = -2.71$ V $E^{\ominus}_{Ag^+/Ag} = +0.8$ V
B metals show a greater tendency to form complex ions than s-block metals	Aluminium forms $Al(OH)_6^{3-}$ ions; similar s-block ions not made
B metals are 'softer' (more easily polarised) and tend to give compounds with a greater degree of covalency than s-block metals	$AlCl_3$ predominantly covalent; KCl ionic
B metals may show more than one oxidation state	Sn^{2+}, Sn^{4+}; Pb^{2+}, Pb^{4+}; Tl^+, Tl^{3+}
B metals show the so-called inert pair effect	B metals at the bottom of their Group tend not to use their outer s electrons in bonding (also, see text)

hence the term 'inert pair effect'. However, these s electrons are not really inert. Rather it seems that the bonds made by the B metals in their higher oxidation state are weaker than the bonds made in their lower oxidation states. Hence, once they are made, the compounds tend to break apart more easily.

Answers

87.1 It is tempting to think that he was 'fiddling' the results. However, if you put yourself in his position you might adopt the same attitude as he did. First, the noble gases had been discovered only very recently (argon was discovered in 1894). Little was known about it except that it was very unreactive. The possibility that another unreactive gas was mixed in with it was not impossible. Secondly, the methods used to determine atomic masses were not always reliable. There was the real possibility of error. Given that the vast majority of the elements fitted his scheme, it must have seemed very likely that the experimental results were at fault.

87.2 In the d block, the 3d, 4d, or 5d orbitals are filling. The properties of the transition elements are a result of them having unfilled d orbitals. However, the IB and IIB elements have full d orbitals. Indeed, these elements have s orbitals filling. This is one reason why some chemists think them misplaced tacked on to the end of the transition elements. However, the chemical properties of the IB metals make it sensible to include them among the transition metals. The IIB metals are best excluded.

87.3 Lustrous ≡ shiny; malleable ≡ able to be beaten into shapes; ductile ≡ able to be drawn into wires.

87.4 (i) Mercury.

(ii) Carbon, either as graphite or diamond.

(iii) Bromine.

87.5 The strongest metal and non-metal are likely to give the most violent reaction, i.e. caesium and fluorine.

UNIT 87 SUMMARY

- Mendeléeff published a table (in 1869) that formed the basis of the modern Periodic Table.
- Moseley showed that the position of an element in the Periodic Table depended on its atomic number.
- The periodic law says that:

The properties of the elements vary in relation to their atomic numbers.

- The Periodic Table splits into four main regions depending on the types of orbital that are being filled. The regions are s block, p block, d block and f block.
- The majority of elements are metals; about 20 are non-metals. Metals are to be found to the left of the zig-zag line in Figure 87.4, and non-metals to the right.
- Metals to the bottom of Group I are the most powerful reducing agents.
- Non-metals to the top of Group VII are the most powerful oxidising agents.
- Metallic nature increases going down any Group.
- Non-metallic nature increases across a Period.
- Inert pair effect:

  B metals have an inner pair of electrons, e.g. $6s^2$ for lead, that tend not to be used in bonding.

88

Periodicity of physical properties

88.1 Periodicity of ionisation energies

The periodic law says that the properties of the elements vary with their atomic numbers. Periodicity is the study of the variation in the properties, which can be both physical and chemical. We have already seen how electron structures and the division between metals and non-metals change in the Periodic Table. Now we shall concentrate on the variation in some other characteristics of the elements. We begin with ionisation energies, or rather return to them. Figure 88.1 repeats the graph in Figure 13.4, which shows the periodicity of ionisation energies of the elements hydrogen to neon, and extends it to include many more elements.

The peaks are always the noble gases, and the troughs the alkali metals (Group I). Notice that the ionisation energies tend to decrease down a Group and to increase across a Period. If you have read Unit 13 you should be able to explain these trends using the ideas that:

(i) Down a Group the shielding of outer by inner electrons overcomes the influence of the increasing nuclear charge. Thus, the outer electron is progressively more easy to remove as we go down a Group.

(ii) Across a Period the reverse is true; increasing nuclear charge has a greater effect than shielding. For this reason the outer electron of an atom is less easily lost as we go across a Period. Shielding is only important once a complete set of orbitals (an electron shell) has been filled.

88.1 Use the data in Appendix B to draw a graph of second ionisation energy against atomic number for the elements hydrogen to potassium. Explain the main differences between your graph and Figure 88.1.

88.2 Periodicity of atomic volume

The atomic volume of an element is the volume occupied by one mole of atoms of the element when it is a solid. The atomic volume is a guide to the size of the atoms. If the radius of an atom increases, then we would expect the atomic volume to increase. Figure 88.2 shows that again there is a series of peaks and

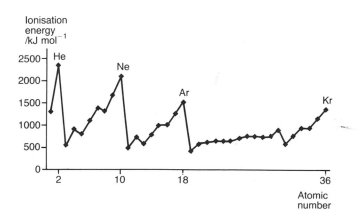

Figure 88.1 *The ionisation energies of the elements hydrogen to krypton*

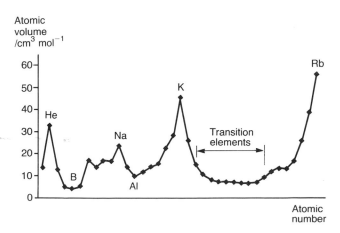

Figure 88.2 *Atomic volumes of the elements hydrogen to rubidium*

troughs, with the peaks appearing at the alkali metals (apart from the first at helium).

> **88.2** Atomic volume is not a direct measure of the size of atoms. Why not?

88.3 Periodicity of atomic radius

The graph of atomic radius plotted against atomic number (Figure 88.3) shows some similarity to that of atomic volume. Again the alkali metals are found at the peaks, but this time the halogens are at the troughs. The contraction we see across a Period occurs for much the same reason as the increase in ionisation energy. With increasing nuclear charge, the electrons are held more tightly by the nucleus. Once a new shell of electrons is made (at the noble gases) there is an expansion of the atomic radius.

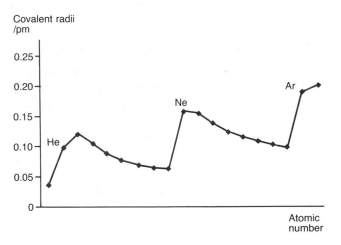

Figure 88.3 Covalent radii of the elements hydrogen to potassium. In the case of the noble gases, van der Waals radii have been plotted

88.4 Periodicity of melting and boiling points

The melting and boiling points (Figure 88.4) tell us something about how strongly the atoms in an element are stuck together. This time the periodicity is rather different to the previous graphs. The atoms that are held together the least strongly appear at the troughs. These are the elements that exist as gases. In particular, the intermolecular forces between atoms of the noble gases are very weak. It is to be expected that these elements should mark the troughs. On the other hand, elements that make giant covalent structures, like carbon, or metallic structures, like the transition elements, appear as peaks. Notice that atoms of the alkali metals

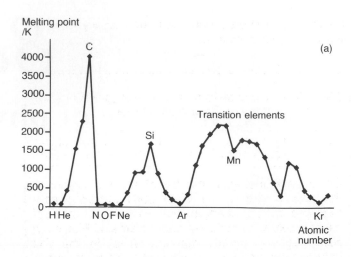

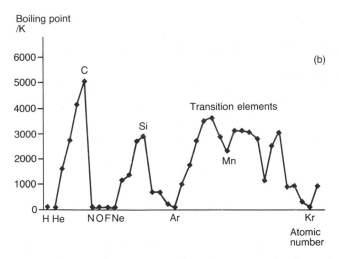

Figure 88.4 (a) Melting points of the elements hydrogen to rubidium. Note that carbon (graphite) and silicon, with giant covalent lattices, have very high melting points. (b) Boiling points of the elements hydrogen to rubidium

are only weakly bound together. This matches with the fact that it is possible to cut samples of potassium or sodium (in Group I) with a knife. The transition metals have high melting and boiling points, and their strength makes them valuable as structural materials.

88.5 Periodicity of valency

The number of moles of hydrogen atoms that will combine with one mole of an element tells us the valency of the element. The valency varies according to the pattern:

Group	I	II	III	IV	V	VI	VII	0
Formula of hydride	LiH	BeH$_2$	BH$_3$	CH$_4$	NH$_3$	H$_2$O	HF	
Valency	1	2	3	4	3	2	1	0

However, you should note that the valency of many elements is not constant.

88.3 The heats of formation of the oxides of the Period sodium to argon are, in kJ mol^{-1}:

Na_2O	MgO	Al_2O_3	SiO_2	P_4O_{10}	SO_3	Cl_2O_7
−416	−602	−1676	−911	−2984	−395	+250

Divide each one by the number of moles of oxygen in the formula of the oxide. The resulting figure gives a measure of the strength with which one mole of oxygen atoms is held by each element. Plot each figure against atomic number. What is the link between the graph (or the figures) and the structures of the oxides?

88.6 Periodicity of electronegativity

Electronegativity is a measure of the tendency of an element to attract electrons to itself. When electronegativity is plotted against atomic number (Figure 88.5), the strongest non-metals, the halogens, appear at the peaks. The alkali metals mark the troughs. This is largely a result of a new shell of electrons starting with the noble gases. The Group I metals have their nuclei

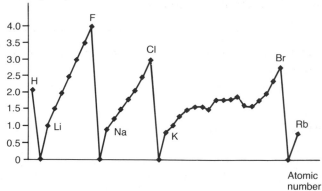

Figure 88.5 *Electronegativities of the elements hydrogen to rubidium. The electronegativities of the noble gases have been set at zero*

quite strongly shielded. Hence they show little tendency to gain new electrons. The electronegativities of the transition elements, like their ionisation energies, do not change greatly one from another.

88.4 Why is it not strictly correct to consider electronegativity values as physical properties of elements?

Answers

88.1 The graph is drawn in Figure 88.6. The alkali metals now come at the peaks. Once they have lost an electron they gain a noble gas structure. Removing an electron from such a structure means breaking in to a more energetically stable set of orbitals. Hence the alkali metal ions M^+ take the place of the noble gases. The pattern of the other elements is similarly shifted.

88.2 Atoms can pack together in different crystal structures. This means that, for example, atoms that pack in a very open structure will have a large atomic volume; slightly larger atoms that pack in a tighter arrangement will have a smaller atomic volume. This is the reverse pattern to their sizes.

88.3 The graph of Figure 88.7 shows that oxygen is most tightly bound to magnesium. This corresponds to the highly ionic nature of magnesium oxide. The weakest bonding is in Cl_2O_7, which happens to be a covalent molecule. Notice though that SiO_2 is closer to MgO than to Cl_2O_7 even though SiO_2 is also covalent. Remember, it is *not* true to say that 'ionic bonds are strong and covalent bonds are weak'. The figures reflect the differences in structure of the substances: MgO, a giant ionic lattice; SiO_2, a giant covalent lattice (see section 96.2); Cl_2O_7, independent molecules. P_4O_{10} has an unusual structure; see section 98.5.

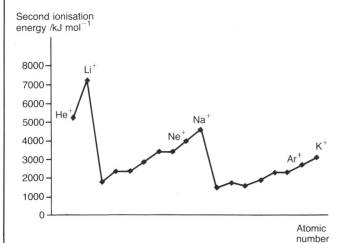

Figure 88.6 *The second ionisation energies of the elements hydrogen to potassium. Note that (i) hydrogen has no second ionisation energy; (ii) the Group I metals are now at the peaks. The ions Li$^+$, Na$^+$, K$^+$ have noble gas electron structures*

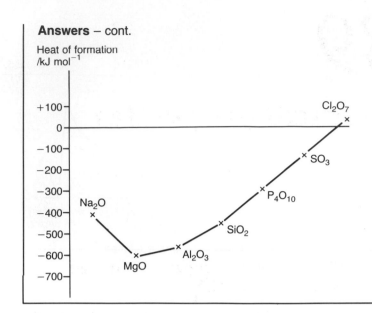

Heat of formation
/kJ mol^{-1}

88.4 We discussed this point in Unit 19. We cannot measure electronegativities because they have been invented by chemists, particularly Linus Pauling. They are not properties of atoms. However, the values have been conjured from measurements we can make, e.g. ionisation potentials.

Figure 88.7 Heats of formation plotted as answer to question 88.3

UNIT 88 SUMMARY

- Periodicity is the study of the variation in chemical and physical properties of the elements.
- Key points:
    - (i) Down a Group,
        shielding of outer by inner electrons overcomes the influence of increasing nuclear charge;
        the outer electron(s) is (are) progressively more easy to remove.
    - (ii) Across a Period,
        increasing nuclear charge has a greater effect than shielding;
        the outer electron(s) is (are) less easily lost.
    - (iii) Shielding is only important once a complete set of orbitals (an electron shell) has been filled.
    - (iv) Ionisation energies decrease down a Group and increase across a Period.
- Periodicity of ionisation energy, atomic volume, etc., are illustrated in Figures 88.1 to 88.7.

89

Periodicity of chemical properties

89.1 How this unit is arranged

In this unit much of the information presented to you is in the form of charts. You will find little discussion of the properties and reactions that are mentioned. The place to look for the detail is in the separate units that follow.

There are three ways in which the periodicity of chemical properties are usually discussed. The first compares the properties of elements going down a Group. The second compares them going across a Period. The third is a comparison of elements that are on a diagonal line traversing two Groups and two Periods. We shall adopt each of these approaches in turn.

89.2 How do properties change down a Group?

We have said a little about this in the previous unit. The key point is that (Table 89.1):

> **Metallic nature increases down a Group.**

Owing to shielding, atoms towards the bottom of a Group lose one or more of their outer electrons more easily than atoms nearer the top. This happens even in the Groups to the right of the Periodic Table, which we normally associate with non-metals. For example, in Group V the non-metals nitrogen and phosphorus are near the top, but by the time we reach bismuth at the bottom, distinct metallic character is present.

Table 89.1. Changes in chemical properties down a Group

Shielding of outer electrons	Metallic nature	Ionic character of compounds	Basic nature of oxides
Increases	Increases	Increases	Increases

89.3 How do properties change across a Period?

Across a Period the number of protons in the nucleus increases, as do the number of electrons. This also happens down a Group; but the key difference is that across a Period a new shell of electrons is not completed until the noble gas at the end of the period is reached. Complete shells of electrons screen the outer electrons from the full attraction of the nucleus. As we go across a Period the efficiency of screening is not so great. Indeed, the pattern is for the increasing nuclear charge to hold the electrons more tightly. (You should know by now that this is the reason why the ionisation energies of the elements increase across a Period.)

As a consequence, the elements become harder to ionise, and they also tend to attract electrons towards them. Thus we see a change from the metals (which give positive ions) on the left of a Period to the non-metals (which give negative ions) on the right. With this change goes a change in the nature of the hydrides, oxides and chlorides of the elements. These changes are summarised in Figures 89.1 to 89.4.

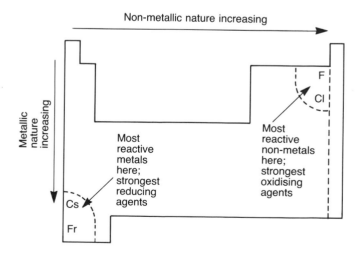

Figure 89.1 *How metallic and non-metallic behaviour change in the Periodic Table*

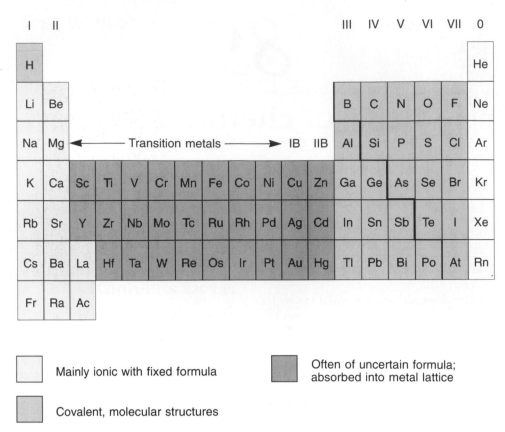

Figure 89.2 *The nature of hydrides of the elements*

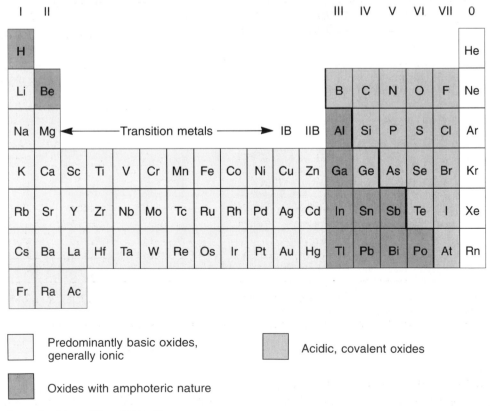

Figure 89.3 *The nature of oxides of the elements*

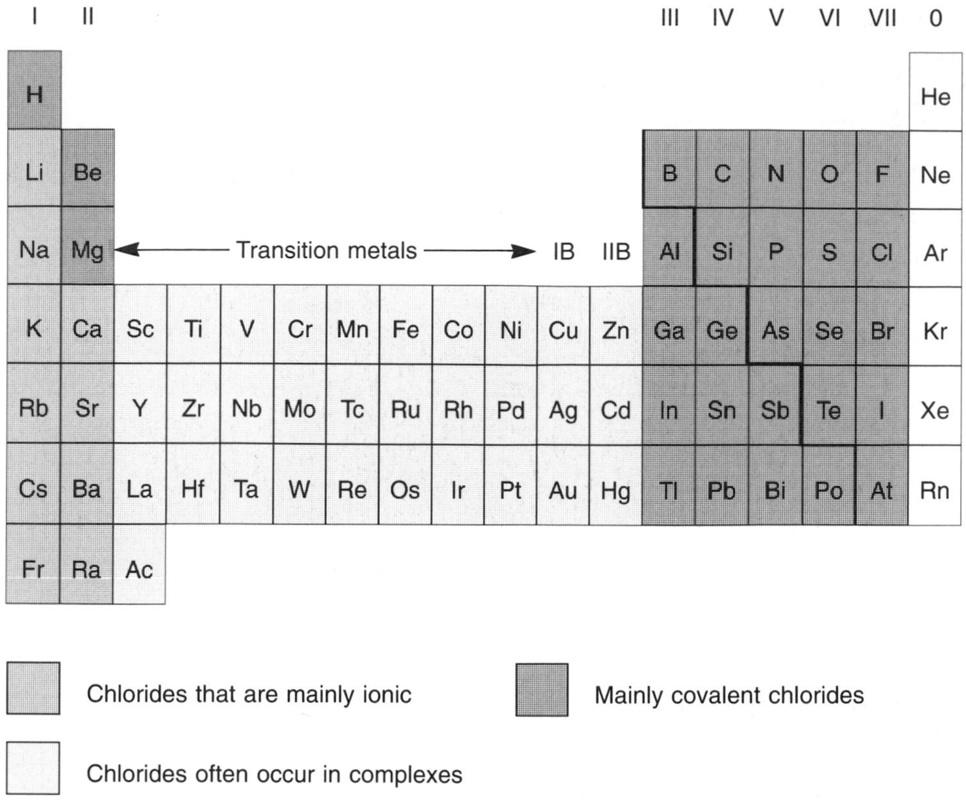

I II   III IV V VI VII 0

H   He

Li Be   B   C   N   O   F   Ne

Na Mg ◄——————— Transition metals ———————► IB IIB Al Si P   S   Cl Ar

K   Ca Sc Ti V   Cr Mn Fe Co Ni Cu Zn Ga Ge As Se Br Kr

Rb Sr Y   Zr Nb Mo Tc Ru Rh Pd Ag Cd In Sn Sb Te I   Xe

Cs Ba La Hf Ta W   Re Os Ir Pt Au Hg Tl Pb Bi Po At Rn

Fr Ra Ac

☐ Chlorides that are mainly ionic   ☐ Mainly covalent chlorides

☐ Chlorides often occur in complexes

Figure 89.4 *The nature of chlorides of the elements*

Table 89.2. The Period lithium to neon*†

Li_2O	BeO	B_2O_3	CO_2	NO_2	O_2	F_2O
Ionic	Covalent/ ionic	Covalent	Covalent	Covalent	Covalent	Covalent
Basic	Amphoteric	Acidic	Acidic	Acidic	Neutral	Acidic
Hydrolysis reaction	Insoluble	Slightly soluble	Hydrolysis reaction	Hydrolysis reaction	Slightly soluble	Hydrolysis reaction
LiOH	$Be(OH)_2$	$B(OH)_3$	H_2CO_3	HNO_3	H_2O	HF
Strong alkali	Amphoteric	Weak acid	Weak acid	Strong acid	Amphoteric‡	Weak acid
LiCl	$BeCl_2$	BCl_3	CCl_4	NCl_3	Cl_2O_7	ClF
Ionic	Covalent	Covalent	Covalent	Covalent	Covalent	Covalent
Dissolves	Dissolves	Hydrolysis	No reaction	Hydrolysis	Hydrolysis	Hydrolysis
$Li^+(aq)$ $Cl^-(aq)$	$Be^{2+}(aq)$ $Cl^-(aq)$	$B(OH)_3$ $H^+(aq)$ $Cl^-(aq)$		$NH_3(aq)$ $HOCl(aq)$	$H^+(aq)$ $ClO_4^-(aq)$	$HF(aq)$ $HOCl(aq)$

*Be careful about taking the labels 'ionic' and 'covalent' too literally. The majority of compounds show a mix of both extreme types of bonding
†The properties listed here are only examples of the reactions undergone by the oxides and chlorides
‡Sometimes water is described as being amphiprotic rather than amphoteric

You might like to look at the way the oxides, hydroxides and chlorides of the elements in the Periods lithium to neon (Table 89.2) and sodium to argon (Table 89.3) compare with one another. (It is likely that your examination syllabus particularly requires you to know about sodium to argon.)

Table 89.3. The Period sodium to argon*

Na_2O	MgO	Al_2O_3	SiO_2	P_4O_{10}	SO_3	Cl_2O_7
Ionic	Ionic	Ionic	Covalent	Covalent	Covalent	Covalent
Basic	Basic	Amphoteric	Acidic	Acidic	Acidic	Acidic
Hydrolysis reaction	Slightly soluble	Insoluble	Hydrolysis reaction	Hydrolysis reaction	Hydrolysis reaction	Hydrolysis reaction
$NaOH$	$Mg(OH)_2$	$Al(OH)_3$	H_2SiO_3	H_3PO_4	H_2SO_4	$HClO_4$
Strong alkali	Weak alkali	Weak alkali	Weak acid	Weak acid	Strong acid	Strong acid
$NaCl$	$MgCl_2$	$AlCl_3$	$SiCl_4$	PCl_3	S_2Cl_2	Cl_2
Ionic	Partly covalent	Covalent	Covalent	Covalent	Covalent	

Dissolve in water $\longleftarrow$ ———— React with water $\rightarrow$ acidic solution ————$\longrightarrow$

$Na^+(aq)$	$Mg^{2+}(aq)$	$Al(OH)_3$	$SiO_2(s)$	$H_3PO_3(aq)$	H_2SO_3	$HOCl(aq)$
$Cl^-(aq)$	$Cl^-(aq)$	$H^+(aq)$	$H^+(aq)$	$H^+(aq)$	$H^+(aq)$	$H^+(aq)$
		$Cl^-(aq)$	$Cl^-(aq)$	$Cl^-(aq)$	$Cl^-(aq)$	$Cl^-(aq)$
					$S(s)$	

*Be careful about taking the labels 'ionic' and 'covalent' too literally. The majority of compounds show a mix of both extreme types of bonding

There are two elements that make life difficult when we want to generalise about the Period lithium to neon (Table 89.2). They are beryllium and boron. (They also tend to give exceptions to the trends in their Groups.) Given their positions in Group II and III respectively, we would expect them to be metals, to give ionic compounds and to give basic oxides. They do none of these things. The main reason is that, if they were to exist as ions, their sizes would be extremely small. The ionic radius of Be^{2+} is 30 pm, and of B^{3+} is only 16 pm. These ions are so small that they represent extremely dense centres of positive charge, and they would polarise any negative ions that approach them. If you can recall Fajans' rules (Table 19.4) you will understand why this leads to covalency.

You can also see changes taking place across the Periods in the way the oxides behave with water. To the left of the Periods the oxides give hydroxides, and to the right they give acids. For example, in the Period lithium to neon we have the change from amphoteric beryllium hydroxide, $Be(OH)_2$, to strongly acidic nitric acid, HNO_3. However, if we show how the atoms are bonded together in hydroxides and acids, we find that they are more similar than we might otherwise expect. Look at Figure 89.5. In each case the hydrogen atoms are bonded to the oxygen atoms. Molecules that we regard as acids tend to give up hydrogen atoms (as hydrogen ions) to water molecules. For example,

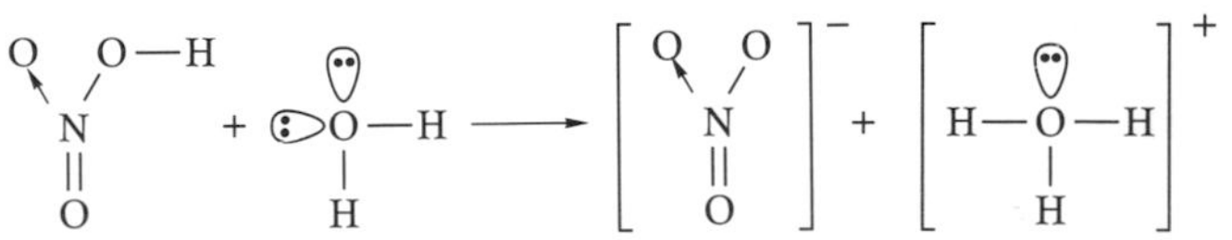

$$HNO_3(aq) + H_2O(l) \longrightarrow NO_3^-(aq) + H_3O^+(aq)$$

The reason why they do this is complicated, but the

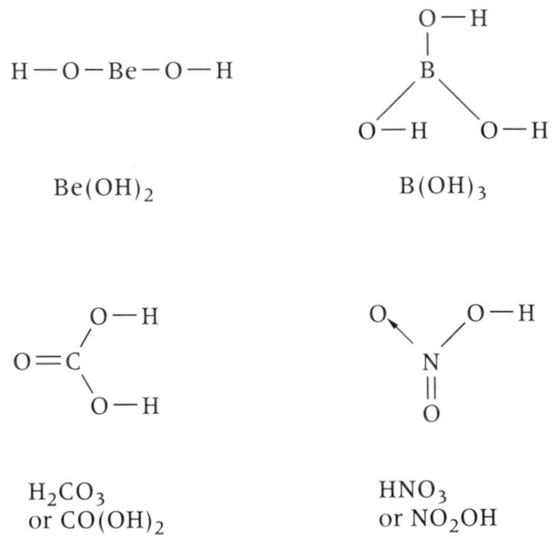

Figure 89.5 Two hydroxides, Be(OH)$_2$ and B(OH)$_3$, and two acids, H$_2$CO$_3$ and HNO$_3$, are structurally similar. They all have OH groups in them

process is certainly helped if the negative ion produced (in this case NO_3^-) is energetically stable. In the nitrate ion and sulphate ion there is a considerable degree of electron delocalisation, which, as we said in Unit 14, leads to just this energetic stabilisation. However, many other things (such as entropy changes and kinetic factors) have to be taken into account if we are to give a really satisfactory explanation of the way in which a molecule with OH groups will behave in water. Fortunately we can leave this task to more advanced books.

The shapes of the chlorides of the elements in the two Periods lithium to neon and sodium to argon are summarised in Figures 89.6 and 89.7.

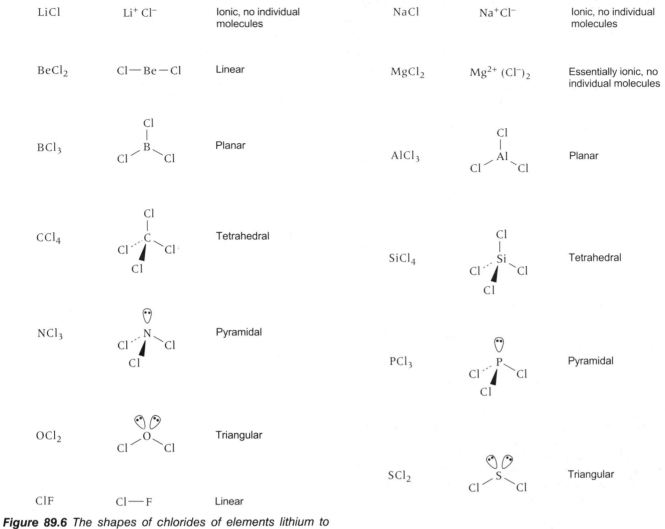

LiCl	Li⁺ Cl⁻	Ionic, no individual molecules
$BeCl_2$	Cl—Be—Cl	Linear
BCl_3		Planar
CCl_4		Tetrahedral
NCl_3		Pyramidal
OCl_2		Triangular
ClF	Cl—F	Linear

Figure 89.6 *The shapes of chlorides of elements lithium to neon*

NaCl	Na⁺Cl⁻	Ionic, no individual molecules
$MgCl_2$	Mg^{2+} $(Cl^-)_2$	Essentially ionic, no individual molecules
$AlCl_3$		Planar
$SiCl_4$		Tetrahedral
PCl_3		Pyramidal
SCl_2		Triangular
Cl_2	Cl—Cl	Linear

Figure 89.7 *The shapes of chlorides of elements sodium to argon*

89.4 Diagonal relationships between some elements

We know that metallic nature increases down a Group, and that non-metallic nature increases across a Period. If we take a diagonal route across the Periodic Table (Figure 89.8), these two trends might be expected to cancel one another out.

Indeed, it is the case that some elements on a diagonal do tend to show similarities. The pairs that we shall consider are (i) lithium and magnesium, (ii) beryllium and aluminium, and (iii) boron and silicon. You will find the similarities summarised in Tables 89.4 to 89.6.

89.1 Give a short explanation of diagonal relationships in the Periodic Table in terms of changes in nuclear charge and shielding of electrons.

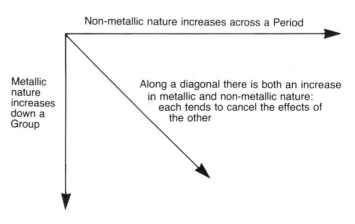

Figure 89.8 *Why neighbouring elements that lie diagonally to one another have similar properties*

Table 89.4. Comparison of the elements lithium and magnesium

Carbonates	Like $MgCO_3$, Li_2CO_3 is decomposed by heat (the other Group I carbonates do not decompose). Both carbonates are insoluble (the other Group I carbonates are soluble)
Chlorides	Both chlorides are hydrated: $LiCl \cdot 2H_2O$, $MgCl_2 \cdot 6H_2O$ (the other Group I chlorides are not hydrated)
Oxides	Both give their normal oxide, Li_2O, MgO, when they burn (the other Group I metals give peroxides, e.g. Na_2O_2)
Hydration	Both Li^+ and Mg^{2+} ions are heavily hydrated in solution (they are both dense centres of charge)

Table 89.5. Comparison of the elements beryllium and aluminium

With nitric acid	Both metals are rendered passive and will not dissolve
With alkali	Both dissolve, giving off hydrogen
Amphoteric nature	Both metals are amphoteric (react with both acid and alkali)
Oxides	Both BeO and Al_2O_3 are amphoteric
Chlorides	Both $BeCl_2$ and $AlCl_3$ are electron deficient; both act as Lewis acids

Table 89.6. Comparison of the elements boron and silicon

Hydrides	Both give a wide variety of unstable hydrides: the boranes and the silanes, e.g. BH_4, B_2H_6, SiH_4, Si_2H_6
Oxides	Both are weakly acidic, e.g. $B_2O_3 \rightarrow H_3BO_3$, $SiO_2 \rightarrow H_2SiO_3$
Chlorides	Both BCl_3 and $SiCl_4$ are volatile and will undergo hydrolysis to give an acidic solution

Answer

89.1 Moving one place to the right across a Period, the increased nuclear charge holds the electrons more tightly to the atom. Moving down one place in a Group, the extra shell of electrons lessens the attraction of the nucleus for the outer electrons. A diagonal move means that these effects tend to compensate for one another.

UNIT 89 SUMMARY

- Down a Group:
  Metallic nature, ionic nature of compounds and basic nature of oxides all increase.
- Across a Period:
  Non-metallic nature, tendency to covalency and acidic nature of oxides all increase.
- Beryllium and boron:
  Are anomalous in their Groups; they are non-metallic, do not give ionic compounds and do not give basic oxides. Their tendency to covalency is explained by the intense polarising power of their very small ions.
- Diagonal relationships are shown by:
  (i) Lithium and magnesium.
  (ii) Beryllium and aluminium.
  (iii) Boron and silicon.
  The similarities are partly due to the cancelling out effect of moving one place to the right along a Period (non-metallic nature increasing) and one place down a Group (metallic nature increasing).

90

Hydrogen and hydrides

90.1 The element

Although air contains almost no free hydrogen, hydrogen atoms are very abundant in Nature. One reason for this is that they are to be found in water, which is widely distributed over the Earth; another is that, combined with carbon, they are found in all living matter. Hydrogen is also to be found in space. It plays a crucial role in stars (such as our Sun) in nuclear fusion reactions. Information about hydrogen is summarised in Table 90.1.

Table 90.1. Information about hydrogen

Relative atomic mass	1.008
Electron structure	1s
Ionisation energy	1312 kJ mol^{-1}
Electron affinity	-72 kJ mol^{-1}
Molecular formula	H_2
Melting point	14 K ($-259°$ C)
Boiling point	20 K ($-253°$ C)
Density at s.t.p.	0.09 g dm^{-3}
Bond energy, H—H	436 kJ mol^{-1}
Bond length, H—H	74 pm
Colourless, odourless, tasteless	

Hydrogen has three isotopes: $^{1}_{1}H$ (protium, H), $^{2}_{1}H$ (deuterium, D) and $^{3}_{1}H$ (tritium, T). Deuterium occurs naturally but only to the extent of 0.015%. Tritium can be made by causing the nuclear reaction:

$$^{2}_{1}H + ^{2}_{1}H \rightarrow ^{3}_{1}H + ^{1}_{1}H$$

Tritium is radioactive, decaying by beta emission.

90.1 What is the equation for the nuclear reaction that takes place when tritium undergoes beta decay? What is produced in the reaction?

90.2 The mass spectrum of a sample of hydrogen gas has a strong peak corresponding to a mass to charge ratio (m/e) value of 2. However, there are other lines at higher m/e values. Why is this? How many peaks would you expect? What would their m/e values be?

90.2 The large-scale extraction of hydrogen

One of the vital uses of hydrogen is in the manufacture of ammonia by the Haber process. We followed the working of the process in Unit 83. Here we shall mention that a very important source of hydrogen for the process is natural gas, methane (Table 90.2). The key idea is to react methane with steam at a high pressure (35 atm) and temperature (800° C) in the presence of a catalyst (nickel). A mixture of carbon monoxide, carbon dioxide and hydrogen results:

$$CH_4(g) + H_2O(g) \rightarrow CO(g) + 3H_2(g)$$
$$CO(g) + H_2O(g) \rightarrow CO_2(g) + H_2(g)$$

The carbon monoxide and carbon dioxide are removed, leaving the hydrogen for further reaction.

Hydrogen can also be obtained from the oil refining industry. It is made in many reactions that involve cracking long-chain hydrocarbons into smaller molecules (see section 86.3), e.g.

$$C_6H_{12}(g) \rightarrow C_6H_6(g) + 3H_2(g)$$

It is possible to make hydrogen by the *Bosch reaction*. The process takes place in three stages.

(i) *Stage 1.* Steam is passed over white hot coke:

$$H_2O(g) + C(s) \rightarrow CO(g) + H_2(g)$$
water gas

(ii) *Stage 2.* The mixture of carbon monoxide and hydrogen, known as water gas, is mixed with

Table 90.2. The manufacture and uses of hydrogen

Manufacture	Uses
From methane	Making ammonia in the Haber
Cracking of hydrocarbons	process
Electrolysis	Production of margarines from
Bosch reaction	vegetable oils
	Welding
	Fuel cells

more steam and passed over an iron catalyst. Only the carbon monoxide reacts:

$$H_2O(g) + CO(g) \rightarrow CO_2(g) + H_2(g)$$

(iii) *Stage 3.* Under pressure the carbon dioxide is dissolved in water, thus leaving the hydrogen available for further use.

A method that is still used where either supplies of methane are limited, or where electricity is relatively cheap, is electrolysis. This can be a direct electrolysis of water (on a suitably large scale), or as a by-product of the electrolysis of brine. The electrolysis of water is not as simple as it might seem. The main problem is that the hydrogen can become contaminated by impurities given off at the cathode. It can also mix with oxygen that diffuses through the electrolyte from the anode. If particularly pure hydrogen is needed, the cathode gas has to be purified.

Apart from the desire to obtain hydrogen, the electrolysis of water is carried out in order to make heavy water, D_2O. This is water that consists of two atoms of deuterium rather than ordinary hydrogen. Heavy water is often used in chemical research.

90.3 Before natural gas can be used to make hydrogen, it has to be purified. Especially, sulphur compounds must not be allowed through. Why?

90.3 The uses of hydrogen

Over 20 million tonnes of hydrogen are produced each year. Almost 50% is used in the Haber process to make ammonia (Table 90.2). In addition to the Haber process, hydrogen finds large-scale use in the hydrogenation of vegetable oils to make margarine. The oils are said to be unsaturated. This means that they contain double bonds between carbon atoms. In the presence of a nickel catalyst, the double bonds take up hydrogen:

$$\begin{array}{c} {>}C{=}C{<} \quad \xrightarrow{H_2} \quad \begin{array}{c} H \ \ H \\ | \ \ \ | \\ {-}C{-}C{-} \\ | \ \ \ | \end{array} \end{array}$$

Hydrogen has also been used in welding. The gas is passed through an electric arc, which splits the molecules into atoms. When the atoms recombine to produce molecules, a great deal of heat is generated. This heat is used to melt and fuse metal surfaces together. A bonus of the process is that because it does not use oxygen, oxidation of the metals is prevented.

One use of hydrogen that has been thoroughly researched in recent years is its potential use as a fuel. When hydrogen burns in air it produces significant amounts of energy, and it has the virtue of being pollution free. Cars have been built that use hydrogen as a fuel rather than petrol. It might be thought that carrying hydrogen around could be dangerous, but it can be

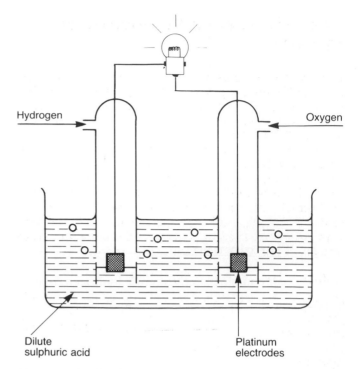

Figure 90.1 *A fuel cell that can be made in the laboratory*

absorbed by a number of metal alloys, which are perfectly safe.

Hydrogen–oxygen fuel cells are useful devices for converting the energy of the reaction

$$2H_2(g) + O_2(g) \rightarrow 2H_2O(l)$$

into electrical energy. You can see a diagram of a fuel cell in Figure 90.1. Hydrogen and oxygen gas are brought together over two electrodes, which can catalyse the reaction between them. The electrolyte can be an acid or an alkali, depending on the type of cell. The virtue of the fuel cell is that it produces electricity with only water as the side product. The cells can operate with high power outputs and efficiencies greater than those of conventional ways of generating electricity, e.g. in oil burning power stations. Fuel cells have been used in space vehicles, where they have the virtue of being reliable, efficient and of relatively small size compared to conventional electric cells.

90.4 The free energy change for the reaction powering a hydrogen–oxygen fuel cell is about $-240\,kJ\,mol^{-1}$. What is the e.m.f. of the cell? (Hint: see Unit 69.)

90.5 Work out the standard enthalpy changes for the reaction:

$$H_2O(g) + C(s) \rightarrow CO(g) + H_2(g)$$

You will need the following information: $\Delta H_f^{\ominus}(H_2O(g)) = -241.8\,kJ\,mol^{-1}$; $\Delta H_f^{\ominus}(CO(g)) = -110.5\,kJ\,mol^{-1}$.

Why is it that air has to be blown through the coke from time to time if the reaction is to be kept going?

90.4 The chemical properties of hydrogen

Free hydrogen atoms are too reactive to exist on their own, and hydrogen gas consists of hydrogen molecules, H_2. Many of the properties of hydrogen have been known for a long time. The most important of them are listed in Table 90.3.

The fact that hydrogen and oxygen make an explosive mixture has had some unfortunate results. On the large scale the destruction of the air ship Hindenburg in 1937 was spectacular. The possibility of explosion exists when hydrogen is made on a large scale in the laboratory, and the experiment should be done with great care. The standard method (Figure 90.2) is to mix dilute sulphuric acid with zinc granules:

$$Zn(s) + 2H^+(aq) \rightarrow Zn^{2+}(aq) + H_2(g)$$

Copper(II) sulphate crystals are usually added. A displacement reaction takes place, which gives the zinc a thin layer of copper metal. An electrochemical cell is set up, which greatly increases the rate of evolution of hydrogen.

Hydrogen from a cylinder can be safely burnt at a jet. It burns with a pale blue flame. If the hydrogen is burnt at the mouth of a glass tube, the flame may appear

Table 90.3. Chemical properties of hydrogen

Forms *hydrides* with many elements:
  hydrides of non-metals are covalent, e.g. CH_4, NH_3, H_2O, HCl
  hydrides of reactive metals are ionic, e.g. Na^+H^-, $Ca^{2+}(H^-)_2$

Forms hydrogen bonds with highly electronegative atoms, e.g. in liquid HF, H_2O

Explodes with oxygen if ignited (gives a 'pop' with a lighted splint):
  $2H_2(g) + O_2(g) \rightarrow 2H_2O(g)$

A reducing agent; will remove the oxygen from many oxides:
  $CuO(s) + H_2(g) \rightarrow Cu(s) + H_2O(g)$

Liberated from acids by many metals:
  $Zn(s) + 2H^+(aq) \rightarrow Zn^{2+}(aq) + H_2(g)$
  This is the basis of the laboratory preparation of the gas

Hydrogen can exist as positive ions, $H^+(aq)$ or $H_3O^+(aq)$, in water. Hydrogen ions are the active agents in aqueous acids

yellow owing to contamination from sodium in the glass. The same reaction occurs as in an explosion:

$$H_2(g) + \tfrac{1}{2}O_2(g) \rightarrow H_2O(g); \quad \Delta H^\ominus = -285.9\,\text{kJ mol}^{-1}$$

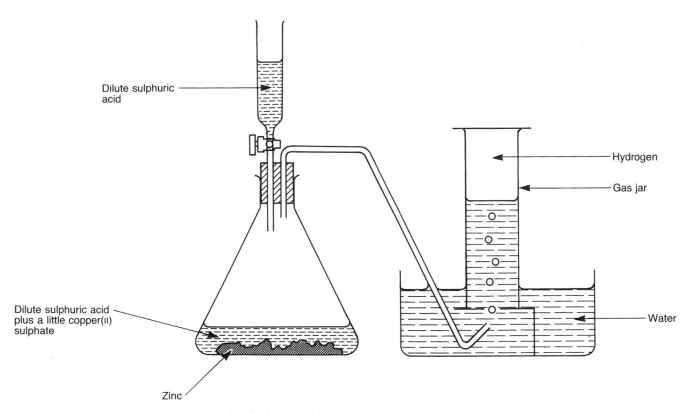

Dilute sulphuric acid

Dilute sulphuric acid plus a little copper(II) sulphate

Zinc

Hydrogen

Gas jar

Water

Figure 90.2 *A simple apparatus for preparing hydrogen. If the gas must be dried before it is collected, it can be passed through concentrated sulphuric acid. However, it must not then be collected over water!*

Figure 90.4 *The chain structure in solid beryllium hydride*

90.6 The reaction between hydrogen and oxygen can be started by ultraviolet light. What does this suggest about the mechanism of the reaction?

90.7 (i) For the most powerful explosion, hydrogen and oxygen should be mixed in a particular proportion. What is it?

(ii) If hydrogen is made from zinc and dilute sulphuric acid in a test tube, there is a loud pop with a lighted splint soon after the reaction starts. However, if there is a delay of some minutes before doing the test, no pop is heard even though many bubbles of gas can be seen. Why does the test not work?

90.8 Copper(II) oxide can be reduced by hydrogen using the apparatus shown in Figure 90.3. The oxide has to be hot to start the reaction, but once it starts it is sufficiently exothermic to keep going without further heating. A student performed the experiment using 2.0 g of the oxide. As soon as all the oxide had changed colour to orange, the hydrogen supply was turned off and air allowed into the tube. When it was cool enough the tube was reweighed and the mass of the product found to be 1.7 g. (The molar masses are: $M(Cu) = 64$ g mol^{-1}; $M(O) = 16$ g mol^{-1}.)

(i) What is the equation for the reaction.

(ii) What should have been the mass of the product?

(iii) What had gone wrong in the experiment?

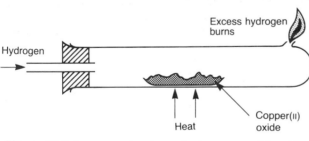

Figure 90.3 *An apparatus for reducing copper(II) oxide with hydrogen*

90.5 Hydride formation with metals

The metals of Group I and calcium, strontium and barium in Group II will combine directly with hydrogen, the outcome being ionic hydrides. Owing to their structures being similar to salts like sodium chloride, they are sometimes known as the *saline hydrides*. A typical reaction is

$$2Na(s) + H_2(g) \rightarrow 2Na^+H^-(s)$$

These hydrides can be used as powerful reducing agents. Especially, they decompose water, sometimes violently:

$$Na^+H^-(s) + H_2O(l) \rightarrow Na^+(aq) + OH^-(aq) + H_2(g)$$

With transition elements, hydrogen behaves somewhat oddly. Although true compounds can occur, for example copper forms CuH, it is more common for hydrogen to be absorbed into the crystal structures of the solids. This can happen because hydrogen molecules are small enough to squeeze into the spaces between the metal atoms; hence they are called the *interstitial hydrides*. They do not have a fixed formula, which is summed up by saying that they are *non-stoichiometric*. The actual formula can vary with temperature and pressure of the hydrogen. For example, palladium hydride can vary in composition between $PdH_{0.1}$ and $PdH_{0.6}$.

The hydrides of beryllium and magnesium are believed to be predominantly covalent, and to exist in chains (Figure 90.4).

90.9 Briefly explain why aluminium hydride, AlH$_3$, would be expected to bond with a hydride ion. (Look back at Unit 14 if you are stuck.)

90.6 Hydride formation with non-metals

Hydrogen reacts readily with highly electronegative elements like fluorine, oxygen and chlorine, but less violently with other elements. It is, for example, very difficult to persuade hydrogen and carbon to react directly; nitrogen and hydrogen require a catalyst to make them react at an appreciable rate (the Haber process, section 83.3). The hydrides are covalent and usually gases at room temperature (Figure 90.5). Water, of course, is an exception owing to the strength of the hydrogen bonding among the molecules. You should look back to Unit 21 if you need to remind yourself about this.

The covalent hydrides are sometimes acidic, sometimes alkaline and sometimes neutral in water. For example, hydrogen chloride, hydrogen bromide and hydrogen iodide are all strong acids in water:

$$HX(aq) + H_2O(l) \rightleftharpoons H_3O^+(aq) + X^-(aq)$$
equilibrium lies far to the right

Ammonia is a weak base:

$$NH_3(aq) + H_2O(l) \rightleftharpoons NH_4^+(aq) + OH^-(aq)$$
equilibrium lies far to the left

Methane, CH$_4$, is only very slightly soluble and does not react with water.

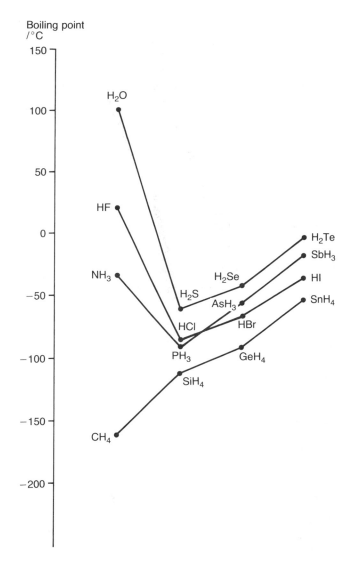

Figure 90.5 *Boiling points for hydrides of elements in Groups IV, V, VI and VII*

90.10 $E^{\ominus}$ for a standard hydrogen electrode (S.H.E.) is defined to be exactly 0 V. This value implies that hydrogen should reduce any redox system that has a positive $E^{\ominus}$. For example, $E^{\ominus}_{Br_2/Br^-} = +1.09\,V$.

(i) What should be the cell reaction if this half-cell is combined with a S.H.E.?

(ii) If hydrogen is bubbled through a solution of bromine, nothing happens. Why might this be?

90.7 Some unusual hydrides

You are almost certain to have used the simpler hydrides, like hydrogen chloride or ammonia, or their solutions in your chemistry course. However, you may

Table 90.4. Some of the hydrides of boron

Name	Formula	Comment
Diborane	B_2H_6	Easily hydrolysed; highly flammable in air
Tetraborane	B_4H_{10}	Rather less reactive than diborane
Hexaborane-10	B_6H_{10}	Similar to tetraborane
Decaborane	$B_{10}H_{14}$	Little reaction with air or water
Icosaborane-16	$B_{20}H_{16}$	Like decaborane

not have met the hydrides of boron before. These hydrides provided a severe shock for theories of chemical bonding when they were first investigated. There are many boron hydrides (Table 90.4), but we shall concentrate on the simplest: diborane, B_2H_6.

If we count the electrons available for bonding we have one each from six hydrogen atoms, and there are three electrons in the valence shell of each boron (B: $1s^2 2s^2 2p$), making 12 electrons in all. The essential problem about diborane is this: the X-ray structure of diborane shows that there are two hydrogen atoms attached to each boron atom, and two more shared between the boron atoms. Therefore it appears that there are eight bonds, for which we should need 16 electrons. Clearly the numbers do not fit. There appear to be too few electrons to account for the number of bonds. A molecule, like diborane, for which this is true is called an *electron deficient* molecule.

One solution to the puzzle is to use molecular orbital theory. Each bridging hydrogen atom is imagined to be held to the boron atom by a bond that stretches between all three atoms. The bond is called a three-centre bond (Figure 90.6). Each one contains two electrons, which, together with the four pairs of electrons in the bonds to the four terminal hydrogen atoms, brings the total to the 12 electrons we have available. The structures of other boranes can be explained in similar fashion, but the larger structures are complicated by bonds between boron atoms as well as between the boron and hydrogen atoms.

A second hydride that is rather unusual is *lithium tetrahydridoaluminate(III)*, $LiAlH_4$. (It has the alternative name lithium aluminium hydride.) This hydride can be made by reacting lithium hydride with aluminium trichloride in ethoxyethane (ether):

$$4LiH + AlCl_3 \rightarrow LiAlH_4 + 3LiCl$$

It is a white ionic solid, which has remarkable powers as a reducing agent. The AlH_4^- ion is the active reducing agent. One of the problems with it is that it will react violently with water, even in small amounts. It finds its main use in organic chemistry. For example, it will reduce acid, aldehyde and ketone groups to alcohols, but leave any double bonds alone. The following conversions are typical:

(i) ethanoic acid to ethanol,

$$CH_3COOH \rightarrow CH_3CH_2OH$$

(a)

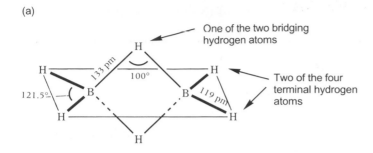

One of the two bridging hydrogen atoms

Two of the four terminal hydrogen atoms

(b)

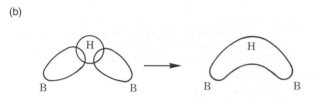

Figure 90.6 *(a) The structure of diborane, B_2H_6. (b) The bridging hydrogen atoms have a 1s orbital, which can overlap with one orbital on each boron atom. A three-centre bond results*

(ii)   ethanal to ethanol,

$$CH_3CHO \rightarrow CH_3CH_2OH$$

(iii)   propanone to propan-2-ol,

$$(CH_3)_2CO \rightarrow (CH_3)_2CH_2OH$$

In addition it will convert nitriles (organic compounds that contain CN groups) to amines, and amides to amines. For example, ethanonitrile changes to ethylamine,

$$CH_3CN \rightarrow CH_3CH_2NH_2$$

and ethanamide changes to ethylamine,

$$CH_3CONH_2 \rightarrow CH_3CH_2NH_2$$

In some reactions sodium tetrahydridoborate(III), $NaBH_4$, behaves similarly to $LiAlH_4$. The main difference between them is that $NaBH_4$ is a less powerful reducing agent.

90.11   Write down the formula of the molecule made when $LiAlH_4$ reacts with CH_2=$CHCOOH$.

90.12 $LiAlH_4$ will reduce chlorides. It is found that 1 mol of it reacts completely with 1 mol of silicon(IV) chloride, $SiCl_4$. Two of the products are 1 mol of lithium chloride and 1 mol of aluminium trichloride. The third product is a gaseous hydride of silicon. Discover its formula by working out the equation for the reaction.

90.13   Suggest a way of making phosphine, PH_3, starting with phosphorus trichloride, PCl_3.

Answers

90.1 $^3_1H \rightarrow {}^3_2He + {}^0_{-1}e$. The product is an isotope of helium.

90.2   A sample of hydrogen gas will contain H_2, HD and D_2 molecules, which have *m/e* values of 2, 3 and 4 respectively. As you would expect, the latter two peaks are extremely small.

90.3   Sulphur is a very effective poison of catalysts, so it has to be removed before it reaches the nickel.

90.4   The equation we need is $\Delta G^\ominus = -zFE^\ominus$. Here, we have

$$-240 \times 10^3 \, J \, mol^{-1} = -2 \times 96\,500 \, C \, mol^{-1} \times E^\ominus$$

which gives $E^\ominus = +1.24$ V.

90.5 $\Delta H^\ominus(reaction) = \Delta H_f^\ominus(CO(g)) - \Delta H_f^\ominus(H_2O(g))$
$$= +131.3 \, kJ \, mol^{-1}$$

The reaction is strongly endothermic. A high temperature is needed to keep the reaction going, so air is blown through to liberate heat by the reaction of coke with oxygen. This is an exothermic reaction no matter whether carbon monoxide or carbon dioxide is produced.

90.6   It goes via a free radical mechanism.

90.7   (i) It is 2 volumes of hydrogen to 1 volume of oxygen. (The same ratio as in the equation.)
(ii) At the start the hydrogen mixes with oxygen already in the air in the tube. This gives an explosive mixture. After some time all the air is driven out and there is no oxygen left to give an explosion. Instead, the hydrogen burns very quietly, but the flame is so small that it is easily missed.

90.8   (i) $CuO(s) + H_2(g) \rightarrow Cu(s) + H_2O(g)$
(ii) The equation tells us that 80 g of oxide should give 64 g of copper; so 2.0 g should produce 1.6 g of copper.
(iii) The result shows that the product has a greater mass than expected. One possibility is that some of the copper(II) oxide was left unreacted. Also, it is likely that by letting air into the tube before it was cold, some of the hot copper reacted with oxygen and changed back to oxide.

90.9   Like aluminium trichloride, there is an empty orbital on the aluminium, which can accept a lone pair of electrons. The hydride ion has the necessary pair of electrons.

90.10   (i) $H_2(g) + Br_2(aq) \rightarrow 2H^+(aq) + 2Br^-(aq)$
(ii) The reaction is kinetically blocked. Largely this is because of the strength of the bond in the hydrogen molecule. In other words, there is a large activation energy to the reaction.

90.11 CH_2=$CHCH_2OH$. The double bond is not affected. The acid group is converted to an alcohol.

90.12 $LiAlH_4 + SiCl_4 \rightarrow LiCl + AlCl_3 + SiH_4$

90.13   One possibility is to react phosphorus trichloride with lithium tetrahydridoaluminate(III):

$$3LiAlH_4 + 4PCl_3 \rightarrow 3LiCl + 3AlCl_3 + 4PH_3$$

UNIT 90 SUMMARY

- Hydrogen:
    (i) Is much less dense than air.
    (ii) Is a good reducing agent.
    (iii) Is highly flammable.
- Laboratory preparation:

    Zinc + dilute sulphuric acid
    + trace copper(II) sulphate

    $$Zn(s) + 2H^+(aq) \rightarrow Zn^{2+}(aq) + H_2(g)$$

- Manufacture:
    (i) From methane at 35 atm, 800°C, Ni catalyst

    $$CH_4(g) + H_2O(g) \rightarrow CO(g) + 3H_2(g)$$
    $$CO(g) + H_2O(g) \rightarrow CO_2(g) + H_2(g)$$

    (ii) Bosch process, reacting steam and carbon or carbon monoxide at high temperature plus Fe catalyst

    $$H_2O(g) + C(s) \rightarrow CO(g) + H_2(g)$$
    $$H_2O(g) + CO(g) \rightarrow CO_2(g) + H_2(g)$$

Reactions

- With oxygen:

    $$2H_2(g) + O_2(g) \rightarrow 2H_2O(g)$$

    This reaction used in fuel cells to generate electricity.
- A reducing agent:

    e.g. $CuO(s) + H_2(g) \rightarrow Cu(s) + H_2O(l)$
- Hydrogenation:
    Converts alkenes to alkanes (Ni catalyst), e.g.

    $$H_2C{=}CH_2 + H_2(g) \rightarrow H_3C{-}CH_3$$

- With reactive metals:
    Ionic hydrides made, e.g.

    $$2Na(s) + H_2(g) \rightarrow 2Na^+H^-(s)$$

- Haber process:
    Used to make ammonia

    $$N_2(g) + 3H_2(g) \rightleftharpoons 2NH_3(g)$$

- With boron:
    Electron deficient hydrides exist, e.g. B_2H_6.

91
Water

91.1 What is special about water?

Water played a crucial part in the origin of life and it still has an essential role in maintaining plant and animal life. Plants depend on water for the transfer of nutrients and for photosynthesis. Owing to the presence of water in cells and body fluids such as blood, human beings are approximately 60% water. Nearly all the processes essential for life depend on reactions that take place in an aqueous solution, be it the division of DNA in a cell, the digestion of foodstuffs in the stomach, or the trans-port of oxygen around the body. Given the importance of water, it is not surprising that men and women can survive very much longer without food than they can without water.

Historically, the availability of water supplies has determined where villages, towns and cities are sited. Nomadic peoples, and animals, may travel hundreds of miles over the course of a year following the seasonal variation in rainfall. A lack of good quality drinking water, and water for sanitation, brings deadly illnesses such as typhoid.

Scenes like this, of flooding in Bangladesh, will become even more common if the worst predictions of the results of global warming come true.

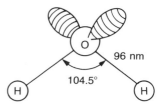

Figure 91.1 *A water molecule has a bond angle of 104.5° and bond length of 96 nm. According to valence bond theory it also has two lone pairs of electrons*

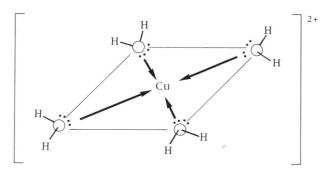

Figure 91.2 *The Cu(H_2O)_4^{2+} ion*

All these factors, and many more, make water a substance of great importance. From a strictly chemical point of view the remarkable thing about water is the amount of hydrogen bonding there is, both in the solid (ice) and in the liquid. If it were not for the fact that hydrogen bonds are of intermediate strength (stronger than van der Waals bonds but weaker than ordinary ionic or covalent bonds) then life as we know it could not exist and the world would be without rivers, lakes or seas.

We have talked about how hydrogen bonding comes about, and the effects it has, in Unit 21. For example, you should know that in many ways water is anomalous when it is compared to the other hydrides of elements in the same Group as oxygen. The solubility of many salts is due in no small measure to the polarity of water molecules. The attraction of the negatively charged oxygen atoms allows them to congregate around positive ions; the positively charged hydrogen atoms can gather round negative ions. This is the process of *hydration*. If you look at Unit 60 you will find that we discussed solubility largely as a competition between the magnitude of the lattice energy of a solid and the hydration energies of its ions or molecules.

According to valence bond theory each water molecule has two lone pairs of electrons (Figure 91.1). Often one of the lone pairs can be used in dative covalent bonding. An important example is the ability of a water molecule to bond to a proton in making the oxonium ion:

$$H_2O(l) \quad + \quad H^+(aq) \longrightarrow H_3O^+(aq)$$

The colour of many transition metal ions in water is due to the ability of water molecules to bond to the ions. Transition metal ions often have d orbitals available, which can be used in dative covalent bonding with water molecules. For example, the colour of copper(II) sulphate in water is due to the presence of $Cu(H_2O)_4^{2+}$ ions (Figure 91.2). Molecules (or other species with lone pairs) that bond with transition metal ions are called *ligands*. This gives us another property of water: its molecules can be ligands.

91.1 What does molecular orbital theory have to say about the presence of lone pairs in a water molecule? (Look back at section 16.8.)

91.2 Some chemical reactions of water

There are two main types of reaction of water. They are reactions with metals and with non-metals. A third type is reactions with compounds.

(a) With metals

The way metals react with water can give us a measure of their reactivities. The most reactive metals are the alkali metals in Group I. Each of them reacts with cold water, giving off hydrogen and leaving an alkaline solution. A typical reaction is

$$2K(s) + 2H_2O(l) \rightarrow 2K^+(aq) + 2OH^-(aq) + H_2(g)$$

The reactivity increases down the Group. The Group II metals react in a similar way but less violently. (Beryllium does not react at all.) However, the hydroxides of the Group II metals are much less soluble than those of Group I.

Some of the transition metals react with hot water or steam to give an oxide. For example, if steam is passed over heated iron, hydrogen is released and an oxide is left:

$$3Fe(s) + 4H_2O(l) \rightarrow Fe_3O_4(s) + 4H_2(g)$$

Water plays an important part in the rusting of iron, the details of which you will find in section 68.2.

(b) With non-metals

The typical reaction with non-metals is for little to happen. Elements such as carbon, sulphur and phosphorus normally do not react with water. Carbon, usually in the guise of coke, when it is white hot will react with steam. The product is called water gas:

$$C(s) + H_2O(g) \rightarrow CO(g) + H_2(g)$$

This is a reaction we have met before, in section 90.2.

The halogens react with water to give an acidic solution. For example,

$$Cl_2(g) + H_2O(l) \rightarrow HOCl(aq) + H^+(aq) + Cl^-(aq)$$

The usefulness of a solution of chlorine lies not so much in its acidity but in the chloric(I) acid, $HOCl(aq)$, otherwise known as hypochlorous acid. This weak acid acts as an oxidising agent. It is responsible for the anti-bacterial action of chlorine water, and for its use as a bleach.

(c) With compounds

We have already discussed why salts will often dissolve in water, some of them suffering hydrolysis. You had best look back at Unit 76 if you have forgotten this. Similarly, in Units 74 and 75 we met many instances where compounds of non-metals react to give acidic or alkaline solutions. Here a few equations might remind you of the possibilities.

Solutions of carbonates are slightly alkaline:

$$CO_3^{2-}(aq) + H_2O(l) \rightleftharpoons HCO_3^-(aq) + OH^-(aq)$$

A solution of ammonia is a weak alkali:

$$NH_3(aq) + H_2O(l) \rightleftharpoons NH_4^+(aq) + OH^-(aq)$$

A solution of carbon dioxide is weakly acidic:

$$CO_2(g) + H_2O(l) \rightleftharpoons H_2CO_3(aq) \rightleftharpoons HCO_3^-(aq) + H^+(aq)$$

Hydrogen chloride gives a strongly acidic solution:

$$HCl(g) + H_2O(l) \rightarrow H_3O^+(aq) + Cl^-(aq)$$

91.2 We have said that some oxides can show the properties of both acids and bases. These are the amphoteric oxides. Water is an amphoteric oxide. It shows this behaviour when it reacts with hydrogen chloride and with ammonia.

(i) Write down equations for the two reactions.

(ii) Identify where water is acting as an acid and where it is a base.

(iii) Which theory of acid and base behaviour have you used?

91.3 Why is it that ethanol, C_2H_5OH, and glucose, $C_6H_{12}O_6$, are both very soluble in water?

91.3 Heavy water

Deuterium, ^2_1H or D for short, is one of the isotopes of hydrogen. It has twice the mass of an ordinary hydrogen atom (protium), but like all isotopes of an element it has the same chemical properties. However, its extra mass makes a deuterium atom react less quickly than ordinary hydrogen, ^1_1H. Deuterium atoms are often used in chemistry to discover the mechanism of a reaction.

Especially, instead of using ordinary water, H_2O, a reaction can be carried out in *heavy water*, D_2O. If hydrogen atoms from water take part in a reaction, they should be found in the products or one of the intermediates. If the mass spectrum of the products or intermediates is taken, then the presence of the heavier deuterium atoms should show up.

It so happens that D_2O does not undergo electrolysis as easily as H_2O. This allows heavy water to be obtained as a product of the electrolysis of water. Apart from its use in the study of chemical reactions, heavy water has been widely used in nuclear reactors. It is a fairly efficient moderator. That is, it can lower the energies of fast neutrons.

91.4 Ammonia gas, NH_3, can easily dissolve in heavy water.

(i) Write down the equation for the equilibrium that is set up.

(ii) If the solution is warmed and the gas produced is dried before being passed into a mass spectrometer, there is a large peak at $m/e = 17$, and a smaller one at 18. What causes the two peaks?

(iii) What does the experiment show about the equilibrium?

91.5 Some pure ordinary water, H_2O, is mixed with heavy water, D_2O. How many different molecules and ions could appear in the mixture once equilibrium Is established? What are their formulae?

91.6 Write down the equation for the reaction of sodium with heavy water.

91.4 The water cycle

It has been estimated that the total volume of water that falls to the Earth's surface each year is about $496\,000\ km^3$, i.e. approaching 500×10^{12} tonne. About one quarter of this precipitation (a general term for water falling as rain, snow, etc.) occurs over land, the rest over the seas. The precipitation is balanced by evaporation of an equal amount of water. (A nice example of the rule that what goes up must come down!) The changes involved in the continuous process of precipitation and evaporation is called the *water cycle* (Figure 91.3).

Evaporation mainly takes place from the sea (about 425×10^{12} tonne each year). On land, much water evaporates from rivers and lakes; but a great deal returns to the atmosphere by the loss of water from the leaves of plants. This is the process of *transpiration*. With the large amount of rain forest that is being cut down in some countries, changes in climate may be caused by

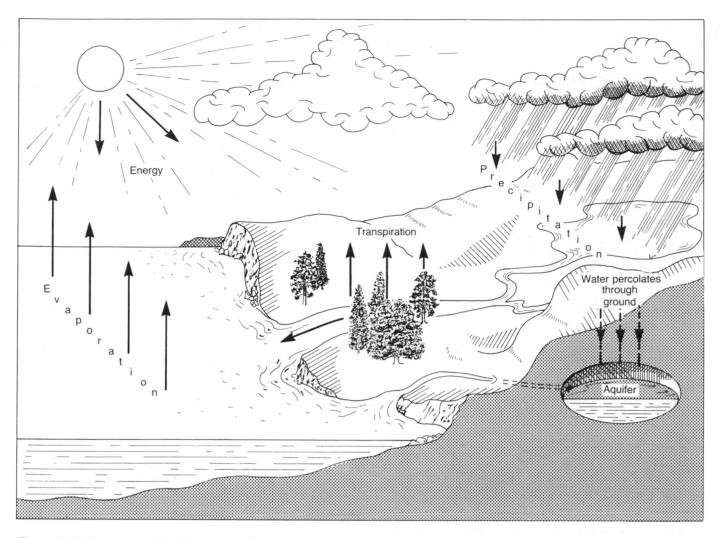

Figure 91.3 *The water cycle. About 875 km³ of water evaporates from oceans, and about 160 km³ from the land. Over 775 km³ of water falls as rain over oceans, more than 260 km³ over land, and 100 km³ flows into the oceans from rivers. Much water is present as clouds and invisible water vapour in the air. (Data taken from: Allaby, M. (1986).* Ecology Facts, *Hamlyn, London)*

the reduction in transpiration. This is in addition to the increase in soil erosion that takes place when forest is cleared. There is, of course, evaporation from all other surfaces that get wet, such as roads, fields and washing hung out to dry. Some precipitation escapes evaporation by finding its way through the surface layers of rock and soil to underground chambers, where it may remain for thousands of years. The chambers are known as *aquifers*, and they are a useful source of water where wells can be sunk sufficiently deeply to reach them.

91.7 At a temperature of 15° C the energy absorbed when water changes from liquid to gas is about 2500 kJ kg⁻¹.

(i) Estimate the energy used in the evaporation of water over land and sea each year.

(ii) When water vapour condenses to rain drops, energy is released. Around 125×10^{12} tonne of precipitation takes place over land each year. How much energy is released?

91.5 Water pollution

When rain drops fall through the atmosphere they dissolve small quantities of gases in the atmosphere. Where there is little air pollution, the gases are mainly nitrogen, oxygen and a little carbon dioxide. Although small in its amount, the carbon dioxide does make the water very slightly acidic owing to the production of weak carbonic acid. Rain falling in a thunderstorm is more acidic than normal. The energy of the lightning is sufficient to dissociate nitrogen molecules, which then combine with oxygen to give oxides such as nitrogen

It is only with the help of charities such as Oxfam that people in many African communities, like the one shown here in Baskare village, Burkina Faso, can gain the resources to have a water well dug.

dioxide. Indeed, rain in a thunderstorm is very dilute nitric acid.

As the rain percolates through soil and rock it dissolves minerals. Depending on the geology of the catchment area for water, reservoirs and aquifers can contain high amounts of dissolved salts. Especially, in regions where there is chalk, $CaCO_3$, the water contains calcium ions and hydrogencarbonate ions owing to the reaction

$$CaCO_3(s) + H^+(aq) \rightarrow Ca^{2+}(aq) + HCO_3^-(aq)$$

Magnesium ions can appear by a similar means where dolomite, a mineral containing $MgCO_3$ and $CaCO_3$, occurs.

Over the last few years the concentrations of sulphate and nitrate ions in water have greatly increased in areas that are intensively farmed. This is partly due to the widespread use of fertilisers such as ammonium sulphate and ammonium nitrate. It is the nitrate that is the main cause of worry because in some circumstances it is able to give rise to nitrite ions. Nitrite ions are known carcinogens. It remains to be seen whether populations that drink water contaminated by nitrate ions will show a higher incidence of cancers in years to come.

The use of machines to spread chemically manufactured fertilisers has led to over-use, and problems with water purity.

In most industrialised areas of the world the air contains significant quantities of the oxides of carbon, nitrogen and sulphur. A major source of nitrogen oxides in the atmosphere is the burning of fuel in inter-

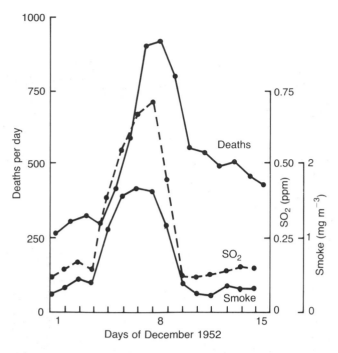

Figure 91.4 *The graphs show how the increased air pollution in a London fog of 1952 caused a marked increase in the death rate. (Taken from: White, I. D. et al. (1987). Environmental Systems, Allen and Unwin, London, figure 24.18)*

nal combustion engines, e.g. in cars and lorries. Carbon dioxide and sulphur dioxide are produced in large quantities by the burning of fossil fuels, e.g. in open coal fires, oil central heating systems and power stations; such burning can cause a health risk (Figure 91.4). The smoke from power stations is usually sent into the air from very tall chimneys. For the people who live close to the power stations this is all to the good; but the acidic oxides can travel hundreds or thousands of miles in the winds higher in the atmosphere. The oxides then give rise to *acid rain* in regions far from the power station. The deaths of huge areas of forest in Germany and parts of Scandinavia has been blamed on power stations in Britain, many of which have inefficient purification systems for the effluent sent to the chimneys.

Acid rain also has a marked effect on the ecology of lakes and rivers. A small change in pH can greatly affect the ability of fish to breed. Some parts of Scotland and Scandinavia have suffered almost complete loss of salmon from rivers where the pH has decreased. Acid rain can also leach minerals from rocks and soil which are unaffected by normal rain water. For example, some inland lakes now have much higher concentrations of aluminium ions than a few years ago.

91.6 Water treatment

As we have seen, water can contain different ions depending on the nature of the area in which it is found. Normally it will also contain small amounts of organic matter, such as particles of clay and decaying vegetation suspended in it. The particles are often of a colloidal size and can be precipitated by the addition of aluminium salts. However, if the addition is not controlled properly, the aluminium can itself end up in the water supply as a pollutant. Larger particles can be removed by passing the water through beds of sand and

These trees in the USA show the characteristic loss of foliage which is the mark of damage by acid rain.

Sprinkler beds contain particles which provide a large surface area upon which bacteria live and digest much of the harmful matter in sewage.

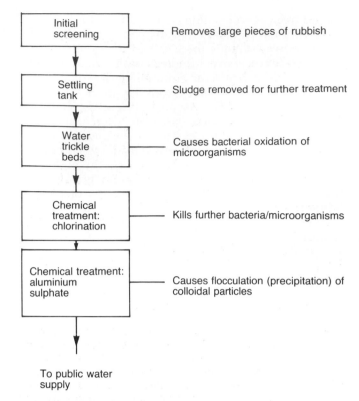

Initial screening	Removes large pieces of rubbish
Settling tank	Sludge removed for further treatment
Water trickle beds	Causes bacterial oxidation of microorganisms
Chemical treatment: chlorination	Kills further bacteria/microorganisms
Chemical treatment: aluminium sulphate	Causes flocculation (precipitation) of colloidal particles

To public water supply

Figure 91.5 *The key stages in the purification of water*

gravel that act as filters (Figure 91.5). Before it is fed to the public water supply, chlorine is added in small quantities in order to oxidise harmful bacteria. In some countries fluoride is also added because there is evidence that it can help prevent tooth decay, particularly among children.

Water that contains calcium and magnesium ions is called *hard water*. When these ions are accompanied by hydrogencarbonate ions, the water contains *temporary hardness*. It is called temporary because the calcium and magnesium ions can be removed from the water by boiling. The hydrogencarbonate ions decompose on heating, producing carbon dioxide and carbonate ions. The carbonate ions give precipitates with the metal ions:

$$2HCO_3^-(aq) \rightarrow CO_3^{2-}(aq) + H_2O(l) + CO_2(g)$$
$$Mg^{2+}(aq) + CO_3^{2-}(aq) \rightarrow MgCO_3(s)$$

When the metal ions are present with chloride or sulphate ions, the water is *permanently hard*. Permanent hardness cannot be removed by boiling.

Hard water has several unwanted properties. In the first place, it gives a scum with soap. Soaps are ionic compounds containing a long hydrocarbon chain with a —COO⁻ group on the end. This group is the remnant of an organic acid that has its hydrogen removed. In a solid soap it is found with a sodium or potassium ion. A typical soap particle is sodium stearate, $C_{17}H_{35}COO^-Na^+$. A particle like this is easily soluble in water, but with calcium or magnesium ions the stearate ion forms an insoluble solid. Not only is this a waste of soap, but the scum it makes is unpleasant. For both

reasons it is desirable to remove these ions from water, a process known as *water softening*. There are several ways that this can be done.

The simplest method is to *add washing soda*, sodium carbonate crystals. The added carbonate ions give a precipitate with the calcium or magnesium ions:

$$Mg^{2+}(aq) + CO_3^{2-}(aq) \rightarrow MgCO_3(s)$$

A more expensive approach is to use *ion exchange*. Here the water is passed through a column of inert material, often beads of a polymer which have ionic groups attached to the surface. There are two types of ion exchange: cation and anion exchange. The first type, which is used in water softening, has negative surface ions to which anions can cling. Figure 91.6 shows you the idea. Initially sodium ions are held to the negative groups, but when water containing calcium ions is poured over the beads, the sodium ions are displaced. The water that comes out contains sodium ions, but it is no longer hard. Of course, eventually all the available sites are full and no more calcium ions can be taken up. When this happens the column is flooded with salt water. The deluge of sodium ions displaces the calcium ions and then the column is ready to be used again.

A similar method to using an ion exchange column is to use *zeolites*. Zeolites are cage-like structures made from silicates (Figure 91.7). They have empty spaces in their structures into which ions can fit. As with ordinary ion exchange, sodium ions are the normal residents in the holes. When hard water passes over the zeolite, the calcium ions displace the sodium ions. The zeolite is regenerated by swamping it with salt water.

There is another reason why magnesium and calcium ions should be removed from water. If hard water is boiled the calcium or magnesium carbonates that are precipitated produce a layer of scale. The presence of scale in a kettle can be annoying, but in hot water pipes in the home or in industry it leads to two problems. First, it takes more energy to heat water in a tank that has a layer of scale on it. This can greatly increase the cost of heating. Secondly, it becomes very much harder for water to pass round the system. This can overload pumps and reduce efficiency. Eventually pipes can become completely blocked or 'furred up'.

The concentration of ions such as Ca^{2+} and Mg^{2+} that cause hardness in water can be found by titration.

Ethylenediaminetetraacetic acid (EDTA; more correctly named 1,2-bis[bis(carboxymethyl)amino]ethane) has an anion with long tentacle-like groups that can wrap round a metal ion. In principle there are six sites that can be used in bonding, but they are not always used. For example, with Ca^{2+} ions only four bonds are made. Figure 91.8 shows you the idea. (The calcium ion makes a complex with the EDTA anion – a process we normally associate with transition metals.) An indicator is added to the sample of water that is to be titrated. The indicator changes colour in the presence of Mg^{2+} or Ca^{2+} ions. Often a solution of Erichrome Black T is used, but this will only work if the water solution is made alkaline. To ensure the solution remains alkaline, a buf-

(a)

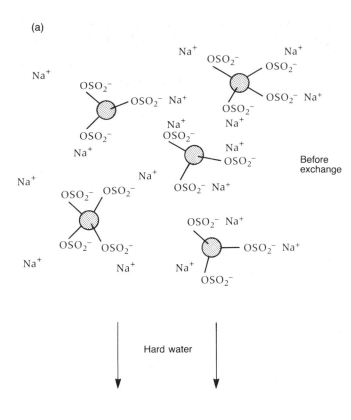

Before exchange

Hard water

(b)

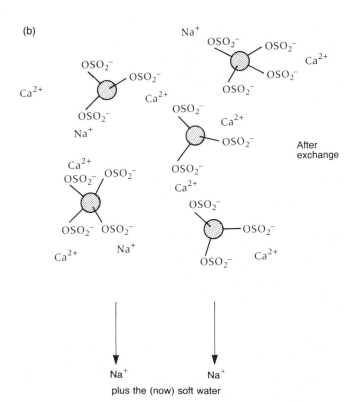

After exchange

Na⁺   Na⁺

plus the (now) soft water

Figure 91.6 *(a) Part of an ion exchange resin. The beads carry SO₃⁻ ions, together with Na⁺ ions close by. (b) After hard water is passed over the resin, most of the Ca²⁺ ions in the water are held on the beads. Na⁺ ions are washed off the beads*

Smaller hole in here

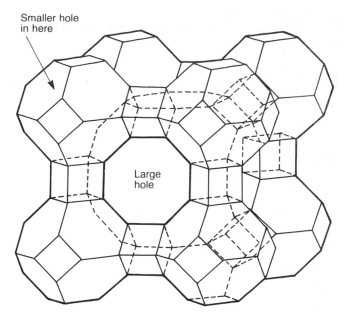

Large hole

Figure 91.7 *The cage structure of zeolite, with formula $Na_{12}(Al_{12}Si_{12}O_{48}) \cdot 27H_2O$. The structure is based on a combination of cubes and octahedra. There is a large hole in the centre, and smaller ones inside the other frameworks, in which atoms, molecules or ions can be trapped. (Adapted from: Wells, A. F. (1950). Structural Inorganic Chemistry, 2nd edn, Clarendon Press, Oxford, figure 23.25)*

(a)

(b)

(c)

Figure 91.8 *How EDTA works. (a) A molecule of EDTA. (b) There are six possible sites that an EDTA⁴⁻ ion can use in bonding. Note that two of these are lone pairs on the nitrogen atoms. (c) The complex made by EDTA⁴⁻ with a Ca²⁺ ion*

fer made from ammonium chloride and concentrated aqueous ammonia is added to the water sample. When the indicator is added to the mixture of buffer and water, a red colour is produced. If a solution of EDTA is run in from a burette, the EDTA anions grab the Mg^{2+} and Ca^{2+} ions and prevent them giving a colour with the indicator. At the endpoint all the ions have been surrounded by EDTA anions, and the colour of the solution changes to blue. If we know the concentration of the EDTA solution, we can work out the concentration of the Mg^{2+} and Ca^{2+} ions in the water. Question 91.10 takes you through the procedure.

91.8 Why is it vital that sulphide and acidic waste is not mixed at a chemical disposal site?

91.9 One way of distinguishing hard from soft water is to mix a little of each with soap solution. When the mixture is shaken, the soft water will give a lather; the hard water will not give a lather until much more soap solution is added.

You have been given a soap solution together with the usual apparatus found in a chemical laboratory. You have also been given three samples of water. Your task is to design an experiment that would allow you to compare (with reasonable accuracy) the degree of hardness of each sample. Briefly explain what you would do.

91.10 25 cm³ of tap water were placed in a conical flask together with an alkaline buffer solution and a few drops of Erichrome Black T indicator. A solution of 0.1 mol dm⁻³ EDTA was run in from a burette until the red-to-blue endpoint was reached. On average, 14.40 cm³ of EDTA were required.

(i) What is the number of moles of EDTA run in at the endpoint?

(ii) How many moles (in total) of Mg^{2+} and Ca^{2+} ions were present in the 25 cm³ of water?

(iii) What was the total concentration, in mol dm⁻³, of the ions?

(iv) Assume that 50% of the ions were Ca^{2+} ions. What mass of chalk, $CaCO_3$, would provide these ions in the water? $M(CaCO_3) = 100$ g mol⁻¹.

(v) Assume the water came from a reservoir holding 10^6 tonnes of water. What mass of $CaCO_3$ had been leached out of the rock structures in the area? Density of water $= 1$ g cm⁻³ or 1 kg dm⁻³; 1 tonne $\equiv 1000$ kg.

91.11 Water that is fit for drinking is called *potable* water. Try a little research of your own to discover the source of potable water in your area. What ions or other chemicals does your local water contain?

91.7 The estimation of the amount of oxygen in water

The amount of dissolved oxygen in water can be found by the Winkler method. It relies on a series of reactions involving manganese ions, iodide ions and oxygen. The outcome of the experiment is a solution of iodine whose concentration reflects the concentration of oxygen in the water. On a small scale the water can be put into a syringe, e.g. a gas syringe, and the nozzle to the syringe sealed with a rubber septum. A little concentrated solution of manganese(II) sulphate is injected into the syringe, followed by an alkaline solution of potassium iodide. When the manganese(II) ions mix with the alkali, an off-white precipitate of manganese(II) hydroxide is produced:

$$Mn^{2+}(aq) + 2OH^-(aq) \rightarrow Mn(OH)_2(s)$$

This hydroxide is sensitive to the presence of oxygen, which converts it into manganese(III) oxide, Mn_2O_3:

$$4Mn(OH)_2(s) + O_2(aq) \rightarrow 2Mn_2O_3(s) + 4H_2O(l)$$

This solid appears as a brown precipitate. Now comes the last part of the process. Manganese(III) is sufficiently oxidising in its nature to convert iodide ions into iodine. However, it will only do so in acidic conditions. Once the precipitation reaction is finished, sulphuric acid is injected into the syringe and the solution takes on the orange-brown colour of triiodide ions, I_3^-. First,

$$Mn_2O_3(s) + 2I^-(aq) + 6H^+(aq) \rightarrow 2Mn^{2+}(aq) + I_2(s) + 3H_2O(l)$$

Then,

$$I_2(s) + I^-(aq) \rightleftharpoons I_3^-(aq)$$

The concentration of iodide ions can be determined by three methods. If only a rough estimate is needed, the colour of the solution can be compared with the colours of pre-prepared solutions of iodine dissolved in potassium iodide. A more accurate method is to determine the concentration of triiodide ions, and hence of iodine, using a colorimeter. Alternatively, a sodium thiosulphate titration can be done (see section 40.3).

If you look back at the equations for the reactions, you will find that

$$1 \text{ mol of } O_2 \equiv 2 \text{ mol } Mn_2O_3 \equiv 2 \text{ mol } I_2$$

Hence if we know the number of moles of iodine produced, we also know the number of moles of oxygen molecules that were in the original water sample.

Answers

91.1 The lone pairs are not to be found in molecular orbital theory. This just goes to show that we should not mistake the results of mathematical theories as a direct picture of 'reality'.

91.2 (i), (ii) $H_2O(l) + HCl(g) \rightarrow H_3O^+(aq) + Cl^-(aq)$
  base
$H_2O(l) + NH_3(aq) \rightleftharpoons NH_4^+(aq) + OH^-(aq)$
  acid

(iii) In the first equation water is a proton acceptor, so using the Brønsted theory it is a base. For the opposite reason, water is a Brønsted acid in the second equation.

91.3 Their solubility is due to hydrogen bonding (Unit 21).

91.4 (i) $D_2O(l) + NH_3(aq) \rightleftharpoons NH_3D^+(aq) + OD^-(aq)$

(ii) As this is an equilibrium the NH_3D^+ ion can react with the OD^- like this:

$NH_3D^+(aq) + OD^-(aq) \rightarrow NH_2D(aq) + HDO(l)$

The NH_2D can be given off as a gas (just like NH_3). This gives rise to the peak at 18 and NH_3 to the peak at 17.

(iii) It is dynamic.

91.5 In ordinary water you might remember that an equilibrium is set up:

$2H_2O(l) \rightleftharpoons H_3O^+(aq) + OH^-(aq)$

Another way of writing this is

$H_2O(l) + H_2O(l) \rightleftharpoons H_3O^+(aq) + OH^-(aq)$

If one of the pair of water molecules is D_2O, we can have

$H_2O(l) + D_2O(l) \rightleftharpoons H_2DO^+(aq) + OH^-(aq)$

or

$H_2O(l) + D_2O(l) \rightleftharpoons D_2HO^+(aq) + OD^-(aq)$

If both molecules are D_2O we have

$2D_2O(l) \rightleftharpoons D_3O^+(aq) + OD^-(aq)$

An OD^- ion can combine with an H^+ ion, or an OH^- ion with a D^+ ion to give HDO molecules. Therefore we can have three different molecules and five different types of ion.

91.6

$2Na(s) + 2D_2O(l) \rightarrow 2Na^+(aq) + 2OD^-(aq) + D_2(g)$

91.7 (i) To convert 500×10^{12} tonne of liquid water to vapour requires

$500 \times 10^{12} \times 10^3 \, kg \times 2500 \, kJ \, kg^{-1} = 1.25 \times 10^{21} \, kJ$

This energy is absorbed and leads to a lowering of temperature.

(ii) $125 \times 10^{12} \times 10^3 \, kg \times 2500 \, kJ \, kg^{-1} = 0.31 \times 10^{21} \, kJ$

This energy is released, and thereby tends to increase the temperature of the atmosphere.

91.8 Sulphides can react with acid to give off very poisonous hydrogen sulphide gas, e.g.

$Na_2S(s) + 2H^+(aq) \rightarrow 2Na^+(aq) + H_2S(g)$

One of the properties of hydrogen sulphide is that it can be smelled in very small quantities; but once someone begins to be poisoned, he or she loses sensitivity to the smell. Accidents have occurred at disposal sites when work people have died as a result of such poisoning.

91.9 The soap solution can be put in a burette and a pipette used to measure out a known volume of each water sample into separate flasks. The soap solution is added in the standard method during a titration. However, instead of looking for a colour change in an indicator, the flask is shaken to see if a permanent lather is produced. Once the lather is achieved the volume of soap solution used is recorded. The harder the water, the more soap solution will be needed.

91.10 (i) There were

$\dfrac{14.4 \, cm^3}{1000 \, cm^3} \times 0.1 \, mol \, dm^{-3} = 14.4 \times 10^{-4} \, mol$

(ii) 14.4×10^{-4} mol in $25 \, cm^3$; the ratio is 1 EDTA anion to each ion.

(iii) We scale up from $25 \, cm^3$ to $1000 \, cm^3$, i.e. multiply by 40. The concentration is $57.6 \times 10^{-3} \, mol \, dm^{-3}$.

(iv) There are 28.8×10^{-3} mol of Ca^{2+} ions in $1 \, dm^3$. Given that 1 mol of Ca^{2+} ions comes from 1 mol of $CaCO_3$, the amount of $CaCO_3$ is also 28.8×10^{-3} mol, i.e. $28.8 \times 10^{-3} \, mol \times 100 \, g \, mol^{-1} = 2.88 \, g$.

(v) 10^6 tonne of water represents 10^9 kg, or $10^9 \, dm^3$. The mass of $CaCO_3$ leached is 2.88×10^9 g, or 2880 tonnes. This provides one more example of the huge scale upon which natural processes take place.

91.11 This will depend on your own locality.

UNIT 91 SUMMARY

- Water:
  (i) Essential for living systems (see Figure 91.3 for the water cycle).
  (ii) A good solvent for ionic and covalent compounds.
  (iii) A good ligand with transition metal ions.
- Hydrogen bonding is responsible for the high melt ing and boiling points of water.

- Hard water
  (i) Contains magnesium ions, Mg^{2+}, and calcium ions, Ca^{2+}, which give precipitates (scum) with soap, and deposits of insoluble magnesium and calcium carbonates after hard water is boiled.
  (ii) Hardness in water can be removed by passing water through an ion exchange column.

- The Winkler method is used to estimate the proportion of dissolved oxygen in water.

Reactions

- Water autoionises:

$$2H_2O(l) \rightarrow H_3O^+(aq) + OH^-(aq)$$

- With reactive metals:
  Hydrogen released, alkaline solution left;

  e.g. $2K(s) + 2H_2O(l) \rightarrow$
  $$2K^+(aq) + 2OH^-(aq) + H_2(g)$$

- With halogens:
  Acidic solutions made;

  e.g. $Cl_2(g) + H_2O(l) \rightarrow$
  $$HOCl(aq) + H^+(aq) + Cl^-(aq)$$

- Hydrolysis reactions:

  e.g. $HCl(g) + H_2O(l) \rightarrow H_3O^+(aq) + Cl^-(aq)$
  $NH_3(g) + H_2O(l) \rightarrow NH_4^+(aq) + OH^-(aq)$
  $CO_3^{2-}(aq) + H_2O(l) \rightarrow$
  $$HCO_3^-(aq) + OH^-(aq)$$

92

Group I

92.1 The nature of the elements

The elements of Group I are all metals. Indeed, from a chemist's point of view they make an excellent set because they have a large number of properties in common. Lithium is the only member of the Group that is not completely typical. Table 92.1 provides you with some of the physical data about the elements and Table 92.2 gives their uses.

The negative values of their standard redox potentials should tell you that they are all good reducing agents. Alternatively, we can say that they are all highly electropositive metals. Indeed, the tendency for them to lose their outermost electron and change into a positive ion is the most important feature of their chemistry. The main reason why they do this is that the outer s electron is very well shielded by the inner electrons. The s electron feels only a fraction of the nuclear charge. As we go down the Group, shielding wins over the effect of the increasing numbers of protons in the nucleus. Caesium, for example, is a much more powerful reducing agent than sodium. The metals are so reactive that in Nature they are always found combined with other elements. Especially, they exist as chlorides, nitrates, sulphates and carbonates.

The exception to the rule that the reducing power of

Table 92.2. Uses of the elements in Group I

Element*	Main uses
Lithium	In small, long-life batteries, e.g. for use in digital watches, calculators and computers In the reducing agents LiH and LiAlH$_4$ Specialist chemicals in a wide range of industries, e.g. making glass, organic chemicals
Sodium	Liquid sodium has been used for heat transfer in nuclear power stations As a reducing agent in the manufacture of some elements, e.g. titanium In alloys In batteries
Potassium	Manufacture of KO$_2$ for oxygen generators

*The other metals have but few uses

the elements increases down the Group appears to be lithium. It has a more negative electrode potential than sodium, potassium or rubidium. The reason for the anomaly lies in the nature of the lithium ion in water. The order of the ionisation energies agrees with our notion that the further you go down the Group, the more easily electrons are lost from the atoms. However,

Table 92.1. Physical properties of the elements in Group I*

Symbol	Lithium Li	Sodium Na	Potassium K	Rubidium Rb	Caesium Cs
Electron structure	(He)2s	(Ne)3s	(Ar)4s	(Kr)5s	(Xe)6s
Electronegativity	1.0	0.9	0.8	0.8	0.7
I.E./kJ mol^{-1}	520	513	419	400	380
Melting point/°C	181	98	63	39	29
Boiling point/°C	1331	890	766	701	685
Atomic radius/pm	123	157	203	216	235
Ionic radius/pm	68	98	133	148	167
Principal oxid. no.	+1	+1	+1	+1	+1
$E^{\ominus}_{M^+/M}$/V	−3.03	−2.71	−2.92	−2.93	−3.08

*The last element in the Group, francium, is omitted. It is not at all common, and its chemistry is of little importance

in solution, the product is a metal ion in water, surrounded by its hydration sphere. Table 92.1 shows that the lithium ion has by far the smallest ion. This results in it being a very dense centre of positive charge. It attracts and holds water molecules to it very strongly. Indeed, its hydration energy is huge: almost $-500\,kJ$ mol^{-1}. (This is about $110\,kJ\,mol^{-1}$ greater than the next largest, for the sodium ion.) The large hydration energy is responsible for the ease with which a lithium ion will be made in solution. The electrode potential of lithium reflects this tendency for lithium to convert into a hydrated ion, rather than its inherent ability as a reducing agent. If lithium takes part in reactions that do not involve water, then it does show less reducing power than the other members of the Group.

It is difficult to convert Group I metal ions into neutral atoms, so if we need to obtain the pure metal we have to use electrolysis. Sodium is by far the most widely used of the metals, and it is made by the electrolysis of sodium hydroxide in the Downs process. You will find details of the process in Unit 85. It was Humphry Davy who in 1807 first isolated pure potassium and sodium by using electrolysis.

The pure metals are silvery white and, apart from lithium, soft and easy to cut. However, they rapidly tarnish in air giving a layer of oxide, peroxide, or sometimes superoxide. They will also react violently with water. For both reasons they are kept under a layer of oil.

92.2   Reactions with oxygen

Lithium oxidises less rapidly than the other metals, but they all give ionic oxides and peroxides. In a plentiful supply of oxygen the reactions can be violent. (You will find more information about different types of oxide in Unit 99.) A typical reaction is:

$$2K(s) + O_2(g) \rightarrow K_2O_2(s)$$

As we should expect with metallic oxides, they are basic; indeed very strongly so. They dissolve in water to give strongly alkaline solutions containing hydroxide ions. For example,

$$Na_2O(s) + H_2O(l) \rightarrow 2Na^+(aq) + 2OH^-(aq)$$
$$2Na_2O_2(s) + 2H_2O(l) \rightarrow 4Na^+(aq) + 4OH^-(aq) + O_2(g)$$

One use of potassium superoxide, KO_2, is for generating oxygen. It has the ability to absorb carbon dioxide, while giving out oxygen at the same time:

$$4KO_2(s) + 2CO_2(g) \rightarrow 2K_2CO_3(s) + 3O_2(g)$$

This property has been made use of in breathing equipment, e.g. for mountaineers, in submarines and in spacecraft.

92.3   Reactions with water

Lithium, sodium and potassium all float on water. Lithium reacts only slowly, but sodium and potassium react more quickly. Hydrogen is given off and the solution remaining is alkaline. You may have seen an experiment in which sodium darts across the surface of water. If it sticks to the side of the container it may even burst into flame along with the hydrogen released. Potassium almost always ignites soon after being placed on water. It too rushes around over the surface. The reactions of rubidium and caesium with water are best not attempted. Explosions result! The reaction for sodium is:

$$2Na(s) + 2H_2O(l) \rightarrow 2Na^+(aq) + 2OH^-(aq) + H_2(g)$$

92.4   The hydroxides

The hydroxides of the Group I metals are among the strongest bases known. They exist as ionic solids and are very soluble in water; except, that is, for lithium hydroxide, which is slightly soluble. (The solubility of LiOH in water is about $130\,g\,dm^{-3}$, the others are greater than $400\,g\,dm^{-3}$.) Lithium hydroxide is also the only one that will convert to an oxide on heating.

Probably you will have used a solution of sodium hydroxide as a source of hydroxide ions. For example, in neutralising acids,

$$OH^-(aq) + H^+(aq) \rightarrow H_2O(l)$$

converting ammonium ions into ammonia,

$$OH^-(aq) + NH_4^+(aq) \rightarrow NH_3(g) + H_2O(l)$$

or precipitating insoluble hydroxides of metals such as iron,

$$Fe^{3+}(aq) + 3OH^-(aq) \rightarrow Fe(OH)_3(s)$$
iron(III) hydroxide

Another use for hydroxides, especially potassium hydroxide, is to absorb carbon dioxide. This happens in the cold, and a carbonate results:

$$2OH^-(aq) + CO_2(g) \rightarrow CO_3^{2-}(aq) + H_2O(l)$$

92.1 Write the equation for the action of heat on LiOH.

92.2 Which indicator would you use for titrating sodium hydroxide solution with (i) dilute hydrochloric acid, (ii) a solution of ethanoic acid?

92.3 Why should an alkali never (or almost never) be put into a burette?

92.5   The carbonates and hydrogencarbonates

The carbonates are all soluble in water, and their hydrogencarbonates exist as solids. The exception once again is lithium, which does not give a hydrogencarbonate. This pair of properties is different to the corresponding compounds of the Group II metals. For example, their carbonates are insoluble, and their hydrogencarbonates only exist in solution.

If you heat one of the carbonates you will not find a great deal happening: they do not decompose, except that is for lithium carbonate. On the other hand, the hydrogencarbonates do decompose, giving off carbon dioxide and water vapour, e.g.

$$2NaHCO_3(s) \rightarrow Na_2CO_3(s) + H_2O(g) + CO_2(g)$$

Sodium carbonate is a useful substance. It may be that you have some in your home. It is sold as washing soda crystals, $Na_2CO_3 \cdot 10H_2O$. In water it gives a slightly alkaline solution owing to salt hydrolysis (see Unit 76):

$$CO_3^{2-}(aq) + H_2O(l) \rightarrow HCO_3^-(aq) + OH^-(aq)$$

In industry sodium carbonate has a much more important role. It is one of the key participants in the chloralkali industry, where it is known as soda or soda ash. It is made in huge quantities by the Solvay process. You will find details in Unit 84. The overall change that takes place is

$$CaCO_3 + 2NaCl \rightarrow Na_2CO_3 + CaCl_2$$

but this hides a number of intermediate steps involving ammonia and carbon dioxide (among other things). The majority of sodium carbonate is used in glass making.

Both the carbonates and hydrogencarbonates react readily with acids, giving off carbon dioxide and water. For example,

$$CO_3^{2-}(s) + 2H^+(aq) \rightarrow CO_2(g) + H_2O(l)$$
$$HCO_3^-(s) + H^+(aq) \rightarrow CO_2(g) + H_2O(l)$$

The ease with which hydrogencarbonates give off carbon dioxide is made use of in fire extinguishers and baking powders. See Unit 95 for details.

92.4 (i) What would you expect to happen if you heated washing soda crystals?

(ii) Why are these crystals used in bath salts (together with a little colouring and perfume)?

92.6 The halides

All the metals give fluorides, chlorides, bromides and iodides. Apart from caesium they have the same crystal structure as sodium chloride: the rock salt structure (Figure 92.1). We have met this structure before in Unit 32. The positive metal ions and the negative halide ions each have a coordination number of 6. The structure of the caesium halides is different. Here the coordination number of the ions is 8. If you look back at Unit 32 you will find that we explained this in terms of the ratio of

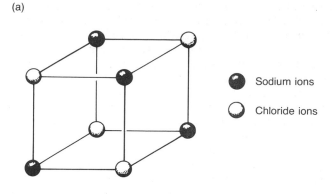

(a)

● Sodium ions

○ Chloride ions

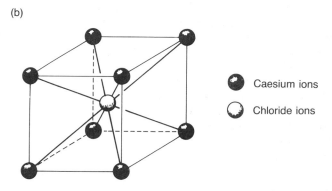

(b)

● Caesium ions

○ Chloride ions

Figure 92.1 (a) The sodium chloride (rock salt) structure and (b) the caesium chloride structure

the radii of the caesium and halide ions. In simple terms, the caesium ion is large enough for eight other ions to fit round it without bumping into one another.

There are many deposits of sodium chloride around the world. Often they are the result of the evaporation of seas and lakes. Rock salt mines are a source of sodium

An age-old method of salt manufacture is to allow salt water to be evaporated by the heat of the sun. This particular scene is near Lanzarote in the Canary Islands.

chloride, which is both an essential mineral for health and an important feedstock for the chemical industry.

No doubt you will know that sodium chloride is soluble in water, and this is typical of the other halides. However, lithium chloride is the only one which is markedly deliquescent, and lithium fluoride dissolves with difficulty (its solubility is only about $2\,\mathrm{g\,dm^{-3}}$). Explaining solubilities is not easy, but in the case of these halides the high energy of hydration of the ions is largely responsible for overcoming the lattice energies.

In the laboratory the halides are useful starting materials for making hydrogen halides, and their acidic solutions. Hydrogen fluoride and hydrogen chloride can be prepared by warming concentrated sulphuric acid with sodium fluoride or chloride. To make hydrogen bromide or iodide, the sodium salt must be reacted with an acid that has less oxidising power than sulphuric acid, usually phosphoric(v) acid:

$$NaCl(s) + H_2SO_4(l) \rightarrow NaHSO_4(s) + HCl(g)$$
$$NaI(s) + H_3PO_4(l) \rightarrow NaH_2PO_4(s) + HI(g)$$

Section 102.1 will give you more information about these reactions.

92.5 Large quantities of rock salt are used on roads in winter. Why?

92.6 Rock salt can either be mined in a conventional way, i.e. by digging it out of the ground, or by another method. What might be the other method?

92.7 The Solvay process is the name of an industrial process that makes use of large quantities of salt. What is made in the process?

92.8 (i) Why is it that lithium salts have a greater degree of covalent character than the other halides of the Group?

(ii) Which of LiCl, LiBr, LiI would you expect to have the most covalent character? (Hint: Fajans' rules.)

92.9 Why is it that although the lithium ion, Li^+, is far smaller than the other metal ions, it moves through a solution less rapidly than the others? (Hint: hydration spheres.)

92.10 Suggest a reason why lithium fluoride has the lowest solubility of the Group I metal halides.

92.7   The nitrates and nitrites

Sodium nitrate, $NaNO_3$, and sodium nitrite, $NaNO_2$, are the most important salts.

(a)   Nitrates

Sodium nitrate can be obtained from deposits in Chile. These deposits of 'Chile saltpetre' contain large amounts of iodates. The latter provide the world's main source of iodine. Sodium nitrate can also be made by neutralising nitric acid with sodium carbonate. The salt is valuable because it is used as a fertiliser. In common with all other nitrates, sodium nitrate is soluble in water, so the nitrate ions are readily available to plants. However, for the same reason much of the nitrate put on soil is washed away by rain. One consequence is that the over-use of nitrate fertilisers is responsible for the pollution of water supplies by nitrate ions.

Chemically, the Group I nitrates are a little different to those of other metals. In particular, when they are heated, they give off oxygen and change into a nitrite, e.g.

$$2KNO_3(s) \rightarrow 2KNO_2(s) + O_2(g)$$

Nitrates of the Group II metals and others, like copper, give off nitrogen dioxide and decompose into an oxide, e.g.

$$2Ca(NO_3)_2(s) \rightarrow 2CaO(s) + 4NO_2(g) + O_2(g)$$

Perhaps we might expect that lithium nitrate might be an exception, and so it is. It decomposes like the majority of nitrates of other Groups.

Most nitrates are energetically stable. However, the nitrogen in a nitrate ion is in a high oxidation state ($+5$) and the ions contain a high percentage of oxygen. With the right chemicals (or wrong, depending on your point of view), the ions will show a considerable ability to act as oxidising agents. Especially, potassium nitrate mixed with sulphur and carbon is used as gun powder. The mixture will explode when ignited by a spark. The explosion is due to the very speedy release of large volumes of nitrogen and carbon dioxide:

$$2KNO_3(s) + S(s) + 3C(s) \rightarrow K_2S(s) + N_2(g) + 3CO_2(g)$$

Sodium or potassium nitrate can be used to make nitric acid in the laboratory. Nitric acid is much more volatile than concentrated sulphuric acid, and we can make it by heating this acid with the nitrate:

$$NaNO_3(s) + H_2SO_4(l) \rightarrow NaHSO_4(s) + HNO_3(g)$$

The main difference between this and the preparation of hydrogen chloride is that the nitric acid fumes must be passed through a Liebig condenser in order to collect the liquid.

(b)   Nitrites

There are two large-scale uses of nitrites. Sodium nitrite is used in the manufacture of dyes. This is a result of its ability to take part in diazotisation reactions with aromatic amines. You will find details of these reactions in Unit 122. The other large-scale use is as an anti-oxidant in foodstuffs. Meat products such as paté, sausages, hamburgers and bacon, as well as raw meat sold in supermarkets, often has sodium nitrite (and sometimes sodium nitrate) added to it. Although the shelf life of the products is increased, it is known that nitrites can be converted into very dangerous carcinogenic compounds in acids. It is no comfort to realise that our stomachs are full of an acidic solution! There are good reasons for not eating too many meat products treated with nitrates and nitrites.

92.11 Write the equation for the decomposition of lithium nitrate, $LiNO_3$.

92.12 Estimate the change in volume when 2 mol of KNO_3 reacts with 1 mol of S and 3 mol of C.

92.8 The sulphates, hydrogen sulphates and sulphites

All the members of the Group give sulphates, SO_4^{2-}, and hydrogensulphates, HSO_4^-. They are all soluble in water. This fact, like the solubility of the carbonates, is a marked difference to the sulphates of Group II.

Sulphites, such as sodium sulphite, Na_2SO_3, are more reactive than either sulphates or hydrogensulphates. For example, if you warm a sulphite with an acid, you will find sulphur dioxide is given off:

$$SO_3^{2-}(aq) + 2H^+(aq) \rightarrow SO_2(g) + H_2O(l)$$

Another reaction that you may come across is the conversion of sodium sulphite into sodium thiosulphate, $Na_2S_2O_3$. This change is performed by boiling a solution of sodium sulphite with powdered sulphur:

$$SO_3^{2-}(aq) + S(s) \rightarrow S_2O_3^{2-}(aq)$$

Sodium thiosulphate is used as 'hypo' in photography. In the laboratory it is used in iodine titrations (see section 40.3). The reaction with iodine is

$$I_2(s) + 2S_2O_3^{2-}(aq) \rightarrow S_4O_6^{2-}(aq) + 2I^-(aq)$$

92.13 How might you show that sulphur dioxide is given off in a reaction?

92.9 The hydrides

All the hydrides of the Group are ionic, with the metal being positive and the hydrogen being negative. As you may know, it is most unusual for hydrogen to make an ion other than the positively charged hydrogen ion, H^+. Indeed, hydride ions are very reactive. In water they immediately react, giving off hydrogen:

$$Li^+H^-(s) + H_2O(l) \rightarrow Li^+(aq) + OH^-(aq) + H_2(g)$$

In organic chemistry, lithium tetrahydridoaluminate(III) (otherwise known as lithium aluminium hydride), $LiAlH_4$, is a very useful reducing agent. It can be made by reacting lithium hydride with aluminium trichloride, and contains Li^+ and AlH_4^- ions. It is the latter ions that supply the hydride ions which perform the reduction. You will find examples of its reactions in the units on organic chemistry, and in Unit 90.

For example, it will reduce amides to amines, $RCONH_2 \rightarrow RCH_2NH_2$. Also, see section 90.7.

92.14 (i) How many electrons does a hydride ion possess?

(ii) What is its electron structure?

(iii) What does this tell you about the statement that all noble gas structures are very stable?

92.15 What is special about aluminium trichloride that allows it to bond with a hydride ion?

Answers

92.1 $2LiOH(s) \rightarrow Li_2O(s) + H_2O(l)$

92.2 Hydrochloric acid is a strong acid. A good strong acid/strong base indicator is screened methyl orange. For ethanoic acid (a weak acid) and a strong base, phenolphthalein is best.

92.3 It gives a deposit of a carbonate with carbon dioxide in the air, which can jam the tap.

92.4 (i) The crystals dissolve in their own water of crystallisation and then steam is given off.

(ii) The carbonate ions give a deposit of $CaCO_3$ with Ca^{2+} ions often found in hard water.

92.5 They lower the freezing point of water, thus preventing ice from forming (see section 65.5).

92.6 Water passed into the deposits will dissolve the salt. After pumping out, the salt is crystallised out.

92.7 Sodium carbonate.

92.8 (i) The small size of Li^+ gives it a huge polarising power. This leads to covalency.

(ii) LiI. The iodide is larger and more polarisable (see Fajans' rules in Unit 19).

92.9 The dense charge of Li^+ attracts several layers of water molecules around it. They increase the effective size of the ion, thus slowing it down. See section 73.4.

92.10 The small size of both the Li^+ and F^- ions leads to a very large lattice energy, which means that the crystal is very hard to break apart.

92.11 $4LiNO_3(s) \rightarrow 2Li_2O(s) + 4NO_2(g) + O_2(g)$

92.12 The volume of the solids will be some tens of cm^3. Four moles of gas at room temperature have a volume of $96\,dm^3$. Hence the expansion is enormous.

92.13 Sulphur dioxide is a reducing agent. A standard test is to show that it turns orange acidified potassium dichromate(VI) solution a green colour; or acidified potassium manganate(VII) from purple to colourless.

92.14 (i) Two. (ii) $1s^2$. (iii) It is wrong.

92.15 $AlCl_3$ is electron deficient. It can accept a pair of electrons into one of its empty orbitals. See Unit 15.

- The Group I metals:
  - (i) Are all good reducing agents (highly electropositive), with reducing power increasing down the Group. (Lithium is an exception, owing to the very small size of the Li^+ ion.)
  - (ii) Make ionic compounds with non-metals.
- Manufacture:
  Isolated by electrolysis of molten salts.

Reactions

- With water:
  Vigorous reaction, hydrogen given off, alkaline solution remains;

  e.g. $2Na(s) + 2H_2O(l) \rightarrow$
  $$2Na^+(aq) + 2OH^-(aq) + H_2(g)$$

- With oxygen:
  Peroxides made;

  e.g. $2K(s) + O_2(g) \rightarrow K_2O_2(s)$

- With halogens:
  Ionic salts made;

  e.g. $2Na(s) + Cl_2(g) \rightarrow 2Na^+Cl^-(s)$

Compounds

- Oxides:
  Basic, dissolve in water to make strong alkalis.
- Hydroxides:
  - (i) Strong alkalis in water; give ammonia with ammonium salts;

    e.g. $NH_4^+(aq) + OH^-(aq) \rightarrow NH_3(g) + H_2O(l)$

  - (ii) Give precipitates with some metal ions;

    e.g. $Fe^{3+}(aq) + 3OH^-(aq) \rightarrow Fe(OH)_3(s)$

- Hydrides:
  Are ionic, give hydrogen with water;

  e.g. $Li^+H^-(s) + H_2O(l) \rightarrow$
  $$Li^+(aq) + OH^-(aq) + H_2(g)$$

- Carbonates:
  - (i) Soluble in water.
  - (ii) Give carbon dioxide with acids;

    e.g. $Na_2CO_3(s) + 2H^+(aq) \rightarrow$
    $$2Na^+(aq) + CO_2(g) + H_2O(l)$$

  - (iii) Do not give carbon dioxide when heated.
- Halides:
  React with concentrated sulphuric acid, e.g. a method of making $HCl(g)$

  $NaCl(s) + H_2SO_4(l) \rightarrow NaHSO_4(s) + HCl(g)$

- Nitrates:
  - (i) Give oxygen but not nitrogen dioxide when heated;

    e.g. $2KNO_3(s) \rightarrow 2KNO_2(s) + O_2(g)$

  - (ii) Can be used to make nitric acid using concentrated sulphuric acid;

    $NaNO_3(s) + H_2SO_4(l) \rightarrow NaHSO_4(s) + HNO_3(l)$

- Sulphates:
  All soluble in water.
- Sulphites:
  Give sulphur dioxide when warmed with acids;

  e.g. $Na_2SO_3(s) + 2H^+(aq) \rightarrow$
  $$2Na^+(aq) + SO_2(g) + H_2O(l)$$

- Thiosulphates:
  - (i) Sodium thiosulphate can be made by boiling a solution of sodium sulphite, Na_2SO_3, with sulphur;

    $SO_3^{2-}(aq) + S(s) \rightarrow S_2O_3^{2-}(aq)$

  - (ii) Thiosulphates deposit sulphur with acids (a common experiment in rates of reactions).
  - (iii) Thiosulphate solutions are used in iodine titrations (iodine decolourised);

    $2S_2O_3^{2-}(aq) + I_2(aq) \rightarrow S_4O_6^{2-}(aq) + 2I^-(aq)$

93

Group II

93.1 The nature of the elements

The elements of Group II are metals. They show the properties we would expect, e.g. they are good reducing agents, they give ionic compounds, their oxides and hydroxides are basic, and they give hydrogen with acids. The alkaline nature of the elements is responsible for them being known as the alkaline earth metals. Their properties and uses are summarised in Tables 93.1 and 93.2. The exception to the common pattern is the first member, beryllium. One reason why beryllium is different is that its electrons are not strongly shielded from its nucleus. The radius of the Be^{2+} ion is extremely small, and it represents a very dense centre of positive charge. Fajans' rules remind us that such an ion would have an immense polarising power. This ability to draw electrons towards itself is responsible for the covalency of many of its compounds. For example, we saw in Unit 17 that beryllium chloride, $BeCl_2$, is a covalent, linear molecule. Beryllium also has a higher electronegativity than the other elements. This tells us that the compounds it makes with non-metals should have less ionic character. Another feature of the chemistry of beryllium is that in solution its compounds tend to suffer from hydrolysis, and some are amphoteric rather than completely basic.

Table 93.2. Uses of the elements in Group II

Element	Main uses
Beryllium	As a moderator in nuclear reactors
Magnesium	In alloys of many kinds—it lends strength with little increase in weight; hence its use in aircraft structures
Calcium	In biological systems it is essential for the healthy growth of bones Its carbonate is used in manufacturing cement, and in the alkali industry
Strontium	Few uses, although its radioactive isotope $^{90}_{38}Sr$ is well known (and feared) because it is produced in nuclear fall-out
Barium	In some alloys with lead and calcium $BaSO_4$ is used in medicine as 'barium meal', which patients swallow—the sulphate is relatively opaque to X-rays so it shows particularly well on X-ray photographs

For example, magnesium is often used in the laboratory to liberate small quantities of hydrogen:

$$Mg(s) + 2H^+(aq) \rightarrow Mg^{2+}(aq) + H_2(g)$$

Table 93.1. Physical properties of the elements in Group II*

Symbol	Beryllium Be	Magnesium Mg	Calcium Ca	Strontium Sr	Barium Ba
Electron structure	$(He)2s^2$	$(Ne)3s^2$	$(Ar)4s^2$	$(Kr)5s^2$	$(Xe)6s^2$
Electronegativity	1.5	1.2	1.0	1.0	0.9
1st I.E./kJ mol^{-1}	899	738	590	550	500
2nd I.E./kJ mol^{-1}	1800	1500	1100	1100	1000
Melting point/°C	1283	650	850	770	710
Boiling point/°C	2477	1117	1492	1367	1637
Atomic radius/pm	106	140	174	191	198
Ionic radius/pm	30	65	94	110	134
Principal oxid. no.	+2	+2	+2	+2	+2
$E^\ominus_{M^{2+}/M}$/V	−1.85	−2.37	−2.87	−2.89	−2.90

*The last element in the Group, radium, is omitted. It is not common, but is famous for its radioactive nature and its discovery by Mme Curie

Even if it is coated with a thin layer of oxide, the metal will still react because the oxide dissolves in the acid. If beryllium is coated with oxide it will not react at all. When it is pure, reaction does take place, especially if it is finely powdered. Hydrogen is given off, but the Be^{2+} ion is heavily hydrated and exists as $Be(H_2O)_4^{2+}$. Here the water molecules act as ligands by bonding to the ion through one of their lone pairs. It is characteristic of beryllium that it gives complexes of the kind we would normally associate with transition metals. The other metals sometimes give complexes, but much less readily than beryllium.

Beryllium will dissolve in alkali. This is something that magnesium and the other metals in the Group will not do:

$$Be(s) + 2OH^-(aq) \rightarrow BeO_2^{2-}(aq) + H_2(g)$$

The product, BeO_2^{2-}, is the beryllate ion, which is better represented in solution as the tetrahydroxoberyllate(II) ion, $Be(OH)_4^{2-}$. However, several other types of ion are usually present as well. In its amphoteric behaviour it resembles aluminium in Group III.

Like the Group I metals, the reactivity of the elements makes it difficult to extract them by chemical means. Magnesium is the most important member of the Group and it is extracted by the electrolysis of magnesium chloride, $MgCl_2$.

Beryllium is hard to extract. Partly this is because its minerals are not widely distributed in Nature. The most common method is to convert minerals such as beryl, $Be_3Al_2Si_6O_{18}$, into beryllium chloride and reduce the chloride with magnesium metal.

Warning: Do not attempt to perform reactions with beryllium compounds. They are intensely poisonous. One way in which they poison is by blocking the reactivity of enzymes and other biologically active systems, e.g. by taking the place of magnesium ions.

93.1 Explain why the ionisation potentials decrease going down the Group.

93.2 Beryllium gives a compound with the following percentage composition: Be, 6.1%; N, 37.8%; Cl, 48%; H, 8.1%. One mole of the compound had a mass of 148 g. $M(Be) = 9\,g\,mol^{-1}$.

(i) What is the molecular formula of the compound?

(ii) In water, 1 mol of the compound reacts with 2 mol of silver ions. Suggest a structural formula for the compound, and explain how the atoms are arranged.

93.3 Do you know of a biologically important molecule in which magnesium atoms are held in position by organic groups?

93.4 A student suggested that calcium should be made if calcium oxide is reacted with aluminium powder. Was the student correct?

You will need to use the following free energies of formation: $\Delta G_f^\circ(CaO) = -604.2$ kJ mol^{-1}; $\Delta G_f^\circ(Al_2O_3) = -1582.4$ kJ mol^{-1}.

93.2 The oxides and hydroxides

Beryllium oxide, BeO, is more like the oxide of aluminium in Group III rather than the oxides of the other elements in Group II. It has a high degree of covalency, which is lacking in the other oxides. It is insoluble in water and it will dissolve only with great difficulty in acids. The reactivity of BeO depends on its treatment. If it is heated to a high temperature (about 800° C) it becomes almost completely inert.

The other oxides will dissolve in water with increasing ease down the Group. The resulting solutions are slightly alkaline owing to reactions between the oxides and water, e.g.

$$MgO(s) + H_2O(l) \rightleftharpoons Mg^{2+}(aq) + 2OH^-(aq)$$
$$CaO(s) + H_2O(l) \rightleftharpoons Ca^{2+}(aq) + 2OH^-(aq)$$

We have to be a little careful here. You may recognise a solution of calcium and hydroxide ions as 'lime water'. Lime water is famous for giving a milky precipitate with carbon dioxide. It is normally treated with much less care than the caustic alkalis of Group I metals such as sodium hydroxide. The reason is not that the hydroxide ions it contains are any the less reactive than those in sodium hydroxide. Rather, there are far fewer of them in solution. Calcium hydroxide, like the other hydroxides in the Group, is only partially soluble in water. You might like to compare the solubilities in Table 93.3. The values tell us that solubility increases down the Group. We can understand this trend if we remember one of the most important factors in explaining trends in solubility: high solubility often correlates with low lattice energy and vice versa. Also, we know that a high lattice energy is given by ionic substances that contain small highly charged ions, and covalent substances with giant molecular lattices. As we go down the Group the size of the metal ion increases. We can claim that this is the chief reason for the lattice energy decreasing, and the solubility increasing. However, a word of warning: if you read Unit 60 you will find that we have to take entropy changes into account if we are to give a thorough explanation of solubilities.

Table 93.3. The solubilities of the Group II hydroxides

Hydroxide	Solubility at 25° C/mol per 100 g water
$Be(OH)_2$	Highly insoluble
$Mg(OH)_2$	2.0×10^{-5}
$Ca(OH)_2$	1.5×10^{-3}
$Sr(OH)_2$	3.4×10^{-3}
$Ba(OH)_2$	1.5×10^{-2}

93.5 Write the equation for the reaction between carbon dioxide and calcium hydroxide solution.

93.6 Apart from lattice energy, what other energy change is important in accounting for solubilities?

93.7 0.2 g of magnesium ribbon was placed in a crucible and heated with the lid on until the magnesium began to burn fiercely. At the end of the experiment there was 0.3 g of a white powder left. Show that this result does *not* agree with the equation

$$2Mg(s) + O_2(g) \rightarrow 2MgO(s)$$

What might have gone wrong?

93.8 Suggest other ways of writing the formulae $BeCl_2 \cdot 4H_2O$ and $MgCl_2 \cdot 6H_2O$.

93.9 Beryllium makes the complex ion BeF_4^{2-}. Predict the shape of this ion, and suggest the types of orbital used in bonding by the beryllium atom.

93.3   The halides

The chlorides are much the most common of the halides, so we shall concentrate on them.

We have discussed the bonding in beryllium chloride before in Units 15 and 17. The important things to know about this chloride is that isolated molecules are linear, while the solid is composed of chains of linked molecules (Figure 93.1). The bonding is very much like that in aluminium trichloride. Lone pairs on some of the chlorine atoms are used in bonding with empty orbitals on the beryllium atoms. The chlorides of the other metals have a greater degree of ionic character. Unlike the Group I chlorides, they easily react with water to give hydrates (Table 93.4). Anhydrous calcium chloride is deliquescent and widely used as a drying agent.

The elements all give fluorides, bromides and iodides as well as chlorides. They are all soluble in water, but the fluorides are much less soluble than the others; e.g. the solubility of magnesium fluoride is little more than 10^{-4} mol per 100 g of water at 18° C. We shall seek an explanation of this in section 93.6.

Cl—Be—Cl

Figure 93.1 *The structures of single BeCl₂ molecules and the chains that exist in solid BeCl₂*

Table 93.4. The chlorides of Group II

Chloride		Solubility at 18°C*
Anhydrous	Hydrated	/mol per 100 g water
BeCl₂	BeCl₂·4H₂O	0.90
MgCl₂	MgCl₂·6H₂O	0.59
CaCl₂	CaCl₂·6H₂O	0.66
SrCl₂	SrCl₂·6H₂O	0.32
BaCl₂	BaCl₂·2H₂O	0.18

*The figures give the solubilities of the anhydrous salts, apart from beryllium chloride

93.4   The carbonates and hydrogencarbonates

The Group II carbonates are different to those of the alkali metals of Group I in two major respects. First, they are only very slightly soluble in water, with the solubility decreasing down the Group. Secondly, they are decomposed by heat, giving off carbon dioxide and leaving an oxide. For example,

$$MgCO_3(s) \rightarrow MgO(s) + CO_2(g)$$

The ease of decomposition decreases down the Group. These properties are summarised in Table 93.5.

Table 93.5. Solubility and decomposition temperature of the carbonates*

Carbonate	Solubility at 25° C /mol per 100 g water	Decomposition temperature/° C
MgCO₃	1.5×10^{-4}	400
CaCO₃	1.3×10^{-5}	900
SrCO₃	7.4×10^{-6}	1280
BaCO₃	9.1×10^{-6}	1360

*Data for BeCO₃ are not available

Calcium carbonate in the form of chalk or limestone is particularly important, both in Nature and in industry. You can find information about them in Unit 84.

Chalk is one Group II carbonate (CaCO₃) that is widely found in nature.

Concrete is one of the world's most common building materials. Unfortunately it is rare for concrete to be used in such an attractive way as in the Law Courts building in Liverpool shown here.

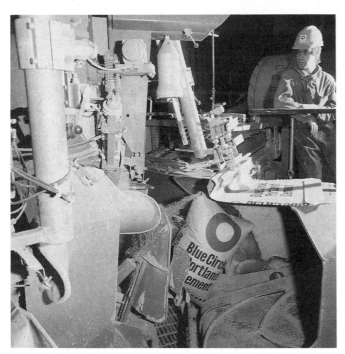

Cement powder being loaded into bags.

One way of explaining the trend in decomposition is to relate it to the size of the metal ion. The smaller the ion, the greater is its polarising power. This suggests that a beryllium ion would have the greatest ability to distort carbonate ions, pulling electron density from the oxygen atoms towards it. Likewise, barium ions would have the least tendency to distort carbonate ions. In addition, if we look at the products of the reaction, we know that carbon dioxide is given off in each case. This is a common factor, so it is unlikely to give us help in explaining the decompositions. However, the metal

oxide is different for each reaction. Given that a large lattice energy is encouraged by small highly charged ions close together, we should not be surprised to find that the lattice energies of the oxides follow the order

$$BeO > MgO > CaO > SrO > BaO$$

largest lattice energy   smallest lattice energy

Therefore, with $BeCO_3$ we have carbonate ions already highly distorted, and a product of the reaction that is energetically very favoured. Both factors encourage decomposition of the carbonate. At the other extreme, $BaCO_3$ has carbonate ions that are not highly polarised, and an oxide product that is not so energetically favoured. The other carbonates fall nicely in order between the two extremes. You will find an explanation of the trend in solubilities in section 93.6.

Finally, another observation that marks out the elements of this Group from the alkali metals of Group I is that those metals possess solid hydrogencarbonates. The hydrogencarbonates of the alkaline earth metals only exist in solution.

93.10   Is BeO ionic?

93.11   Strictly, if we are to explain the decomposition of the carbonates we should refer to the free energies of the reactions. Why, in these reactions, will we not go far wrong in concentrating on enthalpy changes rather than free energies?

93.12   What do you understand by the term *dissociation pressure* of calcium carbonate?

93.13   What is the importance of calcium and magnesium carbonates and hydrogencarbonates in making water hard or soft?

93.14   A student mixed two colourless solutions. A white precipitate was produced, and the student said she thought it might be a carbonate. How could you find out if she was right?

93.5 The sulphates

The solubilities of the sulphates decrease down the Group (Table 93.6). We discussed the reasons for this trend in Unit 60. You will also find a brief comment about it in the following section.

Beryllium, magnesium and calcium sulphates are often found as hydrated crystals, e.g. $BeSO_4 \cdot 4H_2O$, $MgSO_4 \cdot 7H_2O$, $CaSO_4 \cdot 2H_2O$. The crystals of magnesium sulphate have a rather unfortunate reputation. They are better known as Epsom salts, and at one time they were widely used as a laxative in medicine. Crystals of $CaSO_4 \cdot 2H_2O$ are found in Nature as the mineral gyp-

Table 93.6. The solubilities of Group II sulphates

Sulphate	Solubility at 25° C/mol per 100 g water
$BeSO_4$	3.79×10^{-1}
$MgSO_4$	1.83×10^{-1}
$CaSO_4$	4.66×10^{-3}
$SrSO_4$	7.11×10^{-5}
$BaSO_4$	9.43×10^{-7}

sum. Anhydrous calcium sulphate also occurs naturally as anhydrite.

If you have had a broken bone set in plaster you can thank gypsum for your recovery. When gypsum is heated to just under 100° C it loses three quarters of its water of crystallisation. The powder remaining is plaster of Paris, which can be represented by the formula $(CaSO_4)_2 \cdot H_2O$. When water is mixed with it, gypsum crystals are produced again, but this time they set to a hard, solid mass. In surgery the dehydrated gypsum is often embedded in bandages which are wrapped around the damaged limb. The speed with which the plaster sets can be controlled by adding other chemicals to it.

We can make use of the insolubility of barium sulphate in the test for a sulphate. In its simplest form, the test is to add barium chloride solution to the suspected sulphate. If sulphate ions are present, a white precipitate of barium sulphate is made:

$$Ba^{2+}(aq) + SO_4^{2-}(aq) \rightarrow BaSO_4(s)$$

However, this way of doing the test is not to be trusted because, for example, if carbonate ions are present there will also be a precipitate (of barium carbonate). To

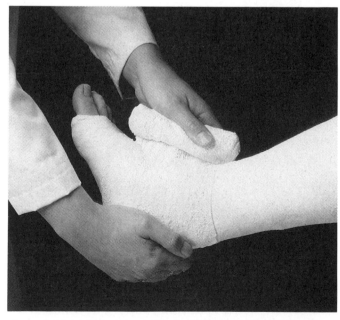

A common sight in hospital casualty departments: bandages impregnated with plaster of Paris are used to set broken bones.

avoid such confusion it is best to add dilute hydrochloric acid to the solution before adding barium chloride.

93.15 Barium phosphate is also insoluble in water. However, a solution of a phosphate will not give a precipitate with barium chloride solution provided dilute acid is added first. What might be the explanation for this? (Hint: weak acids.)

93.16 The solubility products of barium sulphate and barium carbonate are 1×10^{-10} mol² dm⁻⁶ and 5.5×10^{-10} mol² dm⁻⁶ respectively. What would happen, if anything, if (i) solid barium carbonate were boiled with a solution containing sulphate ions, (ii) solid barium sulphate were boiled with a solution containing carbonate ions?

93.6 Explaining trends in solubilities

You should read Unit 60 for a more detailed explanation of trends in solubilities. If you have already read that unit you should know that many of the trends can be explained in terms of a competition between two opposing tendencies:

(i)   The higher the lattice energy, the harder it is to break a crystal apart, so the less likely it is to dissolve.

(ii) The higher the hydration energy of the ions, the greater is the tendency for the crystal to dissolve.

In the case of the relative insolubilities of the fluorides of the Group II elements compared to the other halides, we can claim that this is a result of the higher lattice energies of the fluorides, compared to the chlorides or other halides. Fortunately the claim is backed up by experimental evidence: the lattice energies of the fluorides are between 300 and 600 kJ mol⁻¹ greater than the other halides.

The *decreasing* solubility of the carbonates down the Group (Table 93.7) is not explained by the trend in

Table 93.7. Trends in solubilities in Group II

	Chlorides	Carbonates	Sulphates	Hydroxides
Be	Most soluble	Most soluble	Most soluble	Least soluble
Mg				
Ca	↑	↑	↑	
Sr				↓
Ba	Least soluble	Least soluble	Least soluble	Most soluble

lattice energies, which also decrease down the Group. Rather the trend follows the order of the hydration energies of the metal ions (Be^{2+} having the greatest and Ba^{2+} the least). However, this is another occasion on which we have to take great care. You may have seen in Unit 60 that it is not really respectable to explain solubilities without taking account of entropy changes. Fortunately we do not have to track down all the entropy changes that take place when small portions of these carbonates dissolve, but we can be quite sure that they play a significant part in determining the trend down the Group.

Answers

93.1 The outer electrons are increasingly well shielded from the nucleus by the inner shells of electrons.

93.2 (i) If you follow through the method of Unit 37 you should find that the molecular formula is $BeN_4H_{12}Cl_2$.

(ii) The reaction with silver ions suggests that the 2 mol of chlorine is present as chloride ions. We know that nitrogen and hydrogen are often to be found together in ammonia, so we can guess that the formula is $Be(NH_3)_4Cl_2$, or $[Be(NH_3)_4]^{2+}(Cl^-)_2$. The four ammonia molecules act as ligands. Compare this with some of the complexes in Unit 105.

93.3 Chlorophyll is the compound.

93.4 The equation is

$$3CaO(s) + 2Al(s) \rightarrow 3Ca(s) + Al_2O_3(s)$$
$$\Delta G^{\ominus}(\text{reaction}) = \Delta G_f^{\ominus}(Al_2O_3) - 3\Delta G_f^{\ominus}(CaO)$$
$$= +230.2\, kJ\, mol^{-1}$$

The positive sign tells us that the reaction cannot occur under standard conditions. However, the student would be pleased to know that under non-standard conditions, e.g. at high temperatures, the free energy change is negative and the reaction takes place. This is the method used to extract calcium in industry.

93.5 $CO_2(g) + Ca^{2+}(aq) + 2OH^-(aq) \rightarrow$
$$CaCO_3(s) + H_2O(l)$$

93.6 The solvation energies of the ions.

93.7 The equation shows that 2 mol, i.e. 48 g, of magnesium should give 2 mol, i.e. 80 g, of magnesium oxide. Therefore, 0.2 g of magnesium should give $80\, g \times 0.2\, g/48\, g = 0.33\, g$ of the oxide. Perhaps some of the oxide escaped as smoke, or the magnesium did not completely react, or the magnesium may have made magnesium nitride instead.

93.8 The clue to this is realising that the water molecules can act as ligands: $[Be(H_2O)_4]^{2+}(Cl^-)_2$, $[Mg(H_2O)_6]^{2+}(Cl^-)_2$.

93.9 The electron structure of Be^{2+} is $1s^2$. The ion can make four bonds if it makes use of its empty 2s and 2p orbitals. We can think of them as four sp^3 hybrids. As there are no lone pairs, and only four bond pairs to the four fluorine ions, electron repulsion theory predicts that the BeF_4^{2-} ion should be tetrahedral in shape. It is.

93.10 No. But covalent, as well as ionic, substances have lattice energies.

93.11 One mole of each carbonate gives off 1 mol of gas. Thus the entropy changes for each of the reactions will be approximately the same. Therefore, changes in free energies between the reactions will be governed by the enthalpy changes.

93.12 Look back to section 51.3 for this.

93.13 When slightly acidic rain water percolates through carbonate rocks, some of the rock dissolves. Ca^{2+} and Mg^{2+} ions are dissolved in the water. It is these ions which give scum with soap, or make scale when hard water is boiled.

93.14 Add dilute nitric acid. If it is a carbonate, carbon dioxide would be given off.

93.15 When acid is added to a solution containing phosphate ions, the equilibrium

$$3H^+(aq) + PO_4^{3-}(aq) \rightleftharpoons H_3PO_4(aq)$$

is driven to the right. This reduces the number of free PO_4^{3-} ions that are available to react with the Ba^{2+} ions.

93.16 The values of the solubility products show that $BaSO_4$ is *less soluble* than $BaCO_3$. (The smaller the solubility product, the less soluble is the solid.) Therefore, in (i) there will be no change; but in (ii) over a period of time the free barium ions in the solution will be converted into solid barium carbonate, i.e. the overall change will be

$$BaSO_4(s) + CO_3^{2-}(aq) \rightarrow BaCO_3(s) + SO_4^{2-}(aq)$$

UNIT 93 SUMMARY

- The Group II metals:
    (i) Are all good reducing agents.
    (ii) Make ionic compounds.
    (iii) Make oxides and hydroxides that are less soluble in water than those of Group I.
    (iv) Give hydrogen with water and acids.

    (v) Beryllium is an exception in the Group. Especially, owing to the small size of its ion, Be^{2+} (high polarising power), its compounds are covalent.
- Manufacture:
    Isolated by electrolysis of molten salts.

- Solubilities:
  - (i) Of chlorides, sulphates and carbonates decrease down the Group.
  - (ii) Of hydroxides increase down the Group.

Reactions

- With water:

  Less vigorous reaction than Group I, hydrogen given off, alkaline solution remains;

  e.g. $Ca(s) + 2H_2O(l) \rightarrow Ca(OH)_2(s) + H_2(g)$

  (Magnesium only reacts well with steam.)

- With oxygen:

  Oxides made;

  e.g. $2Mg(s) + O_2(g) \rightarrow 2MgO(s)$

- With halogens:

  Ionic salts made. Beryllium an exception; $BeCl_2$ is covalent.

Compounds

- Oxides:

  Basic, partially soluble in water giving alkaline solutions.

- Hydroxides:

  Calcium hydroxide solution, i.e. lime water, used to test for carbon dioxide ('lime water goes milky');

  $Ca(OH)_2(aq) + CO_2(g) \rightarrow CaCO_3(s) + H_2O(l)$

- Carbonates:

  Only partially soluble in water, give carbon dioxide with acids and when heated;

  e.g. $CaCO_3 + 2H^+(aq) \rightarrow$
  $$2Ca^{2+}(aq) + CO_2(g) + H_2O(l)$$
  $$CaCO_3(s) \rightarrow CaO(s) + CO_2(g)$$

- Halides:
  - (i) Are often hydrated, e.g. $CaCl_2 \cdot 6H_2O$. Anhydrous calcium chloride is used as a drying agent; it is deliquescent.
  - (ii) Barium chloride solution is used to test for sulphates, gives a white precipitate of barium sulphate;

    $Ba^{2+}(aq) + SO_4^{2-}(aq) \rightarrow BaSO_4(s)$

- Nitrates:

  Give oxygen and nitrogen dioxide when heated;

  e.g. $2Ba(NO_3)_2(s) \rightarrow 2BaO(s) + 4NO_2(g) + O_2(g)$

- Sulphates:
  - (i) Are not as soluble as Group I sulphates.
  - (ii) Solubility decreases down the Group.
  - (iii) Gypsum, $CaSO_4 \cdot 2H_2O(l)$ used to make plaster of Paris; $(CaSO_4)_2 \cdot H_2O$.

94
Group III

94.1 The nature of the elements

This Group marks the beginning of the p-block elements. In the following Groups the elements show definite non-metallic nature. For example, carbon, nitrogen and oxygen all tend to gain electrons rather than lose them. (It is the metals that give up electrons to make positive ions.) The compounds of boron are mainly covalent, and it clearly shows non-metallic properties. Going down the Group to the next period we reach aluminium, which is a metal. However, both boron and aluminium are amphoteric. That is, the elements, and their oxides, will react with both acids and alkalis. A summary of their properties and uses is shown in Tables 94.1 and 94.2.

Examples of their amphoteric nature are the reactions of boron and aluminium with alkali. These reactions are often written:

$$2B(s) + 2OH^-(aq) + 2H_2O \rightarrow 2BO_2^-(aq) + 3H_2(g)$$

metaborate
ions

$$2Al(s) + 2OH^-(aq) + 2H_2O \rightarrow 2AlO_2^-(aq) + 3H_2(g)$$

aluminate
ions

Table 94.2. Uses of the elements in Group III

Element*	Main uses
Boron	As a neutron absorber in nuclear reactors In boron nitride, BN, as an extremely hard abrasive material
Aluminium	Widely used in alloys where strength and lightness are needed together, e.g. in aircraft manufacture In making cooking utensils In packaging and cooking foils As $Al_2(SO_4)_3$ in water purification As $Al(OH)_3$ in foam fire extinguishers and as a mordant in dyeing
Gallium	As a semiconductor, e.g. with phosphorus and arsenic in light emitting diodes.

*Indium and thallium have few uses

but the ions are better represented as complex ions, e.g.

$$[Al(OH)_4(H_2O)_2]^- \quad \text{or} \quad Al(OH)_6^{3-}$$

In common with other reactive metals, aluminium will give off hydrogen with hydrochloric or sulphuric

Table 94.1. Physical properties of the elements in Group III*

Symbol	Boron B	Aluminium Al	Gallium Ga	Indium In	Thallium Tl
Electron structure	$(He)2s^2 2p$	$(Ne)3s^2 3p$	$(Ar)3d^{10}4s^2 4p$	$(Kr)4d^{10}5s^2 5p$	$(Xe)4f^{14}5d^{10}6s^2 6p$
Electronegativity	2.0	1.5	1.6	1.7	1.8
1st I.E./kJ mol^{-1}	801	578	580	560	590
2nd I.E./kJ mol^{-1}	2400	1800	2000	1800	2000
3rd I.E./kJ mol^{-1}	3700	1600	3000	2700	2900
Melting point/°C	2027	659	30	256	304
Boiling point/°C	3927	2447	2237	2047	1467
Atomic radius/pm	88	126	126	150	155
Ionic radius/pm	16	45	62	81	95
Principal oxid. no.	+3	+3	+3	+3,+1	+1
$E^{\ominus}_{M^{3+}/M}$/V		−1.66	−0.56	−0.38	+2.18
$E^{\ominus}_{M^+/M}$/V					−0.34

*Boron does not have a simple electrode potential for the change $B^{3+}(aq) + 3e^- \rightleftharpoons B(s)$. The $E^{\ominus}$ values for Al and Ga are in acid solution, the others in basic solution

acids. But it will not react with nitric acid. This acid is said to render the surface of aluminium passive. Amorphous boron does not give hydrogen with sulphuric or nitric acids. Instead, these acids are reduced, e.g.

$$2B(s) + 6HNO_3(aq) \rightarrow 2H_3BO_3(aq) + 6NO_2(g)$$
$$\text{orthoboric}$$
$$\text{acid}$$

Boron, aluminium and gallium all show an oxidation number of +3 in their compounds; but indium and especially thallium prefer an oxidation number of +1. Here we have an example of the so-called inert pair effect, about which we spoke in section 87.5. This refers to the tendency of elements towards the bottom of the B Groups not to lose their outer pair of s electrons.

One of the main reasons why boron does not make ionic compounds lies in the extremely small size of the B^{3+} ion. With an ionic radius of only 16 pm, the ion is such a dense centre of charge that if it did exist it would polarise any neighbouring ion. That is, it would attract electron density towards itself and lead to electrons being shared between the atoms.

Pure boron can be obtained as either a crystalline solid or a fine amorphous powder. When crystalline it is extremely hard, and largely inert. In contrast, the powder will react with non-metals. For example, it gives the oxide B_2O_3, nitride BN and chloride BCl_3. We shall return to a description of boron trichloride later.

Aluminium is far more reactive, although if it is left in air for all but a short time it becomes coated with a layer of aluminium oxide, Al_2O_3. This oxide protects the aluminium from further attack. The layer is so useful that in industry it is purposely increased by an electrolytic process called *anodising*.

The aluminium is made the anode in an electrolyte of sulphuric, phosphoric, or chromic acid. A voltage from 10 to 500 V may be used. The mechanism of anodising is complicated, but the result is a layer of aluminium oxide whose thickness lies between 0.25×10^{-6} and 150×10^{-6} m. A coloured layer can be produced either by adding chemicals to the electrolyte or by immersing the anodised layer in a dye.

Aluminium is widely used in alloys where, like magnesium, it is valued for the strength it can add with little increase in mass. The metal can be rolled into thin sheets that have a wide variety of uses. Especially, cooks will be familiar with it as baking foil. The importance of aluminium as a metal makes it worth extracting from its main ore, bauxite, which also has the formula Al_2O_3. We saw how this is done in Unit 85. It is an indication of how strongly the aluminium is bonded to oxygen that electrolysis has to be used rather than a strictly chemical method.

In recent years aluminium has been under suspicion as a possible cause of Alzheimer's disease. This disease induces senility in relatively young men and women. The disease causes them to lose their memory and they can no longer look after themselves. Needless to say, it is not small particles of the solid that are blamed; rather it is the presence of aluminium ions that can occur in foodstuffs or water supplies. However, as yet the link

Light emitting diodes are used in displays in many types of electrical equipment, including video recorders like that shown here.

between the disease and aluminium has not been proved.

The other elements in the Group are less important than boron or aluminium, so we shall say little about them. However, gallium does have one important use when it is combined with arsenic and phosphorus. Gallium arsenic phosphide gives out light when a voltage is placed across it. The substance is used in light emitting diodes (LEDs).

94.1   You should know that the oxides of non-metals are acidic, and the oxides of metals are basic. Would you expect boron to be more acidic or more basic in its reactions than aluminium?

94.2   What evidence is there from Table 94.1 that thallium has a greater tendency to give Tl^+ ions in solution rather than Tl^{3+} ions?

94.3 $E^{\ominus}_{Al^{3+}/Al} = -1.66$ V; $E^{\ominus}_{SO_4^{2-}/SO_2} = +0.17$ V. What, if anything, would happen if you reacted aluminium with sulphuric acid? If you think a reaction would take place, write down the equation.

94.4   Boron nitride has a layer structure rather like graphite. Remind yourself of the structure of graphite, and then write out the structure of part of a layer of boron nitride.

94.5   Aluminium has a carbide, Al_4C_3. It reacts with water to give a colourless gas, which does not react with bromine. What might be the gas? Write an equation for the reaction.

94.6   The standard free energies of formation of B_2O_3 and MgO are -1194 kJ mol^{-1} and -569 kJ mol^{-1} respectively. Should it be possible to prepare boron by reacting B_2O_3 with magnesium?

94.2 The oxides

All the elements give oxides with the general formula E_2O_3. They can be made by direct reaction with oxygen, but it is more common for B_2O_3 to be made by heating boric acid, H_3BO_3, above $100°$ C:

$$2H_3BO_3(s) \rightarrow B_2O_3(s) + 3H_2O(g)$$

An interesting example of the amphoteric nature of B_2O_3 is shown by its reaction with basic oxides in the *borax bead test*. (Borax is the traditional name of the substance that should really be called disodium tetraborate-10-water, $Na_2B_4O_7 \cdot 10H_2O$. However, for the present we shall continue to refer to it as borax.) In this test a little borax is heated alone, which converts it into boron trioxide, B_2O_3. Then the bead of hot oxide is touched on a sample of a metal oxide, and the two heated in a bunsen flame. Depending on the metal present a bead of a characteristic colour is produced. For example, cobalt oxide gives a blue bead, copper oxide a red bead and iron a green bead. A typical reaction is

$$CoO(s) + B_2O_3(s) \rightarrow Co(BO_2)_2(s)$$

The product is an example of a metaborate. There are several different types of structures for metaborates. They contain chains of boron and oxygen atoms rather than separate BO_2^- ions.

We have said that borax crystals have the empirical formula $Na_2B_4O_7 \cdot 10H_2O$. However, the crystals contain $B_4O_5(OH)_4^{2-}$ ions shown in Figure 94.1. Borax is used in laboratories as a primary standard in titrations. Strong acids convert it into boric acid, H_3BO_3, which is only very weak. The reaction is usually written

$$B_4O_7^{2-}(aq) + 2H^+(aq) + 5H_2O(l) \rightarrow 4H_3BO_3(aq)$$

Boric acid does not itself release hydrogen ions in solution. Rather, it accepts hydroxide ions; thereby an excess of hydrogen ions exists which makes its solution slightly acidic.

Aluminium oxide also shows amphoteric properties, dissolving in both acid and alkali:

$$Al_2O_3(s) + 6H^+(aq) \rightarrow 2Al^{3+}(aq) + 3H_2O(l)$$
$$Al_2O_3(s) + 6OH^-(aq) + 3H_2O(l) \rightarrow 2Al(OH)_6^{3-}(aq)$$

Aluminium oxide occurs in Nature in several minerals, the most important of which is bauxite. This is the mineral from which aluminium is extracted by electrolysis (see section 85.5). Another variety of Al_2O_3 is cor-

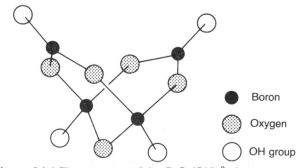

Figure 94.1 *The structure of the $B_4O_5(OH)_4^{2-}$ ion*

undum. This is a particularly hard mineral, and it is used as an abrasive. Corundum can be made in the laboratory by heating powdered aluminium oxide to a high temperature. If you have an eye for beauty you might be more impressed by the minerals in the photos in the colour section.

Like magnesium oxide, aluminium oxide has a very large enthalpy (and free energy) of formation. This suggests that aluminium will remove the oxygen from other metal oxides, which indeed it does. The most famous example of this is the thermit reaction. Here, a combination of powdered magnesium and aluminium is mixed with a metal oxide. The reaction needs energy to make it start, but once begun the reaction is rapid and highly exothermic. For example,

$$Fe_2O_3(s) + 2Al(s) \rightarrow 2Fe(s) + Al_2O_3(s);$$
$$\Delta H^\ominus = -854\,kJ\,mol^{-1}$$

The thermit reaction can be done in the laboratory, but it has found practical use in two rather different ways. A constructive use was when molten iron made by the thermit reaction was once used to plug holes in broken tram lines. A destructive use of the thermit reaction is in incendiary bombs. Those who like to design such weapons regard the thermit reaction as admirable not only because it gives out a lot of heat, but also because it cannot be put out by water.

94.7   Rewrite the equation

$$B_4O_7^{2-}(aq) + 2H^+(aq) + 5H_2O(l) \rightarrow 4H_3BO_3(aq)$$

showing the reaction as one that involves the $B_4O_5(OH)_2^{2-}$ ion instead of $B_4O_7^{2-}$. Does it matter very much which equation is used?

94.8   The endpoint of a borax–acid titration is at about pH = 4. Which indicator would you use in the titration?

94.9   (i) The formula of boric acid does not represent its structure. What is its formula?

(ii) The formula could be written in a different way. What is it?

(iii) How might you classify 'boric acid' if this formula were used?

94.10   Why can the thermit reaction not be stopped by water like an ordinary fire?

94.11   Hot aluminium oxide will catalyse the decomposition of ethanol, C_2H_5OH. A colourless gas is given off, which decolourises bromine water. What are the products of the reaction?

94.3   The hydroxides

The hydroxide of boron is $B(OH)_3$, but earlier we wrote its formula as H_3BO_3 and called it boric acid (Figure

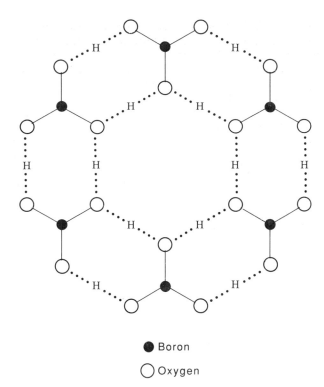

● Boron

○ Oxygen

Figure 94.2 *The structure of boric acid, H₃BO₃, has planar BO₃ groups linked by hydrogen bonds*

94.2). This just goes to show that the structures of many acids and alkalis are similar. (You should look back at Unit 74 to remind yourself about the structures of acids such as HNO₃ and H₂SO₄.)

If you add hydroxide ions to a solution of an aluminium salt, you will find that a gelatinous white precipitate of aluminium hydroxide, Al(OH)₃, is produced:

$$Al^{3+}(aq) + 3OH^-(aq) \rightarrow Al(OH)_3(s)$$

The precipitate will dissolve in both acid and alkali:

$$Al(OH)_3(s) + 3H^+(aq) \rightarrow Al^{3+}(aq) + 3H_2O(l)$$
$$Al(OH)_3(s) + 3OH^-(aq) \rightarrow Al(OH)_6^{3-}(aq)$$

Aluminium hydroxide has two important uses. Fire extinguishers can be made that contain aluminium sulphate together with the usual carbonate and acid. When the extinguisher is used, aluminium hydroxide is produced:

$$2Al^{3+}(aq) + 3CO_3^{2-}(aq) + H_2O(l) \rightarrow$$
$$2Al(OH)_3(s) + 3CO_2(g)$$

The hydroxide has the virtue of stabilising the foam, which covers the fire. A second use is in dyeing. Particles of the hydroxide are able to attach themselves to the fibres in cloth. The hydroxide is able to absorb dye molecules much more efficiently than the fibres alone. A substance which, like aluminium hydroxide, assists in the dyeing of a fabric is called a *mordant*.

94.12 Why is it that solutions of many aluminium salts are acidic in water? (Hint: see Unit 76.)

94.4 The halides

The elements all give halides with the general formula EX₃. For example, BF₃, AlCl₃, GaBr₃. However, there are two complicating factors. The first, which we shall come to in a moment, is that some of the halides dimerise, e.g. aluminium trichloride gives Al₂Cl₆. The second, which we shall largely ignore, is that below aluminium other formulae are found, e.g. GaI₂, which often hide some complicated structures.

Boron trichloride and aluminium trichloride are famous for the bonding they contain, and for the shapes of their molecules. You will find details about them in Unit 17. However, in Figure 94.3 you can see the shapes of the isolated molecules, and of the dimer that aluminium trichloride makes.

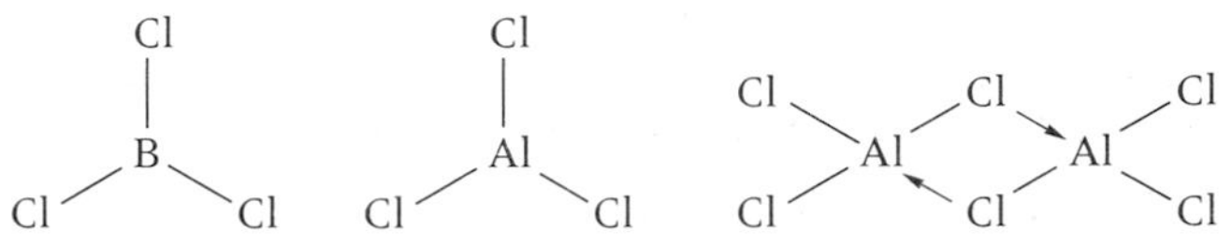

Figure 94.3 *The shapes of BCl₃, AlCl₃ and the dimer Al₂Cl₆*

Both BCl₃ and AlCl₃ are prone to hydrolysis. This is characteristic of many chlorides, but very different to tetrachloromethane, CCl₄, which will not hydrolyse. The reason for the difference lies in the electron deficient nature of BCl₃ and AlCl₃. Both boron and aluminium have more p orbitals in their outer shell than are used in bonding to the three chlorine atoms. We can imagine that the hydrolysis reaction starts by a water molecule donating one of its pairs to the empty orbital (Figure 94.4). This disrupts the arrangement of the electron clouds in the molecules, and the reaction has begun. We can write the equation in two ways. The first is a simple version, the second rather more accurate:

$$AlCl_3(s) + 3H_2O(l) \rightarrow Al(OH)_3(aq) + 3HCl(aq)$$
$$AlCl_3(s) + 6H_2O(l) \rightarrow Al(H_2O)_6^{3+}(aq) + 3Cl^-(aq)$$

Tetrachloromethane cannot react in this way because it has no empty orbitals with the right range of energies to accept a lone pair from a water molecule.

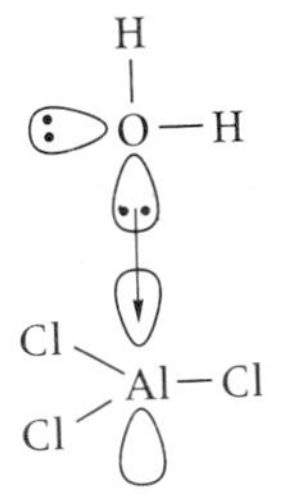

Figure 94.4 *We can think of a reaction between H₂O and AlCl₃ starting when a lone pair on the water molecule overlaps with an empty 2p orbital on the aluminium atom*

94.13 Gallium gives a chloride whose molar mass in the vapour state is approximately 352 g mol^{-1}. What is the likely formula of the chloride? $M(Ga) = 69.7$ g mol^{-1}; $M(Cl) = 35.5$ g mol^{-1}.

94.14 A solution of a thallium compound had some sodium chloride solution added to it. A white insoluble chloride of thallium was made, which was found to contain about 15% of chlorine. What is the likely formula of the chloride? $M(Tl) = 204.4$ g mol^{-1}.

94.15 A student suggested making aluminium trichloride by this recipe:

(i) Add sodium hydroxide solution to aluminium sulphate solution.

(ii) Filter off the precipitate.

(iii) Convert the hydroxide into a chloride by adding hydrochloric acid.

(iv) Evaporate the solution to leave crystals of $AlCl_3$.

Why will this method not work? What method must be used to make the chloride?

94.16 Explain why (i) BF_3 can react with ammonia to give a compound of formula BF_3NH_3; (ii) 1 mol of $AlCl_3$ will combine with 1 mol of ethoxyethane, $(C_2H_5)_2O$.

94.17 $AlCl_3$ is used in the Friedel–Crafts reaction in organic chemistry. Look up this reaction (if you need to) and give an example of it; explain what the connection is between the example and *Lewis acids*.

Some crystals, like that of chrome alum shown here, can be grown to impressive sizes. See Unit 94 for information about alums.

to be too few electrons to explain the number of bonds present. For example, diborane, B_2H_6, has the structure shown in Figure 94.5a. We have discussed this compound in section 90.7. Here we shall just say that the two hydrogen atoms that bridge between the boron atoms are involved in three-centre bonds. These bonds consist of orbitals that contain two electrons in the nor-

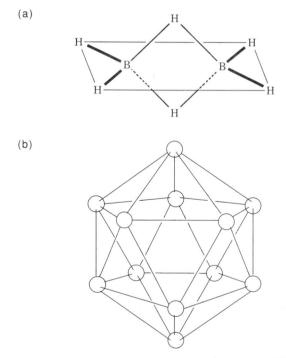

Figure 94.5 (a) The structure of diborane, B_2H_6. (b) The $B_{12}H_{12}^{2-}$ ion has the shape of an icosahedron. Only the boron atoms are shown. (Adapted from: Muetterties, E. L. (1967). The Chemistry of Boron and its Compounds, Wiley, New York, p. 39)

94.5 The sulphates

Aluminium sulphate and the alums are the most widely known of the sulphates. You may come across aluminium sulphate as a white crystalline material, $Al_2(SO_4)_3 \cdot 18H_2O$, but more interesting is the mixed sulphate of aluminium and potassium known as potash alum, $KAl(SO_4)_2 \cdot 12H_2O$. The crystals of this alum are octahedral and translucent. They are isomorphic with the violet crystals of chrome alum, $KCr(SO_4)_2 \cdot 12H_2O$. That is, the two varieties have exactly the same crystal form. Indeed a crystal of potash alum in a saturated solution of chrome alum will grow a layer of chrome alum over it. Other alums include ammonia alum (ammonium aluminium sulphate), $NH_4Al(SO_4)_2 \cdot 12H_2O$, and iron alum (iron(III) ammonium sulphate), $NH_4Fe(SO_4)_2 \cdot 12H_2O$.

94.6 The hydrides

Boron makes a large number of hydrides. The bonding in them is unusual, because in many cases there appear

Table 94.3. Differences between NaBH₄ and LiAlH₄

Class of organic compound	Reaction with	
	NaBH₄	LiAlH₄
Aldehyde	Reduced	Reduced
Ketone	Reduced	Reduced
Acid	No effect	Reduced
Ester	No effect	Reduced
Amide	No effect	Reduced
Nitrile	No effect	Reduced

mal way, but the orbitals spread over three atoms rather than two. Boron and hydrogen can also give a large number of ionic hydrides with remarkable structures, e.g. $B_{12}H_{12}{}^{2-}$ has the shape of an icosahedron (Figure 94.5b).

Some of the hydrides made by boron and aluminium have useful properties. In particular, sodium tetra-hydridoborate(III), $NaBH_4$, and lithium tetrahydrido-aluminate(III), $LiAlH_4$, are widely used in organic chemistry as reducing agents (Table 94.3). These hydrides contain $BH_4{}^-$ and $AlH_4{}^-$ ions respectively. We can think of them as sources of hydride ions, H^-, which are powerful reducing agents. However, $LiAlH_4$ is more powerful than $NaBH_4$. See Unit 90 and the units on organic chemistry for more information on them.

Answers

94.1 Being nearer the top of the Group, we would expect boron to show more non-metallic nature than aluminium; so it should be more acidic in nature than aluminium. (It is.)

94.2 $E^{\ominus}_{Tl^+/Tl}$ is more negative than $E^{\ominus}_{Tl^{3+}/Tl}$, which suggests that thallium has a greater tendency to turn into Tl^+ in solution.

94.3 The aluminium is capable of reducing the acid to sulphur dioxide. The two half-equations are

$$Al^{3+}(aq) + 3e^- \rightarrow Al(s)$$
$$SO_4{}^{2-}(aq) + 2e^- + 4H^+(aq) \rightarrow SO_2(g) + 2H_2O(l)$$

They can be combined by taking three times the second and subtracting twice the first:

$$2Al(s) + 3SO_4{}^{2-}(aq) + 12H^+(aq) \rightarrow$$
$$3SO_2(g) + 6H_2O(l) + 2Al^{3+}(aq)$$

94.4 The structure is shown in Figure 94.6.

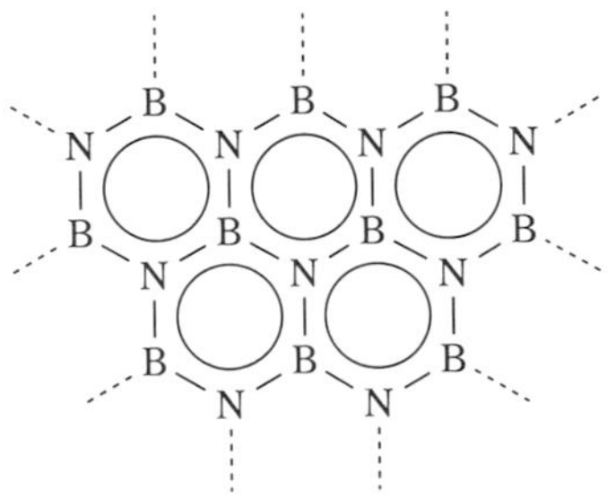

Figure 94.6 Part of a layer of boron nitride. The circles in the ring stand for delocalised π electrons. The structure of the compound is similar to that of graphite shown in Figure 95.2. However, there are marked differences in properties. For example, boron nitride is white and does not conduct electricity

94.5 It is methane. The equation is

$$Al_4C_3(s) + 12H_2O(l) \rightarrow 4Al(OH)_3(s) + 3CH_4(g)$$

94.6 The equation is

$$B_2O_3(s) + 3Mg(s) \rightarrow 2B(s) + 3MgO(s)$$

The free energy change is

$$3\Delta G^{\ominus}_f(MgO) - \Delta G^{\ominus}_f(B_2O_3) = -513 \, kJ \, mol^{-1}$$

The negative sign shows us that the reaction should be spontaneous. As usual, though, the two reactants have to be heated to start the reaction.

94.7 $B_4O_5(OH)_2{}^{2-}(aq) + 2H^+(aq) + 4H_2O(l) \rightarrow$
$$4H_3BO_3(aq)$$

No, it does not; the molar ratio of the ion to hydrogen ions is the same in both cases.

94.8 Screened methyl orange.

94.9 (i) H_3BO_3.

(ii) Boric acid could be written $B(OH)_3$.

(iii) It looks like a hydroxide.

94.10 In an ordinary fire the reactions are kept going by the material reacting with oxygen in the air. Water not only cools the fire, but also keeps oxygen out. In the thermit reaction, the oxygen involved is in the metal oxide, so keeping atmospheric oxygen out has no effect. (Also, at the temperatures of the thermit reaction (above 1000°C), water will give off hydrogen with aluminium.)

94.11 Ethene and water (see section 112.2).

94.12 The $Al(H_2O)_6{}^{3+}$ ion can lose protons (see section 75.5).

94.13 You should guess that the chloride might be $GaCl$, $GaCl_2$, $GaCl_3$, or perhaps Ga_2Cl_6. By trial and error you should find that Ga_2Cl_6 is correct. You might have guessed this by analogy with Al_2Cl_6 and because the molar mass is high.

94.14 15% is not a high proportion of chlorine. Also, if you read the first section you will find that there is a hint that thallium tends to give compounds in which its oxidation number is +1. The chloride is $TlCl$.

94.15 Aluminium chloride is hydrolysed by water so it cannot be crystallised from solution. The chloride is made by direct combination of aluminium and chlorine.

UNIT 94 SUMMARY

- The Group III metals:
  - (i) Show non-metallic character, e.g. BCl_3 and $AlCl_3$ are covalent.
  - (ii) Boron and aluminium are amphoteric, i.e. the elements and their oxides will react with both acids and alkalis.
  - (iii) B^{3+} ion is extremely small, thus leading to covalency (Fajans' rules).
- Manufacture:
  Aluminium isolated by electrolysis of molten bauxite, Al_2O_3.

Reactions

- With oxygen:
  - (i) Oxides made;

    e.g. $2Al(s) + 3O_2(g) \rightarrow 2Al_2O_3(s)$

  - (ii) Aluminium powder is used in the thermit reaction to remove oxygen from (i.e. reduce) iron(III) oxide;

    $Fe_2O_3(s) + 2Al(s) \rightarrow 2Fe(s) + Al_2O_3(s)$

- With halogens:
  Aluminium makes aluminium trichloride, $AlCl_3$, which occurs as dimers, Al_2Cl_6;

  $2Al(s) + 3Cl_2(g) \rightarrow Al_2Cl_6(s)$

Compounds

- Oxides:
  - (i) Basic, but insoluble in water.
  - (ii) Amphoteric nature shown by B_2O_3 and Al_2O_3;

e.g. $Al_2O_3(s) + 6H^+(aq) \rightarrow 2Al^{3+}(aq) + 3H_2O(l)$

$Al_2O_3(s) + 6OH^-(aq) + 3H_2O(l) \rightarrow 2Al(OH)_6{}^{3-}(aq)$

- Hydroxides:
  - (i) The hydroxide of boron, $B(OH)_3$, is better classed as boric acid, H_3BO_3.
  - (ii) Gelatinous white $Al(OH)_3$ is insoluble in water. It is made by adding alkali to a solution containing Al^{3+} ions, but dissolves in excess alkali.
- Hydrides:
  - (i) Boron forms many electron deficient hydrides, e.g. B_2H_6. These hydrides contain hydrogen atoms bridging between boron atoms.
  - (ii) Sodium tetrahydridoborate(III), $NaBH_4$, and lithium tetrahydridoaluminate(III), $LiAlH_4$, are used as reducing agents in organic chemistry; $LiAlH_4$ is the more vigorous.
- Halides:
  - (i) Aluminium trichloride, $AlCl_3$, dimerises, forming Al_2Cl_6 with a mix of covalent and coordinate bonding.
  - (ii) The halides are easily hydrolysed;

    e.g. $AlCl_3(s) + 3H_2O(l) \rightarrow Al(OH)_3(aq) + 3HCl(aq)$

- Sulphates:
  Potash alum, $KAl(SO_4)_2 \cdot 12H_2O$, is isomorphic with chrome alum, $KCr(SO_4)_2 \cdot 12H_2O$.

95

Carbon

95.1 Why is carbon important?

The normal valency of carbon is four. It is most important that you can explain why it is that carbon can make four covalent bonds. Do read Unit 17 if you have to remind yourself of the explanation.

The properties and uses of carbon are listed in Tables 95.1 and 95.2. Carbon is an especially important element because it shows the property of *catenation*. This means that its atoms can join together to make chains. This property is shown in the millions of carbon compounds that are either naturally occurring or have been made for the first time in laboratories. Indeed, there are so many of them that they are best studied as a separate branch of chemistry: organic chemistry. In this unit we shall summarise the chemistry of the simple carbon compounds, e.g. the gases carbon monoxide and carbon dioxide, and the carbonates. However, the main reason why carbon is one of the most important of elements is that life is based upon it. The molecules that are responsible for the growth and development of living organisms, such as vitamins, proteins, enzymes, hormones and DNA, are structures built from chains of carbon atoms. Owing to the key part that carbon plays in maintaining life, it is as well that you know about the *carbon cycle*. The cycle is represented in Figure 95.1.

The atmosphere acts as a reservoir of carbon dioxide. During photosynthesis plants convert the gas into sugars, and then into more complicated molecules. Animals that consume the plants then incorporate many of the carbon compounds into their own structures, or use them as a source of energy. Carbon dioxide is released back into the atmosphere in many ways. Some is breathed out by animals as a breakdown product of respiration, and some is released when organic matter decays. (You count as organic matter!) Another source of carbon dioxide in the atmosphere is the burning of fuels such as wood, peat, coal and oil. The majority of carbon dioxide released in this way is caused by humans: to keep warm, to cook food, to run cars and lorries, and to use in industry.

In previous centuries the proportion of carbon dioxide in the atmosphere has increased fairly slowly; but in the last 20 years the increase has been much more rapid. It appears that humans are releasing more carbon dioxide through the burning of fuels than can be absorbed by plants or dissolved in the oceans. One consequence of this build-up of carbon dioxide is the *greenhouse effect*. Visible and ultraviolet light from the Sun strikes the Earth's surface as photons of relatively short wavelength. Radiation is emitted from the Earth's surface in the infrared region of the spectrum, and it is of a much longer wavelength. Carbon dioxide is capable of absorbing a significant amount of this radiation, so the more of the gas present, the less energy escapes into space. The result is that the temperature of the atmosphere rises. The average temperature of the atmosphere does appear to be rising, and it may have marked effects on the climate. The precise influence is hard to analyse because the weather system is so complicated; but if the rise continues, dramatic changes are sure to take place. For example, a small rise in the average

Table 95.1. Information about carbon

Relative atomic mass, $A_r(C)$	12
Electron structure	$1s^2 2s^2 2p^2$
Ionisation energy	1086 kJ mol^{-1}
Electronegativity	2.5
Bond energy (C—C, average)	346 kJ mol^{-1}
Bond length (C—C, average)	154 pm
Two allotropes: diamond and graphite	

Table 95.2. Uses of carbon

Diamond	Gem stone
	In industry for cutting, milling and drilling
Graphite	Reducing agent in the extraction of metals
	Electrodes in electrolytic extraction of elements
	Moderator in nuclear reactors
	In very high strength carbon fibres
Activated charcoal	To adsorb gases and other chemicals
Charcoal	As a fuel

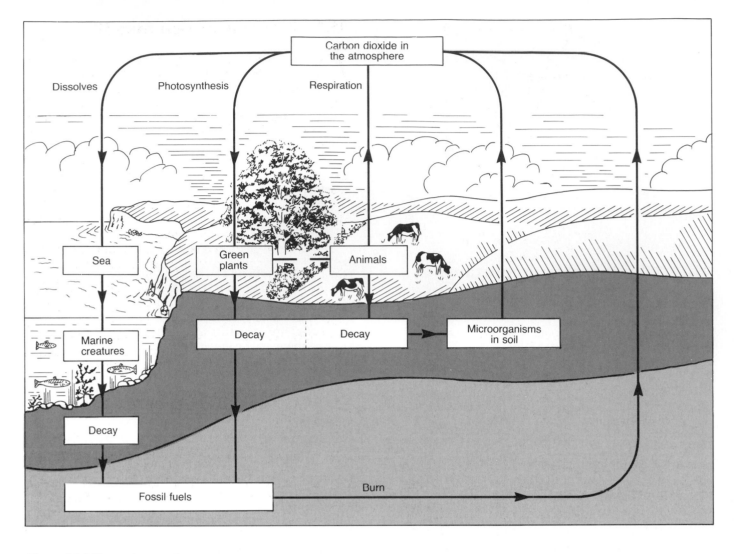

Figure 95.1 *The carbon cycle*

temperature of the polar regions will cause a massive melting of the ice caps. Given that the majority of the water on Earth is locked up as ice in these caps, its release into the oceans will cause a rise of sea level of many feet. Cities like London or New York will be flooded. Perhaps of more importance is that regions of the world in which food is grown may turn into deserts; or alternatively they may become swamped by rain levels now associated with the tropics.

Carbon possesses two famous allotropes: diamond and graphite (Figure 95.2). You will find details of them in Units 32 and 57. Contrary to appearances, graphite is the more thermodynamically stable allotrope of carbon. This is in spite of the fact that graphite is much softer than diamond, and that it burns much more easily. The difference in properties is due to the difference in bonding. Diamond has its carbon atoms joined in an interlocking network of tetrahedra. It is very hard to break diamond apart, either physically or chemically, because of the vast number of bonds that have to be broken. On the other hand graphite has a layer structure in which the atoms are joined in interlocking networks of hexagons. The hexagonal layers can slide over one another.

The two allotropes of carbon, diamond and graphite, have very different properties and values.

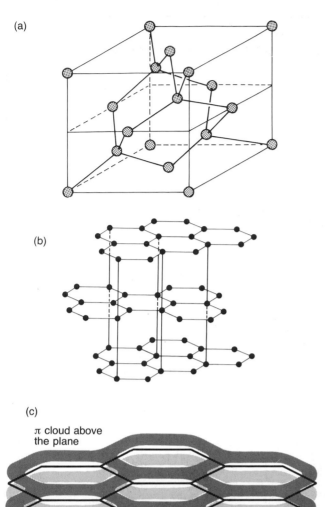

(a)

(b)

(c)

π cloud above
the plane

π cloud below
the plane

Figure 95.2 *Structures of (a) diamond and (b) graphite. In graphite there is a delocalised π cloud of electrons parallel to the planes of carbon atoms*

95.1 What is one explanation of the bonding in methane, CH_4?

95.2 What happens to the infrared energy absorbed by molecules in the atmosphere?

95.3 (i) What type of bonding holds the layers together in graphite?

(ii) Why will graphite conduct electricity well in a direction parallel to the planes of hexagons, but not at all well in a direction perpendicular to the planes?

(iii) What is the name we give to crystals, or substances, that show different properties in different directions?

95.2 Where is carbon found?

Carbon is found as both graphite and diamond in underground deposits. The amount of graphite is very small and not economic to extract. However, there is a large demand for graphite and it can be made by passing electricity through coke at 2700°C. Carbon in the form of charcoal can be made by heating wood in a limited supply of air. It has been of great importance to humans over thousands of years. Especially, once it was found that iron ore could be converted into iron and steel by reacting it with carbon, vast areas of forest and woodland were cut down in order to make charcoal. The charcoal industry was responsible for transforming the ecology of many European countries.

The value of diamonds is such that mining companies will shift enormous amounts of earth in the expectation of finding them. Native diamonds have to be cut and polished before they make good gem stones. The smaller, less well formed, specimens can be used as industrial diamonds, e.g. in cutting and milling machines. However, it is now common for industrial grade diamonds to be made artificially by heating graphite at high pressures and temperatures.

95.3 Carbon dioxide and carbon monoxide

Carbon monoxide and carbon dioxide are the only two gaseous oxides in Group IV. Carbon dioxide is denser than air; carbon monoxide is less dense. Both can be produced when organic matter burns, depending on the amount of oxygen present. For example, if coke burns in a plentiful supply of air,

$$C(s) + O_2(g) \rightarrow CO_2(g)$$

but in a limited supply of air,

$$2C(s) + O_2(g) \rightarrow 2CO(g)$$

Both these reactions are exothermic and take place in furnaces used in reducing metal ores. Generally in such places it is desirable to make carbon monoxide because it is a reducing agent and will, for example, remove the oxygen from a metal oxide. Units 50 and 85 are the places to find out more about this.

The two gases also occur in the exhausts from car and lorry engines. Many countries now have laws that regulate the amount of carbon monoxide that should be emitted from these engines.

Carbon dioxide is used in many fire extinguishers (Table 95.3). There are two types. In water based extinguishers a sealed bottle of acid is immersed in a solution of a hydrogencarbonate. When the plunger on

Table 95.3. Uses of carbon dioxide

In fire extinguishers
Making fizzy drinks, e.g. champagne, lemonade, colas
Dry ice, solid CO_2, is used in show business for making mists

---

the extinguisher is struck, the bottle bursts, allowing the acid and hydrogencarbonate to mix. Carbon dioxide is given off in a watery spray (or in a foam if other chemicals have been added).

A second type contains pure carbon dioxide under pressure. The gas given out covers the fire and prevents oxygen in the air from reacting with the hot materials. If you use this type of extinguisher you will find that the gas given out is very cold. This is a well known effect in thermodynamics (the Joule–Kelvin effect) and takes place when a gas expands from a region of high to a region of low pressure. If the gas cools sufficiently, it will make crystals of CO_2, i.e. solid CO_2, also known as dry ice.

In the laboratory carbon monoxide is best made by dehydrating methanoic acid or sodium methanoate with concentrated sulphuric acid:

$$HCOOH - H_2O \rightarrow CO$$

You may have seen carbon dioxide made in the traditional way by reacting dilute hydrochloric acid with marble chips:

$$CaCO_3(s) + 2H^+(aq) \rightarrow Ca^{2+}(aq) + CO_2(g) + H_2O(l)$$

Both CO and CO_2 are poisonous, odourless, colourless gases. Carbon dioxide is heavier than air. Carbon monoxide acts as a poison by acting as a ligand on the iron atom in haemoglobin. Once it has combined with the iron it tends to stay there, so it prevents oxygen molecules being taken up. The blood of victims of carbon monoxide poisoning is a brilliant red-pink colour rather than the dull red of normal blood. Carbon monoxide is the dangerous component of coal gas, and is produced when, for example, the flues of coal or oil burners become blocked. Carbon dioxide kills people by entering the lungs and preventing sufficient oxygen being absorbed.

You should know how to test for carbon dioxide: it gives a white precipitate with lime water ('lime water goes milky'). The cloudiness is due to a precipitate of calcium carbonate:

The cold vapour escaping from dry ice (solid carbon dioxide) is responsible for the clouds of 'mist' seen at many rock concerts.

$$Ca(OH)_2(aq) + CO_2(g) \rightarrow CaCO_3(s) + H_2O(l)$$

This reaction demonstrates the acidic nature of carbon dioxide. Indeed, a solution of carbon dioxide in water behaves as a weak acid:

$$CO_2(g) + H_2O(l) \rightarrow H_2CO_3(aq)$$
$$H_2CO_3(aq) \rightleftharpoons H^+(aq) + HCO_3^-(aq)$$
$$HCO_3^-(aq) \rightleftharpoons H^+(aq) + CO_3^{2-}(aq)$$

The weakness of the acid is also a reason why solutions of carbonates are often alkaline. For example, if we dissolve sodium carbonate in water, its ions separate, and hydroxide ions are released:

$$CO_3^{2-}(aq) + H_2O(l) \rightarrow HCO_3^-(aq) + OH^-(aq)$$

You will find out more about this type of salt hydrolysis in Unit 76.

Although carbon monoxide has no acidic or basic properties in terms of its ability to donate or accept protons, it can behave as a Lewis base. This, you might remember, defines a base as a species that can donate a pair of electrons to another atom. It shows this property particularly well with transition metals, making a large number of carbonyl complexes, e.g. $Ni(CO)_4$, $Fe(CO)_5$, $Cr(CO)_6$ (Figure 95.3). One of the interesting things about the carbonyls is that the transition metal is in its zeroth oxidation state. In some, more complicated, carbonyls, the carbon monoxide molecule can bridge between two, or sometimes three, metal atoms, e.g. in $Fe_2(CO)_9$. One reason why nickel carbonyl is of interest is that it is used in the extraction of nickel; see section 85.3.

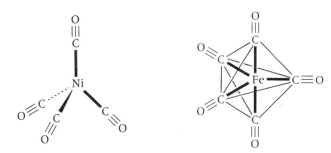

Ni(CO)$_4$: tetrahedral   Fe(CO)$_5$: trigonal bipyramid

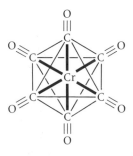

Cr(CO)$_6$: octahedral

Figure 95.3 The structures of three transition metal carbonyl compounds

95.4 Every year some people die by being in a room containing a faulty heater that uses coal, gas, or oil. How might the deaths occur?

95.5 Describe the bonding in CO and CO_2. What is the shape of the CO_2 molecule?

95.6 Why does the precipitate redissolve if excess carbon dioxide is passed through lime water?

95.7 If you had a mixture of carbon monoxide and carbon dioxide, how would you find out the relative proportions of the two gases in the mixture?

95.8 In the carbonyls of Figure 95.3, each carbonyl donates two electrons to the transition metal ion.

(i) How many electrons does each of the metals have in its zeroth oxidation state?

(ii) How many electrons in total are donated by the carbon monoxide molecules?

(iii) What is special about the total number of electrons that the metals now have?

(iv) Predict the formula of molybdenum carbonyl. (The atomic number of molybdenum, Mo, is 42.)

These limestone figures at Wells Cathedral in Somerset show the unfortunate effects of hundreds of years of atmospheric pollution.

95.4 The importance of carbonates

Carbonates (Table 95.4), together with silicates, are some of the most important geological minerals. Calcium carbonate, $CaCO_3$, occurs in a variety of limestones, including chalk. One of the characteristics of limestone is that it is dissolved by acids, even weakly acidic rain water. Among other things this is why limestone deposits are often riddled by caves, and why water that comes from aquifers in chalky areas is hard. We can write the reaction that takes place as

$$CaCO_3(s) + H^+(aq) \rightarrow Ca^{2+}(aq) + HCO_3^-(aq)$$

Chalk has been made from the shells of tiny sea creatures and the bodies of algae. It is a lot softer than other limestones, some of which have been subjected to high temperatures and pressures during changes in the Earth's crust.

Table 95.4. Carbonate minerals

Name	Formula	Comment
Calcite	$CaCO_3$	A pure variety is Iceland spar. It shows the property of double refraction
Aragonite	$CaCO_3$	Similar to calcite, but with a different crystal structure. Gradually changes into calcite
Dolomite	$CaCO_3 \cdot MgCO_3$	More resistant to acid than other limestones. Found as mountains, e.g. 'the Dolomites' in Italy
Siderite	$FeCO_3$	An important ore of iron
Malachite	$CuCO_3 \cdot Cu(OH)_2$	A copper ore with a characteristic green colour

95.9 Explain why a solution containing $Ca^{2+}(aq)$ and $HCO_3^-(aq)$ ions is said to be hard.

95.10 What is wrong with this statement: 'water is often hard because it has calcium carbonate dissolved in it'?

95.11 In recent years limestone rocks have been dumped (on purpose, and legally) in a number of lakes in Scandinavia, Germany and Great Britain. What might be the reason for this?

95.12 Why are there no mountains made of the Group I metal carbonates, e.g. Na_2CO_3, K_2CO_3?

95.13 What is wrong with this description of an

experiment: 'a little nitric acid was added to a solution of barium carbonate and the carbon dioxide given off turned lime water milky'?

95.14 Explain why it is that a solution of washing soda crystals (i) is alkaline, (ii) gives hydrogen if it is heated with aluminium.

be made by reacting sulphur vapour with carbon at very high temperatures:

$$C(s) + 2S(s) \rightarrow CS_2(l); \qquad \Delta H_f^\ominus = +88\,kJ\,mol^{-1}$$

The enthalpy change is strongly positive, so we can call carbon disulphide an endothermic compound. With this piece of knowledge you should expect that it would be quite likely to break apart easily, which it does. It is explosive in its reactions with a number of chemicals.

95.5   Carbon disulphide

Carbon disulphide, CS_2, is a poisonous, volatile liquid with a most unpleasant smell. It has a marked ability to dissolve sulphur and rubber. It can be used to make crystals of rhombic sulphur by dissolving powdered sulphur in it, and allowing the liquid to evaporate slowly. However, it is better to use an organic solvent such as methylbenzene for the purpose. Carbon disulphide can

95.15 A mixture of CS_2 vapour and nitrogen monoxide, NO, gives a vivid blue flame when ignited. A yellow solid is left afterwards. Predict the products of the reaction, and write the equation.

95.16 Why, given its endothermic heat of formation, can you *expect* that CS_2 would decompose easily but you cannot be certain?

Answers

95.1   sp³ hybridisation. See Unit 17 for details.

95.2   It makes the molecules vibrate more rapidly.

95.3   (i) Van der Waals bonding.

(ii) If you look at Figure 95.2c you will see that the delocalised orbitals in graphite lie parallel to the planes. Electrons can move along the planes making use of these orbitals. Conduction is good parallel to the planes. There is no overlap of orbitals perpendicular to the planes, so conduction is poor in this direction.

(iii) They are anisotropic.

95.4   If the flue to the heater becomes blocked, or there is insufficient ventilation, the fumes of CO and CO_2 build up in the room. Unless the person lies on the floor, it is the CO that kills.

95.5   There is a triple bond in CO, and two double bonds in CO_2 (see section 14.2). CO_2 is linear.

95.6   As the gas dissolves, the solution becomes acidic. The insoluble $CaCO_3$ is converted into soluble calcium hydrogencarbonate, $Ca(HCO_3)_2$.

95.7   One method is to measure their total volume, and then pass them through a solution of an alkali. (Potassium hydroxide is very good.) The CO_2 will be absorbed, so by measuring the residual volume of CO their proportions can be found.

95.8   (i) Ni has 28, Fe has 26 and Cr has 24. (These are the same as their atomic numbers.)

(ii) Each donates two electrons, so 8 go to Ni, 10 to Fe and 12 to Cr.

(iii) The totals are 36 for each metal. This is the number of electrons belonging to the noble gas krypton. It is common for the transition metals to make carbonyls with a total number of electrons equal to that of a noble gas.

(iv) The noble gas following molybdenum is xenon. This has 54 electrons. The difference of 12 electrons can be made up from six CO molecules. The formula is $Mo(CO)_6$.

95.9   The Ca^{2+} ions give precipitates (scum) with soaps. On heating the HCO_3^- ions revert to CO_3^{2-}, which gives a precipitate with Ca^{2+} ions. (This is the scale/fur in kettles and boilers.)

95.10   It contains only extremely small amounts of $CaCO_3$ through the solid dissolving in water of its own accord. Mainly it contains the products of the reaction between $CaCO_3$ and slightly acidic rain water.

95.11   Many lakes and waterways have become acidic owing to the effects of acid rain. The limestone is used to neutralise the acid. The treatment can be successful, but is not always so.

95.12   They are all soluble, so even if they were made millions of years ago, they would have soon dissolved away when it rained!

95.13   Barium carbonate is almost insoluble in water, so the solution could not have contained this carbonate. See section 93.4.

95.14   (i) Salt hydrolysis is responsible.

(ii) Aluminium is amphoteric and will give hydrogen with solutions containing H^+ or OH^- ions (see section 94.1).

95.15   You should have predicted that the nitrogen is likely to be released as N_2, and that the solid is sulphur. The equation is

$$2NO(g) + CS_2 \rightarrow 2S(s) + N_2(g) + CO_2(g)$$

95.16   We know that endothermic compounds are energetically unstable; but they may be *kinetically* stable. That is, the reaction may have a large activation energy.

- The element:
    - (i) Carbon is a non-metal.
    - (ii) Carbon atoms show the property of catenation, i.e. they can bond with each other to form chains.
    - (iii) Carbon compounds form the basis of biologically active molecules.
    - (iv) Carbon has two allotropes: diamond and graphite.
    - (v) Graphite is the most energetically stable of the two.
- Manufacture:
    Charcoal (graphite) is made by heating wood in the absence of air; diamond occurs naturally, but can be made by heating graphite at a high temperature and pressure.

Reactions

- A reducing agent:
    Will reduce many metal oxides, especially useful in the extraction of iron in the blast furnace.
- With water:
    Takes part in the Bosch reaction: see Unit 90.

- With oxygen:
    Produces carbon monoxide or carbon dioxide depending on the quantity of oxygen present;

    $$C(s) + O_2(g) \rightarrow CO_2(g)$$
    $$2C(s) + O_2(g) \rightarrow 2CO(g)$$

Compounds

- Oxides:
    - (i) Carbon dioxide is acidic in water;

        $$CO_2(g) + H_2O(l) \rightarrow HCO_3^-(aq) + H_3O^+(l)$$

    - (ii) Carbon dioxide turns lime water milky.
    - (iii) Carbon monoxide is a neutral oxide; but it can act as a ligand to transition metal ions, e.g. $Fe(CO)_5$.
- Hydrides:
    A vast number of hydrides exist (CH_4, C_2H_6, etc.), which are the subject of organic chemistry.
- Halides:
    Tetrachloromethane, CCl_4, is used as a solvent. Unlike many other chlorides of non-metals it will not undergo hydrolysis. See Unit 96.

96

Group IV

96.1 The nature of the elements

Carbon and silicon show many of the properties that are characteristic of non-metals; but as we move down the Group, the metallic nature of the elements increases. For example, carbon and silicon give acidic oxides, whereas those of germanium, tin and lead are amphoteric, although some lead oxides are definitely basic. There is a change in the nature of the bonding down the Group as well. Covalency dominates until, as we would expect with metals, tin and lead make some ionic compounds.

The physical properties and uses are listed in Tables 96.1 and 96.2.

The normal valency of the elements is four; but apart from carbon, the elements can make more than four bonds. This is because they make use of a set of d orbitals in bonding. The d orbitals are not those listed in Table 96.1 (which are full of electrons). Rather they are the empty d orbitals of the outer electron shell, e.g. the 3d set for silicon and the 4d set for germanium (Figure 96.1). The availability of d orbitals is responsible for the ability of the elements, except carbon, to make complex ions such as SiF_6^{2-}. Here silicon can provide a maximum of four unpaired electrons by making use of one electron in each of the 3s and 3p orbitals. To make the extra two bonds, two of the 4d orbitals are used. In fact the bonds in the SiF_6^{2-} ion are a mixture of 3s, 3p and 4d orbitals. If we use the theory of hybridisation, which

we discussed in Unit 17, we can describe the bonding orbitals as a set of d^2sp^3 hybrids.

Another feature of the chemistry of the Group is that some carbon compounds are less reactive than the corresponding compounds of the other members of the Group. For example, tetrachloromethane, CCl_4, will not react with water at all easily, whereas silicon tetrachloride, $SiCl_4$, will. As you will see in section 96.5, this too is a result of the elements (except carbon) making use of an empty set of d orbitals.

Table 96.2. Uses of the elements in Group IV

Element	Main uses
Silicon	As a semiconductor in transistors and other electronic components
	In silicone waxes and polymers
	As silicates, in the ceramics and glass industries
Germanium	As a semiconductor
Tin	In alloys
	As a protective coating for iron or steel
Lead	In alloys and water pipes (this use is much in decline)
	In petrol additives (also in decline)
	In car batteries

Table 96.1. Physical properties of the elements in Group IV

Symbol	Carbon C	Silicon Si	Germanium Ge	Tin Sn	Lead Pb
Electron structure	$(He)2s^22p^2$	$(Ne)3s^23p^2$	$(Ar)3d^{10}4s^24p^2$	$(Kr)4d^{10}5s^25p^2$	$(Xe)5d^{10}6s^26p^2$
Electronegativity	2.5	1.8	1.8	1.8	1.8
I.E./kJ mol^{-1}	1086	786	760	710	720
Melting point/°C	3550 diamond	1410	940	232 white	328
Boiling point/°C	4830 diamond	2680	2830	2690 white	1751
Atomic radius/pm	77	117	122	140	154
Principal oxid. no.	+4	+4	+2, +4	+2, +4	+2, +4

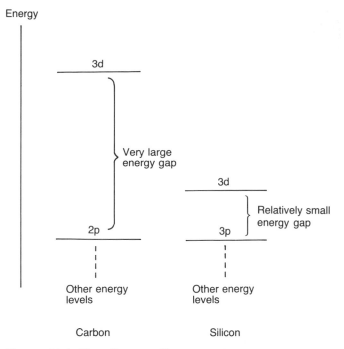

Figure 96.1 *The diagram illustrates the large difference between the 2p and 3d energy levels for carbon. The carbon 3d orbitals are so high in energy that they cannot be used in bonding. The 3p to 3d gap for silicon is relatively small. Silicon can use the 3d orbitals in bonding. (Note: the diagram is not to scale)*

96.1 (i) Which theory other than valence bond theory could be used to explain the bonding in SiF_6^{2-}?

(ii) Predict the shape of this ion.

96.2   The importance of silicon

Silicon is one of the most abundant of elements, being the essential ingredient of a large number of minerals that make up the Earth's surface. Sand, clay and the stuff we called 'earth' is mainly a mixture of compounds of silicon and oxygen. The majority of the compounds consist of silicates, in which silicon and oxygen make giant three-dimensional structures. They are all based on the same basic building block made from one silicon atom and four oxygen atoms at the corners of a tetrahedron. Figure 96.2 illustrates the more important structures, and examples of minerals in which they occur.

The name 'silica' covers an entire group of minerals, which have the general formula SiO_2, the most common of which is quartz. Quartz is a framework silicate with SiO_4 tetrahedra arranged in spirals. The spirals can turn in a clockwise or anticlockwise direction – a feature that results in there being two mirror image, optically active, varieties of quartz.

In recent years silicon and silicates have been studied from a new point of view. It has been suggested that the first primitive forms of life on Earth may have been based on silicon rather than carbon. This suggestion cannot be discounted out of hand because the silicates show that these minerals have some ability to replicate and develop their structures into long chains. However, clearly carbon won the competition (if such it was) owing to its ability to make a wider variety of more flexible structures.

Silicates are also used as the basis of ceramics. Ceramics are much better electrical and thermal insulators than metals, and have greater rigidity, hardness and temperature stability than organic polymers. Following a great deal of research, new ceramic materials have been developed that contain, for example, borides, carbides and nitrides. They have found an increasing number of uses, e.g. glasses for covering solar panels, parts of turbines and internal combustion engines, and refractory brick linings for high temperature furnaces.

96.2   Talc, $Mg_3Si_4O_{10}(OH)_2$, is the softest mineral. It has a smooth, greasy touch. It is used in talcum powder. Muscovite, $KAl_2(Si_3Al)O_{10}(OH)_2$, is one of the micas, which split into thin layers extremely easily. Both minerals have layer structures. Which type of force holds the layers together?

96.3   The extraction of the elements

(a)   Silicon

Silicon is the most important element used in semiconductors such as transistors and diodes. Without a supply of pure silicon it would be impossible to make the range of high quality computers, calculators, telephones, radios, etc., upon which modern industrial societies have come to rely. Crude silicon can be obtained by reducing silica (sand) with coke:

$$SiO_2(s) + 2C(s) \rightarrow Si(s) + 2CO(g)$$

The product is contaminated by silicon carbide, SiC_2. This carbide is also known as carborundum. It is extremely hard. The bits of steel drills are often impregnated with carborundum in order to reduce the wear on them.

Much better quality silicon is produced by reducing silicon tetrachloride with hydrogen:

$$SiCl_4(l) + 2H_2(g) \rightarrow Si(s) + 4HCl(g)$$

However, even silicon made in this way must be subjected to further purification if it is to be used in making semiconductors. The method used is called zone refining (see Figure 96.3). A cylinder of impure silicon is drawn very slowly (no faster than 1 cm per hour) through a small furnace. The key to understanding zone refining is to think about it as a variety of fractional

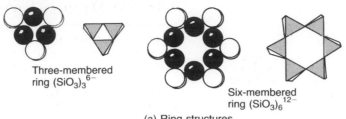

Three-membered
ring $(SiO_3)_3^{6-}$

Six-membered
ring $(SiO_3)_6^{12-}$

(a) Ring structures

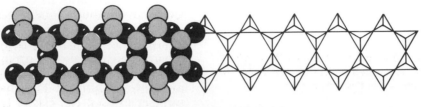

(b) Double chain structure $(Si_4O_{11})_n^{6-}$

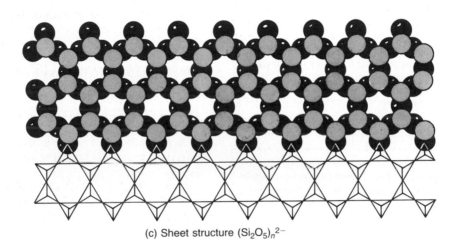

(c) Sheet structure $(Si_2O_5)_n^{2-}$

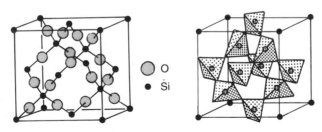

O   O

● Si

(d) Framework structure (SiO_2)

Figure 96.2 *The structures of four types of silicates. (Taken from: Sorrel, C. A. and Sandstrom, E. F. (1977).* The Rocks and Minerals of the World, *Collins, London, p. 157)*

Examples

Topaz	$Al_2SiO_4(OH)_2F_2$	*Tetrahedral units*
Beryl	$Be_3Al_2Si_6O_{18}$	*Six-membered rings*
Tremolite	$Ca_2Mg_5Si_8O_{22}(OH)_2F_2$	*Double chains*
Talc	$Mg_3Si_4O_{10}(OH)_2$	*Sheets*
Quartz	SiO_2	*Framework*

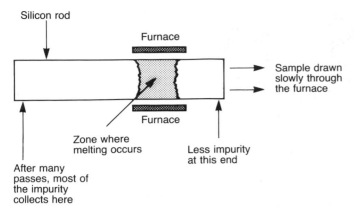

Silicon rod

Furnace

Sample drawn slowly through the furnace

Furnace

Zone where melting occurs

Less impurity at this end

After many passes, most of the impurity collects here

Figure 96.3 *Zone refining. The sample is passed through the furnace many times. The impurities collect at one end of the sample rod. (Note: the furnace is cylindrical)*

crystallisation. Inside the furnace the solid melts; and as the rod moves through it, the part that emerges begins to cool. It so happens that the pure silicon crystallises from the solution (of molten silicon plus impurities) more easily than the impurities. Therefore the liquid in the furnace tends to retain the impurities and the rod leaving the furnace is of higher purity than before. The best results (impurities less than $10^{-6}\%$) are obtained by zone refining the rod many times.

(b) *Germanium*

Germanium is isolated as a side product from zinc and copper refining. Impure germanium compounds are converted into the tetrachloride by reacting them with hydrogen chloride. Then the chloride is hydrolysed with water and the resulting oxide reduced by hydrogen or carbon. Germanium for the semiconductor industry will also be zone refined.

(c) *Tin*

The chief mineral containing tin is cassiterite, SnO_2. It can be reduced by coke in much the same way as in the iron blast furnace. Impurities, e.g. lead, can be removed by electrolysis.

(d) *Lead*

Lead is found mainly as galena, its sulphide mineral PbS. Essentially, the sulphide is converted into an oxide and then reduced with carbon. You will find details of the method in Unit 85.

96.3 Write equations for the conversion of $GeCl_4$ into GeO_2 by water, and the reduction of the oxide by hydrogen.

96.4 The hydrides

All the elements give hydrides. Carbon, of course, gives an immense number, but silicon and germanium also

Table 96.3. The major hydrides of Group IV

Element	Hydrides
Carbon	CH_4 (methane), C_2H_2 (ethyne), C_2H_4 (ethene), C_2H_6 (ethane), C_3H_8, and many, many more
Silicon	SiH_4 (silane), Si_2H_6, Si_3H_8, . . ., Si_6H_{14}
Germanium	GeH_4 (germane), Ge_2H_6, Ge_3H_8, . . ., Ge_9H_{20}
Tin	SnH_4 (stannane), Sn_2H_6
Lead	PbH_4 (plumbane)

show a wide variety (see Table 96.3). The geometries of the hydrides follow those of methane, and are based on a tetrahedral arrangement around the central atom.

The carbon hydrides will not ignite in air unless a flame is put to them. Apart from silane, SiH_4, the silicon hydrides are less well behaved. For example, Si_3H_8 is spontaneously flammable in air:

$$Si_3H_8(l) + 5O_2(g) \rightarrow 3SiO_2(s) + 4H_2O(l)$$

Like the carbon hydrides, silicon hydrides are not hydrolysed by water alone. However, traces of alkali will convert them into hydrated silica, $SiO_2 \cdot nH_2O$, and hydrogen gas. Carbon hydrides are not hydrolysed by alkali.

The hydrides of tin and lead are not very important and we shall ignore them.

96.4 What is a simple explanation for the tetrahedral shape of the hydrides?

96.5 What is the polarity of (i) a carbon–hydrogen bond, (ii) a carbon–silicon bond? (See Table 96.1 for electronegativity values; the electronegativity of hydrogen is 2.1.)

96.5 The halides

The halides are listed in Table 96.4. Once again there is a tendency for the elements to make four bonds, and

Table 96.4. The major halides of Group IV

Element	Typical halides*	Complex halides
Carbon†	CCl_4	None; d orbitals are needed
Silicon	$SiCl_4$	SiF_6^{2-}
Germanium	$GeCl_4$, $GeCl_2$	GeF_6^{2-}, $GeCl_6^{2-}$
Tin	$SnCl_4$, $SnCl_2$	SnF_6^{2-}, $SnCl_4^{2-}$, $SnCl_6^{2-}$
Lead	$PbCl_4$, $PbCl_2$	$PbCl_4^{2-}$, $PbCl_6^{2-}$

*For the simple halides only the chlorides are shown. Similar compounds are given with fluorine, bromine and iodine; except that $PbBr_4$ and PbI_4 do not exist
†As in the case of the hydrides, carbon forms many halides, e.g. C_2Cl_6, C_2F_4. Silicon has a similar, but less marked, tendency to make a variety of halides

with a tetrahedral arrangement. As with the hydrides, there is a marked difference in the hydrolysis reactions of tetrachloromethane and silicon tetrachloride. CCl_4 will not react with water, but $SiCl_4$ is immediately converted into silica:

$$SiCl_4(l) + 2H_2O(l) \rightarrow SiO_2(s) + 4H^+(aq) + 4Cl^-(aq)$$

We can explain the difference by assuming that one of the lone pairs on a water molecule can overlap with one of the empty 3d orbitals on the silicon atom. If you look back at Unit 16 you will find that we made the point that efficient overlap can only occur between orbitals that have not only the right shape (or symmetry) but also similar energies. (An orbital is not merely a region of three-dimensional space, it has the added dimension of energy.) The 3d orbitals of carbon are much higher in energy than those of silicon, so bonding cannot occur between them and a water molecule. Once electron density is fed into the silicon atom, the chlorine atoms can detach themselves by converting into chloride ions, and the $SiCl_4$ is destroyed.

The metallic nature of tin is shown by the way it dissolves in hydrochloric acid, giving off hydrogen. The solution that remains contains tin(II), Sn^{2+}, ions. These ions are good reducing agents, a property that is made use of in organic chemistry in the reduction of nitrobenzene to phenylamine (see section 122.2 for details).

The halides of lead show some interesting characteristics. Lead(IV) chloride, $PbCl_4$, decomposes very easily into lead(II) chloride and chlorine:

$$PbCl_4(s) \rightarrow PbCl_2(s) + Cl_2(g)$$

The corresponding bromides and iodides are too unstable to have any life of their own. We can relate these observations to the tendency of lead(IV) compounds to act as oxidising agents. For example, we can interpret the equation we have just looked at as an example where two changes take place:

(i) lead(IV) takes electrons
$$Pb^{4+} + 2e^- \rightarrow Pb^{2+}$$

(ii) chloride loses electrons
$$2Cl^- - 2e^- \rightarrow Cl_2$$

Given that iodide ions and bromide ions are less resistant to oxidation than chloride ions, we can appreciate why PbI_4 and $PbBr_4$ do not exist.

The reason why lead(IV) acts as an oxidising agent can be related to the inert pair effect. The B metals at the bottom of their Group have a pair of s electrons (6s in the case of lead), which tend not to be used in bonding. In the higher oxidation state, lead(IV), these electrons are involved in bonding; in the lower oxidation state, lead(II), they are not used.

On the other hand, from silicon downwards the elements all give one or more complex ions, e.g. SiF_6^{2-}. In these octahedral complexes the elements do make use of their outer s electrons, e.g. the 6s, 6p and 6d orbitals are used by lead.

Lead(II) chloride is very insoluble in water and you

Table 96.5. Soluble and insoluble salts of lead(II)

Salts	Colour	Comment
Soluble		
Lead(II) nitrate, $Pb(NO_3)_2$	Colourless	
Lead(II) ethanoate, $Pb(CH_3COO)_2$	Colourless	
Insoluble		
Lead(II) hydroxide, $Pb(OH)_2$	White	Dissolves in excess alkali
Lead(II) sulphate, $PbSO_4$	White	
Lead(II) chloride, $PbCl_2$	White	Soluble in hot water
Lead(II) iodide, PbI_2	Orange	Soluble in hot water
Lead(II) sulphide, PbS	Black	Soluble in hot nitric acid. Converted to white $PbSO_4$ by hydrogen peroxide
Lead(II) chromate(VI), $PbCrO_4$	Yellow	Soluble in ammonia and ethanoic acid solution

will see it precipitate as a white solid if you add chloride ions to a solution of lead(II) nitrate. Similarly, lead(II) iodide is insoluble in water, but this substance is a beautiful orange-yellow colour. Both of them are more soluble in hot water than in cold. In fact the majority of lead(II) salts are insoluble in water (see Table 96.5).

Lead shows the ability to make a complex with chloride ions. If ice cold concentrated hydrochloric acid has lead(IV) oxide added to it, yellow hexachloroplumbate(IV) ions, $PbCl_6^{2-}$, are made:

$$PbO_2(s) + 6Cl^-(aq) + 4H^+(aq) \rightarrow PbCl_6^{2-}(aq) + 2H_2O(l)$$

Lead(II) also gives an analogous complex ion.

Tetrachloroplumbate(II) can be made by dissolving lead(II) oxide in concentrated hydrochloric acid:

$$PbO(s) + 4Cl^-(aq) + 2H^+(aq) \rightarrow PbCl_4^{2-}(aq) + 2H_2O(l)$$

96.6 Why is it reasonable to think that the release of chloride ions from $SiCl_4$ is a favourable process?

96.7 One problem with making crystals of tin(II) chloride is that they are easily hydrolysed by warm water. Instead of dissolving tin in hydrochloric acid, how else might the anhydrous chloride be made?

96.8 Use the redox potentials below to decide if Sn^{2+} ions will (i) reduce Fe^{3+} to Fe^{2+}; (ii) reduce Fe^{2+} to Fe.

(iii) Is it possible to convert Sn^{2+} ions to Sn^{4+} ions using $Cr_2O_7^{2-}$ ions? In each case, if you think a reaction is possible, write the equation.

$E^{\ominus}_{Sn^{2+}/Sn} = -0.14\,V$; $E^{\ominus}_{Sn^{4+}/Sn^{2+}} = +0.15\,V$;

$E_{Fe^{3+}/Fe^{2+}}^{\ominus} = +0.77\,V; E_{Fe^{2+}/Fe}^{\ominus} = -0.44\,V;$
$E_{Cr_2O_7^{2-}/Cr^{3+}}^{\ominus} = +1.33\,V.$

96.9 Explain why we know that lead(IV) is acting as an oxidising agent in the change

$PbCl_4(s) \rightarrow PbCl_2(s) + Cl_2(g)$

96.10 A student was given a colourless solution and was told to add a little sodium chloride solution to it. A cloudy white precipitate was made. What conclusion should the student draw about the nature of the colourless solution?

96.6 The oxides

The oxides of Group IV are shown in Table 96.6. The oxides of carbon and silicon are predominantly covalent, but the chief oxide of silicon, SiO_2, unlike the small gaseous molecules CO and CO_2, has a giant molecular structure that is better represented by the formula $(SiO_2)_n$. GeO_2 is found with two different crystal structures and SnO_2 with three. The most important of them is the rutile structure, which we saw in Figure 32.10. This is also the structure of PbO_2. The melting points of these oxides show what a great difference there is in the bonding when compared with that in carbon dioxide.

The oxides of silicon and lead are more important than those of germanium and tin, so for the most part we will concentrate on the former rather than the latter.

Carbon and silicon oxides are definitely acidic. For example, silica behaves like carbon dioxide when it reacts with an alkali. The alkali has to be hot and concentrated if it is in solution; alternatively the silica can be heated with pellets of potassium hydroxide or sodium hydroxide, e.g.

$SiO_2(s) + 2OH^-(aq) \rightarrow SiO_3^{2-}(aq) + H_2O(l)$

Depending on whether KOH or NaOH is used, the product will be potassium or sodium silicate.

Silica will also react with metal carbonates, giving off carbon dioxide:

$SiO_2(s) + Na_2CO_3(s) \rightarrow Na_2SiO_3(s) + CO_2(g)$

Silicates are not destroyed by hydrogen ions as are carbonates. Rather, if you were to add dilute acid to

Table 96.6. The oxides of Group IV

Element	Oxides	Nature	Melting point/°C
Carbon	CO, CO_2	Acidic	CO_2: -56
Silicon	SiO_2	Acidic	SiO_2: 1610
Germanium	GeO, GeO_2	Amphoteric	GeO_2: 1116
Tin	SnO, SnO_2	Amphoteric	SnO_2: 1127
Lead	PbO, PbO_2, Pb_3O_4	Basic/amphoteric	PbO_2: decomposes at 300°C

sodium silicate solution you would find the solution turning to a gel. It is colloidal and an example of hydrated silica, $SiO_2 \cdot nH_2O$, often with $n = 2$. If the product is heated carefully it loses much of its water and gives a lumpy solid. The solid is *silica gel*, which is widely used as a drying agent. You may, for example, see small sachets of it in boxes containing electrical equipment such as videos and radios. It has the ability to lose the water it absorbs if it is heated, so the gel can be recycled.

Solutions of sodium silicate are sold as *water glass*. Water glass was once used for preserving eggs (a layer of insoluble calcium silicate is made, which keeps out air and bacteria). The solution can also be used to grow 'crystal gardens'.

The oxygen can be removed from silica by reacting it with a reactive metal like magnesium. It is possible to do this in the laboratory by heating a mixture of *dry* sand and magnesium powder in a *dry* boiling tube. The reaction is exothermic and takes place readily. Two reactions can occur, depending on the proportions of silica and magnesium:

$SiO_2(s) + 2Mg(s) \rightarrow 2MgO(s) + Si(s)$
$SiO_2(s) + 4Mg(s) \rightarrow 2MgO(s) + 2Mg_2Si(s)$

The second gives magnesium silicide, Mg_2Si. The silicon made is amorphous (i.e. without a consistent crystal structure), and if it is made as a very fine powder it is very reactive. For example, it will readily revert to SiO_2 with oxygen, and combine directly with fluorine and chlorine, e.g.

$Si(s) + 2F_2(g) \rightarrow SiF_4(l)$

It will also react with steam, giving off hydrogen:

$Si(s) + 2H_2O(g) \rightarrow SiO_2(s) + 2H_2(g)$

You may notice that this reaction is one that is also performed by many metals.

As we descend the Group, the basic character of the oxides increases. The oxides of germanium and tin are amphoteric rather than distinctly basic. For example, tin(II) oxide will dissolve in both acid and alkali:

$SnO(s) + 2H^+(aq) \rightarrow Sn^{2+}(aq) + H_2O(l)$
$SnO(s) + 4OH^-(aq) + H_2O(l) \rightarrow Sn(OH)_6^{4-}(aq)$

(a) Glass

Silica, in the form of sand, has one very important use: it is used to make glass. Glasses of one kind or another are found in various geological deposits. Glass has been made by humans for around 6000 years. Although glass gives every appearance of being a solid, its structure is best regarded as belonging to a supercooled liquid containing a mixture of silicates, e.g. $CaSiO_3$, Na_2SiO_3. There are tetrahedra of silicon and four oxygen atoms bonded together, but the arrangement is not regular, as it should be in a true solid. Indeed, we have mentioned before that over long periods of time glass will flow (see section 23.5).

The essential ingredients for making glass are silica, SiO_2, and an alkaline substance, although a wide variety of chemicals can be added to give glasses of

The manufacture of plate glass by the float process was invented by the Pilkington company in the UK.

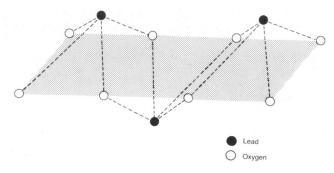

Figure 96.4 *The layer structure of lead(II) oxide, PbO*

different characteristics, e.g. colour and strength. A typical mix consists of sand, limestone, sodium carbonate, metal oxides such as PbO and MgO, and coke. In modern processes, recycled glass is also added. Molten glass is formed if the mixture is heated to about 1500° C. The way the glass is treated depends on its eventual use. It can be blown by hand to make fine articles, or blown automatically to make mass produced items such as milk bottles. Much of the glass used for shop, office and house windows is made by a float process. Here the liquid glass flows onto a bath of liquid tin. The tin gives a smooth surface for the glass, and by heating the glass from above and below a more uniform structure to the glass is obtained. The float process gives very high quality plate glass.

(b) Oxides of lead

There are three oxides of lead: lead(II) oxide, PbO; lead(IV) oxide, PbO_2; and Pb_3O_4. This last oxide is best considered as a combination of the other two, i.e. $2PbO\cdot PbO_2$. For this reason it can be called dilead(II) lead(IV) oxide. However, a more friendly name is 'red lead'.

Lead(II) oxide

Lead(II) oxide, also known as litharge, is a yellow-orange solid that is easily reduced, e.g. by heating with carbon. It is amphoteric:

$$PbO(s) + 2H^+(aq) \rightarrow Pb^{2+}(aq) + H_2O(l)$$
$$PbO(s) + 2OH^-(aq) + H_2O(l) \rightarrow Pb(OH)_4{}^{2-}(aq)$$

The crystal structure of the oxide shows that it has a considerable amount of covalent character. The repeating unit in the crystal is a square-based pyramid with lead at one apex, and four oxygen atoms at the other corners. You can see the pattern in Figure 96.4.

Lead(IV) oxide

Lead(IV) oxide is a dark brown solid. Given that lead is showing its highest oxidation state in this oxide, we might expect it to behave as an oxidising agent; and so it does. Here are two examples, involving hot concentrated hydrochloric acid and sulphur dioxide:

$$PbO_2(s) + 4HCl(aq) \rightarrow PbCl_2(s) + 2H_2O(l) + Cl_2(g)$$
$$PbO_2(s) + SO_2(g) \rightarrow PbSO_4(s)$$

When the oxide is heated it also shows a tendency to convert to the lower oxidation state. It is one of the few oxides that give off oxygen on heating:

$$2PbO_2(s) \rightarrow 2PbO(s) + O_2(g)$$

It also gives off oxygen if it is heated with concentrated sulphuric acid:

$$2PbO_2(s) + 2H_2SO_4(aq) \rightarrow 2PbSO_4(s) + 2H_2O(l) + O_2(g)$$

Unlike PbO the oxide will not simply dissolve in an acid, although it will dissolve in concentrated alkali:

$$PbO_2(s) + 2OH^-(aq) + 2H_2O(l) \rightarrow Pb(OH)_6{}^{2-}(aq)$$

Dilead(II) lead(IV) oxide

This oxide, Pb_3O_4, can be made by heating lead(II) oxide in air. Its crystal structure shows that it is predominantly covalent and composed of a combination of two types of lead atom. One type is bonded to six oxygen atoms, the other to three oxygen atoms. However, as far as its chemistry is concerned, it sometimes behaves like a mixture of PbO and PbO_2. For example, if Pb_3O_4 is heated we find that oxygen is given off:

$$2Pb_3O_4(s) \rightarrow 6PbO(s) + O_2(g)$$

We can imagine that this occurs in a different way:

$$2(2PbO\cdot PbO_2) \equiv 4PbO + 2PbO_2$$

Of this combination only the PbO_2 is affected by heat:

$$2PbO_2(s) \rightarrow 2PbO(s) + O_2(g)$$

The overall change is the same as the first equation.

This pattern is not always maintained. For example, when it reacts with sulphuric acid, oxygen is given off, but the other main product is lead(II) sulphate. On its own, lead(II) oxide will not react with the acid. See the answer to question 96.13 for the reason. With hot dilute nitric acid, Pb_3O_4 gives a precipitate of lead(IV) oxide:

$$Pb_3O_4(s) + 4HNO_3(aq) \rightarrow$$
$$PbO_2(s) + 2Pb(NO_3)_2(aq) + 2H_2O(l)$$

This is one way of making PbO_2.

96.11 Silica will react in a similar way with metal sulphates (and phosphates) as it does with carbonates, giving a solid residue and a gas. 1 mol of silica reacts with 1 mol of sodium sulphate, Na_2SO_4. Write the equation for the reaction.

96.12 Why does powdered silicon react much more quickly with steam than a lump of silicon?

96.13 If you were to attempt to dissolve lead(II) oxide in acid, which acid would you choose?

96.14 Explain why the reactions with hydrochloric acid and sulphur dioxide show the oxidising nature of PbO_2.

96.15 Predict the result of reacting Pb_3O_4 with concentrated hydrochloric acid.

96.7 Organic compounds of the elements

As we said at the beginning of this unit, we shall not concern ourselves here with the organic chemistry of carbon. On the other hand, the other elements all give compounds with carbon or carbon compounds that we can classify as organic in nature. We shall consider just a few examples.

(a) Organo-silicon compounds

Among the most important organic compounds of silicon are the silicones. A typical silicone has a chain of repeating units like this:

$$-\underset{\underset{CH_3}{|}}{\overset{\overset{CH_3}{|}}{Si}}-O-\underset{\underset{CH_3}{|}}{\overset{\overset{CH_3}{|}}{Si}}-O-\underset{\underset{CH_3}{|}}{\overset{\overset{CH_3}{|}}{Si}}-O-$$

$$-Si(CH_3)_2-O-Si(CH_3)_2-O-Si(CH_3)_2-O-$$

Typically the chains may be built of several thousand monomer units. In this chain the monomer is $-Si(CH_3)_2-O-$. They can be made by the hydrolysis of organo-silicon compounds such as $Si(CH_3)_2Cl_2$, or in general SiR_2Cl_2 where R is an organic group.

The reason why silicones are useful depends on the fact that they are kinetically and energetically stable.

Their sheer size together with the non-polar organic side chains make them immiscible with water. They are widely used as sealants and greases, where their water repellent properties are extremely useful. (It may be that a burette you use may have its tap lubricated with silicone grease.) Their ability to repel water also makes them a suitable material for treating fabrics to be used in rain wear. Others are used in waxes and polishes. Some silicones exist as crosslinked polymers. These can be hard, and resistant to wear. They find uses in insulating electrical equipment.

By changing the side chains, and using a mixture of different starting materials, the length and nature of the chain can be changed. For example, when $Si(CH_3)_2Cl_2$ loses its two chlorine atoms in hydrolysis there are two sites per molecule that can be used in bonding. This allows a chain to grow using oxygen atoms to link the silicon atoms. If we were to use $Si(CH_3)_3Cl$ as a starting material, it could not make a chain any longer than two units long. There is only one chlorine to lose, so each molecule can make but one link to another.

On the other hand, this molecule has a use if we mix it with $Si(CH_3)_2Cl_2$. When the $Si(CH_3)_3Cl$ reacts, it will stop the polymer chain growing. For example, we could get

$$CH_3-\underset{\underset{CH_3}{|}}{\overset{\overset{CH_3}{|}}{Si}}-O-\underset{\underset{CH_3}{|}}{\overset{\overset{CH_3}{|}}{Si}}-O-\underset{\underset{CH_3}{|}}{\overset{\overset{CH_3}{|}}{Si}}-O-\underset{\underset{CH_3}{|}}{\overset{\overset{CH_3}{|}}{Si}}-CH_3$$

$$Si(CH_3)_3-O-Si(CH_3)_2-O-Si(CH_3)_2-O-Si(CH_3)_3$$

By changing the proportions of $Si(CH_3)_3Cl$ and $Si(CH_3)_2Cl_2$ we can control the length of the chains.

On the other hand, if we use $SiCH_3Cl_3$ as a starting material the chain can grow in three places. This gives the opportunity for crosslinking to take place. For example, we might obtain

$$-O-\underset{\underset{O}{|}}{\overset{\overset{CH_3}{|}}{Si}}-O-\underset{\underset{O}{|}}{\overset{\overset{CH_3}{|}}{Si}}-O-$$
$$-O-\underset{\underset{CH_3}{|}}{\overset{\overset{|}{}}{Si}}-O-\underset{\underset{CH_3}{|}}{\overset{\overset{|}{}}{Si}}-O-$$

Some specialist adhesives are made by mixing two silicone-based materials, which react together when mixed to give a crosslinked polymer.

(b) Organo-lead compounds

The most famous, or infamous, organic compound of lead is tetraethyl-lead(IV), $Pb(C_2H_5)_4$. We have discussed this compound before for two reasons. First, it has been used as a petrol additive. It allows petrol to burn more smoothly, but at the expense of polluting the world with tens of thousands of tonnes of lead each year. Secondly, and related to its use in petrol, is its ability to break into free radicals. You will find more details about both these matters in Units 81 and 86.

96.16 What would be the structure of the silicone molecule made from $Si(CH_3)_3Cl$?

96.17 One type of silicone adhesive uses a polymer that has two OH groups left on the ends of the chain, e.g.

$HO—Si(CH_3)_2—[O—Si(CH_3)_2]_n—O—Si(CH_3)_2—OH$

It has been found that this kind of polymer reacts easily with an organo-silicon molecule that contains ethanoate groups, CH_3COO. Which of these three molecules:

$CH_3COOSi(CH_3)_3$
$(CH_3COO)_2Si(CH_3)_2$
$(CH_3COO)_3SiCH_3$

would you choose to add to the hydroxy polymer in order to make a strong adhesive? Briefly explain your choice.

96.8 **Sulphides**

You will find a list of the sulphides of the Group in Table 96.7. You are unlikely to come across many of them.

Table 96.7. The sulphides of Group IV

Element	*Sulphides*
Carbon	CS_2
Silicon	SiS_2
Germanium	GeS, GeS_2
Tin	SnS, SnS_2
Lead	PbS, PbS_2

We described some of the properties of carbon disulphide in the previous unit. Here, only tin(II) sulphide and lead(II) sulphide need concern us. They are precipitated from solutions by hydrogen sulphide. Tin(II) sulphide, SnS, is brown, but rapidly converts into yellow tin(IV) sulphide, SnS_2. Lead(II) sulphide, PbS, is black. In the days when many paints contained white pigments based on lead compounds, the paints would be discoloured by hydrogen sulphide present in the air of industrial towns and cities. The white colour could be recovered by washing with hydrogen peroxide solution. This has the ability to oxidise the sulphide to sulphate:

$$PbS(s) + 4H_2O_2(aq) \rightarrow PbSO_4(s) + 4H_2O(l)$$
black   white

Answers

96.1 (i) Molecular orbital theory. (ii) Octahedral.

96.2 Van der Waals forces.

96.3 $GeCl_4(l) + 2H_2O(l) \rightarrow GeO_2(s) + 4HCl(aq)$
$GeO_2(s) + 2H_2(g) \rightarrow Ge(s) + 2H_2O(l)$

96.4 Electron repulsion theory says that if there are four bonds and no lone pairs, the repulsions will be minimised if the bonds are at the tetrahedral angle.

96.5 (i) The hydrogen is slightly positive and the carbon slightly negative.
(ii) The silicon is slightly positive and the hydrogen atoms slightly negative.

96.6 Chloride ions, Cl^-, are kinetically and energetically stable in water and they can be solvated by water molecules, so they making good leaving groups in the reaction.

96.7 React tin with hydrogen chloride gas. This is the same approach needed to make anhydrous iron(II) chloride.

96.8 (i) If Sn^{2+} ions are to act as reducing agents they must gain electrons. This means they will convert into tin atoms, so we look at the values of $E^{\ominus}_{Sn^{2+}/Sn}$ and $E^{\ominus}_{Fe^{3+}/Fe^{2+}}$ to decide. $E^{\ominus}_{Sn^{2+}/Sn}$ is much more negative than $E^{\ominus}_{Fe^{3+}/Fe^{2+}}$ so the reduction should take place. The reaction is
$Sn^{2+}(aq) + 2Fe^{3+}(aq) \rightarrow Sn^{4+}(aq) + 2Fe^{2+}(aq)$
(ii) This time $E^{\ominus}_{Fe^{2+}/Fe}$ is more negative than $E^{\ominus}_{Sn^{2+}/Sn}$ so the reduction will not take place.

(iii) $E^{\ominus}_{Cr_2O_7^{2-}/Cr^{3+}}$ is much more positive than $E^{\ominus}_{Sn^{4+}/Sn^{2+}}$ so the oxidation should take place. The reaction is
$Cr_2O_7^{2-}(aq) + 3Sn^{2+}(aq) + 14H^+(aq) \rightarrow$
$\qquad\qquad 2Cr^{3+}(aq) + 3Sn^{4+}(aq) + 7H_2O(l)$
It helps to remember that $Cr_2O_7^{2-}$ ions are 'six-electron oxidising agents' (see section 41.7) so 1 mol of $Cr_2O_7^{2-}$ will react with 3 mol of Sn^{2+}.

96.9 It gains electrons.

96.10 First, the student should realise that, because sodium salts are soluble in water, it is the chloride ions that are giving the precipitate. There might be Pb^{2+} ions in the solution, but equally there might be Ag^+ ions. One way to distinguish them is to warm the solution: $PbCl_2$ dissolves in hot water. Alternatively, $AgCl$ dissolves in ammonia solution.

96.11 $SiO_2(s) + Na_2SO_4(s) \rightarrow Na_2SiO_3(s) + SO_3(g)$

96.12 It has a greater surface area.

96.13 Nitric acid. This is because the lead(II) nitrate made is soluble in water. If sulphuric acid is used, an insoluble layer of $PbSO_4$ clings to the surface of the powder and stops the reaction. Likewise, $PbCl_2$ prevents the reaction with dilute hydrochloric acid.

96.14 The lead changes from lead(IV) to lead(II). This represents a gain of electrons, and therefore an oxidation of the other reactant.

96.15 We would expect the acid to be oxidised, with chlorine given off, and the oxide to be converted into $PbCl_2$:

$$Pb_3O_4(s) + 8HCl(aq) \rightarrow 3PbCl_2(s) + 4H_2O(l) + Cl_2(g)$$

96.16 The structure is

$$CH_3-\underset{\underset{CH_3}{|}}{\overset{\overset{CH_3}{|}}{Si}}-O-\underset{\underset{CH_3}{|}}{\overset{\overset{CH_3}{|}}{Si}}-CH_3$$

$$Si(CH_3)_3-O-Si(CH_3)_3$$

96.17 $(CH_3COO)_3SiCH_3$. This should give the cross-linking necessary for strength. (Compare $SiCH_3Cl_3$ above.)

UNIT 96 SUMMARY

- The Group IV elements:
  - (i) Metallic nature increases down the Group: silicon is a non-metal; silicon and germanium are semiconductors; tin and lead show definite metallic properties.
  - (ii) Silicon is found combined with oxygen in a wide range of rocks and minerals (silicates) whose structures are based upon SiO_4 tetrahedra.
  - (iii) Silicon is of key importance in making semiconductors for the electronics and computer industries.
- Extraction:
  - (i) Silicon is obtained by reducing silicon tetrachloride with hydrogen;

    $$SiCl_4(l) + H_2(g) \rightarrow Si(s) + 4HCl(g)$$

  - (ii) Tin is isolated by the reduction of SnO_2 with carbon (coke).
  - (iii) Lead is made from its sulphide, galena;

    $$2PbS(s) + 3O_2(g) \rightarrow 2PbO(s) + 2SO_2(g)$$
    $$PbO(s) + C(s) \rightarrow Pb(s) + CO(g)$$

Reactions

- With oxygen:
  Oxides are formed;

  e.g. $Si(s) + O_2(g) \rightarrow SiO_2(s)$

- With halogens:
  Chlorides are made;

  e.g. $Si(s) + 2Cl_2(g) \rightarrow SiCl_4(l)$

Compounds

- Oxides:
  - (i) The oxides are predominantly covalent.
  - (ii) Silica, SiO_2, will undergo hydrolysis with alkali, showing an acidic nature expected of a non-metal oxide. GeO, SnO and PbO are amphoteric;

    e.g. $PbO(s) + 2H^+(aq) \rightarrow Pb^{2+}(aq) + H_2O(l)$
    $PbO(s) + 2OH^-(aq) + H_2O(l) \rightarrow$
    $$Pb(OH)_4{}^{2-}(aq)$$

  - (iii) Lead(IV) oxide, PbO_2, and dilead(II) lead(IV) oxide, Pb_3O_4, show oxidising properties;

    e.g. $PbO_2(s) + 4HCl(aq) \rightarrow$
    $$PbCl_2(s) + H_2O(l) + Cl_2(g)$$

    Pb_3O_4 behaves like a combination of lead(II) and lead(IV) oxides (PbO and PbO_2).

  - (iv) Silica can be reduced by reacting with magnesium;

    $$SiO_2(s) + 2Mg(s) \rightarrow Si(s) + 2MgO(s)$$

    Other oxides can be reduced by carbon;

    $$PbO(s) + C(s) \rightarrow Pb(s) + CO(g)$$

- Chlorides:
  Silicon tetrachloride (like tetrachloromethane) is a volatile liquid at room temperature and pressure. It will undergo hydrolysis (tetrachloromethane will not). This is the result of silicon having 3d orbitals that are much lower in energy than those of carbon and can be used in reactions.
- Sulphides:
  Tin(II) sulphide, SnS (brown), and lead(II) sulphide, PbS (black), are precipitated by hydrogen sulphide from solutions containing Sn^{2+} or Pb^{2+} ions.
- Organic:
  All the elements make a range of compounds with organic groups.
  - (i) Tetraethyl-lead(IV), $Pb(C_2H_5)_4$, is used as anti-knock in petrol.
  - (ii) The silicones contain chains of $-Si(CH_3)_2-O-$ groups, and are used as lubricants and adhesives.

97

Nitrogen

97.1 The element

The properties of nitrogen are summarised in Table 97.1. Nitrogen makes up almost 78% of the atmosphere, and is therefore the most common gas. The fact that there is so much nitrogen about tells us something about its chemistry. That is, it is particularly unreactive. Most of the oxygen and other gases that may have been produced when the Earth was being made combined with other elements to give us the solid and liquid matter that we see around us. For example, nearly all the hydrogen combined with oxygen to make the vast quantities of water on and around the Earth's surface. Nitrogen is different because there is a triple bond between the atoms in a nitrogen molecule. This gives the molecule a very high bond strength, so it is correspondingly hard to break apart. (The bond energy of N_2 is over twice as large as either H_2 or O_2.) However, nitrogen will react directly with some metals. For example, burning magnesium will continue to burn in nitrogen, giving white magnesium nitride, Mg_3N_2:

$$3Mg(s) + N_2(g) \rightarrow Mg_3N_2(s)$$

Nitrides are ionic, containing the N^{3-} ion.

Although nitrogen molecules are almost inert, nitrogen atoms can bond readily to many other atoms, mainly the non-metals hydrogen, oxygen and carbon. It shows a valency of three, e.g. in ammonia, NH_3, and amines like methylamine, CH_3NH_2. The most important compounds of nitrogen are in living things, e.g. amino acids and proteins. Among other things, the high electronegativity of nitrogen allows it to take part in hydrogen bonding and, for example, help to keep the strands of DNA together. The main classes of organic compounds that contain nitrogen are amines, amides, amino acids and proteins, and nitriles. You will find each type discussed in the units on organic chemistry.

97.2 Nitrogen, ammonia and fertilisers

The extraction and uses of nitrogen are given in Table 97.2. Nitrogen is isolated by the liquefaction of air – a process that we discussed in Unit 83. The conversion of nitrogen into ammonia through the equilibrium reaction

$$N_2(g) + 3H_2(g) \rightleftharpoons 2NH_3(g)$$

is one of the most important industrial chemical processes in the world. This is because ammonia is a feedstock for the manufacture of nitric acid and fertilisers. You can find details about the industrial processes concerned in Unit 83.

The most important fertilisers are urea, $(NH_2)_2CO$, ammonium sulphate, $(NH_4)_2SO_4$, ammonium nitrate, NH_4NO_3, and monoammonium phosphate, $NH_4H_2PO_4$. These are either used on their own or in conjunction with other chemicals, often to make NPK fertilisers. These are designed to release nitrogen (N), phosphorus (P) and potassium (K) into the soil.

Urea is made by the direct combination of ammonia

Table 97.1. Information about nitrogen

Relative atomic mass, $A_r(N)$	14
Electron structure	$1s^2 2s^2 2p^3$
Ionisation energy	1402 kJ mol^{-1}
Electronegativity	3.0
Molecular formula	N_2
Melting point	63 K ($-210°$C)
Boiling point	77 K ($-196°$C)
Bond energy	945 kJ mol^{-1}
Bond length	110 pm
Colourless, odourless, tasteless	

Table 97.2. The extraction and uses of nitrogen

Extraction	Uses
Liquefaction of air Nitrogen fixation in the Haber process	Provides an inert atmosphere for food storage and metal working When liquid, as a refrigerant, e.g. for frozen foods Manufacture of ammonia, and then fertilisers, nitric acid and other nitrogen-containing chemicals

with carbon dioxide at 180°C and a pressure of around 140 atm:

$$2NH_3(g) + CO_2(g) \rightarrow NH_2CONH_2(g) + H_2O(g)$$

Ammonium nitrate is made on an industrial scale by continuously mixing nitric acid and ammonia. If the proportions are kept properly balanced, the resulting solution will crystallise if it is dropped through a tower against a counter-current of air. The granules of the nitrate are removed, mixed with drying agents and packed ready for distribution.

Ammonium sulphate can be made in a similar manner, but using sulphuric acid of course.

97.1 If ammonium nitrate decomposes rapidly, the reaction

$$2NH_4NO_3(s) \rightarrow 2N_2(g) + O_2(g) + 4H_2O(g)$$

takes place.

(i) Calculate the enthalpy change for the reaction. The heats of formation of $NH_4NO_3(s)$ and $H_2O(g)$ are $-365.6\,kJ\,mol^{-1}$ and $-241.8\,kJ\,mol^{-1}$ respectively.

(ii) Explain why this is the basis for the use of the 'fertiliser' as an explosive.

97.3 The place of nitrogen in Nature

Over the course of millions of years of evolution, some microorganisms have developed methods of capturing nitrogen from the atmosphere and converting it into useful compounds. This process is called *nitrogen fixation*. The microorganisms manage to fix nitrogen through a complicated system of biochemical reactions involving an enzyme, nitrogenase, and adenosine triphosphate (ATP). Nitrogenase has active sites in which iron and molybdenum atoms are involved. It first converts nitrogen into ammonia, which is then used to make more complicated molecules.

Some plants fix nitrogen by use of *Rhizobium* bacteria, which are often found in the root system. Peas and other legume plants are particularly good at fixing nitrogen in this way. The virtue of this system is that such plants provide their own fertiliser. The only large-scale method that chemists have developed to fix nitrogen also involves ammonia; it is the Haber process, which we discussed earlier. However, the Haber process is extremely inefficient when compared to the way nitrogenase works. Nonetheless the demand for ammonia has remained high, largely because of the part that the gas plays in the manufacture of nitrogenous fertilisers. The increasing demand for efficiency in farming has meant that farmers have often been persuaded to take two or more crops each year from a given piece of land. By itself the land cannot supply the nutrients that the crops need, so fertilisers have to be added to the soil. Unfortunately, much of the fertiliser is wasted because rain washes it out of the soil before it can be used by the crops. Recently, it has been found that some water supplies have been contaminated by nitrogenous fertilisers. This has had the effect of encouraging the uncontrolled growth of algae and other water plants. It has also meant that nitrate ions have collected in drinking water. There is some evidence that links nitrate ions with cancer in humans, so for several environmental reasons the over-use of nitrogenous fertilisers is being discouraged.

Nitrogen is also fixed by chemical reactions in the atmosphere, particularly in thunderstorms. Here the energy released by lightning converts nitrogen into nitrogen oxides and nitric acid, which falls to earth in the rain. However, the proportion of nitrogen in the atmosphere is not continually decreasing because that lost by fixation is replaced by the decomposition of nitrogenous compounds by bacteria. This process is called bacterial denitrification. The balance between the removal and return of nitrogen to the atmosphere gives us the nitrogen cycle of Figure 97.1.

97.4 Ammonia and hydrazine

(a) Ammonia

We have discussed the bonding in ammonia in Units 14 and 17. Here we shall just summarise the key points. Ammonia is a tetrahedral molecule, with three bond pairs and one lone pair (Figure 97.2). The lone pair is responsible for decreasing the H—N—H bond angles to 107° (as compared with the perfect tetrahedral angle of 109°28'). The lone pair is involved in most of the reactions of ammonia and of amines. It can be used in coordinate bonding, for example in combining with a hydrogen ion to make an ammonium ion:

$$H^+ + NH_3 \longrightarrow NH_4^+$$

Similarly it is responsible for ammonia's ability to act as a ligand with transition metal ions, e.g.

$$Cu^{2+}(aq) + 4NH_3(aq) \rightarrow [Cu(NH_3)_4]^{2+}(aq)$$
tetraamminecopper(II)
(royal blue colour)

Ammonia is extremely soluble in water, but its solution is only a weak alkali. Although it is sometimes labelled ammonium hydroxide, it is better called aqueous ammonia. By far the majority of ammonia molecules in water remain intact. Only a few react with the solvent:

$$NH_3(aq) + H_2O(l) \rightleftharpoons NH_4^+(aq) + OH^-(aq)$$

There are many salts that contain ammonium ions, e.g. ammonium chloride, NH_4Cl, and those in the previous section, which we found were used as fertilisers. Ammonium chloride is famous for changing

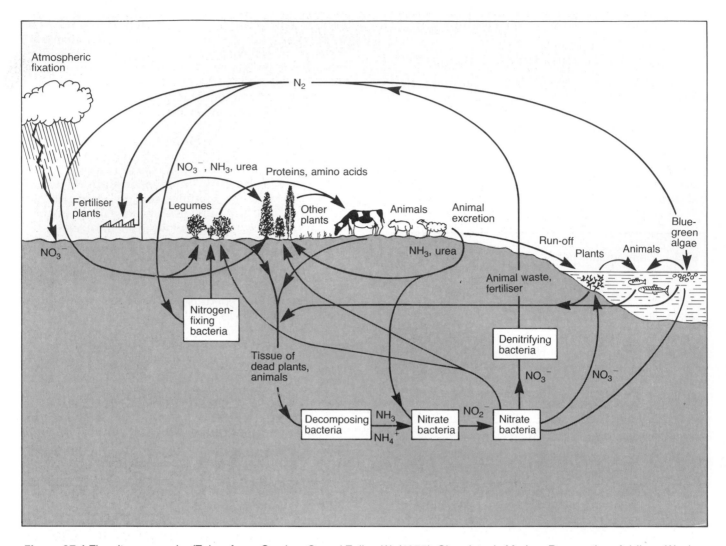

Figure 97.1 *The nitrogen cycle. (Taken from: Gordon, G. and Zoller, W. (1975).* Chemistry in Modern Perspective, *Addison-Wesley, Reading, MA, figure 11.5)*

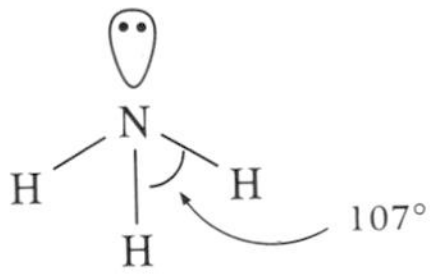

Ammonia, NH_3

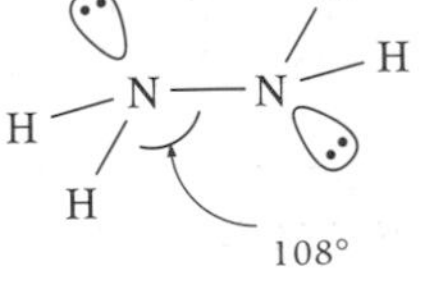

Hydrazine, N_2H_4

Figure 97.2 *The structures of ammonia and hydrazine*

from solid to gas, and vice versa, without liquefying. This is often taken as an example of sublimation, but it is something of a cheat because the vapour above hot ammonium chloride consists of a mixture of ammonia and hydrogen chloride. In other words, ammonium chloride decomposes on heating:

$$NH_4Cl(s) \rightleftharpoons NH_3(g) + HCl(g)$$

Ammonia is a reducing agent. This is shown by the way it will remove oxygen from a number of metal oxides. A good example is the reduction of hot copper(II) oxide:

$$3CuO(s) + 2NH_3(g) \rightarrow 3Cu(s) + N_2(g) + 3H_2O(l)$$

(b)   *Hydrazine*

Hydrazine has the formula N_2H_4 (Figure 97.2). It is a gas at room temperature, but it readily dissolves in water, and its solution can be used to show many of its reactions. In particular, it is a powerful reducing agent; for

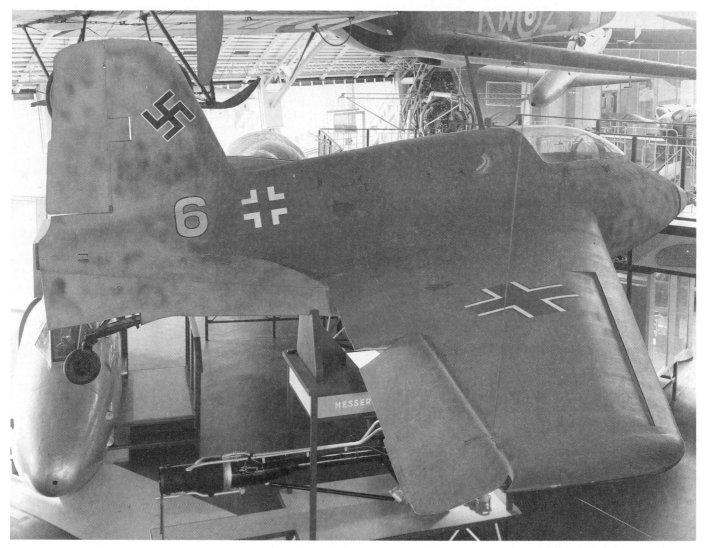

This Messerschmitt ME163 Komet was one of the early jet planes developed in Germany towards the close of the Second World War. These planes used hydrazine as a fuel. They had the unfortunate habit of exploding and killing the pilot.

example, if you drop a little aqueous hydrazine into a test tube containing silver nitrate solution you will see a silver mirror on the glass:

$$4Ag^+(aq) + N_2H_4(aq) \rightarrow 4Ag(s) + N_2(g) + 4H^+(aq)$$

Hydrazine, together with hydrogen peroxide, has been used as a rocket fuel. We might expect a vigorous reaction when a good reducing agent meets a good oxidising agent, but in this case the reaction is particularly violent:

$$N_2H_4(l) + 2H_2O_2(l) \rightarrow N_2(g) + 4H_2O(g);$$
$$\Delta H^\ominus(298\,K) = -642\,kJ\,mol^{-1}$$

One reason for this is that hydrazine is an endothermic compound, i.e. it has a positive heat of formation, so when it breaks up into its elements heat should be given out. (Even though its hydrogen is converted into water in the reaction, this only adds to the exothermic nature of the reaction.) In addition, the products of the reaction are gases, which give an explosive edge to the process.

97.2 Why does silver chloride dissolve in aqueous ammonia?

97.3 (i) Why is ammonia a polar molecule?

(ii) Why does it show a large deviation from ideal gas behaviour?

(iii) Why does ammonia have no difficulty in dissolving in water?

97.4 Sketch an apparatus that would allow you to pass ammonia over hot copper(II) oxide, and collect the water as well as the nitrogen given off.

97.5 Suggest a reason why you would expect hydrazine to dissolve easily in acids.

97.6 What would you expect to see if hydrazine solution were added to copper(II) sulphate solution?

97.5 Nitrogen oxides

The oxides of nitrogen that we shall consider are listed in Table 97.3. The oxides show the wide range of oxidation numbers that can be assigned to nitrogen, ranging from +1 in N_2O to +5 in N_2O_5. However, all the oxides are essentially covalent compounds. Except for N_2O_5, which is an unstable colourless solid, they are all gases at room temperature.

Dinitrogen oxide, N_2O, has something of a chequered history. It has been widely used as an anaesthetic. Many operations, and visits to the dentist, have been rendered (almost) painless by its use. It is also known as laughing gas, owing to its ability to induce laughter in people when it is breathed in small quantities. Among certain groups of people in Victorian times this property of the gas was used as a source of entertainment, which we would now describe as drug abuse. Less harmful is the use of the gas as a propellant in producing whipped cream from a can. Dinitrogen oxide also happens to be one of the few colourless gases that will support combustion.

Nitrogen monoxide, NO, is an odd-electron molecule (see section 16.7). Alternatively, we can say that it is a free radical. In either event, the odd electron makes the gas highly reactive. Nitrogen monoxide will not survive long in air because it reacts with oxygen (which has two unpaired electrons) to give brown fumes of nitrogen dioxide:

$$2NO(g) + O_2(g) \rightarrow 2NO_2(g)$$
$$\text{colourless} \qquad \text{brown}$$

Nitrogen dioxide, NO_2, is also an odd-electron molecule, but it is not given to reacting with oxygen. However, it will combine with itself to give dinitrogen tetraoxide:

$$2NO_2(g) \rightleftharpoons N_2O_4(g)$$

This is an equilibrium reaction that is influenced by both pressure and temperature. If you want to remind yourself about the details of the equilibrium, look at section 53.6. Nitrogen dioxide will often support combustion. This happens if the temperature reaches above 150°C when the gas decomposes:

$$2NO_2(g) \rightarrow 2NO(g) + O_2(g)$$

It is the presence of the oxygen that aids the combustion. Many hot metals will even decompose the nitrogen monoxide, giving nitrogen as one of the products.

Dinitrogen tetraoxide, N_2O_4, solidifies at −11.2°C, and is colourless. As the temperature increases, the proportion of brown NO_2 increases and the colour of the liquid darkens. At around 140°C the mixture is almost entirely NO_2.

Dinitrogen trioxide, N_2O_3, has no significant claim to fame. It melts at a little below −100°C, and promptly decomposes. Such is life.

Dinitrogen pentaoxide, N_2O_5, is made when concentrated nitric acid is dehydrated. Therefore we can regard it as the anhydride of nitric acid. However, its formula N_2O_5 belies the nature of the solid, which is ionic and consists of nitryl cations, NO_2^+, and nitrate, NO_3^-, ions.

97.7 Apart from N_2O, which other colourless gas supports combustion?

97.8 Write an equation for the reaction of burning magnesium with nitrogen dioxide.

97.6 Nitrogen oxides and air pollution

Collectively, nitrogen oxides are often given the symbol NO_x. Two of them, nitrogen monoxide and nitrogen dioxide, are of special importance in environmental

Table 97.3. The oxides of nitrogen

Name	Formula	Preparation
Dinitrogen oxide (nitrous oxide)	N_2O	Heat ammonium nitrate $NH_4NO_3(s) \rightarrow N_2O(g) + 2H_2O(g)$
Nitrogen monoxide (nitric oxide)	NO	Add copper turnings to 5 mol dm^{-3} nitric acid $3Cu(s) + 8HNO_3(aq) \rightarrow 3Cu(NO_3)_2(aq) + 2NO(g) + 4H_2O(l)$
Nitrogen dioxide	NO_2	Heat lead(II) nitrate $2Pb(NO_3)_2(s) \rightarrow 2PbO(s) + 4NO_2(g) + O_2(g)$
Dinitrogen tetraoxide	N_2O_4	Cool nitrogen dioxide $N_2O_4(g) \underset{\text{low temp}}{\overset{\text{high temp}}{\rightleftharpoons}} 2NO_2(g)$
Dinitrogen trioxide	N_2O_3	Combine NO and NO_2 $NO(g) + NO_2(g) \rightarrow N_2O_3(g)$
Dinitrogen pentaoxide	N_2O_5	Dehydrate nitric acid with P_2O_5 $2HNO_3(l) - H_2O(l) \rightarrow N_2O_5(s)$

chemistry. They are responsible for a considerable amount of air pollution. They are given off in car exhaust fumes and when fossil fuels are burned, as well as being made during thunderstorms. In each case nitrogen monoxide is made when nitrogen and oxygen in air react at a high temperature:

$$N_2(g) + O_2(g) \rightarrow 2NO(g)$$

and then the nitrogen monoxide goes on to make nitrogen dioxide. Of the two, it is the nitrogen dioxide that is the more unpleasant. Not only does it give rise to acid rain, it also takes part in photochemical reactions, which results in the photochemical smogs enveloping cities such as Los Angeles. The photons making up sunlight can cause the NO_2 molecules to fall apart, releasing oxygen atoms:

$$NO_2(g) \xrightarrow{hf} NO(g) + O(g)$$

Oxygen atoms are extremely reactive and will attack other molecules, especially those in car exhaust fumes. For example, hydrocarbons that escape combustion can be changed into aldehydes. The oxygen atoms can also react with oxygen molecules to make trioxygen (ozone, O_3).

Nitrogen dioxide is also involved in the production of one of the most unpleasant of all chemicals present in photochemical smog. It is known as PAN (peroxyacetyl nitrate):

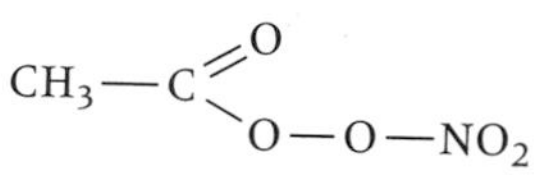

Between them, PAN, the aldehydes, ozone and nitrogen dioxide make a most unpleasant mix. They irritate eyes and lungs, and generally make breathing difficult.

97.7   The chemistry of nitric acid

You will find the industrial manufacture of nitric acid described in Unit 83.

Nitric acid, HNO_3, can be made in the laboratory by warming sodium nitrate with concentrated sulphuric acid. Nitric acid is more volatile than sulphuric acid, and is given off as a vapour:

$$NaNO_3(s) + H_2SO_4(l) \rightarrow NaHSO_4(s) + HNO_3(g)$$

The vapour condenses to a yellow fuming liquid ('fuming nitric acid'). It takes on this colour owing to nitrogen dioxide made during the reaction being absorbed by the acid. It is colourless when pure. On heating it decomposes at high temperatures, giving off steam, oxygen and nitrogen dioxide.

Nitric acid is both a very strong acid in water, and a vigorous oxidising agent.

In water, as we would expect for a strong acid, the equilibrium

$$HNO_3(aq) + H_2O(l) \rightarrow NO_3^-(aq) + H_3O^+(aq)$$

lies far to the right. The acid is a good source of hydro-gen ions and shows many of the properties of acids, e.g. it neutralises a base, and gives carbon dioxide with carbonates or hydrogencarbonates. However, it does not always give hydrogen with metals. The reason is that its oxidising power wins over its action as an acid. In fact it is only with extremely dilute nitric acid that magnesium will behave in its typical manner and give off hydrogen.

(a)   *The oxidising power of nitric acid*

To appreciate fully the varieties of reaction that can occur, you will benefit from reading about redox reactions in Unit 41. Here is a list of the major half-reactions that may take place:

$$2NO_3^-(aq) + 4H^+(aq) + 2e^- \rightarrow 2NO_2(g) + 2H_2O(l);$$
$$E^\ominus = +0.80\,V$$
$$NO_3^-(aq) + 4H^+(aq) + 3e^- \rightarrow 2NO(g) + 2H_2O(l);$$
$$E^\ominus = +0.95\,V$$
$$NO_3^-(aq) + 10H^+(aq) + 8e^- \rightarrow NH_4^+(aq) + 3H_2O(l);$$
$$E^\ominus = +0.88\,V$$

With metals

Given that $E^\ominus_{Cu^{2+}/Cu} = +0.34\,V$, we can see that copper will reduce nitric acid. (Copper has the more negative redox potential.) In principle each of the three half-reactions could take place; in practice the first predominates if concentrated nitric acid is used. The overall reaction is

$$Cu(s) + 4H^+(aq) + 2NO_3^-(aq) \rightarrow$$
$$Cu^{2+}(aq) + 2NO_2(g) + 2H_2O(l)$$

On the other hand, if the concentrated acid is diluted by about 50%, nitrogen monoxide is released:

$$3Cu(s) + 8H^+(aq) + 2NO_3^-(aq) \rightarrow$$
$$3Cu^{2+}(aq) + 2NO(g) + 4H_2O(l)$$

With non-metals

The pattern is for concentrated nitric acid to oxidise a non-metal to a high oxidation state, and give off brown fumes of nitrogen dioxide at the same time. Fuming nitric acid gives the most vigorous reaction. Be warned: these reactions can be violent, and nitrogen dioxide is very poisonous. Typical reactions are

$$S(s) + 6HNO_3(l) \rightarrow H_2SO_4(aq) + 6NO_2(g) + 2H_2O(l)$$
$$I_2(s) + 10HNO_3(l) \rightarrow 2HIO_3(aq) + 10NO_2(g) + 4H_2O(l)$$
$$P(s) + 5HNO_3(l) \rightarrow H_3PO_4(aq) + 5NO_2(g) + H_2O(l)$$

97.9 Sketch an apparatus that you could use to collect liquid nitric acid made in the reaction between $NaNO_3$ and H_2SO_4.

97.10 Explain how the equation

$$Cu(s) + 4H^+(aq) + 2NO_3^-(aq) \rightarrow$$
$$Cu^{2+}(aq) + 2NO_2(g) + 2H_2O(l)$$

was obtained from the half-equations given earlier in this section.

97.11 Zinc is a sufficiently good reducing agent ($E^{\ominus}_{Zn^{2+}/Zn} = -0.76$ V) to give ammonium ions in its reaction with nitric acid. Write the equation for the reaction.

97.12 A reasonably safe reaction using fuming nitric acid is to drop it onto warm sawdust. What do you think happens?

97.13 Nitric acid will also oxidise ions in solution. For example Fe^{2+} to Fe^{3+}, and sulphite, SO_3^{2-}, to sulphate, SO_4^{2-}.

(i) With Fe^{2+}, nitrogen monoxide is given off. Write the equation. Why would you still see brown fumes evolved?

(ii) With SO_3^{2-}, nitrogen dioxide is given off. Write the equation.

97.8 Nitrates

There are five things you should know about nitrates.

(i) There is a rule in chemistry that says that all nitrates are soluble. This is the reason why, for example, if you want a solution of silver ions you will use a solution of silver nitrate; most other silver salts are insoluble in water. (For once this is a rule that has no exceptions.)

(ii) Nitrates of Group I metals decompose in a different way to other nitrates. They give off oxygen and turn into nitrites, e.g.

$$2KNO_3(s) \rightarrow 2KNO_2(s) + O_2(g)$$

(iii) Nearly all other nitrates give off oxygen and nitrogen dioxide when they are heated. The metal oxide is left. For example,

$$2Pb(NO_3)_2(s) \rightarrow 2PbO(s) + 4NO_2(g) + O_2(g)$$
$$2Cu(NO_3)_2(s) \rightarrow 2CuO(s) + 4NO_2(g) + O_2(g)$$

You should take care if you heat lead(II) nitrate. It crackles and the crystals fly into pieces once they start to give off the gases. Copper(II) nitrate is less tiresome, but it contains a large amount of water of crystallisation. When it is heated it might seem to melt; actually it is dissolving in the water of crystallisation. Reactions like these are best done in a fume cupboard, owing to the danger of inhaling nitrogen dioxide.

(iv) Two exceptions to the last pattern are the nitrates of mercury and silver. These metals have oxides that are themselves destroyed by heat and they decompose to mercury and silver respectively.

(v) Ammonium nitrate decomposes to give dinitrogen oxide; see Table 97.3.

97.9 Nitrous acid and nitrites

Nitrous acid has the formula HNO_2, but it can also be thought of as a mixture of nitrite ions, NO_2^-, and

hydrogen ions in water. The acid cannot exist in its own right free from water. It appears as a pale blue solution when a nitrite is dissolved in dilute hydrochloric acid. In order to prevent the solution decomposing it must be kept below 5°C. The main use of the acid is in organic chemistry where, among other things, it is used in making dyes. You can find details in Unit 122.

Unlike nitric acid, nitrous acid tends to react as either an oxidising or a reducing agent. As a reducing agent, it will decolourise acidified potassium manganate(VII) solution:

$$2MnO_4^-(aq) + 6H^+(aq) + 5NO_2^-(aq) \rightarrow$$
purple
$$2Mn^{2+}(aq) + 5NO_3^-(aq) + 3H_2O(l)$$
colourless

As an oxidising agent it will, for example, oxidise Fe^{2+} to Fe^{3+}:

$$2Fe^{2+}(aq) + 4H^+(aq) + 2NO_2^-(aq) \rightarrow$$
$$2Fe^{3+}(aq) + 2NO(g) + 2H_2O(l)$$

We can think of these two reactions in terms of the redox reactions:

(i) as an oxidising agent

$$2HNO_2(aq) + 2H^+(aq) + 2e^- \rightarrow$$
$$2NO(g) + 2H_2O(l); \quad E^{\ominus} = +0.99 \text{ V}$$

(ii) as a reducing agent

$$NO_3^-(aq) + 3H^+(aq) + 2e^- \rightarrow$$
$$HNO_2(aq) + H_2O(l); \quad E^{\ominus} = +0.94 \text{ V}$$

In the first reaction, the acid is acting as an oxidising agent, taking electrons from the chemical with which it reacts. The second shows what will happen if the acid is oxidised by another chemical: it is converted into nitrate ions. The acid will behave as a reducing agent if it becomes involved with a half-reaction that has a redox potential more positive than +0.94 V. This is the case with acidified manganate(VII) ions, for which $E^{\ominus}_{MnO_4^-/Mn^{2+}} = +1.51$ V. However, $E^{\ominus}_{Fe^{3+}/Fe^{2+}} = +0.77$ V, so in this case nitrous acid will be the oxidising agent and Fe^{2+} will be the reducing agent.

97.14 Nitrous acid is able to oxidise sulphite ions, SO_3^{2-}, to sulphate, SO_4^{2-}. Write the equation for the reaction.

97.10 Nitrogen halides

The most important halides are NF_3 (a gas) and NCl_3 (an oily liquid). Nitrogen trifluoride is almost inert, but the trichloride is easily hydrolysed:

$$NCl_3(l) + 3H_2O(l) \rightarrow NH_3(aq) + 3HOCl(aq)$$

NCl_3 can be made by passing chlorine into a solution of ammonium chloride:

$$NH_4Cl(aq) + 3Cl_2(g) \rightarrow NCl_3(l) + 4HCl(aq)$$

It is not a good idea to try this reaction because NCl_3 has an unfortunate tendency to explode.

Both NBr_3 and NI_3 can be obtained in a similar way but combined with ammonia, as in $NI_3 \cdot NH_3$. The latter compound is also explosive.

Answers

97.1 (i) Heat of reaction
$$= \Delta 4H_f^{\ominus}(H_2O(g)) - \Delta H_f^{\ominus}(NH_4NO_3)$$
$$= -761.6 \, kJ \, mol^{-1}$$

(ii) Not only is the reaction highly exothermic, the change is from a solid of negligible volume to a mixture of gases of large volume. It is the sudden, violent, expansion of the gases that represents the explosion.

97.2 The soluble complex ion $Ag(NH_3)_2{}^+$ is made (see section 53.8).

97.3 (i) The nitrogen pulls electron density towards itself, and away from the hydrogen atoms. Thus the nitrogen carries a slight negative charge, and the hydrogen atoms a slight positive charge. The separation of charge gives rise to the dipole moment.

(ii) Owing to the dipole moment, there are strong intermolecular forces between the molecules.

(iii) It can hydrogen bond with water molecules.

97.4 Figure 97.3 shows the sketch.

97.5 Hydrazine has two lone pairs. These can be protonated, in much the same way as the lone pair on the nitrogen in ammonia is protonated in acid.

97.6 A copper mirror is produced and the blue colour of the solution fades.

97.7 Oxygen.

97.8 $4Mg(s) + 2NO_2(g) \rightarrow 4MgO(s) + N_2(g)$

97.9 A simple distillation apparatus using all glass apparatus should be used, as in Figure 115.5, except that the vessel containing the nitric acid distillate should be cooled.

97.10 We can produce the equation by noting that the conversion
$$Cu(s) \rightarrow Cu^{2+}(aq) + 2e^-$$
represents the loss of 2 mol of electrons. The first nitric acid half-equation involves the gain of 2 mol of electrons. Hence, all we have to do is to add the two half-equations.

97.11 $4Zn(s) + 10H^+(aq) + NO_3{}^-(aq) \rightarrow$
$$4Zn^{2+}(aq) + 3H_2O(l) + NH_4{}^+(aq)$$

97.12 The sawdust is oxidised, with brown fumes of nitrogen dioxide being given off. It may burst into flame.

97.13 (i) $3Fe^{2+}(aq) + 4H^+(aq) + NO_3{}^-(aq) \rightarrow$
$$3Fe^{3+}(aq) + 2H_2O(l) + NO(g)$$
The nitrogen monoxide will react with oxygen in the air to give brown nitrogen dioxide.

(ii) $SO_3{}^{2-}(aq) + 2H^+(aq) + 2NO_3{}^-(aq) \rightarrow$
$$SO_4{}^{2-}(aq) + H_2O(l) + 2NO_2(g)$$

97.14 $SO_3{}^{2-}(aq) + 2H^+(aq) + 2NO_2{}^-(aq) \rightarrow$
$$SO_4{}^{2-}(aq) + 2NO(g) + H_2O(l)$$

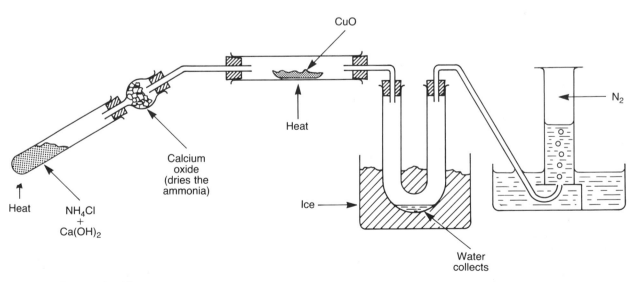

Figure 97.3 *One method in answer to question 97.4*

- Nitrogen:
  - (i) Makes up almost 78% of the atmosphere, and is the most common gas.
  - (ii) Shows very few reactions, mainly as a result of the very high $N\equiv N$ bond strength.
  - (iii) Is used in the Haber process to make ammonia;

    $$N_2(g) + 3H_2(g) \rightleftharpoons 2NH_3(g)$$

    Conditions: pressures between 15 and 30 MPa (150 and 300 atm), temperature around 500°C, iron catalyst with a promoter (e.g. aluminium oxide, zirconium oxide, potassium oxide).
  - (iv) Is fixed by microorganisms, i.e. converted direct from the atmosphere into biologically active compounds. (Also, see the nitrogen cycle of Figure 97.1.)
- Isolation:
  By the liquefaction of air followed by fractional distillation.

Reactions

- With burning magnesium:
  An ionic nitride is made;

  $$3Mg(s) + N_2(g) \rightarrow Mg_3N_2(s)$$

- With hydrogen:
  Makes ammonia in the Haber process;

  $$N_2(g) + 3H_2(g) \rightleftharpoons 2NH_3(g)$$

Compounds

- Oxides:
  - (i) Dinitrogen oxide ('laughing gas'), N_2O, supports combustion.
  - (ii) Nitrogen monoxide, NO, is a neutral oxide that readily combines with oxygen making brown fumes of nitrogen dioxide;

    $$2NO(g) + O_2(g) \rightarrow 2NO_2(g)$$

  - (iii) Nitrogen dioxide (brown) is an acidic oxide. It exists in equilibrium with dinitrogen tetra-oxide (colourless);

    $$2NO_2(g) \rightleftharpoons N_2O_4(g)$$

  - (iv) Nitrogen oxides are unpleasant atmospheric pollutants.
- Hydrides:
  Ammonia, NH_3
  - (i) Has a lone pair of electrons, which can be protonated by hydrogen ions to give the ammonium ion;

    $$NH_3(aq) + H^+(aq) \rightarrow NH_4^+(aq)$$

    Ammonium ions are found in a variety of ammonium salts. These give off ammonia when warmed with alkali.
  - (ii) Owing to its lone pair, an ammonia molecule can act as a ligand with transition metal ions, e.g. in making $Cu(NH_3)_4^{2+}$.
  - (iii) Ammonia is a reducing agent;

    e.g. $3CuO(s) + 2NH_3(g) \rightarrow$
    $$3Cu(s) + N_2(g) + 3H_2O(l)$$

  Hydrazine, N_2H_4
  - (i) Is an endothermic compound and a vigorous reducing agent;

    e.g. $4Ag^+(aq) + N_2H_4(aq) \rightarrow$
    $$4Ag(s) + N_2(g) + 4H^+(aq)$$

  - (ii) With hydrogen peroxide, it is used as a rocket fuel.
- Nitric acid, HNO_3:
  - (i) Made by warming a mixture of concentrated sulphuric acid and sodium nitrate;

    $$NaNO_3(s) + H_2SO_4(l) \rightarrow NaHSO_4(s) + HNO_3(g)$$

  - (ii) Is a strong acid in water;

    $$HNO_3(l) + H_2O(l) \rightarrow H_3O^+(l) + NO_3^-(aq)$$

    and will, for example, give CO_2 with carbonates.
  - (iii) Shows oxidising ability when concentrated;

    e.g. $S(s) + 6HNO_3(l) \rightarrow$
    $$H_2SO_4(aq) + 6NO_2(g) + 2H_2O(l)$$
    $$Cu(s) + 4H^+(aq) + 2NO_3^-(aq) \rightarrow$$
    $$Cu^{2+}(aq) + 2NO_2(g) + 2H_2O(l)$$

    Often, brown fumes of nitrogen dioxide are produced in its oxidation reactions.
  - (iv) 50% nitric acid liberates nitrogen monoxide with copper;

    $$3Cu(s) + 8H^+(aq) + 2NO_3^-(aq) \rightarrow$$
    $$3Cu^{2+}(aq) + 2NO(g) + 4H_2O(l)$$

- Nitrates:
  - (i) All nitrates are soluble in water. They are widely used as fertilisers.
  - (ii) Group I nitrates give off oxygen but not nitrogen dioxide when they are heated;

    e.g. $2KNO_3(s) \rightarrow 2KNO_2(s) + O_2(g)$

  - (iii) Heavy metal nitrates give off both oxygen and nitrogen dioxide when heated;

    e.g. $2Pb(NO_3)_2(s) \rightarrow$
    $$2PbO(s) + 4NO_2(g) + O_2(g)$$

  - (iv) Silver and mercury nitrates decompose to the metal rather than oxide.
  - (v) Ammonium nitrate decomposes to give dinitrogen oxide;

    $$2NH_4NO_3(s) \rightarrow 2N_2O(g) + 4H_2O(g)$$

- Nitrous acid, HNO_2:

    Can act as both an oxidising and reducing agent, e.g. oxidises Fe^{2+} to Fe^{3+}, and will reduce (decolourise) potassium manganate(VII) solution.

- Halides:

    Nitrogen trichloride, NCl_3, is easily hydrolysed;

    $$NCl_3(l) + 3H_2O(l) \rightarrow NH_3(aq) + 3HOCl(aq)$$

98

Group V

98.1 The nature of the elements

In this unit we shall see how the properties of the elements in Group V change as we go down the Group. You will not find details of the chemistry of nitrogen here; the previous unit is the place to find out about nitrogen.

Nitrogen and phosphorus show the typical properties of non-metals. For example, they are poor conductors of heat and electricity and give acidic oxides. Their compounds are predominantly covalent. However, bismuth at the bottom of the Group shows definite metallic properties, e.g. it conducts electricity and makes salts such as bismuth nitrate, $Bi(NO_3)_3$. This trend is typical of the behaviour of elements in all the Groups of the Periodic Table: metallic nature increases going down a Group. This is also reflected in the electronegativities of the elements. Nitrogen has the greatest tendency to attract electrons, antimony and bismuth the least. The properties and uses of the elements are summed up in Tables 98.1 and 98.2.

As you might expect then, the trend down the Group is a move from covalent bonding to ionic bonding, with the majority of compounds showing a mix of each. That is, many of the bonds made by the elements are polar.

Phosphorus, arsenic and antimony have allotropes. However, only the allotropes of phosphorus (red and

Table 98.2. Uses of the elements in Group V

Element	Main uses
Nitrogen	Manufacture of ammonia, fertilisers and nitric acid
Phosphorus	In phosphoric acid and phosphate fertilisers Rust preventatives Matches, flares, fireworks and explosives
Arsenic	In alloys The manufacture of semiconductors, e.g. light emitting diodes Agricultural insecticides
Antimony	Treatment of parasitic diseases, in insecticides Antimony sulphide is used in fireworks
Bismuth	In alloys * In medicines

*Alloys of bismuth often have melting points between 50 and 250° C. This makes them useful in fire safety devices. For example, part of the valve in a water sprinkler can be made of a bismuth alloy. When a fire causes the temperature to rise the valve will open automatically as the alloy melts

white phosphorus) are of importance. We have met these in Unit 57, so look there if you want to remind yourself about them.

Table 98.1. Physical properties of the elements in Group V

Symbol	Nitrogen N	Phosphorus P	Arsenic As	Antimony Sb	Bismuth Bi
Electron structure	$(He)2s^22p^3$	$(Ne)3s^23p^3$	$(Ar)3d^{10}4s^24p^3$	$(Kr)4d^{10}5s^25p^3$	$(Xe)5d^{10}6s^26p^3$
Electronegativity	3.0	2.1	2.0	1.9	1.9
I.E./kJ mol^{-1}	1402	1012	950	830	700
Melting point/° C	−210	597 red 44 white	817	630	272
Boiling point/° C	−204	431 red 281 white	613	1637	1559
Atomic radius/pm	74	110	121	141	152
Principal oxid. no.	+3,+5	+3,+5	+3,+5	+3,+5	+3*

* Bismuth will show an oxidation number of +5, but it is not common

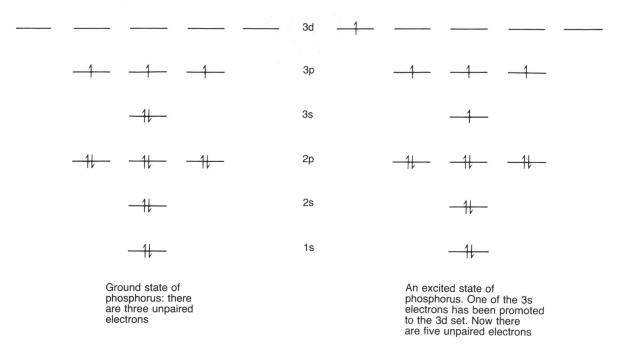

Figure 98.1 *One way of explaining how phosphorus can make three or five covalent bonds*

Phosphorus and the other members of the Group can make use of d orbitals in their bonding. This is because the energy of these orbitals is not a lot greater than those of the other valence electrons. For example, phosphorus can make use of its 3s, 3p and the empty 3d orbitals. If we imagine that the pair of 3s electrons can be parted, and that one of the five valence electrons is promoted to a 3d orbital, then we would expect that phosphorus could make a maximum of five single covalent bonds (see Figure 98.1). If the 3s electrons are not split, then phosphorus should make three covalent bonds. Indeed, three and five are the favoured valencies of the Group V elements (Figure 98.2). The exception is nitrogen, whose 3d orbitals are far too high in energy to be used in bonding.

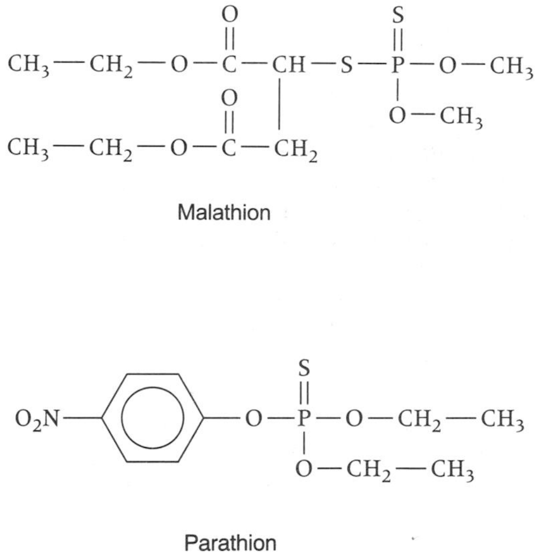

> **98.1** Does the polarisability of the atoms increase or decrease going down the Group? Briefly explain your answer.

98.2 The extraction of the elements

(a) *Phosphorus*

Fluorapatite, $CaF_2 \cdot 3Ca_3(PO_4)_2$, is the chief ore of phosphorus. The phosphorus is removed from the ore by first grinding it to remove large lumps, and then heating it in a mixture with sand (silica, SiO_2) and coke (carbon). The reactions take place at around 1500°C. This very high temperature is reached by using an electric arc furnace (see Figure 98.3). Be careful that you do not think of this as an electrolysis. Current is not passed

Malathion

Parathion

Figure 98.2 *Two organic phosphorus insecticides in which phosphorus is showing a valency of five*

through the chemicals, it is only used to heat them. The key reaction is

$$2Ca_3(PO_4)_2 + 6SiO_2 + 10C \rightarrow 6CaSiO_3 + 10CO + P_4$$

The phosphorus vapour is allowed to escape from the furnace and is condensed by passing it into water spraying towers. The product is white phosphorus, which can be sold as it stands or first converted into red phosphorus. The conversion is done by heating white phosphorus at 400°C for some hours (or at a lower temperature for a longer time).

The chief reason for extracting the phosphorus is to convert it into phosphoric(v) acid, H_3PO_4 (Figure 98.3). This is done by burning red phosphorus in excess air, which produces a dusty cloud of phosphorus(v) oxide (tetraphosphorus decaoxide):

$$P_4(s) + 5O_2(g) \rightarrow P_4O_{10}(s)$$

The fumes are sprayed with water, which produces the acid:

$$P_4O_{10}(s) + 6H_2O(l) \rightarrow 4H_3PO_4(aq)$$

Phosphoric(v) acid has many uses, among the most important of which is the production of superphosphates – see section 98.6 below.

(b) Arsenic, antimony and bismuth

Each of these elements is often found as its sulphide or oxide in deposits of metal ores, especially those of lead and copper. They are extracted from the waste products of the copper or lead extraction processes. We need not worry about the details.

98.3 The hydrides

Just as nitrogen makes ammonia, NH_3, so the other elements give similar hydrides (Table 98.3). They all have the same pyramidal shape, although their bond angles differ from that of ammonia. All the hydrides have unpleasant smells, but AsH_3 and SbH_3 are additionally dangerous because they are extremely poisonous. Phosphine can be made by warming white phosphorus with sodium hydroxide solution. It is necessary to exclude air from the apparatus because, although pure PH_3 will not burn unless ignited, it is often contaminated with P_2H_4, which is spontaneously flammable. The equation is

$$P_4(s) + 3OH^-(aq) + 3H_2O(l) \rightarrow PH_3(g) + 3H_2PO_2^-(aq)$$

Like ammonia, phosphine can accept a proton onto its lone pair and give the phosphonium ion, PH_4^+, and it

Table 98.3. The major hydrides of Group V*

Hydride	Formula	Bond angle	Melting point/° C	Boiling point/° C
Ammonia	NH_3	106°45′	−78	−33.5
Phosphine	PH_3	94°	−136	−87
Arsine	AsH_3	91°30′	−114	−55
Stibine	SbH_3	91°30′	−88	−18

*Bismuthine, BiH_3, is very difficult to prepare and decomposes extremely easily. Nitrogen also forms hydrazine, N_2H_4. There is an analogous hydride of phosphorus, P_2H_4

will combine with hydrogen iodide to make phosphonium iodide, PH_4I. Like the analogous ammonium salts it is ionic.

However, phosphine will not accept protons as readily as ammonia, but as if to compensate it is a much more powerful reducing agent than ammonia. For example, if phosphine is bubbled through silver nitrate solution, silver metal is released:

$$PH_3(g) + 6Ag^+(aq) + 3H_2O(l) \rightarrow$$
$$H_3PO_3(aq) + 6Ag(s) + 6H^+(aq)$$

One of the characteristics of the hydrides of arsenic, antimony and bismuth is that they are decomposed by heat (AsH_3 the least and BiH_3 the most easily). This property has been used in an important test for arsenic called Marsh's test. The substance thought to contain arsenic is placed in a mixture of zinc and sulphuric acid. The hydrogen given off converts any arsenic into AsH_3. The gases from the reaction flask are led through a heated glass tube. If AsH_3 is present, a black deposit of solid arsenic will be seen. The reason why this test has been widely used is the unfortunate tendency of some people to murder their acquaintances by poisoning them with arsenic. Marsh's test has been a cornerstone

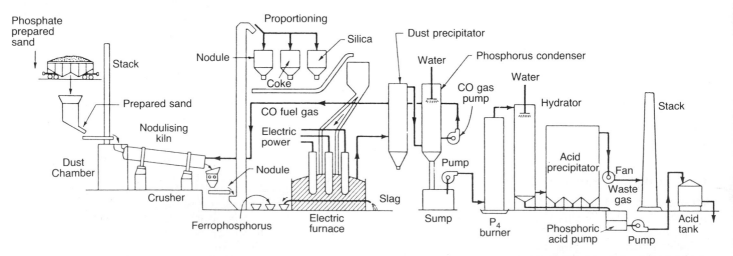

Figure 98.3 An outline of the manufacture of phosphoric(v) acid. In order to produce 1 tonne of 85% H_3PO_4, the following materials and utilities are needed: phosphate rock (35.6% P_2O_5), 1000 kg; silica rock, 320 kg; coke, 377 kg; iron, depends on ferrophosphorus requirements; electricity, 13 840 MJ; direct labour (est.), 0.5–1 work hours. (Taken from Austin, G. T. (1984). Shreve's Chemical Process Industries, 5th edn, McGraw-Hill, New York)

of forensic science. However, fashions in murder change and the use of arsenic is a little behind the times now.

98.2 Look at the values of the bond angles in Table 98.3.

(i) Why do the molecules all have a tetrahedral shape?

(ii) Explain the change in the bond angle going from NH_3 to AsH_3. (Hint: look back at Unit 17, and also take notice of the electronegativities of the elements.)

98.3 (i) Briefly describe how you would make and collect phosphine.

(ii) What is special about the changes in oxidation number of phosphorus in the reaction?

98.4 When phosphine burns in air, phosphorus(V) oxide, P_4O_{10}, is produced. What else do you think is made? Write the equation for the reaction.

98.5 Phosphine will also reduce Cu^{2+} ions, but this time phosphoric(V) acid, H_3PO_4, is formed rather than phosphonic acid, H_3PO_3. Write the equation for the reaction.

98.6 Why is it likely that phosphine and chlorine would react? Predict what will be made in the reaction.

98.7 What would you expect to happen if (i) NH_4Cl and (ii) PH_4I are warmed with sodium hydroxide solution?

98.8 Predict the outcome of heating PH_4I.

98.4 The halides and oxohalides

The major halides that the elements make are listed in Table 98.4. You can see that they all give trihalides and all but bismuth give one or more pentahalides. Of these the phosphorus chlorides are by far the most important, and we shall use them as examples.

Table 98.4. The principal halides of Group V

Element	Trihalides	Pentahalides
Nitrogen	NF_3, NCl_3, NBr_3, NI_3	None
Phosphorus	PF_3, PCl_3, PBr_3, PI_3	PF_5, PCl_5, PBr_5
Arsenic	AsF_3, $AsCl_3$, $AsBr_3$, AsI_3	AsF_5
Antimony	SbF_3, $SbCl_3$, $SbBr_3$, SbI_3	SbF_5, $SbCl_5$
Bismuth	BiF_3, $BiCl_3$, $BiBr_3$, BiI_3	BiF_5

(a) The trihalides

The trihalides are covalent and can be made by reacting the element and halogen directly together. For example, to make phosphorus trichloride, chlorine can be passed over molten white phosphorus:

$$P_4(l) + 6Cl_2(g) \rightarrow 4PCl_3(l)$$

The product is a fuming oily liquid. The reason why it fumes in air is that it reacts with water:

$$PCl_3(l) + 3H_2O(l) \rightarrow H_3PO_3(aq) + 3HCl(aq)$$

When the water is present as water vapour, the two acids form misty fumes.

Phosphorus trichloride can be used to prepare one of the oxochlorides of phosphorus: phosphorus trichloride oxide, $POCl_3$. This is done by warming the liquid with potassium chlorate(V):

$$3PCl_3(l) + KClO_3(s) \rightarrow 3POCl_3(l) + KCl(s)$$

$POCl_3$ is another unpleasant fuming liquid, but in industry it is widely used in the manufacture of organic phosphate esters. You will find it is often produced when PCl_3 or PCl_5 react.

Although bismuth trichloride is not as well known as phosphorus trichloride, we have had reason to refer to it elsewhere (see section 53.2). The reason is that it too is hydrolysed by water, but an equilibrium is set up:

$$\underset{\text{colourless}}{BiCl_3(aq)} + H_2O(l) \rightleftharpoons \underset{\text{white}}{BiOCl(s)} + 2HCl(aq)$$

The bismuth(III) chloride oxide, $BiOCl$, is a white insoluble powder. The equilibrium can be shifted to one side or the other by adding water or concentrated hydrochloric acid.

(b) Pentahalides

Phosphorus pentachloride is the most well known of the pentahalides. It is a pale yellow solid, whose structure belies its formula, PCl_5. In fact the solid consists of pairs of PCl_4^+ and PCl_6^- ions. The former is tetrahedral in shape, the latter octahedral. You can see the shapes in Figure 98.4.

If you do a quick count of the electrons that must be involved in bonding in PCl_6^- or PCl_5 you will find that the 3s and 3p orbitals are insufficient to accommodate them all. As we said earlier, the 3d orbitals of phosphorus can also be used in bonding. If we assume

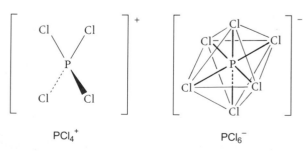

Figure 98.4 The two types of ion present in solid PCl_5

hybridisation to occur, the octagonal shape can be obtained by using a set of d^2sp^3 hybrids, and the trigonal bipyramid using dsp^3 hybrids.

To make the chloride, you can pass chlorine into an ice-cooled glass container into which phosphorus trichloride is dropped:

$$PCl_3(l) + Cl_2(g) \rightarrow PCl_5(s)$$

Phosphorus pentachloride also fumes in air. It too reacts with water. If water is in short supply the reaction is

$$PCl_5(s) + H_2O(l) \rightarrow POCl_3(l) + 2HCl(aq)$$

If water is in excess,

$$PCl_5(s) + 4H_2O(l) \rightarrow H_3PO_4(aq) + 5HCl(g)$$

Indeed, in the units on organic chemistry you will find that we shall speak of a use for PCl_5 as a test for the presence of OH groups. If OH groups are present, white fumes of HCl will be given off. At the same time the OH group is replaced by Cl. For example,

$$CH_3-CH_2-OH + PCl_5 \longrightarrow CH_3-CH_2-Cl + POCl_3 + HCl$$

$$C_2H_5OH + PCl_5 \longrightarrow C_2H_5Cl + POCl_3 + HCl$$
ethanol

$$CH_3-C\!\!\begin{smallmatrix}O\\OH\end{smallmatrix} + PCl_5 \longrightarrow CH_3-C\!\!\begin{smallmatrix}O\\Cl\end{smallmatrix} + POCl_3 + HCl$$

$$CH_3COOH + PCl_5 \longrightarrow CH_3COCl + POCl_3 + HCl$$
ethanoic
acid

It will also react with aldehydes and ketones, but no HCl is given off:

$$CH_3-C\!\!\begin{smallmatrix}O\\H\end{smallmatrix} + PCl_5 \longrightarrow CH_3-C\!\!\begin{smallmatrix}Cl\\|\\H\end{smallmatrix}\!\!-Cl + POCl_3$$

$$CH_3CHO + PCl_5 \longrightarrow CH_3CHCl_2 + POCl_3$$
ethanal

Phosphorus pentachloride has one other facet to its nature that we have looked at before (see section 52.6). If you were to measure the molar mass of its vapour you would find that, depending on the temperature, it appears to be anywhere between 104 and 208 g mol^{-1}. The reason is that it dissociates when heated:

$$PCl_5(g) \rightleftharpoons PCl_3(g) + Cl_2(g)$$

The pentachlorides of the other elements in the Group behave similarly, but they dissociate even more readily.

Before we leave this section, we should note the greater reluctance of the elements below phosphorus to form pentahalides. This has been ascribed to the inert pair effect. The heavier B metals show a reluctance to use their outer s electrons in bonding. For example,

these are the 6s electrons for bismuth. If these electrons are not used, the elements will make molecules with only three bonds.

98.9 Sketch an apparatus that you could use to make PCl_3. What precautions would you have to take (i) on account of the phosphorus, (ii) to stop the product being hydrolysed.

98.10 Predict the shape of (i) PCl_3, (ii) $POCl_3$ and draw dot-and-cross diagrams to represent the bonding in them.

98.11 In the gaseous state, PCl_5 can exist as individual, covalent, molecules. In one sentence explain why the shape of the molecule is a trigonal bipyramid, as shown in Figure 98.5.

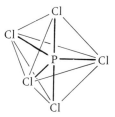

Figure 98.5 *An isolated PCl_5 molecule*

98.12 Suggest a reason why nitrogen does not give NF_5 or NCl_5.

98.13 The ability of PCl_5 to replace OH groups by chlorine atoms extends to its reaction with sulphuric and nitric acids. Draw the structures of these two acids (see Figure 74.1 if you are stuck), and draw diagrams of the molecules that are left after the reaction.

98.5 The oxides

The oxides of the elements are shown in Table 98.5. We have dealt with the oxides of nitrogen in the previous unit, so we shall ignore them here. The oxides of phosphorus were once given the formulae P_2O_3 and P_2O_5 (hence their old names of phosphorus trioxide and

Table 98.5. The oxides of Group V

Element	Oxides	Trend
Nitrogen	N_2O, NO, N_2O_3, NO_2, N_2O_4, N_2O_5	Acidic
Phosphorus	P_4O_6, P_4O_{10}	
Arsenic	As_4O_6, As_4O_{10}	Amphoteric
Antimony	Sb_4O_6, Sb_4O_{10}	
Bismuth	Bi_2O_3	Basic

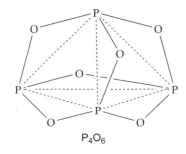

P_4O_6

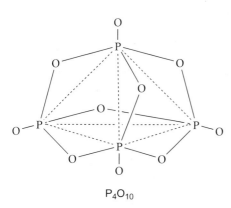

P_4O_{10}

Figure 98.6 *The structures of P_4O_6 and P_4O_{10}*

phosphorus pentoxide) but this was before their structures had been found using X-ray diffraction. The basic building block of both of them is a tetrahedron of phosphorus atoms (Figure 98.6). The oxygen atoms in phosphorus(III) oxide, P_4O_6, act as bridges between the corners of the tetrahedron, as if the edges had been plucked outwards. In phosphorus(V) oxide, P_4O_{10}, each phosphorus atom is also bonded to an extra oxygen atom.

The oxides of phosphorus react vigorously with water to give acidic solutions:

$$P_4O_6(s) + 6H_2O(l) \rightarrow 4H_3PO_3$$
phosphonic acid

$$P_4O_{10}(s) + 6H_2O(l) \rightarrow 4H_3PO_4$$
phosphoric(V) acid

Phosphorus(V) oxide has such an affinity for water that it finds use as a dehydrating agent. For example, it will dehydrate nitric acid, releasing the acid anhydride, N_2O_5:

$$P_4O_{10}(s) + 4HNO_3(l) \rightarrow 4HPO_3(l) \quad + 2N_2O_5(s)$$
metaphosphoric acid

It can also be a useful reagent in organic chemistry, for example in converting amides into nitriles.

Given that we know that nitrogen and phosphorus are definitely non-metallic in character, we should expect them to give acidic oxides, However, as we go down the Group, we know that metallic nature increases, and this should give a greater tendency for the oxides to be basic. In fact the oxides of arsenic and antimony show amphoteric properties, i.e. partly acidic and partly basic properties. For example, arsenic(III) oxide will dissolve in both alkali and concentrated acid:

$$As_4O_6(s) + 12OH^-(aq) \rightarrow 4AsO_3^{3-}(aq) + 6H_2O(l)$$
$$As_4O_6(s) + 12HCl(aq) \rightarrow 4AsCl_3(aq) + 6H_2O(l)$$

The oxide of bismuth is definitely basic in nature. A good example is the way it obeys the simple rule that a base plus an acid gives a salt plus water:

$$Bi_2O_3(s) + 6HNO_3(aq) \rightarrow Bi(NO_3)_2(aq) + 3H_2O(l)$$

Indeed, bismuth is alone in the Group in giving a stable nitrate, sulphate and carbonate. In this respect it is showing the typical behaviour of a metal.

98.14 Use your knowledge of bonding to suggest how each of the phosphorus atoms can bond to an extra oxygen atom in P_4O_{10}.

98.6 The oxoacids

The oxoacids of phosphorus are by far the most important, and there is a remarkable range of them. Table 98.6 lists their names and formulae. The structure of phos-

Table 98.6. The oxoacids of phosphorus

Name	Formula	Bonding
Phosphoric(V) (orthophosphoric) acid	H_3PO_4	HO, HO—P=O, HO
Phosphonic (phosphorus) acid	H_3PO_3	HO, H—P=O, HO
Phosphinic (hypophosphorus) acid	H_3PO_2	H, H—P=O, HO
Phosphinous acid	H_3PO	H, H—P, HO
Pyrophosphoric acid	$H_4P_2O_7$	HO—P(=O)(OH)—O—P(=O)(OH)—OH
Tripolyphosphoric acid	$H_5P_3O_{10}$	HO—P(=O)(OH)—O—P(=O)(OH)—O—P(=O)(OH)—OH

Table 98.6. (cont.)

Name	Formula	Bonding
Tetrapolyphosphoric acid	$H_6P_4O_{13}$	
Pyrophosphonic acid	$H_4P_2O_5$	
Metaphosphoric acid	HPO_3	
Trimetaphosphoric acid	$H_3P_3O_9$	

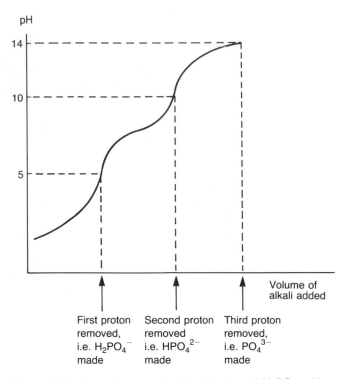

Figure 98.8 *The pH curve for the titration of H_3PO_4 with a strong alkali (e.g. NaOH). The endpoints for the removal of successive protons are not sharp. This is characteristic of a weak acid. (Compare the weak acid part of Figure 76.1)*

phoric(V) acid is shown in Figure 98.7. Fortunately you will not have to remember any but the first two or three. However, this should not prevent you from admiring the structures of the others.

Figure 98.7 *The structure of phosphoric(V) acid, H_3PO_4, is a distorted tetrahedron*

The first point to understand about these acids is that the hydrogen atoms directly attached to oxygen atoms can usually be lost from the molecule, but those bonded to the phosphorus atoms cannot. This means that, for example, phosphoric(V) acid is a triprotic (or tribasic) acid. In a titration with alkali, the pH changes as shown in Figure 98.8.

The graph shows that the first hydrogen is completely removed at pH = 5, the second at pH = 10, and the third at pH = 13.

The salts made by crystallising the solution at each stage are sodium dihydrogenphosphate(V), NaH_2PO_4, disodium hydrogenphosphate(V), Na_2HPO_4, and trisodium phosphate(V), Na_3PO_4. Salts like NaH_2PO_4 and Na_2HPO_4 that contain hydrogen are often known as acid salts. A salt like Na_3PO_4 *can* be categorised as a neutral salt. However, these names should not be taken too literally. If you were to dissolve a sample of each of

them in water, you would find that NaH_2PO_4 does give an acidic solution owing to the equilibrium

$$H_2PO_4^-(aq) + H_2O(l) \rightleftharpoons HPO_4^{2-}(aq) + H_3O^+(aq)$$

lying over to the right-hand side. However, with Na_2HPO_4 the equilibrium is

$$HPO_4^{2-}(aq) + H_2O(l) \rightleftharpoons H_2PO_4^-(aq) + OH^-(aq)$$

That is, the solution of this 'acid' salt is really slightly alkaline. Likewise, a solution of 'neutral' Na_3PO_4 is also alkaline:

$$PO_4^{3-}(aq) + H_2O(l) \rightleftharpoons HPO_4^{2-}(aq) + OH^-(aq)$$

98.7 The uses of phosphate salts

Some phosphate salts have great importance in agriculture. They are a source of phosphorus, which is essential for the healthy growth of plants. The element is used, for example, in building the phosphate groups of DNA, and the biochemically important substance adenosine triphosphate (ATP). Originally, bone ash was used as a source of phosphate, but the ash suffers the same problem as the mineral fluorapatite: the phosphates they contain are released into the soil only very slowly. In the 1830s and 1840s it was discovered that the rate of release was greatly improved by treating the ash or rock with sulphuric acid. This remains the method used to treat fluorapatite today. The more soluble phosphate salts that can be made in the process are known as superphosphates (Figure 98.9).

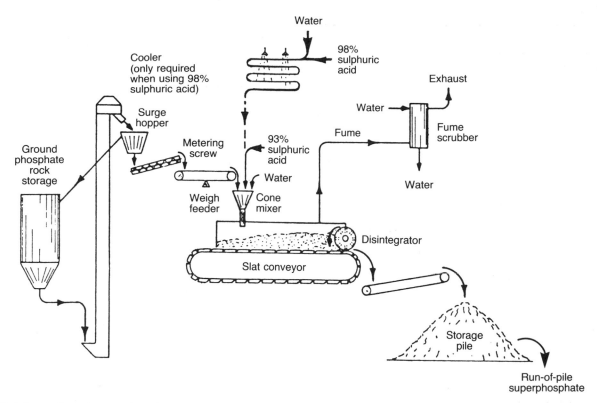

Figure 98.9 *An outline of superphosphate manufacture. (Taken from: Austin, G. T. (1984). Shreve's Chemical Process Industries, 5th edn, McGraw-Hill, New York, figure 16.3, p. 275)*

The fluorapatite rock is ground and mixed with concentrated sulphuric acid. It takes about an hour for the mixture to pass along a slowly moving conveyor, by which time the crucial reactions have finished. The most important of them is

$$Ca_3(PO_4)_2(s) + 2H_2SO_4(l) + 4H_2O(l) \rightarrow$$
$$Ca(H_2PO_4)_2(s) + 2CaSO_4(s)$$

At the end of the conveyor, the product is broken into small pieces and is ready for distribution.

The mixture of the much more soluble calcium dihydrogenphosphate(v), $Ca(H_2PO_4)_2$, and anhydrite, $CaSO_4$, is sold as 'superphosphate'. The market for phosphate fertiliser is far larger than that for elemental phosphorus. About 90% of fluorapatite goes in fertiliser manufacture, and only 10% in extracting phosphorus.

The calcium fluoride in the mineral releases hydrogen fluoride with the sulphuric acid, which itself reacts to give various side products. The gaseous waste products are removed by washing with water in the fume scrubber.

Superphosphate is also made by reacting fluorapatite, or other phosphate rock, with phosphoric(v) acid:

$$Ca_3(PO_4)_2(s) + 4H_3PO_4(l) \rightarrow 3Ca(H_2PO_4)_2(s)$$

Cooks might be particularly interested in some salts of phosphorus because they are ingredients of baking powders. For example hydrated calcium dihydrogenphosphate(v), $Ca(H_2PO_4)_2 \cdot H_2O$, mixed with sodium hydrogencarbonate and starch makes a good baking powder. When the temperature is high enough the salt and the hydrogencarbonate give off carbon dioxide:

$$3Ca(H_2PO_4)_2(s) + 8NaHCO_3(s) \rightarrow$$
$$Ca_3(PO_4)_2(s) + 4Na_2HPO_4(s) + 8H_2O(l) + 8CO_2(g)$$

Other acid phosphorus salts used in baking powder are NaH_2PO_4, KH_2PO_4 and $Na_2H_2P_2O_7$.

Phosphoric(v) acid is useful in the preparation of the hydrogen halides, especially HBr and HI. The acid has little oxidising power, and is not very volatile. When it is heated with potassium bromide or iodide the hydrogen halide is released uncontaminated with the impurities that result from using sulphuric acid. (Also, see section 102.1.)

98.15 The product $Ca(H_2PO_4)_2$ is known as triple superphosphate. Why does it command a higher price than ordinary superphosphate?

98.16 Classify each of the acids in Table 98.6 according to how many protons can (in principle) be lost from their molecules.

98.8 The sulphides

Phosphorus and the other members of the Group give a number of sulphides (Table 98.7). Sulphides of phos-

Table 98.7. Some sulphides of Group V

Element	Sulphides
Nitrogen	N_4S_4
Phosphorus	P_4S_3, P_4S_5, P_4S_7, P_4S_{10}
Arsenic	As_4S_4, As_4S_3, As_4S_6, As_2S_5, As_2S_3
Antimony	Sb_2S_5, Sb_2S_3
Bismuth	Bi_2S_3

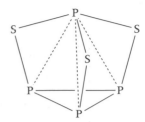

Figure 98.10 *The structure of P_4S_3*

phorus, particularly P_4S_3 (Figure 98.10), have been used in making matches. To make the match head the sulphide is mixed with an oxidising agent, e.g. potassium chlorate(v), and a little ground glass, and the mixture is bound together with a glue. This combination gives a 'strike anywhere' match. The match boxes these come in usually have a strip of sand paper along the side. The friction generated by rubbing the match head along the paper is enough to make the sulphide and chlorate(v) react. In safety matches, the sulphide in the head is an antimony sulphide, Sb_2S_3. This has to be rubbed on a strip that is a mixture of red phosphorus, ground glass and glue.

The sulphides As_2S_3 (yellow), Sb_2S_3 (orange) and Bi_2S_3 (brown) all have characteristic colours, and they are all highly insoluble in water. Both features allow the presence of these elements in solution to be detected by passing hydrogen sulphide through the solution.

98.17   Why does the precipitation of these sulphides depend on the hydrogen ion concentration? (Hint: see section 64.5.)

Answers

98.1   It increases. Going down the Group, the atoms become larger and more polarisable ('squashy') because the outer electrons are increasingly well shielded from the nucleus.

98.2   (i) This arrangement minimises the repulsions between one lone pair and three bond pairs.

(ii) The nitrogen atom pulls its lone pair of electrons very strongly towards itself. (Note its high electronegativity.) This makes the lone pair a very dense centre of negative charge, which in turn repels the bond pairs strongly. This tends to reduce the H—N—H bond angles. By comparison the lone pair on the antimony atom is much less tightly held to the atom. The diffuse charge means that the bond pairs of electrons are less strongly repelled, so the bond angle is greater. Given that arsenic and antimony have identical electronegativities, we might expect the bond angle to be similar. Phosphorus is in between nitrogen and arsenic, and the bond angle is correspondingly intermediate in value.

98.3   (i) The main thing that you should realise is that air must be removed from the apparatus before the alkali and phosphorus are allowed to meet. This can be done by passing carbon dioxide, or nitrogen, through the flask and then allowing alkali to pour into the reaction flask. The phosphine can be collected in a gas jar over water in the time honoured manner.

(ii) The phosphorus changes from 0 to -3 (in PH_3) and $+1$ (in $H_2PO_2^-$). This is a disproportionation reaction.

98.4   Water.

$$4PH_3(g) + 8O_2(g) \rightarrow P_4O_{10}(s) + 6H_2O(l)$$

98.5 $PH_3(g) + 4Cu^{2+}(aq) + 4H_2O(l) \rightarrow$
$$H_3PO_4(aq) + 4Cu(s) + 8H^+(aq)$$

98.6   Chlorine is a good oxidising agent, so when it meets a good reducing agent a vigorous reaction can be expected. The chlorine removes the hydrogen, giving fumes of hydrogen chloride. The phosphorus and chlorine give phosphorus trichloride. (You might not have known which chloride phosphorus makes, but you should have predicted that *a* chloride would be made.) The reaction is

$$PH_3(g) + 3Cl_2(g) \rightarrow PCl_3(l) + 3HCl(g)$$

98.7   (i) Ammonia is given off.

(ii) By analogy, phosphine is given off.

98.8   It dissociates:

$$PH_4I(s) \rightleftharpoons PH_3(g) + HI(g)$$

(Compare NH_4Cl.)

98.9   The apparatus is sketched in Figure 98.11.

98.10   The diagrams are shown in Figure 98.12.

98.11   This shape minimises the repulsions.

98.12   If it is to make these compounds it would have to use the 3d set of orbitals. These are far too high in energy for them to be used by nitrogen, and others in the same Period. Also, see section 96.5.

98.13   The diagrams are drawn in Figure 98.13.

98.14   Each of the phosphorus atoms in P_4O_6 is bonded to three oxygen atoms. This leaves a lone pair over on each phosphorus, which can give a coordinate bond with an oxygen atom.

98.15   It contains no anhydrite, $CaSO_4$, so each kilogram is richer in fertiliser.

Answers – cont.

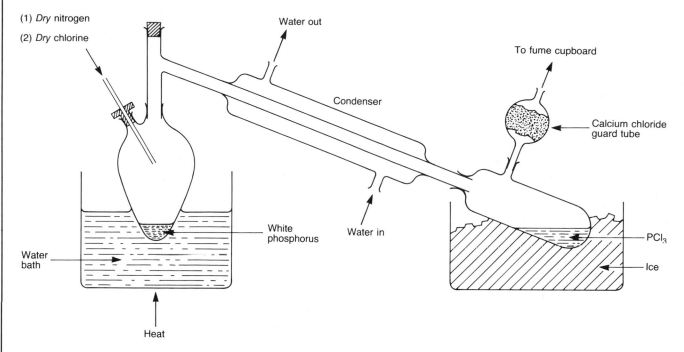

(1) *Dry* nitrogen
(2) *Dry* chlorine

Water out

To fume cupboard

Condenser

Calcium chloride
guard tube

Water in

White
phosphorus

Water
bath

PCl₃

Ice

Heat

Figure 98.11 *One apparatus for preparing PCl₃. Dry nitrogen (or CO₂) is passed through the apparatus to remove air. The CaCl₂ guard tube prevents moisture in the air entering the apparatus*

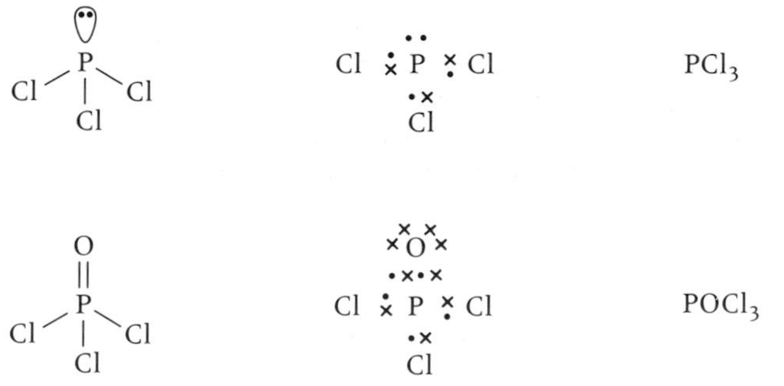

PCl_3

$POCl_3$

SO_2Cl_2

Sulphur dichloride dioxide

NO_2Cl

Nitryl chloride

Figure 98.12 *Diagrams of PCl₃ and POCl₃*

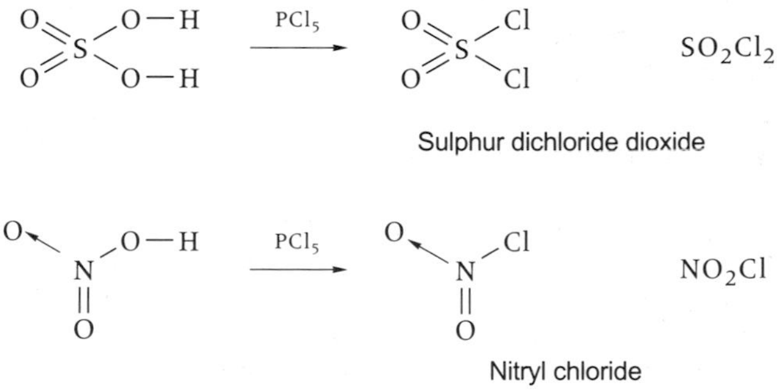

Figure 98.13 *Diagrams for answer to question 98.13*

98.16 We only count the hydrogen atoms attached to oxygen atoms. Going down the list in order: 3, 2, 1, 1, 4, 5, 6, 2, 1, 3.

98.17 This is because of the different solubility products of the sulphides. Addition of hydrogen ions suppresses the production of S^{2-} ions, so in acidic solution $[S^{2-}]$ is very low and only the most insoluble sulphides will precipitate.

UNIT 98 SUMMARY

- The Group V elements:
    - (i) Metallic nature increases down the Group.
    - (ii) Phosphorus and following elements can make use of d orbitals in bonding, and show several valencies.
    - (iii) Phosphorus (red and white), arsenic and antimony have allotropes.

- Extraction:
    - (i) Phosphorus is extracted by heating fluorapatite, $CaF_2 \cdot 3Ca_3(PO_4)_2$, with sand (silica, SiO_2) and coke (carbon) in an electric arc furnace;

    $$2Ca_3(PO_4)_2 + 6SiO_2 + 10C \rightarrow$$
    $$6CaSiO_3 + 10CO + P_4$$

(ii) The other elements are extracted by reducing their oxides or sulphides.

Reactions

- With oxygen:
  Oxides made, e.g. phosphorus burns in oxygen;

$$P_4(s) + 5O_2(g) \rightarrow P_4O_{10}(s)$$

- With halogens:
  Form halides;

$$e.g. \ P_4(s) + 6Cl_2(g) \rightarrow 4PCl_3(l)$$
$$PCl_3(l) + Cl_2(g) \rightarrow PCl_5(l)$$

Compounds

- Oxides:
  (i) Phosphorus makes phosphorus(III) oxide, P_4O_6, and phosphorus(V) oxide, P_4O_{10}. Both are acidic in water;

$$e.g. \ P_4O_{10}(s) + 6H_2O(l) \rightarrow 4H_3PO_4$$

  (ii) Phosphorus(V) oxide is used as a dehydrating agent.
  (iii) Oxides of arsenic and antimony are amphoteric; that of bismuth is basic.

- Oxoacids:
  (i) Phosphorus makes a series of oxoacids (see Table 98.6), of which phosphoric(V) acid, H_3PO_4, is the most important.
  (ii) Owing to its only very weak oxidising nature it is used to prepare hydrogen bromide and hydrogen iodide from their sodium salts. (Concentrated sulphuric acid oxidises HBr and HI.)
- Hydrides:
  (i) Phosphine, PH_3, is made by reacting white phosphorus with sodium hydroxide solution in an inert atmosphere.
  (ii) Arsine, AsH_3, is used in Marsh's test for arsenic.
- Halides:
  (i) All give trihalides and pentahalides (except bismuth).
  (ii) Phosphorus trichloride, PCl_3, is an oily liquid.
  (iii) Phosphorus pentachloride, PCl_5, a pale yellow solid, is composed of a combination of PCl_4^+ and PCl_6^- ions.
  (iv) Both PCl_3 and PCl_5 are easily hydrolysed;

$$e.g. \ PCl_5(s) + 4H_2O(l) \rightarrow H_3PO_4(aq) + 5HCl(g)$$

99

Oxygen and oxides

99.1 The element

Oxygen exists as diatomic molecules, O_2, and constitutes approximately 20.8% of air. It has three isotopes. $^{16}_8O$ is the main one, with an abundance of nearly 99.8%. The others, are $^{17}_8O$ (0.04%) and $^{18}_8O$ (0.2%). The most important use of oxygen is in keeping animals, including the human variety, alive through its activity in respiration. Its slight solubility in water is essential to fish and other aquatic life.

Another virtue of oxygen is that it supports combustion, especially of organic substances such as coal, oil and wood. We rely on the combustion of fuels to provide heat, electricity and transportation.

Information about oxygen is summarised in Table 99.1. Oxygen has an allotrope: trioxygen or ozone, O_3. The importance of ozone has become widely recognised. It is present in the upper atmosphere where it absorbs a good deal of ultraviolet radiation from the Sun. The ozone layer has suffered damage owing to the widespread use of chlorofluorocarbons (CFCs) in aerosols.

Table 99.1. Information about oxygen

Relative atomic mass, $A_r(O)$	16
Electron structure	$1s^2 2s^2 2p^4$
Ionisation energy	1314 kJ mol^{-1}
First electron affinity	$-141.4 \text{ kJ mol}^{-1}$
Second electron affinity	$+790.8 \text{ kJ mol}^{-1}$
Molecular formula	O_2
Bond energy	497 kJ mol^{-1}
Bond length	121 pm
Melting point	54 K ($-219°$ C)
Boiling point	90 K ($-183°$ C)
Solubility* in water at s.t.p.	$8.6 \text{ cm}^3 \text{ dm}^{-3}$
Colourless, odourless, tasteless	

*This is the solubility of oxygen from air in contact with water. Pure oxygen dissolves to the extent of about $49 \text{ cm}^3 \text{ dm}^{-3}$ at s.t.p.

99.2 The large-scale extraction of oxygen

Oxygen is obtained by the fractional distillation of air. This is achieved by forcing air under pressure through a nozzle. The air expands into a region of lower pressure, which cools the gas. (This is known as the Joule–Thomson effect.) The cooled air circulates around the expansion tubes, cooling the air even more. Eventually a point is reached when the air condenses to a liquid. It consists of a mixture of nitrogen, oxygen and noble gases. The mixture can be separated by allowing the liquid to increase in temperature. Oxygen boils less easily than the other gases, so it evaporates last of all. Once the other gases have evaporated, liquid oxygen can be stored in insulated containers, or allowed to vaporise and kept in cylinders under pressure.

Some manufacturing plants operate a more elaborate system, which allows both nitrogen and oxygen to be liquefied. It is rather more difficult to effect the separation of the two liquids from liquid air rather than to allow the nitrogen to escape. Although they use a more tricky process, such plants have the benefit of being able to change the proportions of the two gases in order to meet varying demand.

Liquid oxygen is pale blue in colour and, owing to an oxygen molecule having two unpaired electrons, it is strongly paramagnetic. Molecular orbital theory, but not valence bond theory, has a neat explanation of this.

> **99.1** You might be surprised to discover that even reactive metals like sodium do not react with liquid oxygen. What explanation (from reaction kinetics) can you give for this lack of reactivity?

99.3 The uses of oxygen

The uses of oxygen are shown in Table 99.2. Oxygen is of immense importance in medicine. Patients with respiratory problems can be given an oxygen enriched

Table 99.2. The extraction and uses of oxygen

Extraction	Uses
Liquefaction of air	Medical uses in aiding respiration Breathing by divers, astronauts, etc. Oxy-acetylene welding Rocket fuels

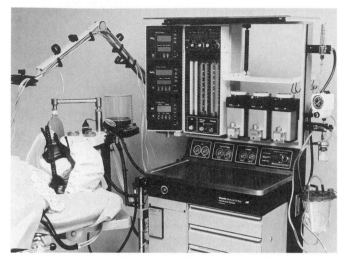

A supply of pure oxygen is essential to the survival of some patients in hospital. The volume of gas given to the patient has to be carefully monitored using equipment like that shown here.

Oxy-acetylene welding is an essential part of many manufacturing processes.

Liquid oxygen is one component of the fuel that lifts the Space Shuttle into orbit around the Earth.

atmosphere to breathe. In operations, the supply of oxygen and anaesthetic have to be carefully balanced.

Divers need a supply of oxygen, but not too much. Often they breathe oxygen greatly diluted with helium, a mixture that results in their voices taking on a high pitched squeaky sound.

Oxygen has found much use in rocketry and space flight. Apart from the needs of the astronauts, liquid oxygen is widely used as the oxidant in the fuel mixture.

Owing to the highly exothermic nature of the reaction, oxygen and ethyne (acetylene) are used in oxy-acetylene welding:

$$C_2H_2(g) + \tfrac{5}{2}O_2(g) \rightarrow 2CO_2(g) + H_2O(g);$$
$$\Delta H^\ominus = -1255.6 \, \text{kJ mol}^{-1}$$

99.2 Hydrazine, N_2H_4, is known as an endothermic compound because its heat of formation is positive: $\Delta H_f^\ominus = +50.6 \, \text{kJ mol}^{-1}$. It reacts with oxygen according to the equation

$$N_2H_4(g) + O_2(g) \rightarrow N_2(g) + 2H_2O(g)$$

At 25°C hydrazine is a liquid. Neglecting the heat needed to convert it into a gas, estimate the enthalpy change in the reaction.

99.4 The place of oxygen in Nature

Oxygen is constantly being used up. Living things use it to generate energy through respiration. Humans and other mammals take in oxygen through their lungs, where it passes into the blood and combines with haemoglobin. Haemoglobin is a protein that has an iron atom held by a ring of carbon atoms called a haem group (Figure 99.1). An oxygen molecule can bond to the iron sufficiently strongly that it can be carried round the body, but not so strongly that the oxygen cannot be removed. Indeed, the oxygen is given up to the various tissues as the blood passes through the labyrinth of capillaries.

A simple way in which we can represent the chemical change of respiration is

$$C_6H_{12}O_6 + 6O_2 \rightarrow 6CO_2 + 6H_2O;$$
$$\Delta H^\ominus = -2820 \text{ kJ mol}^{-1}$$

Here we have used glucose to represent the conversion of carbohydrate into carbon dioxide and water. These two products of respiration are breathed out. The energy released is essential to keep us alive. Respiration is a rather important type of combustion reaction.

Oxygen is also used up by more obvious types of combustion. Burning coal, peat, or wood in a fire, or oil in central heating boilers, converts the carbon and other elements into oxides. Oxides of nitrogen and carbon are responsible for producing acid rain. Carbon dioxide plays a large part in the greenhouse effect. For the present we shall concentrate on the explanation of how it is that the proportion of oxygen in the atmosphere stays constant. The reason is that oxygen is constantly being returned to the atmosphere by the process of *photosynthesis*, which occurs in green plants.

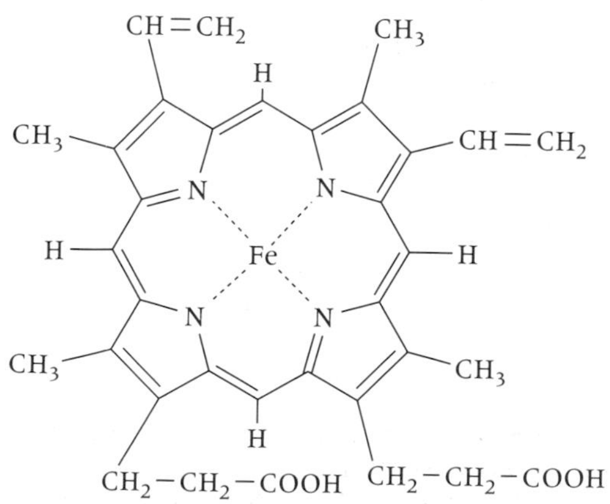

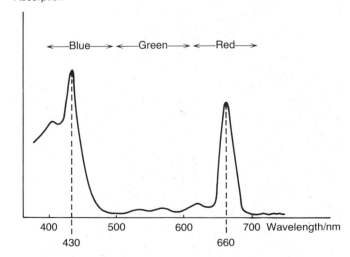

Figure 99.1 A single haem group. In haemoglobin, haem groups are attached to a protein molecule

Photosynthesis takes place through a complicated series of reactions in which *light energy* from the Sun and *chlorophyll* are essential components. Chlorophyll exists in two varieties called chlorophyll *a* (Figure 99.2) and chlorophyll *b*. Both are large organic molecules that contain magnesium.

The visible spectrum of chlorophyll (Figure 99.3) confirms that the molecule absorbs strongly in the red and blue-violet regions of the visible spectrum. The light that is not absorbed is, of course, primarily in the green region. When photons with the right wavelength strike chlorophyll, the energy is given to an electron, which is able to escape and start a highly complicated chain of reactions. The conversion of carbon dioxide

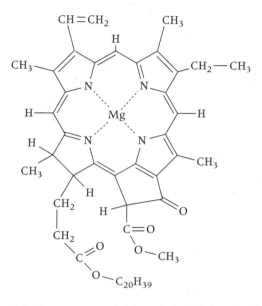

Figure 99.2 The structure of chlorophyll a. Notice the similarity to the structure of haem (Figure 99.1)

Absorption

[Figure 99.3: Absorption spectrum with wavelength axis marked 400, 500, 600, 700 Wavelength/nm; dashed lines at 430 and 660; regions labelled Blue, Green, Red]

Figure 99.3 The spectrum of chlorophyll a. The spectrum of chlorophyll b is similar, with peaks near 450 and 640 nm. There is strong absorption in the blue and red regions of visible light

and water into carbohydrate takes place according to the equation

$$6CO_2 + 6H_2O + \text{light energy} \rightarrow C_6H_{12}O_6 + 6O_2$$

The overall change in photosynthesis is the reverse of respiration.

99.3 Carbon monoxide and cyanide ions are highly poisonous. A sign that someone has been poisoned by either of these two chemicals is that the person's blood changes colour. They both have the effect of preventing oxygen being taken up by the blood stream. What might happen to haemoglobin to stop oxygen being taken up?

99.5 Ozone (trioxygen)

Ozone, O_3, has a triangular shape (Figure 99.4). Both bonds have the same length, about 128 nm, so they must have the same degree of single and double bond character. We know that there are six valence electrons on each oxygen atom. We must account for all of them in describing the bonding. If we use valence bond theory, we claim that each bond involves a set of resonance hybrids in which one of the bonds is a double bond, and the other a coordinate bond. You must remember that the actual structure does not swap between the resonance structures. Rather, the pictures suggest that each bond has partly the nature of a double bond, and partly that of a single bond.

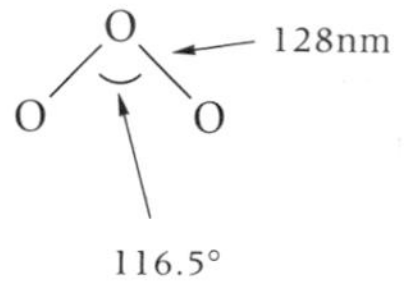

The shape of an ozone molecule

Two of the resonance hybrids for ozone

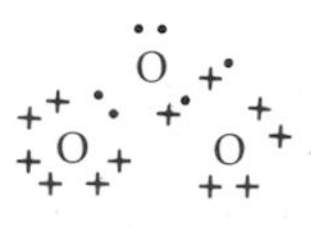

Dot-and-cross diagram for the hybrid on the left

Figure 99.4 *The structure of an ozone (trioxygen) molecule*

In the laboratory, ozone can be made by passing oxygen through a strong electric field. An equilibrium is set up:

$$3O_2(g) \rightleftharpoons 2O_3(g)$$

Ozone is only a metastable allotrope, always having a tendency to convert back into oxygen. It is a vigorous oxidising agent, often reacting with the evolution of oxygen. For example,

$$2I^-(aq) + H_2O(l) + O_3(g) \rightarrow I_2(s) + 2OH^-(aq) + O_2(g)$$
$$S^{2-}(aq) + 2O_3(g) \rightarrow SO_4^{2-}(aq) + O_2(g)$$

Some 30 km above the Earth's surface oxygen molecules can be split apart by ultraviolet light from the Sun. Some of the atoms join with other oxygen molecules to make ozone:

$$O_2(g) + O(g) \rightarrow O_3(g)$$

This reaction has been of extreme importance in the maintenance of life on Earth. The reason is that ozone has the ability to absorb ultraviolet radiation. This is radiation of short wavelength, and therefore of high energy. Other molecules in the atmosphere also absorb electromagnetic radiation, but none of them absorb ultraviolet radiation to a useful extent. If too much ultraviolet radiation reaches Earth, two things happen.

First, the energy balance of the Earth will be upset. The radiation that is reflected from the Earth's surface, which includes rocks, trees and animals, etc., has a longer wavelength than the incoming radiation. The reflected radiation tends to be in the infrared region of the spectrum, which is easily absorbed by molecules in the atmosphere (Figure 99.5). That is, the reflected energy is trapped in the Earth's atmosphere; less than 10% escapes into space. This is known as the *greenhouse effect*. (The glass in a greenhouse allows higher energy radiation through, but tends to trap the reflected radiation. This ensures that the temperature in the greenhouse rises.) If ultraviolet radiation reaches Earth in large amounts, the likely outcome is an increase in the temperature of the atmosphere. Carbon dioxide is one of the substances responsible for the greenhouse effect, but methane and chlorofluorocarbons are also involved. Over the last few hundred years the amount of carbon dioxide in the atmosphere has increased by 25%, and that of methane by 200%. There is no doubt that the temperature of the atmosphere has been rising. This, together with an increase in the radiation reaching Earth owing to the depletion of the ozone layer, is extremely worrying. The effects on the climate could be disastrous. Among other things, patterns of rainfall and agriculture would change drastically. Some people think that such changes are already happening.

Secondly, increased ultraviolet radiation will bring changes in the behaviour of cells in living tissue. One unpleasant outcome will be an increase in the incidence of skin cancers. Such an increase has already occurred in some countries. This has been a result of more people taking holidays during which they lie in the sun for many hours each day. The longer the time spent in the sun, the more ultraviolet radiation hits the skin, and the

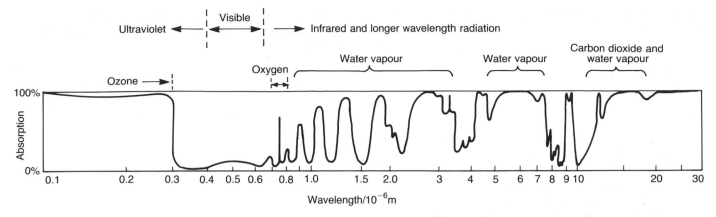

Figure 99.5 *The peaks on this graph show those parts of the spectrum which are absorbed by oxygen, ozone, water vapour and carbon dioxide. (Adapted from: White, I. D. et al. (1987).* Environmental Systems, *Allen and Unwin, London, figure 3.6, p. 47)*

more likely is it that changes in the skin take place.

In 1985 researchers noticed that a gap in the ozone layer appeared above the Antarctic. This hole in the ozone layer was at first thought to be just one of the many fluctuations that take place in the Earth's atmosphere from time to time. However, it has been shown that it is mainly due to the presence of chlorofluorocarbons (CFCs) reacting with ozone. CFCs are gases that are easily liquefied, and in most conditions they are unreactive molecules. Both properties allow them to be used in aerosols, for example in deodorant sprays, paint sprays and furniture polishes, in refrigerators and so on. The problem is that they *do* react with ozone. When the CFCs diffuse into the ozone layer the ozone is removed faster than it can be replaced. The dangers caused by CFCs have now been recognised, and many manufacturers are finding different chemicals to use instead. However, it is certain that the decline in their use will have come too late for many future sufferers from skin cancer. Whether it has come too late to prevent irreversible changes in climatic conditions is less clear.

99.4 Why can molecules in the atmosphere absorb infrared radiation? What happens to the energy supplied by the radiation?

99.6 Ozonolysis

Ozone has a use in organic chemistry. In the right conditions it has the effect of breaking apart molecules with double bonds. The method is known as *ozonolysis*. Essentially, the alkene reacts with ozone to give an unstable ozonide. If the ozonide is treated with dilute acid it breaks up. We can show the result of an ozonolysis experiment on an alkene like this:

As you can see, an oxygen atom is joined on to each carbon atom. The molecules made will be aldehydes or ketones. In Unit 118 you will discover how the nature of these types of molecule can be determined. Once this is done the point of ozonolysis becomes clear: it allows us to find out where the double bond was in the original molecule.

99.5 An alkene was used in an ozonolysis experiment. One mole of the alkene was found to react with four moles of ozone. What does this tell you about the alkene?

99.7 There are four types of oxide

The one chemical property that dominates the chemistry of oxygen is its ability to combine with both metals and non-metals to make oxides (Table 99.3). Oxides can be of four types: *neutral, basic, acidic,* or *amphoteric.* Amphoteric oxides can show both basic and acidic properties. Not many oxides are neutral: nitrous oxide and carbon monoxide are examples.

As is often the case we tend to think of acidity and basicity in relation to water. An oxide will be thought to be basic if it reacts with hydrogen ions to give water. On

Table 99.3. The formulae of some oxides in the Periodic Table*

Group I	Group II	Group III	Group IV	Group V	Group VI	Group VII
Li_2O	BeO	B_2O_3	CO_2	N_2O_5	O_2	F_2O
Na_2O	MgO	Al_2O_3	SiO_2	P_4O_{10}	SO_2	Cl_2O
K_2O	CaO	Ga_2O_3	GeO_2	As_2O_3	SeO_2	Br_2O
Rb_2O	SrO	In_2O_3	SnO	Sb_2O_3	TeO_2	I_2O_5
Cs_2O	BaO	Tl_2O	PbO	Bi_2O_3	PoO_2	

*Not all oxides are shown. Especially, nitrogen has many oxides. There are no noble gas oxides

the other hand, an oxide is acidic if it reacts with water liberating hydrogen ions.

The best examples of *basic oxides* are the oxides of Groups I and II. For example,

$$MgO(s) + 2H^+(aq) \rightarrow Mg^{2+}(aq) + H_2O(l)$$

An alternative way of looking at this is to say that

$$base + acid \rightarrow a\ salt + water$$

Acidic oxides are mainly found among the oxides of non-metals. For example,

$$SO_3(g) + H_2O(l) \rightarrow SO_4^{2-}(aq) + 2H^+(aq)$$

In addition, if an oxide reacts with hydroxide ions, we would say it showed acidic properties.

Aluminium oxide is an example of an *amphoteric oxide*. It shows basic properties by reacting with hydrogen ions in the usual way:

$$Al_2O_3(s) + 6H^+(aq) \rightarrow 2Al^{3+}(aq) + 3H_2O(l)$$

However, it also dissolves in alkali in a reaction that we can write as

$$Al_2O_3(s) + 6OH^-(aq) + 3H_2O(l) \rightarrow 2Al(OH)_6^{3-}(aq)$$

or

$$Al_2O_3(s) + 2OH^-(aq) + 5H_2O(l) \rightarrow$$
$$2[Al(OH)_4(H_2O)_2]^-(aq)$$

Sometimes the product ions are written as the aluminate ion, AlO_2^-.

Many non-metals also exist in combination with oxygen as ions; for example sulphate ions, SO_4^{2-}, and nitrate ions, NO_3^-. Such ions are called *oxoanions*. Oxoanions are often thermodynamically and kinetically stable. As such they exist in salts and minerals of many types, and in solutions of acids such as sulphuric and nitric acids. It is wise to know something of the bonding in these ions. Make sure you have read Unit 14, which explains the key points.

99.6 Zinc oxide, ZnO, is amphoteric. It dissolves in alkali to give the ion $Zn(OH)_4^{2-}$.

(i) What is the equation for the reaction?

(ii) What is the equation for the reaction of the oxide with hydrogen ions?

99.7 Oxides and peroxides are not the only types of compound that metals can make with oxygen. *Superoxides* are also possible.

Here is some information that you should use to work out the formula of sodium superoxide: 1.15 g of sodium reacted with oxygen at a high pressure gives 2.75 g of a white powder, sodium superoxide.

(i) What mass of oxygen was combined with the sodium?

(ii) How many moles of oxygen *atoms* is this?

(iii) How many moles of sodium were used?

(iv) What is the ratio of the moles of the elements?

(v) What is the empirical formula of the compound?

(vi) What is the likely formula of the oxide?

(vii) What is the formula of the superoxide ion?

99.8 Typical basic oxides

Group I and II metals combine directly with oxygen to give basic oxides. Especially, sodium and potassium have to be kept under oil in order to stop them converting into oxides. The reactivity of Group II metals is less marked, but a coating of oxide will soon give the otherwise shiny metal surfaces a dull grey appearance. Some, like magnesium, burn violently in air. Typical reactions are

Group I $4Na(s) + O_2(g) \rightarrow 2Na_2O(s)$
Group II $2Mg(s) + O_2(g) \rightarrow 2MgO(s)$

However, if the Group I metals are heated with oxygen they make *peroxides*, which contain O_2^{2-} ions:

$$2Na(s) + O_2(g) \rightarrow Na_2O_2(s)$$
$$\text{sodium peroxide}$$

The Group II oxides can be converted into peroxides by heating them in oxygen; for example,

$$2BaO(s) + O_2(g) \rightarrow 2BaO_2(s)$$

The oxides are generally ionic, although, as is often the case in its reactions, beryllium at the top of Group II is an exception. Beryllium oxide, BeO, is covalent.

The Group I oxides dissolve in water to give strongly alkaline solutions:

$$Na_2O(s) + H_2O(l) \rightarrow 2Na^+(aq) + 2OH^-(aq)$$

The Group II oxides react less vigorously with water giving hydroxides that are only partially soluble in water and weakly alkaline. For example,

$$MgO(s) + H_2O(l) \rightleftharpoons Mg(OH)_2(s)$$
$$Mg(OH)_2(s) \rightleftharpoons Mg^{2+}(aq) + 2OH^-(aq)$$

The solubility product of magnesium hydroxide is small, which shows that there are very few hydroxide ions in solution. However, the solubilities of the hydroxides do increase down the Group.

99.9 Typical acidic oxides

The oxides of sulphur and phosphorus are typical acidic oxides in that they all react with water to give acidic solutions; e.g.

$$SO_2(g) + H_2O(l) \rightarrow SO_3^{2-}(aq) + H^+(aq)$$
$$\text{sulphurous acid}$$
$$SO_3(g) + H_2O(l) \rightarrow SO_4^{2-}(aq) + 2H^+(aq)$$
$$\text{sulphuric acid}$$
$$P_4O_{10}(s) + 4H_2O(l) \rightarrow$$
$$H_3PO_4(aq) \rightleftharpoons 2PO_4^{3-}(aq) + 6H^+(aq)$$
$$\text{phosphoric(v) acid}$$

Nitrogen dioxide, NO_2, is acidic, but nitrogen monoxide, NO, and dinitrogen oxide, N_2O, are neutral:

$$2NO_2(g) + H_2O(l) \rightarrow NO_2^-(aq) + NO_3^-(aq) + 2H^+(aq)$$

The resulting solution is a mixture of nitrous, HNO_2, and nitric, HNO_3, acids.

Some metals can give acidic oxides. Usually they are transition metals, and have oxides that have a particularly high proportion of oxygen in them. An example is manganese(VII) oxide,

$$Mn_2O_7(s) + H_2O(l) \rightarrow 2MnO_4^-(aq) + 2H^+(aq)$$

The solution contains the purple manganate(VII) ion.

99.8 Oxides that react with water to give acids can be called *acid anhydrides*. It is possible to work out the formula of an anhydride by subtracting the formula of water from the formula of the acid. For example, sulphur trioxide is the anhydride of sulphuric acid: $H_2SO_4 - H_2O = SO_3$. For an acid like hypochlorous acid, HClO, we have to write the change as $2HClO - H_2O = Cl_2O$. The anhydride is chlorine monoxide. What is the anhydride of (i) carbonic acid, H_2CO_3; (ii) perchloric acid, $HClO_4$; (iii) nitric acid, HNO_3?

99.9 At one time, before the poisonous properties of lead were realised, paints contained lead compounds. Especially, lead(II) carbonate, $PbCO_3$, was used as a white pigment. Unfortunately, owing to air often being polluted by acids and sulphides, the pigment was converted to black lead(II) sulphide, PbS. A dilute solution of hydrogen peroxide can be used to remove the black discoloration by oxidising the sulphide to sulphate. What is the equation for the reaction?

99.10 Peroxides

The most important peroxide is hydrogen peroxide, H_2O_2 (Figure 99.6). When pure, it is a colourless liquid, but it is too dangerous to use in this form in the laboratory. Instead, it is kept in solution with water. Often the concentrations of the solutions are given as a number of 'volumes'. This gives a measure of the volume of oxygen that would be given off from a given volume of solution. For example, a 100 volume solution would give approximately $100\ cm^3$ of oxygen from $1\ cm^3$ of solution.

The equation for the decomposition is

$$2H_2O_2(aq) \rightarrow 2H_2O(l) + O_2(g)$$

and, as we found in Unit 77, it can be catalysed by a variety of substances, especially manganese(IV) oxide. The easiest way of making oxygen in the laboratory is to add this oxide to hydrogen peroxide solution.

A solution of hydrogen peroxide is very slightly acidic through the reaction

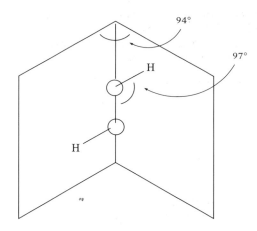

Figure 99.6 *One way of explaining the shape of an H_2O_2 molecule. The hydrogen atoms would lie on the pages of a partly opened book, and the oxygen atoms along the spine*

$$H_2O_2(aq) + H_2O(l) \rightleftharpoons H_3O^+(aq) + HO_2^-(aq)$$

It is possible to regard peroxides as salts of hydrogen peroxide.

Apart from its use in making oxygen, hydrogen peroxide is a useful oxidising agent. At one time it was widely used as a crude type of bleach for hair. This gave rise to women (usually) who had their hair bleached being called 'peroxide blondes'. However, as far as more mundane matters of chemistry are concerned, hydrogen peroxide can react in two ways. In acid it gives rise to a redox potential of $+1.77\ V$, but in alkali the potential can drop to as low as $+0.22\ V$. This behaviour means that:

In acid, hydrogen peroxide is an oxidising agent.

However, in alkaline conditions some reagents are able to oxidise hydrogen peroxide because they have a redox potential greater than $+0.22\ V$. Therefore:

In alkali, hydrogen peroxide is a reducing agent.

Here are some examples.

As an oxidising agent

$$2I^-(aq) + H_2O_2(aq) + 2H^+(aq) \rightarrow I_2(s) + 2H_2O(l)$$
iodide   iodine

$$2Fe^{2+}(aq) + H_2O_2(aq) + 2H^+(aq) \rightarrow 2Fe^{3+}(aq) + 2H_2O(l)$$
iron(II)   iron(III)

As a reducing agent

$$2Fe(CN)_6^{3-}(aq) + H_2O_2(aq) + 2OH^-(aq) \rightarrow$$
hexacyanoferrate(III) $2Fe(CN)_6^{4-}(aq) + 2H_2O(l) + O_2(g)$
             hexacyanoferrate(II)

$$2MnO_4^-(aq) + 5H_2O_2(aq) + 6H^+(aq) \rightarrow$$
manganate(VII) $2Mn^{2+}(aq) + 8H_2O(l) + 5O_2(g)$
           manganese(II)

99.10 Hydrogen peroxide can be made by adding a peroxide to dilute acid. For example,

$$BaO_2(s) + H_2SO_4(aq) \rightarrow BaSO_4(s) + H_2O_2(aq)$$
$$Na_2O_2(s) + H_2SO_4(aq) \rightarrow$$
$$2Na^+(aq) + SO_4{}^{2-}(aq) + H_2O_2(aq)$$

If you were making hydrogen peroxide by one of these methods, which one would you choose? (Look carefully at the state symbols for the products.)

99.11 Hydrogen peroxide can be oxidised by chlorine.

(i) What ion does chlorine turn into when it has been reduced?

(ii) What is the equation for the reaction?

99.12 Design an apparatus that could be used to make oxygen from hydrogen peroxide solution. Your apparatus should allow the oxygen to be given off in controlled amounts.

99.13 At one time oxygen was made by heating barium oxide in air. Provided the temperature was not too high, barium peroxide was made. The product was then heated much more strongly, which caused the peroxide to decompose. We can show the reactions as involving an equilibrium

$$2BaO(s) + O_2(g) \rightleftharpoons 2BaO_2(s)$$

Some time later it was discovered that oxygen could be made from the peroxide more easily by adjusting the pressure in the apparatus. What changes in pressure would you make during the reaction?

Answers

99.1 At the temperature of liquid oxygen (less than 90 K) the sodium and oxygen have insufficient energy to get over the energy barrier.

99.2

$$\Delta H^{\ominus}(\text{reaction}) = 2\Delta H_f^{\ominus}(H_2O) - \Delta H_f^{\ominus}(N_2H_4)$$
$$= -534.2 \text{ kJ mol}^{-1}$$

99.3 Carbon monoxide and cyanide ions are able to bond more strongly to the iron atom in haemoglobin than can oxygen. If you were to breathe in carbon monoxide for any length of time, your haemoglobin would become saturated with carbon monoxide molecules rather than oxygen. As a result all the body processes that rely on a supply of oxygen gradually cease. Death is the result.

99.4 Infrared radiation has the right amount of energy to excite the vibrations of molecules (see Unit 24). The energy increases the vibrational energy of the molecules.

99.5 One mole of ozone is absorbed for each double bond, so there are four double bonds in the alkene.

99.6 (i) $ZnO(s) + 2OH^-(aq) + H_2O(l) \rightarrow Zn(OH)_4{}^{2-}(aq)$
(ii) $ZnO(s) + 2H^+(aq) \rightarrow Zn^{2+}(aq) + H_2O(l)$

99.7 (i) $2.75 \text{ g} - 1.15 \text{ g} = 1.60 \text{ g}$.
(ii) $1.60 \text{ g}/16 \text{ g mol}^{-1} = 0.1 \text{ mol of oxygen atoms}$.
(iii) $1.15 \text{ g}/23 \text{ g mol}^{-1} = 0.05 \text{ mol of sodium atoms}$.
(iv) 2 mol oxygen to 1 mol sodium.
(v) NaO_2.
(vi) Usually we need to know the relative molecular mass in order to work out the molecular formula from the empirical formula. In this case we have the simplest whole number ratio of the atoms. If NaO_2 is not the correct formula, we would have to consider Na_2O_4,

Na_3O_6, etc. All these are most unlikely, so we are left with NaO_2 as the molecular formula.
(vii) Because sodium always forms the Na^+ ion, the superoxide ion has the formula $O_2{}^-$.

99.8 (i) CO_2, carbon dioxide
(ii) $2HClO_4 - H_2O = Cl_2O_7$, dichlorine heptaoxide
(iii) $2HNO_3 - H_2O = N_2O_5$, dinitrogen pentaoxide.

99.9
$$\underset{\text{sulphide}}{S^{2-}(aq)} + 2H_2O_2(aq) + 4H^+(aq) \rightarrow \underset{\text{sulphate}}{SO_4{}^{2-}(aq)} + 4H_2O(l)$$

99.10 The advantage of the first reaction is that the barium sulphate is highly insoluble in water and can be filtered off. This leaves a fairly pure solution of hydrogen peroxide behind. To set against this is that the insoluble sulphate can form on the surface of the barium peroxide. This will prevent the acid reacting, so bringing the reaction to a halt before all the peroxide is used up. In the second reaction all the barium peroxide can be used up because the products are all soluble. However, the disadvantage is that the solution will be contaminated with sodium and sulphate ions.

99.11 (i) Chloride ion, Cl^-.
(ii) $\underset{\text{chlorine}}{Cl_2(g)} + H_2O_2(aq) + 2OH^-(aq) \rightarrow$
$$\underset{\text{chloride}}{2Cl^-(aq)} + 2H_2O(l) + O_2(g)$$

99.12 The apparatus of Figure 90.2 would be suitable.

99.13 If we use Le Chatelier's principle, we can see that the decomposition of the peroxide is favoured by a *low* pressure. Therefore, after the peroxide is made, the pressure in the apparatus should be reduced.

UNIT 99 SUMMARY

- Oxygen:
  - (i) Has two allotropes, O_2, and trioxygen (ozone), O_3.
  - (ii) Is essential to life.
  - (iii) Is released into the atmosphere through photosynthesis in green plants.
  - (iv) Is slightly soluble in water, essential for aquatic life.
  - (v) Is an oxidising agent.
  - (vi) Supports combustion.
- Laboratory preparation:
  By the decomposition of hydrogen peroxide using manganese(IV) oxide as a catalyst;

  $$2H_2O_2(aq) \rightarrow 2H_2O(l) + O_2(g)$$

- Isolation:
  By the liquefaction of air followed by fractional distillation.

Reactions

- With hydrogen:

  $$2H_2(g) + O_2(g) \rightarrow 2H_2O(g)$$

  This reaction is used in fuel cells to generate electricity.
- With metals and non-metals:
  Oxides made, e.g. magnesium, carbon and sulphur burn in oxygen;

  $$2Mg(s) + O_2(g) \rightarrow 2MgO(s)$$
  $$C(s) + O_2(g) \rightarrow CO_2(g)$$
  $$S(s) + O_2(g) \rightarrow SO_2(g)$$

- Combustion reactions:
  Among the most important is the burning of organic matter, e.g. wood, coal, oil and natural gas. Carbon dioxide and water vapour are released.
- Photosynthesis:
  The conversion of carbon dioxide and water into carbohydrate and oxygen using light energy, and chlorophyll as a catalyst;

  $$6CO_2 + 6H_2O + \text{light energy} \rightarrow C_6H_{12}O_6 + 6O_2$$

- Respiration:
  A variety of combustion reaction that releases energy for use by living systems;

  $$C_6H_{12}O_6 + 6O_2 \rightarrow 6CO_2 + 6H_2O + \text{energy}$$

Compounds

- Oxides:
  - (i) The oxides of metals are basic, e.g. CaO.

Group I oxides dissolve to give highly alkaline solutions;

e.g. $Na_2O(s) + H_2O(l) \rightarrow$
$$2Na^+(aq) + 2OH^-(aq)$$

- (ii) The oxides of non-metals are acidic;

  e.g. $SO_3(g) + H_2O(l) \rightarrow SO_4^{2-}(aq) + 2H^+(aq)$

- (iii) Some oxides are neutral, e.g. CO.
- (iv) Some oxides are amphoteric, e.g. Al_2O_3 dissolves in both acid and alkali;

  $$Al_2O_3(s) + 6H^+(aq) \rightarrow 2Al^{3+}(aq) + 3H_2O(l)$$
  $$Al_2O_3(s) + 6OH^-(aq) + 3H_2O(l) \rightarrow$$
  $$2Al(OH)_6^{3-}(aq)$$

- Water:
  An extremely abundant oxide that is essential for life, and a good solvent.
- Oxoanions:
  Occur in a wide range of compounds, e.g. sulphate, SO_4^{2-}, nitrate, NO_3^-, carbonate, CO_3^{2-}, manganate(VII), MnO_4^-, chlorate(VII), ClO_4^-.
- Peroxides:
  - (i) Examples are barium peroxide, BaO_2, sodium peroxide, Na_2O_2, and hydrogen peroxide, H_2O_2. All are vigorous oxidising agents and decompose, giving off oxygen.
  - (ii) Hydrogen peroxide is used to prepare oxygen;

    $$2H_2O_2(aq) \rightarrow 2H_2O(l) + O_2(g)$$

  - (iii) H_2O_2 acts as an oxidising agent in acidic conditions, and as a reducing agent in alkaline conditions.
- Trioxygen (ozone)
  - (i) The ozone layer in the upper atmosphere controls the amount of ultraviolet light reaching the Earth's surface. The layer is in danger of being destroyed by chlorofluorocarbons.
  - (ii) Trioxygen is an oxidising agent;

    e.g. $S^{2-}(aq) + O_3(g) \rightarrow SO_4^{2-}(aq) + O_2(g)$

- Ozonolysis:
  Is important in organic chemistry. Alkenes are converted into mixtures of aldehydes and/or ketones; e.g.

100

Group VI

100.1 The nature of the elements

You will find a summary of the important physical properties of the elements in Table 100.1, and their uses are shown in Table 100.2. We have discussed the chemistry of oxygen in some detail in the previous unit. Here we shall concentrate on the other elements in the Group, especially sulphur. As is common with the first member of a Group, the properties of oxygen are not necessarily typical of the Group as a whole. For example, oxygen is the only gas among them, and has by far the highest electronegativity. Oxygen tends to bond with nearly every element in the Periodic Table, which cannot be said of the other members of the Group. Similarly, oxygen will make ionic compounds with the most reactive metals, whereas unless they react with a Group I metal, the sulphides, selenides, etc., are normally covalent (even though they may be polar). There is the usual tendency for metallic nature to increase going down the Group. However, polonium, right at the bottom of the Group, which displays the most metallic nature, is hardly a pleasant element to study: it is radioactive. One indication of its metallic nature is that it makes an ionic oxide, PoO_2. Polonium is the only B metal in the Group.

Sulphur and selenium possess allotropes, with those of sulphur being the most important. The structure of, and relationship between, rhombic and monoclinic sulphur was discussed in Unit 57. You should look there for details about them. However, just to remind you,

Table 100.2. Uses of the elements in Group VI

Element	Main uses
Oxygen	Aids to breathing
Sulphur	Sulphuric acid manufacture
	Fertilisers
	Explosives, dyes, polymers, detergents and myriad other chemical processes
Selenium	In some alloys
	The manufacture of light sensitive resistors and semiconductors
	Medicines
Tellurium	Alloys
	Semiconductors
	In photocopiers
Polonium	Little or no commercial use

both allotropes are built of puckered rings of eight sulphur atoms. (Selenium has a similar structure.)

Another major difference to oxygen comes about through the use of d orbitals in bonding. In many sulphur compounds the bonds to sulphur are shorter than expected. This suggests a degree of double bond character, which can occur if sulphur makes use of its empty 3d orbitals as well as its s and p orbitals. Similarly, apart from oxygen the elements can make up to six covalent bonds by using d orbitals, e.g. sulphur in sulphates and sulphur hexafluoride, SF_6. (The empty 3d orbitals

Table 100.1. Physical properties of the elements in Group VI

Symbol	Oxygen O	Sulphur S	Selenium Se	Tellurium Te	Polonium Po
Electron structure	$(He)2s^2 2p^4$	$(Ne)3s^2 3p^4$	$(Ar)3d^{10}4s^2 4p^4$	$(Kr)4d^{10}5s^2 5p^4$	$(Xe)5d^{10}6s^2 6p^4$
Electronegativity	3.5	2.5	2.4	2.1	
I.E./$kJ\,mol^{-1}$	1314	1000	950	830	700
Melting point/°C	−219	114.5	217	450	254
Boiling point/°C	−183	444.6	685	1390	962
Atomic radius/pm	74	104	117	137	164
Principal oxid. no.	−2	−2,+4,+6	−2,+4,+6	−2,+4,+6	+2,+4

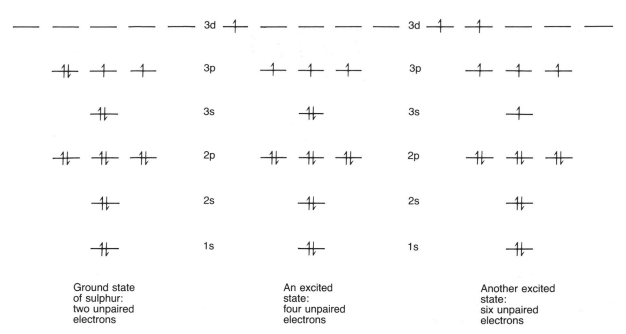

Figure 100.1 *Ways in which sulphur can have two, four or six unpaired electrons, and then make two, four or six covalent bonds*

belonging to oxygen are too high in energy to be used in bonding.)

In fact there are three possibilities for sulphur in using its valence electrons together with its 3d orbitals. Let us assume that electrons can be promoted between the 3s, 3p and 3d sets while keeping the order of energies of the orbitals constant (this is not really possible; see section 17.4 on hybridisation). Then sulphur can have two, four, or six unpaired electrons (Figure 100.1). This is reflected in the common oxidation states of the element (Figure 100.2), and especially in the nature of its halides (see Tables 100.1 and 100.5). The other members of Group VI also have d orbitals available for them to use in bonding.

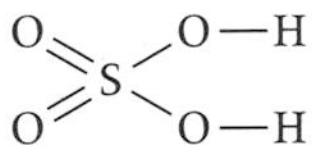

$$
\begin{array}{c}
O \diagdown \quad \diagup O-H \\
S \\
O \diagup \quad \diagdown O-H
\end{array}
$$

Figure 100.2 *The structure of sulphuric acid shows sulphur making six bonds to the four oxygen atoms. This number of bonds occurs when sulphur uses its 3d orbitals in bonding*

100.1   What is wrong with this statement: 'oxygen forms the oxide ion, O^{2-}, in many compounds in order that it can gain the noble gas structure of neon'?

100.2   One isotope of polonium, $^{218}_{84}Po$, decays by alpha emission with a half-life of about 3 min. What is the product of its decay?

100.2   The action of heat on sulphur

If you heat sulphur in a test tube fairly quickly, you will find that the yellow solid changes into a pale orange, mobile (i.e. runny) liquid. This happens at around 128° C. Even at this temperature some of the S_8 rings are breaking apart. Although the change from rhombic to monoclinic sulphur takes place at 96.5° C (the transition temperature, see Unit 57), you cannot see this change take place, and in any case rapid heating does not allow an equilibrium to be set up between the two allotropes. On further heating the colour darkens, and the liquid becomes more viscous. The increase in viscosity is due to many S_8 rings breaking apart, and the sulphur atoms joining to make long chains up to 10^6 atoms long. The chains become entangled and cannot slide over one another at all easily. Viscosity is a maximum near 200° C. Eventually the liquid becomes almost black in colour and mobile again as many chains break apart to give smaller molecules. Although sulphur will not boil until 444° C, fumes of sulphur will be given off. If you do this experiment, take care: the hot fumes may well catch fire, giving a blue flame and pungent fumes of sulphur dioxide. If you pour the almost black liquid into cold water, an orange/brown solid is produced. This solid is called *plastic sulphur*, and is elastic. Plastic sulphur is *not* an allotrope of sulphur. It consists of a tangle of long chains of sulphur atoms. Gradually the solid loses its rubbery feel; it becomes hard and takes on the yellow colour of rhombic sulphur to which all of it eventually converts. The key stages in the process are shown in Figure 100.3.

100.3   The extraction of sulphur

Sulphur is found in many minerals, especially combined with B metals like copper, tin, mercury and lead. Sulphur can be obtained as a by-product of the extraction of these metals from their ores. However, this is a costly and involved process. Some deposits of oil and

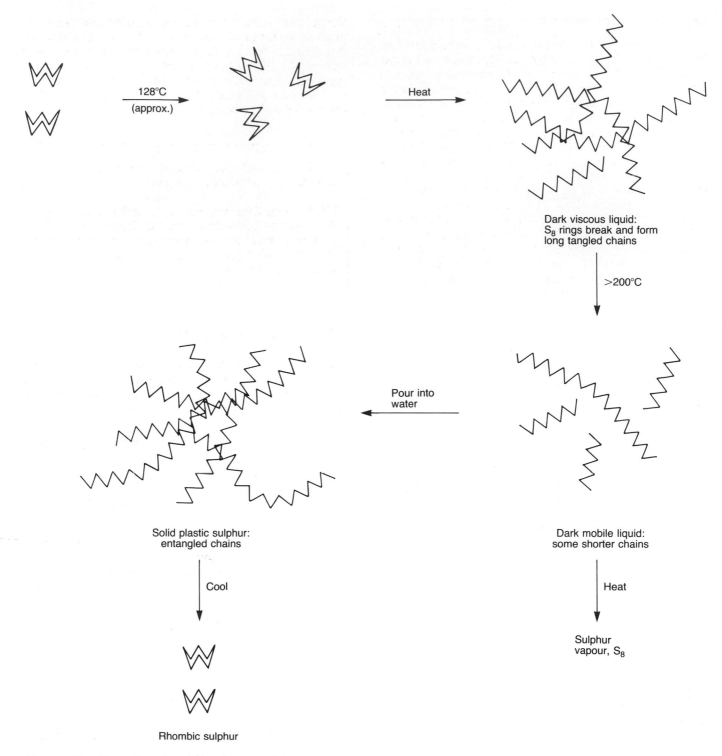

128°C (approx.)

Heat

Dark viscous liquid: S₈ rings break and form long tangled chains

>200°C

Pour into water

Solid plastic sulphur: entangled chains

Dark mobile liquid: some shorter chains

Cool

Heat

Rhombic sulphur

Sulphur vapour, S₈

Figure 100.3 *The action of rapid heating on sulphur*

natural gas contain a significant amount of sulphur, particularly in the form of hydrogen sulphide. Owing to the necessity of removing all traces of sulphur compounds before the oil and gas can be converted into anything useful, some oil refineries are able to sell the sulphur they have to extract. It is far easier to extract sulphur directly out of the ground using a method

invented by Hermann Frasch about 100 years ago. In certain parts of the world, especially USA, Poland and Iraq, there are huge deposits of elemental sulphur buried underground. Often the sulphur is mixed with other minerals, and although it is possible to dig the mixture out and separate the sulphur, Frasch developed a better method.

(a) The Frasch process

This is shown in Figure 100.4. A hole is drilled down to the rock containing the sulphur, which is usually less than 1000 m down. Then a set of concentric tubes is placed down the hole. The lower part of the widest tube has holes through it and the bottom of the tube terminates in a metal surround drilled with holes. This grid is connected to the second of the three concentric pipes. Superheated water at about 160°C is pumped through the outer tube. Even though the water cools as it enters the surrounding rock, it is sufficiently hot to melt the sulphur. Molten sulphur mixed with some water is forced through the lower grid by the pressure of the superheated water and rises up the inner tube. About half-way up the second tube, the third tube ends and feeds hot compressed air into the mixture of molten sulphur and water. This makes the mixture froth, and the bubbles rise up the tube, taking the sulphur with them. On the surface the air is allowed to escape and the sulphur is stored, usually as a liquid. It is common practice to store and transport sulphur as a liquid, even though there is a cost in providing the heat necessary to stop it solidifying. This is largely a matter of convenience; liquids are easier to pump from one place to another than are solids. Also, there are no problems of the escape of fine particles of sulphur dust.

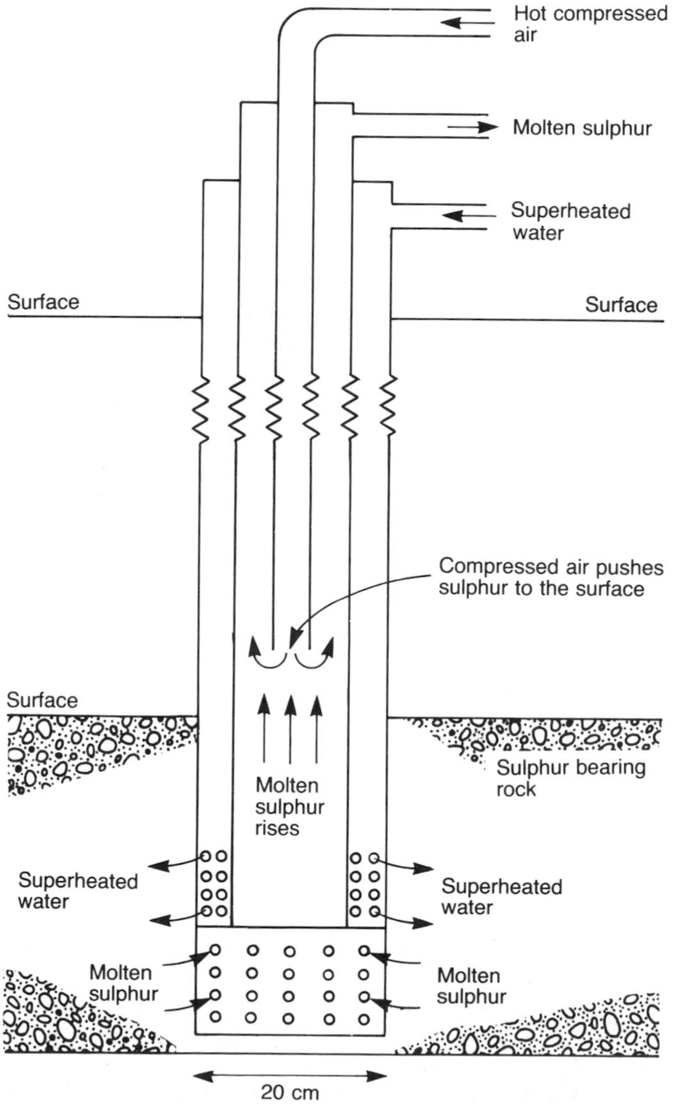

100.3 Why is it necessary to remove sulphur compounds from oil and natural gas?

100.4 A second hole has to be drilled into the sulphur deposit some way from the Frasch pipes. The purpose of this hole is to pump out waste water. Why does the Frasch process come to a halt if the excess water is not removed?

100.4 Sulphuric acid

The majority of sulphur is used to make sulphuric acid. You will find the method of manufacture described in Unit 83. The uses of sulphuric acid are shown in Table 100.3. Sulphuric acid has four different aspects to its chemistry:

(i)   in dilute solution it behaves like a typical strong acid;
(ii) when concentrated it shows some oxidising ability;
(iii) when concentrated it is a good dehydrating agent;
(iv) it is a sulphonating agent in organic chemistry.

Table 100.3. The uses of sulphuric acid

Making superphosphate fertiliser
Making ammonium sulphate fertiliser
Processing of metal ores
Manufacture of detergents
Manufacture of paper
Manufacture of Rayon and other polymers
Manufacture of paints and pigments
Electrolyte in heavy duty batteries
Industrial treatment of metals
Laboratory reagent

Figure 100.4 The Frasch process for extracting sulphur

(a) As a strong acid

Sulphuric acid is regarded as a strong acid in water. It dissociates in two stages:

$$H_2SO_4(aq) + H_2O(l) \rightarrow HSO_4^-(aq) + H_3O^+(aq)$$
$$HSO_4^-(aq) + H_2O(l) \rightleftharpoons SO_4^{2-}(aq) + H_3O^+(aq)$$

Only the first dissociation is complete; the second is partial.

When it is *dilute* it shows all the usual properties of

acids. For example, it will give hydrogen with metals above hydrogen in the activity series, e.g.

$$Zn(s) + H_2SO_4(aq) \rightarrow ZnSO_4(aq) + H_2(g)$$

or

$$Zn(s) + 2H_3O^+(aq) \rightarrow Zn^{2+}(aq) + H_2(g) + 2H_2O(l)$$

When reacting in this way, the sulphate ion remains intact.

A mixture of concentrated nitric and sulphuric acids is used as a *nitrating mixture* in organic chemistry. When the two acids are mixed an equilibrium is set up, the active ingredient of which is the nitryl cation (nitronium ion), NO_2^+:

$$HNO_3 + 2H_2SO_4 \rightarrow NO_2^+ + H_3O^+ + 2HSO_4^-$$

The nitryl cation is a sufficiently good electrophile to attack a benzene ring.

(b)   *As an oxidising agent*

The acid shows its oxidising nature when it is *concentrated*. For example, it cannot be used to prepare hydrogen bromide from sodium bromide. This is because it can oxidise the hydrogen bromide released:

$$2HBr(g) + H_2SO_4(l) \rightarrow Br_2(l) + SO_2(g) + 2H_2O(l)$$

(You will find details of this and similar reactions in Unit 101 on the halogens.)

The oxidising reaction is a feature of the sulphate ion, in which sulphur has the high oxidation state (or number) of +6. We have said before that high oxidation states lead to oxidation reactions because the element involved tends to take electrons and revert to a lower oxidation state.

A variety of possibilities exist for the way it reacts, of which the most important follow:

(i)   Sulphur dioxide given off

$$SO_4^{2-}(aq) + 4H^+(aq) + 2e^- \rightarrow H_2SO_3(aq) + H_2O(l)$$

or

$$SO_4^{2-}(aq) + 4H^+(aq) + 2e^- \rightarrow SO_2(g) + 2H_2O(l)$$

(ii)   Sulphur deposited

$$SO_4^{2-}(aq) + 8H^+(aq) + 6e^- \rightarrow S(s) + 4H_2O(l)$$

(iii)   Hydrogen sulphide given off

$$SO_4^{2-}(aq) + 10H^+(aq) + 8e^- \rightarrow H_2S(g) + 4H_2O(l)$$

If you look at the number of electrons involved in each of these changes, you should realise that it is harder to remove six or eight electrons from an atom or ion than to remove two. The first change, where SO_2 is given off, will occur unless the acid is reacting with something that can easily lose electrons. That is, we expect sulphur or hydrogen sulphide only when there is a good reducing agent reacting with the acid. Hydrogen bromide and hydrogen iodide are two of the more common substances that will give these changes; see section 102.1.

Normally the hot, concentrated acid reacts as in the following examples:

$$Cu(s) + 2H_2SO_4(aq) \rightarrow CuSO_4(aq) + 2H_2O(l) + SO_2(g)$$
$$S(s) + 2H_2SO_4(aq) \rightarrow 2H_2O(l) + 3SO_2(g)$$
$$C(s) + 2H_2SO_4(aq) \rightarrow CO_2(g) + 2H_2O(l) + 2SO_2(g)$$

(c)   *As a dehydrating agent*

Concentrated sulphuric acid will remove the elements of water from a wide variety of organic compounds. A reaction that you might see performed is the addition of a few drops of the concentrated acid to sugar (glucose). The sugar becomes very hot and froths, leaving a black mass of carbon:

$$C_6H_{12}O_6 - 6H_2O \rightarrow 6C$$

In organic chemistry the acid is used to convert alcohols into alkenes (and vice versa; see section 112.2).

Its dehydrating action on methanoic acid can be used as a method of preparing carbon monoxide:

$$\underset{\substack{\text{methanoic}\\\text{acid}}}{HCOOH} - H_2O \rightarrow CO$$

In practice the acid is added to sodium methanoate; methanoic acid is produced when they mix.

If ethanedioic acid (oxalic acid) is used, at about 60°C a mixture of carbon monoxide and carbon dioxide is released:

$$(COOH)_2 - H_2O \rightarrow CO + CO_2$$

Sulphuric acid has also been used for drying gases that are resistant to oxidation, e.g. chlorine.

The black mass in the tube is carbon left from the dehydration of sucrose by concentrated sulphuric acid.

(d)   *Sulphonation*

Please look at section 113.6 for information about this.

100.5 Apart from those listed above, what other properties are typical of strong acids?

100.6 Which of nitric acid and sulphuric acid is the stronger Brønsted acid, i.e. which donates a proton to the other?

100.7 Here is some information about redox potentials: $E^{\ominus}_{Cu^{2+}/Cu} = +0.34\,V$; $E^{\ominus}_{Zn^{2+}/Zn} = -0.76\,V$; $E^{\ominus}_{SO_4^{2-}/H_2SO_3} = +0.17\,V$; $E^{\ominus}_{SO_4^{2-}/H_2S} = +0.3\,V$; $E^{\ominus}_{SO_4^{2-}/S} = +0.36\,V$.

(i) Why do these figures show that copper should *not* be oxidised by sulphuric acid, under standard conditions?

(ii) What is an explanation of why the figures do not apply to the reaction as it is carried out in the laboratory?

(iii) What would you predict to happen if zinc reacts with concentrated sulphuric acid?

100.5 The hydrides

Of the hydrides of Group VI (Table 100.4), water is by far the most important, and is not typical of the others. By now you should know that water is anomalous in being a liquid at room temperature; the other hydrides are gases. Hydrogen bonding is responsible. Read Unit 21 for more information about this.

Hydrogen sulphide is very poisonous. If you have smelled the gas (it has the smell of rotten eggs) and (presumably) survived, it is because your nose is very sensitive to the gas. The time to worry is when the gas is present but you can no longer smell it: death will follow soon afterwards. The gas can be made by mixing hydrochloric acid with a metal sulphide, often iron(II) sulphide:

$$FeS(s) + 2HCl(aq) \rightarrow FeCl_2(aq) + H_2S(g)$$

Unlike water, hydrogen sulphide will burn in air (with a pale blue flame):

$$2H_2S(g) + 3O_2(g) \rightarrow 2H_2O(g) + 2SO_2(g)$$

A useful property of hydrogen sulphide is that it

Table 100.4. The major hydrides of Group VI

Hydride	Formula	Bond angle	Melting point/°C	Boiling point/°C
Water	H_2O	104°30′	0	100
Hydrogen sulphide	H_2S	93°	−85.5	−60.3
Hydrogen selenide	H_2Se	91°	−65.7	−42
Hydrogen telluride	H_2Te	89°30′	−49	−2

releases sulphide ions when it dissolves in water. Two equilibria are involved:

$$H_2S(aq) + H_2O(l) \rightleftharpoons H_3O^+(aq) + HS^-(aq)$$
$$HS^-(aq) + H_2O(l) \rightleftharpoons H_3O^+(aq) + S^{2-}(aq)$$

Transition metals and B metals often give precipitates with sulphide ions, some of which are coloured (e.g. MnS is pink), although many are black (e.g. PbS, CuS). The presence (or absence) of precipitates allows us to identify metal ions by passing the gas into their solution. You will find more details in Unit 64.

Again unlike water, but like ammonia, hydrogen sulphide can be a good reducing agent. For example it performs the simple tests that we expect of a reducing agent. In particular, it will decolourise acidified potassium manganate(VII) solution, and turn an orange solution of potassium dichromate(VI) green:

$$5H_2S(aq) + 2MnO_4^-(aq) + 6H^+(aq) \rightarrow$$
purple
$$5S(s) + 2Mn^{2+}(aq) + 8H_2O(l)$$
yellow   colourless

$$3H_2S(aq) + Cr_2O_7^{2-}(aq) + 8H^+(aq) \rightarrow$$
orange
$$3S(s) + 2Cr^{3+}(aq) + 7H_2O(l)$$
yellow   green

Chlorine is also a good oxidising agent and will remove the hydrogen from hydrogen sulphide:

$$Cl_2(g) + H_2S(g) \rightarrow 2HCl(g) + S(s)$$

100.8 Sketch the way the boiling points of the Group VI hydrides change going down the Group, and explain the appearance of the graph.

100.9 Suggest an explanation of the way the bond angles in the hydrides change down the Group. (Hint: it will help if you have read Unit 17 first.)

100.10 Different sulphides are precipitated at different times depending on the concentration of sulphide ions in the solution, which can be controlled by passing H_2S into an acid or alkaline solution. Explain how the concentration of sulphide ions changes with the pH of the solution.

100.11 The redox reaction involving the reducing power of hydrogen sulphide is

$$S(s) + 2H^+(aq) + 2e^- \rightarrow H_2S(aq); \quad E^{\ominus}_{S/H_2S} = +0.14\,V$$

Two other half-equations are

$$Fe^{3+}(aq) + e^- \rightarrow Fe^{2+}(aq); \quad E^{\ominus}_{Fe^{3+}/Fe^{2+}} = +0.77\,V$$
$$Br_2(l) + 2e^- \rightarrow 2Br^-(aq); \quad E^{\ominus}_{Br_2/Br^-} = +1.07\,V$$

Under standard conditions:

(i) Will hydrogen sulphide react with iron(III) ions?

(ii) Will it react with bromine?

If you decide a reaction should take place, write the equation.

Table 100.5. The principal halides of Group VI

Element	Fluorides	Chlorides	Bromides	Iodides
Oxygen	F_2O	Cl_2O, ClO_2, Cl_2O_7	Br_2O, BrO_2	I_2O_5
Sulphur	SF_4, SF_6	S_2Cl_2, SCl_2	S_2Br_2	
Selenium	SeF_4, SeF_6	Se_2Cl_2	$Se_2Br_2, SeBr_4$	
Tellurium	TeF_4, TeF_6	$TeCl_2, TeCl_4$	$TeBr_2, TeBr_4$	TeI_4
Polonium		$PoCl_2, PoCl_4$	$PoBr_2, PoBr_4$	PoI_4

100.6   The halides and oxohalides

You will find the main halides listed in Table 100.5. There are two or three features of their chemistry that we shall quickly look at. First, notice that fluorine brings out the highest oxidation state, as in SF_6, SeF_6 and TeF_6. The fact that these atoms can actually make six bonds is due to their use of d orbitals in bonding.

Secondly, the sulphur halides tend to hydrolyse easily. (In section 40.4 we used this type of reaction to show how the formula of a chloride could be discovered.) Sulphur hexafluoride, SF_6, is an exception. The sulphur–fluorine bonds in this molecule are very strong, and shorter than normal. Both features suggest that sulphur is using its 3d orbitals in providing a degree of double bond character.

Disulphur dichloride, S_2Cl_2, is the chloride made when sulphur is warmed in dry chlorine:

$$2S(s) + Cl_2(g) \rightarrow S_2Cl_2(l)$$

Of the oxohalides, the most important are those of sulphur, especially sulphur dichloride oxide (thionyl chloride), $SOCl_2$, and sulphur dichloride dioxide (sulphuryl chloride), SO_2Cl_2 (Figure 100.5). The former is a colourless liquid that is easily hydrolysed:

$$SOCl_2(l) + 2H_2O(l) \rightarrow H_2SO_3(aq) + 2HCl(aq)$$

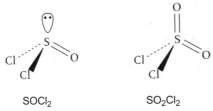

Figure 100.5 *Both $SOCl_2$ and SO_2Cl_2 have structures based on a tetrahedron (although both are distorted)*

Its main use comes in organic chemistry, where it can be used to replace the OH group of an acid. For example,

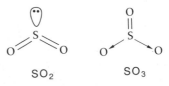

$$CH_3COOH(l) + SOCl_2(l) \rightarrow CH_3COCl(l) + SO_2(g) + HCl(g)$$

100.12   Predict the shape of SF_6 and (harder) SCl_4 and $TeCl_4$.

100.13   Why might SF_6 not hydrolyse easily? (There are two possible reasons; one to do with energy, the other to do with steric hindrance.)

100.14   If you look at section 116.4 you will find that $SOCl_2$ is not the only reagent that reacts with the OH group of an organic compound. However, it has an advantage over its main rival. What is the rival, and what is the advantage?

100.7   The oxides

Sulphur dioxide and sulphur trioxide are gases; the others are solids (although those of selenium are volatile). The oxides are shown in Table 100.6. In an earlier unit we saw how SO_2 and SO_3 can be made; the first by burning sulphur in air, the second by reacting sulphur dioxide with oxygen in the presence of a catalyst.

$$S(s) + O_2(g) \rightarrow SO_2(g)$$
$$2SO_2(g) + O_2(g) \rightarrow 2SO_3(g)$$

They are both highly soluble in water, with the reaction between SO_3 and water being explosive. The structure and bonding of the two gases are illustrated in Figure 100.6.

Table 100.6. The principal oxides of Group VI

Element	Oxides
Sulphur	SO_2, SO_3
Selenium	SeO_2, SeO_3
Tellurium	TeO, TeO_2, TeO_3
Polonium	PoO, PoO_2

Sulphur dioxide is the anhydride of sulphurous acid, H_2SO_3. This acid is made in small amounts when the gas dissolves in water. A solution of sulphur dioxide contains many more hydrated sulphur dioxide molecules and hydrogen sulphite ions, HSO_3^{2-}, than H_2SO_3 molecules. Sulphur trioxide is the anhydride of sulphuric acid:

$$SO_2(g) + H_2O(l) \rightarrow H_2SO_3(aq)$$
$$SO_3(g) + H_2O(l) \rightarrow H_2SO_4(aq)$$

Figure 100.6 *The structures of SO_2 and SO_3 molecules*

A solution of sulphur dioxide is a good reducing medium and will decolourise acidified potassium manganate(VII) and acidified potassium dichromate(VI) solutions.

100.15 Sulphur trioxide can be made on a small scale by passing dry oxygen and dry sulphur dioxide over a hot catalyst (platinum is good if you can afford it). The sulphur trioxide can be collected as thin needle-like crystals by cooling the collection flask with a mixture of salt and ice.

(i) What would you use to dry the gases?

(ii) Why must the apparatus be thoroughly dried before use?

(iii) Sketch an apparatus that you could use to make and collect a sample of sulphur trioxide. Assume that you are provided with a supply of sulphur dioxide and oxygen.

100.16 Sulphur dioxide and hydrogen sulphide react together according to the equation

$$2H_2S(g) + SO_2(g) \rightarrow 3S(s) + 2H_2O(l)$$

We have said that both gases are reducing agents. Which is the more powerful reducing agent of the two? (Hint: oxidation numbers.)

100.8 Sulphites, sulphates and other oxoanions

(a) *Sulphites*

Sulphites contain the SO_3^{2-} ion. Most sulphites are soluble in water, and act as reducing agents. When they are warmed with acid, sulphur dioxide is given off, e.g.

$$SO_3^{2-}(aq) + 2H^+(aq) \rightarrow SO_2(g) + H_2O(l)$$

However, there are one or two sulphites that are insoluble. In particular, if you add silver nitrate solution to a sulphite in water, you will find that a white precipitate is produced:

$$2Ag^+(aq) + SO_3^{2-}(aq) \rightarrow Ag_2SO_3(s)$$

This is one reason why the silver nitrate test for a chloride should always be done after adding dilute nitric acid. Indeed, it is best to warm the acidified solution first, as this will drive sulphur dioxide from the solution.

(b) *Sulphates*

Sulphates contain the SO_4^{2-} ion. The sulphates of Group II metals tend to be insoluble; for example, calcium sulphate, $CaSO_4$, and barium sulphate, $BaSO_4$. The latter is the precipitate made when barium chloride solution is used to test for a sulphate. Other sulphates are often soluble. Probably the one you know best is copper(II) sulphate, which as the crystals $CuSO_4{\cdot}5H_2O$ or in solution is a deep blue colour. (However, when anhydrous it is white.)

Sulphates usually decompose when heated to a sufficiently high temperature. The usual pattern is for them to change into an oxide, and give off SO_3, e.g.

$$\underset{\text{iron(III) sulphate}}{Fe_2(SO_4)_3} \rightarrow Fe_2O_3(s) + 3SO_3(g)$$

(c) *Thiosulphates*

Thiosulphates contain the ion $S_2O_3^{2-}$. The structure of the ion is like that of a sulphate ion, except that one of the oxygen atoms is replaced by a sulphur atom. It is possible to use a solution of a sulphite to make thiosulphate ions. All you need do is boil a solution of sodium sulphite with powdered sulphur:

$$SO_3^{2-}(aq) + S(s) \rightarrow S_2O_3^{2-}(aq)$$

Sodium thiosulphate solution is widely used as a fixing agent in photography. It has the ability to dissolve the silver salts that have not been affected by light. In the chemistry laboratory, thiosulphate solutions are used in iodine titrations, the key reaction being

$$I_2(aq) + 2S_2O_3^{2-}(aq) \rightarrow 2I^-(aq) + S_4O_6^{2-}(aq)$$

You will find details in section 40.3.

The reaction between thiosulphate ions and hydrogen ions is a useful one to show the essentials of chemical kinetics; see section 79.2.

(d) *Peroxodisulphates*

These ions have the formula $S_2O_8^{2-}$ and are found in salts such as $K_2S_2O_8$. They are oxidising agents, and behave according to the half-equation

$$S_2O_8^{2-}(aq) + 2e^- \rightarrow 2SO_4^{2-}(aq)$$

They will, for example, oxidise iodide to iodine and iron(II) to iron(III).

Peroxodisulphates can be thought of as salts of peroxodisulphuric(VI) acid, $H_2S_2O_8$. This acid can be made by the electrolysis of cold sulphuric acid and is one of a number of more unusual acids that sulphur will make. The structures of three oxoacids of sulphur are shown in Figure 100.7.

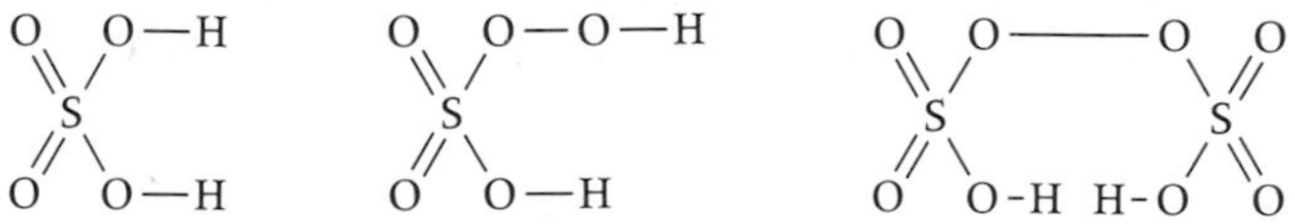

H_2SO_4 Sulphuric acid | H_2SO_5 Caro's acid | $H_2S_2O_8$ Peroxodisulphuric(VI) acid

Figure 100.7 *Three oxoacids of sulphur*

100.17 When a sulphite acts as a reducing agent it reacts according to the equation

$$SO_3^{2-}(aq) + H_2O(l) \rightarrow SO_4^{2-}(aq) + 2H^+(aq) + 2e^-$$

Write the equations for the reaction of a sulphite with MnO_4^- and with $Cr_2O_7^{2-}$ ions in acid solution.

100.18 Suggest two ways of dehydrating $CuSO_4 \cdot 5H_2O$.

100.19 Write equations for the reactions of $S_2O_8^{2-}$ ions with (i) iodide ions, (ii) iron(II) ions.

Table 100.7. Some sulphide ores

Metal	Ore
Copper	Copper pyrites, $CuFeS_2$ (also known as fool's gold)
	Chalcocite, Cu_2S
	Covellite, CuS
Lead	Galena, PbS
Mercury	Cinnabar, HgS
Nickel	Pentlandite, $Fe_9Ni_9S_{16}$
Zinc	Zinc blende, ZnS

100.9   The sulphides

The sulphides of Group I metals are ionic, e.g. $(Na^+)_2S^{2-}$. The sulphides of other metals, especially the B metals, are covalent to a greater or lesser extent. Some metals are found in Nature in combination with sulphur (see Table 100.7), and these are often used as the starting material for the extraction of the metals. You will find details in Unit 85. The crystal structure of zinc blende, ZnS, is important in crystal chemistry (see Unit 32).

100.20 What would happen if you warmed dilute hydrochloric acid with sodium sulphide?

100.21 A student needed a solution of a sulphide, and shook some copper(II) sulphide with water.

(i) Why would the student not be successful in making a solution in this way?

(ii) What is a better way of making a solution that contains sulphide ions?

Answers

100.1 Look back at section 46.2. It is energetically very unfavourable for an isolated oxygen atom to become an oxide ion. The production of oxide ions is a result of the high lattice energies of the ionic solids in which it is found.

100.2 Alpha decay gives us the element two places to the left in the Periodic Table (atomic number decreases by two), and reduces the mass number by four. This gives us $^{214}_{82}Pb$.

100.3 Sulphur compounds easily poison the catalysts used in many industrial processes.

100.4 If the water remains, the pressure builds up underground and stops the transfer of superheated water into the sulphur deposit.

100.5 They give carbon dioxide with carbonates, and neutralise a base to give a salt plus water.

100.6 Sulphuric acid molecules are converted into HSO_4^- ions. Therefore they donate protons to the nitric acid. Sulphuric acid is the stronger Brønsted acid.

100.7 (i) Because $E^{\ominus}_{Cu^{2+}/Cu}$ is more positive than $E^{\ominus}_{SO_4^{2-}/H_2SO_3}$ and $E^{\ominus}_{SO_4^{2-}/H_2S}$ it means that Cu^{2+} ions are better oxidising agents than SO_4^{2-} ions. Therefore the acid should not oxidise Cu to Cu^{2+}. Although $E^{\ominus}_{SO_4^{2-}/S}$ is more positive than $E^{\ominus}_{Cu^{2+}/Cu}$ it would appear that copper should be oxidised to Cu^{2+} and the acid reduced to H_2S.

However, the difference between the two e.m.f.s is so small that the reaction is unlikely to occur at an appreciable rate (see section 70.3).

(ii) Standard conditions refer to acid of concentration $1\ mol\ dm^{-3}$. This is very different to concentrated acid. Similarly, the acid is used hot, not at 25°C. Non-standard conditions can radically alter redox potentials.

(iii) Zinc is a good reducing agent (this is shown by the negative value of $E^{\ominus}_{Zn^{2+}/Zn}$). It is good enough to reduce sulphuric acid to sulphuric dioxide, and even hydrogen sulphide.

100.8 See Figure 90.5. Apart from water, which is anomalous owing to the large amount of hydrogen bonding, the trend is as one would expect: the boiling points increase as the mass of the molecules increases.

100.9 The V shape can be regarded as tetrahedral if the two lone pairs of the molecules are included. Oxygen, with the greatest electronegativity of the Group, has lone pairs that are the most compact and dense centres of negative charge. These have the effect of forcing the bond pairs closer together than in the other hydrides. As the lone pairs become more spread out, their repulsive nature decreases.

100.10 If we add acid (H^+) to a solution in which the two equilibria are set up, each of them will move to the left, so as to reduce the acid concentration. Thus, adding acid reduces the concentration of sulphide ions. Adding

Answers – cont.

alkali works in the opposite direction. Thus, the lower the pH, the greater $[S^{2-}(aq)]$ and vice versa.

100.11 (i) Yes, Fe^{3+} ions will oxidise H_2S to S:
$$2Fe^{3+}(aq) + H_2S(aq) \rightarrow 2Fe^{2+}(aq) + S(s) + 2H^+(aq)$$

(ii) Yes:
$$Br_2(l) + H_2S(aq) \rightarrow 2Br^-(aq) + S(s) + 2H^+(aq)$$

100.12 The shapes are shown in Figure 100.8.

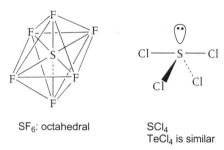

SF₆: octahedral

SCl₄
TeCl₄ is similar

Figure 100.8 *The shapes of SF₆, SCl₄ and TeCl₄ molecules. Note the lone pair in SCl₄ (and TeCl₄)*

100.13 One possibility is that the S—F bonds are too strong to break easily. Secondly, the fluorine atoms surround the central sulphur atom so well that an attacking water molecule (or hydroxide ion) cannot get its electrons near enough to the sulphur to start making a bond.

100.14 PCl_5. The side products from $SOCl_2$ are gases and easily removed. PCl_5 gives the non-volatile $POCl_3$.

100.15 (i) Concentrated sulphuric acid works well.

(ii) The oxides will make tiny drops of acid in the apparatus. SO_3 will not solidify properly if it is damp.

(iii) Figure 100.9 shows the apparatus.

100.16 The oxidation number of sulphur in SO_2 is +4, and in H_2S is −2. In both cases the sulphur is left with an oxidation number of 0. The SO_2 gains electrons, the H_2S loses them; therefore H_2S is the reducing agent.

100.17 The equations are:
$$5SO_3^{2-}(aq) + 2MnO_4^-(aq) + 6H^+(aq) \rightarrow$$
$$5SO_4^{2-}(aq) + 2Mn^{2+}(aq) + 3H_2O(l)$$
$$3SO_3^{2-}(aq) + Cr_2O_7^{2-}(aq) + 8H^+(aq) \rightarrow$$
$$3SO_4^{2-}(aq) + 2Cr^{3+}(aq) + 4H_2O(l)$$

100.18 Heating it. Leaving it with concentrated sulphuric acid.

100.19 (i) $S_2O_8^{2-}(aq) + 2I^-(aq) \rightarrow 2SO_4^{2-}(aq) + I_2(s)$

(ii) $S_2O_8^{2-}(aq) + 2Fe^{2+}(aq) \rightarrow 2SO_4^{2-}(aq) + 2Fe^{3+}(aq)$

100.20 Hydrogen sulphide is given off.

100.21 (i) Like many sulphides, CuS is insoluble in water.

(ii) Bubble H_2S through water.

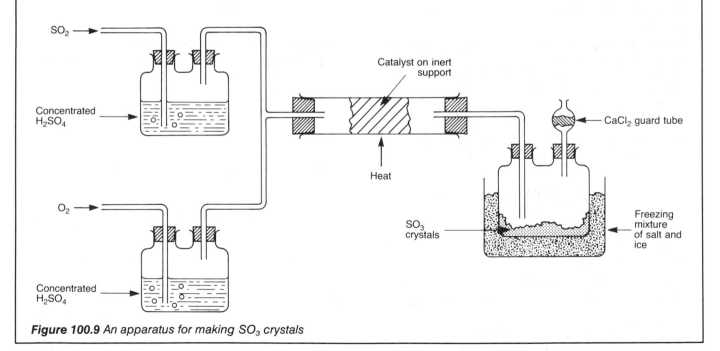

Figure 100.9 *An apparatus for making SO₃ crystals*

- The Group VI elements:
  - (i) Predominantly non-metallic, e.g. make acidic oxides.
  - (ii) Sulphur and selenium have allotropes.
  - (iii) Elements below oxygen can make use of d orbitals in bonding.
  - (iv) Sulphur is the most important element in the Group, and its chemistry is summarised here.
- Sulphur:
  - (i) Has the two allotropes, rhombic and monoclinic sulphur, both existing as S_8 rings.
  - (ii) Can make use of d orbitals in bonding, thus showing more than the single valence of 2 shown by oxygen.
  - (iii) Is extracted by the Frasch process.

Reactions

- With oxygen:
  - (i) Sulphur burns to make sulphur dioxide;

    $$S(s) + O_2(g) \rightarrow SO_2(g)$$

  - (ii) Sulphur trioxide is made by passing oxygen and sulphur dioxide over heated platinum gauze (a catalyst);

    $$2SO_2(g) + O_2(g) \rightarrow 2SO_3(g)$$

Compounds

- Sulphuric acid:
  - (i) In dilute solution behaves like a typical strong acid;

    $$H_2SO_4(aq) + H_2O(l) \rightarrow HSO_4^{2-}(aq) + H_3O^+(aq)$$
    $$HSO_4^{2-}(aq) + H_2O(l) \rightleftharpoons SO_4^{2-}(aq) + H_3O^+(aq)$$

  - (ii) When concentrated, shows oxidising ability. Signs of oxidation are sulphur deposited, and sulphur dioxide or hydrogen sulphide given off;

    $$2HBr(g) + H_2SO_4(l) \rightarrow$$
    $$Br_2(l) + SO_2(g) + H_2O(l)$$

  - (iii) When concentrated, the acid is a good dehydrating agent; e.g. dehydrates sugar, copper(II) sulphate crystals.
  - (iv) In organic chemistry it is used as: a sulphonating agent; a nitrating agent with concentrated nitric acid;

    $$HNO_3 + 2H_2SO_4 \rightarrow NO_2^+ + H_3O^+ + 2HSO_4^-$$

    The nitryl cation, NO_2^+, is the active species.
- Oxides:
  - (i) Sulphur dioxide and sulphur trioxide are acidic oxides.
  - (ii) Sulphur trioxide is the anhydride of sulphuric acid;

    $$SO_3(g) + H_2O(l) \rightarrow 2H^+(aq) + SO_4^{2-}(aq)$$
- Oxoanions:
  - (i) Sulphites (e.g. Na_2SO_3), thiosulphates (e.g. $Na_2S_2O_3$) and sulphates (e.g. $CaSO_4$) are common.
  - (ii) Sulphites: give off SO_2 when warmed with acids; act as reducing agents, e,g, decolourise acidified potassium manganate(VII) solution.
  - (iii) Thiosulphates: give a deposit of sulphur with acids; decolourise iodine solutions;

    $$I_2(aq) + 2S_2O_3^{2-}(aq) \rightarrow 2I^-(aq) + S_4O_6^{2-}(aq)$$

  - (iv) Sulphates: give off sulphur trioxide when heated strongly; Group II sulphates tend to be insoluble in water.
  - (v) Peroxodisulphates, e.g. $K_2S_2O_8$, are oxidising agents;

    $$\text{e.g. } 2I^-(aq) + S_2O_8^{2-}(aq) \rightarrow 2SO_4^{2-}(aq) + I_2(s)$$
- Halides:
  - (i) Disulphur dichloride, S_2Cl_2, is made when sulphur is warmed with chlorine. It is easily hydrolysed.
  - (ii) Sulphur hexafluoride, SF_6, is octahedral in shape. The bonding involves sulphur's 3d orbitals.
- Oxohalides:
  Sulphur dichloride oxide (thionyl chloride), $SOCl_2$, is used in organic chemistry to replace an OH group by Cl.
- Hydrogen sulphide:
  - (i) Is made by warming a sulphide with dilute hydrochloric acid;

    $$\text{e.g. } FeS(s) + 2HCl(aq) \rightarrow FeCl_2(aq) + H_2S(g)$$

  - (ii) Is highly poisonous.
  - (iii) Burns in air;

    $$2H_2S(g) + 3O_2(g) \rightarrow 2H_2O(g) + 2SO_2(g)$$

  - (iv) Is slightly soluble in water, giving an acidic solution;

    $$H_2S(aq) + H_2O(l) \rightleftharpoons H_3O^+(aq) + HS^-(aq)$$
    $$HS^-(aq) + H_2O(l) \rightleftharpoons H_3O^+(aq) + S^{2-}(aq)$$

  - (v) In solution it is used as a source of sulphide ions in testing for metal ions that give insoluble sulphides;

    $$\text{e.g. } Cu^{2+}(aq) + S^{2-}(aq) \rightarrow CuS(s)$$

  - (vi) Is a reducing agent;

    $$\text{e.g. } 3H_2S(aq) + Cr_2O_7^{2-}(aq) + 8H^+(aq) \rightarrow$$
    $$3S(s) + 2Cr^{3+}(aq) + 7H_2O(l)$$

101

Group VII: the halogens

101.1 The nature of the elements

The elements of Group VII, fluorine, chlorine, bromine, iodine and astatine, are known as the halogens. Many years ago this name was derived from two Greek words meaning 'to make sea salt'. Indeed, all the halogens react with Group I metals to make ionic salts such as sodium fluoride, NaF, sodium chloride, NaCl, potassium bromide, KBr, and so on.

The halogens are all covalent and they show just the sort of trends in their properties that we should expect from their position in the Periodic Table. Rather more importantly, they have a number of extremely useful properties that make them valuable commodities. Table 101.1 lists some basic information about the elements, and Table 101.2 shows their uses. Astatine has only been prepared in small amounts in the course of nuclear reactions, and it is radioactive. There is little need for you to know about its chemistry, and data on this element are omitted from the tables.

101.2 Discovery and extraction of the halogens

(a) Fluorine

During 1813 and 1814, Humphry Davy performed a set of experiments in which he proved the presence of a new element in a number of different compounds; he called the element fluorine. Unfortunately he was unable to collect a sample of it. This was achieved in 1886 by the French chemist Henri Moissan. Moissan electrolysed liquid anhydrous hydrogen fluoride with potassium hydrogenfluoride, KHF_2, dissolved in it. This discovery continues to be the basis of the commercial extraction of fluorine.

The fact that electrolysis has to be used to extract fluorine from its compounds tells us that this element is highly reactive. Once it reacts with another element it tends to give very strong bonds, which are hard to break. You might like to check your knowledge of the reactivity of elements by looking at section 66.9.

(b) Chlorine

Chlorine was first isolated by Scheele in 1774, but its nature as an element was established by Davy. Indeed,

Table 101.1. Physical properties of the halogens

Formula	Fluorine F$_2$	Chlorine Cl$_2$	Bromine Br$_2$	Iodine I$_2$
Electron structure	$(He)2s^22p^5$	$(Ne)3s^23p^5$	$(AR)3d^{10}4s^24p^5$	$(Kr)3d^{10}5s^25p^5$
Electronegativity	4.0	3.0	2.8	2.5
Bond energy/kJ mol^{-1}	158	242	193	151
Melting point/°C	−220	−101	−7	114
Boiling point/°C	−188	−34	58	183
I.E./kJ mol^{-1}	1681	1251	1140	1010
E.A./kJ mol^{-1}	361	388	365	332
Atomic radius/pm	64	99	111	128
Ionic radius/pm	133	181	196	219
$E^{\ominus}_{X_2/X^-}$/V	2.87	1.36	1.09	0.54
Oxidising power	Most			Least
Common oxid. no.	−1	−1,+1,+5,+7	−1,+1,+5,+7	−1,+1,+5,+7

Table 101.2. Uses of the halogens and their compounds

Halogen	Uses
Fluorine	In chlorofluorocarbons (CFCs), used as aerosol propellants, refrigerants and as foaming agents in polymers CFCs are also used in artificial blood PTFE (polytetrafluoroethene): lubricant, non-stick cooking pans Production of UF_6 in uranium purification Fluoridation of water supplies using, e.g., NaF Tin(II) fluoride (SnF_2) used in 'fluoride toothpastes' In hydrofluoric acid (HF), an etching agent for glass
Chlorine	Also used in CFCs and many organic chemicals, e.g. PVC (polyvinyl chloride, polychlorothene) Solutions of chlorine or bleaching powder are used as household and commercial bleaches Drinking water is treated with chlorine to kill bacteria Unfortunately, chlorine has been used in poison gases, e.g. Cl_2 itself and mustard gas ($ClCH_2CH_2SCH_2CH_2Cl$)
Bromine	In 1,2-dibromoethane ($BrCH_2CH_2Br$), as a petrol additive Bromochloromethane (CH_2ClBr) is used in fire extinguishers Silver bromide is widely used in photographic film
Iodine	A solution of iodine in alcohol can be used as a mild antiseptic Iodine is an essential part of our diet

in 1810 it was Davy who gave chlorine its name (this being taken from a Greek word meaning pale green). Before it received its new name, chlorine was called oxymuriatic acid. One reason for this name was that, until Davy's experiments were accepted, many chemists believed the gas to be a compound of oxygen and another element.

Chlorine is obtained on an industrial scale by the electrolysis of brine. You will find details of the method in section 84.2. Notice that, like fluorine, the use of electrolysis to isolate the element is an indication of its great reactivity.

(c) Bromine

Bromine, whose name means 'stinking', was discovered by Antoine Balard in 1826. (Balard was not a particularly famous chemist: a nasty comment once made about him was that, rather than Balard discovering bromine, it was bromine that discovered Balard!) He discovered the element by accident in the course of some experiments on sea water. He noticed the yellow-orange colour that bromine gives when it was made by treating sea water with chlorine. The reaction is due to the oxidation of bromide ions by chlorine:

$$2Br^-(aq) + Cl_2(aq) \rightarrow Br_2(aq) + 2Cl^-(aq)$$

This reaction remains at the heart of the commercial isolation of bromine. One method (there are several variations) is as follows:

(i) Sulphuric acid is added to sea water before chlorine gas is passed through the solution. This liberates the bromine according to the equation above.

(ii) Bromine evaporates from sea water fairly easily, so it is removed by blowing air, or sometimes steam, through the solution.

(iii) The bromine vapour reacts with another reagent, often sulphur dioxide, which together with water converts it into bromide ions again:

$$Br_2(g) + SO_2(g) + 2H_2O(l) \rightarrow$$
$$2Br^-(aq) + SO_4^{2-}(aq) + 4H^+(aq)$$

But this time the solution that collects is very much more concentrated than the original sea water.

(iv) This solution is also treated with a mixture of chlorine and steam. The bromine produced can be separated from the water with which it is mixed by fractional distillation.

(d) Iodine

Deposits of sodium nitrate ('Chile saltpetre') often contain sodium iodate, $NaIO_3$, as an impurity. If the degree of contamination is high, it can be economic to convert the iodate to iodine. The extraction takes place in three stages:

(i) The rock containing the nitrate and iodate is crushed and dissolved in hot water. On cooling, the sodium nitrate crystallises first, leaving the solution richer in sodium iodate. This is a large-scale example of fractional crystallisation; see section 59.2. After the solid is separated, part of the solution is treated with sodium hydrogensulphite, $NaHSO_3$:

$$IO_3^-(aq) + 3HSO_3^-(aq) \rightarrow I^-(aq) + 3HSO_4^{2-}(aq)$$

(ii) Now more of the solution is added. This brings about a reaction we have seen in the unit on titrations:

$$IO_3^-(aq) + 5I^-(aq) + 6H^+(aq) \rightarrow 3I_2(s) + 3H_2O(l)$$

(iii) The iodine can be filtered from the solution and, if necessary, purified by allowing it to sublime.

101.1 Sea water contains around 65×10^{-3} g of bromide ions in $1\,dm^3$. If all the bromide ions are converted into bromine, how many dm^3 of sea water are needed to produce 1 kg of bromine?

101.2 Deposits of brine in the USA sometimes contain significant amounts of iodide ions. Suggest a way of converting the iodide content of the brine into iodine.

101.3 What is meant by sublimation?

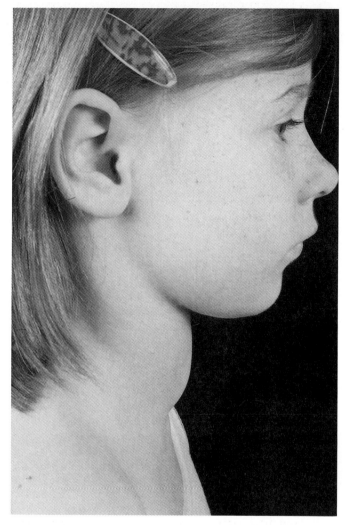

The swelling in this girl's neck is an indication of goitre, which resulted from a lack of iodine in her diet.

101.3 The laboratory preparation of the halogens

All the halogens are poisonous, and you should take great care if you do reactions with them. They should only be prepared in a fume cupboard. Fluorine is difficult to make, so we shall ignore its preparation. The others can be made by oxidising the appropriate hydrogen halide, HCl, HBr, or HI. The usual oxidising agent is manganese(IV) oxide.

(a) Chlorine

The most convenient source of hydrogen chloride is concentrated hydrochloric acid. If the acid and manganese(IV) oxide are warmed together, chlorine is given off. If you want pure chlorine, the gas can be passed first through water and then through concentrated sulphuric acid. The water dissolves any fumes of

hydrochloric acid, and the sulphuric acid dries the gas. The reaction is:

$$MnO_2(s) + 4HCl(aq) \rightarrow MnCl_2(aq) + Cl_2(g) + 2H_2O(l)$$

or

$$2Cl^-(aq) + 4H^+(aq) + MnO_2(s) \rightarrow$$
$$Mn^{2+}(aq) + Cl_2(g) + 2H_2O(l)$$

If you want a quick, but not too pure, supply of chlorine it is possible to drop concentrated hydrochloric acid on to crystals of potassium manganate(VII):

$$2KMnO_4(s) + 16HCl(aq) \rightarrow$$
$$2KCl(aq) + 2MnCl_2(aq) + 5Cl_2(g) + 8H_2O(l)$$

(b) Bromine

Solutions of hydrogen bromide are not generally available, so it must be made during the preparation. Hydrogen bromide is produced when, for example, potassium bromide is mixed with concentrated sulphuric acid. However, if manganese(IV) oxide is mixed with the acid, before the hydrogen bromide can escape it is oxidised and bromine is made. Overall the reaction is much the same as that for the preparation of chlorine: just substitute bromine for chlorine in the equation.

(c) Iodine

Iodine can be made by a similar method to bromine. The main difference is that iodine is a solid at room temperature.

101.4 By working out the changes in oxidation numbers, or states, of the reactants, explain why the reaction of hydrochloric acid with MnO_2 or $KMnO_4$ is an oxidation of the hydrogen chloride.

101.5 Is bromine a gas or liquid at room temperature (about 20° C)? Sketch an apparatus that you could use to prepare bromine.

101.6 Sketch an apparatus that you could use to make and collect iodine in the laboratory.

101.4 The reactivity of the halogens

By reactivity we mean the ease with which the halogens react. This is largely a matter of rates of reaction, which in turn is related to the activation energies of the reactions. Fluorine has the lowest bond energy. A collision involving a fluorine molecule needs less energy to break the molecule apart than does the same collision with, say, a chlorine molecule. As a result fluorine reacts more readily than chlorine. Indeed, fluorine is the most reactive of all the halogens. Reactivity decreases down the Group.

Even if a reaction takes place very quickly this does not mean that the product is energetically stable. As a rule this will only happen if the reaction is strongly exothermic (see Unit 50). In fact fluorine often takes part in reactions that are highly exothermic. This is a result of the strength of the bonds that fluorine makes with many atoms, in both covalent and ionic compounds. For example, the energy of a carbon–fluorine bond in CF_4 is about $485\,kJ\,mol^{-1}$, while that of the carbon–hydrogen bond in CH_4 is $435\,kJ\,mol^{-1}$. You should know that carbon–hydrogen bonds are very hard to break; carbon–fluorine bonds are even harder to break! The lattice energy of sodium fluoride (an ionic solid) is $915\,kJ\,mol^{-1}$, which is over $100\,kJ\,mol^{-1}$ greater than that of sodium chloride. Not only does fluorine react very quickly with sodium, it gives a highly exothermic reaction as well.

101.7 Look back at Unit 46, and note the order of ionic radii in Table 101.1. Explain why sodium fluoride has a high lattice energy.

101.5 The halogens are oxidising agents

In this section we shall think about three types of reactions that show that the halogens are oxidising agents.

(a) Displacement reactions

Fluorine has the largest electronegativity of all the elements, but the values for the other halogens show that they all have a tendency to accept electrons. The only type of negative ion they all give is the halide ion, X^-. (From now on we shall often use the symbol X_2 to stand for the halogen molecules, F_2, Cl_2, etc., and X^- for the halide ions, F^-, Cl^-, etc.) This should not surprise us given that they have one electron less than the nearest noble gas. By taking up one extra electron they will complete the octet of their outermost p orbitals. We know that when elements react they have a tendency to adopt noble gas electron structures. (However, make sure you try question 46.2 about oxygen and oxide ions before you take this observation as a *reason* why the halogens all give halide ions.) If an element takes one or more electrons from another, then it is acting as an oxidising agent. The values of the standard electrode potentials in Table 101.1 show that in solution the oxidising power decreases from fluorine to iodine.

We can see the relative oxidising powers in a series of simple test tube experiments by adding the halogens to solutions of the halides. The results are listed in Table 101.3. The most convenient way to carry out the reac-

Table 101.3. Displacement reactions of the halogens*

| | Halogen added | | |
Solution	Chlorine	Bromine	Iodine
Chloride, Cl^-	No reaction	No reaction	No reaction
Bromide, Br^-	Yellow-orange Br_2 released	No reaction	No reaction
Iodide, I^-	Black-dark brown I_2 released	Black-dark brown I_2 released	No reaction

*The results show that:
 (i) Chlorine displaces bromine and iodine
  $Cl_2(aq) + 2Br^-(aq) \rightarrow 2Cl^-(aq) + Br_2(aq)$
  $Cl_2(aq) + 2I^-(aq) \rightarrow 2Cl^-(aq) + I_2(s)$
 (ii) Bromine displaces iodine
  $Br_2(aq) + 2I^-(aq) \rightarrow 2Br^-(aq) + I_2(s)$
 (iii) Iodine displaces neither of the other two

tions is to use the halogens as solutions in water. Although iodine is almost insoluble in water, it will dissolve in a solution of an iodide such as potassium iodide. The soluble triiodide ion, I_3^-, is made:

$$I_2(s) + I^-(aq) \rightleftharpoons I_3^-(aq)$$

One of the problems with doing the tests is deciding if a reaction has taken place. For example, a solution containing a small amount of bromine can appear almost colourless. Similarly, a very dilute solution of iodine in potassium iodide can look like a solution of bromine. One way of distinguishing the possibilities is to add a little 1,1,1-trichloroethane to the reaction mixture. Any bromine or iodine present will dissolve more readily in the purely covalent organic liquid than in water. If bromine is present, the organic layer will become orange-yellow; if iodine is present, it will become purple.

(b) With alkali metals

The alkali metals towards the bottom of Group I can react violently with fluorine and chlorine. However, it is safe to burn sodium in chlorine. The sodium burns quietly, giving off clouds of white fumes. The fumes are particles of sodium chloride (salt). We can show the reaction in two ways:

$$2Na(s) + Cl_2(g) \rightarrow 2NaCl(s)$$
$$2Na(s) + Cl_2(g) \rightarrow 2Na^+Cl^-(s)$$

The second is useful because it shows that the reaction involves a sodium atom losing an electron and a chlorine atom gaining it. Given that oxidation is the loss of electrons, the sodium has been oxidised by the chlorine. Sodium chloride is one of the most typical ionic solids.

A word of warning: you need to know something about thermochemistry if you are to provide an explanation of why two elements react to give an ionic substance of a particular formula. You might like to read Unit 46 to remind yourself about this.

Not all metals give ionic compounds with the halogens. In particular, Group III metals and B metals in general give covalent compounds. When the metal can have more than one oxidation state, the higher state is brought out by fluorine.

(c) With hydrocarbons

If a lighted candle is put into chlorine the wax continues to burn, but with a very smoky flame. Together with a lot of soot, white fumes of hydrogen chloride are released. The chlorine removes the hydrogen from the hydrocarbon (an oxidation) and leaves carbon behind, e.g.

$$C_{10}H_{22}(s) + 11Cl_2(g) \rightarrow 10C(s) + 22HCl(g)$$

If you look at Unit 113 you will find that the reaction of chlorine with some hydrocarbons can be explosive, especially in the presence of ultraviolet light. The reaction can take place by a free radical mechanism, which almost always leads to a very rapid change.

101.8 Explain why the displacement reactions of the halogens are best regarded as redox reactions.

101.9 If you did have a supply of fluorine, what would happen if fluorine was added to the solutions listed in Table 101.3?

101.10 Is it true that iodine will not displace *any* other halogen?

101.11 The standard redox potential for a half-cell containing Fe^{3+} and Fe^{2+} ions is $E^{\ominus}_{Fe^{3+}/Fe^{2+}}$ = +0.77 V. Which of the halogens will oxidise Fe^{3+} to Fe^{2+} in solution?

101.6 Reactions with water and alkali

(a) With water

Both fluorine and chlorine are able to oxidise water. Fluorine can give a mixture of oxygen and trioxygen (ozone), e.g. for oxygen

$$2F_2(g) + 2H_2O(l) \rightarrow O_2(g) + 4HF(aq)$$

Chlorine, which is not such a powerful oxidising agent, does not release oxygen. Instead, a solution containing a mixture of hydrochloric and chloric(I) acids is produced (the old name for chloric(I) acid is hypochlorous acid):

$$Cl_2(g) + H_2O(l) \rightarrow \underset{\text{hydrochloric acid}}{H^+(aq) + Cl^-(aq)} + \underset{\text{chloric(I) acid}}{H^+(aq) + ClO^-(aq)}$$

Chlorate(I) ions, ClO^-, in a solution of chlorine are

responsible for its bleaching action. For example, coloured organic materials, like grass or some clothing dyes, are decolourised if they are put into chlorine water. The chlorate(I) ion is able to lose its oxygen fairly readily, which is used in the oxidation process.

Chlorine water, and bleach, will give off bubbles of oxygen if they are left in sunlight, owing to the decomposition of the chlorate(I) ions:

$$2ClO^-(aq) \rightarrow 2Cl^-(aq) + O_2(g)$$

Iodine is so insoluble in water that its solution has no oxidising power.

(b) With alkali

There are two types of change depending on the temperature of the alkali:

(i) Cold dilute alkali. The change is summarised in the general equation

$$X_2(g) + 2OH^-(aq) \rightarrow X^-(aq) + XO^-(aq) + H_2O(l)$$

e.g.

$$Cl_2(g) + 2OH^-(aq) \rightarrow Cl^-(aq) + ClO^-(aq) + H_2O(l)$$

(ii) Hot concentrated alkali. The equation is

$$3X_2(g) + 6OH^-(aq) \rightarrow \\ 5X^-(aq) + XO_3^-(aq) + 3H_2O(l)$$

e.g.

$$3Cl_2(g) + 6OH^-(aq) \rightarrow \\ 5Cl^-(aq) + ClO_3^-(aq) + 3H_2O(l)$$

The main difference between the two is that, with cold dilute alkali, chlorate(I), ClO^-, ions are produced; while with hot concentrated alkali, chlorate(V), ClO_3^-, ions are made.

101.12 Write a balanced equation for the reaction between fluorine and water that gives off trioxygen, O_3, rather than oxygen.

101.13 Bromine in water does not give appreciable quantities of bromate(I), BrO^-, ions. However, in sunlight bromine water will slowly give off oxygen. Write an equation for the reaction (bromide ions are also liberated).

101.14 Explain why the reactions of chlorine with alkali can be regarded as disproportionations. (Hint: see section 105.2b if you are unsure.)

101.7 Halide ions

Often, when a halogen reacts, each atom gains an electron to give a halide ion. It is useful to be able to perform tests to distinguish between chloride, bromide and iodide ions. The simplest test involves adding silver

Table 101.4. The silver nitrate test for halide ions

	Addition of silver nitrate in presence of nitric acid	Addition of ammonia solution
Chloride	White precipitate of silver chloride $Ag^+(aq)+Cl^-(aq)\rightarrow AgCl(s)$	Precipitate dissolves; clear solution
Bromide	Pale yellow precipitate of silver bromide $Ag^+(aq)+Br^-(aq)\rightarrow AgBr(s)$	Precipitate partly dissolves
Iodide	Yellow precipitate of silver iodide $Ag^+(aq)+I^-(aq)\rightarrow AgI(s)$	Precipitate does not dissolve

nitrate solution to a solution of the halide. This should be done in the presence of dilute nitric acid; otherwise other ions may give a precipitate. Silver ions react with halide ions to give precipitates. These in turn can be identified by their colour, or by their reaction with ammonia solution. The solubility of silver halides (especially silver chloride) in ammonia solution is due to the formation of the diamminesilver(I) ion, $Ag(NH_3)_2^+$, which is soluble in water:

$$AgCl(s) + 2NH_3(aq) \rightarrow Ag(NH_3)_2^+(aq) + Cl^-(aq)$$

Table 101.4 summarises the tests for you.

Answers

101.1   Volume needed
$$=\frac{1\,kg}{65\times 10^{-6}\,kg}\times 1\,dm^3 \approx 15\,000\,dm^3$$
This should give you an idea of the huge scale upon which the industry works.

101.2   See stages (i), (ii) and (iii) for extracting bromine from sea water.

101.3   It is the conversion of a solid directly into a gas (or vice versa).

101.4   The oxidation numbers of manganese in $KMnO_4$ and MnO_2 are $+7$ and $+4$, and of chlorine in HCl and Cl_2 are -1 and 0, respectively. The manganese gains electrons while the HCl loses them. Therefore the HCl has been oxidised and the manganese reduced.

101.5   It is a liquid. This means that the preparation apparatus must condense the bromine. A distillation apparatus will work.

101.6   The apparatus is shown in Figure 101.1.

101.7   A small radius means that the ions in the crystal are closer together. This leads to greater attractions in the crystal, and to higher lattice energies. Also, see the work on the Born–Mayer equation in section 46.3.

101.8   When, for example, chlorine converts iodide ions into iodine, the iodide ions lose electrons, and the chlorine atoms gain electrons. That is, the chlorine oxidises the iodide (and the iodide reduces the chlorine).

101.9   Being the most powerful oxidising agent of all the halogens, fluorine would displace chlorine, bromine and iodine from their respective solutions.

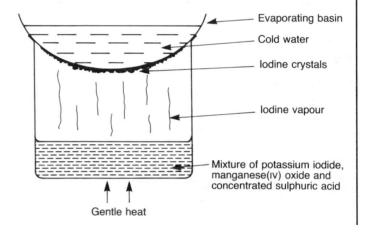

Figure 101.1 A simple way of preparing iodine

101.10   No. It would displace astatine from solution.

101.11   Fluorine, chlorine and bromine. Their $E^\ominus$ values are more positive than $E^\ominus_{Fe^{3+}/Fe^{2+}}$, so they are the stronger oxidising agents.

101.12 $3F_2(g) + 3H_2O(l) \rightarrow O_3(g) + 6HF(aq)$

101.13
$$2Br_2(aq) + 2H_2O(l) \rightarrow 4Br^-(aq) + O_2(g) + 4H^+(aq)$$

101.14   One oxidation state converts into two different oxidation states during the reaction. This is what is meant by disproportionation. See the next unit for more details.

UNIT 101 SUMMARY

- The Group VII elements:
    - (i) Are all oxidising agents, fluorine being the most vigorous.
    - (ii) Exist as diatomic molecules.
    - (iii) Form halides with metals, e.g. K^+I^-, $AlCl_3$.
    - (iv) Reactivity follows the order $F_2 > Cl_2 > Br_2 > I_2$.
- Laboratory preparation:
    Chlorine is made by

(i) Warming concentrated hydrochloric acid and manganese(IV) oxide;

$$MnO_2(s) + 4HCl(aq) \rightarrow$$
$$MnCl_2(aq) + Cl_2(g) + 2H_2O(l)$$

(ii) Dropping concentrated hydrochloric acid onto crystals of potassium manganate(VII). Dry chlorine is obtained by passing the gas through concentrated sulphuric acid.

Bromine and iodine are made in a similar way to chlorine except that HBr(aq) and HI(aq) are made *in situ*, and the methods of collection are different.

- Extraction:
  (i) Fluorine by the electrolysis of a mixture of liquid anhydrous hydrogen fluoride and potassium hydrogenfluoride, KHF_2.
  (ii) Chlorine by the electrolysis of brine.
  (iii) Bromine from bromide ions in sea water.
  (iv) Iodine from sodium iodate(V) found in sodium nitrate deposits.

Reactions

- Displacement reactions:
  Each will displace the halogen below it from solution;
  e.g. $Cl_2(aq) + 2Br^-(aq) \rightarrow 2Cl^-(aq) + Br_2(aq)$

- With metals:
  Salts are made;
  e.g. $2Na(s) + Cl_2(g) \rightarrow 2Na^+Cl^-(s)$
  $2Fe(s) + 3Cl_2(g) \rightarrow 2FeCl_3(s)$

- With a burning hydrocarbon:
  The hydrogen is removed by chlorine;
  e.g. $CH_4(g) + 2Cl_2(g) \rightarrow C(s) + 4HCl(g)$
  Also, see Unit 111.

- With water:
  (i) Fluorine oxidises water;
  $$2F_2(g) + 2H_2O(l) \rightarrow O_2(g) + 4HF(aq)$$

  (ii) Chlorine and bromine give acidic solutions in water;
  $$Cl_2(g) + H_2O(l) \rightarrow$$
  $$H^+(aq) + Cl^-(aq) + H^+(aq) + ClO^-(aq)$$

  The solution acts as a bleach.

  (iii) Iodine is almost insoluble in water, but it does dissolve in solutions containing iodide ions. The soluble triiodide ion is made;
  $$I_2(aq) + I^-(aq) \rightleftharpoons I_3^-(aq)$$

- With alkali, disproportionation occurs:
  (i) Cold dilute alkali gives a mixture of halide and halate(I) ions;
  $$e.g. \; Cl_2(g) + 2OH^-(aq) \rightarrow$$
  $$Cl^-(aq) + ClO^-(aq) + H_2O(l)$$
  chloride   chlorate(I)

  (ii) Hot concentrated alkali gives a mixture of halide and halate(V);
  $$e.g. \; 3Cl_2(g) + 6OH^-(aq) \rightarrow$$
  $$5Cl^-(aq) + ClO_3^-(aq) + H_2O(l)$$
  chloride   chlorate(V)

Compounds of the halogens

102.1 The hydrogen halides

The hydrogen halides are hydrogen fluoride, HF, hydrogen chloride, HCl, hydrogen bromide, HBr, and hydrogen iodide, HI. Information about them is listed in Table 102.1. Although they all have properties in common, hydrogen fluoride is not completely typical (Figure 102.1). As you will find, this is largely a result of its very high bond energy, and the highly polar nature of the molecule.

To explain thoroughly the high bond energy is a difficult task. We can see from the values of the dipole moment in Table 102.1 that hydrogen fluoride has a considerable degree of ionic character, with the fluorine atom pulling electron density towards it. Indeed, hydrogen fluoride molecules are strongly attracted to each other by hydrogen bonds (Figure 102.2). The hydrogen bonds between hydrogen and fluorine atoms in hydrogen fluoride are the strongest known. In solution with water, hydrogen fluoride is only a weak acid, whereas all the other hydrogen halides are strong acids. That is, the equilibrium

$$HF(aq) + H_2O(l) \rightleftharpoons H_3O^+(aq) + F^-(aq)$$

lies well to the left. Although we need to take account of both enthalpy and entropy changes to explain the dissociation of weak acids, here there is much truth in the observation that links the strength of the bond in HF with its lack of dissociation.

Table 102.1. Information about the hydrogen halides

	HF	HCl	HBr	HI
Melting point/°C	−83	−114	−87	−51
Boiling point/°C	20	−85	−67	−35
Bond energy /kJ mol^{-1}	560	431	366	299
Dipole moment /D	1.91	1.05	0.8	0.42
Acidity constant /mol dm^{-3}	5.6×10^{-4}	← very large →		

However, you should remember that acidity is not an absolute concept. A chemical can be an acid in one situation and a base in another. Especially, very concentrated or pure hydrogen fluoride is an extremely strong acid.

For example, we normally think of nitric acid as a strong acid because it acts as a proton donor with water:

$$HNO_3(aq) + H_2O(l) \rightarrow H_3O^+(aq) + NO_3^-(aq)$$

But in liquid hydrogen fluoride,

$$HNO_3 + HF \rightarrow H_2NO_3^+ + F^-$$

Here the HNO$_3$ accepts the proton, and HF donates it.

Hydrogen fluoride and hydrogen chloride can be prepared by reacting the appropriate sodium or calcium halide with sulphuric acid. For example, hydrogen chloride is given off if sodium chloride is warmed with concentrated sulphuric acid:

$$NaCl(s) + H_2SO_4(l) \rightarrow NaHSO_4(s) + HCl(g)$$

If you see this reaction carried out you will find that the gas gives white fumes in air. The fumes are small droplets of concentrated hydrochloric acid made when the gas mixes with water vapour in air.

It would be nice to think that exactly the same method could be used to make hydrogen bromide and hydrogen iodide. Unfortunately it cannot. This is because sulphuric acid has a significant oxidising power; it can oxidise HBr and HI as soon as they are made:

$$2HBr(g) + H_2SO_4(l) \rightarrow Br_2(l) + SO_2(g) + 2H_2O(l)$$
$$2HI(g) + H_2SO_4(l) \rightarrow I_2(s) + SO_2(g) + 2H_2O(l)$$
$$6HI(g) + H_2SO_4(l) \rightarrow 3I_2(s) + S(s) + 4H_2O(l)$$
$$8HI(g) + H_2SO_4(l) \rightarrow 4I_2(s) + H_2S(s) + 4H_2O(l)$$

The way round the problem is to use an acid with very little oxidising power. Phosphoric(v) acid, H_3PO_4, is the usual one to choose. For example, on warming,

$$NaI(s) + H_3PO_4(l) \rightarrow NaH_2PO_4(s) + HI(g)$$

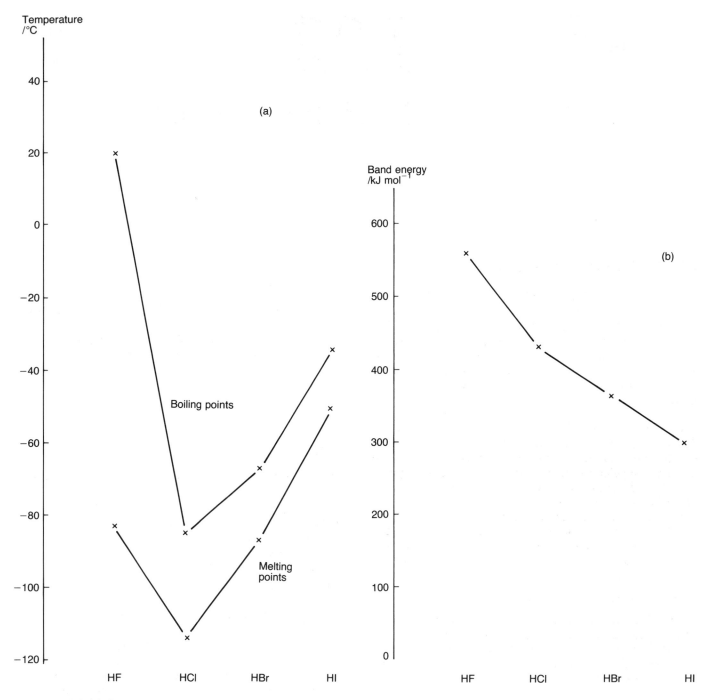

Figure 102.1 *(a) Graphs showing the melting and boiling points of the hydrogen halides. (b) The bond energies of the hydrogen halides*

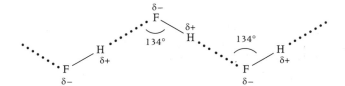

Figure 102.2 *Liquid hydrogen fluoride is very strongly hydrogen bonded, with many of the molecules in zig-zag chains. Solid hydrogen fluoride has a six-membered ring structure, (HF)₆*

An alternative method for HBr and HI is to use a combination of the halogen and red phosphorus in water. We can think of the reaction taking place in two stages. In the case of iodine, first

$$3I_2(s) + 2P(s) \rightarrow 2PI_3(s)$$

then

$$PI_3(s) + 3H_2O(l) \rightarrow H_3PO_3(aq) + 3HI(g)$$

Bromine behaves similarly.

Compounds of the halogens 655

102.1 We can see that the trend is for the polarity and the degree of hydrogen bonding between the molecules to decrease from HF to HI. If necessary, look back to Unit 19 and remind yourself of Fajans' rules. Then explain why you would expect hydrogen iodide to be less polar than, say, hydrogen chloride.

102.2 If the relative molecular mass of hydrogen fluoride vapour is measured at various temperatures, a graph like that shown in Figure 102.3 is obtained. At around 31°C, the relative molecular mass appears to be 40; above 60°C it is nearly 20. What is the explanation for these observations? $A_r(H) = 1$, $A_r(F) = 19$.

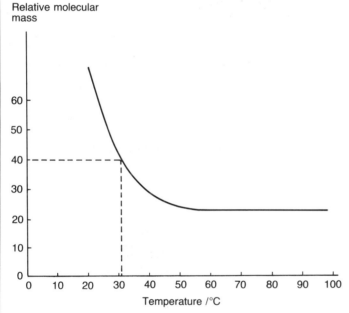

Figure 102.3 *How the relative molecular mass of hydrogen fluoride changes with temperature*

102.3 Only two of the hydrogen halides will dissociate if you put a red hot wire into them. Which are they? What would you see?

102.4 Write an equation for the production of hydrogen fluoride from calcium fluoride, CaF_2 (also known as the mineral fluorspar), and concentrated sulphuric acid. Why is this not a good reaction to carry out in a test tube?

102.5 What are the oxidation numbers of sulphur in H_2SO_4 and in the products of its reactions with the hydrogen halides? Which is the better reducing agent, HBr or HI? Explain.

102.6 The hydrogen halides are very soluble in water. Imagine you had a flask in which you were about to prepare hydrogen chloride. Your aim is to dissolve the gas in water in order to make a solution of hydrochloric acid. You must not put the outlet tube from the flask directly into the water because the water will suck back into the flask. An explosion could result when the water and concentrated acid meet. Suggest three ways of safely making the acid solution. (Hint: Figure 102.4 shows you diagrams of pieces of apparatus you might use.)

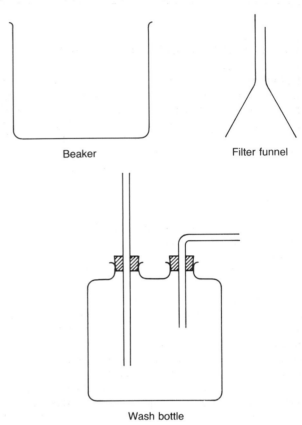

Figure 102.4 *Apparatus that might be used to dissolve hydrogen chloride in water*

102.7 Hydrogen and chlorine will explode if they are mixed in sunlight. What would be the product, and what is the likely mechanism for the reaction?

102.8 Hydrogen chloride reacts with ammonia.

(i) What is made?

(ii) Describe the bonding in the product.

(iii) Describe the reaction in terms of Lewis acids and bases.

(Hint: look back at Units 15 and 74 for help.)

102.9 (Try this question only if you have studied the properties of alkenes; Unit 112.) The halogens and hydrogen halides will react with alkenes such as ethene, C_2H_4, and propene, C_3H_6.

(i) What is made if (a) bromine, (b) hydrogen bromide reacts with these alkenes.

(ii) What is bromine used to test for in organic chemistry?

Table 102.2. Inter-halogen compounds

ClF	ClF$_3$	ClF$_5$	IF$_7$
BrF	BrF$_3$	BrF$_5$	
BrCl	ICl$_3$	IF$_5$	
BrI	IF$_3$		
ICl			

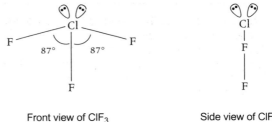

Front view of ClF$_3$ Side view of ClF$_3$

102.2 Inter-halogen compounds

The halogens make a number of inter-halogen compounds, some of which are listed in Table 102.2. Many of the compounds can be made by direct combination of the elements. For example, fluorine and chlorine will react at around 200° C to give chlorine fluoride, ClF; iodine and excess chlorine will combine at room temperature to make iodine trichloride, ICl$_3$:

$$F_2(g) + Cl_2(g) \rightarrow 2ClF(g)$$
$$I_2(s) + 3Cl_2(g) \rightarrow 2ICl_3(l)$$

Given their chemical similarity, perhaps we should expect the halogens to make compounds like BrCl and ICl. However, it is quite noticeable that fluorine gives compounds in which the other halogen has a much higher oxidation number than usual. For example, in iodine pentafluoride, IF$_5$, we have $Ox(I) = +5$. This ability of fluorine to bring out high oxidation states in elements with which it combines is another way of saying that fluorine is a very powerful oxidising agent. Nonetheless, the fact that five fluorine atoms can bond to an iodine atom needs some explanation.

Iodine has the electron structure I:(Kr)4d^{10}5s^{2}5p^5. It appears that only one of the 5p orbitals has space for an additional electron. (We can imagine that this electron is provided by the chlorine atom in, say, ICl.) However, accommodating five extra electrons in IF$_5$ is not feasible unless we assume other orbitals are used in the bonding. We have seen this type of situation before in section 17.4. There we found that we could use the theory of hybridisation to explain how carbon might make four bonds to other atoms. In the case of IF$_5$ we have seven electrons from the iodine valence shell, and one electron from each fluorine to make the five bonds. Thus we need six orbitals to cope with 12 electrons. They are obtained if we assume that not only the 5s and 5p orbitals are involved in bonding, but also some of the 5d orbitals are used. In this way we can build up a set of six d^{2}sp^3 hybrid orbitals. (You might like to look at Table 17.5, where you will find that a number of different combinations of orbitals were employed to describe the shapes of a wide range of molecules and ions.) The shape of the IF$_5$ molecule is illustrated in Figure 102.5.

Iodine molecules have the distinction of being able to combine with iodide ions to give triiodide ions, I$_3^-$. These ions are made whenever iodine is produced in a solution containing iodide ions. They are soluble in water, and deep brown in colour. If we wish to use a solution of iodine, it is always made by dissolving iodine crystals in a solution of iodide ions. An equilibrium,

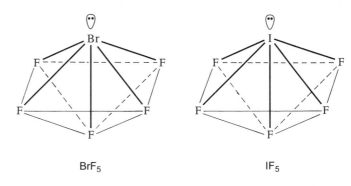

BrF$_5$ IF$_5$

BrF$_5$ and IF$_5$ have similar structures

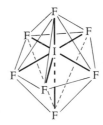

IF$_7$ has the shape of a pentagonal bipyr

Figure 102.5 *The shapes of inter-halogen fluorides*

lying far to the right-hand side, exists in the solution:

$$I_2(s) + I^-(aq) \rightleftharpoons I_3^-(aq)$$

The solution acts as a source of iodine because the equilibrium shifts to the left and replaces those iodine molecules that may be removed. This reaction is used in iodine titrations (see section 40.3).

102.10 Figure 102.5 shows that ClF$_3$ has a distorted T-shape. Explain why two lone pairs of electrons are shown in the diagram, and why this might lead to distortion of the shape.

102.11 Predict the shapes of the molecules in each of the four columns in Table 102.2.

102.12 From the nature of the bonding in the inter-halogen compounds, would you expect them to have high or low melting and boiling points?

102.3 Metal halides

Here you will find a brief survey of metal halides. Unit 89 will tell you about the solubilities and the hydrolysis of halides.

In previous units (e.g. Units 89, 92 and 93) we discussed the nature of bonding in substances like sodium chloride. We established that the Group I metals give predominantly ionic compounds with the halogens. Apart from beryllium and magnesium, the metals of Group II also follow this pattern. For example, beryllium gives a chloride whose formula $BeCl_2$ belies its structure, which is a chain of linked $BeCl_2$ units (Figure 102.6).

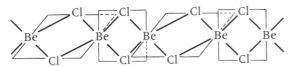

Figure 102.6 *The structure of solid beryllium chloride, $BeCl_2$*

The halides of Group III are predominantly covalent. The structure of aluminium chloride is one that you should know. It was described in section 15.1. In the solid it has a chain structure like that of $BeCl_2$; but in the vapour, or in some organic liquids, it exists as dimers.

Uranium hexafluoride, UF_6, has been used in the process of separating the isotope ^{235}U from ^{238}U. (^{235}U has been used as a fuel for nuclear power stations and in nuclear weapons.) Because $^{235}UF_6$ is lighter than $^{238}UF_6$, it diffuses more rapidly than the latter. This difference between the two fluorides is the basis of the separation.

The halogens give a variety of complexes with transition and B metals where the halide ions can act as

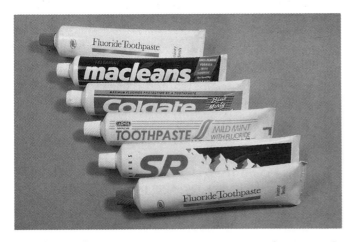

Fluoride toothpastes are now very common, and can contain a number of different metal fluorides e.g. sodium fluoride.

Solutions of hydrogen fluoride have the ability to dissolve glass. This property is often used to etch car windows with the registration number as a security measure.

ligands. Structures of the complexes vary; e.g. $[Cr(H_2O)_4Cl_2]^-$, $CuCl_4^{2-}$, $PbCl_6^{4-}$. You will find more information about such complexes in Unit 106.

A reaction in which a complex ion of silicon is made is the etching of glass by hydrofluoric acid solution. Glass contains a mixture of calcium and sodium silicates. Calcium silicate reacts according to the equation

$$Ca^{2+}SiO_3^{2-}(s) + 6HF(aq) \rightarrow Ca^{2+}(aq) + SiF_6^{2-}(aq) + 3H_2O(l)$$

The products are all soluble in water, so we see the glass being dissolved. However, if only a little of the acid is put on the glass, the surface takes on a frosted look. This is made use of by motorists who have their car windows etched with the car registration number. The idea is that this will help dissuade thieves from making off with the car and disguising it by merely changing the number plates. It is likely that glassware you use in the laboratory has been etched using the same method.

Apart from silver fluoride, the silver halides are notable for their insolubility in water. This property is used in the silver nitrate test for the halides.

102.13 What does the ease with which silver fluoride dissolves in water suggest about the bonding in silver fluoride?

102.14 Explain why a chlorine atom is able to give a coordinate bond with a neighbouring beryllium atom. (Hint: do you know what an electron deficient compound is?)

102.4 Making metal halides

For many halides it is usually possible to combine the metal and halogen directly. However, this is not always the case, and sometimes precautions have to be taken. These are not only ones of safety, but of keeping moisture out of the apparatus. Although the ionic halides do not suffer from the complaint of hydrolysis, many of the predominantly covalent halides do react with water. Figure 102.7 shows you one way of making a small amount of aluminium trichloride. Notice that the chlorine is dried before it passes into the reaction tube, and that the collection bottle has a tube containing anhydrous calcium chloride attached. This allows air and excess chlorine to escape from the apparatus, but it prevents moisture entering. If moisture does get to the solid, hydrolysis takes place:

$$AlCl_3(s) + 3H_2O(l) \rightarrow Al(OH)_3(s) + 3HCl(aq)$$

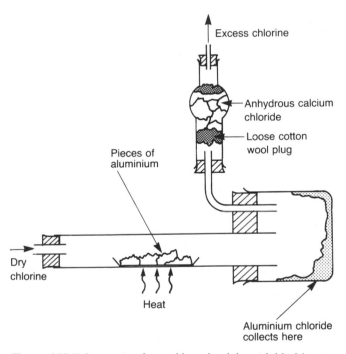

Figure 102.7 *Apparatus for making aluminium trichloride*

102.15 Iron(III) chloride and iron(II) chloride both suffer from hydrolysis. The former can be made by passing chlorine over hot iron. Why can the latter not be made in this way? What other gaseous source of chlorine might be used instead?

102.16 A student reacted iron(II) sulphide with dilute hydrochloric acid in a fume cupboard. The reaction was

$$FeS(s) + 2HCl(aq) \rightarrow Fe^{2+}(aq) + 2Cl^-(aq) + H_2S(g)$$

After the reaction was over the student decided to make some iron(II) chloride crystals by evaporating the solution.

(i) Why was it necessary to do the reaction in a fume cupboard?

(ii) Why did the student fail to make crystals of iron(II) chloride?

102.5 Compounds containing oxygen and halogens

There are many molecules and ions that contain both oxygen and halogens. The general name for ions that contain oxygen and another element is *oxoanions*. There is a list in Table 102.3, and two are shown in Figure 102.8. Of the molecules, chlorine dioxide, ClO_2, is the most useful beyond the laboratory. At room

Table 102.3. Molecules and ions containing both oxygen and halogens*

Molecules			
Fluorine	*Chlorine*	*Bromine*	*Iodine*
OF_2	Cl_2O	Br_2O	I_2O_5
O_2F_2	ClO_2	BrO_2	
	Cl_2O_6	BrO_3	
	Cl_2O_7		

Oxoanions†		
Chlorine	*Bromine*	*Iodine*
Chlorate(I), ClO^-	Bromate(I), BrO^-	Iodate(I), IO^-
Chlorate(V), ClO_3^-	Bromate(V), BrO_3^-	Iodate(V), IO_3^-
Chlorate(VII), ClO_4^-		

Acids		
Modern name	*Old name*	*Formula*
Chloric(I) acid	Hypochlorous acid	$HClO$
Chloric(V) acid	Chloric acid	$HClO_3$
Chloric(VII) acid	Perchloric acid	$HClO_4$

*Only the more common molecules and ions are listed
†Fluorine does not form oxoanions

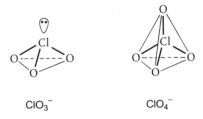

Figure 102.8 *Both the chlorate(V), ClO_3^-, and chlorate(VII), ClO_4^-, ions are tetrahedral*

temperature it is a dark yellow gas. Below its boiling point of 11°C it exists as an oily red liquid. It is an extremely powerful oxidising agent and, in spite of the dangers associated with it, the gas has been used in treating water supplies to kill bacteria, and in bleaching wood pulp prior to paper making.

Some of the oxoanions you may already know about: many of them are used as oxidising agents, and some show up in disproportionation reactions. We shall consider both types of reaction now. First, we shall stick to stating what happens, mainly using the chlorine oxoanions as examples; later, if you are interested, you can use your knowledge of redox potentials to explain the reactions.

(a) *Oxidation reactions*

The usual way to employ the oxoanions is in acidic solution. The pattern of reaction is:

chlorate(I) → chloride + loss of oxygen
chlorate(V) → chloride + loss of oxygen
chlorate(VII) → very little oxidising power

(Note: anhydrous chloric(VII) acid, $HClO_4$, is an extremely vigorous oxidising agent.)

Bromates and iodates behave similarly. Here are some examples.

Chlorate(I) oxidises iron(II) to iron(III)

Chlorate(I) ions combine with hydrogen ions to give chloric(I) acid (also known as hypochlorous acid). This is a weak acid, so it mainly reacts as HClO molecules:

$$HClO(aq) + 2Fe^{2+}(aq) + H^+(aq) \rightarrow$$
$$Cl^-(aq) + 2Fe^{3+}(aq) + H_2O(l)$$

However, a useful source of free chlorate(I) ions is bleaching powder. This is calcium chlorate(I), $Ca(OCl)_2$. (It is also known as calcium hypochlorite). Bleaching powder has a long history. It was one of the first bleaches that was effective in bleaching cotton. The use of bleaching powder allowed cotton to be dyed to a consistently high quality, and therefore gave manufacturers who used it a great advantage over those who did not.

Chlorate(V) oxidises sulphite ions to sulphate

The reaction is

$$ClO_3^-(aq) + 3SO_3^{2-}(aq) \rightarrow Cl^-(aq) + 3SO_4^{2-}(aq)$$

This is a relatively harmless oxidation reaction. Others can be very dangerous. Especially, a dry mixture of a chlorate(V) and, for example, carbon or sulphur will explode.

Chlorate(VII) ions have limited oxidising action

For example, a solution of a chlorate(VII) will not oxidise iron(II) or sulphite ions, SO_3^{2-}. In fact, these reactions are energetically favourable. The reason why they do not take place is a matter of kinetics. The activation energy of the reactions is very large. This is partly due to the number of covalent bonds that have to be broken if the ClO_4^- ion is to break apart.

(b) *Disproportionation reactions*

A disproportionation reaction is one where an element with a single oxidation number (or state) reacts to give two or more different oxidation numbers. An example that you should remember is when one of the halogens reacts with alkali. We shall use bromine as our example, but chlorine and iodine react similarly.

With cold dilute alkali

The reaction is

$$Br_2(l) + 2OH^-(aq) \rightarrow Br^-(aq) + OBr^-(aq) + H_2O(l)$$
$$0 \qquad\qquad\qquad -1 \qquad +1$$

i.e.

bromine → bromide + bromate(I)

where here and in the following the oxidation number of bromine is written below the symbol.

With hot concentrated alkali

The reaction is

$$3Br_2(l) + 6OH^-(aq) \rightarrow 5Br^-(aq) + BrO_3^-(aq) + 3H_2O(l)$$
$$0 \qquad\qquad\qquad -1 \qquad +5$$

i.e.

bromine → bromide + bromate(V)

It is possible to prepare salts of some of the oxoanions in this way. For example, if you were to bubble chlorine through hot concentrated potassium hydroxide solution, the solution would contain a mixture of potassium chloride and potassium chlorate(V). Once you allowed the solution to cool, the potassium chlorate(V) would crystallise first because it has a much smaller solubility in water than potassium chloride. This property can be used in fractional crystallisation (see section 59.2). However, you would be disappointed if you used dilute alkali and attempted to crystallise potassium chlorate(I). Once you started to evaporate the solution, the chlorate(I) ions would decompose.

Heating a chlorate(V)

Potassium chlorate(V) behaves in two ways depending on the temperature at which it is heated. At around its melting point (367°C), chlorate(VII) and chloride ions are made:

$$4ClO_3^- \rightarrow 3ClO_4^- + Cl^-$$

This is another example of disproportionation. However, at a higher temperature, the chlorate(VII) ions decompose giving off oxygen:

$$ClO_4^- \rightarrow Cl^- + 2O_2$$

102.17 A solution of bleaching powder contains chlorate(I) ions. The solution will easily convert iodide ions, I^-, into iodine, I_2. Write an equation for the reaction.

102.18 Chlorate(v) ions will also oxidise iron(II) to iron(III) and iodide to iodine. Write equations for these reactions. Assume they take place with acid present.

102.6 Explaining redox reactions using redox potentials

It is essential that you understand the work on electrode and redox potentials in Units 66 and 70 if you are to follow the explanations in this section.

We can see how the oxidising powers of the oxoanions vary by looking at their redox potentials. These are gathered together in Table 102.4. You will see that some of the oxoanions are shown as their acids; e.g HClO rather than OCl^-. Where this happens, we assume that the reactions take place in acidic solutions. First, you can see from the fairly high positive values of the redox potentials that all the oxoanions are oxidising agents. We shall work through some of the more important reactions.

(a) Iodate(v)–iodide reaction

The data show that iodate(v) ions will oxidise another species whose redox potential is less than 0.91 V. In particular, if we look at the iodine–iodide reaction, this has a redox potential of +0.54 V. Therefore, if we mix iodate(v) and iodide ions, the iodide ions will be oxidised to iodine. The reactions will take place in the directions shown:

$$IO_3^-(aq) + 6H^+(aq) + 6e^- \rightarrow I^-(aq) + 3H_2O(l)$$
$$2I^-(aq) \rightarrow I_2(s) + 2e^-$$

We can combine the equations by taking three times the second and adding it to the first. This gives us

$$IO_3^-(aq) + 5I^-(aq) + 6H^+(aq) \rightarrow 3I_2(s) + 3H_2O(l)$$

The value of this reaction is in performing titrations designed to determine the concentration of iodide in a solution. You will find details of the method in Unit 40.

(b) Chlorate(v)–chloride reaction

Because $E^\ominus_{ClO_3^-/Cl^-} = +1.45\,V$ and $E^\ominus_{Cl_2/Cl^-} = +1.36\,V$, chlorate(v) ions should oxidise chloride ions to chlorine. However, the difference between the two potentials is far less than the rule of thumb of 0.6 V that we used in Unit 70. In fact, the reaction will take place, but only if the chlorate(v) is heated with concentrated hydrochloric acid. As well as chlorine being released, chlorine dioxide is released. The mixture of the two gases was investigated by Humphry Davy, who called the mixture euchlorine. Needless to say, investigating the properties of this mixture of gases caused him, and other chemists, much bother.

(c) Disproportionation in halogen–alkali reactions

We have already discussed this type of reaction in Unit 71. If you have forgotten how to decide if an element or ion will take part in a disproportionation reaction, do consult that unit.

102.7 The halogen oxoacids

There are no oxoacids of fluorine. The oxoacids of chlorine are the most important and are listed in Table 102.5.

We have mentioned some of their properties in the previous sections, particularly that chloric(I) acid is a good oxidising agent (like chlorate(I) ions). It is made when chlorine dissolves in water:

$$Cl_2(g) + H_2O(l) \rightarrow HOCl(aq) + Cl^-(aq) + H^+(aq)$$

It is responsible for the bleaching action of chlorine water.

Each of the oxoacids has its own particular anhydride. The anhydride is a molecule that reacts to give the acid when water is added to it. Alternatively,

Table 102.4. Standard redox potentials for halogen oxoanions

Ion	Half-reaction	$E^\ominus/V$
In acid solution		
Chlorate(I)	$2HClO(aq) + 2H^+(aq) + 2e^- \rightarrow Cl_2(g) + 2H_2O(l)$	+1.63
Chlorate(v)	$ClO_3^-(aq) + 6H^+(aq) + 6e^- \rightarrow Cl^-(aq) + 3H_2O(l)$	+1.45
Chlorate(vII)	$ClO_4^-(aq) + 8H^+(aq) + 8e^- \rightarrow Cl^-(aq) + 4H_2O(l)$	+1.39
Bromate(I)	$2HBrO(aq) + 2H^+(aq) + 2e^- \rightarrow Br_2(l) + 2H_2O(l)$	+1.57
Bromate(v)	$BrO_3^-(aq) + 6H^+(aq) + 6e^- \rightarrow Br^-(aq) + 3H_2O(l)$	+1.93
Iodate(I)	$2HIO(aq) + 2H^+(aq) + 2e^- \rightarrow I_2(aq) + 2H_2O(l)$	+1.45
Iodate(v)	$IO_3^-(aq) + 6H^+(aq) + 6e^- \rightarrow I^-(aq) + 3H_2O(l)$	+0.91
Chloride	$Cl_2(g) + 2e^- \rightarrow 2Cl^-(aq)$	+1.36
Bromide	$Br_2(l) + 2e^- \rightarrow 2Br^-(aq)$	+1.09
Iodide	$I_2(aq) + 2e^- \rightarrow 2I^-(aq)$	+0.54

Table 102.5. The oxoacids of chlorine*

	Formula	Oxidising power	Acid strength	Thermal stability
		Properties		
Chloric(I) acid (hypochlorous acid)	HClO	Strong ↑	Weak ↓	Poor ↓
Chloric(V) acid (chloric acid)	HClO$_3$			
Chloric(VII) acid (perchloric acid)	HClO$_4$	Weak	Strong	Good

*Bromic(V) acid is very strong; bromic(VII) acid does not exist. Iodine forms an analogous set of acids to those of chlorine, but with weaker oxidising power

we can think of the anhydride as the species that is left when the elements of water are removed from the acid. For example, the anhydride of chloric(I) acid is dichlorine oxide, Cl_2O:

$$2HClO - H_2O \rightarrow Cl_2O$$

Similarly we predict the anhydrides of the other oxoacids to be:

$$2HClO_3 - H_2O \rightarrow Cl_2O_5$$
$$2HClO_4 - H_2O \rightarrow Cl_2O_7$$

Of these, dichlorine pentaoxide, Cl_2O_5, does not exist; but dichlorine heptaoxide, Cl_2O_7, does exist. In spite of its unlikely looking formula, it is energetically and kinetically stable at room temperature and exists as a colourless, viscous liquid. It can be made by reacting chloric(VII) acid with a powerful dehydrating agent, phosphorus(V) oxide.

Answers

102.1 Fajans' rules tell us that a covalent bond is favoured if the cation has a small ionic radius and the anion a large ionic radius. A large anion is more easily polarised. An iodide ion is larger than a chloride ion, so would be more readily polarised by a hydrogen ion. This is one reason why hydrogen iodide is less polar than hydrogen chloride.

102.2 Hydrogen fluoride is hydrogen bonded not only in the liquid state, but also in the vapour. You should have worked out that $M_r(HF) = 20$. At the lower temperature, it seems likely that the molecules are going around in pairs, or dimers, i.e. $(HF)_2$. This would give the observed relative molecular mass of 40. As the temperature increases, the collisions of the particles are so energetic that the dimers break up to give single HF molecules, for which $M_r(HF) = 20$.

102.3 Hydrogen bromide and hydrogen iodide have weaker bonds than the others, so these will be more likely to decompose. You would see the pale yellow and purple colours of bromine and iodine vapours, respectively.

102.4 $CaF_2(s) + H_2SO_4(l) \rightarrow CaSO_4(s) + 2HF(l)$

Any moisture will encourage the hydrogen fluoride to etch the glass.

102.5 The oxidation numbers are:

Element	Substance	Oxidation number
Sulphur	H$_2$SO$_4$	+6
	SO$_2$	+4
	S	0
	H$_2$S	−2
Bromine	Br$_2$	0
Iodine	I$_2$	0

Iodine converts sulphur from oxidation number +6 to −2, whereas bromine can only manage the conversion +6 to +4. Iodine is the better reducing agent because it

can supply electrons more readily. Alternatively, you can say that HI ions are more easily oxidised than HBr.

102.6 The arrangement of the apparatus is shown in Figure 102.9.

102.7 Hydrogen chloride is made. A free radical mechanism (see Unit 81).

102.8 (i) Ammonium chloride:

$HCl(g) + NH_3(g) \rightarrow NH_4Cl(s)$

(ii) See section 15.1

(iii) See section 74.4.

102.9 (i) (a) Ethene makes 1,2-dibromoethane:

Propene makes 1,2-dibromopropane:

(b) Ethene makes bromoethane:

Propene makes 2-bromopropane:

HBr obeys Markovnikoff's rule (section 112.4) when it adds to an alkene (provided free radicals are not involved).

(ii) Unsaturated compounds, in particular double and triple bonds in alkenes and alkynes.

Answers – cont.

(a)   (b)

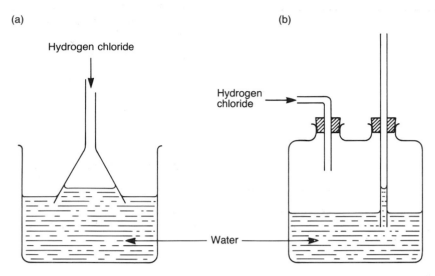

Figure 102.9 *Diagrams for answer to question 102.6. (a) When the water level rises inside the funnel, the level outside falls. Eventually the level falls below the lip of the funnel. This releases the pressure inside the funnel and the water drops back to its original level. (b) Here there is no chance of the water level reaching the inlet tube. If the pressure increases too much, the water rises up the second tube*

102.10 Each fluorine will provide one of the electrons in the three bonding pairs. Chlorine has seven electrons in its valence shell. This makes 10 electrons in total. Six of them are used in bonding, so there are four left over. These make the two lone pairs. Electron repulsion theory tells us that the lone pairs will repel the bonding pairs, thereby distorting the molecule.

102.11 The first column will be linear, the second T-shaped, the third square-based pyramid, and IF_7 is a pentagonal bipyramid.

102.12 They are all covalent, and do not give giant structures, so they should have low melting and boiling points; they do.

102.13 The bonding is mainly ionic. The fluoride ion is far less polarisable than the other halides, so AgF will not be so covalent as the other silver halides.

102.14 You will find an explanation of this in section 15.1. Briefly, a beryllium atom has an empty 2p orbital, which can accept a pair of electrons from chlorine.

102.15 Chlorine will oxidise the iron(II) to iron(III). Hydrogen chloride is used to make iron(II) chloride from iron.

102.16 (i) H_2S not only has an awful smell, it is highly poisonous.

(ii) The chloride is hydrolysed by water. It will turn into iron(III) hydroxide.

102.17
$$ClO^-(aq) + 2I^-(aq) + 2H^+(aq) \rightarrow Cl^-(aq) + I_2(s) + H_2O(l)$$

102.18 The key to doing this is to spot that the chlorine changes from oxidation number +5 to −1. This is an effective gain of six electrons. Hence the $6I^-$ and $6Fe^{2+}$ appearing in the equations:

$$ClO_3^-(aq) + 6I^-(aq) + 6H^+(aq) \rightarrow$$
$$Cl^-(aq) + 3I_2(s) + 3H_2O(l)$$
$$ClO_3^-(aq) + 6Fe^{2+}(aq) + 6H^+(aq) \rightarrow$$
$$Cl^-(aq) + 6Fe^{3+}(aq) + 3H_2O(l)$$

UNIT 102 SUMMARY

Hydrogen halides

- Preparation:
  (i) Hydrogen fluoride and chloride: warm the sodium halide with concentrated sulphuric acid;

  e.g. $NaCl(s) + H_2SO_4(l) \rightarrow NaHSO_4(s) + HCl(g)$

  (ii) Hydrogen bromide and iodide: react the sodium halide with phosphoric acid;

  e.g. $NaI(s) + H_3PO_4(l) \rightarrow NaH_2PO_4(s) + HI(g)$

Properties

- Hydrogen fluoride:
  (i) Strongly hydrogen bonded.

(ii) Aqueous HF is only a weak acid.
- With water:
  Hydrogen chloride, bromide and iodide are strongly acidic;

  e.g. $HBr(g) + H_2O(l) \rightarrow H_3O^+(l) + Br^-(aq)$

- With ammonia:
  Salts made;

  e.g. $NH_3(g) + HCl(g) \rightarrow NH_4Cl(s)$

Interhalogen compounds

A number of interhalogen compounds exist, e.g. ICl, IF_5; the latter shows fluorine's ability to bring out the highest oxidation state of an element.

Metal halides

(i) Can be made by direct reaction between the metal and halogen, or in some cases the hydrogen halide;

  e.g. iron(III) chloride
  $$2Fe(s) + 3Cl_2(g) \rightarrow 2FeCl_3(s)$$
  iron(II) chloride
  $$Fe(s) + 2HCl(g) \rightarrow FeCl_2(s) + H_2(g)$$

(ii) Some halides are easily hydrolysed, e.g. $AlCl_3$, $FeCl_2$, $FeCl_3$.

Properties

- Nature of halides:
  (i) Group 1 halides are ionic solids, e.g. Na^+Cl^-.
  (ii) The chlorides of beryllium, boron and aluminium ($BeCl_2$, BCl_3, $AlCl_3$) are covalent. $AlCl_3$ dimerises, making Al_2Cl_6.
- Hydrolysis:
  Some covalent halides are easily hydrolysed, e.g. $AlCl_3$, $FeCl_2$, $FeCl_3$. (Therefore, require dry conditions in their preparations.)
- With transition metal ions:
  Halide ions can act as ligands, e.g. $[Cr(H_2O)_4Cl_2]^+$.

Compounds containing oxygen and halogens

Many molecules containing both oxygen and a halogen exist. See Table 102.3.

- Oxoanions:
  (i) Most important are halate(I), XO^-, halate(V), XO_3^-, and in the case of chlorine only, chlorate(VII), ClO_4^-.
  (ii) Chlorate(I) (as HOCl, hypochlorous acid) is responsible for the bleaching ability of household bleaches.
  (iii) Pattern of reactions is

  chlorate(I) $\rightarrow$ chloride + loss of oxygen
  chlorate(V) $\rightarrow$ chloride + loss of oxygen
  chlorate(VII) $\rightarrow$ very little oxidising power

  e.g. $HClO(aq) + 2Fe^{2+}(aq) + H^+(aq) \rightarrow$
  $$Cl^-(aq) + 2Fe^{3+}(aq) + H_2O(l)$$
  $ClO_3^-(aq) + 3SO_3^{2-}(aq) \rightarrow$
  $$Cl^-(aq) + 3SO_4^{2-}(aq)$$
  $IO_3^-(aq) + 5I^-(aq) + 6H^+(aq) \rightarrow$
  $$3I_2(aq) + 3H_2O(l)$$

- Disproportionation reactions:
  (i) Occur when an element with a single oxidation number (or state) reacts to give two or more different oxidation numbers.
  (ii) Take place when halogens react with alkali. With cold dilute alkali;

  $Br_2(l) + 2OH^-(aq) \rightarrow$
  $$Br^-(aq) + OBr^-(aq) + H_2O(l)$$

  With hot concentrated alkali;

  $3Br_2(l) + 6OH^-(aq) \rightarrow$
  $$5Br^-(aq) + BrO_3^-(aq) + 3H_2O(l)$$

- Heat on a chlorate(V):
  (i) Near its melting point ($367°C$);

  $4KClO_3 \rightarrow 3KClO_4 + KCl$
  (a disproportionation)

  (ii) At higher temperatures;

  $4KClO_3 \rightarrow 4KCl + 6O_2$

- Oxoacids:
  Examples are chloric(I) acid, HClO (hypochlorous acid); chloric(V) acid, $HClO_3$ (chloric acid); chloric(VII) acid, $HClO_4$ (perchloric acid).

103

Pseudohalides and pseudohalogens

103.1 What is a pseudohalide?

A pseudohalide is an ion that behaves like a halide ion. For example, we should expect pseudohalides to

(i) have a negative charge,
(ii) give salts with alkali metals, and
(iii) act as a ligand with transition metal ions.

The most common ones are listed in Table 103.1.

(a) Azides

Group I metals give ionic azides, whereas B metals tend to give covalent azides. One rather important difference between the two classes of azide is that the covalent ones are explosive, whereas the ionic ones are not. Azide ions are linear and can join with organic radicals in much the same way as the halogen atoms, e.g. methyl azide, CH_3N_3.

Hydrazoic acid, HN_3, is a weak acid in water, and can be made by reacting sulphuric acid with sodium azide, $Na^+N_3^-$.

(b) Cyanides

Cyanide ions are famous largely because of their reputation as an extremely dangerous poison. These ions satisfy each of our three conditions:

(i) they have a negative charge;
(ii) Group I metals give ionic salts, e.g. sodium cyanide, Na^+CN^-;
(iii) cyanide ions can act as ligands, e.g. hexacyanoferrate(II), $Fe(CN)_6^{4-}$, and hexacyanoferrate(III), $Fe(CN)_6^{3-}$, ions.

Indeed, it is the ability of cyanide ions to bond with

Table 103.1. The most common pseudohalides

Name	Ion
Azide	N_3^-
Cyanide	CN^-
Thiocyanate	SCN^-

iron ions that is partly responsible for their action as a poison. They bond very strongly to the iron in cytochrome oxidase, and prevent oxygen molecules being taken up by haemoglobin. Cytochrome oxidase is involved in the biochemical cycle controlling respiration.

Hydrogen cyanide, HCN, can be made by reacting sodium cyanide with sulphuric acid. It has a smell of bitter almonds, although it would be a foolish chemist who spent too long in savouring the smell! Like hydrogen fluoride, hydrogen cyanide is a weak acid in water.

(c) Thiocyanates

Thiocyanate ions also complex with transition metal ions. With iron(III) ions, an intense blood red colour is produced. This reaction can be used as a test for iron(III) ions:

$$Fe^{3+}(aq) + SCN^-(aq) \rightarrow FeSCN^{2+}(aq)$$
$$\text{blood red}$$

103.1 Write the equation for the reaction between sulphuric acid and sodium azide.

103.2 Sodium cyanide can be regarded as a salt of a weak acid (HCN) and a strong alkali (NaOH). What would you predict for the pH of a solution of sodium cyanide in water. Briefly explain your answer.

103.3 The stability constants of $FeSCN^{2+}$ and FeF^{2+} are approximately $10^3 \, mol^{-1} \, dm^3$ and $10^5 \, mol^{-1} \, dm^3$ respectively. A solution of FeF^{2+} is colourless. What, if anything, should happen if (i) F^- ions are added to a solution of $FeSCN^{2+}$, (ii) SCN^- ions are added to a solution of FeF^{2+}?

103.2 Pseudohalogens

In the same way that chloride ions have chlorine molecules, Cl_2, as their parents, so many of the pseudohalides have pseudohalogens. Two of our pseudohalides give pseudohalogens. They are:

(i) cyanogen, $(CN)_2$;
(ii) thiocyanogen, $(SCN)_2$.

Both are gases at room temperature.

There is no pseudohalogen matching the azide ion.

Like the true halogens, cyanogen reacts with alkali:

$$(CN)_2(g) + 2OH^-(aq) \rightarrow CN^-(aq) + \underset{\text{cyanate}}{OCN^-(aq)} + H_2O(l)$$

This can be compared with:

$$Cl_2(g) + 2OH^-(aq) \rightarrow Cl^-(aq) + OCl^-(aq) + H_2O(l)$$

103.4 Cyanogen will combine directly with Group I metals. Write the equation for the reaction between sodium and cyanogen.

Answers

103.1 $NaN_3(s) + H_2SO_4(aq) \rightarrow NaHSO_4(aq) + HN_3(g)$

103.2 In water, the equilibrium

$$CN^-(aq) + H_2O(l) \rightleftharpoons HCN(aq) + OH^-(aq)$$

is set up. The solution will be alkaline, i.e. the pH will be rather more than 7. See Unit 76 about salt hydrolysis.

103.3 The stability constants show that FeF^{2+} is energetically more favoured than $FeSCN^{2+}$. Therefore we would expect (i) the blood red colour to clear as thiocyanate ions are displaced by fluoride ions, and (ii) nothing to happen. In the first reaction, kinetics is on our side, and the colour change takes place quickly.

103.4 $2Na(s) + (CN)_2(g) \rightarrow 2NaCN(s)$

UNIT 103 SUMMARY

- Pseudohalides:
  Show some, or all, of the properties of true halides. For example, they
  (i) Have a negative charge.
  (ii) Give salts with alkali metals.
  (iii) Act as ligands with transition metal ions.
  Examples are azide ions, N_3^-; cyanide ions, CN^-; thiocyanate ions, SCN^-.
- Pseudohalogens:
  Are related to pseudohalides, like halides are to halogens. Examples are cyanogen, $(CN)_2$; thiocyanogen, $(SCN)_2$.

104

The noble gases

104.1 What is special about the noble gases?

Sometimes it seems that the most honour goes to those people who are furthest removed from the nasty practicalities of life; so it is with the gases helium, neon, argon, krypton, xenon and radon, whose physical properties are outlined in Table 104.1. These gases were once thought to be so extremely idle that they had to belong to the nobility. The hurly burly of chemical reactions was thought to be above their station in life. Hence they were given the alternative name of the inert gases. However, at least in the case of xenon, it is not so; xenon will react with other elements and compounds.

However, the reactivity of the gases *is* very low, and this has given them a number of uses, which you will find listed in Table 104.2.

The discovery of the gases was a feat in itself (Table 104.3). You may remember from Unit 87 that the presence of one of them caused Mendeléeff some difficulty. Radon was discovered as the product of the radioactive decay of uranium and thorium. Helium was first detected as a new element in the spectrum of the Sun in 1868 (hence its name is connected with *helios*, the Greek word for sun). It was not until some years later that it was collected as a gas given off from radioactive minerals. Argon was discovered by Sir William Ramsay as an impurity in nitrogen from air. The other noble gases were discovered by an analysis of fractions of different boiling points taken from liquid air.

Helium has the lowest boiling point of any element,

Helium is used in small airships like this. (They are more properly called 'dirigibles'.)

and its behaviour as a liquid is most unusual. It will, for example, creep up the walls of its container. This phenomenon is a result of there being two different phases of liquid helium, one of which has an extremely low viscosity.

Table 104.1. Physical properties of the noble gas elements

Symbol	Helium He	Neon Ne	Argon Ar	Krypton Kr	Xenon Xe	Radon Rn
Electron structure	$1s^2$	$(He)2s^22p^6$	$(Ne)3s^23p^6$	$(Ar)3d^{10}4s^24p^6$	$(Kr)4d^{10}5s^25p^6$	$(Xe)4f^{14}6s^26p^6$
I.E./kJ mol^{-1}	2372	2081	1521	1350	1170	1040
Melting point/°C		−249	−189	−157	−112	−71
Boiling point/°C	−269	−249	−186	−152	−109	−62
Atomic radius/pm	120	160	190	200	220	
Principal oxid. no.	0	0	0	0	+2,+4,+6	0

Table 104.2. Uses of the noble gas elements

Element	Main uses
Helium	In air ships Mixed with oxygen, as a breathing gas for divers Used in some lasers Provides an inert atmosphere for welding
Neon	In 'neon lamps'
Argon	In electric light bulbs to prevent oxidation of the filament In Geiger counter tubes
Xenon	In high light intensity photographic flash tubes
Krypton	Also used in electric light bulbs and, like neon, in coloured display lamps
Radon	Radon is an alpha emitter It has been used in radiotherapy (Recently, the natural occurrence of radon in some houses has been linked to a very slight increased risk of cancer among occupants of those houses. The risk is highest in regions where there is much granite)

Table 104.3. The discovery of the noble gases

Gas	Year of discovery	Discovered by
Helium	1895	Sir William Ramsay
Neon	1898	Sir William Ramsay, M. W. Travers
Argon	1894	Sir William Ramsay, Lord Rayleigh
Krypton	1898	Sir William Ramsay, M. W. Travers
Xenon	1898	Sir William Ramsay, M. W. Travers
Radon	1900	Friedrich Dorn

104.1 What is the link between helium and (i) radioactive decay, (ii) the Sun's energy?

104.2 There are three isotopes of radon: $^{219}_{86}$Rn, $^{220}_{86}$Rn, $^{222}_{86}$Rn. Each of them decays by alpha emission with a half-life of about 4 s, 55 s and 4 days respectively. That they are all isotopes of radon was not known at first, and they were given the names actinon, thoron and radon. What are the products of their decay?

104.3 The noble gases are very close to being ideal.

(i) Which of them is the nearest to being an ideal gas?

(ii) Why (or how) is it that the gases liquefy?

104.4 Why are there no electronegativity values listed in Table 104.1?

104.2 Compounds of the noble gases

The first authentic noble gas compound was made by N. Bartlett in 1962. He showed that oxygen and platinum hexafluoride would react together to give an orange substance with the formula $O_2^+PtF_6^-$. The ionisation energy of an oxygen molecule is about 1165 kJ mol^{-1}, a value that Bartlett noticed was almost the same as the first ionisation energy of xenon (1170 kJ mol^{-1}). Not only that, but he expected that the sizes of the O_2^+ and Xe^+ ions would be comparable. This would mean that the lattice energies of $O_2^+PtF_6^-$ and $Xe^+PtF_6^-$ should be similar. In fact, $Xe^+PtF_6^-$ was deposited as an orange solid as soon as Bartlett mixed xenon and the hexafluoride.

Following this discovery, a number of other xenon compounds have been made. For example, XeF_2, XeF_4, XeF_6. The other noble gases remain unreactive.

104.5 Why is the chemistry of radon particularly difficult to investigate?

Answers

104.1 (i) An alpha-particle, $^4_2$He, is the nucleus of a helium atom. Alpha-particles are often given off in nuclear reactions.

(ii) Much of the Sun's energy is given by the fusion of hydrogen nuclei. Helium is a major product of fusion. See Unit 6.

104.2 Alpha decay decreases the mass number by four units and the atomic number by two units. This gives, respectively, $^{215}_{84}$Po, $^{216}_{84}$Po and $^{218}_{84}$Po.

104.3 (i) Helium.

(ii) They liquefy owing to the van der Waals forces that are set up between the atoms. The more electrons an atom has, the greater the forces. Helium has the least number of electrons, so the forces are weakest. Hence intermolecular attractions are the least, and it is closest to being ideal.

104.4 Pauling's electronegativity values tell us about the elements' relative attractions for electrons. Apart from a very few reactions, the noble gases are inert. Thus their electronegativity values are of little practical use.

104.5 It is radioactive, and has a short half-life. Hence it is very difficult to work with.

UNIT 104 SUMMARY

- The Group 0 elements:
  - (i) Have very high ionisation energies.
  - (ii) Are all highly unreactive.
- Isolation:
  By the liquefaction of air followed by fractional distillation.

- Reactions:
  Xenon shows a slight tendency to react, e.g. to make XeF_2, XeF_4, XeF_6, $Xe^+PtF_6^-$.

105

Transition metals

105.1 Transition metals and their electron structures

The transition metals are the block of elements sandwiched between Group II and Group III. There are three rows of them – the first, second and third transition series. For the most part we shall only be concerned with the first series. These are the elements scandium to zinc. However, zinc is not a typical transition metal, and it is best regarded as a B metal. We shall deal with its chemistry in Unit 108.

The graphs in Figure 105.1 will give you a visual impression of how many of the properties in Table 105.1 vary across the series. The electron structures of the transition metals are at the heart of their chemistry. It would be wise to learn them. Especially, notice that

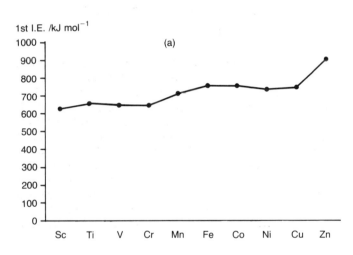

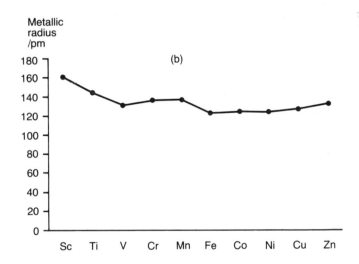

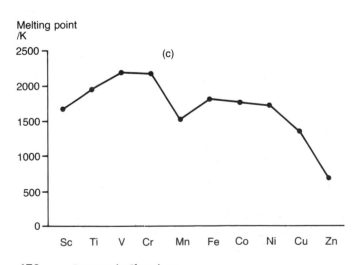

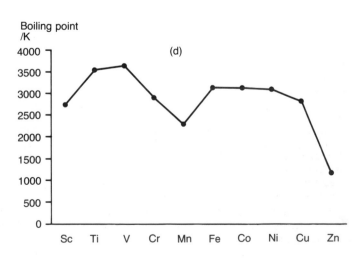

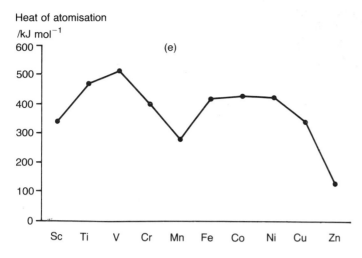

Heat of atomisation
/kJ mol^{-1}

(e)

Figure 105.1 *(a) First ionisation energies. The values are similar to but significantly higher than those of Group I metals. (b) Metallic radii. The similarity in radii is one reason why the metals form alloys with each other. Little distortion of a lattice may occur when one atom is substituted by another. (c) Melting points, (d) boiling points and (e) heats of atomisation. The graphs of melting points, boiling points and heats of atomisation all show that atoms of the elements are all held tightly in their lattices. However, the dips at manganese indicate that the combination of 3d^5 and 4s^2 electrons is less readily involved in metallic bonding. (The reason for this is complicated)*

they all involve the filling of the 3d set of orbitals, shown in Figure 105.2. (The second and third series of transition metals have 4d and 5d orbitals being filled.) For the elements before scandium, the 4s orbital has a lower energy than the 3d orbital. However, as the nuclear charge increases, at scandium the 3d orbitals become lower in energy than the 4s orbital. Partly this is because the 3d orbitals can penetrate into the region of space between the nucleus and the maximum in the 4s probability density. As a result, the 3d electrons can feel

a stronger attraction for the nucleus than the 4s electrons. The order of energies of the orbitals also depends on the interactions between the electrons in the orbitals. These interactions can be very complicated, and you would need to know a lot about quantum theory to understand them. They are responsible for the unusual electron structures of chromium and copper. We might expect chromium to have the structure (Ar)3d^{4}4s^2; instead it is (Ar)3d^{5}4s. The 3d set of orbitals can contain a maximum of 10 electrons, and the 4s orbital a maximum of two electrons. Therefore chromium has both sets of orbitals half-full. This is a result of a quantum effect that gives rise to a contribution to the energy called the *exchange energy*. The same effect lies behind the electron structure of copper. In this case the expected structure is (Ar)3d^{9}4s^2, whereas in reality it is (Ar)3d^{10}4s.

In Table 105.2 you will find a summary of the most important properties of the transition metals. In the following eight sections we shall examine each of them in turn. In fact we have already tackled the electron structures.

105.2 The oxidation states of transition metals

Taken together, the metals in all the transition series are the majority of the elements in the Periodic Table. All but a few of them can be found in two or more oxidation states. This is unlike many other elements, which have one or perhaps two oxidation states at most. You should know that the oxidation state of an atom can be shown by writing it as a Roman numeral in the name, or as a number of positive or negative charges. Do remember that by doing this we do *not* mean that the atom really does exist as an ion. The most common oxidation states are listed in Table 105.3. Actually, if you try hard enough you can make the transition

Table 105.1. The first transition series*†

	Sc	Ti	V	Cr	Mn	Fe	Co	Ni	Cu	Zn
Atomic no.	21	22	23	24	25	26	27	28	29	30
Outer electrons	3d4s^2	3d^{2}4s^2	3d^{3}4s^2	3d^{5}4s	3d^{5}4s^2	3d^{6}4s^2	3d^{7}4s^2	3d^{8}4s^2	3d^{10}4s	3d^{10}4s^2
1st I.E./kJ mol^{-1}	630	660	650	650	720	760	760	740	750	910
Covalent radius/pm	144	132	122	117	117	116	116	115	135	131
Atomic radius/pm	161	145	132	137	137	124	125	125	128	133
Melting point/K	1673	1950	2190	2176	1517	1812	1768	1728	1356	693
Boiling point/K	2750	3550	3650	2915	2314	3160	3150	3110	2855	1181
Sublimation energy/kJ mol^{-1}	305	428.9	458.6	348.9	219.7	351	382.4	371.8	304.6	115.3

*Each of the atoms has the argon gas core, (Ar) = 1s^{2}2s^{2}2p^{6}3s^{2}3p^6. The pattern of orbital filling changes at chromium and copper
†Zinc is included here for comparison; for most purposes it is better regarded as a B metal rather than a member of the transition series

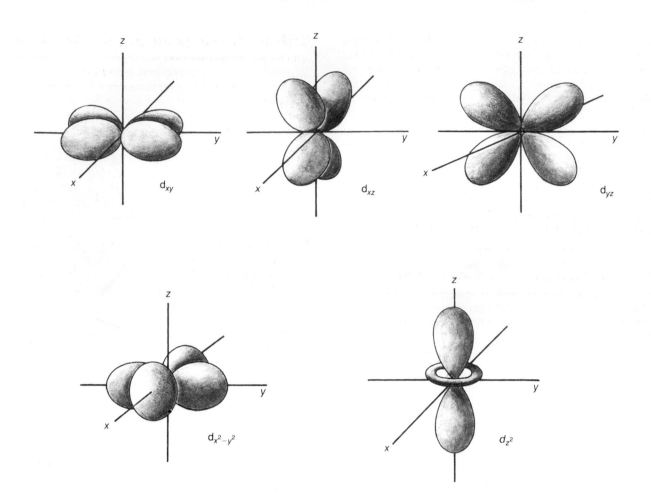

Figure 105.2 *The set of five 3d orbitals*

Table 105.2. Properties of transition metals

Outer electron structures involve d orbitals
Show variable oxidation states
Make complex ions
Paramagnetic compounds
Have coloured compounds
Act as catalysts
Form alloys

Table 105.3. The common oxidation states of the transition metals

Sc	Ti	V	Cr	Mn	Fe	Co	Ni	Cu	Zn
+3	+4	+3	+3	+2	+2	+2	+2	+1	+2
		+4	+6	+4	+3	+3		+2	
		+5		+7					

metals show a wide variety of oxidation states. For example, manganese can be found in all the oxidation states ranging from −3 (in $Mn(NO)_3CO$) to +7 (in $KMnO_4$), but only 0, +2, +4 and +7 are common.

> **105.1** What are the oxidation states of manganese in the following compounds: (i) $MnCl_4{}^{2-}$, (ii) $MnO_4{}^-$, (iii) Mn_2O_3?

105.3 What are complex ions?

From time to time we have met complex ions in previous units. For example, when copper(II) ions mix with aqueous ammonia, the tetraamminecopper(II) ion, $Cu(NH_3)_4{}^{2+}$, is made. This is a typical complex ion as far as its bonding is concerned. An ammonia molecule has a lone pair of electrons that can be donated to the copper ion. A molecule or ion like ammonia that gives a coordinate bond with a transition metal ion is called a *ligand*. A complex ion is made by a number of ligands bonding to the metal ion. A molecule or ion that can give one bond to a metal ion is a *monodentate* ligand. You will find examples of monodentate ligands in Table 105.4.

Usually four or six bonds are made to the ion. However, this does not necessarily mean that four or six ligands are involved. Some ligands have more than one site that can bond to an ion. These are called *polydentate* ligands. An example is ethane-1,2-diamine, $NH_2CH_2CH_2NH_2$, which is drawn in Figure 105.3. This molecule has a lone pair on each of the two nitrogen atoms. Therefore it can make two coordinate bonds with a metal ion.

A ligand like this is a *bidentate* ligand. We only need two ethane-1,2-diamine molecules to take the place of the four single ammonia molecules, and the complex ion $[Cu(NH_2CH_2CH_2NH_2)_2]^{2+}$ results. Ethane-1,2-diamine is given a shorthand symbol, en. This allows us

Table 105.4. Examples of monodentate ligands*

Ligand	Diagram
Water	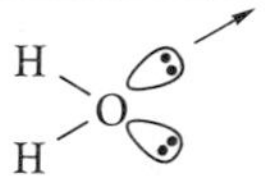
Ammonia	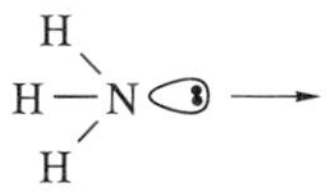
Chloride ion	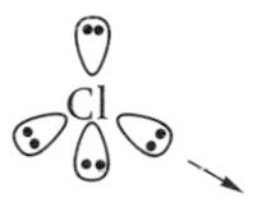
Cyanide ions	
Carbon monoxide	

*They all have at least one lone pair of electrons, which can make a coordinate bond with a transition metal ion

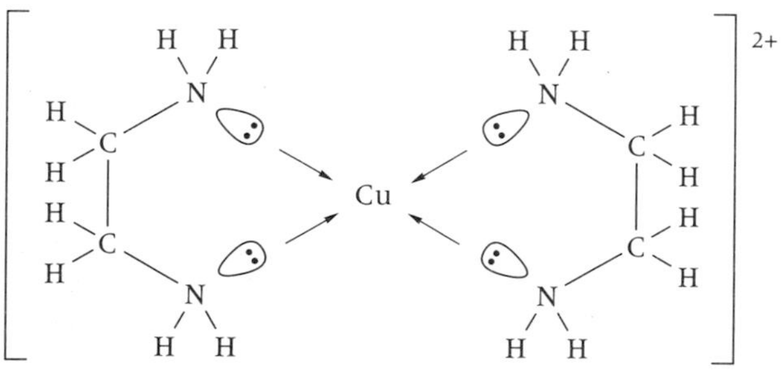

Figure 105.3 *The $[Cu(NH_2CH_2CH_2NH_2)_2]^{2+}$ ion has a copper(II) ion bonded to two ethane-1,2-diamine molecules through the four lone pairs on the nitrogen atoms*

Table 105.5. Examples of polydentate ligands†

Ligand	Diagram
Bidentate	
Ethane-1,2-diamine	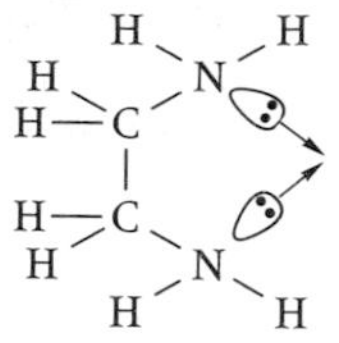
Ethanedioate ion	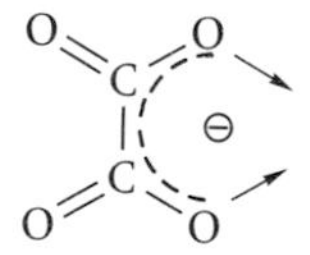
Pentane-2,4-dione (acetylacetonato ion)	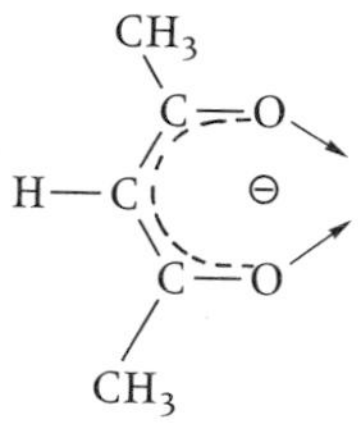
Hexadentate	
Ethylenediaminetetraacetic acid‡ (*marks a site for coordinate bonding)	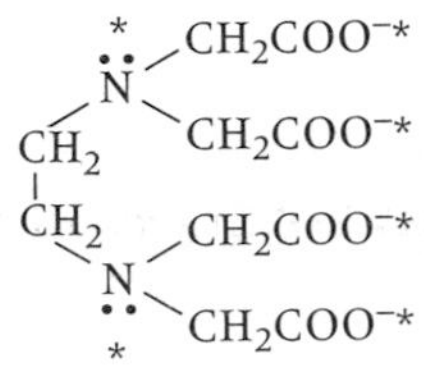

†Polydentate ligands are also called *chelating agents* (chelate is pronounced to rhyme with 'keylate')
‡Actually the anion of EDTA is shown. The acid molecule might be better written H_4EDTA

to write the formula of the complex in a neater way as $Cu(en)_2^{2+}$.

You will find examples of bidentate ligands in Table 105.5. Ligands with three, four, five and six bonding sites are also known. One hexadentate ligand (six bonding sites) called ethylenediaminetetraacetic acid (or 1,2-bis[bis(carboxymethyl)amino]ethane), EDTA, is extremely useful. It is used in analysis for the detection and estimation of the concentrations of metal ions. Not only transition metal ions react with EDTA: see section 91.6.

Ligands that have up to six sites for bonding are common, but monodentate, bidentate and hexadentate ligands are by far the most important. You might like to read panel 105.1 now. This will explain the naming system for transition metal compounds, and give you examples of compounds with a variety of oxidation states and ligands.

Panel 105.1
How to name transition metal compounds

There are two parts to the name of a complex. The first tells us about the ligands attached to the ion; the second gives the name of the metal ion and its oxidation state. The oxidation state tells us the charge that the metal would have if it were on its own as an ion. Here are some examples to show you the pattern:

(i) $Cu(NH_3)_4^{2+}$
Tetraammine ¦ copper(II)
four ammonia ¦ copper ion with +2 charge

(ii) $[Co(NH_3)_4Cl_2]^+$
Dichloro ¦ tetraammine ¦ cobalt(III),
two chlorine ¦ four ammonia ¦ cobalt with +3 charge

(iii) $Pt(NH_2CH_2CH_2NH_2)_2^{2+}$
Bis(ethane-1,2-diamine) ¦ platinum(II)
two ethane-1,2-diamine ¦ platinum with +2 charge

(iv) $[Fe(CN)_6]^{4-}$
Hexacyano ¦ ferrate(II)
six cyanide ¦ iron with +2 charge

Ending *ate* means entire ion is negatively charged.

Some of the rules that are at work in these examples are as follows:

Numbers of ligands

With monodentate ligands		With bidentate ligands	
Prefix	*Number*	*Prefix*	*Number*
Di	2	Bis	2
Tri	3	Tris	3
Tetra	4	Tetrakis	4
Penta	5		
Hexa	6		

Names of ligands

Name in complex	Ligand
Aqua (or aquo)	H_2O
Ammine	NH_3
Ethane-1,2-diamine	$NH_2CH_2CH_2NH_2$
Chloro	Cl^-
Cyano	CN^-
Nitro	NO_2^-

Notice that, when ammonia molecules act as ligands, they are referred to as 'ammine'; the term 'amine' refers to an NH_2 group as a ligand.

Naming the metal

Positive charge on complex	Use standard name with oxidation state in brackets
Negative charge on complex	Use name with *ate* ending and oxidation state in brackets

---

105.2 Carbon monoxide has a lone pair of electrons on the carbon atom that can make a coordinate bond to a transition metal ion. In addition, it has an empty antibonding π orbital. This can overlap with a d orbital of the right symmetry as shown in Figure 105.4.

(i) How would you describe the bond that is made?

Many metals give carbonyls with carbon monoxide, e.g. $Cr(CO)_6$, $Fe(CO)_5$.

Figure 105.4 *Carbon monoxide can make bonds to a transition metal ion: one through a lone pair on the carbon atom, another through an antibonding π orbital (which has four lobes)*

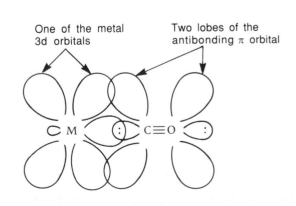

One of the metal 3d orbitals

Two lobes of the antibonding π orbital

(ii) What is the oxidation state of the metals in these compounds? (Hint: is carbon monoxide charged?)

105.3 What are the names of the following complexes or compounds: (i) $[Cu(NH_3)_2(H_2O)_2]^{2+}$; (ii) $[Cr(en)_2Cl_2]^+$; (iii) $K_4[Fe(CN)_6]$?

105.4 What are the formulae of the following complexes:

(i) tris(ethane-1,2-diamine)chromium(III),

(ii) trinitrotriamminecobalt(III),

(iii) chloropentaaquachromium(III)?

105.4 **The shapes of complex ions**

There are three typical shapes for complex ions. They are usually *octahedral, square planar* or *tetrahedral* (Figure 105.5).

(a) *Octahedral complexes*

An octahedral complex results when there are six bonds to the metal ion. This can happen if there are six monodentate ligands, three bidentate ligands, or other combinations that give the six bonds (see Figure 105.6 for examples).

(b) *Square planar complexes*

In a square planar complex there are four bonds to the metal ion. The metal ion and the four atoms attached to it all lie in the same plane (see Figure 105.7 for examples).

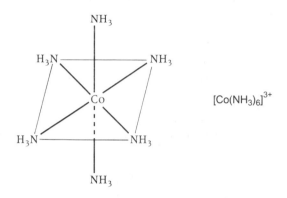

$[Co(NH_3)_6]^{3+}$

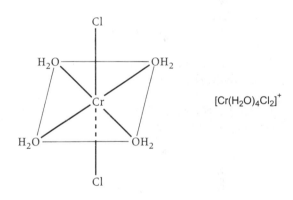

$[Cr(H_2O)_4Cl_2]^+$

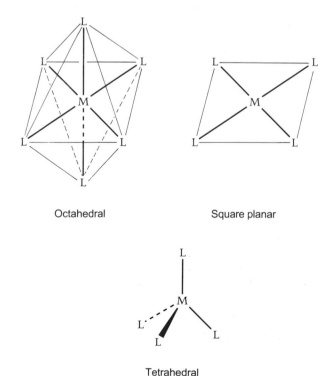

Octahedral   Square planar

Tetrahedral

Figure 105.5 *The three main shapes adopted by transition metal complexes*

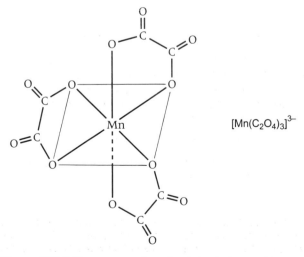

$[Mn(C_2O_4)_3]^{3-}$

Figure 105.6 *Three examples of octahedral complexes*

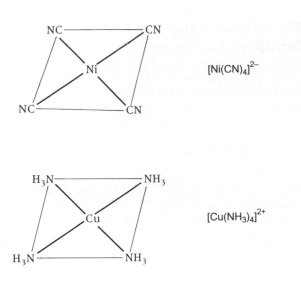

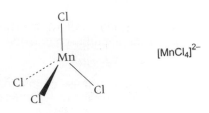
$[MnCl_4]^{2-}$

$[Ni(CN)_4]^{2-}$

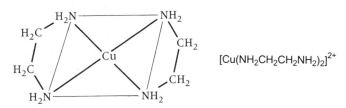

$[Cu(NH_3)_4]^{2+}$

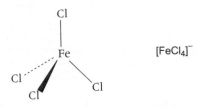
$[FeCl_4]^-$

$[Cu(NH_2CH_2CH_2NH_2)_2]^{2+}$

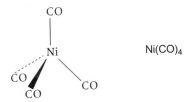
$Ni(CO)_4$

Figure 105.7 *Three square planar complexes*

Figure 105.8 *Three tetrahedral complexes*

(c) Tetrahedral complexes

A tetrahedral complex also has four bonds from the ligands to the metal ion. However, the ligands lie at the corners of a tetrahedron rather than a square (see Figure 105.8 for examples).

In the next unit you will disover that the various shapes of the complexes can produce isomerism. That is, there can be several different complexes with the same formula.

(d) Some unusual complexes

You should not think that the three geometries we have looked at are the only ones. There are linear complexes, complexes with five ligands and complexes that have distorted tetrahedral or octahedral shapes. A most unusual complex is made between iron and the organic molecule cyclopentadiene. Cyclopentadiene can gain an electron, for example by reacting with a sodium atom, to give a negative ion:

$$2C_5H_6 + 2Na \rightarrow 2C_5H_5^- + 2Na^+ + H_2$$

The importance of the ion is that it is cyclic, and the charge is delocalised over the entire ring in a series of π orbitals. Figure 105.9 shows you the idea. Two of these ions can bond to iron, giving the complex ion commonly known as ferrocene, $Fe(C_5H_5)_2$. (Its formal name is di-π-cyclopentadienyliron.) Ferrocene is held together by the bonding between the delocalised π

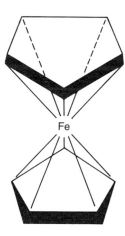

Figure 105.9 *Ferrocene has an iron ion sandwiched between two planar cyclopentadiene, $C_5H_5^-$, ions (shown here as pentagons)*

electrons on the two rings and some of the iron d orbitals. Once ferrocene was discovered, similar compounds were soon prepared, e.g. $Cr(C_5H_5)_2$, $Ni(C_5H_5)_2$.

Much more important than ferrocene are complexes that occur in biochemical systems. Haemoglobin is responsible for transporting oxygen through the blood stream. This large molecule has an iron(II) ion held by bonds to four nitrogen atoms as shown in Figure 105.10. (A further bond can be made to part of a protein chain not shown in the diagram.) The iron is able to bond to an oxygen molecule; but the strength of the

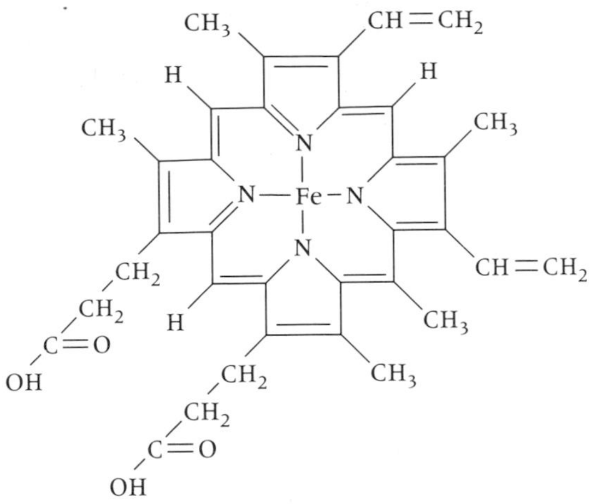

Figure 105.10 The arrangement of atoms in the haem group of haemoglobin

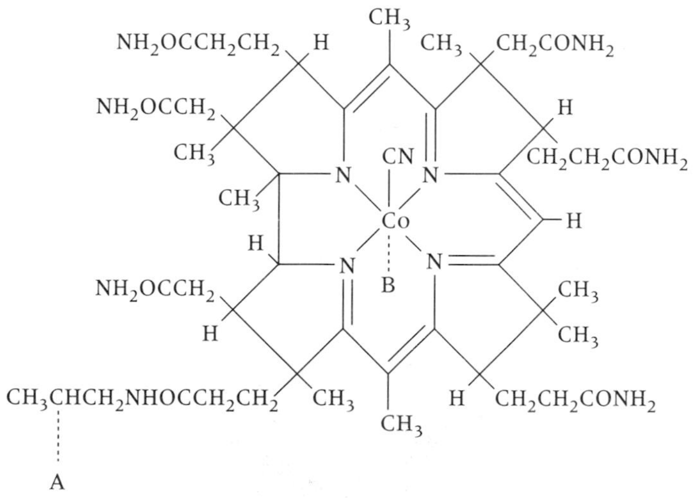

Figure 105.11 Part of the structure of vitamin B_{12}. Points A and B are linked together through a combination of phosphate, sugar and nitrogen ring groups

bond is such that the oxygen is easily removed, thus making it available throughout the body for use in respiration.

Cobalt is found in vitamin B_{12}, which is responsible for maintaining the supply of red blood cells in the liver (Figure 105.11). The cobalt is present as cobalt(III), with five bonds made to nitrogen atoms in different parts of the surrounding structure. The sixth site is where the biochemical activity takes place.

105.5 We are leaving the main work on isomers until the next unit. However, you might like to make a model of a complex ion with the formula $M(NH_3)_2Cl_2$. Use different coloured balls for the metal and the two different ligands. Assume it is square planar. How many different complexes (isomers) can you make? How many can you make if the complex is tetrahedral?

105.5 What happens to the d orbitals in a complex ion?

There are three theories of bonding in transition metal complexes. The first is *crystal field theory*, the second is *ligand field theory* and the third is *valence bond theory*. We shall examine each of them in turn.

(a) Crystal field theory

In crystal field theory we imagine that the ligands act as centres of electric charge. Then we try to work out what effect the field of the ligands has on the electrons in the d orbitals. This was the first theory invented to explain the properties of complexes. Given that it ignores the finer, and even the important, points of bonding, it is remarkably successful. This is what it has to say about the effects on the d orbitals.

First, if you bring negative charges near to electrons, the energies of the electrons rise. This has the effect of increasing the energies of all the d orbitals. Now let us look at the effects on the individual orbitals. Figure 105.2 showed the shapes of the d orbitals. Let us make a diagram of an octahedral complex (Figure 105.12) and show the d_{z^2} and $d_{x^2-y^2}$ orbitals as well. The lobes of the two orbitals point directly at the ligands. Electrons in these orbitals will suffer a significant amount of repulsion. Now look at Figure 105.13, which shows the arrangements of the d_{xy}, d_{xz} and d_{yz} orbitals. These have their lobes pointing between the ligands. As a result, electrons in these orbitals feel less repulsion than those in the other two orbitals. Thus crystal field theory says that the five d orbitals should split into two groups, two of which have a higher energy than the other three.

We say that the d orbitals are split. Owing to the splitting being caused by the electric field of the ligands, it is called the *ligand field splitting* (Figure 105.14). The extent of the splitting is given the symbol Δ_0. The two orbitals with the higher energy are called the e_g set; the three with the lower energy are the t_{2g} set.

(b) Ligand field theory

In ligand field theory we take a more realistic view of the bonding in complexes. We use molecular orbital theory to decide how the orbitals on the metal ion and the ligands interact. As you might expect, this is rather complicated. However, the nice thing is that ligand field theory comes up with the same scheme of splitting as crystal field theory. Figure 105.15 gives an impression of the results. You might be surprised to know that according to ligand field theory the two orbitals in the e_g set are *antibonding* orbitals. If this worries you, read section 16.6 again; antibonding orbitals can be perfectly respectable ones to use in bonding.

Once we have the arrangement of the orbitals, we are left with the task of filling them with electrons. This can be done in detail using ligand field theory, in much the same way as we used for oxygen in section 16.6. Fortunately, we do not need this level of detail, so we shall return to the crystal field picture. This allows us to

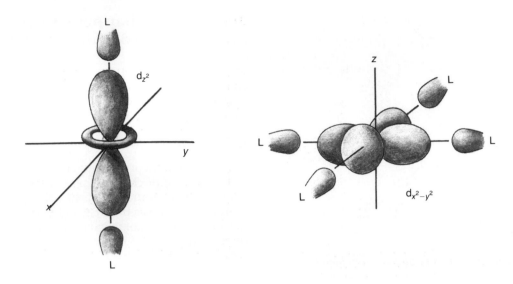

Figure 105.12 *Ligands at the ends of the x, y and z axes have electron density that will interact strongly with electrons in d_{z^2} and $d_{x^2-y^2}$ orbitals*

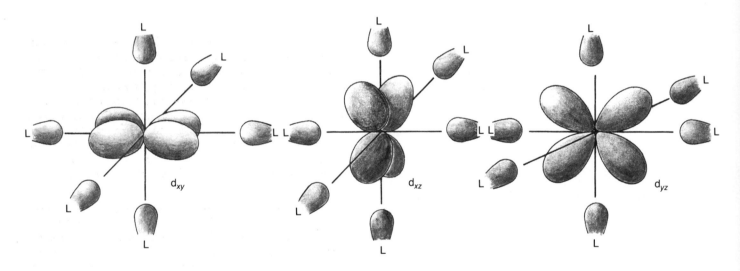

Figure 105.13 *Ligands at the ends of the x, y and z axes have weaker interactions with d_{xy}, d_{yz} and d_{xz} orbitals than with d_{z^2} or $d_{x^2-y^2}$ orbitals*

concentrate on the metal d electrons without getting bogged down in the detail of considering all the other electrons.

Before we do so, you might like to look at the way the d orbitals split in square planar and tetrahedral complexes (Figure 105.16). The pattern is quite different to the octahedral case.

(c) *Valence bond theory*

In valence bond theory the shapes of complex ions are explained by invoking hybridisation. Table 105.6 lists the combinations of orbitals that are needed for the main three geometries. Actually, valence bond theory can give us little help in explaining many properties of transition metal ions, so we shall say no more about it.

105.6 Why are transition metal compounds often paramagnetic?

Most people know that iron is magnetic. In fact, iron shows a very special type of magnetism called ferromagnetism. The two metals following it, cobalt and nickel, also show this type of magnetism.

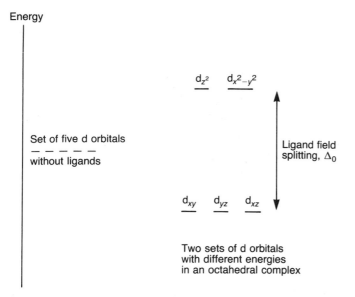

Figure 105.14 *How the energies of the d orbitals change in an octahedral complex*

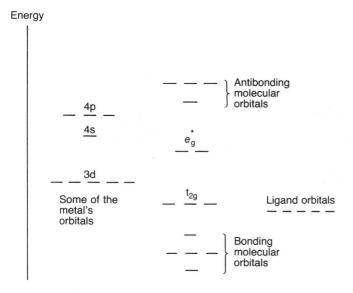

Figure 105.15 *In ligand field theory there is a set of five molecular orbitals that split apart in the same way as do the five d orbitals in crystal field theory. Two (e*$_g^*$ *set) increase in energy, three (t*$_{2g}$ *set) decrease in energy*

Table 105.6. Three types of hybridisation found in transition metal complexes*

Shape of complex	Hybrid orbitals
Tetrahedral	d^3s
Square planar	dsp^2
Octahedral	d^2sp^3

*Also, see Table 17.5

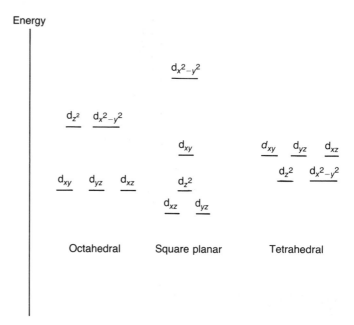

Figure 105.16 *How the d orbitals change their energies in octahedral, square planar and tetrahedral complexes*

Ferromagnetism is a variety of magnetism known as *paramagnetism*. A paramagnetic substance put close to a magnetic field will be attracted into the field. Many transition metal compounds are paramagnetic.

It is the magnetic field associated with the spin of an electron that can give rise to paramagnetism. However, only compounds that have unpaired electrons are paramagnetic. When two electrons occupy the same orbital, they have their spins paired, and we can think of the magnetic fields cancelling out. To sum up:

> **Transition metal compounds are paramagnetic when they have one or more unpaired electrons.**

If we measure the amount of their paramagnetism we can work out how many unpaired electrons are present. This can be done using a *Gouy balance* (Figure 105.17). An empty glass tube is suspended between the jaws of an electromagnet. Then the tube is filled with crystals of the chemical being investigated. The magnet is switched on and, if the substance is paramagnetic, the tube is pulled down into the field. Masses are placed on the balance pan until the balance is zeroed again. The mass used is a measure of the amount of paramagnetism, and therefore the number of unpaired electrons. If we are to understand how these unpaired electrons come about, we need to look carefully at what happens to the d orbitals when a complex ion is made.

The two complexes of iron(II), hexaaquaferrate(II), $[Fe(H_2O)_6]^{2+}$, and hexacyanoferrate(II), $[Fe(CN)_6]^{4-}$,

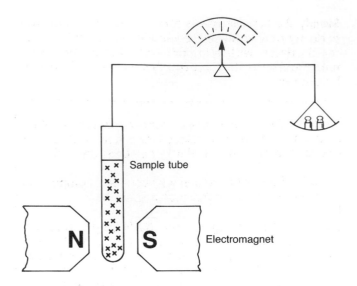

Figure 105.17 *The principle of a Gouy balance. The sample is balanced with the electromagnet off, and again with it on. The difference between the two weighings is a measure of the number of unpaired electrons in the sample*

electron in each of the five orbitals, and one left over. This will go into the orbital of lowest energy. Making sure that we stick to the Pauli principle, the two electrons in the same orbital will have their spins paired. This is the arrangement in Figure 105.18a. You can see that there are four unpaired electrons, just as found from experiment.

In the case of $[Fe(CN)_6]^{4-}$ there are no unpaired electrons. The only way of achieving this, and keeping the energy of the electrons to a minimum, is to put them as three pairs in the t_{2g} set. This is the arrangement of Figure 105.18b. The reason why they take up this arrangement is that cyanide ions have a larger effect on the splitting of the d orbitals than do water molecules. The ligand field splitting is larger in $[Fe(CN)_6]^{4-}$ than in $[Fe(H_2O)_6]^{2+}$. When the gap Δ_0 becomes large, it takes more energy to put an electron into one of the e_g set than it does to pair electrons in the t_{2g} orbitals. For fairly obvious reasons, $[Fe(H_2O)_6]^{2+}$ is called a *high spin* complex, and $[Fe(CN)_6]^{4-}$ a *low spin* complex.

are both octahedral. The first has paramagnetism corresponding to four unpaired electrons. The second is not paramagnetic; it has no unpaired electrons. Now, the electron structure of iron is $(Ar)3d^64s^2$. When it loses two electrons, it is left with the argon core and six 3d electrons. We have to arrange these six electrons on the energy level diagram for an octahedral complex. We know that electrons will go into separate orbitals whenever possible so that the repulsion among them is minimised. If we follow this pattern we shall have one

105.6 An atom with eight d electrons can be written in shorthand as d^8. There is only one way of arranging the electrons in a d^8 octahedral complex. Show the arrangement on a diagram. How many unpaired electrons are there?

105.7 There is a high and a low spin state for a d^7 octahedral complex. Draw diagrams to show them, and the number of unpaired electrons in each case.

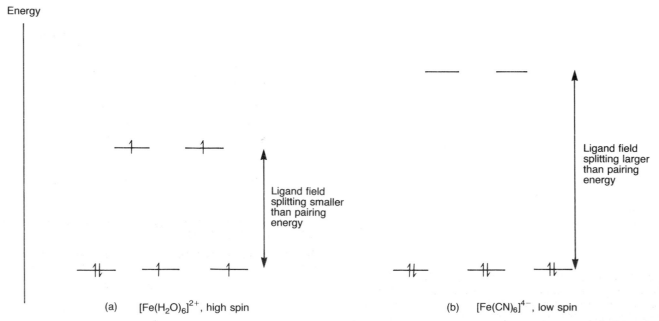

Figure 105.18 *Cyanide ions cause a much larger splitting of the d orbitals than do water molecules. One result is that cyanide complexes are often 'low spin'*

105.7 Why are transition metal compounds often coloured?

Transition metal complexes are coloured because visible light has just about the right energy to excite an electron in the lower set of d orbitals into the higher set. For example, the octahedral complex hexaaquanickel(II), $[Ni(H_2O)_6]^{2+}$, is responsible for the green colour of many nickel(II) salts in water (Figure 105.19). The complex appears green because it absorbs in the blue region of the spectrum (and also in the red). In this case the splitting of the d orbitals corresponds to the energy of blue light.

Actually there is a rule called the Laporte rule which says that transitions between d orbitals should not occur. The reason for this lies in the symmetry of the orbitals, something we spoke about in Units 24 and 25.

Strictly the Laporte rule is correct, but it does not work perfectly here because the vibrations of complexes mix the d orbitals with p orbitals. This changes their symmetry and gives the electrons their chance to be excited by photons.

105.8 Zinc compounds almost always contain zinc(II). Look at the electron structure of zinc and suggest a reason why zinc compounds are white.

105.9 Suggest a reason why copper(I) compounds are unlikely to be highly coloured.

105.10 The complex ion $[Ni(H_2O)_6]^{2+}$ has an absorption peak at a wavelength of around 410 nm. Remind yourself of the connection between wavelength, frequency and energy. What is the ligand field splitting in the complex in joules, and in kJ mol^{-1}?

105.11 When ammonia is added to a solution of $[Ni(H_2O)_6]^{2+}$, the colour changes to dark blue-purple. The absorption peak has moved to around 350 nm. Has the ligand field splitting increased or decreased compared to the aqua complex?

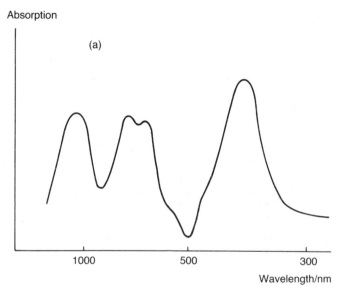

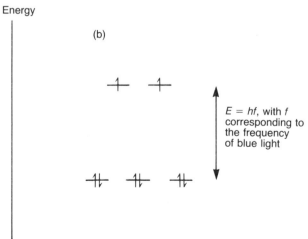

$E = hf$, with f corresponding to the frequency of blue light

Figure 105.19 (a) The spectrum of $[Ni(H_2O)_6]^{2+}$ shows a number of strong absorptions. However, the complex absorbs only weakly around 500 nm, which is the green and blue-green region of the visible spectrum. (b) The ligand field splitting in $[Ni(H_2O)_6]^{2+}$ matches the frequency of light in the blue region of the visible spectrum

105.8 How do the transition metals act as catalysts?

Transition metals can act as catalysts (Table 105.7), for two reasons. First, because they can have several dif-

Table 105.7. Transition metal catalysts

Reaction	Catalyst
The Haber process $N_2(g) + 3H_2(g) \rightleftharpoons 2NH_3(g)$	Iron (and aluminium oxide)
Sulphuric acid manufacture $2SO_2(g) + O_2(g) \rightleftharpoons 2SO_3(g)$	Platinum and vanadium(v) oxide
Nitric acid manufacture $4NH_3(g) + 5O_2(g) \rightleftharpoons 4NO(g) + 6H_2O(g)$	Platinum
Hydrogenation of alkenes	Nickel
Laboratory preparation of oxygen $2H_2O_2(aq) \rightarrow 2H_2O(l) + O_2(g)$	Manganese(IV) oxide
Polymerisation of alkenes, e.g.	Ziegler–Natta catalysts $Al(C_2H_5)_3/TiCl_3$

ferent oxidation states, they can take part in electron transfer reactions. One example that demonstrates this is the effect of adding iron(II) or iron(III) ions to a mixture of peroxodisulphate(VI) ions, $S_2O_8^{2-}$, and iodide ions. Normally the reaction:

$$S_2O_8^{2-}(aq) + 2I^-(aq) \rightarrow 2SO_4^{2-}(aq) + I_2(s) \qquad (A)$$

takes place at a convenient rate, which we can follow by one of the usual methods. If a little iron(III) sulphate solution is added, the rate of production of iodine increases markedly. This is because the Fe^{3+} ions react with the iodide ions:

$$2Fe^{3+}(aq) + 2I^-(aq) \rightarrow 2Fe^{2+}(aq) + I_2(s) \qquad (B)$$

If this were all that happened, we would not be dealing with true catalysis: the Fe^{3+} ions appear to be used up in the reaction. However, the Fe^{2+} ions are oxidised back to Fe^{3+} by $S_2O_8^{2-}$ ions:

$$2Fe^{2+}(aq) + S_2O_8^{2-}(aq) \rightarrow 2Fe^{3+}(aq) + 2SO_4^{2-}(aq) \quad (C)$$

If you combine equations (B) and (C) you will obtain the original equation, (A). The Fe^{3+} ions have taken part in the reaction, but they are regenerated. The key point is that by giving a different path by which electron transfer can take place, the ions have provided a route with a lower activation energy.

Transition metals and their ions can also act as catalysts by providing sites at which reactions can take place. They can bond to a wide range of ions and molecules, e.g. those with lone pairs or π electrons. Also they show the ability to make different numbers of bonds, often four or six, but sometimes two, three and five as well. They can show catalytic behaviour when dissolved in solution or as solids, i.e. as homogeneous or heterogeneous catalysts.

An important example of a catalytic reaction in solution is the Ziegler–Natta polymerisation of alkenes (Figure 105.20). A typical Ziegler–Natta catalyst is a mixture of an organic aluminium compound, e.g. $Al(C_2H_5)_3$, and titanium(III) chloride. One of the ethyl groups from the aluminium compound bonds to the titanium ion; but being a transition metal ion, it has further sites available for bonding. The π cloud of an alkene can fill one of these sites, with the result that the alkene and ethyl group are held very close to one another. For reasons that we do not have to consider, the alkene inserts itself between the ethyl group and the titanium ion. Now there is room for a further alkene to bond to the titanium ion, which in turn swaps its position. In this way a hydrocarbon chain that originally had only two carbon atoms in it now has six of them joined together. Provided the supply of alkene molecules is kept up, the chain will grow to great length. In other words, a polymer is made. The discovery of this

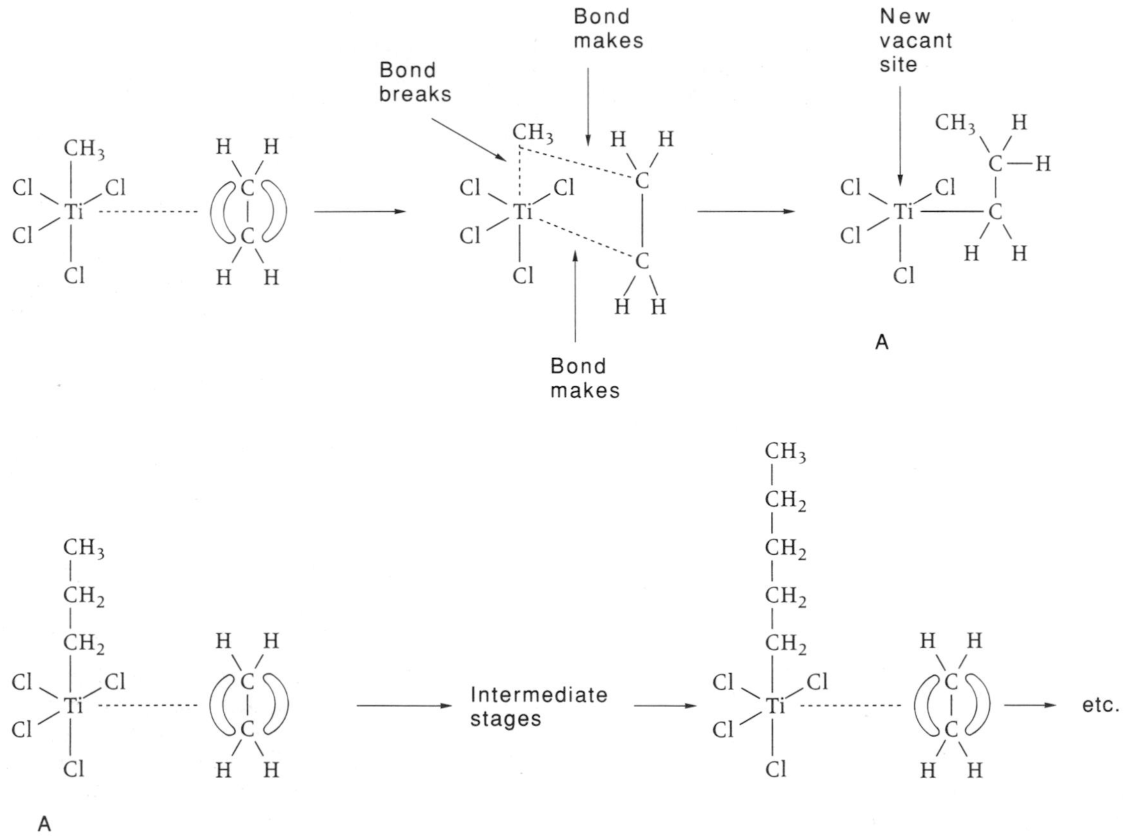

Figure 105.20 *These diagrams show you the type of change that takes place with Ziegler–Natta catalysts. The first step is an alkene bonding to a titanium complex through its π cloud of electrons. Then there is a rearrangement of the atoms, which leaves a complex, A, that can repeat the process. The result is growth of the polymer chain*

type of reaction by Ziegler and Natta has revolutionised the manufacture of polymers, partly because the method can give polymers of consistent quality.

105.9 Transition metals and alloys

The transition metals are generally hard, and difficult to melt. They are also strong, but they can be brittle. (A strong metal bar can support a heavy load hung from its mid-point. However, if the bar is dropped on to a hard surface and splits into pieces, the metal is brittle.) One way of making metals less brittle is to mix them with other atoms to make an *alloy*. The transition metals are very good at making alloys with transition, and other, metals as well as some non-metals (see Table 105.8).

Table 105.8. Examples of alloys containing transition metals

Name	Composition	Use
Steel*	Iron with small amounts of C and Mn	Car bodies, bridges, etc., where strength but greater ductility and malleability than iron is needed
Stainless steel	Iron with up to 35% Cr; Ni may also be present	Steel with great resistance to corrosion, e.g. knives, sinks, industrial pipes, jet engines
Coinage metal	Cu 75%, Ni 25% Cu 95.5%, Zn 1.5%, Sn 3%	'Silver' coins 'Copper' coins
Brass	Various proportions of copper and zinc, e.g. Cu 70%, Zn 30%	Door knobs, ornaments, bullet cases
Duralumin	Aluminium with up to 5% Cu, less than 1% Mn, Mg and Si	In aircraft, owing to combination of strength with lightness
Monel	About 66% Ni, 33% Cu, with some Fe, Mn, Si and C	Equipment that must be resistant to corrosion, e.g. in steam turbines
Nichrome	Ni 60%, Fe 20%, Cr 20%	Wires for electrical equipment

*There are many different types of steel, with properties depending on the quantity of the added elements. These can include V, Cr, Mn, Co, Ni, C, Si, P and S

Answers

105.1 (i) Mn(II), (ii) Mn(VII), (iii) Mn(III).

105.2 (i) It is a type of π bond ($d\pi$).

(ii) Carbon monoxide is neutral. As the carbonyls are uncharged, the metals can have no charge. They are in the oxidation state zero.

105.3 (i) Diamminediaquacopper(II).

(ii) Dichlorobis(ethane-1,2-diamine)chromium(III).

(iii) Potassium hexacyanoferrate(II).

105.4 (i) $[Cr(NH_2CH_2CH_2NH_2)_3]^{3+}$.

(ii) $Co(NO_2)_3(NH_3)_3$.

(iii) $[Cr(H_2O)_5Cl]^{2+}$.

105.5 There are two isomers in the square planar arrangement. No isomers are possible if the complex is tetrahedral.

105.6 See Figure 105.19b. $[Ni(H_2O)_6]^{2+}$ is such a complex.

105.7 The diagrams are shown in Figure 105.21.

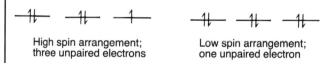

High spin arrangement; three unpaired electrons   Low spin arrangement; one unpaired electron

Figure 105.21 Diagrams for answer to question 105.7

105.8 Zinc(II) has ten d electrons. Therefore transitions between the d orbitals cannot take place: they are all full, and you cannot have more than two electrons in an orbital.

105.9 Copper(I), like zinc(II), has a full set of d orbitals, so transitions between them cannot occur. However, there can be complicated interactions that produce colour in compounds, even though a set of d orbitals is full; for example, copper(I) oxide is orange.

105.10 $c = f\lambda$ and $E = hf$, so

$$E = \frac{6.626 \times 10^{-34}\,J\,s \times 2.998 \times 10^{8}\,m\,s^{-1}}{410 \times 10^{-9}\,m}$$

i.e. $E = 4.85 \times 10^{-19}\,J$, which converts to $292\,kJ\,mol^{-1}$.

105.11 It has increased. We can work this out because the absorption has moved from 410 nm to a lower wavelength, and lower wavelength corresponds to increasing energy. The spectrochemical series puts ligands in order of their ability to increase the splitting of d orbitals. The series is

$CN^- > NO_2^- > en > NH_3 > H_2O$
Greatest splitting $> OH^- > F^- > Cl^- > Br^- > I^-$
                       Least splitting

UNIT 105 SUMMARY

- The transition elements:
    - (i) Are all metals.
    - (ii) Have similar ionisation energies.
    - (iii) Form alloys.
    - (iv) Have a set of d orbitals filling with electrons.
    - (v) Give coloured compounds.
    - (vi) May have paramagnetic compounds.
    - (vii) Make complex ions.
    - (viii) Show a range of different oxidation states.
    - (ix) Often act as catalysts.
- Ligands:
    - (i) Monodentate ligands make one bond per molecule/ion to a metal ion, e.g. NH_3, Cl^-.
    - (ii) Bidentate ligands make two bonds per molecule/ion to a metal ion, e.g. ethane-1,2-diamine, $NH_2CH_2CH_2NH_2$.
    - (iii) EDTA is a hexadentate ligand (six bonding sites).

- Shapes of complexes:
    Mainly octahedral, square planar or tetrahedral.
- Paramagnetism:
    Is due to the presence of unpaired electrons in the d orbitals.
- Colour:
    Is often a result of electrons moving between the d orbitals.
- Catalytic action:
    - (i) Is due to the availability of d orbitals for making bonds, and ability to transfer electrons when changing oxidation state.
    - (ii) Ziegler–Natta catalysts are used in the polymerisation of alkenes. They are mixtures of an organic aluminium compound, e.g. $Al(C_2H_5)_3$, and titanium(III) chloride, $TiCl_3$.

106

More about complex ions

106.1 Isomerism in complex ions

An ion or molecule has isomers if it can exist in two or more different forms with the same formula. Isomers frequently occur in organic compounds, but one of the fascinating things about transition metal chemistry is that many complex ions have isomers. The study of isomerism in complexes was one of the main interests of the German chemist Alfred Werner. We owe much of our understanding of transition metal complexes to him. In his book *New Ideas on Inorganic Chemistry* published in 1911 he discussed the main types of isomerism. We shall follow some of his examples, although we shall change the names that he used to their modern versions.

(a) *Ionisation isomerism*

This is the first type of isomerism we shall look at. Werner says:

> A peculiar type of isomerism ... is that in which compounds of the same composition yield different ions in solution.

His example was two compounds with the formula $Co(NH_3)_5BrSO_4$. Werner noted that in solution:

> The one is a reddish-violet colour, and gives in a freshly prepared solution no reaction for bromide ions, but does for the sulphate ion. On the other hand, the solution of the second salt gives the reactions of bromide ion, but not of the sulphate ion.

Well, so much for the facts. How can we explain them? The difference is due to the way the bromide and sulphate are bonded in the compounds. In both of them the five ammonia molecules are ligands. This leaves space for one other ligand. In one of the complexes the bromide is the ligand, in the other it is the sulphate ion. The two possibilities are shown in Figure 106.1. In the first one the bromide is strongly bonded to the cobalt ion. We say that the ammonia molecules and the bromide ion are in the *coordination sphere*. The sulphate ion is held in the crystal with its two negative charges just balancing the two positive charges on the complex.

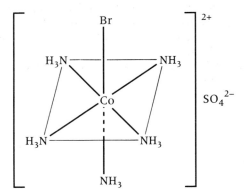

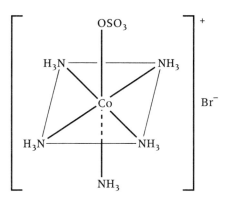

Figure 106.1 *The two isomers of Co(NH₃)₅BrSO₄*

When the solid is put in water the sulphate ion floats away, but the bromide is left attached to the cobalt. Therefore, if we test for the presence of sulphate ions we find a positive result, but a test for free bromide ions is negative. With the second variety of the salt, the opposite is true. The sulphate ion is bonded to the cobalt, and free bromide ions can be found in the solution. (The sulphate ion is now in the coordination sphere with the ammonia molecules.)

(b) *Hydrate isomerism*

This is the second type of isomerism that we need to know about. Werner comments:

The most beautiful example of hydrate isomerism is that furnished by chromium chloride. Three different hydrates of the formula $CrCl_3 + 6H_2O$ are known. One is greenish-blue, which dissolves giving a blue-violet solution, and two are green, which dissolve yielding green solutions.

The difference between the three hydrates is shown by writing their formulae as:

$[Cr(H_2O)_6]^{3+}(Cl^-)_3$ blue-violet in solution
$[Cr(H_2O)_5Cl]^{2+}(Cl^-)_2$ green in solution
$[Cr(H_2O)_4Cl_2]^+Cl^-$ darker green in solution

You should understand why it is that one mole of each of them give three, two and one moles of chloride ions in solution respectively. When they are crystallised from solution the second takes up one mole, and the third two moles of water of crystallisation. This gives them all the overall formula $CrCl_3 \cdot 6H_2O$.

(c)   *Geometrical isomerism*

This is the third type of isomerism. Werner points out that a complex ion like $[Co(NH_3)_5Cl]^{2+}$ has no isomers. However, two isomers of $[Co(NH_3)_4Cl_2]^+$ can be identified:

Up to the present, and in spite of much careful work, it has been found impossible to obtain three isomers. These facts are best explained by assuming that the groups are placed around the central atom at the corners of a regular octahedron.

At this point you should do your best to make models of the structures that we are going to draw. It can be hard to visualise a shape drawn on paper; it is much easier to see it in three dimensions.

The ion $[Co(NH_3)_5Cl]^{2+}$ is an octahedral complex. Because of the symmetry of the octahedron all the diagrams in Figure 106.2 represent the same complex. A model will soon convince you of this.

If an ammonia molecule is replaced by a chloride ion we have $[Co(NH_3)_4Cl_2]^+$. This time there are two possible structures. In one of them the four ammonia molecules and the cobalt ion lie in a plane. The chloride ions lie above and below the plane, in the axial positions. In the second, a plane containing the cobalt ion also contains three ammonia molecules and one of the chloride ions. This leaves the remaining ammonia molecule and chloride ion in the axial positions. The two arrangements are different, as illustrated in Figure 106.3. They are geometrical isomers.

Square planar complexes can also have geometrical isomers. Figure 106.4 shows two isomers that have the formula $Pt(NH_3)_2Cl_2$. When the two chlorides (or ammonia molecules) are on the same side of the square, they are said to make the *cis isomer*. When they are opposite one another, we have the *trans isomer*.

(d)   *Optical isomerism*

This is the final type of isomerism we shall meet here. If you are to understand optical isomers, you must know what is meant by polarised light. If you have not yet

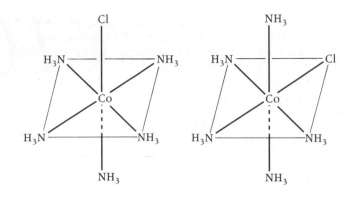

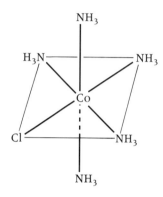

Figure 106.2 *Although these diagrams might seem to show different complex ions with the formula $[Co(NH_3)_5Cl]^{2+}$, they are in fact identical*

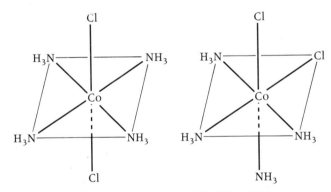

Figure 106.3 *Two (true) isomers of $[Co(NH_3)_4Cl_2]^+$*

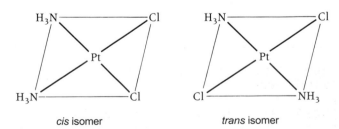

     cis isomer   *trans* isomer

Figure 106.4 *The* cis *and* trans *isomers of $Pt(NH_3)_2Cl_2$*

done so, read Units 24 and 110, which respectively tell you about light in general and polarised light in particular.

An optical isomer rotates the plane of polarised light.

If it rotates the plane to the left, it is called an *l* form (*l* for *laevo*, left). If it rotates it to the right, it is called a *d* form (*d* for *dextro*, right). A test to discover if a molecule or ion is optically active is to see if it can be superimposed on its mirror image. Examples that occur in transition metal chemistry are octahedral complexes involving a bidentate ligand such as ethane-1,2-diamine. The three ligand molecules can attach themselves in two ways, shown in Figure 106.5, which resemble aeroplane propellers. You should make models of the two structures to convince yourself that they are different. It is impossible to match the positions of all the atoms at once.

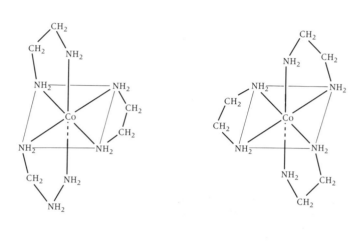

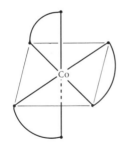 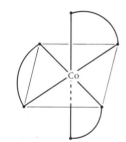

Figure 106.5 *Two isomers of* $[Co(NH_2CH_2CH_2NH_2)_3]^{3+}$, *which are mirror images of each other, are known. They are optical isomers. The top two diagrams show the arrangement of the atoms; the bottom two use curved lines to represent the ligands*

106.1 How would you test a solution of a complex to see if it contained free sulphate ions?

106.2 There are two isomers with the formula $Pt(NH_3)_4Br_2SO_4$. In both isomers the ammonia molecules are firmly bonded to the platinum ion. By measuring the depression of freezing point of water it is found that 1 mol of one isomer has a depression corresponding to the presence of three moles of particles. The other isomer gives two moles of particles. Explain the results and write formulae that

show clearly the make-up of the coordination spheres.

106.3 A crystal of a complex was discovered to have chromium, water and bromine present in the ratio 1 mol:6 mol:3 mol. The solution gave a pale yellow precipitate with silver nitrate solution. On analysis it was found that 10 g of the crystal gave 9.40 g of silver bromide.

(i) What is the mass of 1 mol of the crystal? $M(Cr) = 52$ g mol^{-1} and $M(Br) = 80$ g mol^{-1}.

(ii) How many moles of the crystal were taken?

(iii) The molar mass of silver bromide is 188 g mol^{-1}. How many moles of silver bromide were made in the reaction?

(iv) How many moles of bromide ion are given by 1 mol of crystal?

(v) What does your result tell you about the nature of the complex in the solution?

(vi) Draw a diagram showing the geometry of the complex ion.

106.4 Are there any isomers of $Co(NH_3)_3Cl_3$?

106.5 Are there any isomers of dichloroethane-1,2-diaminenickel(II), $Ni(NH_2CH_2CH_2NH_2)Cl_2$?

106.6 There are *three* isomers of the complex dichlorobis(ethane-1,2-diamine)cobalt(III), which has the formula $Co(NH_2CH_2CH_2NH_2)_2Cl_2$. Two of them are optically active, and one is not. Make models of the three isomers and draw diagrams of them.

(i) Which two are mirror images of each other?

(ii) Why is the third not optically active?

106.7 The formula of a complex ion can sometimes be found using a colorimeter. Here is an example. A solution of nickel(II) sulphate, $NiSO_4$, is green, but if ammonia is added the solution goes blue. Ten tubes each containing 20 cm^3 of 0.05 mol dm^{-3} solution of $NiSO_4$ had 0.4 mol dm^{-3} aqueous ammonia added. The first tube had 2 cm^3 of the ammonia solution added, the second 4 cm^3, and so on. Distilled water was added to bring the total volume in each tube to 50 cm^3. Each tube was then placed in a colorimeter and the reading recorded. The results are shown in Figure 106.6.

(i) What colour filter should be used in the colorimeter?

(ii) Why does the reading on the colorimeter go down?

(iii) Explain why the graph comes to a minimum, and then stays horizontal.

(iv) How many moles of Ni^{2+} ions were there in each tube?

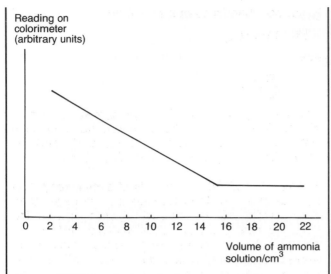

Reading on colorimeter (arbitrary units)

Volume of ammonia solution/cm³

Figure 106.6 *Graph of results for question 106.7*

(v) How many moles of ammonia give the minimum reading?

(vi) What is the formula of the complex?

(vii) Why was the total volume in each tube kept constant?

106.2 Stability constants of complex ions

The colour of a solution of copper(II) sulphate is due to the presence of hydrated Cu^{2+} ions. A standard test for the presence of these ions is to add dilute ammonia solution, which produces a beautiful clear royal blue solution. You should know that the colour change is a result of the production of the tetraamminecopper(II) ion, $Cu(NH_3)_4^{2+}$. This change is an example of an equilibrium reaction. The equilibrium set up is

$$Cu(H_2O)_4^{2+}(aq) + 4NH_3(aq) \rightleftharpoons Cu(NH_3)_4^{2+}(aq) + 4H_2O(l)$$

Given that the reaction is done in water, whose concentration remains constant, we can write the equilibrium constant as

$$K_c = \frac{[Cu(NH_3)_4^{2+}(aq)]}{[Cu(H_2O)_4^{2+}(aq)][NH_3(aq)]^4}$$
$$= 1.2 \times 10^{13} \, dm^{12} \, mol^{-4}$$

This is the value of K_c at 25°C.

If chloride ions are added (e.g. from concentrated hydrochloric acid) to copper(II) ions in solution, a yellow colour is produced. The equilibrium set up is

$$Cu(H_2O)_4^{2+}(aq) + 4Cl^-(aq) \rightleftharpoons CuCl_4^{2-}(aq) + 4H_2O(l)$$
yellow

for which

$$K_c = \frac{[CuCl_4^{2-}(aq)]}{[Cu(H_2O)_4^{2+}(aq)][Cl^-(aq)]^4}$$
$$= 4.2 \times 10^5 \, dm^{12} \, mol^{-4}$$

The two equilibrium constants are called *stability constants*. If you compare the values of two stability constants (Table 106.1) they tell you what should be the outcome of a reaction in which there is a competition between two ligands.

For example, the stability constant of $Cu(NH_3)_4^{2+}$ is greater than that of $CuCl_4^{2-}$. This means that if chloride ions and ammonia molecules are in the same solution, the blue ammine complex will form in preference to the chloride complex.

However, be careful if you actually do some test tube experiments. Side reactions can make things difficult. For example, if you add ammonia solution to copper(II) ions in concentrated hydrochloric acid, the ammonia molecules will be protonated immediately by the hydrogen ions. You would have to add a great deal of ammonia solution before there are free ammonia molecules able to give the complex. Also, note that we are talking about energetic stability here. In some cases it might take a long time for one ligand to displace another. Once again, thermodynamics tells us nothing about rates.

Owing to stability constants often being large, unwieldy numbers, it is very common for them to be quoted as logarithms. For example, you may find the stability constant of $Cu(NH_3)_4^{2+}$ quoted as

$$\lg K_c = \lg(1.2 \times 10^{13}) = 13.08$$

Table 106.1 gives you examples of stability constants.

Table 106.1. Values of stability constants

Ion with water as ligands	Complex*	Lg(stability constant)
Fe^{3+}	$[Fe(H_2O)_5F]^{2-}$	5.3
Fe^{3+}	$[Fe(H_2O)_5(SCN)]^{2-}$	3.0
Co^{2+}	$[Co(NH_3)_6]^{2+}$	4.4
Ni^{2+}	$[Ni(NH_3)_6]^{2+}$	8.6
Ni^{2+}	$[Ni(en)_3]^{2+}$	18.3
Cu^{2+}	$[Cu(NH_3)_4]^{2+}$	13.1
Cu^{2+}	$CuCl_4^{2-}$	5.6
Zn^{2+}	$[Zn(NH_3)_4]^{2+}$	8.7

*The two complexes with iron(III) will also be accompanied by other ions such as $[Fe(H_2O)_4F_2]^-$; en = ethane-1,2-diamine

106.8 The stability constant for the complex $[Cu(NH_2CH_2CH_2NH_2)_2]^{2+}$ is $1.1 \times 10^{20} \, dm^6 \, mol^{-2}$. What will happen, if anything, if ethane-1,2-diamine is added to a solution of $Cu(NH_3)_4^{2+}$?

106.9 Table 106.1 shows you that thiocyanate ions, SCN^-, and fluoride ions, F^-, can act as ligands with iron(III) ions. The thiocyanate complex is blood red and the fluoride complex is colourless. A solution containing iron(III) ions in water is split into two portions. To the first is added a solution of thiocyanate ions, followed after a few minutes by a solution of fluoride ions. To the second the same solutions are added, but in the reverse order. Describe what you would expect to see in these reactions.

106.10 Would you expect ammonia to displace ethane-1,2-diamine molecules from a Ni(en)$_3^{2+}$ complex?

106.11 A complex ion like $\{Co(NH_3)_6\}^{2+}$ can be made by adding ammonia to a solution containing $\{Co(H_2O)_6\}^{2+}$ ions. (Here we use the curly brackets to avoid confusion with the square brackets meaning 'concentration of' in the equations for K.) If only a small amount of ammonia is added at a time, the water molecules can be replaced one at a time. The first complex will be $\{Co(H_2O)_5NH_3\}^{2+}$, the second $\{Co(H_2O)_4(NH_3)_2\}^{2+}$, and so on. For each complex there will be a corresponding equilibrium:

$$\{Co(H_2O)_6\}^{2+}(aq) + NH_3(aq) \rightleftharpoons$$
$$\{Co(H_2O)_5NH_3\}^{2+}(aq) + H_2O(l)$$

$$K_1 = \frac{[\{Co(H_2O)_5NH_3\}^{2+}(aq)]}{[\{Co(H_2O)_6\}^{2+}(aq)][NH_3(aq)]}$$

We can write the $\{Co(H_2O)_6\}^{2+}$ ion as M for short, and the ammonia molecules as A. This is to simplify the equations we are going to write. For example we can put

$$M + A \rightleftharpoons MA$$

and

$$K_1 = \frac{[MA]}{[M][A]}$$

In a similar fashion we can write the production of $\{Co(H_2O)_4(NH_3)_2\}^{2+}$ from $\{Co(H_2O)_5NH_3\}^{2+}$ by the equation

$$MA + A \rightleftharpoons MA_2$$

with

$$K_2 = \frac{[MA_2]}{[MA][A]}$$

(i) Write down the corresponding equations for the production of the other ions, $\{Co(H_2O)_3(NH_3)_3\}^{2+}$, ..., $\{Co(NH_3)_6\}^{2+}$. Call the equilibrium constants K_3, ..., K_6.

(ii) We can represent $\{Co(NH_3)_6\}^{2+}$ being made directly from $\{Co(H_2O)_6\}^{2+}$ by the equation

$$\{Co(H_2O)_6\}^{2+}(aq) + 6NH_3(aq) \rightleftharpoons$$
$$\{Co(NH_3)_6\}^{2+}(aq) + 6H_2O(l)$$

and

$$K_T = \frac{[\{Co(NH_3)_6\}^{2+}(aq)]}{[\{Co(H_2O)_6\}^{2+}(aq)][NH_3(aq)]^6}$$

In our shorthand way these become

$$M + 6A \rightleftharpoons MA_6$$

and

$$K_T = \frac{[MA_6]}{[M][A]^6}$$

Explain why or show that $K_T = K_1 \times K_2 \times K_3 \times K_4 \times K_5 \times K_6$. You have just demonstrated the connection between the *overall stability constant*, K_T, and the *step-wise stability constants*, K_1, K_2, ..., K_6.

106.12 The stability constants of complexes made from bidentate or tridentate ligands are larger than those made with monodentate ligands of a similar type. For example, look at the values for Ni(NH$_3$)$_6^{2+}$ and Ni(en)$_3^{2+}$ in Table 106.1. The problem is to explain this effect, called the *chelate effect*. The bonding in the two complexes both involve the lone pairs on the nitrogen atoms. Therefore we would expect the bond strengths to be similar, and the enthalpy changes of formation of the two complexes to be similar (which they are). Given that stability constants are related to free energy changes through the usual equation $\Delta G^\ominus = -RT \ln K$, and that $\Delta G^\ominus = \Delta H^\ominus - T\Delta S^\ominus$ we must look to the entropy term to explain the differences between the stability constants.

(i) If one monodentate ligand from a solution displaces one water molecule from a complex, would you expect there to be a large entropy change?

(ii) A bidentate ligand can displace two water molecules from a complex. How does the entropy change compare with that in (i)?

You should now have an idea of how the chelate effect is related to entropy changes.

106.3 Complexes and redox potentials

In the previous unit we saw that iron(III) ions could oxidise iodide ions to iodine:

$$2Fe^{3+}(aq) + 2I^-(aq) \rightarrow 2Fe^{2+}(aq) + I_2(aq)$$

This fits with our knowledge of the redox potentials, $E^\ominus_{Fe^{3+}/Fe^{2+}} = +0.77\,V$, $E^\ominus_{I_2/I^-} = +0.54\,V$. That is, because $E^\ominus_{Fe^{3+}/Fe^{2+}}$ is more positive than $E^\ominus_{I_2/I^-}$, we know that Fe^{3+} ions are stronger oxidising agents than iodine molecules. (Or, if you prefer, I$^-$ is a stronger reducing agent than Fe^{2+}.) It is simple to do this reaction in a test tube, but if a solution of the complex hexacyanoferrate(III), Fe(CN)$_6^{3-}$, is added to iodide ions, no iodine is produced. This is in spite of the fact that the iron is in the same oxidation state as before. The redox potential shows that this is expected; $E^\ominus_{Fe(CN)_6^{3-}/Fe(CN)_6^{4-}} = +0.36\,V$. The potential is not more positive than $E^\ominus_{I_2/I^-}$ so the oxidation of iodide ions does not occur. We say that the cyanide ions have stabilised the higher oxidation state of iron, i.e. it is less likely to act as an oxidising agent and revert to iron(II). It is important that you realise that we are talking about energetics here, not kinetics. It is common for a complex of a transition metal ion to have a different redox potential to that of the uncomplexed ion. (In fact the ion will be complexed by water molecules, but we often neglect water molecules as ligands.) When ligands gather round a metal ion, they upset the energies of the orbitals; it is this rearrangement that changes the oxidising power.

There is another effect that can occur, but it is not so much one of energetics as kinetics. When an ion is complexed, the ligands can prevent electrons being transferred so readily between the metal ion and a neighbouring ion or molecule. There can be a considerable increase in the energy barrier, which stops the anticipated reaction taking place. This can happen even though the reaction is energetically still favourable.

You will find further examples where transition metal ions are involved in redox reactions in Unit 70.

Before we leave this unit we shall compare the redox potentials $E^\ominus_{Cr^{3+}/Cr^{2+}} = -0.41\,V$ and $E^\ominus_{Mn^{3+}/Mn^{2+}} = +1.51\,V$. The two figures tell us that Cr^{2+} is a much better reducing agent than Mn^{2+}, i.e. Cr^{2+} ions will give up electrons and change into Cr^{3+} more readily than Mn^{2+} will change into Mn^{3+}. It seems that Mn^{2+} are more energetically stable than Cr^{3+}. It so happens that Mn^{2+} ions are the only ones out of the four that have half-filled sets of 3d orbitals. (The 3d structures are Cr^{3+}, $3d^3$; Cr^{2+}, $3d^4$; Mn^{3+}, $3d^4$; Mn^{2+}, $3d^5$.) We have noted before (section 105.1) that half-filling a set of 3d orbitals appears to be energetically favourable, and so it appears here. The change between Cr^{3+} and Cr^{2+} does not involve disrupting a half-filled set of 3d orbitals, and this correlates with the greater reducing power of Cr^{2+} than Mn^{2+}.

Answers

106.1 The solution should be acidified with a little hydrochloric acid and then barium chloride solution added. A white precipitate should appear with a sulphate.

106.2 The isomers are $[Pt(NH_3)_4Br_2]^{2+}SO_4^{2-}$ and $[Pt(NH_3)_4SO_4]^{2+}(Br^-)_2$. One mole of the first gives 2 mol of particles, $[Pt(NH_3)_4Br_2]^{2+}$ and SO_4^{2-}. One mole of the second gives 3 mol of particles: 1 mol of $[Pt(NH_3)_4SO_4]^{2+}$ and 2 mol of Br^-.

106.3 (i) 400 g.

(ii) 1/40 or 0.025 mol.

(iii) $9.40\,g/188\,g\,mol^{-1} = 0.05$ mol.

(iv) The ratio is 1 mol of crystal to 2 mol of bromide.

(v) If only 2 mol of bromide are available in solution, it means that 1 mol of bromide is attached to the chromium as a ligand.

(vi) The remaining water molecule and two bromides are held in the crystal structure.

106.4 No.

106.5 No.

106.6 (i) These are shown in Figure 106.7.

(ii) The third isomer has both a plane and a centre of symmetry. Therefore it cannot be optically active.

106.7 (i) The complex is blue, so it absorbs in the red; a red filter should be used.

(ii) As more of the complex is made, more light is absorbed. (You may sometimes see graphs with absorbance on the vertical scale; in this case the graph will be the other way up.)

(iii) As ammonia is added, more complex is produced and the amount of light passing through the solution decreases. When all the Ni^{2+} ions have been complexed, adding more ammonia will not change the colour, nor the reading on the colorimeter.

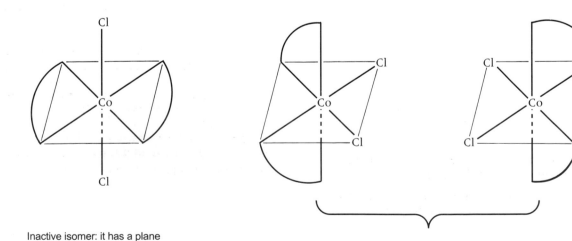

Inactive isomer: it has a plane of symmetry

Mirror images: these are optically active

Figure 106.7 The isomers of $[Co(NH_2CH_2CH_2NH_2)_2Cl_2]^+$. The curved lines stand for $NH_2CH_2CH_2NH_2$

Answers – cont.

(iv) There are

$$\frac{20}{1000} \text{ dm}^3 \times 0.05 \text{ mol dm}^{-3} = 10^{-3} \text{ mol of Ni}^{2+}$$

(v) The minimum comes at 15 cm³ ammonia solution added. This contains

$$\frac{15}{1000} \text{ dm}^3 \times 0.4 \text{ mol dm}^{-3} = 6 \times 10^{-3} \text{ mol of NH}_3$$

(vi) The ratio fits the formula $Ni(NH_3)_6^{2+}$.

(vii) If the volume were not constant, the concentration of the reagents would vary with the tube they were in. The colorimeter reading would change because of the dilution effect of adding more water to each tube, rather than because of the amount of complex made. Note: The shape of the graph can change depending on how an experiment like this is carried out. Sometimes the amount of metal ion solution and ligand solution are both changed. In such cases the graph has a V-shape.

106.8 The ethane-1,2-diamine complex has the higher stability complex, so the ammonia will be displaced to give $[Cu(NH_2CH_2CH_2NH_2)_2]^{2+}$.

106.9 In the first reaction, a blood red colour appears as the thiocyanate ions join with the Fe^{3+} ions. When the fluoride ions are added, the red colour disappears. This is because the complex with the fluoride ions has the larger of the two stability constants. In the second case, the red colour does not appear. The fluoride ions immediately complex with the Fe^{3+} ions. The thiocyanate ions cannot displace the fluoride ions.

106.10 No. The values in Table 106.1 show that $[Ni(en)_3]^{2+}$ is far more stable than $[Ni(NH_3)_6]^{2+}$.

106.11 (i)

$$MA_2 + A \rightleftharpoons MA_3; \quad K_3 = [MA_3]/([MA_2][A])$$
$$MA_3 + A \rightleftharpoons MA_4; \quad K_4 = [MA_4]/([MA_3][A])$$
$$MA_4 + A \rightleftharpoons MA_5; \quad K_5 = [MA_5]/([MA_4][A])$$
$$MA_5 + A \rightleftharpoons MA_6; \quad K_6 = [MA_6]/([MA_5][A])$$

(ii) The simplest way of doing this is to multiply the six K values together. You will find that $[MA]$, $[MA_2]$, ..., $[MA_5]$ cancel out, leaving the correct result for the overall stability constant.

106.12 (i) No.

(ii) It will be larger. The bidentate ligand will lose its ability to wander through the solution, and to rotate more or less randomly. This leads to a decrease in entropy. However, by releasing *two* molecules into the solution, they gain the ability to move and rotate with much less restriction than they had in the complex. This leads to an increase in entropy. The increase is larger than the decrease. It is this increase in entropy that contributes to the larger stability constants of complexes of bidentate and tridentate ligands.

UNIT 106 SUMMARY

- Isomerism:
  An ion or molecule has isomers if it can exist in two or more different forms with the same formula.
- Types of isomerism in complex ions:
  (i) Ionisation isomerism

  e.g. $[Co(NH_3)_5Br]^{2+}SO_4^{2-}$, bromide ion in the coordination sphere (gives test for sulphate, not bromide);

  $[Co(NH_3)_5SO_4]^+Br^-$, sulphate ion in the coordination sphere (gives test for bromide, not sulphate).

  (ii) Hydrate isomerism

  e.g. $[Cr(H_2O)_6]^{3+}(Cl^-)_3$, blue-violet in solution (no chloride ions in hydration sphere);

  $[Cr(H_2O)_5Cl]^{2+}(Cl^-)_2$, green in solution (one chloride ion in hydration sphere);

  $[Cr(H_2O)_4Cl_2]^+Cl^-$, darker green in solution (two chloride ions in hydration sphere).

- Geometrical isomerism:
  Different isomers are possible owing to the different ways the ligands are arranged in space. See the figures in the unit for help.
- Optical isomerism:
  This variety of isomerism occurs in octahedral complexes involving a bidentate ligand such as ethane-1,2-diamine. See Figure 106.5, which shows two isomers that cannot be superimposed upon one another.
- Stability constants:
  (i) A variety of equilibrium constant that gives a guide to the energetic stability of a complex.
  (ii) In a mixture of different ligands and a transition metal ion, the complex with the greater stability constant should be made.
- Redox potentials:
  When a metal forms a complex, its redox potential can change markedly, e.g. $E^{\ominus}_{Fe(CN)_6^{3-}/Fe(CN)_6^{4-}} = +0.36$ V compared to $E^{\ominus}_{Fe^{3+}/Fe^{2+}} = +0.77$ V.

Chromium, manganese and iron

107.1 The chemistry of chromium

In this section you will find a summary of the chemistry of chromium, much of which has appeared in previous units, e.g. those on redox potentials and equilibria. However, some of the reactions may be new to you. The two most important oxidation states of chromium are chromium(VI) and chromium(III).

(a) Chromium(VI) compounds

As you should expect, chromium(VI) compounds show oxidising properties. Especially, dichromate(VI) ions, $Cr_2O_7^{2-}$, are used to oxidise other compounds. The ion reacts best in acid conditions (see below) and the colour change from orange to green is a sure sign of the reaction taking place. The half-equation for the process is

$$Cr_2O_7^{2-}(aq) + 14H^+(aq) + 6e^- \rightarrow 2Cr^{3+}(aq) + 7H_2O(l)$$
orange   green

Acidified sodium dichromate(VI) solution is used in organic chemistry to oxidise alcohols to aldehydes and acids. For example, it will cause the changes

$CH_3CH_2OH \rightarrow CH_3CHO \rightarrow CH_3COOH$
ethanol   ethanal   ethanoic acid

($Na_2Cr_2O_7$ is used rather than $K_2Cr_2O_7$ owing to its greater solubility in the reaction mixture.)

We have said that $Cr_2O_7^{2-}$ ions react best in acid conditions. There are two reasons for this. The first is that the redox potential is highly dependent on the concentration of acid (see section 67.2). However, equally important is the fact that if the solution is not acidified, the number of $Cr_2O_7^{2-}$ ions decreases markedly. This is due to the equilibrium

$$Cr_2O_7^{2-}(aq) + H_2O(l) \rightleftharpoons 2CrO_4^{2-}(aq) + 2H^+(aq)$$
orange   yellow

Especially, if alkali is added nearly all the orange

dichromate(VI) ions are converted into yellow chromate(VI) ions, CrO_4^{2-}.

Chromium(VI) oxide, CrO_3, can be made by reacting sodium dichromate(VI) with concentrated sulphuric acid:

$$Cr_2O_7^{2-}(aq) + 2H^+(aq) \rightarrow 2CrO_3(s) + H_2O(l)$$
red crystals

The oxide is precipitated as red needle-like crystals. The oxide is a powerful oxidising agent, especially with organic chemicals. It is the anhydride of chromic(VI) acid, H_2CrO_4. In laboratories, this acid is used for cleaning glassware when all other methods of cleaning have failed.

There are many salts of chromic(VI) acid. Often they are highly coloured and insoluble in water. Examples are yellow lead(II) chromate(VI), $PbCrO_4$, and red silver chromate(VI), Ag_2CrO_4. The latter is used to indicate the endpoint when chlorides are titrated with silver nitrate.

(b) Chromium(III) compounds

Apart from complexes such as $CrCl_3 \cdot 6H_2O$, which we met in the previous unit, chromium(III) oxide, Cr_2O_3, is the most important compound. It is a bulky green powder, which can be made by heating ammonium dichromate(VI):

$$(NH_4)_2Cr_2O_7(s) \rightarrow Cr_2O_3(s) + 4H_2O(g) + N_2(g)$$

Once the reaction starts, it will keep going at a reasonable rate. Also, a small amount of $(NH_4)_2Cr_2O_7$ gives a much larger volume of Cr_2O_3. Both features of the reaction lead to it being used in indoor fireworks.

107.1 Why should you expect chromium(VI) compounds to show oxidising properties?

107.2 A student dipped two small pieces of filter paper into acidified potassium dichromate(VI) solution. One of them was held in the mouth of a test tube containing sulphur dioxide, and the other piece in a test tube containing hydrogen sulphide. In both cases

---

the paper turned green. The two half-reactions involving the gases are

$$SO_2(aq) + 2H_2O(l) \rightarrow SO_4^{2-}(aq) + 4H^+(aq) + 2e^-$$
$$H_2S(aq) \rightarrow S(s) + 2H^+(aq) + 2e^-$$

(i) Write the equations for the reactions that took place.

(ii) Describe what the student might have seen if she had shaken a *solution* of acidified potassium dichromate(VI) with hydrogen sulphide gas.

107.3 Use the simplest definitions of oxidation and reduction to explain why the changes ethanol → ethanal → ethanoic acid represent oxidations of the organic molecules.

107.4 Briefly explain why adding alkali to the dichromate(VI)/chromate(VI) equilibrium produces a strong yellow colour.

107.5 Explain how you would discover the concentration of a solution of chloride in a silver nitrate titration.

107.6 In alkaline solution, hydrogen peroxide has $E^\ominus = +0.87\,V$. Will hydrogen peroxide oxidise chromium(III) hydroxide, $Cr(OH)_3$, to chromate(VI)? $E^\ominus_{CrO_4^{2-}/Cr(OH)_3} = -0.13\,V$.

107.7 (i) What is the oxidation state of chromium in CrO_2Cl_2.

(ii) Predict the shape of this molecule.

(Hint: assume that only the six outer electrons of chromium are used in bonding.)

107.2 The chemistry of manganese

(a) Manganese(VII)

You are likely to have met the highest oxidation state of manganese in potassium manganate(VII), $KMnO_4$. It is often used as an oxidising agent. Like potassium dichromate(VI), it gives a definite colour change when it reacts. In acidic conditions the change is from purple to (nearly) colourless:

$$MnO_4^-(aq) + 5e^- + 8H^+(aq) \rightarrow Mn^{2+}(aq) + 4H_2O(l);$$
purple   colourless
$$E^\ominus = +1.51\,V$$

In alkali the change is from purple to a clear solution and a black precipitate of manganese(IV) oxide:

$$MnO_4^-(aq) + 3e^- + 4H^+(aq) \rightarrow MnO_2(s) + 2H_2O(l);$$
purple   black
$$E^\ominus = +1.70\,V$$

Acidified potassium manganate(VII) solution can be used to test for sulphur dioxide and hydrogen sulphide. More importantly, it is used in titrations to discover the concentrations of solutions of ethanedioates (oxalates), $C_2O_4^{2-}$. In organic chemistry it is used to oxidise alkenes to diols: for example,

$$C_2H_4 \rightarrow CH_2OHCH_2OH$$
ethene   ethane-1,2-diol

Potassium manganate(VII) can be made in the laboratory by heating manganese(IV) oxide with a strong oxidising agent, usually potassium chlorate(V). The oxide and chlorate(V) are mixed with pellets of potassium hydroxide, and the three heated together. Gradually the green colour of manganate(VI) ions appears:

$$3MnO_2(s) + 6KOH(s) + KClO_3(s) \rightarrow$$
$$3K_2MnO_4(s) + KCl(s) + 3H_2O(l)$$
green

Once the reaction mixture cools, the products are boiled with water, and the green colour is quickly replaced by the purple colour of manganate(VII) ions:

$$3MnO_4^{2-}(aq) + 2H_2O(l) \rightarrow$$
manganate(VI) $MnO_4^-(aq) + MnO_2(s) + 4OH^-(aq)$
green   manganate(VII)
                         purple

After the solid is filtered off, shiny black crystals of $KMnO_4$ crystallise from the solution.

(b) Manganese(IV)

Manganese(IV) oxide is the most common compound showing this oxidation state of manganese. It is a black powder, which can be used as an oxidising agent. For example, a steady supply of chlorine can be made by warming it with concentrated hydrochloric acid:

$$MnO_2(s) + 4HCl(aq) \rightarrow MnCl_2(aq) + 2H_2O(l) + Cl_2(g)$$

It is also used as a catalyst, particularly for releasing oxygen from hydrogen peroxide:

$$2H_2O_2(aq) \xrightarrow{MnO_2\ catalyst} 2H_2O(l) + O_2(g)$$

The standard way of making the oxide is to oxidise manganese(II) hydroxide with a solution of sodium chlorate(I):

$$Mn(OH)_2(s) + OCl^-(aq) \rightarrow MnO_2(s) + H_2O(l) + Cl^-(aq)$$

(c) Manganese(II)

This is the oxidation state of manganese that shows little or no oxidising power. Manganese(II) ions occur as $Mn(H_2O)_6^{2+}$ complexes in water, which give a pale pink colour to the solution. However, you are unlikely to spot this colour under the conditions used in your laboratory. Some manganese(II) compounds are insoluble in water, in particular white manganese(II) hydroxide, $Mn(OH)_2$, and pink manganese(II) sulphide, MnS. The precipitation of the sulphide is used as a test for the presence of Mn^{2+} ions. However, the chloride, sulphate and nitrate are all soluble.

107.8 Which is the stronger oxidising agent, manganate(VII) ions in acid, or alkaline conditions?

107.9 (i) Write down the equation for the reaction between ethanedioate and manganate(VII) ions. Carbon dioxide is given off in the reaction.

(ii) Why is an indicator not needed in a titration between these two chemicals?

107.10 How would you classify the conversion of manganate(VI) ions into manganate(VII) ions and manganese(IV) oxide?

107.11 Write down the equation for the reaction between iron(II) ions, Fe^{2+}, and manganate(VII) ions in acidic solution.

107. 12 What is the connection between potassium manganate(VII) and chlorine gas?

107.13 The solubility product of manganese(II) sulphide is $1.4 \times 10^{-15}\,mol^2\,dm^{-6}$. What is the solubility of MnS in water?

107.14 When $Mn(OH)_2$ is made by adding an alkali to a solution containing Mn^{2+} ions, the precipitate quickly darkens, and eventually goes black. What might be the chemical giving the black colour, and how is it made?

107.3   The chemistry of iron

The key process in the manufacture of iron is the reduction by carbon and carbon monoxide of iron(III) oxide, which occurs in the minerals haematite and limonite. We have discussed the process, and the uses of the metal, in Unit 85.

Here we shall mention the most important reactions of compounds of its two common oxidation states, iron(III) and iron(II).

(a)   Iron(III)

Iron(III) chloride is made by the direct combination of the elements. It is subject to hydrolysis, so the apparatus and chemicals must be kept dry:

$$2Fe(s) + 3Cl_2(g) \rightarrow 2FeCl_3(s)$$

Like aluminium trichloride, it can form dimers.

Iron(III) chloride (or bromide) is made as an intermediate when benzene rings are chlorinated (or brominated) in the presence of iron powder; see section 113.6.

Of all iron(III) compounds, iron(III) oxide is the most common. In its hydrated form it is well known (even to non-chemists) as rust, $Fe_2O_3 \cdot nH_2O$. The way in which rust is formed is complicated; you will find details in section 68.2.

(b)   Iron(II)

Iron(II) sulphate, $FeSO_4$, is the iron(II) compound that you are most likely to meet in the laboratory. It is a useful source of Fe^{2+} ions. However, in order to prevent it being oxidised, the salt must be dissolved in water containing dilute sulphuric acid. Solutions of iron(II) ammonium sulphate, $FeSO_4 \cdot (NH_4)_2SO_4 \cdot 6H_2O$, are used as primary standards in volumetric analysis. The Fe^{2+} ions released by the salt will react with solutions of oxidising agents such as potassium manganate(VII) and potassium dichromate(VI).

Iron(II) chloride cannot be made from iron and chlorine directly; instead dry hydrogen chloride is passed over heated iron:

$$Fe(s) + 2HCl(g) \rightarrow FeCl_2(s) + H_2(g)$$

(c)   Tests for iron(II) and iron(III)

A simple test is to add hydroxide ions to solutions of the ions. Iron(II) hydroxide will appear as a gelatinous pale green precipitate, whereas iron(III) hydroxide has a definite orange-brown colour:

$$Fe^{2+}(aq) + 2OH^-(aq) \rightarrow Fe(OH)_2(s)$$
pale green

$$Fe^{3+}(aq) + 3OH^-(aq) \rightarrow Fe(OH)_3(s)$$
orange-brown

An even simpler method is to add a few drops of potassium thiocyanate solution. Thiocyanate ions give an intense blood red colour with Fe^{3+}, but not with Fe^{2+} ions:

$$Fe^{3+}(aq) + SCN^-(aq) \rightarrow FeSCN^{2+}(aq)$$
blood red

A more involved test, and one that has quite a history, makes use of the two complex ions hexacyanoferrate(II), $Fe(CN)_6^{4-}$, and hexacyanoferrate(III), $Fe(CN)_6^{3-}$. If the former is added to a solution of iron(III) ions, a dark blue precipitate is produced. The precipitate is known as Prussian blue. If the latter is added to a solution of iron(II) ions, another dark blue precipitate is made. This precipitate is called Turnbull's blue:

$$Fe(CN)_6^{4-}(aq) + Fe^{3+}(aq) \rightarrow \text{Prussian blue}$$
hexacyanoferrate(II)

$$Fe(CN)_6^{3-}(aq) + Fe^{2+}(aq) \rightarrow \text{Turnbull's blue}$$
hexacyanoferrate(III)

It is now known that the chemical structures of both precipitates are identical, which, for simplicity, we shall write as $FeFe(CN)_6^-$.

107.15 Briefly describe the apparatus that you might use to make $FeCl_3$.

107.16 Solutions of iron(III) salts are often acidic. Suggest a reason why this is so. You may assume that $Fe(H_2O)_6^{3+}$ ions are present. (Hint: see section 75.5.)

107.17 Iron(III) ions are fairly good oxidising agents. ($E^{\ominus}_{Fe^{3+}/Fe^{2+}} = +0.77\,V$.)

(i) Will they oxidise sulphide ions? ($E^{\ominus}_{S/S^{2-}} = -0.48\,V$.)

(ii) Will they oxidise Sn^{2+} ions? ($E^{\ominus}_{Sn^{4+}/Sn^{2+}} = +0.15\,V$.)

(iii) Will they convert Cu into Cu^{2+} ions? ($E^{\ominus}_{Cu^{2+}/Cu} = +0.34\,V$.)

(iv) Will they change Ce^{3+} into Ce^{4+}? ($E^{\ominus}_{Ce^{4+}/Ce^{3+}} = +1.77\,V$.)

(v) In each case, if you decide there will be a reaction, write the equation.

107.18 $E^{\ominus}_{Fe^{2+}/Fe} = -0.44\,V$. Should iron give hydrogen with acids? Comment on your answer.

107.19 What appears to be the oxidation state of the iron atoms in $FeFe(CN)_6^-$? Explain your answer.

Answers

107.1 Ions in high oxidation states tend to take electrons from another chemical and return to a lower oxidation state. As it loses electrons, the other chemical must have been oxidised.

107.2 (i) $Cr_2O_7^{2-}(aq) + 3SO_2(aq) + 2H^+(aq) \rightarrow$
$$2Cr^{3+}(aq) + 3SO_4^{2-}(aq) + H_2O(l)$$
$Cr_2O_7^{2-}(aq) + 3H_2S(aq) + 8H^+(aq) \rightarrow$
$$2Cr^{3+}(aq) + 3S(s) + 7H_2O(l)$$

(ii) She would have seen the green colour together with a yellow precipitate of sulphur.

107.3 Ethanol to ethanal involves the loss of hydrogen atoms, and ethanol to ethanoic acid the gain of oxygen; these are both features of oxidation.

107.4 Adding alkali will remove hydrogen ions from the solution. As Le Chatelier's principle says, the equilibrium will move to the right, i.e. in the direction that tends to replace the hydrogen ions.

107.5 See section 40.4, which describes this type of titration.

107.6 Yes. Hydrogen peroxide has an $E^{\ominus}$ considerably more positive than $E^{\ominus}_{CrO_4^{2-}/Cr(OH)_3}$, so the peroxide will act as the oxidising agent. This reaction can be used to make chromate(VI) in the laboratory.

107.7 (i) $Ox(Cr) + 2 \times Ox(O) + 2 \times Ox(Cl) = 0$ and so $Ox(Cr) - 4 - 2 = 0$, which gives $Ox(Cr) = +6$.

(ii) Tetrahedral.

107.8 The alkaline solution: it has the more positive redox potential.

107.9 (i) $2MnO_4^-(aq) + 5C_2O_4^{2-}(aq) + 16H^+(aq) \rightarrow$
$$2Mn^{2+}(aq) + 10CO_2(g) + 8H_2O(l)$$

(ii) The manganate(VII) solution changes colour, from purple to colourless, while there are ethanedioate ions with which it can react. As soon as the latter are used up, the solution stays purple; hence the solution acts as its own indicator.

107.10 It is a disproportionation (see section 71.3).

107.11 $MnO_4^-(aq) + 5Fe^{2+}(aq) + 8H^+(aq) \rightarrow$
$$Mn^{2+}(aq) + 5Fe^{3+}(aq) + 4H_2O(l)$$

107.12 Chlorine can be made by dropping concentrated hydrochloric acid onto crystals of potassium manganate(VII).

107.13 $MnS(s) \rightarrow Mn^{2+}(aq) + S^{2-}(aq)$

The equation tells us that $[Mn^{2+}(aq)] = [S^{2-}(aq)]$, so the formula for the solubility product, $K_{sp} = [Mn^{2+}(aq)] \times [S^{2-}(aq)]$ becomes $K_{sp} = [Mn^{2+}(aq)]^2$. Therefore, $[Mn^{2+}(aq)]^2 = 1.4 \times 10^{-15}\,mol^2\,dm^{-6}$ and $[Mn^{2+}(aq)] = 3.7 \times 10^{-8}\,mol\,dm^{-3}$. To obtain 3.7×10^{-8} mol of Mn^{2+} ions in $1\,dm^3$, an equal number of moles of MnS must have dissolved, i.e. the solubility is $3.7 \times 10^{-8}\,mol\,dm^{-3}$.

107.14 The black colour is due to manganese(IV) oxide, MnO_2. It is made by the $Mn(OH)_2$ being oxidised by oxygen in the air.

107.15 An apparatus similar to the one used to make $AlCl_3$ can be used; see Figure 40.1.

107.16 The complex ion can act as a proton donor:
$$Fe(H_2O)_6^{3+}(aq) + H_2O(l) \rightarrow$$
$$[Fe(H_2O)_5OH]^{3+}(aq) + H_3O^+(l)$$

107.17 (i), (ii) and (iii) Yes, because $E^{\ominus}_{Fe^{3+}/Fe^{2+}}$ is the more positive in each case.

(iv) No; rather, Ce^{4+} oxidises Fe^{2+}. This is the basis of the redox titration described in section 70.4.

(v) The reactions are:
$$S^{2-}(aq) + 2Fe^{3+}(aq) \rightarrow S(s) + 2Fe^{2+}(aq)$$
$$Sn^{2+}(aq) + 2Fe^{3+}(aq) \rightarrow Sn^{4+}(aq) + 2Fe^{2+}(aq)$$
$$Cu(s) + 2Fe^{3+}(aq) \rightarrow Cu^{2+}(aq) + 2Fe^{2+}(aq)$$

The last reaction is used to etch copper-clad electronics circuit boards.

107.18 Yes. However, concentrated nitric acid renders iron passive. Even so, the layer that causes the reaction to stop is easily lost, so if the mixture is shaken the reaction continues.

107.19 Each cyanide carries a charge of -1. The two iron atoms take up five negative charges, so their average oxidation number is $+2.5$. In practice, there is a mixture of Fe^{2+} and Fe^{3+} ions.

Chromium

- Chromium(VI) compounds are oxidising agents, e.g. dichromate(VI) ions, $Cr_2O_7{}^{2-}$, in acid solution:

$$Cr_2O_7{}^{2-}(aq) + 14H^+(aq) + 6e^- \rightarrow$$
orange

$$2Cr^{3+}(aq) + 7H_2O(l)$$
green

- Chromium(VI) oxide, CrO_3, made by reacting sodium dichromate(VI) with concentrated sulphuric acid:

$$Cr_2O_7{}^{2-}(aq) + 2H^+(aq) \rightarrow 2CrO_3(s) + H_2O(l)$$

Manganese

- Potassium manganate(VII), $KMnO_4$, is a common oxidising agent.
  - (i) In acid,

$$MnO_4{}^-(aq) + 5e^- + 8H^+(aq) \rightarrow$$
purple

$$Mn^{2+}(aq) + 4H_2O(l); \; E^\ominus = +1.51\,V$$
colourless

  - (ii) In alkali,

$$MnO_4{}^-(aq) + 3e^- + 4H^+(aq) \rightarrow$$
purple

$$MnO_2(s) + 2H_2O(l); \; E^\ominus = +1.70\,V$$
black

- Potassium manganate(VII) is made by first heating manganese(IV) oxide with potassium chlorate(V) and potassium hydroxide pellets, then boiling with water.
- Manganese(IV) oxide, MnO_2, is:
  - (i) Used as a catalyst with hydrogen peroxide to make oxygen.
  - (ii) Warmed with concentrated hydrochloric acid to make chlorine.
- Manganese(II) ions can be identified by the salmon pink precipitate (MnS) made with sulphide ions.

Iron

- Iron is magnetic.
- It is extracted in the blast furnace from minerals containing iron(III) oxide, Fe_2O_3, and is the major component of steel.
- Iron(III) chloride, $FeCl_3$, is made by heating iron with dry chlorine; the chloride is easily hydrolysed.
- Hydrated iron(III) oxide is better known as rust.
- Iron(II) chloride, $FeCl_2$, is made by warming iron with dry hydrogen chloride. It too is easily hydrolysed.
- Iron(II) ammonium sulphate, $FeSO_4(NH_4)_2SO_4 \cdot 6H_2O$, is used as a primary standard in volumetric analysis.
- Fe^{2+} ions give a gelatinous pale green precipitate (of $Fe(OH)_2$), and Fe^{3+} a gelatinous brown precipitate (of $Fe(OH)_3$), with hydroxide ions.
- Fe^{3+} ions (not Fe^{2+}) give a blood red colour with thiocyanate ions, SCN^-.

108

Group IIB

108.1 The nature of the elements

The elements of Group IIB are zinc, cadmium and mercury. Their properties and uses are shown in Tables 108.1 and 108.2. They follow the transition metals, and as such they have a complete set of ten d electrons as well as a pair of outer s electrons. Mercury also has a set of fourteen f electrons. The differences between the second and third ionisation energies of these metals and the transition metals that precede them are not markedly different. However, unlike the transition elements, none of them show oxidation states higher than +2. Mercury is unusual in that it shows two oxidation states, mercury(I) and mercury(II), but the lower state is found in the dimercury(I) ion, Hg_2^{2+}, in which two mercury atoms are joined by a metal–metal bond.

Table 108.1. Physical properties of the elements

Symbol	Zinc Zn	Cadmium Cd	Mercury Hg
Electron structure	$(Ar)3d^{10}4s^2$	$(Kr)4d^{10}5s^2$	$(Xe)4f^{14}5d^{10}6s^2$
Electro-negativity	1.6	1.7	1.9
1st I.E. /kJ mol^{-1}	910	870	1010
2nd I.E. /kJ mol^{-1}	1700	1600	1800
3rd I.E. /kJ mol^{-1}	3800	3600	3300
Melting point/° C	423	321	−39
Boiling point/° C	908	765	357
Atomic radius/pm	131	148	148
Ionic radius/pm	74	97	110
Principal oxid. no.	+2	+2	+1,+2
$E^{\ominus}_{M^{2+}/M}$/V	−0.76	−0.40	−0.79*

*The $E^{\ominus}$ of mercury is for $Hg_2^{2+}(aq)+2e^-\rightarrow 2Hg(l)$

Table 108.2. Uses of the elements

Element	Main uses
Zinc	In alloys such as brass, e.g. 70% Cu, 30% Zn For galvanising iron (see section 68.3) In batteries
Cadmium	In cadmium sulphide photocells It can also be used for protecting metals in much the same way as zinc
Mercury	In thermometers In amalgams for dental fillings In various agricultural chemicals and pharmaceuticals

(Mercury(I) and mercury(II) compounds were once labelled mercurous and mercuric respectively.)

They also show the ability to give complex ions. For example, $Zn(CN)_4^{2-}$, $Cd(CN)_4^{2-}$, $Hg(CN)_4^{2-}$. You will find examples of other complexes later on.

The melting and boiling points of the metals are much lower than their transition metal neighbours. The melting point of mercury is so low that it is the only liquid metal at room temperature. The ease with which they vaporise is made use of in the methods used for their production. The chief ores of the metals are their sulphides. Zinc blende, ZnS, is roasted with air and changes into its oxide. In turn the oxide is reduced using coke (as a cheap variety of carbon).

The method for extracting mercury is similar to that of zinc, except that it is not necessary to heat mercury(II) oxide with carbon: it decomposes into mercury and oxygen at 300°C. The reactions are

$$2HgS(s) + 3O_2(g) \rightarrow 2HgO(s) + 2SO_2(g)$$
$$2HgO(s) \rightarrow 2Hg(l) + O_2(g)$$

Overall the reaction is

$$HgS(s) + O_2(g) \rightarrow Hg(l) + SO_2(g)$$

Mercury and its compounds are dangerous. The vapour pressure of mercury is high enough for the liquid to be a health hazard if it is spilled. Like many of

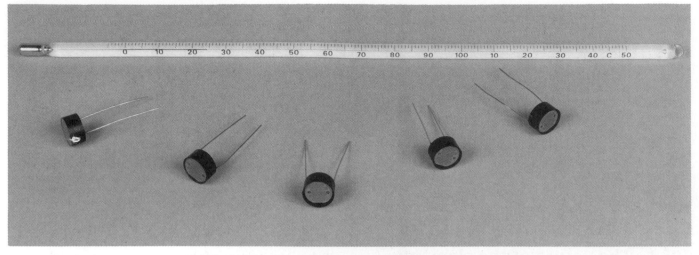

Two uses of the B metals cadmium (in light dependent resistors) and mercury (in thermometers).

the 'heavy metals' (which includes cadmium), mercury causes brain damage. It is far more dangerous as a vapour, when it can be taken in through the lungs, than it is as a liquid. One of the problems with mercury is that often the poisoning is discovered only after symptoms appear. At this stage, antidotes will not work.

If you come across liquid mercury, treat it with caution. One way of removing the danger it presents is to cover the droplets with powdered sulphur. Even at room temperature, the two elements will combine to give mercury(II) sulphide, HgS.

108.1 Why is it that, unlike the transition metals, the complexes of Zn, Cd and Hg are normally colourless. (Hint: electron structures.)

108.2 Zinc blende is often contaminated by significant proportions of cadmium. In the course of extracting zinc, cadmium can also be isolated. A difference in one of the physical properties of the elements makes the separation possible? What is the property?

108.3 This method of extracting metals is one that we discussed in Unit 85 from the point of view of thermodynamics. What are the free energy changes of the reactions:

(i) $2ZnS(s) + 3O_2(g) \rightarrow 2ZnO(s) + 2SO_2(g)$

(ii) $ZnO(s) + C(s) \rightarrow Zn(s) + CO(g)$

The standard free energies of formation of ZnS, ZnO, SO_2 and CO are -205.4, -318.2, -300.4 and -137.3, all in units of kJ mol^{-1}.
Comment on your answers.

108.4 $E^{\ominus}_{Zn^{2+}/Zn} = -0.76\,V$, $E^{\ominus}_{Cd^{2+}/Cd} = -0.40\,V$ and $E^{\ominus}_{Hg^{2+}/Hg} = +0.85\,V$. Which, if any, of the metals will give hydrogen with acids?

108.2 The oxides and hydroxides

Zinc oxide will dissolve in both acid and alkali:

$$ZnO(s) + 2H^+(aq) \rightarrow Zn^{2+}(aq) + H_2O(l)$$
$$ZnO(s) + 2OH^-(aq) + H_2O(l) \rightarrow Zn(OH)_4^{2-}(aq)$$

This shows us the amphoteric nature of zinc. Cadmium oxide and mercury(II) oxide are essentially basic, which illustrates the usual tendency of metallic nature increasing down a group.

Zinc oxide is white when it is cold, a property that has given it a use as a pigment in paints. However, it changes colour when hot to a pale yellow. This is due to changes in the structure of the lattice.

You can make zinc hydroxide, $Zn(OH)_2$, by adding dilute alkali to a solution of a zinc(II) salt. It appears as a gelatinous white precipitate:

$$Zn^{2+}(aq) + 2OH^-(aq) \rightarrow Zn(OH)_2(s)$$

The precipitate will dissolve in excess alkali to give soluble $Zn(OH)_4^{2-}(aq)$ ions.

Zinc hydroxide is also precipitated with aqueous ammonia, and it will dissolve in excess ammonia. However, this time the complex ion $Zn(NH_3)_4^{2+}$ is made. This reaction allows us to distinguish a solution containing zinc ions from one containing aluminium ions (see Table 108.3).

Table 108.3. How to tell apart solutions of zinc and aluminium ions

Ion	With sodium hydroxide	With aqueous ammonia
Zn^{2+}	White precipitate; dissolves in excess	White precipitate; dissolves in excess
Al^{3+}	White precipitate; dissolves in excess	White precipitate; does not dissolve*

*Actually it dissolves to a very slight extent, but unless you added a vast amount of ammonia you would see no change in the precipitate

The reactions of mercury(I) and mercury(II) ions with alkali are markedly different to those of zinc and cadmium. A mercury(II) solution eventually gives a yellow precipitate of mercury(II) oxide:

$$Hg^{2+}(aq) + 2OH^-(aq) \rightarrow HgO(s) + H_2O(l)$$
$$\text{yellow}$$

(The difference between the yellow and red varieties of HgO is simply one of the size of the particles.)

Dimercury(I) ions give a black precipitate of mercury(I) oxide:

$$Hg_2^{2+}(aq) + 2OH^-(aq) \rightarrow Hg_2O(s) + H_2O(l)$$
$$\text{black}$$

108.5 Predict the shape of the $Zn(NH_3)_4^{2+}$ ion. Suggest which orbitals might be used in making the bonds.

108.6 A student was given a solution of cadmium sulphate, $CdSO_4$, and split it into two parts. To one part he added dilute sodium hydroxide, to the other aqueous ammonia. What do you think he saw?

108.3   The halides

The halides of zinc and cadmium are mainly ionic, while those of mercury are covalent. The mercury halides are insoluble in water, whereas zinc fluoride is the only zinc halide that is insoluble. Mercury(II) chloride is sufficiently covalent to have a crystal structure in which individual molecules can be identified. The only halide that is of significant use in the laboratory is dimercury(I) chloride, Hg_2Cl_2. This is because it is one of the ingredients of the calomel cell. It consists of mercury in contact with dimercury(I) chloride and has a stable e.m.f. of $E^{\ominus} = +0.789\,V$ at $25\,°C$. We said a little about this cell in section 66.6. Look there for more information if necessary. The cell reaction is

$$Hg_2Cl_2(s) + 2e^- \rightleftharpoons 2Hg(l) + 2Cl^-(aq)$$

108.7 Zinc(II) chloride is extremely deliquescent. Briefly explain how you would attempt to make this chloride in the laboratory.

108.8 What might be the reason for the insolubility of zinc(II) fluoride?

108.9 Dimercury(I) iodide, Hg_2I_2, is a greenish colour and is precipitated if iodide ions are added to a solution of dimercury(I) sulphate. Likewise, the red mercury(II) iodide, HgI_2, is precipitated from a solution of mercury(II) sulphate. However, both precipitates dissolve in excess iodide solution. What might be the reason for this?

108.4   The sulphates, nitrates and carbonates

(a) Sulphates

All the sulphates are soluble in water and appear as hydrated crystals, e.g. $ZnSO_4 \cdot 7H_2O$, $HgSO_4 \cdot 6H_2O$. They are useful as sources of the metal ions in solution. However, when they dissolve in water, the solution becomes acidic. This is a result of hydrolysis. For example,

$$[Zn(H_2O)_4]^{2+}(aq) \rightarrow [Zn(H_2O)_2OH]^{2+}(aq) + H^+(aq)$$

(b) Nitrates

The nitrates of zinc and cadmium decompose in the normal way for heavy metal nitrates, i.e. they give off nitrogen dioxide and oxygen, with the metal oxide being left behind. In the case of mercury(II) nitrate, the metal is left:

$$Hg(NO_3)_2(s) \rightarrow Hg(l) + 2NO_2(g) + O_2(g)$$

(c) Carbonates

Zinc carbonate is the only carbonate of much interest. It occurs as a mineral calamine and at one time was widely used in calamine lotion, a pink suspension of the carbonate, for treating spots and other skin ailments. Zinc carbonate is insoluble in water, but shows the usual properties of carbonates.

108.10 Which other metal ion (not from Group IIB) behaves like $[Zn(H_2O)_4]^{2+}$ in giving an acidic solution?

108.11 Why is mercury(II) nitrate a slight exception to the normal pattern of decomposition by heat?

108.12 What are the 'usual properties' of carbonates mentioned above?

108.13 If a solution of a hydrogencarbonate is added to a solution containing Zn^{2+} ions, why is carbon dioxide given off? (Zinc carbonate is also precipitated.)

108.5   The sulphides

Each of the sulphides(ZnS, CdS, HgS) is highly insoluble in water. (There is no mercury(I) sulphide.) They also have distinctive colours (Table 108.4). We can make use of both features in qualitative analysis. If hydrogen sulphide, or a solution of the gas in water, is added to a solution containing the ions, the sulphides will precipitate under different conditions.

Both zinc sulphide and mercury(II) sulphide can exist in two crystal structures. For ZnS the structures are

Table 108.4. The sulphides of zinc, cadmium and mercury

Formula	Colour	Solubility product /mol² dm⁻⁶
ZnS	Cream	1.6×10^{-24}
CdS	Yellow	8.0×10^{-27}
HgS	Black	1.6×10^{-52}

called by the names of the minerals in which the sulphide is found: zinc blende and wurtzite. We discussed the structures in Unit 32.

Mercury(II) sulphide has red and black forms. Although the black variety is precipitated in solution, the red type (cinnabar) is the more energetically stable form. If you find a bottle of HgS in your laboratory, the contents will be red in colour.

The sulphides are somewhat unusual in that they can be made by directly combining the metal and sulphur.

The reaction between zinc and sulphur can be violent, while that between mercury and sulphur is mild. As we noted earlier, this reaction is useful for clearing mercury spillages.

108.14 Use Table 108.4 to answer these questions:

(i) Which sulphide is the most insoluble?

(ii) ZnS or CdS can be precipitated from a solution containing a mixture of Zn^{2+} and Cd^{2+} by controlling the pH of the solution. Explain what you would do to separate the zinc from the cadmium. (Hint: look back at Unit 64.)

108.15 $E^{\ominus}_{Hg^{2+}/Hg_2^{2+}} = 0.91\,V$, $E^{\ominus}_{Hg_2^{2+}/Hg} = 0.79\,V$ and $E^{\ominus}_{Sn^{4+}/Sn^{2+}} = 0.15\,V$. What, if anything, would you expect to happen if a solution containing (i) Hg^{2+} ions, (ii) Hg_2^{2+} ions were individually mixed with a solution of tin(II) chloride?

Answers

108.1 In the transition metals the colours are caused by electrons moving between d orbitals. For the IIB metals their d orbitals are full, so the transitions cannot take place.

108.2 Their boiling points.

108.3 (i) $\Delta G^{\ominus}$(reaction)
$= 2\Delta G^{\ominus}_f(ZnO) + 2\Delta G^{\ominus}_f(SO_2) - 2\Delta G^{\ominus}_f(ZnS)$
$= -826.4\,kJ\,mol^{-1}$

(ii) $\Delta G^{\ominus}$ (reaction) $= \Delta G^{\ominus}_f(CO) - \Delta G^{\ominus}_f(ZnO)$
$= +180.9\,kJ\,mol^{-1}$

The first reaction is spontaneous even at 25° C, but the second is not. The second reaction only becomes spontaneous at high temperatures (above 1000° C). However, the reaction also takes place by carbon monoxide reducing the oxide.

108.4 Owing to their negative $E^{\ominus}$ values, zinc and cadmium will give hydrogen, but mercury will not. However, with oxidising acids other reactions can take place.

108.5 Tetrahedral. Use is made of the outer s and p orbitals.

108.6 You are expected to predict that cadmium behaves like zinc; which it does. With both sodium hydroxide and ammonia solution, a precipitate is produced first, which then dissolves.

108.7 Pass dry hydrogen chloride over hot zinc. See Figure 40.1 for a suitable apparatus.

108.8 Its high lattice energy.

108.9 A complex ion is produced. Its formula is HgI_4^{2-}. Notice that this is a mercury(II) complex. Dimercury(I) complexes do not occur. If a complex can be made in a reaction, it is common for dimercury(I) to change its

oxidation state, giving the mercury(II) complex. In fact, disproportionation reactions occur; e.g.

$Hg_2^{2+}(aq) + 4I^-(aq) \rightarrow HgI_4^{2-}(aq) + Hg(l)$

108.10 Aluminium.

108.11 Because HgO is easily decomposed by heat, the product of the reaction is liquid mercury rather than the oxide.

108.12 They are decomposed by heat or acid, giving off carbon dioxide.

108.13 We said that solutions containing Zn^{2+} ions are slightly acidic. Hence with the hydrogencarbonate, carbon dioxide is given off.

108.14 (i) Mercury(II) sulphide. (It has the smallest solubility product.)

(ii) The solubility product of ZnS is larger than that of CdS. If the acid is present, the concentration of free sulphide ions is reduced (see section 64.5), and $[S^{2-}(aq)]$ becomes too low for ZnS to be precipitated. However, CdS will be precipitated. Thus, add a little acid to the solution before adding the sulphide solution. Filter off the precipitate of CdS. Then add a little alkali to the filtrate to neutralise the acid. Now ZnS will precipitate.

108.15 The $E^{\ominus}$ values for the mercury ions are more positive than $E^{\ominus}_{Sn^{4+}/Sn^{2+}}$, so both Hg^{2+} and Hg_2^{2+} are able to oxidise Sn^{2+} to Sn^{4+}. Alternatively we can say that $E^{\ominus}_{Sn^{4+}/Sn^{2+}}$ is more negative than the other $E^{\ominus}$ values, so the Sn^{2+} ions can reduce the mercury ions. Notice that Hg^{2+} can be reduced to Hg_2^{2+}; also Hg_2^{2+} can be reduced to Hg. So mercury is the final product in both reactions. The reactions are

$Hg_2^{2+}(aq) + Sn^{2+}(aq) \rightarrow 2Hg(l) + Sn^{4+}(aq)$
$Hg^{2+}(aq) + Sn^{2+}(aq) \rightarrow Hg(l) + Sn^{4+}(aq)$

UNIT 108 SUMMARY

- The B metals:
    - (i) Have a complete set of ten d electrons as well as a pair of outer s electrons.
    - (ii) Tend not to use their outer pair of s electrons in bonding, hence the name 'inert pair'.
    - (iii) Can form complex ions.
    - (iv) Have much lower melting and boiling points than transition metals.
    - (v) Mercury shows two oxidation states, mercury(I) and mercury(II).
- Extraction:
    All are extracted from their sulphides; first by roasting in air to give the oxide, then by reduction with carbon to the metal. Mercury(II) oxide decomposes directly to mercury and oxygen.

Compounds

- Oxides:
    Zinc oxide is amphoteric;

    $ZnO(s) + 2H^+(aq) \rightarrow Zn^{2+}(aq) + H_2O(l)$
    $ZnO(s) + 2OH^-(aq) + H_2O(l) \rightarrow Zn(OH)_4^{2-}(aq)$

    Cadmium and mercury oxides are basic.

- Halides:
    Dimercury(I) chloride, Hg_2Cl_2, is used in calomel cells.
- Sulphates:
    All are soluble in water and make hydrated crystals, e.g. $ZnSO_4\cdot7H_2O$.
- Nitrates:
    All give off NO_2 and O_2. Zinc and cadmium leave their oxides, mercury(II) nitrate leaves liquid mercury.
- Sulphides:
    All are insoluble, and allow identification of the metal ions: ZnS cream, CdS yellow, HgS black.
- Zinc ions:
    Zn^{2+} ions give a white gelatinous precipitate (of $Zn(OH)_2$) with hydroxide ions. The precipitate is soluble in both excess alkali and aqueous ammonia. $Al(OH)_3$ is similar in appearance to $Zn(OH)_2$, but is not soluble in aqueous ammonia.

ORGANIC
CHEMISTRY

109

Organic chemistry

109.1 What is organic chemistry?

The common feature of organic chemicals is that they all contain the element carbon. We do not usually count small molecules like carbon dioxide or carbon monoxide as organic chemicals, but this is just a matter of convenience, not a strict rule. Living things are largely made of organic chemicals, as are a vast number of other substances that we use or see around us everyday. The range of properties and appearance of organic chemicals is bewilderingly large. Table 109.1 lists just a few examples. You can see that they contain carbon combined with one or more other elements, especially hydrogen, oxygen and nitrogen. Notice also that in many of the molecules the carbon atoms are joined together. The ability of atoms to join together is called *catenation*. Other elements can catenate, e.g. sulphur, silicon, boron and some metals; but no other element is able to make as wide a variety of chains as carbon.

109.1 One of the early ideas about chemicals was that they split into two types. *Organic* chemicals were thought to be found only in living things. *Inorganic* chemicals were those that were found with non-living things. For a long time it was widely believed that organic chemicals were different to inorganic chemicals because they had a special 'life force' within them.

As with many forces, the life force was invisible, and hard to detect. How would you show that the life force is not the reason why organic chemicals are different to inorganic chemicals?

109.2 What are organic chemicals like?

The overwhelming majority of them are covalent. This has several consequences. In the first place many have fairly low melting and boiling points. Examples are gases like methane and propane, or liquids like ethanol and octane. However, some are solids at room temperature, and have high melting points. This can happen for a number of reasons. In plastics it is because the molecules are extremely long, heavy polymers. In other cases it is a result of some ionic bonding being present, e.g. amino acids such as glycine.

The second effect of covalency is that organic reactions tend to be slow. In a typical inorganic reaction, ions are present. Reactions between ions are extremely fast, taking a fraction of a second to complete. Where there are no ions, covalent bonds have to be broken. This takes a fair amount of energy, and quite a lot of time. For example, the preparation of an organic chemical may take several hours of continuous heating. It is very common for a number of different reactions to take place at the same time. For this reason an important part of an organic reaction is the separation of the main product from the less important, unwanted, products.

109.3 The main types of organic chemical

We can think of a typical organic molecule as having carbon atoms joined together in a chain. Most of the carbon atoms will also have hydrogen atoms attached. The chain of carbon atoms makes up the *backbone* of the molecule. The *hydrocarbons* have a carbon backbone with only hydrogen atoms joined to it, but in other types of molecule somewhere along the chain there will be an atom, or group of atoms, that gives rise to the major properties of the molecule. For example, in the alcohols there is a backbone with an OH group attached (Figure 109.1). Among other things the presence of this group is responsible for many alcohols being completely miscible with water, so allowing people to become intoxicated (to a greater or lesser extent) by drinking ethanol in beer, wine, whisky, etc. A group like OH is called the *functional group* of the molecule.

Molecules with the same functional group can be classified together into a *homologous series* (Table 109.2). Owing to them having the same functional group, the

Table 109.1. Examples of organic chemicals

Substance	Diagram	Use
Ethanoic acid	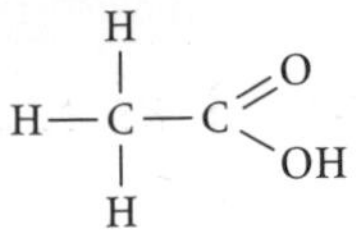	Vinegar
Ethanol	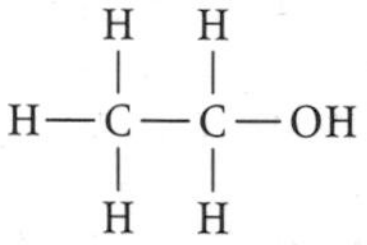	Alcoholic drinks
Glucose	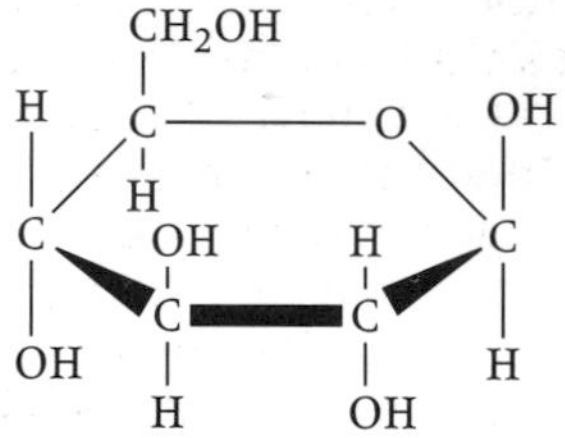	A sugar
Glycine	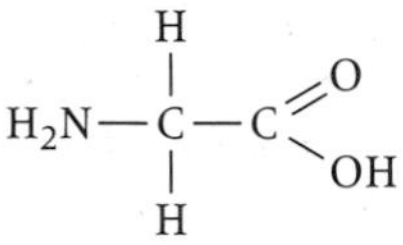	An amino acid, found in proteins
Nylon-6,6	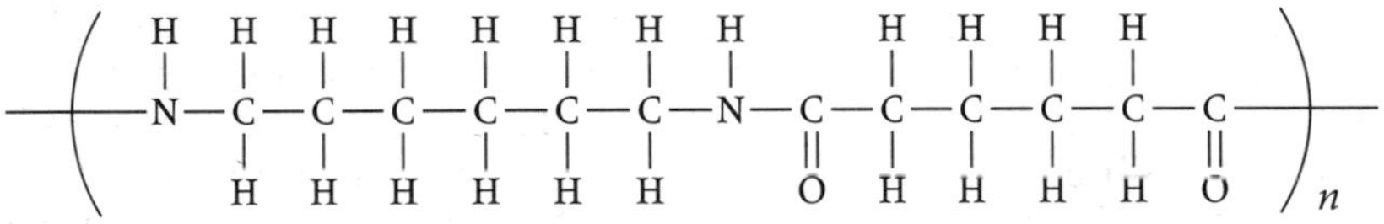	Clothing, ropes
Octane	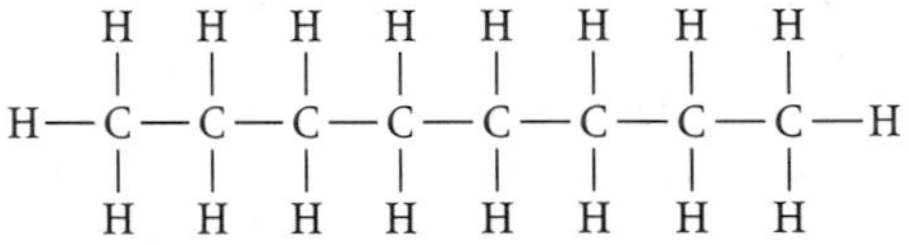	Petrol
Poly(ethene)	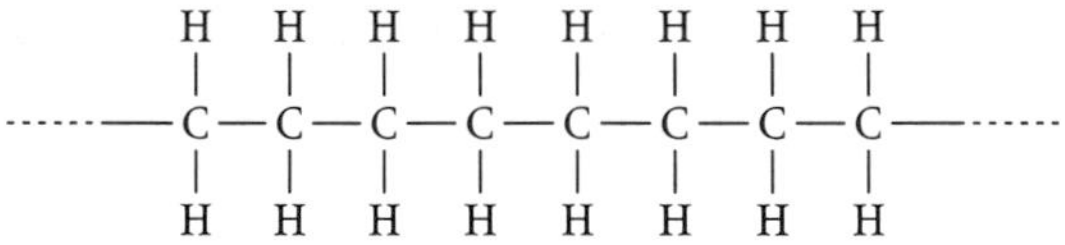	Polythene packaging
Progesterone	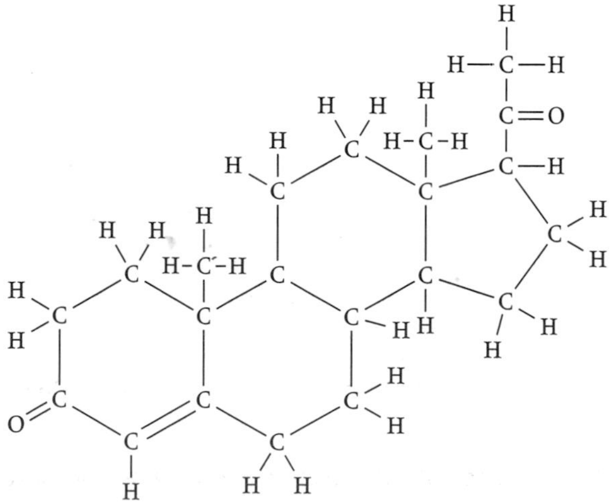	Sex hormone, building block of DNA

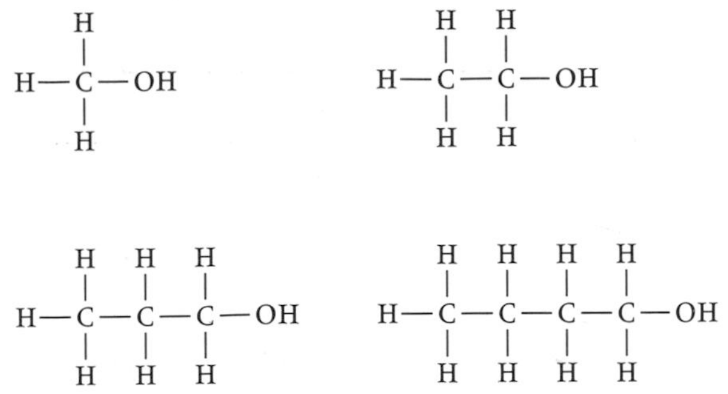

Figure 109.1 *All organic compounds have a 'carbon backbone', i.e. a series of carbon atoms joined together. Here there are four alcohols with, respectively, carbon backbones of one, two, three and four atoms. The OH in each molecule is the functional group of alcohols*

Table 109.2. Homologous series

The members of a homologous series have
  the same functional group,
  similar chemical properties,
  the same general formula,
  gradually changing physical properties

chemical properties of the various members of a homologous series are very similar. They also have the same general formula. For example, the alkanes, methane, CH_4, ethane, C_2H_6, and propane, C_3H_8, all fit the general formula C_nH_{2n+2} with $n = 1$, 2 and 3. The members of a homologous series do differ in their physical properties such as melting and boiling points.

There are millions of different organic chemicals. If we had to study each of them as a completely separate type of substance, life for an organic chemist would be almost intolerable. Fortunately the presence of functional groups allows us to classify the millions of individual molecules into a smaller number of homologous series. You will be pleased to discover that you are

expected to know something of only a small fraction of the total. Even so, you will be able to predict the properties of many thousands of compounds by studying just one or two in each series.

109.4 The tetrahedral arrangement around carbon atoms

If you look carefully at the diagrams in Table 109.1 you will find that the carbon atoms always have four bonds. This fact, together with our knowledge of the bonding habits of other atoms, can help us to establish the arrangement of the atoms in a molecule. For example, suppose we find that a compound has the formula CH_4O. At first sight it appears impossible to say how the atoms might be arranged. However, we can start by writing down the carbon atom and putting four bonds around it. Oxygen always makes two bonds, so we could try joining the oxygen directly to the carbon atom by a double bond. (This is shown in Figure 109.2.) If we do this, there are only two bonds left but four hydrogen atoms to fit. This arrangement will not work. Instead, we can try putting a single bond between the carbon and oxygen atoms. If we do this there is one bond left on the oxygen and three around the carbon. Therefore we have room to fit the four hydrogen atoms. The resulting molecule is the alcohol, methanol, CH_3OH.

In this example, and in most other areas of organic chemistry, you will find it a great help to make models of the molecules that you see in only two dimensions on paper. Do make models whenever possible. You will find it helps you to visualise the shapes much more easily, and to appreciate how the shape of a molecule can influence the way it reacts.

The feature that dominates the shapes of organic molecules is the tetrahedral arrangement around a carbon atom, which has four single bonds. (You will find an explanation of why carbon adopts this geometry in Unit 17.) We often draw the carbon backbone in a

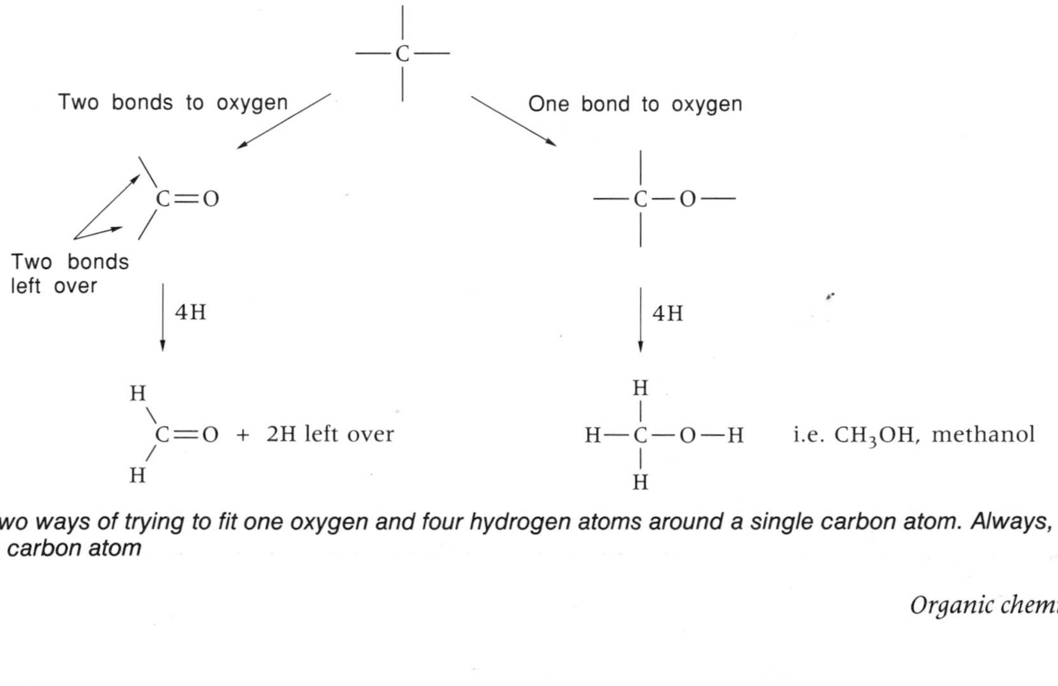

Figure 109.2 *Two ways of trying to fit one oxygen and four hydrogen atoms around a single carbon atom. Always, there must be four bonds to a carbon atom*

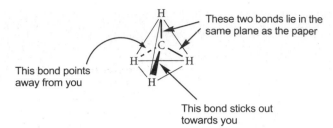

Figure 109.3 *Both diagrams represent the molecule butane,* C_4H_{10}. *The one on the left shows only which atoms are bonded together. It gives the impression that the carbon backbone makes a straight line. Actually, the chain adopts a zig-zag pattern as in the second diagram*

Figure 109.4 *The convention for showing the shape of an organic molecule*

Figure 109.5 *The shape of a methanol molecule,* CH_3OH

This model of methane shows clearly the tetrahedral arrangement of bonds around the carbon atom.

you, and narrowing down to a point as it gets further away.) The second bond is shown by a broken line.

If you have to draw a diagram for a molecule with a functional group, like the OH in methanol, CH_3OH, it makes sense to put the group on one of the bonds in the plane of the paper, like Figure 109.5.

109.2 Draw diagrams of the molecules CH_3Cl and CH_2ClBr using the convention for showing the bonds in Figure 109.5.

109.3 By obeying the rule that carbon atoms always have four bonds, see if you can discover a molecule that has the same formula as benzene, C_6H_6, but does not have a ring of carbon atoms.

109.4 Look at the diagram of progesterone in Table 109.1. Does the diagram give you a realistic impression of the structure of the molecule?

---

molecule as a straight line, but really it is a zig-zag shape. The two ways of drawing the molecules are shown in Figure 109.3.

You may well imagine that it can be quite tricky to show the shapes of molecules on paper. There is a convention for drawing them. Let us take methane, CH_4, as an example. With the help of a model you will see that two of the hydrogen atoms and the carbon atom all lie in the same plane. We draw them as shown in Figure 109.4. These atoms are imagined to lie in the same plane as the paper at which you are looking. Arranged in this way, one of the remaining hydrogen atoms points outwards, towards your eye; the other points backwards, behind the plane of the paper. The one that points towards you is shown by a black wedge in the figure. (It is supposed to represent the perspective of a bond, which would appear thicker at the end nearer to

109.5 When is the arrangement of atoms around carbon not tetrahedral?

One answer is: when a carbon atom forms double or triple bonds to another atom. The most familiar examples are the hydrocarbons, ethene, C_2H_4, and ethyne, C_2H_2. We discussed the bonding in these two molecules in section 16.8, so look there if you have forgotten the explanation of why they are flat or *planar*.

Another planar molecule is benzene, C_6H_6 (Figure 109.6). The history of this molecule is interesting because its structure was a puzzle to chemists for many years. Essentially the puzzle was to explain how six carbon atoms and six hydrogen atoms could be fitted

Two models showing the shapes of ethane and ethene. Note that the double bond in ethene causes the molecule to be planar.

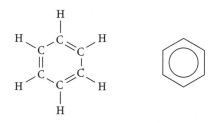

Figure 109.6 *Two ways of representing a benzene ring. The first shows all the atoms labelled, together with the correct number of single and double bonds. The second is a shorthand version. The hydrogen atoms are left out (we have to imagine they are present). A carbon atom is assumed to be at the corners of the hexagon, and the ring in the middle stands for the delocalised π electrons*

together and give a molecule that was not at all reactive. It is possible to fit the atoms together to give compounds containing double or triple bonds; but this would make benzene resemble an alkene (like ethene) or alkyne (like ethyne). Alkenes and alkynes are very reactive, but benzene is not. The solution to part of the puzzle was provided by Friedrich Kekulé in 1865. His notion was that benzene consisted of six carbon atoms in a hexagon, and joined by alternating single and double bonds. Again, you will find details of the modern theory of bonding in benzene in Unit 16. We now think it a mistake to identify individual single and double bonds. Rather, six of the electrons are *delocalised* around the ring in π bonds.

One of the most interesting carbon molecules made in recent years is given the name of 'buckminsterfullerene'. It has the formula C_{60}, and is made from interlocking hexagonal and pentagonal rings of carbon atoms (see Figure 109.7). It was first made in 1985 as a result of the action of a laser beam on a sample of graphite. Such molecules are now thought to exist even in chimney soot or candle smoke. (You may be able to

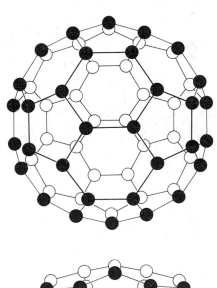

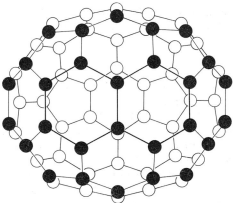

Figure 109.7 *Two bucky balls. On the top is C_{60}, with C_{70} below*

envisage the structure of C_{60} more easily by looking at a football, whose surface may well also be composed of a similar set of interlocking hexagons and pentagons.) Incidentally, the molecule is named after the American

Organic chemistry 709

architect Buckminster Fuller, who had a tendency to design geometrical structures of great complexity. C_{60} has now been outclassed by the discovery of C_{70}. These, and similar, large carbon molecules, are sometimes referred to as 'bucky balls'.

109.6 Baeyer's strain theory

In the early days of organic chemistry it was known that carbon atoms could join to give fairly long chains, e.g. in the alkanes. However, apart from benzene, few compounds were known which had rings of carbon atoms. In 1885 A. Baeyer suggested a reason why this might be so. He believed that ring compounds would be flat and because of this the angles between the carbon atoms would be different to the normal tetrahedral bond angle (109°28′). The idea was that the greater the departure from the tetrahedral angle, the weaker the bonding would be, and the more strain there would be in the ring. Some rings would be so strained that they would not survive. The essence of his theory has survived, and it is called *Baeyer's strain theory*:

> **Rings that have bond angles different to the tetrahedral angle will suffer from strain; if the strain is too great, the ring may break.**

You can see how this theory works in the case of cyclopropane, C_3H_6, by looking at Figure 109.8.

If we had a simple molecule like methane, we could explain the bonding using hybridisation (see section 17.4), in which case we would assume that there were four tetrahedral sp^3 orbitals. We know that the strongest σ bond between two carbon atoms will come with the greatest amount of overlap between the sp^3 orbitals. In cyclopropane, the overlap between such hybrid orbitals is not a maximum because they are pointing slightly away from each other. Therefore, the argument goes, the bond between the carbon atoms is weaker than in an ordinary alkane.

Baeyer's strain theory makes sense, but we have to be careful in using it. Especially, Baeyer's assumption that rings are flat is often false. One important example of this is cyclohexane, C_6H_{12}. If the molecule were flat, we

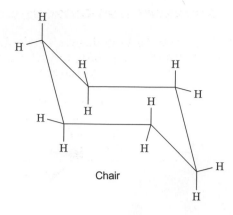

Chair

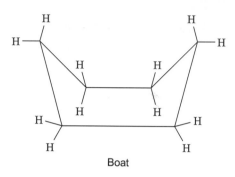

Boat

Figure 109.9 *The chair and boat forms of cyclohexane, C_6H_{12}*

would expect it to have the same hexagonal shape as benzene. However, cyclohexane does not have the incentive of delocalised bonding to make it adopt this planar structure. Instead, the ring puckers. The molecule can bend in two ways, which allow the bond angles to reach the tetrahedral angle (Figure 109.9). The two shapes are called the *boat* and *chair conformations* of cyclohexane. At room temperature a bottle of cyclohexane contains an equilibrium mixture of both conformations, but the chair is energetically more stable and makes up over 99.9% of the liquid. Conformations are varieties of the same molecule that can be changed into one another by rotating parts of the molecule around the carbon–carbon bonds.

If you build a model of cyclohexane you should discover that a suitable combination of rotations (or twists) will convert the chair into the boat, and vice versa.

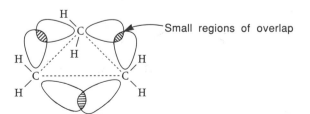

Figure 109.8 *The carbon atoms are at 120° to each other in cyclopropane. We can imagine the bonds to be made by overlapping sp^3 orbitals. However, because of the angle the orbitals make with one another, the overlap is slight. This makes for weak bonds*

> **109.5** If you were a chemist who put your trust in electron repulsion theory to explain molecular shapes, how would you account for the strain in cyclopropane?
>
> **109.6** This is a list of heats of formation (all in kJ mol^{-1}) of propane, cyclopropane and hexane: −199, −104, +55. However, the values have been mixed up. Your task is to put them in the right order.

109.7 Important homologous series

Table 109.3 lists the 16 homologous series that you need to know about. The key thing to spot is the functional group of each series. This group is responsible for giving the molecules in the series their particular properties. For example, alkenes will react with bromine whereas alkanes will not. This is due to the presence of a double bond in alkenes. Similarly, alcohols will react with sodium, but alkenes and alkanes

Table 109.3. Sixteen homologous series

Series	Functional group	General formula	Typical member
Alkane	Hydrocarbon chain	C_nH_{2n+2}	Ethane 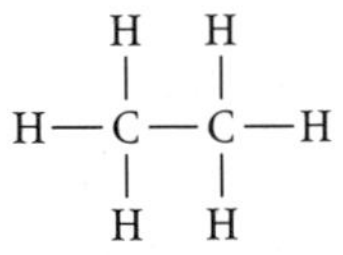
Alkene	Double bond 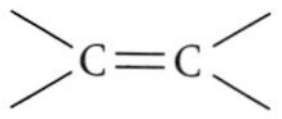 $>C=C<$	C_nH_{2n}	Ethene 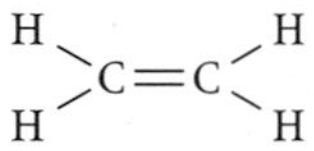
Alkyne	Triple bond $-C\equiv C-$	C_nH_n	Ethyne
Halogenoalkane	Halogen atom	RX (X=Cl,Br,I)	Chloroethane 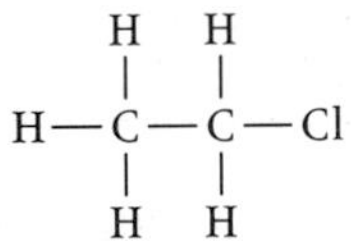
Nitrile	Cyanide group $-C\equiv N$	RCN	Ethanenitrile 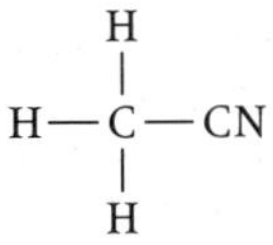
Alcohol	OH group	ROH	Ethanol 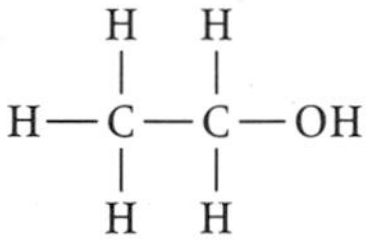
Aldehyde	CHO group 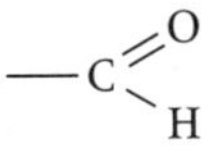	RCHO	Ethanal 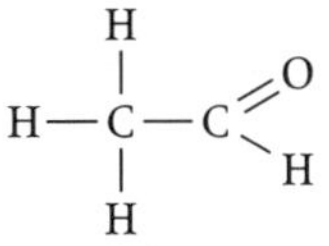
Ketone	CO group 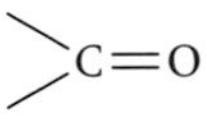 $>C=O$	RR'CO	Propanone 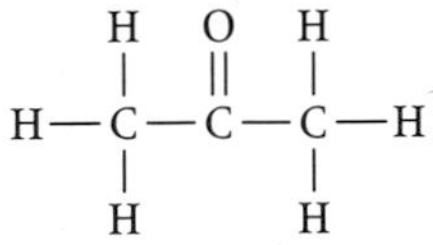
Acid	COOH group 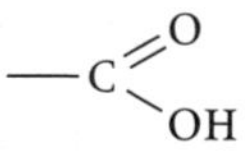	RCOOH	Ethanoic acid 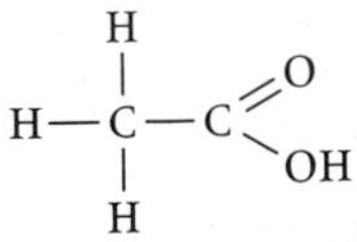

Table 109.3– cont.

Series	Functional group	General formula	Typical member
Ester	COOC group	RCOOR′	Ethyl ethanoate
Ether	COC group	ROR′	Ethoxyethane
Amine	NH_2 group	RNH_2	Ethylamine
Amide	$CONH_2$ group	$RCONH_2$	Ethanamide
Acid chloride	COCl group	RCOCl	Ethanoyl chloride
Arene	Benzene ring	C_6H_6	Benzene
Substituted arene	Benzene ring plus another functional group	Many examples	Phenol
			Chlorobenzene
			Phenylamine

Table 109.4. Four organic radicals

Alcohol	Formula	ROH with R as	Name of R
Methanol	CH_3OH	CH_3	Methyl
Ethanol	CH_3CH_2OH	CH_3CH_2	Ethyl
Propan-1-ol	$CH_3CH_2CH_2OH$	$CH_3CH_2CH_2$	Propyl
Butan-1-ol	$CH_3CH_2CH_2CH_2OH$	$CH_3CH_2CH_2CH_2$	Butyl

will not. It is the OH group in alcohols that is responsible for the reaction.

It can be tedious to write out the entire formula of an organic molecule, so we use a shorthand. We show the carbon atoms and their accompanying hydrogen atoms by the letter R. For example, each member of the homologous series of alcohols can be represented by ROH. Table 109.4 shows you the scheme. There are four common hydrocarbon groups listed in the table, called methyl, ethyl, propyl and butyl. Each of these is an example of an organic radical (notice that we use the symbol R for the radicals). It can be that a more complicated molecule might have two or more different radicals. For example, the alcohol 3-methylhexan-3-ol has methyl, ethyl and propyl radicals in the molecule as well as an OH group. We could show the structure in three ways:

$$CH_3-CH_2-CH_2-\underset{\underset{CH_3}{|}}{\overset{\overset{OH}{|}}{C}}-CH_2-CH_3$$

$$\text{or} \quad R-\underset{\underset{R'}{|}}{\overset{\overset{OH}{|}}{C}}-R'' \quad \text{or even} \quad RR'R''COH$$

R, R' and R'' stand for the three radicals.

109.7 The first three members of the homologous series of chloroalkanes are chloromethane, CH_3Cl, chloroethane, C_2H_5Cl, and 1-chloropropane, C_3H_7Cl.

(i) What is the general formula of the series?

(ii) Chloromethane can be converted into methanol, CH_3OH, by boiling with an alkali. Write down the formulae of the alcohols that would be made from the other two molecules and predict their names.

(iii) Make a model of 1-chloropropane. Now see if you can make another molecule with the same formula, but not identical to the first one. Draw out their structures on paper.

109.8 Naming organic compounds

The key to naming organic molecules is to spot the longest carbon chain, and the functional group or groups. The root name of the molecule depends on the carbon chain according to the system of Table 109.5.

For example, the molecule CH_3CH_2OH has the functional group OH on the end of a chain whose root is ethane (two carbon atoms in the chain). We name the molecule ethanol. When there are three carbon atoms in the chain, the OH group could be on one of the end carbon atoms or on the middle one. We use a numbering system to distinguish them:

$$\overset{3}{C}H_3-\overset{2}{C}H_2-\overset{1}{C}H_2-OH \qquad \overset{3}{C}H_3-\underset{\underset{OH}{|}}{\overset{2}{C}H}-\overset{1}{C}H_3$$

propan-1-ol   propan-2-ol

It does not matter if we write the structure of the first one like this

$$HO-\overset{1}{C}H_2-\overset{2}{C}H_2-\overset{3}{C}H_3$$

we still call it propan-1-ol rather than propan-3-ol. This is a general rule: we use the lowest numbers possible in assigning the name.

When there are two or more functional groups in the same molecule and there is a conflict about which suffix should come first, we put them in alphabetical order. For example, we name the molecule

$$H-\underset{\underset{H}{|}}{\overset{\overset{Cl}{|}}{\overset{2}{C}}}-\underset{\underset{H}{|}}{\overset{\overset{Br}{|}}{\overset{1}{C}}}-H$$

as 1-bromo-2-chloroethane rather than 1-chloro-2-bromoethane.

109.8 Draw the structures of: (i) butan-2-ol; (ii) 2-amino-1-chlorobutane; (iii) propanenitrile; (iv) 2-chloropropanoic acid; (v) 1,1,2-tribromoethane; (vi) ethane-1,2-diol; (vii) buta-1,2-diene; (viii) methoxyethane. Hint: consult Table 109.3.

109.9 Give the names of the following molecules:
(i) $CH_3CHClCH_2OH$;   (ii) $CH_3CHBrCH_2Cl$;
(iii) $(CH_3)_2CHCH_3$;   (iv) CH_3COCH_3;
(v) $HCOOH$;   (vi) CH_3CH_2COCl;
(vii) $CH_3CH_2CH_2OH$;   (viii) $CH_3CHNH_2CH_2OH$.
Hint: consult Table 109.5.

109.9 Chain isomerism

There are several different varieties of isomerism in organic chemicals. Three of them are chain isomerism,

Table 109.5. Naming organic compounds

Longest carbon chain	Root name	Side chain
One carbon atom	Methane	Methyl
Two carbon atoms	Ethane	Ethyl
Three carbon atoms	Propane	Propyl
Four carbon atoms	Butane	Butyl
Five carbon atoms	Pentane	Pentyl
Six carbon atoms	Hexane	Hexyl

Functional group	Naming system		Examples
Alkene	suffix	-ene	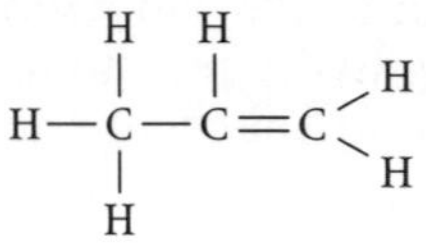 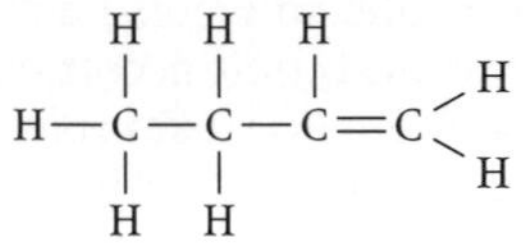 Propene But-1-ene
Alkyne	suffix	-yne	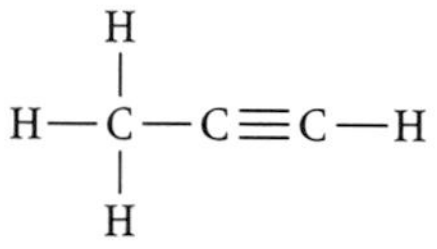 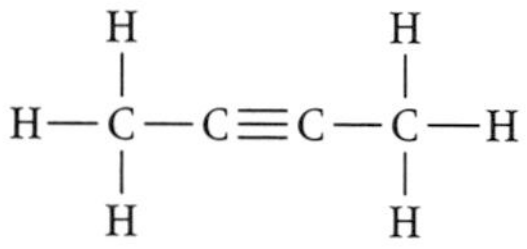 Propyne But-2-yne
Halogenoalkane	prefix	chloro-	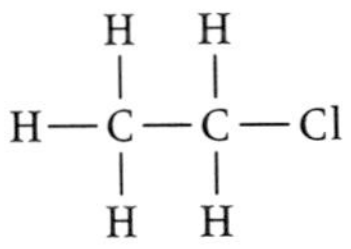 Chloroethane
		bromo-	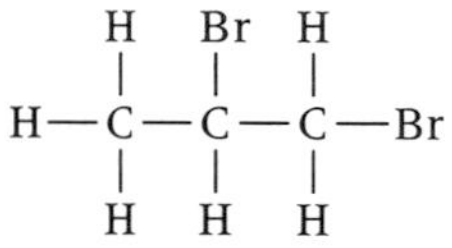 1,2-Dibromopropane
		iodo-	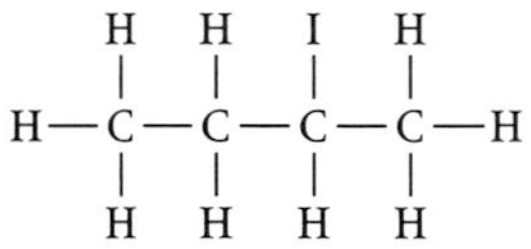 2-Iodobutane
Nitrile	suffix	-nitrile	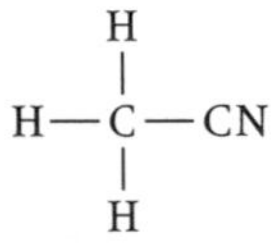 Ethanenitrile
Alcohol	suffix	-ol	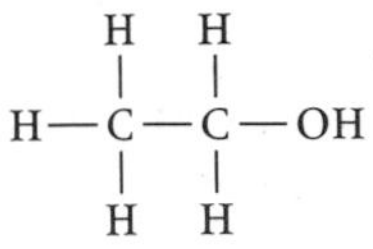 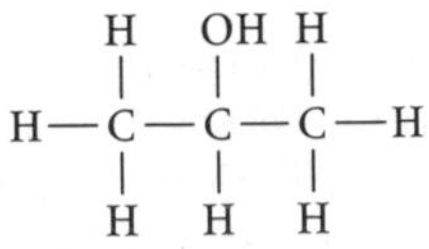 Ethanol Propan-2-ol

Table 109.5 – cont.

Functional group	Naming system		Examples
Aldehyde	suffix	-al	Ethanal Butanal
Ketone	suffix	-one	Butanone Pentan-2-one
Acid	suffix	-oic acid	Ethanoic acid Propanoic acid
Ether	prefix	alkoxy-	Methoxyethane
Amine	suffix	-amine	Ethylamine
Amide	suffix	-amide	Ethanamide
Acid chloride	suffix	-oyl chloride	Ethanoyl chloride
Arene	varies with arene		

functional group isomerism and optical isomerism. You will find the last one explained in the following unit. We shall deal with the other two here and in the next section.

The formula of an organic chemical does not necessarily tell you the arrangement of the atoms. You can see this if you build models of an alkane with the formula C_4H_{10}. There are two possibilities, shown in Figure 109.10. The two molecules are chain isomers.

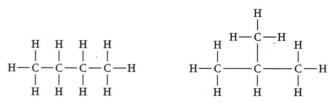

Figure 109.10 Both molecules have the formula C_4H_{10}, but different structures. They are chain isomers

> **Chain isomers have the same molecular formula and functional groups but a different arrangement of the atoms in space.**

One problem we have is deciding how to name isomers. The method we use is as follows. First, identify the longest chain of carbon atoms. This gives us the root name of the compound. Secondly, give each carbon atom in the chain a number from 1 upwards and identify the *side chains* attached to the longest chain. Finally, write down the name of the compound in the order: number, then side chain, then root name. There are examples in Table 109.6. Notice that 2-methylpentane might be named 3-methylpentane if you started numbering from the other end. However, as we have already said, we always use the numbering system that gives the lowest number in the name.

Chain isomers have the same chemical properties, but their physical properties such as density and boiling point differ. The variations in physical properties are often due to changes in intermolecular forces. We spoke about this in Unit 20. If you have understood (and remembered) that work, you should be able to explain why, for example, the boiling point of pentane is about 26 °C higher than its isomer 2,2-dimethylpropane.

On a diagram there are several ways of showing the arrangements of the atoms in an organic molecule. For example, both of the diagrams in Figure 109.11 show the same molecule, 1-bromo-2-chloroethane. At first sight it may seem that the diagrams are of different molecules, but if you make a model of the molecule you will discover why this is not so. By rotating the end groups you should be able to reproduce the arrangements shown in the figure. There is an important point of chemistry here. We say that there is *free rotation* around carbon–carbon single bonds. There is experimental evidence to show that real molecules do rotate in this way.

However, there is *no* free rotation about carbon–carbon double bonds. The π bond in an alkene (or alkyne) prevents the rotation. Thus the two molecules in Figure 109.12 really are different. They are geometric isomers called *cis* and *trans* isomers; but more of this in Unit 112.

Table 109.6. Examples of naming alkanes

Diagram	Name
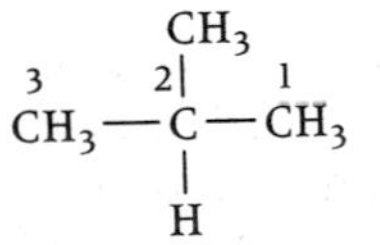 $\overset{3}{CH_3}-\overset{2}{\underset{\underset{H}{\vert}}{\overset{\overset{CH_3}{\vert}}{C}}}-\overset{1}{CH_3}$	2-Methylpropane
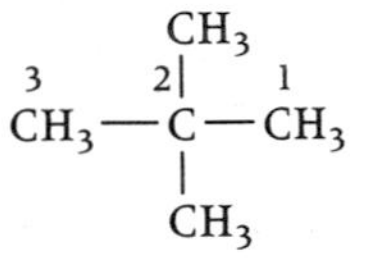 $\overset{3}{CH_3}-\overset{2}{\underset{\underset{CH_3}{\vert}}{\overset{\overset{CH_3}{\vert}}{C}}}-\overset{1}{CH_3}$	2,2-Dimethylpropane
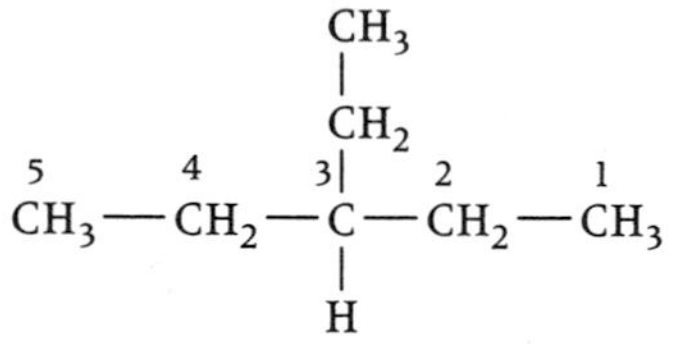 $\overset{5}{CH_3}-\overset{4}{CH_2}-\overset{3}{\underset{\underset{H}{\vert}}{\overset{\overset{CH_3}{\overset{\vert}{\underset{\vert}{CH_2}}}}{C}}}-\overset{2}{CH_2}-\overset{1}{CH_3}$	3-Ethylpentane
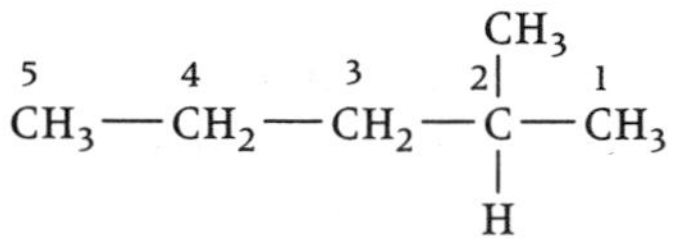 $\overset{5}{CH_3}-\overset{4}{CH_2}-\overset{3}{CH_2}-\overset{2}{\underset{\underset{H}{\vert}}{\overset{\overset{CH_3}{\vert}}{C}}}-\overset{1}{CH_3}$	2-Methylpentane
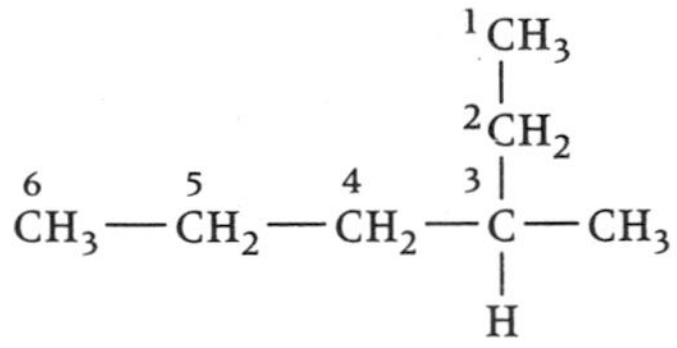 $\overset{6}{CH_3}-\overset{5}{CH_2}-\overset{4}{CH_2}-\overset{3}{\underset{\underset{H}{\vert}}{\overset{\overset{\overset{1}{CH_3}}{\overset{\vert}{\underset{\vert}{\overset{2}{CH_2}}}}}{C}}}-CH_3$	3-Methylhexane*

*Not 2-ethylpentane because the former uses the longest carbon chain

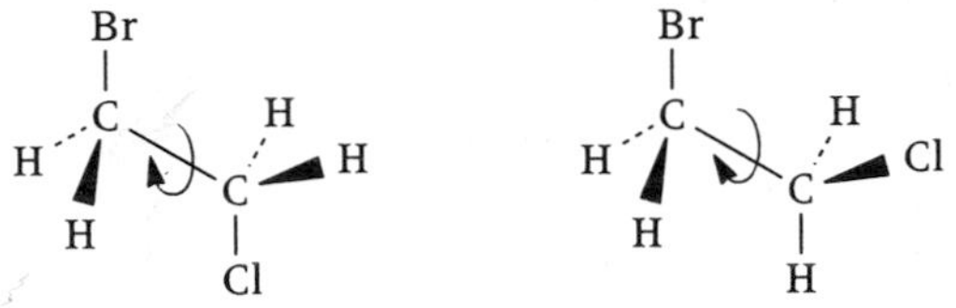

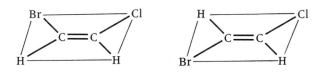

Figure 109.11 *These diagrams may appear to show different molecules. However, they are the same: 1-bromo-2-chloroethane. Each can change into the other form because of free rotation around a carbon–carbon single bond (shown by the arrow)*

cis-1-Bromo-2-chloroethene   trans-1-Bromo-2-chloroethene

Figure 109.12 *Two molecules that are different. Rotation about a carbon–carbon double bond cannot take place*

109.10 How many alcohols can you find with the formula C_4H_9OH:

(i) if the four carbon atoms are joined in one long chain;

(ii) if the carbon atoms do not have to be in one chain?

109.11 Why is the boiling point of pentane about 26 °C higher than that of its isomer, 2,2-dimethylpropane?

109.10 Functional group isomerism

A definition of this variety of isomerism is:

> **Functional group isomers have the same molecular formula but different functional groups.**

If we attempt to make models or draw structures of molecules with the formula C_2H_6O, we have two possibilities:

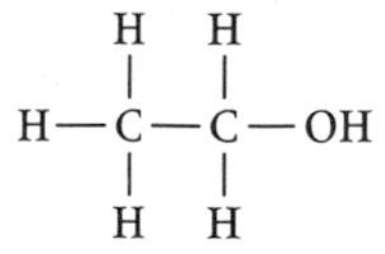

ethanol 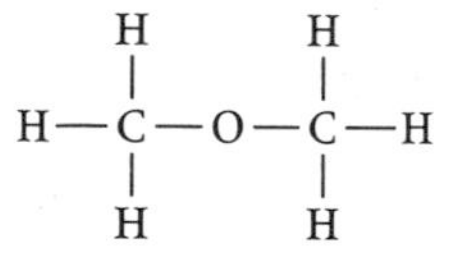   methoxymethane

The first is an alcohol, the second an ether.

109.12 Make a model of benzene, or draw a diagram of it on paper. Now replace one of the hydrogen atoms by a chlorine atom to give a molecule like that shown in Figure 109.13.

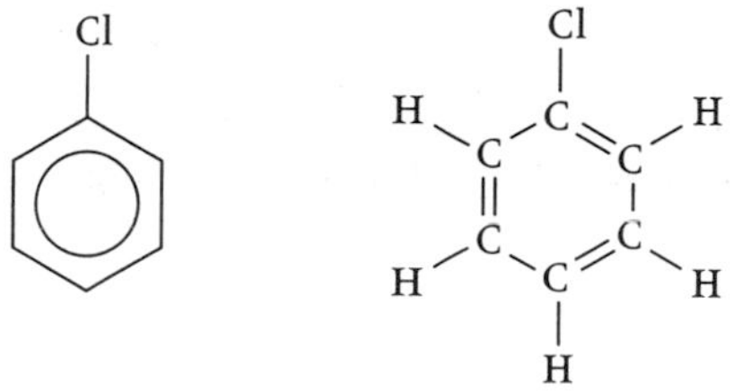

Figure 109.13 *Two ways of representing chlorobenzene*

(i) What is the formula of the molecule?

(ii) Briefly explain why there is only one variety of this molecule.

(iii) Now replace another hydrogen by a chlorine, giving four hydrogen and two chlorine atoms attached to the ring. You should be able to make three different molecules of this type. Draw diagrams of them.

(iv) How many different molecules can you make from a benzene ring with three hydrogen and three chlorine atoms?

109.13 Is it always true that molecules with the general formula C_nH_{2n} are alkenes?

109.14 Draw structures of isomers with formula $C_3H_6O_2$. You will find that it is possible to have more than one functional group in each molecule. Do not include isomers with double bonds between carbon atoms.

109.11 What happens in organic reactions?

This section gives you a brief summary of key ideas that you will find mentioned in later units. You must look at them for details of particular reactions.

In section 109.1 we said that most organic chemicals are covalent and that their reactions are correspondingly slow. When bonds break, they can do so in two ways. If one of the atoms takes both electrons in the bond, the change is called *heterolysis*. If both atoms take one electron each, the change is an example of *homolysis*:

You will meet examples of both in later units.

Often the reagents that we use in organic reactions are of two types. They seek out centres of either positive charge or negative charge:

> **Ions or molecules that are attracted to positive charges are called *nucleophiles*.**
>
> **Those that are attracted to negative charges are *electrophiles*.**

Nucleophiles generally have one or more lone pairs and/or a negative charge. Electrophiles normally have a positive charge or a partial positive charge. To be precise, nucleophiles and electrophiles should be capable of forming a covalent bond when they react. For this reason, we discount species such as sodium ions, Na^+, as electrophiles. Examples are shown in Table 109.7.

From time to time you will find that we attempt to explain how organic reactions take place by referring to the roles played by electrophiles and nucleophiles.

109.12 Organic analysis

In order to identify the nature of an organic compound, we need to know which elements it contains, their pro-

Table 109.7. Electrophiles and nucleophiles

Electrophiles	Nucleophiles
Nitryl cation, NO_2^+	Hydroxide ions, OH^-
Benzenediazonium ions, 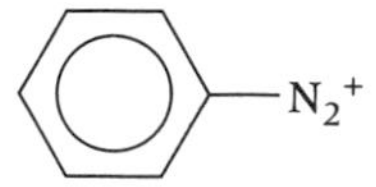	Water molecules, H_2O
	Cyanide ions, CN^-
Bromine cation, Br^+	
Acyl cation, CH_3CO^+	

portions and how they are arranged into the various functional groups. The problem of finding which elements are present is the task of *qualitative analysis*. Some methods of qualitative analysis are relatively simple. You will find them described in Appendix E. More complicated methods involve spectrometry, e.g. mass spectrometry. The task of finding out the nature of the functional groups can also be done chemically or by spectrometry. Examples of tests that indicate the presence of specific functional groups are described in the units that follow. If you have read the units on spectrometry, you will also be aware of how infrared spectrometry can achieve similar results.

Quantitative analysis is used to determine the percentage composition of a compound. Mass spectroscopy is one method by which the molar mass can be found. There are standard methods of analysing compounds to find out the percentages of the elements present. These methods are often performed automatically by companies that specialise in such work. We shall ignore methods of quantitative analysis, but you will find that we sometimes make use of the information that they provide.

Lastly, if you have a sample of a compound whose identity you think you know, a simple way of confirming its identity is to do a melting point test. Methods of performing a melting point determination are described in Unit 58.

Answers

109.1 The German chemist Wohler found one way. In 1828 he took ammonium cyanate, NH_4CNO, which everybody agreed was an inorganic chemical, and heated it. The product was urea, $CO(NH_2)_2$, a substance known to occur in living things, and therefore an organic chemical. It is extremely hard to explain how a life force can enter into a compound simply by heating it. Eventually chemists gave up the life force theory.

You might notice that Wohler's experiment did not *prove* that the life force was absent; only that the theory was not capable of explaining the result of his experiments.

109.2 Figure 109.14 shows the diagrams.

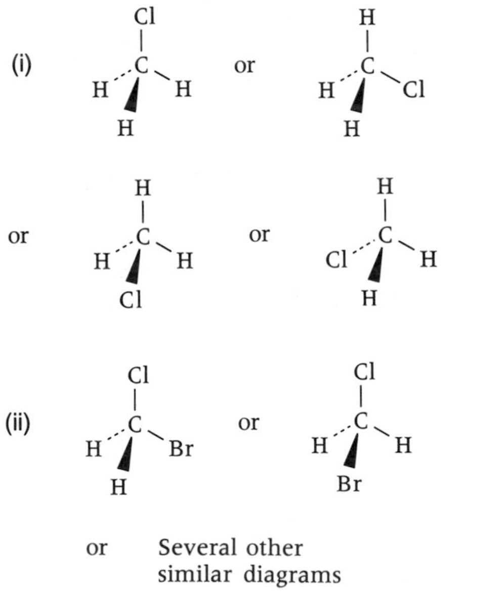

Figure 109.14 Diagrams for answer to question 109.2

109.3 One possibility is $CH_2{=}CH{-}C{\equiv}C{-}CH{=}CH_2$.

109.4 The diagram gives the impression that the molecule is flat; it is not. The rings are puckered.

109.5 You would explain that with a bond angle of 60° in cyclopropane the bonding pairs of electrons are brought closer together than they would be if the bond angle were 109°. This will increase electron repulsion, so there will be greater strain.

109.6 We know that cyclopropane will be the least energetically favoured molecule, so this will have the endothermic heat of formation ($+55\ kJ\ mol^{-1}$). The heats of formation of molecules with six carbons in them will usually be greater than those with three carbon atoms. (There are more bonds to be made, so more energy to be released.) This tells us that the value of $-104\ kJ\ mol^{-1}$ belongs to propane and $-165\ kJ\ mol^{-1}$ to hexane.

109.7 (i) $C_nH_{2n+1}Cl$.

(ii) C_2H_5OH, ethanol; C_3H_7OH, propanol.

(iii) The structures are shown in Figure 109.15.

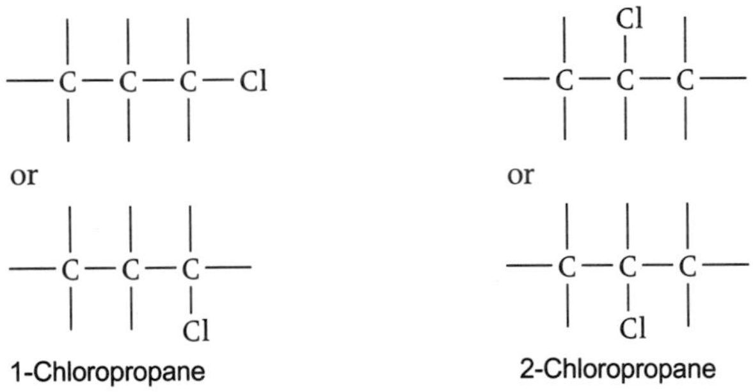

1-Chloropropane   2-Chloropropane

Figure 109.15 Diagrams for answer to question 109.7. The two different arrangements for C_3H_7Cl. Note that there are several ways of drawing each structure, but these are not different molecules

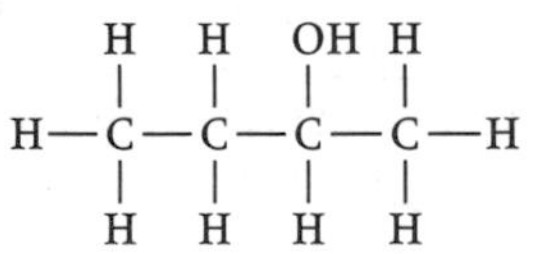

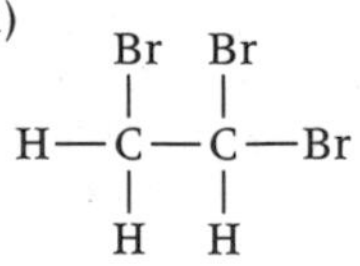

109.8

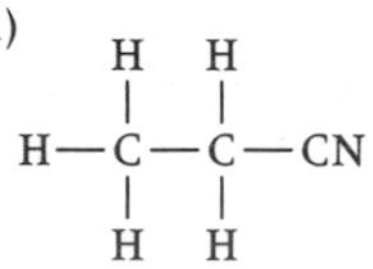

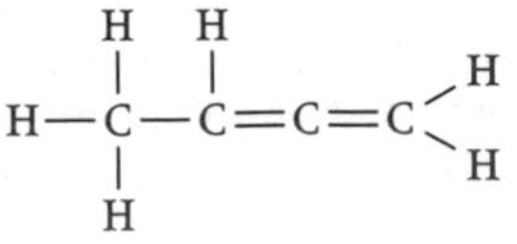

Answers – cont.

(v)

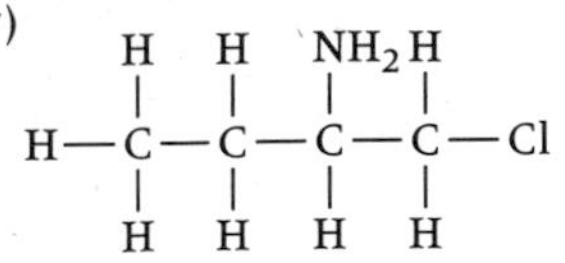

H   H   NH$_2$H
| | | |
H—C—C—C—C—Cl
| | | |
H   H   H   H

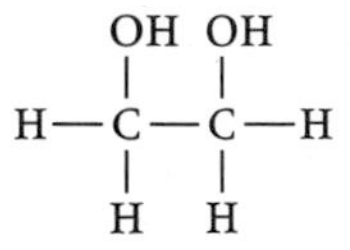

(vi)

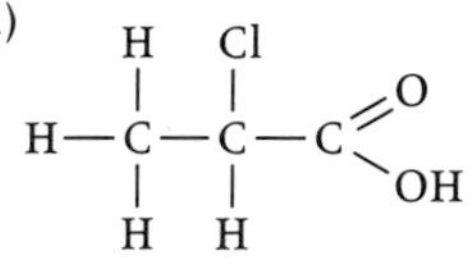

H   Cl
| | O
H—C—C—C⟨
| | OH
H   H

(vii)

OH OH
| |
H—C—C—H
| |
H   H

(viii)

H   H   H
| | |
H—C—O—C—C—H
| | |
H   H   H

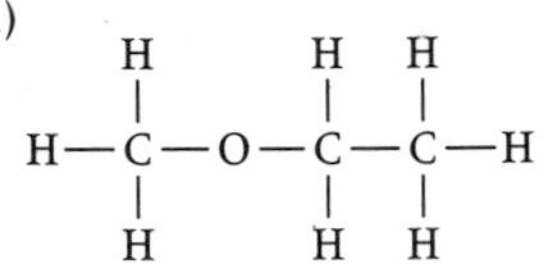

109.9 (i) 2-chloropropan-1-ol; (ii) 2-bromo-1-chloropropane; (iii) 2-methylpropane; (iv) propanone; (v) methanoic acid; (vi) propanoyl chloride; (vii) propan-1-ol; (viii) 2-aminopropan-1-ol.

109.10 (i) There are two of them. If you have found more, you have made the mistake of thinking that, say, butan-2-ol and 'butan-3-ol' are different. They are not, and we choose the lowest number for the name:

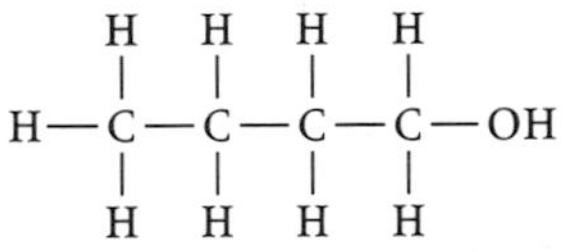

H   H   H   H
| | | |
H—C—C—C—C—OH
| | | |
H   H   H   H

butan-1-ol

H   H   OH H
| | | |
H—C—C—C—C—H
| | | |
H   H   H   H

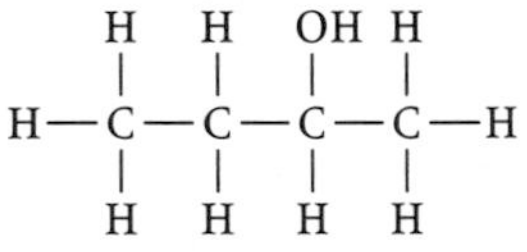

butan-2-ol

(ii) Now there are four isomers: butan-1-ol and butan-2-ol as before, plus

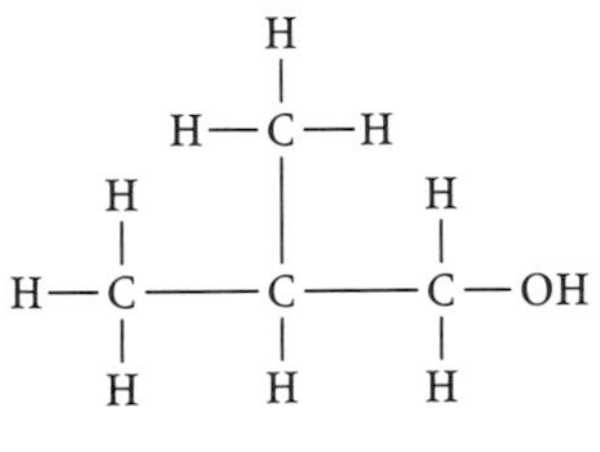

H
|
H—C—H
H | H
| | |
H—C — C — C—OH
| | |
H   H   H

2-methylpropan-1-ol

H
|
H—C—H
H | H
| | |
H—C — C — C—H
| | |
H   H   H
        |
        OH

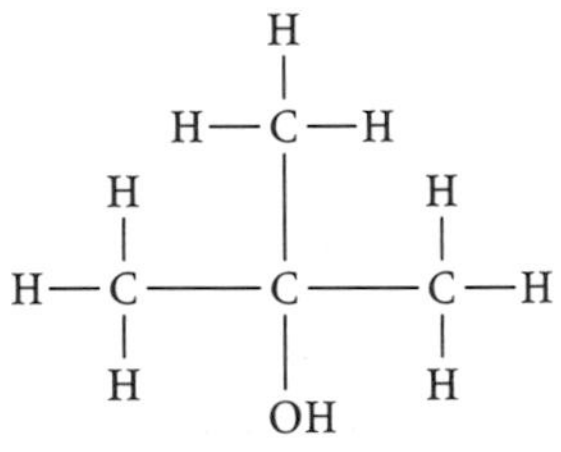

2-methylpropan-2-ol

109.11 The van der Waals forces between two molecules of pentane are stronger than between two molecules of 2,2-dimethylpropane. This is because propane molecules can come into closer contact than the others. 2,2-Dimethylpropane molecules are (roughly) spherical, and only have small area of contact with one another.

109.12 (i) C$_6$H$_5$Cl.

(ii) Owing to the symmetrical nature of benzene, it does not matter where you put the chlorine, the molecule is always the same shape.

(iii) Figure 109.16 shows the three different molecules.

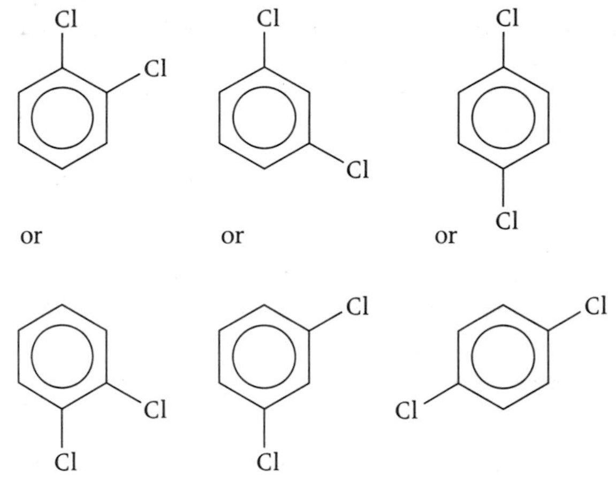

1,2-Dichloro-benzene   1,3-Dichloro-benzene   1,4-Dichloro-benzene

Figure 109.16 Diagrams for answer to question 109.12(iii). Note that there are several ways of drawing each structure, but they are not different. Make models if you do not understand this

(iv) Figure 109.17 shows the molecules that are possible.

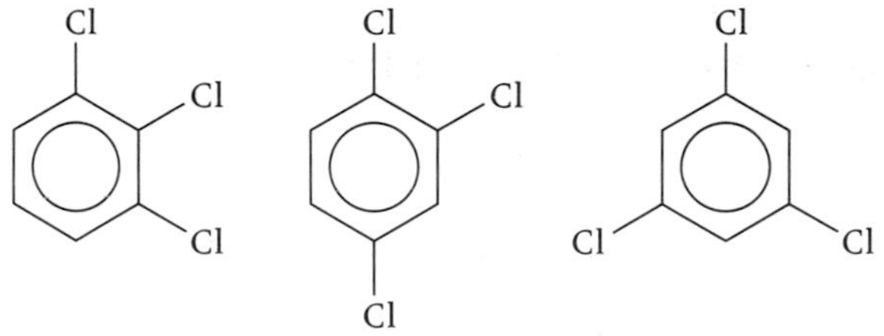

1,2,3-Trichloro-benzene   1,2,4-Trichloro-benzene   1,3,5-Trichloro-benzene

Figure 109.17 Diagrams for answer to question 109.12(iv). Three possibilities for C$_6$H$_3$Cl$_3$. They are the only three; again models should convince you of this

109.13 No. Cycloalkanes have the same general formula.

109.14

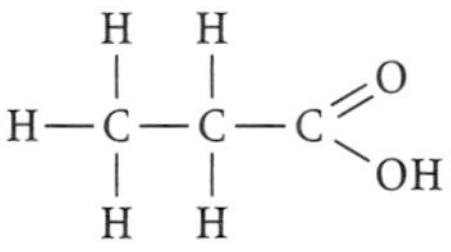

H   H
| | O
H—C—C—C⟨
| | OH
H   H

CH$_3$CH$_2$COOH
propanoic acid

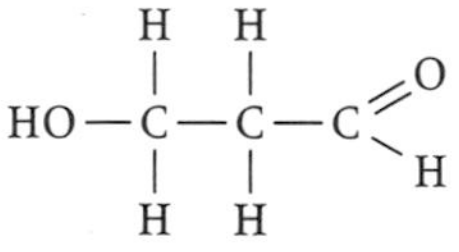

H   H
| | O
HO—C—C—C⟨
| | H
H   H

CH$_2$OHCH$_2$CHO
3-hydroxypropanal

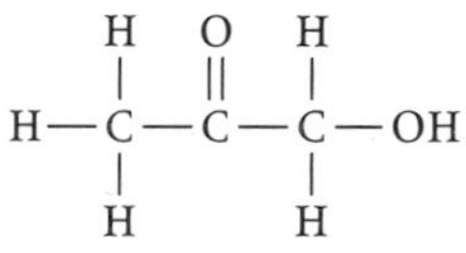

H   O   H
| || |
H—C—C—C—OH
| |
H   H

CH$_3$COCH$_2$OH
hydroxypropanone

- Organic chemistry is the study of carbon compounds.
- Organic compounds:
    - (i) Are mainly covalent.
    - (ii) Always have four bonds to each carbon atom.
    - (iii) Where there are four single bonds there is normally a tetrahedral arrangement of the groups bonded to the carbon atom.
    - (iv) Around single bonds there is free rotation of the groups.
- Baeyer's strain theory says that:

    Rings that have bond angles different to the tetrahedral angle will suffer from strain; if the strain is too great the ring may break.

    Exceptions to the theory include compounds, such as benzene, with delocalised electrons.
- Homologous series:

    The members of a series have
    - (i) The same functional group.
    - (ii) Similar chemical properties.
    - (iii) The same general formula.
    - (iv) Gradually changing physical properties.
- Naming systems

    The root name is taken from the name of the longest carbon chain. See Table 109.3.

- Isomerism:
    - (i) Chain isomers have the same molecular formula and functional groups but a different arrangement of the atoms in space.
    - (ii) Functional group isomers have the same molecular formula but different functional groups.
    - (iii) Geometric isomers are *cis* and *trans* isomers of alkenes, which occur owing to the lack of free rotation about double bonds.
- Bond breaking occurs through:
    - (i) Heterolysis

        $$A \overset{x}{\cdot} B \longrightarrow A^+ + B \overset{x}{\cdot}{}^-$$
        (ions)
    - (ii) Homolysis

        $$A \overset{\cdot}{x} B \longrightarrow A\cdot + B x$$
        (radicals)
- Reactions often involve:
    - (i) Nucleophiles, which seek out centres of positive charge.
    - (ii) Electrophiles, which seek out centres of negative charge.
- Analysis:
    - (i) Qualitative analysis finds the *type* of each element or group present.
    - (ii) Quantitative analysis finds the *amount* of each element or group present.

110

Optical activity

110.1   What is optical activity?

To understand optical activity you need to know about plane polarised light, so this is the first thing we shall explain. In Unit 24 we discovered that light waves consist of constantly changing electric and magnetic fields. These fields are always at right angles to the direction of travel of the light. If you look at Figure 110.1 you will see a diagram showing the electric field of one light wave travelling towards someone's eye. If you imagine that you could see the light waves coming towards you, you would see the electric field as a vertical line; the field can only move in one plane. We say that the light is *plane polarised*.

Light that comes from an ordinary lamp is not plane polarised. The electric fields of the light waves from the lamp are arranged in many different planes. However, if we pass the light through a piece of polaroid, it becomes plane polarised (Figure 110.2). We can place a second piece of polaroid in the path of the light. If we rotate this piece so that it is at right angles to the first one, then no light gets through. When the two pieces of polaroid are

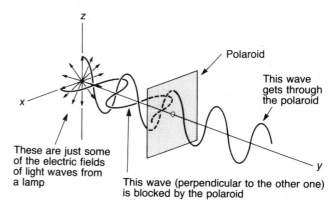

Figure 110.2 *Light can be polarised by passing it through a piece of polaroid. (Light waves with their electric fields somewhere between the two extremes in the diagram will pass through the polaroid, but with reduced intensity)*

lined up in the same direction, the maximum amount of light gets through.

It was the French physicist Jean Biot who, in 1812, discovered that some substances have the ability to *rotate the plane of polarised light*. To see what this means, look at Figure 110.3. In the first diagram we have plane polarised light passing through water. When the light emerges from the tube, it remains polarised in the same plane as when it entered the tube. In the second diagram the light passes through a solution of an optically active chemical (we need not worry about its nature). When the light comes out of the tube, it has been rotated through an angle, in this case 30°. To allow the maximum amount of light through the second piece of polaroid, we would have to rotate it through 30°. As we look at the beam of light the polaroid has to be rotated to the right (clockwise direction). We say that the chemical in the tube is *dextrorotatory*.

The third tube in the figure has a chemical in it that rotates the plane of polarised light to the left (anticlockwise) by 12°. This chemical is *laevorotatory*.

There is a more modern way of talking about optically active molecules. We say that they are *chiral*. (The 'ch' is pronounced like a 'k'.) Thus we can talk about chiral molecules that are dextro- or laevorotatory.

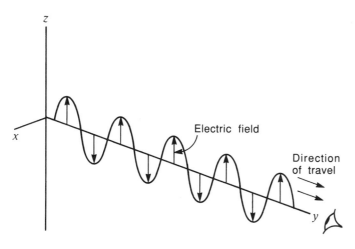

Figure 110.1 *This diagram illustrates a light wave travelling in the y direction. The electric field oscillates up and down, always parallel to the z axis. The field stays in the yz plane: the wave is plane polarised*

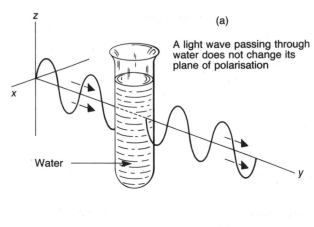

(a)

A light wave passing through water does not change its plane of polarisation

Water

(b)

This light wave has had its plane of polarisation turned through 30°

30°

Optically active solution

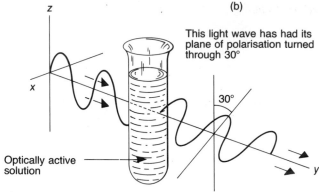

(a)   (b)   (c)

30°   12°

Summary views of results of three experiments

Figure 110.3 *Examples of rotation of the plane of polarised light. (a) Through water the electric field of the light wave is not rotated. (b) This solution has rotated the plane 30° to the right. (c) This solution has rotated the plane 12° to the left. Cases (a) and (b) are illustrated*

110.2 Polarimeters

In the laboratory, optical activity is investigated using a *polarimeter* (Figure 110.4). The apparatus has a source of light, which in sophisticated polarimeters will be a sodium lamp. The light passes through the *polariser*, and then into a tube containing the solution under test. Finally the light passes into the *analyser*. The angle of rotation given by a solution depends on a number of things: concentration, temperature and wavelength of the light. The *specific rotation* [α] at 20°C and using yellow light from sodium (the D lines) is given by

$$[\alpha]_D^{20} = \frac{\text{measured rotation}}{\text{length of tube (dm)} \times \text{concentration (g cm}^{-3})}$$

A positive or negative sign is used to show the sign of the rotation. For example, the specific rotation of one variety of tartaric acid (2,3-dihydroxybutanedioic acid) has $[\alpha]_D^{20} = +14.4°$. The plus sign tells us that the acid is dextrorotatory.

One further complication with specific rotations is that their values depend on the solvent used in the polarimeter tube.

110.1   You may have thought about this point. Suppose you find that a chemical requires the analyser to be rotated by 180°. With this result you cannot tell whether the plane polarised light has been rotated to the left or to the right. What other experiments would you do to discover whether the chemical was dextrorotatory or laevorotatory?

110.2   In practice, if you use a polarimeter you should not move the analyser to find the position of maximum brightness; rather you should move it to find the position of maximum darkness.

(i) Why?

(ii) Suppose that you find the position of maximum darkness by rotating the analyser clockwise through +110°. What angle has the plane of polarised light turned through?

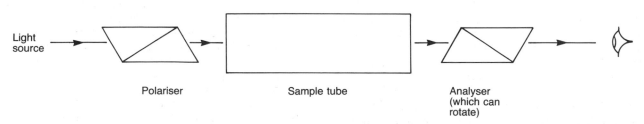

Light source

Polariser   Sample tube   Analyser (which can rotate)

Figure 110.4 *The three key parts of a polarimeter. The polariser and analyser are made of quartz in the best polarimeters. Simpler ones use sheets of polaroid instead. The analyser is connected to a scale, which allows the rotation to be measured*

Sodium ammonium tartrate

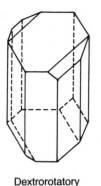

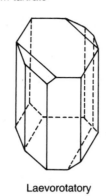

Dextrorotatory

Laevorotatory

Tartaric acid

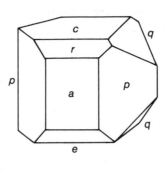

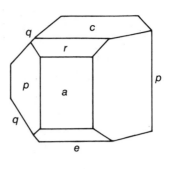

Dextrorotatory

Laevorotatory

Figure 110.5 *Diagrams of the mirror image forms of crystals discovered by Pasteur*

110.3 Why are some substances chiral?

Louis Pasteur was one of the first scientists to make a study of optical activity. In 1848 he made a crucial observation when he was working on the salts of tartaric acid. (The modern name of this acid is 2,3-dihydroxybutanedioic acid.) He crystallised a solution of sodium ammonium tartrate, and spotted that there were two types of crystal present. He picked out the two types and showed that they differed in the arrangements of the faces. Later he drew diagrams of two types of crystalline tartaric acid (see Figure 110.5).

The key thing about the crystals is that, in each pair, the two varieties are *mirror images* of one another. Pasteur describes the sodium ammonium salts in this way:

> I carefully separated the crystals which were hemihedral to the right from those hemihedral to the left, and examined their solutions separately in the polarising apparatus. I then saw with no less surprise than pleasure that the crystals hemihedral to the right deviated the plane of polarisation to the right, and that those hemihedral to the left deviated to the left.

It was clear that the two crystal structures must be the result of different molecular structures. At the time the hunt began for an explanation of how molecules could exist in two mirror image forms. The simplest solution to the problem was given by J. A. le Bel and (independently) by J. H. van't Hoff in 1874. They proposed that chiral molecules were based on a tetrahedral structure. Figure 110.6 shows you that if we have *four different atoms or groups* arranged in a tetrahedron, then the molecule will have two mirror image forms, called *enantiomers* or *optical isomers*. A famous example of such a molecule is lactic acid (2-hydroxypropanoic acid).

At this stage you should make models of lactic acid and see for yourself what the enantiomers look like.

The carbon atom at the centre of the four different groups is called an *asymmetric carbon atom*. It is tempting to think that any substance that has an asymmetric carbon atom will be chiral; but this is not so. Neither is it

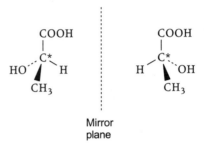

Mirror plane

Figure 110.6 *The two mirror image forms of lactic acid (2-hydroxypropanoic acid). The stars mark the asymmetric carbon atoms. Two mirror image forms cannot be superimposed on each other in such a way that the positions of all the groups match. Note: We treat the CH_3, OH and COOH as single groups; we do not worry about how the hydrogen atoms are arranged in CH_3, or the oxygen and hydrogen atoms in COOH. For this reason we do not have to show the acid groups as COOH and HOOC in the two diagrams*

true that if a substance is chiral then it must contain an asymmetric carbon atom. Since 1874 we have tightened up the conditions that must be satisfied by a molecule if it is to be chiral. We can use this rule:

> **A molecule will be chiral if it has neither a plane nor a centre of symmetry.**

To understand this, you need to know a little more about the symmetry of molecules. A *plane of symmetry* is fairly easy to understand. A cube has many such planes, some of which are shown in Figure 110.7. If there is a plane of symmetry, the arrangement of the object on one side of the plane is exactly the same as on the other.

A *centre of symmetry* can be a little harder to spot. Here we take a line out from the centre until it reaches some part of the figure; for example the line OA in Figure 110.7. Then we take another line of equal length but in exactly the opposite direction, like OB in the figure. If there are two identical spots at the end of the line, then the figure has a centre of symmetry. A cube has a centre of symmetry.

Models of the two mirror image forms of lactic acid. Also, see the colour section.

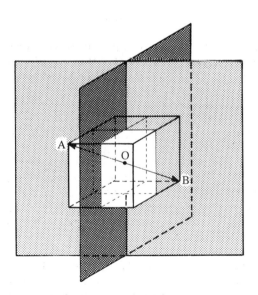

Figure 110.7 *Two planes of symmetry of a cube. These cut the cube in half across four of the sides. Other planes cut across the diagonals. The centre, O, is a centre of symmetry*

Now look at Figure 110.8, where the shapes of a number of different molecules are drawn. Make sure you can understand where the planes and centres of symmetry are to be found. All of these molecules have at least one plane of symmetry or a centre of symmetry. As a consequence, none of them are chiral.

However, the molecules in Figure 110.9 have neither a plane nor a centre of symmetry so they are both chiral. Notice that they do not both have an asymmetric

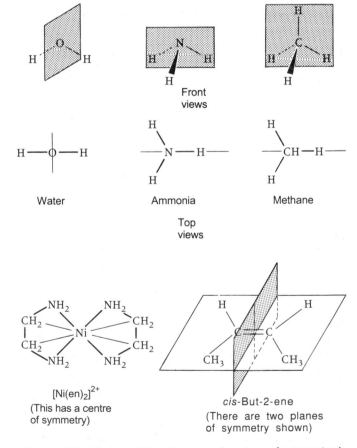

Figure 110.8 *Some of the planes and centres of symmetry in five molecules*

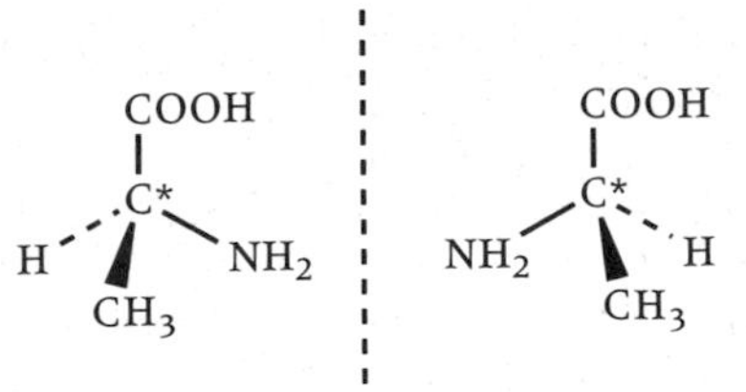

Mirror image forms of alanine, CH₃CH(NH₂)COOH. The stars mark the asymmetric carbon atoms

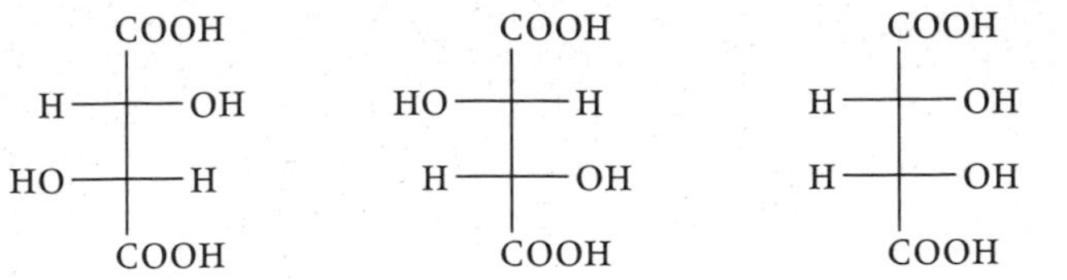

(+)-Tartaric acid
(chiral)

(−)-Tartaric acid
(chiral)

meso-Tartaric acid
(not chiral)

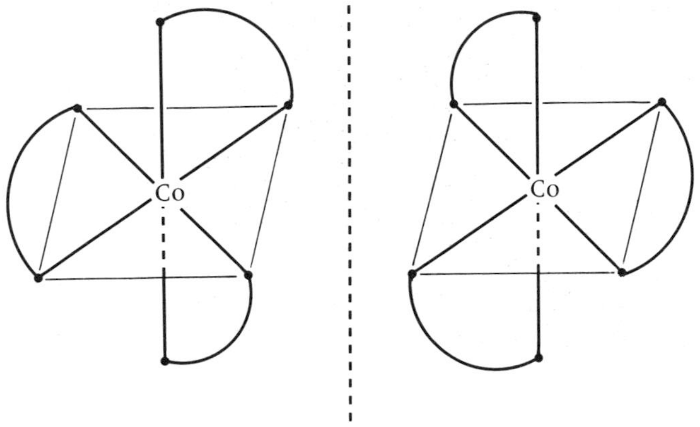

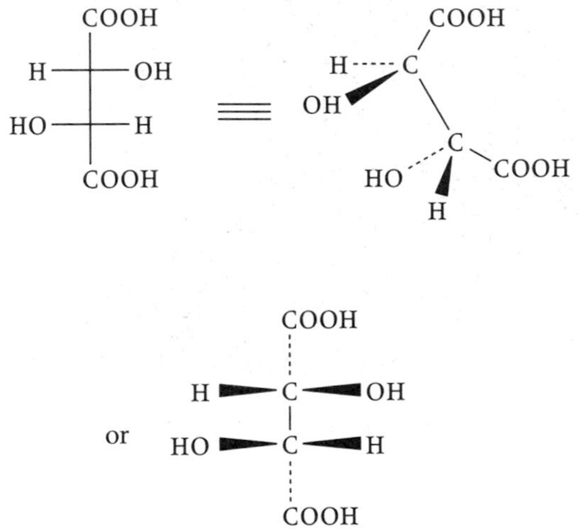

Mirror image forms of tris(ethane-1,2-diamine)cobalt(III),

[Co(NH₂CH₂CH₂NH₂)₃]³⁺. The curved lines

represent NH₂CH₂CH₂NH₂

Figure 110.9 *Diagrams of two molecules whose mirror images cannot be superimposed upon each other. They are chiral molecules. Lactic acid (Figure 110.6) is also chiral*

Figure 110.10 *The three types of tartaric acid. These diagrams of tartaric acid are a shorthand way of showing their structures. The horizontal bonds point towards you, out of the paper, as shown in the lower diagram*

carbon atom. The asymmetric carbon atoms are marked with a star.

110.4 More about tartaric acid

We shall now return to tartaric acid, so that we can discover some further subtle points about optical activity. Until le Bel's and van't Hoff's ideas became accepted, it was a great puzzle to explain why there were *three* types of tartaric acid. Two of them were chiral, but the third was not. The three types can be called (+)-tartaric acid, (−)-tartaric acid and *meso*-tartaric acid. The structures of these molecules are shown in Figure 110.10.

For the first time you can see that each of the three molecules has two asymmetric carbon atoms. However, the arrangement of the groups around each asymmetric carbon atom is different. The (+) and (−) forms are enantiomers. Neither molecule has a plane or a centre of symmetry; but the *meso* form does have a plane of symmetry. This is the reason why it is not chiral. Thus *meso*-tartaric acid *is* one of the three isomers of tartaric acid; but it is not an enantiomer. It is called a *diastereoisomer*.

Sometimes it is said that the *meso* form is not chiral because of *internal compensation*. Here the idea is that the arrangement around one of the asymmetric carbon atoms would cause the plane of polarised light to be rotated to the right, whereas the arrangement around the other asymmetric carbon would rotate by an equal amount to the left. The net result is that there would be no overall rotation.

There is another reason why there was so much confusion over the tartaric acids in Pasteur's time. Solutions known to contain the acid were found to be optically inactive; but they did *not* contain *meso*-tartaric acid. The solutions were inactive because they contained equal amounts of the (+) and (−) isomers. Solutions like this contain a *racemic mixture* of the two enantiomers. We can also say that the solutions are inactive because they are *externally compensated*.

110.3 Decide whether the following molecules can be optically active. In some cases you might need to make a model of the molecule. Which of them, if any, have *meso* forms? Copy each formula and mark the asymmetric carbon atom(s) with a star.

(i) Glycine

$$H_2N-\underset{\underset{H}{|}}{\overset{\overset{H}{|}}{C}}-C\overset{\nearrow O}{\underset{\searrow O-H}{}}$$

(ii) *cis*-But-2-ene

$$\underset{H}{\overset{CH_3}{}}C=C\underset{H}{\overset{CH_3}{}}$$

(iii) *trans*-But-2-ene

$$\underset{CH_3}{\overset{H}{}}C=C\underset{H}{\overset{CH_3}{}}$$

(iv) 2-Methylbutan-1-ol

$$CH_3-\underset{\underset{H}{|}}{\overset{\overset{H}{|}}{C}}-\underset{\underset{H}{|}}{\overset{\overset{CH_3}{|}}{C}}-\underset{\underset{H}{|}}{\overset{\overset{H}{|}}{C}}-OH$$

(v) *trans*-Butenedioic acid

$$\underset{H}{\overset{COOH}{}}C=C\underset{H}{\overset{COOH}{}}$$

(vi) 2,3-Dibromobutane
(Take care with this one!)

$$CH_3CH_2BrCH_2BrCH_3$$

(vii) Dichlorobis(ethane-1,2-diamine)chromium(III), $[Cr(en)_2Cl_2]^+$

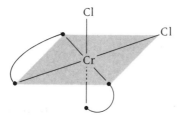

(viii) Hex-2,3-diene
(It is best to make a model of this molecule.)

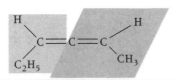

110.5   Properties of enantiomers and racemic mixtures

Enantiomers have identical physical properties, e.g. the melting point of (+)- and (−)-tartaric acids are both 170 °C. Often they have chemical properties in common, e.g. both the chiral forms of lactic acid will react

with phosphorus pentachloride. However, sometimes their reactions do differ. Usually this happens when the chemical with which they react is itself chiral, or has some particular symmetry requirement such as the active site in an enzyme. We can use the difference in reactivity to separate the components from a racemic mixture, a process known as the *resolution* of a racemic mixture.

A racemic mixture sometimes crystallises into a mixture of two different crystals. As Pasteur found, this happens with sodium ammonium tartrate. A mixture like this behaves as a eutectic (see section 58.3). However, sometimes two enantiomers combine to give a single type of crystal (a racemate), and sometimes they make solid solutions. Fortunately we need not worry about the detail of such matters.

110.6   Resolution of racemic mixtures and diastereoisomers

Separating a racemic mixture into its components is known as resolving the mixture. A common method of resolving a racemic mixture is to react the mixture with an optically active substance. One method of doing this is to pass the mixture down a column packed with polymer beads having an optically active group on their surface. (This method is like column chromatography.) The idea is that a (+) isomer on the beads might attract the (+) isomer of the racemic mixture more strongly than the (−) isomer, and the (−) isomer could be collected at the bottom of the column. (It could be the other way round, i.e. the (−) isomer from the mixture might stick to the beads more strongly. Note that (+) and (−) isomers do *not* behave like positive and negative charges, where opposites always attract.)

Another method is to mix the racemic mixture physically with a second chiral substance. If we call the isomers in the racemic mixture (+)R and (−)R, and the substance we add is (+)A, then two compounds are possible: (+)R(+)A and (−)R(+)A. If we choose A carefully, the compounds may have different solubilities and we can separate them by fractional crystallisation. The final task is to decompose the separated crystals to release the original isomers in the racemic mixture.

Compounds like (+)R(+)A and (−)R(+)A will be chiral, but they will not be enantiomers. Like *meso*-tartaric acid, they are diastereoisomers.

110.7   The configurations of optical isomers

If you look back at Figure 110.10, you might wonder how we know that the (+) and (−) isomers have those particular arrangements of the atoms in space. The evidence is complicated, and relies partly on X-ray diffraction, and partly on synthetic techniques. We shall not go into the details of the methods used; but you might find it useful to know something of the results.

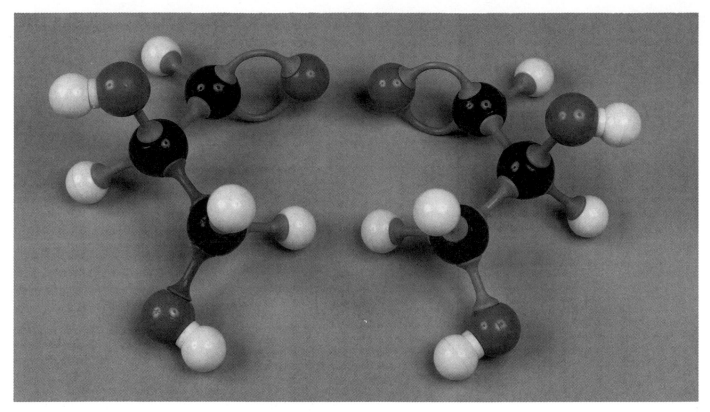

Models of the two mirror image forms of glyceraldehyde.

If you make models of 2,3-dihydroxypropanal (glyceraldehyde), $CH_2OHCHOHCHO$, you will find the two optical isomers in the photo.

We know that one isomer will be dextrorotatory and the other laevorotatory, but for the present let us ignore their optical activity. Even if they did not affect plane polarised light, we can tell from the models that there are two versions of the molecule. There is a special notation that we can use to label them. The first step in the method (invented in 1956 by Cahn, Ingold and Prelog) is to put the groups attached to the asymmetric carbon atom in an order of priority. The order is:

$$I, Br, Cl, F, OH, NH_2, COOH, CHO, CH_2OH,$$
$$CN, C_6H_5, CH_2R, CH_2, CH_3, H$$

Most   Least
important   important

(Actually this is only part of a much longer list.) The order of priority is based on the order of atomic mass.

The next step is to look at models of the molecules and arrange them so that the group with the lowest priority is pointing directly away from you. You should do this for the optical isomers of 2,3-dihydroxypropanal.

You should find that if you trace the groups round in their order of importance, in one case you will move in a clockwise direction:

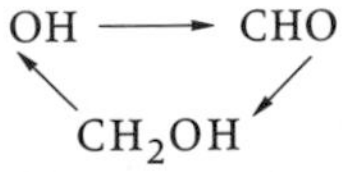

$$OH \longrightarrow CHO$$
$$\searrow \quad \swarrow$$
$$CH_2OH$$

This follows the order of importance in the list. We say that this isomer has a *rectus configuration*, and we give it the symbol *R*. On the other hand, the second isomer has its groups in the order:

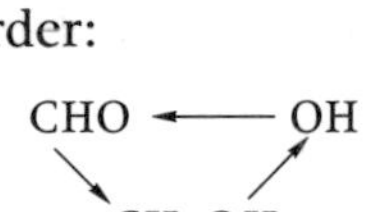

$$CHO \longleftarrow OH$$
$$\searrow \quad \swarrow$$
$$CH_2OH$$

If we follow the groups round in the order of their importance, we move in an *anticlockwise* direction. This isomer has a *sinister configuration*, and is labelled with an *S*.

Thus, we have two configurations of 2,3-dihydroxypropanal. Notice that we have not said which is dextrorotatory and which is laevorotatory. This is a matter for experiment, which shows that *R*-2,3-dihydroxypropanal is dextrorotatory. If we wish to show all this information, we should write the names of the two optical isomers as *R*-(+)-2,3-dihydroxypropanal and *S*-(−)-2,3-dihydroxypropanal. Please be sure you understand that for other molecules an *R* configuration can belong to a laevorotatory isomer, and *S* configuration to a dextrorotatory isomer.

110.4 In our work on mechanisms in Unit 81, we discovered that halogenoalkanes can undergo nucleophilic attack in two ways. They react by an S_N1 or by an S_N2 mechanism. In this case we shall suppose that 2-iodobutane reacts with hydroxide ions. By working through the following questions, you should gain a better understanding of the background to these two mechanisms.

(i) Draw a diagram, or make a model, of 2-iodobutane. Explain why the molecule will be chiral. Establish the order of priority of each of the groups and decide whether your molecule is *rectus* or *sinister*.

(ii) Assume that the molecule reacts with hydroxide ions by an S_N1 mechanism. Draw another diagram, or make a second model, of the product and discover whether the product is *rectus* or *sinister*.

(iii) Can you draw any conclusions about what might happen to the configurations of the products of iodoalkanes that react by an S_N1 mechanism?

(iv) Now assume that the 2-iodobutane reacts via an S_N2 route. Describe the structure of the intermediate formed during the reaction.

(v) Two products of the hydrolysis are possible. What are they? Describe their configurations.

(vi) Are the products chiral? Briefly explain.

(vii) Would you expect the final solution to be optically active? Briefly explain.

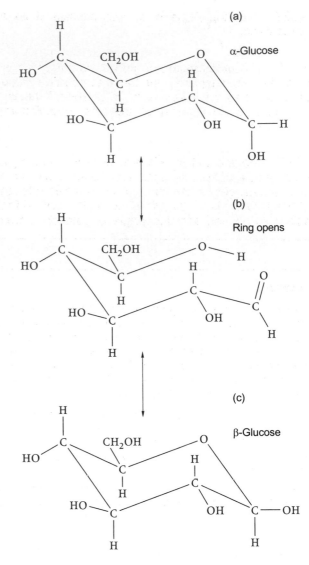

Figure 110.11 How α-glucose and β-glucose are related

110.8 The mutarotation of glucose

Studies of optical activity have given a great deal of information about many types of reaction. We shall look at one example, which concerns glucose molecules.

Glucose, $C_6H_{12}O_6$, is a sugar based on a six-membered ring. You can see the structure of the ring in Figure 110.11a. This version of glucose is called α-glucose. It was known as long ago as 1846 that a fresh solution of α-glucose had a specific rotation of +111°. However, if the solution is left for some time, the rotation eventually falls to +52.5°. This decrease in optical rotation is called the *mutarotation* of glucose.

When the final solution is analysed we find that there are two types of molecule present. Both have the formula $C_6H_{12}O_6$, and both have a ring structure. The rings differ in only one respect: the arrangement of one of the pairs of H and OH groups. This isomer of α-glucose is called β-glucose (Figure 110.11c). When pure, it has a specific rotation of +19.2°. Thus, whenever a molecule of α-glucose changes into β-glucose, the optical rotation

of the solution will decrease. Eventually an equilibrium is established between the isomers, and then the specific rotation remains constant at +52.5°.

We believe that the reason why the mutarotation takes place is that the ring of α-glucose spontaneously breaks from time to time to give the open chain molecule shown in Figure 110.11b. This molecule can remake a ring structure, but sometimes the orientation of a hydrogen atom and an OH group is opposite to that of the original. This alternative structure is β-glucose.

110.5 This question asks you to work out the percentage of α-glucose and β-glucose in an equilibrium mixture of the two. Call the fraction of α-glucose in the mixture x.

(i) What are the specific rotations of α-glucose and β-glucose?

(ii) What does the fraction x of α-glucose contribute to the rotation of the mixture?

(iii) What does the fraction $1-x$ of β-glucose contribute to the rotation of the mixture?

(iv) Given that the actual rotation of the mixture is

+52.5°, find the value of x, and express it as a percentage.

110.6 The mutarotation of glucose is catalysed by acid. Let us suppose that the mutarotation takes place twice as fast in acid as in water alone. What will be the specific rotation of the final (equilibrium) solution?

110.7 Imagine that you place a solution containing a pure sample of an optical isomer of a substance in a polarimeter. Let us assume that the specific rotation of the solution is +100°. You add a little acid to the solution and find that the angle of rotation slowly decreases. After 20 minutes you find the solution has no optical rotation. Some time later you find that the angle of rotation is nearly −50°.

After some research in the library you find that the original isomer has been converted into its mirror image. You also discover that the reaction is first order in the original isomer.

(i) How much of the original isomer was present in the solution after 20 minutes?

(ii) How many minutes after the start of the reaction would it take for the rotation to reach −50°?

(iii) Estimate how long it would take for the rotation to reach −100°.

Answers

110.1 You could change the concentration of the solution. For example, by halving the concentration the rotation would be halved. If the substance were dextrorotatory, the rotation would be +90°; if it were laevorotatory, the rotation would be −90°. (If you had halved the concentration, you would not be likely to confuse −90° with +270°. Why not?)

110.2 (i) It is easier to see if *any* light is getting through than it is to tell if the *maximum* amount of light is entering your eye. That is, it is easier to see a difference between light and dark rather than between two lights of slightly different brightness.

(ii) To reach maximum darkness the analyser must be at right angles to the plane of polarisation; so the angle must be 90° greater than the true rotation. The latter is +20°.

110.3 (i) There is no asymmetric carbon atom in glycine: it is inactive.

(ii), (iii) Both molecules have planes of symmetry, so neither is chiral. One plane contains the four carbon and two hydrogen atoms; another is perpendicular to this plane, and cuts the molecule in two:

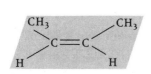

(iv) This molecule has an asymmetric carbon atom, and it has neither a plane nor a centre of symmetry; therefore it is chiral:

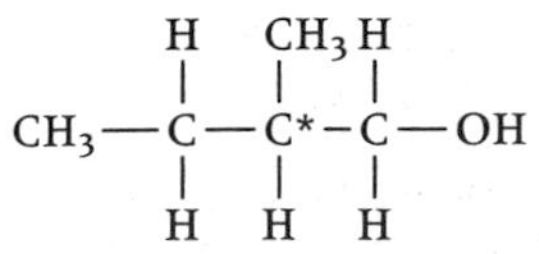

$$CH_3 - \overset{\overset{\displaystyle H}{|}}{\underset{\underset{\displaystyle H}{|}}{C}} - \overset{\overset{\displaystyle CH_3}{|}}{\underset{\underset{\displaystyle H}{|}}{C^*}} - \overset{\overset{\displaystyle H}{|}}{\underset{\underset{\displaystyle H}{|}}{C}} - OH$$

(v) As in (ii) and (iii) there is a plane of symmetry so the molecule is inactive.

(vi) As far as its optical activity is concerned, this molecule is in the same league as tartaric acid. There are two asymmetric carbon atoms, so like tartaric acid there are *three* isomers, two chiral enantiomers and an inactive diastereoisomer:

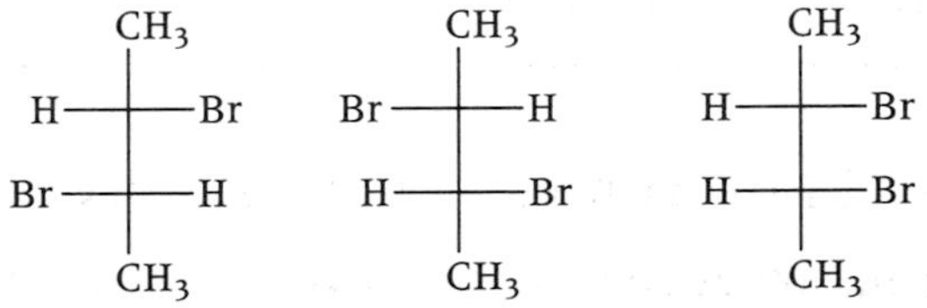

This example has two asymmetric carbon atoms with identical groups attached. If the groups were not identical, there would be more than three isomers possible.

(vii) You may be surprised to find that this molecule does have optical isomers:

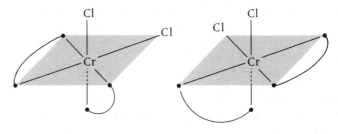

(viii) This is an example of a molecule that does *not* have an asymmetric carbon atom. However, its twisted shape means that it has neither a plane nor a centre of symmetry. Therefore it is entitled to be chiral:

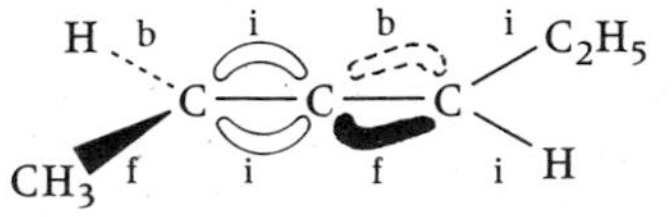

i   bonds in same plane as the paper
f   bonds in front of the plane of the paper
b   bonds behind the plane of the paper

Answers – cont.

110.4 (i)

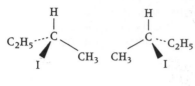

S form   **R** form

(Note: C_2H_5 is the same as CH_2R with $R = CH_3$.)

(ii) The products of the reaction are:

from the *S* isomer   from the *R* isomer

With the hydrogen atom pointing away:

R form   *S* form

(iii) In an extreme case there can be (almost) complete inversion of the configuration during the S_N1 reaction, i.e. *R* changes to *S*, and vice versa.

(iv) The intermediate is a planar (flat) carbocation (see section 81.5).

(v) The hydroxide ions can attack the carbocation from either side. The product should be a 50% mixture of *R* and *S* molecules.

(vi) Yes, the individual molecules are chiral. They have an asymmetric carbon atom (or we can say that they have neither a plane nor a centre of symmetry).

(vii) It should be inactive because equal proportions of the two enantiomers are made. We obtain a racemic mixture.

110.5 (i) +111° and +19.2° respectively.

(ii) 111*x*.

(iii) 19.2(1 − *x*).

(iv) 111*x* + 19.2(1 − *x*) = 52.5. On solving this you will find *x* = 0.36, i.e. 36%.

110.6 The equilibrium value will still be +52.5°. The rate of a reaction has no effect on the final position of equilibrium.

110.7 (i) 50% is left. To see why this is so, imagine that we started with 100 molecules. If 50 of them convert to the other enantiomer, we have 50 of each enantiomer. One will contribute +50° to the rotation, the other −50°; net result, 0° rotation.

(ii) It will take another 20 minutes, i.e. 40 minutes after the start. This is because a rotation of −50° will be obtained by the second enantiomer contributing −75°, and the original enantiomer +25°. A rotation of +25° represents a quarter of the original enantiomer. As 20 minutes is the half-life, it will take another 20 minutes to change from a half to a quarter of the original enantiomer.

(iii) −100° represents complete conversion of the first enantiomer into the second. A rule of thumb is that, for practical purposes, this takes about five or six half-lives, i.e. about 2 hours.

UNIT 110 SUMMARY

- Optically active, or chiral, molecules:
    (i) Rotate the plane of polarised light.
    (ii) Have neither a plane nor a centre of symmetry.
    (iii) Usually have at least one carbon atom with four different groups attached (an asymmetric carbon atom).
    (iv) Have two mirror image forms (enantiomers).
    (v) Enantiomers cannot be completely superimposed on each other.
- Rotation of the plane of polarised light is measured with a polarimeter.
- A racemic mixture consists of equal proportions of two enantiomers and does not rotate the plane of polarised light.
- Diastereoisomers contain at least two asymmetric carbon atoms (chiral centres); the rotation of one centre cancels out the other.

- Resolution:
    (i) Is the separation of enantiomers or diastereoisomers.
    (ii) May be achieved by reacting with another chiral substance, e.g. on an ion exchange column.
- Configuration:
    (i) Refers to the precise three-dimensional arrangement of groups in space.
    (ii) Is given the label *R* or *S* depending on how the arrangement compares with the two forms of 2,3-dihydroxypropanal (glyceraldehyde).
    (iii) The symbols *R* and *S* do *not* give the sign of the optical rotation (+) or (−).
- Mutarotation of glucose:
    This is the name given to the spontaneous change of a solution of α-glucose into β-glucose; the two forms having opposite signs of rotation.

111
Alkanes

111.1 The main types of hydrocarbon

Hydrocarbons contain only carbon and hydrogen atoms. The simplest ones have single bonds only. These are the *alkanes* (Table 111.1). If a hydrocarbon has one or more double bonds, it is an *alkene*; and if it has triple bonds, it is an *alkyne*. Hydrocarbons can also contain benzene rings. These are the *arenes*. We shall investigate each of these homologous series in turn. However, the reactions of arenes can be different to the others and we shall put them in a separate unit.

111.1 (i) Plot graphs showing how the melting and boiling points of the alkanes of Table 111.1 vary as the number of carbon atoms in the chain increases.

(ii) Explain the trend in the two graphs.

(iii) Which of the alkanes is a liquid at room temperature and pressure?

111.2 The importance of alkanes

The alkanes are particularly important to us owing to their use as fuels. The first member of the series, methane, has long been known as 'marsh gas' (rotting organic matter trapped in stagnant water gives off the gas) and 'fire damp' in coal mines. It is also produced from decaying animal dung, and from rubbish buried underground. Methane is trapped in huge quantities underground in areas where oil is found. The source of the gas and oil is the same: they are the result of the decay of marine life from some millions of years ago. Methane can be pumped directly from the deposits via pipelines into homes and factories, where it is more often known as 'natural gas'. Methane is odourless, so for reasons of safety, traces of a foul smelling chemical are usually added to natural gas.

Many hydrocarbons are extracted by the distillation of crude oil in an oil refinery. (You will find details of the process in Unit 86.) The lighter, lower boiling point hydrocarbons are collected from the top of the distillation column, and the heavier ones from the various stages lower down. The mixtures of hydrocarbons removed at each stage are called *fractions* (Table 111.2). Each fraction has its own set of uses.

111.2 Land that has been used as a site for rubbish dumps has sometimes been used later as building land. There have been cases of explosions occurring in houses built on such land. What causes the explosions?

111.3 The reactions of alkanes

The alkanes have few chemical properties (Table 111.3). We have already mentioned the main one in connection with methane: they all burn in air or oxygen. The general equation for the reaction is

$$C_xH_y + (x+y/4)O_2 \rightarrow xCO_2 + (y/2)H_2O$$

Actually this fits the burning of any hydrocarbon, not just alkanes. For example, with pentane, C_5H_{12}, we have $x=5$, $y=12$, $(x+y/4)=8$ and

$$C_5H_{12} + 8O_2 \rightarrow 5CO_2 + 6H_2O$$

In the past this type of reaction was used to discover the formula of hydrocarbons. You might like to try question 111.3 to see how the method worked.

The second type of reaction they perform is a *substitution* reaction with halogens. We have found previously that covalent bonds can be broken in two ways:

$$X{:}Y \rightarrow X^+ + Y{:}^- \text{ or } X{:}^- + Y^+ \qquad \text{Heterolysis}$$
$$X{:}Y \rightarrow X{\cdot} + Y{\cdot} \qquad \text{Homolysis}$$

In homolysis, free radicals are made, and the reaction with halogens shows all the signs of free radical reactions. For example, the reaction between an alkane and chlorine is explosive if light (especially ultraviolet light) shines on the mixture. If the mixture is kept in the dark,

Table 111.1. The alkanes, general formula C_nH_{2n+2}

Name	Formula	Structure	Melting point/°C	Boiling point/°C
Methane	CH_4		-182	-161
Ethane	C_2H_6		-172	-89
Propane	C_3H_8		-188	-42
Butane	C_4H_{10}		-138	0
Pentane	C_5H_{12}		-130	36
Hexane	C_6H_{14}		-95	69

Table 111.2. Products of the fractional distillation of oil

Name of fraction	Boiling range/°C	Use
Gases	<30	Source of propane and butane for fuels; feedstock for chemical industry
Gasoline	30–75	Petrol manufacture
Naphtha	75–190	Feedstock for chemical industry
Kerosene	190–250	Aircraft fuel, central heating boiler fuel
Gas oil	250–350	Diesel fuel, central heating boiler fuel
Waxes, tars, heavy oils, asphalt	>350	Polishes, lubricants, specialised fuels, e.g. for power stations

Table 111.3. Chemical properties of the alkanes

Burn with oxygen
e.g. $CH_4(g) + 2O_2(g) \rightarrow CO_2(g) + 2H_2O(l)$

Substitution with halogens (free radical mechanism)
e.g. $C_2H_6(g) + Cl_2(g) \rightarrow C_2H_5Cl + HCl(g)$
then $C_2H_5Cl \rightarrow C_2H_4Cl_2 \rightarrow C_2H_3Cl_3$, etc.

no reaction takes place. In section 81.4 we found that a free radical reaction takes place in three stages:

(i) Initiation; e.g.

$Cl_2 \rightarrow 2Cl\cdot$

This is the stage that is caused by ultraviolet light. The energy of the photons in the light must be sufficient to break the bond between the halogen atoms.

(ii) Propagation; e.g.

$$CH_4 + Cl\cdot \rightarrow CH_3\cdot + HCl$$

In the propagation stage, a radical may be used up, but another one takes its place.

(iii) Termination; e.g.

$$CH_3\cdot + Cl\cdot \rightarrow CH_3Cl$$

Here, radicals are removed from the reaction.

Please do look at Unit 81 to check that you understand how free radical reactions work.

111.3 Before the days of mass spectroscopy and other methods of analysis, one of the ways in which the formula of a hydrocarbon was found was by burning it in oxygen. If the volume of hydrocarbon, oxygen, carbon dioxide and steam is known then, with the help of Avogadro's theory, the formula can be discovered. Here are the results of an experiment in which a hydrocarbon was burned in an excess of oxygen in a closed container.

Volume of hydrocarbon used $= 20\,cm^3$
Volume of oxygen used $= 150\,cm^3$

Both these readings were taken at room temperature and pressure.

An electric spark was passed through the mixture, which caused the hydrocarbon to burn to carbon dioxide and steam. The gases were allowed to cool to room temperature and pressure.

The total volume of gas in the apparatus was then $110\,cm^3$.

To find the volume of carbon dioxide present, the tube was opened under potassium hydroxide solution. Again at room temperature and pressure, the volume of gas left in the tube was $50\,cm^3$.

Now answer these questions:

(i) Why do we not have to worry about the volume of water produced if we measure volumes at room temperature and pressure?

(ii) What is the gas left at the end of the experiment?

(iii) What volume of oxygen actually reacted with the hydrocarbon?

(iv) What was the volume of carbon dioxide produced in the reaction?

(v) What is the ratio of the volumes of hydrocarbon, oxygen and carbon dioxide?

(vi) Using Avogadro's theory, what is the ratio of the number of moles of each gas used?

(vii) What is the value of x and y in the equation for the burning of a hydrocarbon in oxygen?

(viii) Hence, write down the formula of the hydrocarbon.

111.4 When methane reacts with chlorine the prod-

---

ucts vary with the amount of chlorine in the mixture. For example, if very little chlorine is used, the main product is chloromethane, CH_3Cl; but if a lot of chlorine is used, the product is mainly tetrachloromethane, CCl_4.

(i) Why does the amount of chlorine in the mixture have these effects?

(ii) If roughly equal amounts of chlorine and methane were mixed, what would you expect to find at the end of the reaction?

111.4 How to prepare alkanes

Making alkanes in the laboratory is not often done; but if you were really stuck without a badly needed alkane, here are some methods you might use.

First, a halogenoalkane can be reduced by hydrogen prepared by dripping ethanol onto zinc that has been reacted with a little copper(II) sulphate. The zinc gains a slight coating of copper, and between them they are powerful enough to liberate hydrogen from ethanol. This goes on to attack the halogenoalkane; for example,

$$\underset{\text{iodoethane}}{C_2H_5I} + Zn + 2H^+ \rightarrow \underset{\text{ethane}}{C_2H_6} + HI + Zn^{2+}$$

A second method was invented by the German chemist Wurtz, and as a consequence it is known as Wurtz's reaction. Here an iodoalkane is reacted with sodium; for example,

$$\underset{\text{iodomethane}}{2CH_3I} + 2Na \rightarrow \underset{\text{ethane}}{C_2H_6} + 2NaI$$

The sodium is kept in dry ether and the iodomethane dropped into the ether. The reaction is somewhat exothermic, so a condenser has to be placed on the reaction flask. This has cold water running through the outer jacket, which cools and condenses iodomethane and ether vapour rising into the inner tube. The ethane passes through the condenser and can be collected over water. The apparatus is shown in Figure 111.1.

A third method is to heat the sodium or calcium salt of an organic acid with soda lime (a mixture of sodium and calcium hydroxides). Overall the reaction brings about the loss of carbon dioxide. It is an example of a *decarboxylation* reaction (also see section 113.2); for example,

$$\underset{\substack{\text{sodium} \\ \text{ethanoate}}}{CH_3COONa(s)} + NaOH(s) \rightarrow \underset{\text{methane}}{CH_4(g)} + Na_2CO_3(s)$$

These three methods are a little old fashioned. A more sophisticated method makes use of a *Grignard reaction*. The French inventor of the reaction, Victor Grignard, received the Nobel Prize for his work in 1912. He discovered an entirely new series of reactions. They are all based on the use of magnesium combined with an organic radical and a halogen atom. A typical

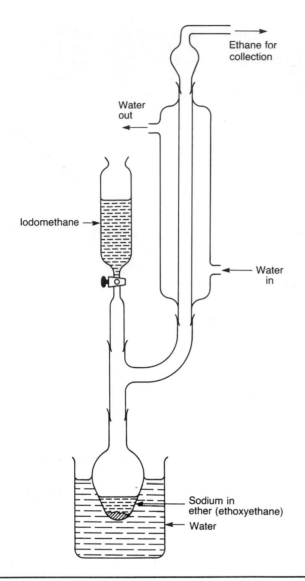

Ethane for collection

Water out

Iodomethane →

Water in

Sodium in ether (ethoxyethane)

Water

Figure 111.1 *An apparatus for preparing ethane by the Wurtz method*

example is CH_3MgBr, methylmagnesium bromide. If this Grignard reagent reacts with water, the organic radical joins with a hydrogen atom from a water molecule:

$$CH_3MgBr + H_2O \rightarrow CH_4 + Mg(OH)Br$$
methane

Similarly, if we wanted to make ethane, we would use ethylmagnesium iodide:

$$C_2H_5MgI + H_2O \rightarrow C_2H_6 + Mg(OH)I$$

111.5 Which alkanes would be made if (i) iodoethane, (ii) 2-iodopropane were used in the Wurtz reaction?

111.6 A student wanted to make propane by the Wurtz reaction. He decided that a mixture of iodomethane and iodoethane would be needed. Was he correct? What would be made in the reaction?

111.7 Write down the formula of the Grignard reagent that you would use to make butane.

Answers

111.1 (i) The graphs are shown in Figure 111.2.

(ii) The melting and boiling points increase as the number of carbon atoms in the chain increases. This is because the intermolecular forces increase with increasing relative molecular mass (see section 20.3).

(iii) Pentane and hexane.

111.2 Methane is released when the rubbish decomposes under the ground. It can happen that the methane escapes into houses built over the site. The explosion is the rapid burning of methane if, for example, a spark starts the reaction.

111.3 (i) The water condenses. The volume of liquid water is insignificant compared to when it is a gas. (For 1 mol, $18 \, cm^3$ compared to over $24\,000 \, cm^3$.)

(ii) Unreacted oxygen.

(iii) We started with $150 \, cm^3$, and $50 \, cm^3$ is left over. Therefore, $100 \, cm^3$ of oxygen were used.

(iv) The volume reduced from $110 \, cm^3$ to $50 \, cm^3$ with the potassium hydroxide, so $60 \, cm^3$ of carbon dioxide were made.

(v), (vi) The ratio is $20 \, cm^3$:$100 \, cm^3$:$60 \, cm^3$, i.e. the ratio of the moles of each is 1:5:3.

(vii) From the number of moles of carbon dioxide we get $x = 3$. From the number of moles of oxygen, we get $x + y/4 = 5$. This means $y = 8$.

(viii) C_3H_8, propane.

111.4 (i) With only a little chlorine, the number of chlorine radicals are greatly outnumbered by the methane molecules. When the methyl radicals are made in the propagation step, they react with chlorine radicals to make chloromethane. This mops up the chlorine radicals quickly, so that very few are left to make dichloromethane, etc.

(ii) Now there are many chlorine radicals in the mixture. Dichloromethane, CH_2Cl_2, can be made:

$$CH_3Cl + Cl\cdot \rightarrow CH_2Cl\cdot + HCl$$
$$CH_2Cl\cdot + Cl\cdot \rightarrow CH_2Cl_2$$

Similarly, CH_2Cl_2 can go on to make $CHCl_3$, and this to make CCl_4. Thus we would expect to find a mixture of all the four products.

Answers – cont.

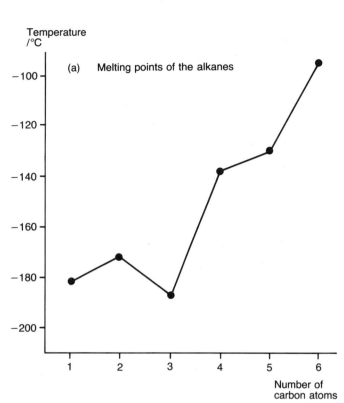

Temperature /°C

(a)   Melting points of the alkanes

Temperature /°C

(b) Boiling points of the alkanes

Figure 111.2 *Graphs for answer to question 111.1(i)*

111.5   (i) Butane, C_4H_{10}.

(ii) If we remove the iodine atoms from two molecules, and join them together, we would have a product that looks like this:

$$
\begin{array}{cc}
CH_3 & CH_3 \\
| & | \\
H-C-C-H \\
| & | \\
CH_3 & CH_3
\end{array}
$$

However, if we stretch the molecule out to show the longest carbon chain, we have

$$
\begin{array}{cc}
& H & H \\
& | & | \\
CH_3-C-C-CH_3 \\
& | & | \\
& CH_3 & CH_3
\end{array}
$$

i.e. 2,3-dimethylbutane.

111.6   Certainly some propane would be made, but also there would be ethane and butane as side products:

$CH_3I + C_2H_5I \rightarrow CH_3{-}C_2H_5$ i.e. C_3H_8, propane
$CH_3I + CH_3I \rightarrow CH_3{-}CH_3$ i.e. C_2H_6, ethane
$C_2H_5I + C_2H_5I \rightarrow C_2H_5{-}C_2H_5$ i.e. C_4H_{10}, butane

The side products would make the yield of the reaction low, and make the isolation and purification of the propane difficult. In other words, this is not a good way to make propane. Likewise, the Wurtz reaction is not useful to make any alkane with an odd number of carbon atoms.

111.7   Butane is C_4H_{10}, so we would use C_4H_9MgI (or Br or Cl in place of the I).

- Alkanes:
  Have the general formula C_nH_{2n+2}; e.g. methane, CH_4; ethane, C_2H_6.
- Preparation:
  (i) Reduction of halogenoalkanes by zinc in ethanol plus trace of copper(II) sulphate;

  e.g. C_2H_5I + Zn + 2H$^+$ → C_2H_6 + HI + Zn^{2+}
  iodoethane   ethane

  (ii) Sodium in dry ether plus an iodoalkane;

  e.g. $2CH_3I$ + 2Na → C_2H_6 + 2NaI
  iodomethane   ethane

  (iii) Decarboxylation of a salt of an organic acid by heating with soda lime;

  e.g. CH_3COONa + NaOH → CH_4 + Na_2CO_3
  sodium ethanoate   methane

  (iv) Reacting a Grignard reagent with water;

  e.g. $CH_3MgBr + H_2O → CH_4 + Mg(OH)Br$
  methane

Reactions

- Burn in air/oxygen:

  $$C_xH_y + (x + y/4)O_2 \rightarrow xCO_2 + (y/2)H_2O$$

- Free radical reaction with chlorine in presence of ultraviolet light:

  e.g. $CH_4 \xrightarrow{Cl_2,\ hf} CH_3Cl, CH_2Cl_2, CHCl_3, CCl_4 + HCl$

- Free radical reactions:
  Three main stages are
  (i) Initiation.
  (ii) Propagation.
  (iii) Termination.

112

Alkenes and alkynes

112.1 The structure of alkenes

Alkenes contain one or more double bonds. They contain fewer hydrogen atoms than an alkane with the same number of carbon atoms. An alkane is saturated with hydrogen atoms; an alkene is said to be *unsaturated*. A double bond is really the combination of a σ and a π bond between two carbon atoms (Figure 112.1). In Unit 16 we discovered how these bonds are made from the overlap of s and p orbitals. A π bond is not as strong as a σ bond, and is more easily broken. This is the reason why alkenes are more reactive than alkanes. However, the bond is sufficiently strong to stop the two ends of the molecule rotating. Shortly you will see that this results in many alkenes having isomers not found with alkanes.

The simplest homologous series contains only one double bond along a hydrocarbon chain. The structures of the early members of the series are shown in Table 112.1.

You can tell if a compound contains a double bond by

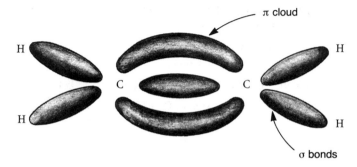

Figure 112.1 *An alkene has a double bond. This is a combination of a σ and a π bond*

the -*ene* ending to its name. In the first few members of the series there is no choice in where the double bond can be placed; but in the later members, we have to show the position of the double bond using a numbering system. As with alkanes, we find the longest carbon chain and number the carbon atoms along it. We then

Table 112.1. The alkenes, general formula C_nH_{2n}

Name	Formula	Structure	Melting point/°C	Boiling point/°C
Ethene	C_2H_4	H⟍C=C⟋H / H⟋ ⟍H	−169	−104
Propene	C_3H_6	CH_3⟍C=C⟋H / H⟋ ⟍H	−185	−48
But-1-ene	C_4H_8	CH_3—CH_2⟍C=C⟋H / H⟋ ⟍H	−185	−6
Pent-1-ene	C_5H_{10}	CH_3—CH_2—CH_2⟍C=C⟋H / H⟋ ⟍H	−165	29

write the lowest number of the two carbon atoms that have the double bond. Table 112.1 shows you the structure of but-1-ene. The '1' in the name tells you that the double bond is between the first and second carbon atoms. This molecule has an isomer, but-2-ene, in which the double bond runs between the second and third carbon atoms.

In fact there are *two* different types of but-2-ene. They are shown in Figure 112.2. They are known as *cis*- and *trans*-but-2-ene. The difference is in the arrangement of the hydrogen and methyl groups. In the *trans* isomer the two methyl groups are on opposite sides of the double bond (*trans* means across); in the *cis* isomer they are on the same side (*cis* means similar or same). *Cis* and *trans* isomers are sometimes called *geometric* isomers.

If there were free rotation about a double bond, these isomers could not exist. We have evidence that they do exist from various measurements. Nuclear magnetic resonance spectroscopy shows that the arrangements of the hydrogen atoms are different in the two isomers; infrared (vibrational) spectroscopy shows them up as well. However, one of the simplest pieces of evidence comes from measurements of dipole moments.

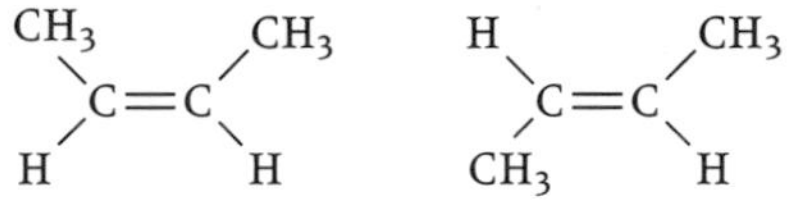

$$
\underset{\text{cis–But–2–ene}}{\overset{CH_3}{\underset{H}{>}}C=C\overset{CH_3}{\underset{H}{<}}} \qquad \underset{\text{trans–But–2–ene}}{\overset{H}{\underset{CH_3}{>}}C=C\overset{CH_3}{\underset{H}{<}}}
$$

Figure 112.2 *The two versions of but-2-ene. There is no rotation of the end groups round a double bond, so the molecules cannot change into one another*

112.1 Three different molecules with the formula $C_2H_2Cl_2$ are known. Two of them have dipole moments, and one does not. Draw structures of the molecules. (Models may help.)

112.2 A student said that the reason why alkenes are unsaturated is that they can be made by removing water from alcohols. Was she correct?

112.2 Ways of making alkenes

We shall examine two ways of making an alkene. The first is heating a halogenoalkane with a concentrated solution of potassium hydroxide in alcohol (called alcoholic potassium hydroxide). For example, if we wanted to make ethene we would warm a mixture of chloroethane, potassium hydroxide and ethanol. The reaction that takes place is called a *dehydrohalogenation* reaction. The name tells us what happens: a hydrogen and a halogen atom are removed. For example,

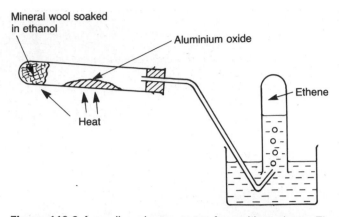

Figure 112.3 *A small-scale apparatus for making ethene. The aluminium oxide must be heated strongly and the alcohol only very slightly. (Great care must be taken to avoid the water sucking back into the reaction tube)*

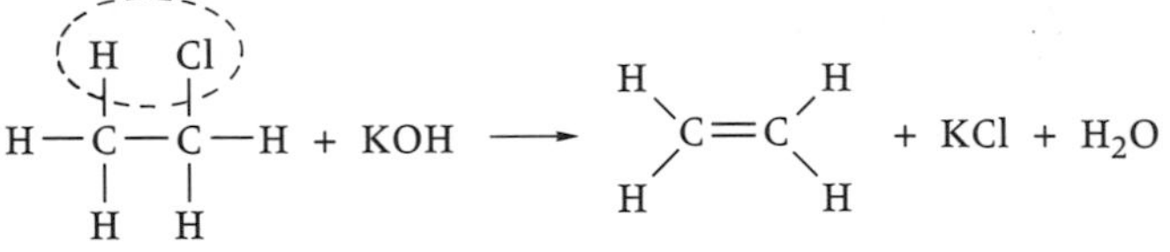

$$
\underset{\text{(H Cl)}}{H-\overset{H}{\underset{H}{C}}-\overset{Cl}{\underset{H}{C}}-H} + KOH \longrightarrow \overset{H}{\underset{H}{>}}C=C\overset{H}{\underset{H}{<}} + KCl + H_2O
$$

$$C_2H_5Cl + KOH \rightarrow C_2H_4 + KCl + H_2O$$

The second method of making an alkene is to *dehydrate* an alcohol. There are two ways of doing this. The simplest is to pass the alcohol vapour over a hot catalyst, often aluminium oxide. You might be able to do this in the laboratory using the apparatus shown in Figure 112.3. If we use ethanol, C_2H_5OH, we can write the reaction as

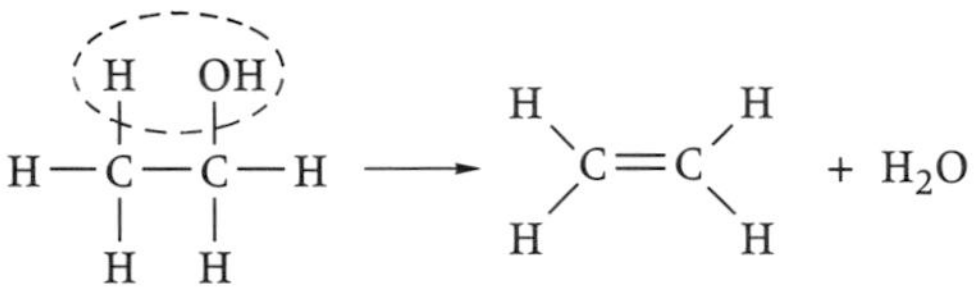

$$
\underset{\text{(H OH)}}{H-\overset{H}{\underset{H}{C}}-\overset{OH}{\underset{H}{C}}-H} \longrightarrow \overset{H}{\underset{H}{>}}C=C\overset{H}{\underset{H}{<}} + H_2O
$$

$$\underset{\text{ethanol}}{C_2H_5OH(g)} \rightarrow \underset{\text{ethene}}{C_2H_4(g)} + H_2O(g)$$

An alternative method is to use concentrated sulphuric acid. If ethanol is used, the mixture must be heated to about 180°C, but the temperature needed to dehydrate other alcohols is often less than this. Concentrated sulphuric acid has a strong attraction for water. A simple view of the reaction is that the acid acts as a dehydrating agent:

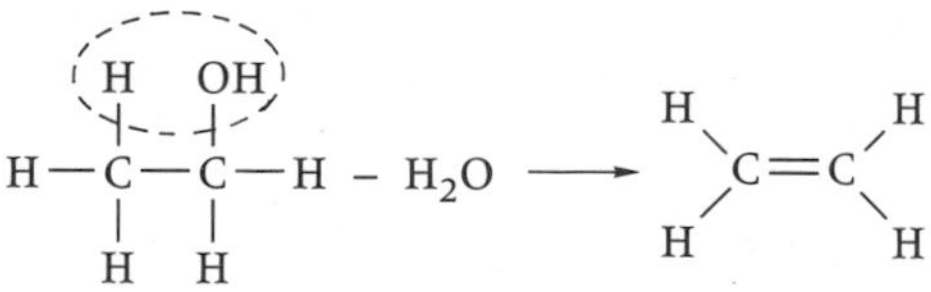

$$
\underset{\text{(H OH)}}{H-\overset{H}{\underset{H}{C}}-\overset{OH}{\underset{H}{C}}-H} - H_2O \longrightarrow \overset{H}{\underset{H}{>}}C=C\overset{H}{\underset{H}{<}}
$$

$$C_2H_5OH - H_2O \rightarrow C_2H_4$$

However, the mechanism of the reaction is more complicated than this simple idea suggests. You can read about the mechanism in panel 112.1 and Figure 112.4 if you wish.

Panel 112.1

The mechanism of the dehydration of alcohols by concentrated sulphuric acid

Like water, alcohols have an OH group. This brings with it two lone pairs, and just as water molecules can be protonated:

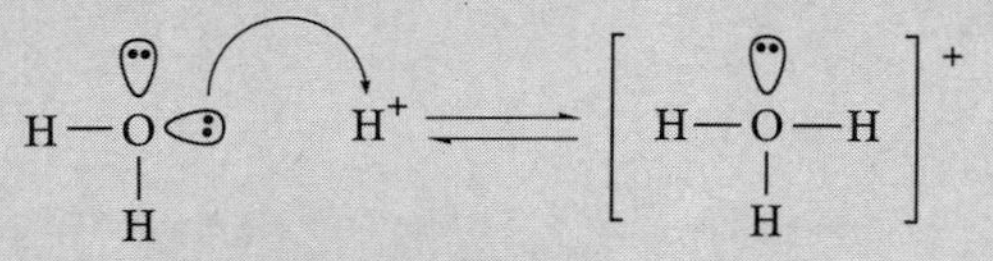

$$H_2O(l) + H^+(aq) \rightleftharpoons H_3O^+(aq)$$

so too can ethanol molecules be protonated:

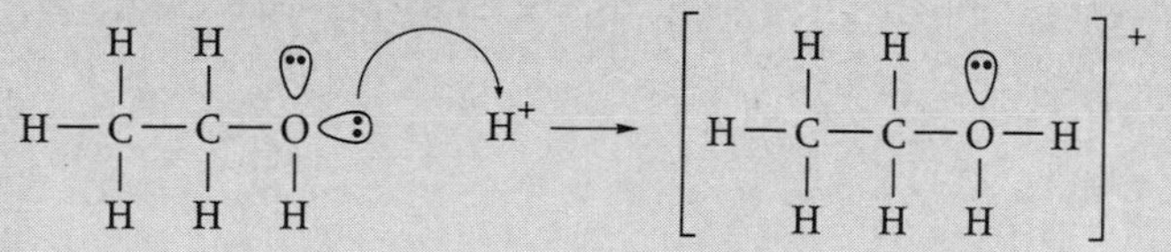

It is believed that the water attached to the carbon atom is lost, leaving a carbocation behind. The positive charge on the end carbon atom attracts electrons towards it. If you build a scale model of ethanol, you will find that the hydrogen atoms on the end methyl group are very close to the carbon atom holding the charge. It takes very little movement for the pair of electrons making the bond to one of the hydrogen atoms to be attracted to the positive carbon. If the pair of electrons are drawn away, the hydrogen atom is left just as a bare proton, i.e. hydrogen ion, H^+. We can show these changes on a diagram (Figure 112.4).

According to this mechanism, the sulphuric acid donates a proton, which is later released back to the acid. Therefore, it suffers no permanent change during the reaction. It is a catalyst.

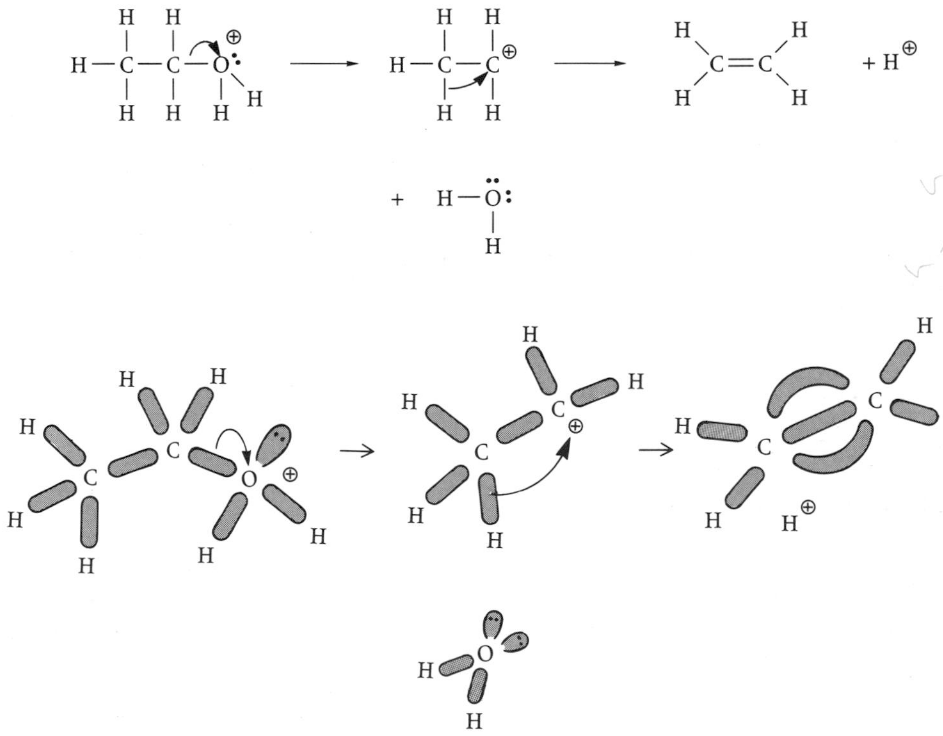

Figure 112.4 *Two ways of representing (i) the loss of a water molecule from a protonated ethanol molecule followed by (ii) rearrangement of the carbonium ion to give an alkene. In the bottom set, the σ bonds are shown, together with the final π bond. In both cases, the curly arrows indicate the movement of a pair of electrons from one place to another. In order to draw attention to the positive charges, we have drawn circles around the + signs.*

112.3 The oxidation of alkenes

The easiest way of oxidising an alkene is to burn it. Like alkanes, alkenes burn to give carbon dioxide and water. However, alkenes contain more carbon in proportion to hydrogen than do alkanes. This means that they need relatively more oxygen to burn completely than do alkanes. If you burn an alkene in the laboratory you will find that it gives a smoky flame because there is insufficient oxygen in the neighbourhood of the flame to turn all the carbon into carbon dioxide.

A more interesting way to oxidise an alkene is to react it with an alkaline solution of potassium manganate(VII), a solution known as *Baeyer's reagent*. The π

bond breaks and is replaced by two OH groups. For example,

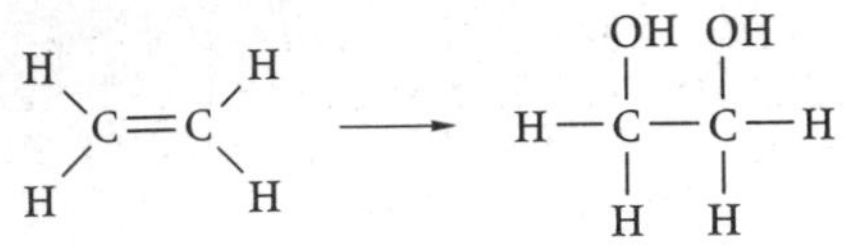

The product, ethane-1,2-diol, is better known as glycol (or ethylene glycol). It is used in huge quantities (in colder regions of the world) as antifreeze for cars.

If you carry out this reaction you will see the purple colour of the solution fade. It is replaced by a cloudy brown colour caused by the precipitation of manganese(IV) oxide.

112.3   What will be made if but-2-ene,

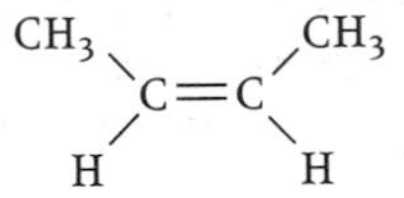

is reacted with Baeyer's reagent?

112.4   (i) Write down the balanced chemical equation for the reaction of alkaline potassium manganate(VII) with ethene. You might like to use these two half-equations:

$$MnO_4^-(aq) + 4H^+(aq) + 3e^- \rightarrow MnO_2(s) + 2H_2O(l)$$

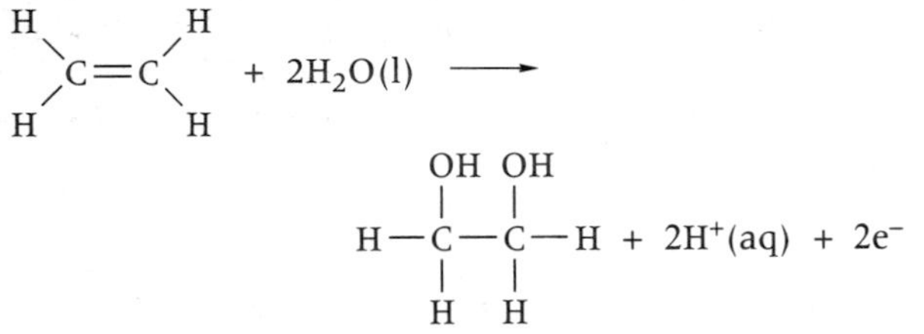

(ii) Explain why these equations show that the reaction is an oxidation of the alkene.

112.4   The addition reactions of alkenes

The most important reactions of alkenes are *addition* reactions. In the simpler addition reactions, the double bond is replaced by two single bonds to two new atoms. We can show the general rule for addition in this way:

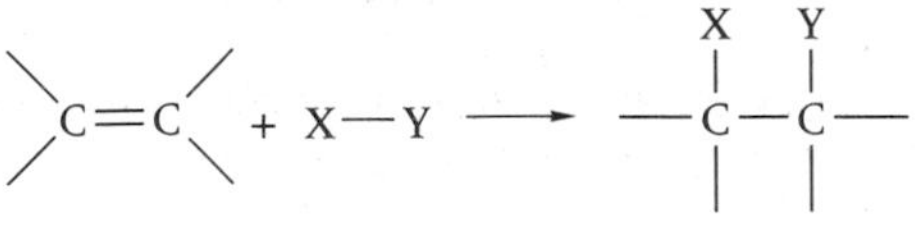

Table 112.2 lists the addition reactions that you need to know. Three other types of addition reaction are *polymerisation*, *ozonolysis* and the *oxo reaction*. We shall leave these until later.

Table 112.2. The addition reactions of alkenes

Reaction with	X — Y	Conditions
Hydrogen 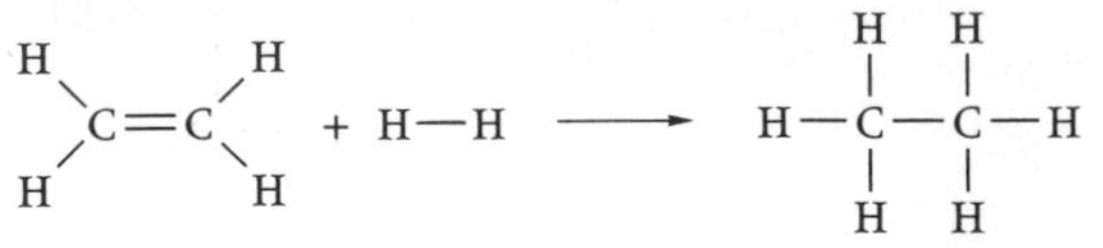	H — H	Heat with a Ni catalyst at about 140 °C

| Bromine 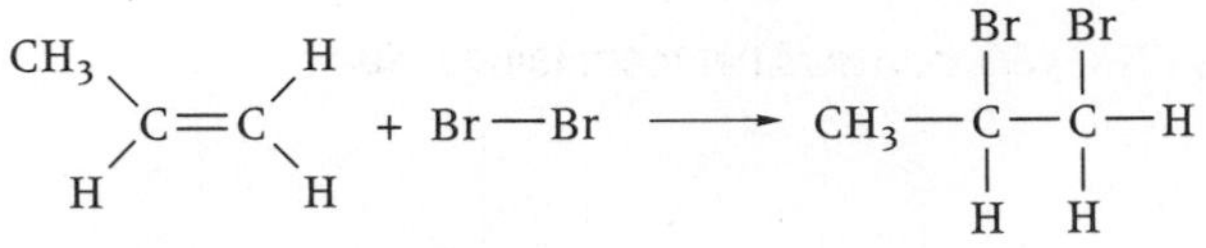 | Br — Br | Shake the alkene with bromine in the cold |

| Hydrogen chloride 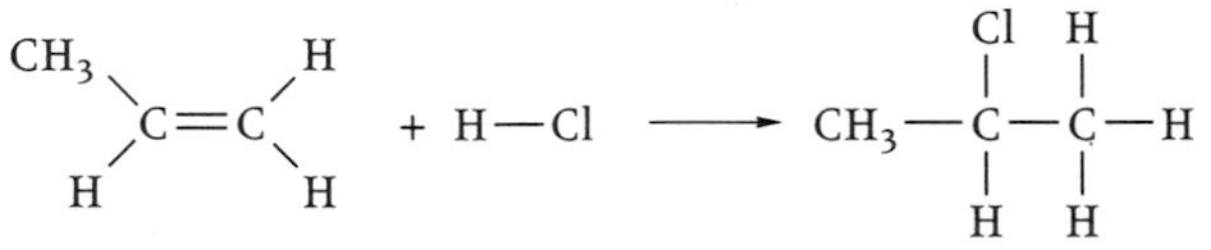 | H — Cl | Direct reaction |

| Water 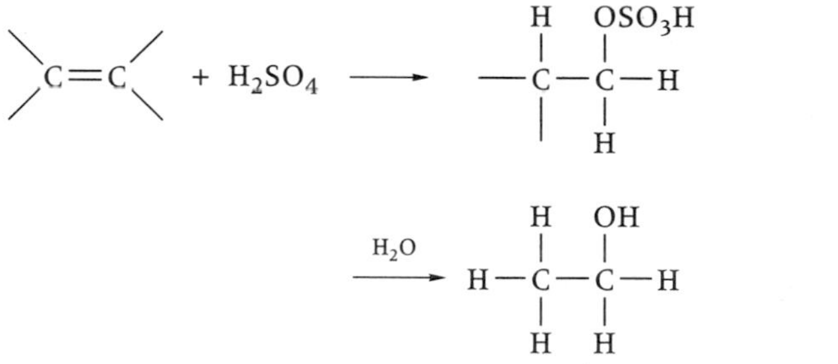 | H — OH | Pass alkene into concentrated sulphuric acid, and then dilute carefully with water |

(a)   Hydrogenation

The addition of hydrogen across a double bond is called hydrogenation. Nickel is often used as a catalyst, although other transition metals will work as well. Hydrogenation is widely used in the manufacture of margarine from unsaturated fats (Figure 112.5), and in the oil industry to convert alkenes into alkanes.

(b)   Test for unsaturation

The reaction of an alkene with bromine is particularly important as it acts as a simple test for an alkene. Bromine is a rich red-brown colour, which changes to colourless when it adds to an alkene. Bromine water can be used instead of pure bromine, but in either case, if you do this test, take care. Bromine is dangerous, and the products of the addition reactions with alkenes are often carcinogenic. If you are interested, you will find the mechanism of the reaction in panel 112.2.

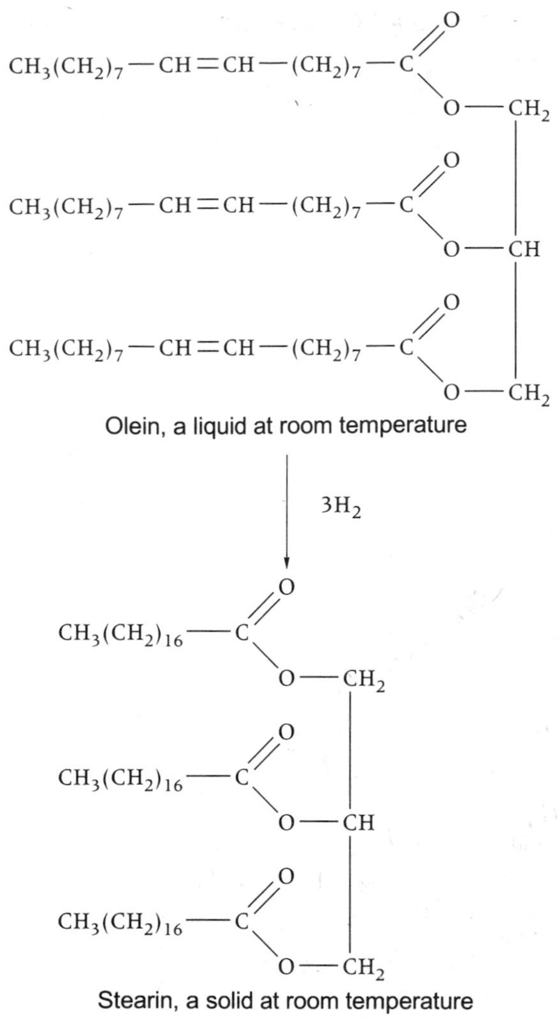

$$CH_3(CH_2)_7-CH=CH-(CH_2)_7-C \overset{\displaystyle O}{\underset{\displaystyle O-CH_2}{}}$$

$$CH_3(CH_2)_7-CH=CH-(CH_2)_7-C \overset{\displaystyle O}{\underset{\displaystyle O-CH}{}}$$

$$CH_3(CH_2)_7-CH=CH-(CH_2)_7-C \overset{\displaystyle O}{\underset{\displaystyle O-CH_2}{}}$$

Olein, a liquid at room temperature

$\downarrow$ $3H_2$

$$CH_3(CH_2)_{16}-C \overset{\displaystyle O}{\underset{\displaystyle O-CH_2}{}}$$

$$CH_3(CH_2)_{16}-C \overset{\displaystyle O}{\underset{\displaystyle O-CH}{}}$$

$$CH_3(CH_2)_{16}-C \overset{\displaystyle O}{\underset{\displaystyle O-CH_2}{}}$$

Stearin, a solid at room temperature

Figure 112.5 *The conversion of an unsaturated fat into a saturated fat*

Panel 112.2

The mechanism for the addition of bromine to an alkene

First, let us think about experimental evidence concerning addition reactions. If, say, ethene is passed into a solution that contains chloride ions but no bromine, there is no reaction. This shows that chloride ions cannot join to ethene directly. However, if bromine is added to the solution, a reaction does occur. If you were to analyse the products you would find that, along with 1,2-dibromoethane, there was a considerable amount of 1-bromo-2-chloroethane. This tells us something important about the way the reaction takes place. Chloride ions are negatively charged, and we know that they will react rapidly with positive ions. Can it be that, during the course of the reaction with bromine, positive ions are made? We believe they are.

If you were a bromine molecule approaching the π bond of an alkene, the first thing you would feel

---

would be the negative charge of the electrons forming the bond. The electrons round the bromine atom nearest to the π bond would be slightly repelled. This atom is left with a slight positive charge while the other end of the molecule gains a slight negative charge. We say that the bromine molecule is *polarised* (see Unit 19):

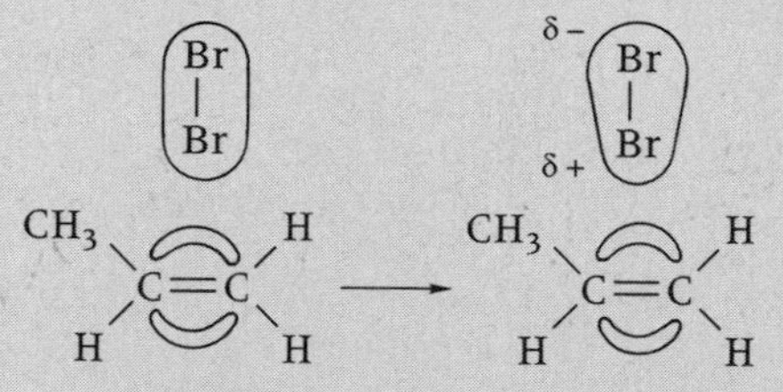

Once this process begins, the positive end of the bromine molecule becomes a centre of attraction for the π electrons. This pair of electrons makes a bond to the bromine atom, while at the same time the electrons making the bond between the bromine atoms gather around the second bromine atom:

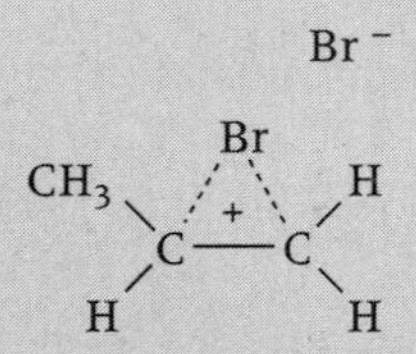

When this stage is complete, we have an intermediate that carries a positive charge; a carbocation has been made. Also, a bromide ion is left floating around. The carbocation will change so that it has the lowest energy. In our example this is the first of the ions shown below:

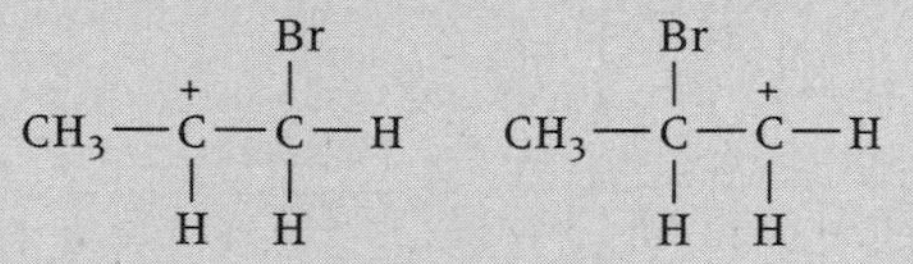

The reason for this lies in the ability of methyl groups (CH_3) to feed electrons into a centre of positive charge. We say they show a *positive inductive effect*. The transfer of electron density towards the carbon atom carrying the charge means that the amount of positive charge on that atom is reduced. Similarly, the regions on the methyl group that lose the electron density gain a slight positive charge. In this way the positive charge is spread, rather than concentrated in one place. We made the point in section 14.7 that such a spread of charge leads to a more energetically stable state.

The bromide ions in the solution can attack the carbocation to give the final product:

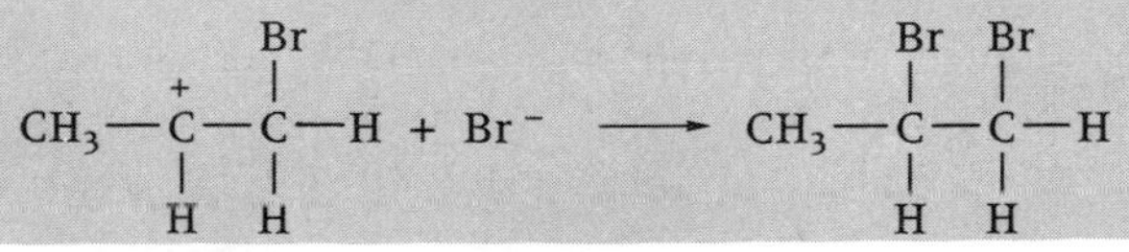

(In practice, both the attack by bromide ion and the rearrangement of the charge may take place together.)

If the solution is flooded with chloride ions, then one of them is more likely than a bromide ion to collide and join to the carbocation. In this way we have explained how 1-bromo-2-chloroethane is made from ethene:

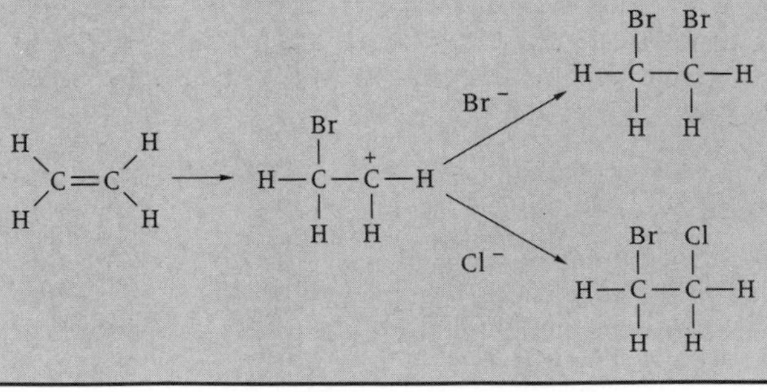

(c) Hydrogen chloride

The addition of hydrogen chloride to alkenes has been widely investigated. Unlike hydrogen or bromine, hydrogen chloride is an unsymmetrical molecule. With ethene the product is the same no matter which end of the molecule the chlorine joins, but with other alkenes the outcome can be two different molecules. For example, with propene, the outcome is either 1-chloropropane, $CH_3CH_2CH_2Cl$, or 2-chloropropane, $CH_3CHClCH_3$:

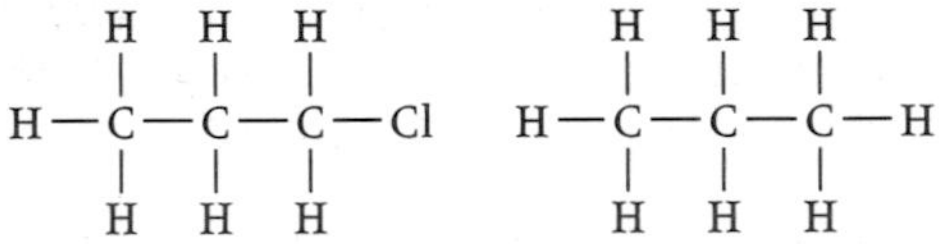

If you were to do the reaction, you would find that the product is almost entirely 2-chloropropane. In 1869 the Russian chemist Vladimir Markovnikoff discovered the rule that predicts the correct product of the reaction. *Markovnikoff's rule* says that:

> **When hydrogen chloride adds to a double bond the hydrogen joins to the carbon that already has the most hydrogen atoms bonded to it.**

In fact the rule applies equally to the addition of hydrogen bromide and hydrogen iodide. However, there are a number of conditions that must be met. For example, no free radicals are involved (which can happen if peroxides are present) and only addition to hydrocarbons is involved. Here are some examples that illustrate the rule.

(i) Addition of HCl to but-1-ene

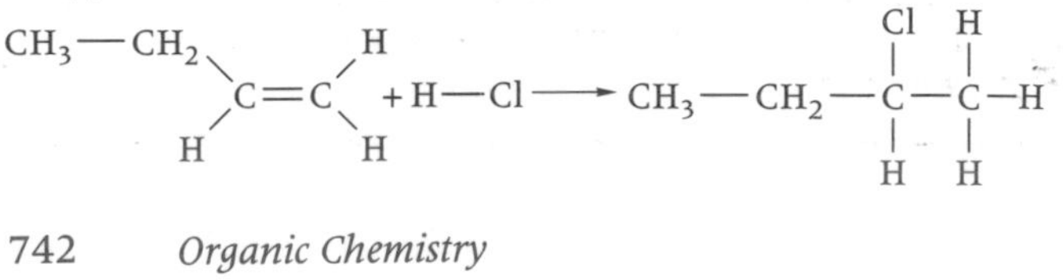

(ii) Addition of HI to pent-1-ene

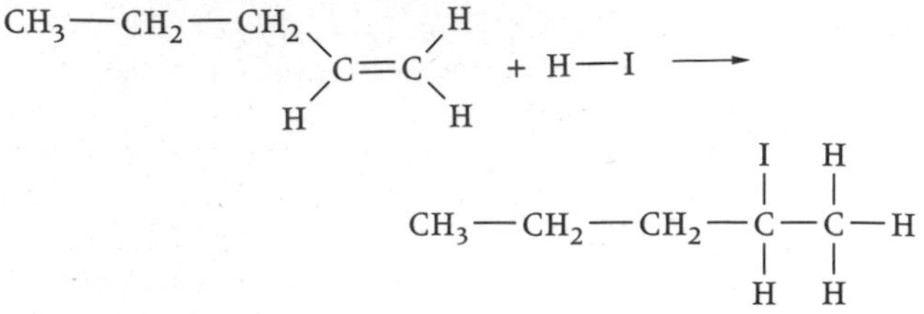

The reason why the addition takes place in this way is explained in panel 112.3 and Table 112.3.

Panel 112.3

The mechanism for the addition of hydrogen halides to alkenes

If you have understood the mechanism for the addition of bromine, you will see obvious similarities with this mechanism. We shall use the addition of hydrogen chloride to ethene as our first example. Hydrogen chloride is already polarised, the chlorine being slightly negatively and the hydrogen slightly positively charged. If the hydrogen end of the molecule approaches the π cloud of electrons, the π bond begins to break, and the hydrogen chloride molecule becomes even more polarised. This bond breaks, leaving the negatively charged chloride ion and a carbocation. A chloride ion will combine with the carbocation to give the final product:

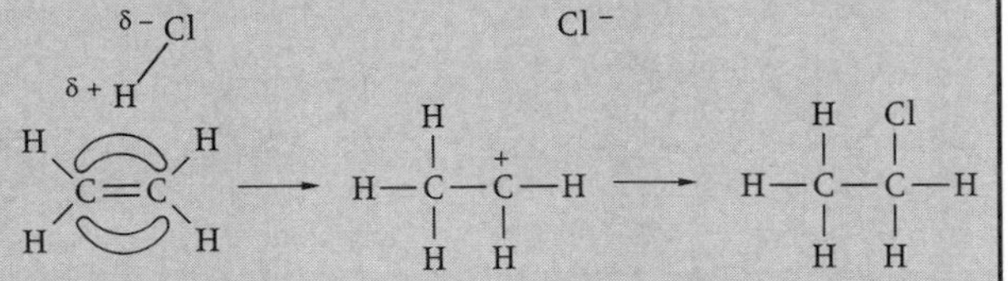

Now let us see what might happen if hydrogen chloride adds to propene. There are two different carbocations that can be made depending on how the hydrogen attaches itself:

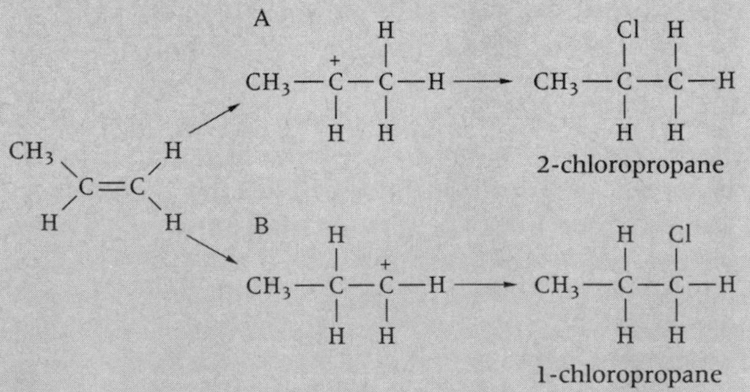

We know that the intermediate A is more favoured than the second (B) because, as Markovnikoff discovered, the product is 2-chloropropane. This suggests that the carbocation A is made more readily than the other (B). The more methyl groups there are attached to the carbon atom with the positive charge, the greater is the gain in energetic stability. This gives us the order of energetic stability shown in Table 112.3.

Now we can see why Markovnikoff's rule works. The carbocation that is made during the initial stage of the reaction is the most energetically stable one. All other things being equal, this will be the one that gives the most methyl groups attached to the carbon atom carrying the positive charge. You can see the effects of this in the pattern of results shown for the addition of HCl to propene above.

Table 112.3. Relative energetic stability of carbocations

Diagram	Name	Energetic stability
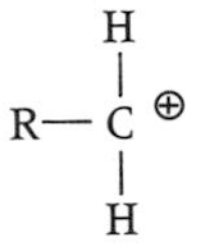	Primary carbocation	Least stable
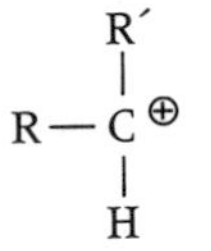	Secondary carbocation	
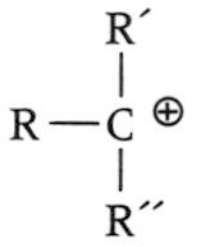	Tertiary carbocation	Most stable

(d) Addition of water

In some ways you can think of this as the reverse of the preparation of an alkene from an alcohol. The first step in the reaction is to absorb the alkene in concentrated sulphuric acid. This is when the addition takes place:

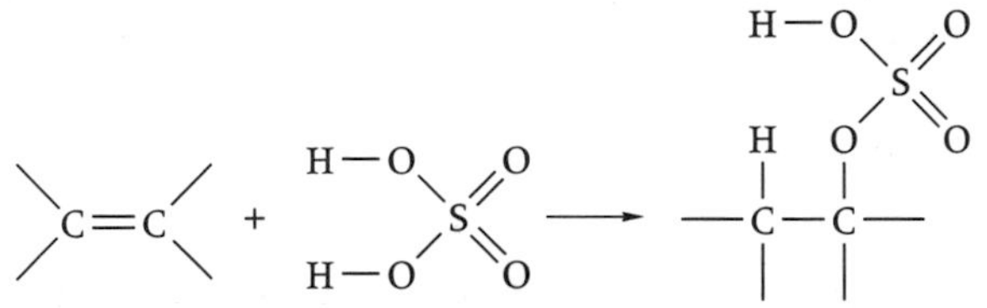

The second step is to warm the reaction mixture with water. An OH group from a water molecule displaces the hydrogensulphate and a molecule of sulphuric acid is released:

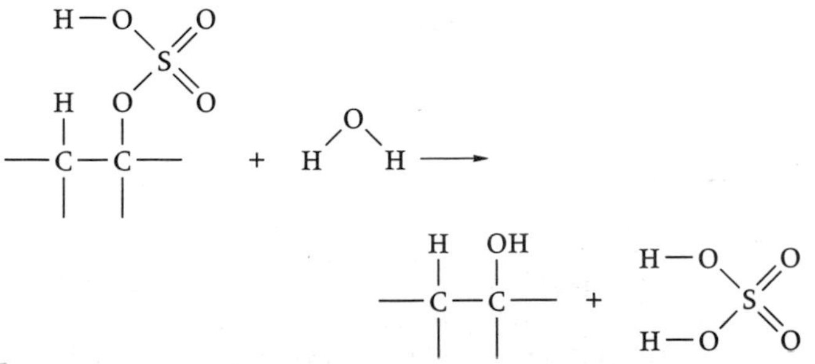

112.5 You should realise that, in a hydrogenation reaction, one double bond takes up one molecule of hydrogen, two double bonds take up two molecules, and so on. This fact can be used to discover the number of double bonds in a molecule.

Here is an example. The mass spectrum of a hydrocarbon showed a parent ion that corresponded to a molar mass of $82\,g\,mol^{-1}$. $2.03\,g$ of the hydrocarbon were found to absorb approximately $600\,cm^3$ of hydrogen (measured at room temperature and pressure).

(i) How many moles of hydrogen gas were absorbed?

(ii) How many moles of the hydrocarbon were used?

(iii) What is the ratio of the moles of hydrocarbon to moles of hydrogen?

(iv) How many double bonds does one molecule of the hydrocarbon contain?

(v) What might be the structure of the hydrocarbon? (Hint: think about alkanes that have a similar formula, and do not worry if you go round in circles.)

112.6 The addition of sulphuric acid to an alkene obeys Markovnikoff's rule. Predict the structure of the alcohols that will be made by the addition of the acid to (i) propene, (ii) but-1-ene, (iii) but-2-ene, followed by reaction with water.

112.7 What would you expect to be made if you passed ethene into a solution containing a mixture of bromine, chloride ions and nitrate ions?

112.8 Is it true that a methyl group is polarisable? Briefly explain your answer. If you need to do so, look at Unit 19 to help you decide your answer to this question.

112.9 Predict, and explain, the most likely product for the addition of hydrogen chloride to 2-methyl-but-2-ene.

112.10 Draw the structure of the alcohol you would expect to obtain if 2-methylpropene were absorbed in concentrated sulphuric acid and then reacted with water.

112.11 If bromine water is used to test for an alkene, the water contains HOBr. What will be made if this molecule adds across the double bond in ethene? Draw the structure.

112.12 Here is a description of the reaction of ethene with bromine: 'During the reaction the double bond breaks and bromine adds on to the molecule.'

This is not quite right. What is wrong with the description?

112.5 Ozonolysis

In an ozonolysis reaction, ozone (trioxygen), O_3, is usually bubbled through a solution of the alkene in an organic solvent. The ozone adds across a double bond in a rather unusual manner:

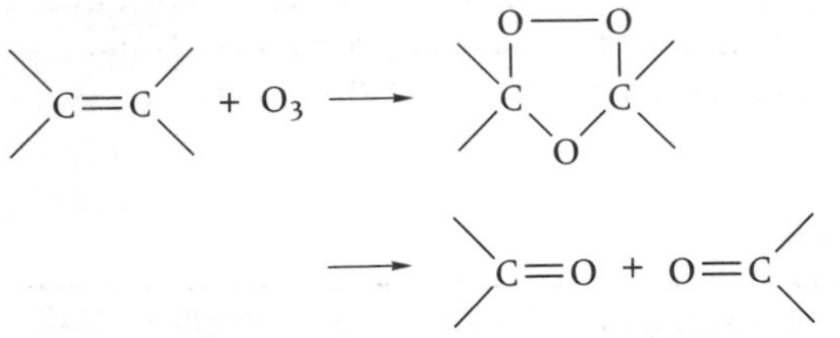

The molecule made in the reaction is an *ozonide*. Ozonides are not particularly stable (either energetically or kinetically). If the ozonide is warmed with a combination of zinc and water, the molecule breaks up. Oxygen atoms bond to each of the carbon atoms that had the double bond between them. The products are aldehydes or ketones (or mixtures of both). Here is an example in which an alkene was found to give propanone and methanal as the products:

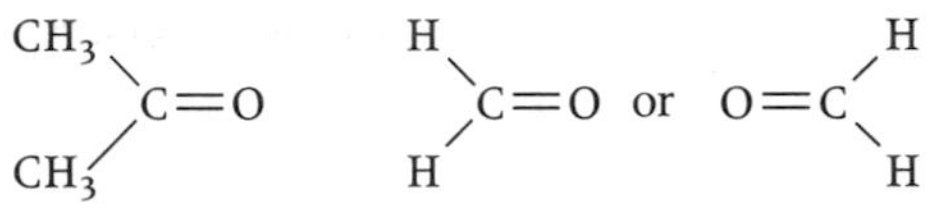

propanone   methanal

What was the original molecule? All we have to do is to take away the oxygen atoms and join the carbon atoms by a double bond. This gives us:

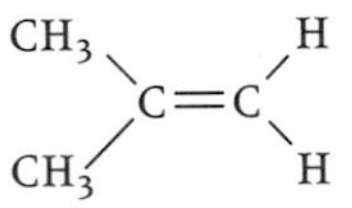

2-methylpropene

Ozonolysis gives us a method of discovering which groups are attached to each end of the double bond. Needless to say the method only works if we can identify the aldehydes or ketones made during the reaction. You will find in Unit 118 that this can usually be done fairly easily.

112.13 After an ozonolysis experiment, the only product was ethanal, CH_3CHO. The chemist who did the experiment correctly claimed that there were two different structures for the starting material. What were they?

(To answer this question you may need to make molecular models of ethanal, remove the oxygen and try to join the remaining parts together.)

112.6 Polymerisation

Alkenes undergo *addition polymerisation*. You can get an idea of why this happens by looking at Figure 112.6.

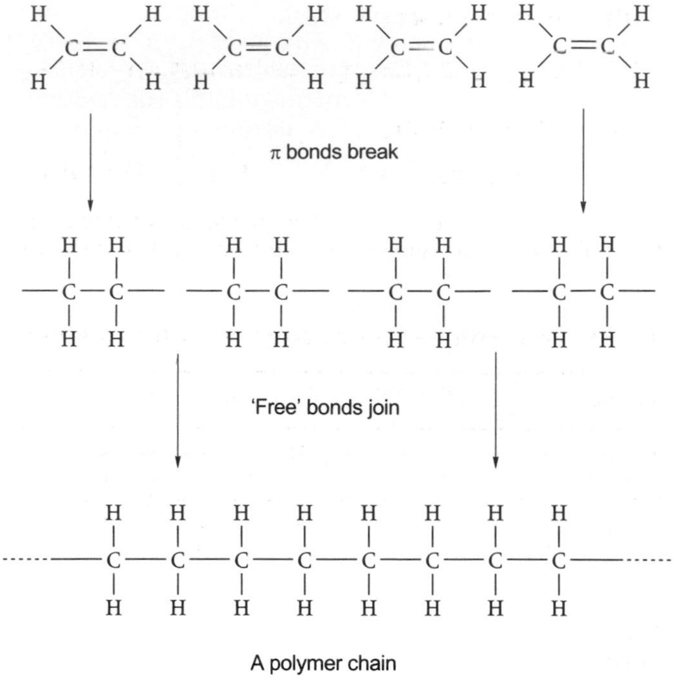

Figure 112.6 *An extremely simplified version of how ethene molecules are converted into poly(ethene) (polythene)*

Here each of the three neighbouring ethene molecules has by some happy chance had its double bond broken. We have arranged things so that the bond breaks homolytically, i.e. one electron stays with each of the two carbon atoms. Next we have shown the unpaired electrons recombining to give a bond pair, but this time the bond is between carbon atoms on different molecules. If this happens, a chain of molecules is made. In a typical polymer chain, many tens of thousands of alkene molecules may join.

Ethene molecules are the *monomer* units that combine to give the *polymer*. In this case the polymer is polyethene, or polythene for short.

In reality, of course, alkenes will not sit neatly arranged breaking their π bonds for no good reason. Persuasion in the form of high pressures and an extra reactant are often needed. The chemical added is the *initiator*. As their name suggests, initiators take part in the initiation step of a free radical reaction. Organic peroxides make good initiators in a free radical polymerisation. These molecules have two short hydrocarbon chains, which we shall write as R, joined by a peroxide link. Thus we can show an organic peroxide as ROOR. When they are heated, the peroxide bond breaks homolytically, producing free radicals.

(i) Initiation step

$$ROOR \rightarrow 2RO\cdot$$

These radicals attack the alkene.

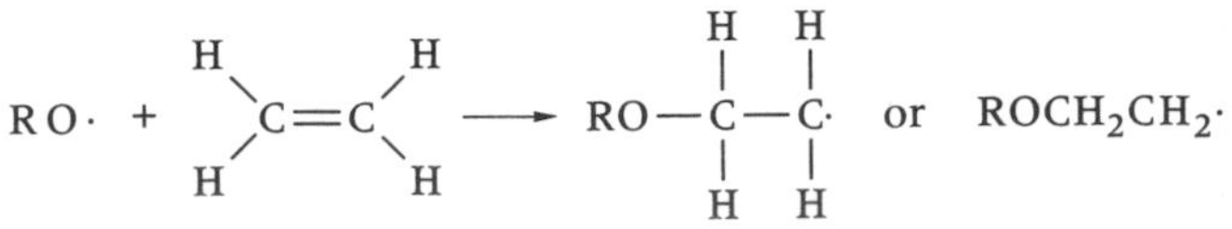

(ii) Propagation step

$$ROCH_2CH_2\cdot \ + \ CH_2{=}CH_2 \longrightarrow ROCH_2CH_2CH_2CH_2\cdot$$

(iii) Termination step

$$ROCH_2(CH_2)_nCH_2\cdot \ + \ RO\cdot \longrightarrow ROCH_2(CH_2)_nCH_2OR$$

As you can see, it is during the propagation stage that the polymer chain grows. At some stage the reaction terminates, leaving polymer chains of varying length, but often many thousands of monomer units long. A typical *average* molar mass would be 20 000 to 30 000 g mol^{-1}. Note that we must speak of the average molar mass. The product of a polymerisation experiment consists of a mixture of chains of different lengths.

A remarkable variety of alkenes have been used in addition polymerisation experiments. Table 112.4 lists examples of the more important ones.

Table 112.4. Some alkenes and their addition polymers

Monomer	Section of polymer	Short formula	Name
Ethene 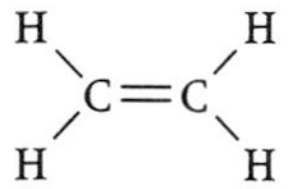	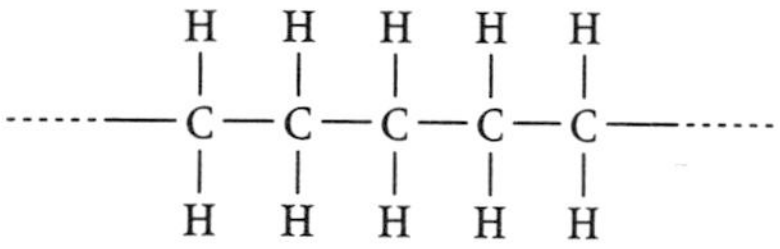	$+CH_2+_n$	Polyethene (polythene)
Chloroethene (old name: vinyl chloride) 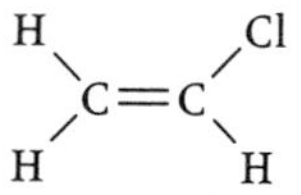	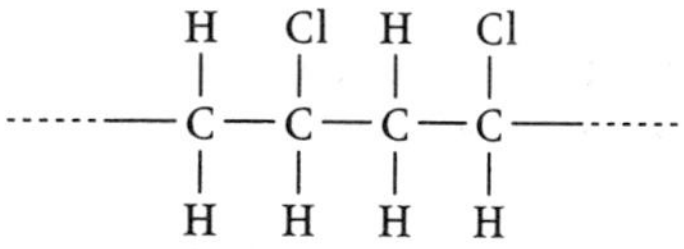	$+CH_2CHCl+_n$	Poly(chloroethene) (polyvinyl chloride, PVC)
Styrene 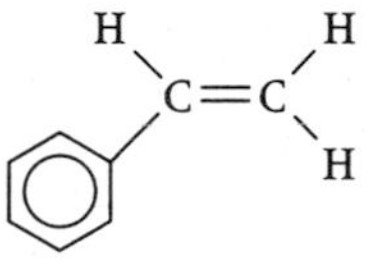	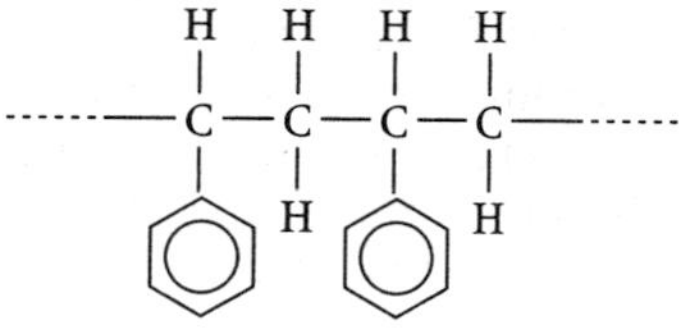	$+CHC_6H_5CH_2+_n$	Polystyrene
Tetrafluoroethene 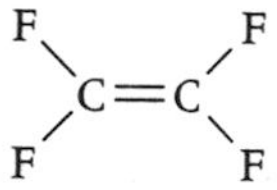	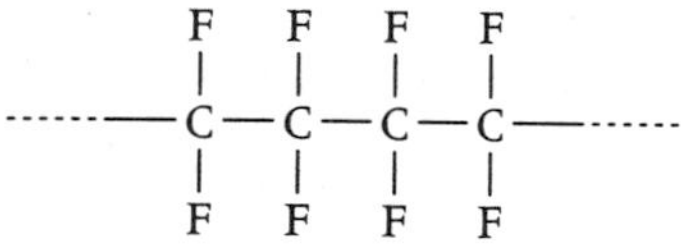	$+CF_2+_n$	Poly(tetrafluoroethene) (PTFE, Teflon)

112.14 What was the monomer that would produce this polymer:

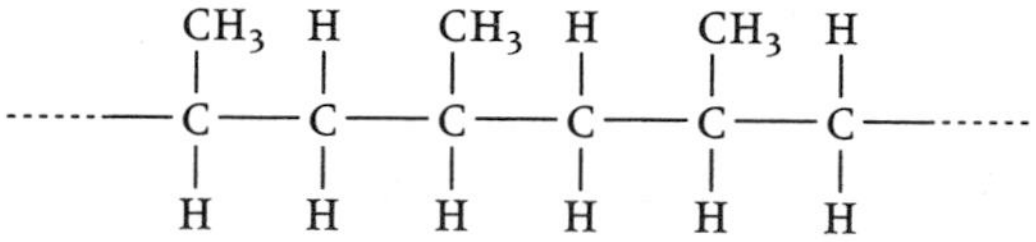

112.15 The molecule 2-methylbuta-1,3-diene,

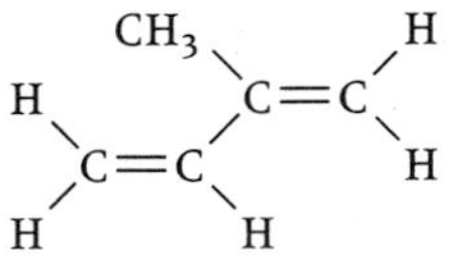

is otherwise known as isoprene. It is one of the basic building blocks of rubber and a number of other natural products. Try to discover how isoprene units can combine to make the two substances shown in Figure 112.7.

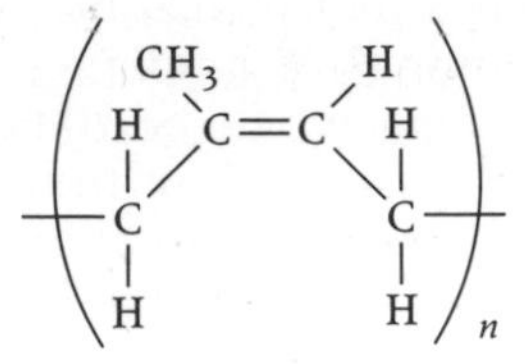

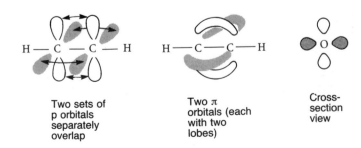

The repeat unit of natural rubber

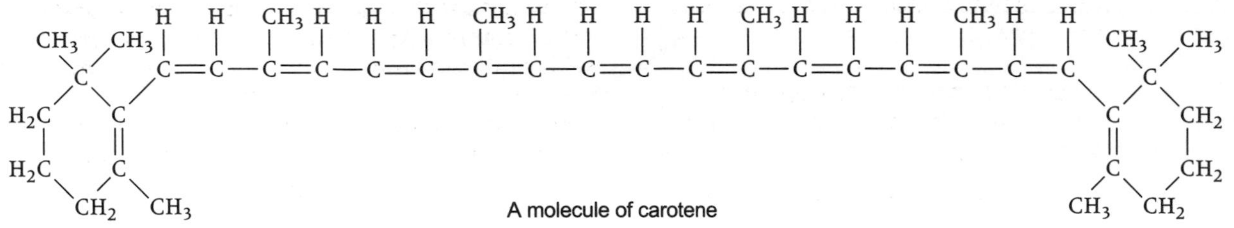

A molecule of carotene

Figure 112.7 *Diagrams for question 112.5*

112.7 The oxo reaction

This is the name given to a number of different reactions that have great importance in the chemical industry. The variety we are concerned with is the conversion of alkenes into aldehydes, which in turn may be converted into alcohols. This is achieved by reacting an alkene with a mixture of carbon monoxide and hydrogen in the presence of a catalyst. This is also called *hydroformylation*. Often the catalyst is a carbonyl compound of a transition metal, e.g. $Co(CO)_4$. The reaction is carried out at high pressures, around 200 atm, and high temperatures, around 200 °C.

One reason why the reaction is important is that it is one of the few ways of directly increasing the number of carbon atoms in a compound, For example,

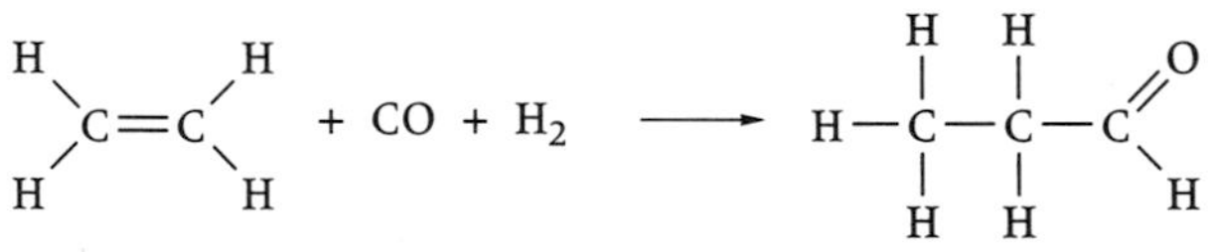

$$CH_2{=}CH_2(g) + CO(g) + H_2(g) \rightarrow CH_3CH_2CHO(g)$$
propanal

followed by

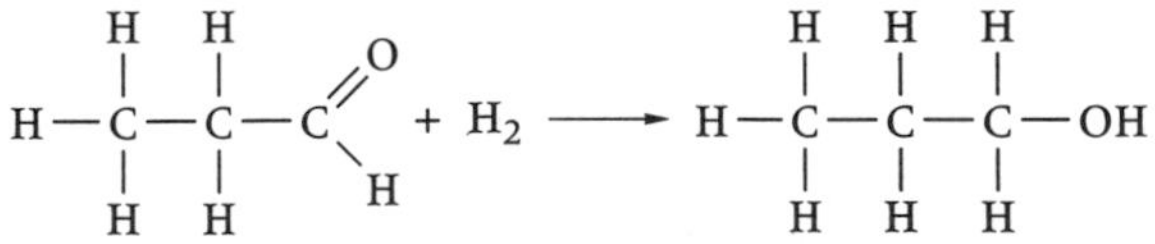

$$CH_3CH_2CHO(g) + H_2(g) \rightarrow CH_3CH_2CH_2OH(g)$$
propan-1-ol

> **112.16** Predict the result of using the alkene, propene, in the oxo process.

112.8 The structure and preparation of alkynes

Alkynes have at least one triple bond. A triple bond is a combination of one σ and two π bonds. We described how these bonds are made in section 16.8.

The simplest alkyne is ethyne, C_2H_2 (Figure 112.8). This gas is easily prepared. All you have to do is add water to a metal carbide such as calcium dicarbide, CaC_2:

$$CaC_2(s) + 2H_2O(l) \rightarrow Ca(OH)_2(s) + C_2H_2(g)$$

Few people use this reaction now, but at one time it was widely used in acetylene lamps (acetylene is the old name for ethyne). In essence, the lamps were very simple: calcium dicarbide was placed in the bottom of a tin and water slowly dripped onto it. The gas escaped through a narrow opening, where it was lit. The gas burns with a vivid flame, and the lamps were used as lights on bicycles and horse-drawn carriages, as well as the early motorcars. Unfortunately, there were many

Two sets of p orbitals separately overlap

Two π orbitals (each with two lobes)

Cross-section view

Figure 112.8 *Diagram showing the bonding in ethyne, [C₂H₂ (or H—C≡C—H). The triple bond is made from one σ and two π bonds*

accidents because a mixture of ethyne and air (oxygen) will explode if the gases are mixed in the right proportions.

112.17 Are alkynes unsaturated?

112.18 (i) Write down the equation for ethyne burning in oxygen.

(ii) What proportions of ethyne and oxygen will give a highly explosive mixture?

(iii) What do you think went wrong with acetylene lamps to make them explode?

112.19 (i) The standard heat of combustion of ethyne at $25\,^{\circ}C$ is approximately $-2600\,kJ\,mol^{-1}$. Comment on this value.

(ii) Why is ethyne (acetylene) used in oxy-acetylene welding?

112.9 The properties of alkynes

We have already mentioned the fact that alkynes will burn, but given their structure, we might expect their properties to be similar to those of alkenes, and this is often the case.

Especially, alkynes give addition reactions. For example, they will decolourise bromine or bromine water. However, their reactions are often more violent than those of alkenes. For example, chlorine will explode with ethyne:

$$H-C\equiv C \quad H + Cl_2 \longrightarrow \begin{array}{c} Cl \quad Cl \\ | \quad | \\ H-C-C-H \\ | \quad | \\ Cl \quad Cl \end{array}$$

1,1,2,2-tetrachloroethane

Ethyne will polymerise. If it is heated to $400\,^{\circ}C$ the product of the reaction is benzene:

$$3C_2H_2(g)\rightarrow C_6H_6(g)$$

More interesting than this is a fairly recent discovery that, if ethyne is polymerised in the presence of iodine, then the polymer that is made will conduct electricity. This is a remarkable thing. Totally covalent materials are almost always insulators.

However, not all properties of alkynes are like those of alkenes. In particular, one reaction that only alkynes will show is that they give precipitates with solutions of silver or copper salts. For example, ethyne gives a white precipitate of silver dicarbide (silver acetylide), AgC_2, with a combination of silver nitrate and ammonia solutions. With a solution containing copper(I) chloride and ammonia, a red precipitate, CuC_2, is produced. *Warning*: Do not attempt these reactions. If the precipitates dry out they can explode unpredictably!

The reaction of ethyne with Baeyer's reagent is a little different to that of ethene. Oxidation takes place, but the product is not an alcohol. Instead, ethanedioic acid (oxalic acid) is produced:

$$H-C\equiv C-H \longrightarrow \begin{array}{c} O \qquad\qquad O \\ \diagdown \qquad \diagup \\ C=C \\ \diagup \qquad \diagdown \\ HO \qquad\qquad OH \end{array}$$

112.20 Ethyne can be prepared by using 1,2-dibromoethane as a starting material. What else would you use in the preparation? (Hint: look back at methods of preparing alkenes.)

112.21 Predict the result of reacting hydrogen chloride with ethyne. Give the name of the product and draw the structure. (Hint: you can treat ethyne as a molecule with two π bonds, and remember Markovnikoff's rule.)

Answers

112.1

trans-1,2-dichloroethene has no dipole

cis-1,2-dichloroethene has a dipole

1,1-dichloroethene has a dipole

112.2 No. In organic chemistry unsaturation has nothing to do with water. The name means that an unsaturated compound could contain more hydrogen atoms than it has at present.

112.3 The result is butan-2,3-diol

$$\begin{array}{c} OH \quad OH \\ | \qquad | \\ CH_3-C-C-CH_3 \\ | \qquad | \\ H \qquad H \end{array}$$

Answers – cont.

112.4 (i)

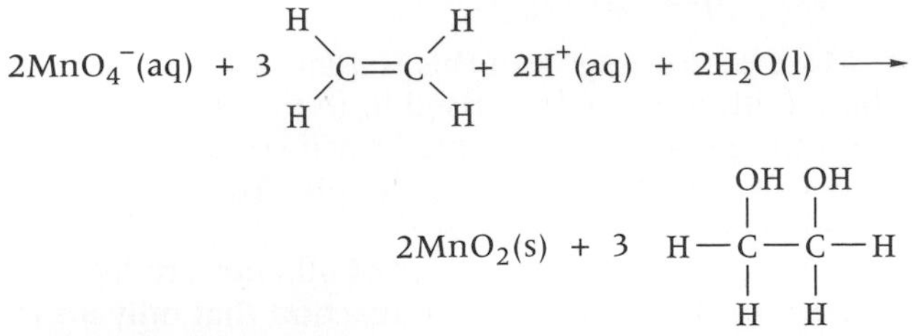

$$2MnO_4^-(aq) + 3\ \underset{H}{\overset{H}{>}}C=C\overset{H}{\underset{H}{<}} + 2H^+(aq) + 2H_2O(l) \longrightarrow$$

$$2MnO_2(s) + 3\ \ H-\underset{\underset{H}{|}}{\overset{\overset{OH}{|}}{C}}-\underset{\underset{H}{|}}{\overset{\overset{OH}{|}}{C}}-H$$

(ii) The half-equations show that the manganate(VII) ion gains electrons, so it is acting as an oxidising agent. Alternatively, we can say that, because the alkene has gained oxygen, it has been oxidised.

112.5
(i) One mole of gas occupies approximately 24 dm³ at room temperature and pressure, so we have 600 cm³/24 000 cm³ mol⁻¹ = 0.025 mol hydrogen.

(ii) 2.03 g/81 g mol⁻¹ = 0.025 mol of hydrocarbon.

(iii) 1 mol hydrogen to 1 mol hydrocarbon.

(iv) One.

(v) You have to work out that, with a molar mass of 82 g mol⁻¹, the only possibility for the formula of the hydrocarbon is C_6H_{10}. (If there were less than six carbon atoms there would be too many hydrogen atoms to fit; e.g. C_5H_{22} is impossible. If there were more than six carbon atoms, the molar mass would be too large.) The molecule is *cyclohexene*:

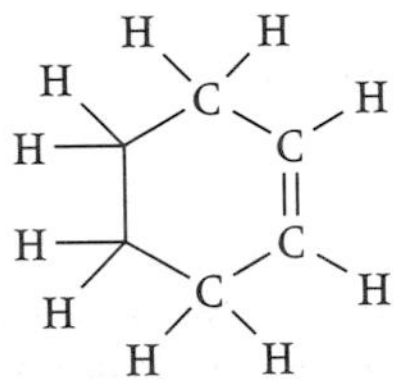

This is a useful alkene, being a liquid at room temperature. You may use it in your laboratory work.

112.6 (i) Propan-2-ol

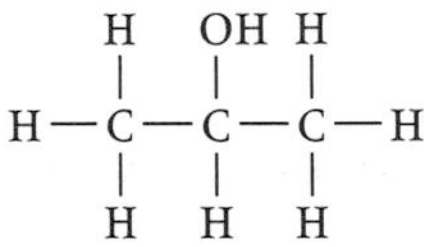

(ii) Butan-2-ol

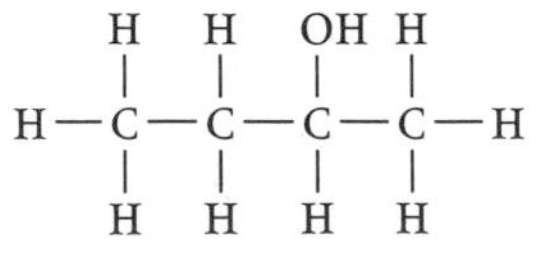

(iii) Same as (ii).

112.7 There should be three products:

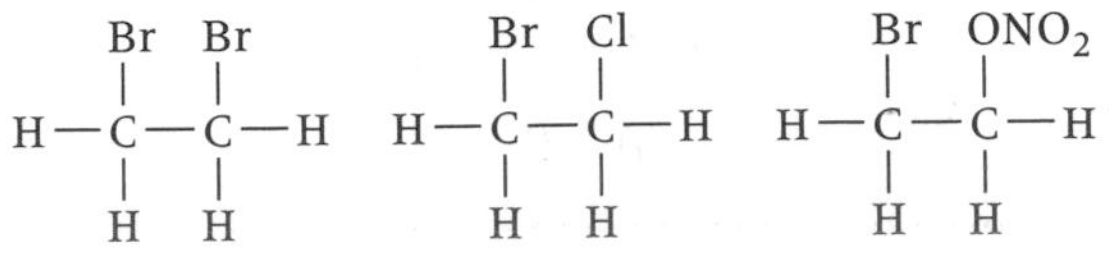

112.8 Yes, the movement of charge that we have described as the inductive effect is another way of describing polarisation.

112.9 The most favourable carbocation intermediate is the one with the most methyl groups attached to the carbon carrying the positive charge. There are two alternatives:

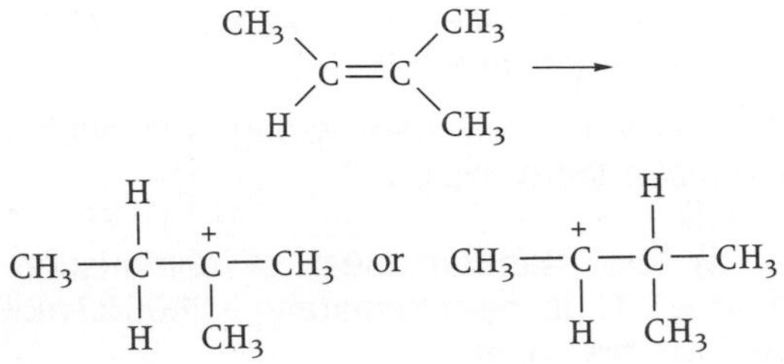

The first one is the most favoured, so the product will be 2-chloro-2-methylbutane:

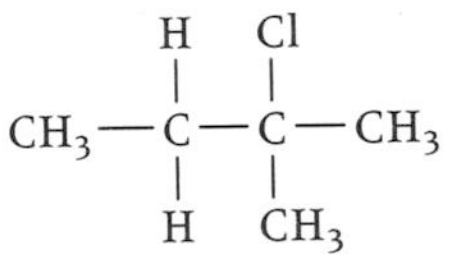

112.10 2-Methylpropan-2-ol:

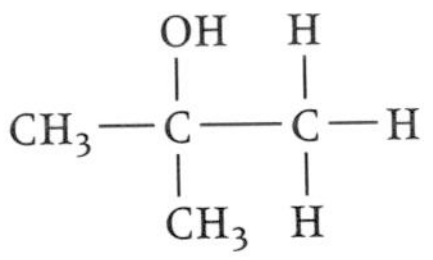

112.11 The product is CH_2BrCH_2OH:

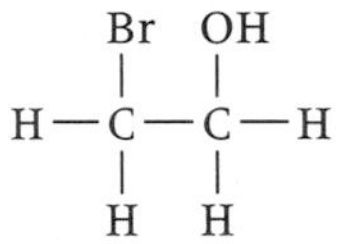

sometimes known as ethylene bromohydrin, but more properly called 2-bromoethan-1-ol.

112.12 We say that alkenes contain a double bond, but strictly *a* double bond does not exist. There are two bonds, the σ and the π. Only the π bond breaks. So the sentence should say that the π bond breaks rather than 'the double bond'.

112.13 The original alkene could have been either of the following:

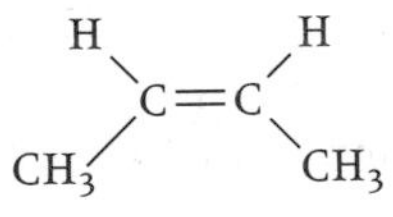

cis-but-2-ene 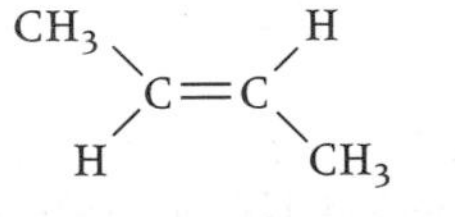   *trans*-but-2-ene

112.14 Propene:

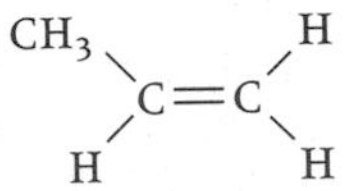

112.15 The isoprene units from which the structures are built are shown in Figure 112.9. (Carotene is the

Answers – cont.

Assuming the double bonds in isoprene break, we have

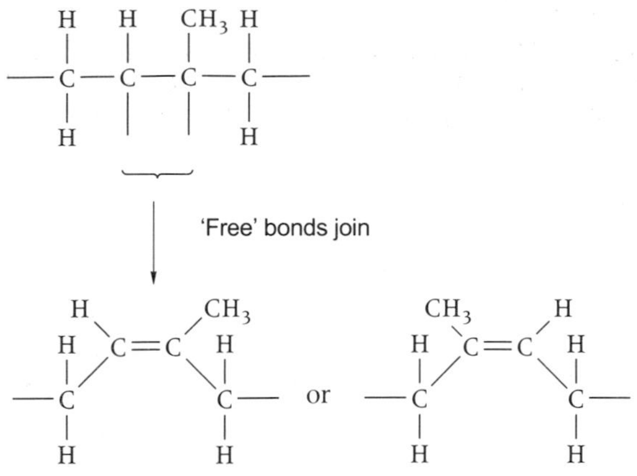

'Free' bonds join

The 'free' bonds on the ends can join with similar fragments to make the polymer

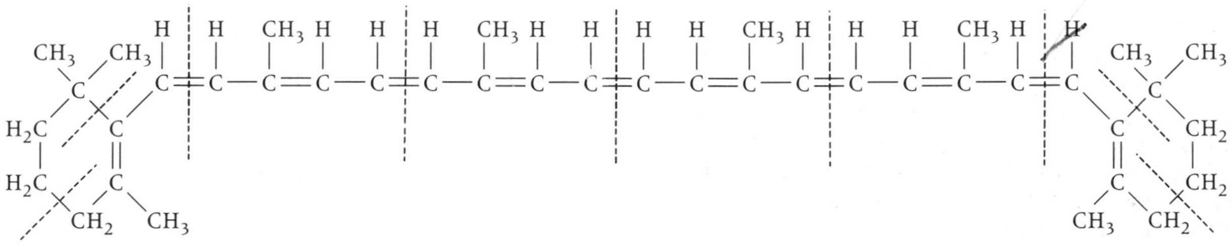

The lines through the bonds show where the isoprene units fit into the structure of carotene

Figure 112.9 *Diagrams for answer to question 112.15*

substance that gives carrots, and other things, their colour.) Isoprene is the basic building block of a large number of naturally occurring compounds.

112.16 The reaction is:

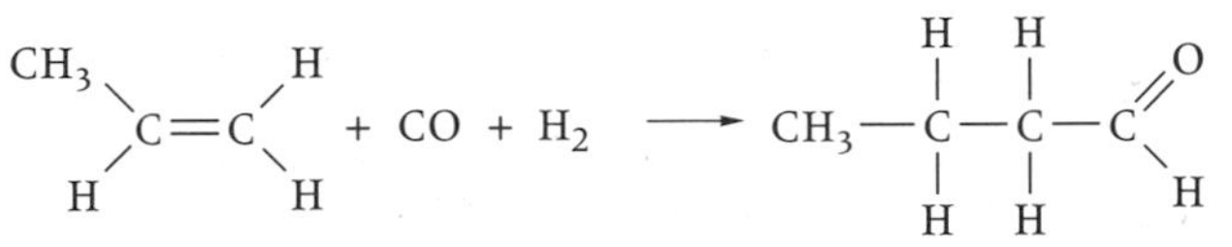

$CH_3CHCH_2(g) + CO(g) + H_2(g) \rightarrow CH_3CH_2CH_2CHO(g)$
butanal

followed by:

$CH_3CH_2CH_2CHO(g) + H_2(g) \rightarrow CH_3CH_2CH_2CH_2OH(g)$
butan-1-ol

112.17 Yes. They contain fewer hydrogen atoms than the corresponding alkane.

112.18 (i) $2C_2H_2(g) + 5O_2(g) \rightarrow 4CO_2(g) + 2H_2O(g)$

(ii) Two volumes of ethyne to five volumes of oxygen.

(iii) People attempted to light the ethyne before all the air was swept out of the burner, or the burner leaked and air could get in to where the calcium dicarbide was stored.

112.19 (i) The value shows that under standard conditions the reaction is extremely exothermic. In part, this is because ethyne is an endothermic compound, i.e. it has a positive heat of formation.

(ii) The exothermic nature of the reaction allows the very high temperatures needed to melt metals to be reached.

112.20 Alcoholic potassium hydroxide.

112.21 After the first molecule of HCl adds, we have

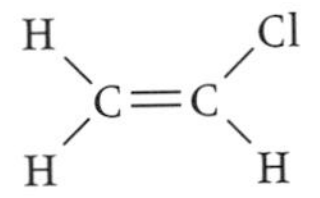

Following Markovnikoff's rule, the next molecule of HCl gives 1,1-dichloroethane:

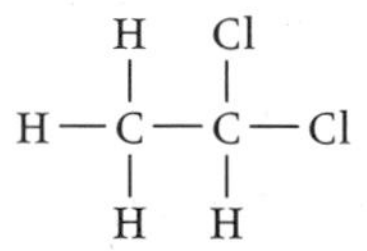

Alkenes

(i) Have the general formula C_nH_{2n}; e.g. ethene, C_2H_4; propene, C_3H_6.

(ii) Have one or more double bonds (unsaturated).

(iii) May have *cis* and *trans* isomers.

- Preparation:

  (i) Heat halogenoalkane with alcoholic potassium hydroxide;

  $$e.g.\ C_2H_5Cl + KOH \rightarrow C_2H_4 + KCl + H_2O$$

  (ii) Heat an alcohol with concentrated sulphuric acid (dehydration reaction);

  $$e.g.\ C_2H_5OH(g) \xrightarrow{180\,°C} C_2H_4(g) + H_2O(g)$$
  ethanol   ethene

  (iii) Pass vapour of an alcohol over hot aluminium oxide (another dehydration reaction).

Reactions

- Oxidation:

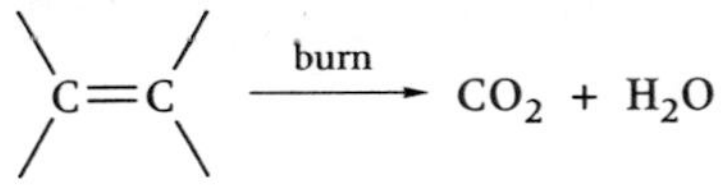

  $$\overset{\diagdown}{\underset{\diagup}{C}}=\overset{\diagup}{\underset{\diagdown}{C}} \xrightarrow{burn} CO_2 + H_2O \quad \text{sooty flame}$$

- Addition reactions:

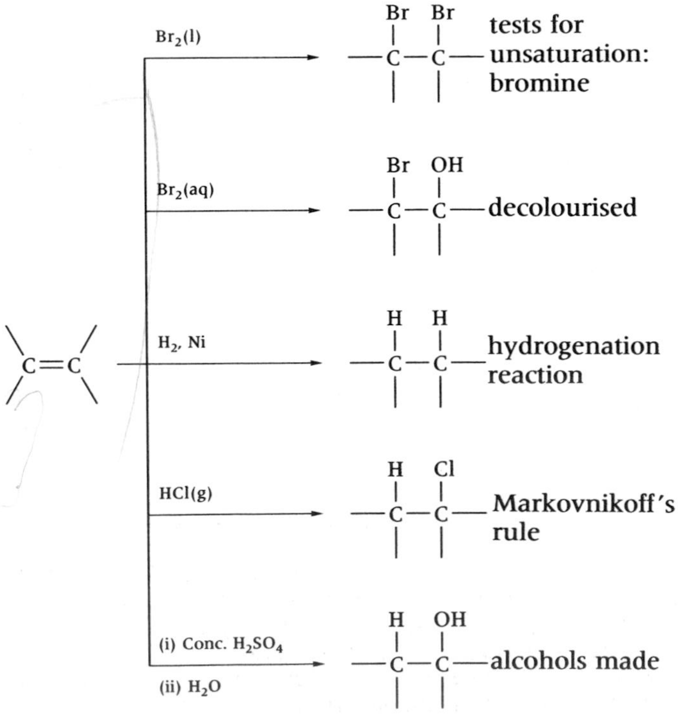

$$\overset{\diagdown}{\underset{\diagup}{C}}=\overset{\diagup}{\underset{\diagdown}{C}}$$

reagent	product	note
$Br_2(l)$	$-\overset{Br}{\underset{\mid}{C}}-\overset{Br}{\underset{\mid}{C}}-$	tests for unsaturation: bromine
$Br_2(aq)$	$-\overset{Br}{\underset{\mid}{C}}-\overset{OH}{\underset{\mid}{C}}-$	decolourised
H_2, Ni	$-\overset{H}{\underset{\mid}{C}}-\overset{H}{\underset{\mid}{C}}-$	hydrogenation reaction
HCl(g)	$-\overset{H}{\underset{\mid}{C}}-\overset{Cl}{\underset{\mid}{C}}-$	Markovnikoff's rule
(i) Conc. H_2SO_4 (ii) H_2O	$-\overset{H}{\underset{\mid}{C}}-\overset{OH}{\underset{\mid}{C}}-$	alcohols made

- Ozonolysis:

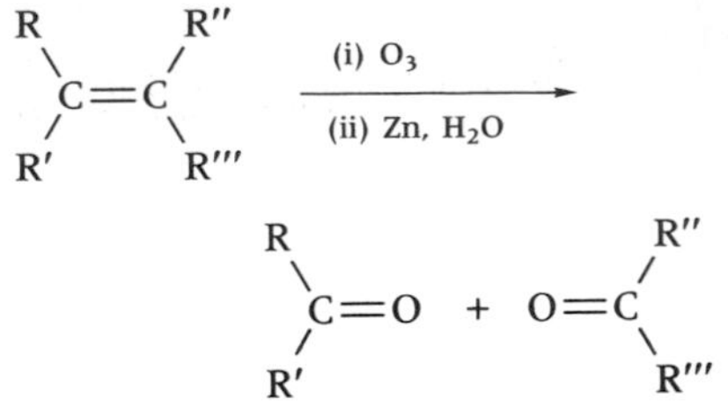

$$\overset{R}{\underset{R'}{>}}C=C\overset{R''}{\underset{R'''}{<}} \xrightarrow[\text{(ii) Zn, }H_2O]{\text{(i) }O_3}$$

$$\overset{R}{\underset{R'}{>}}C=O \ + \ O=C\overset{R''}{\underset{R'''}{<}}$$

used to identify alkenes

- Polymerisation:

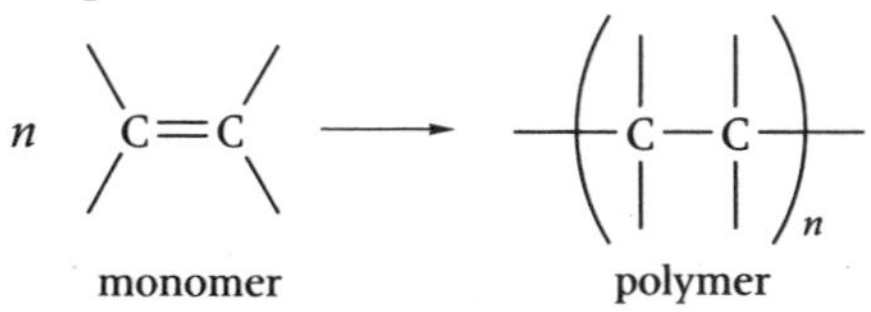

$$n \ \overset{\diagdown}{\underset{\diagup}{C}}=\overset{\diagup}{\underset{\diagdown}{C}} \longrightarrow \left(\!\!\begin{array}{c} | \ \ | \\ -C-C- \\ | \ \ | \end{array}\!\!\right)_n$$
monomer   polymer

initiated by free radicals

- Oxo reaction:

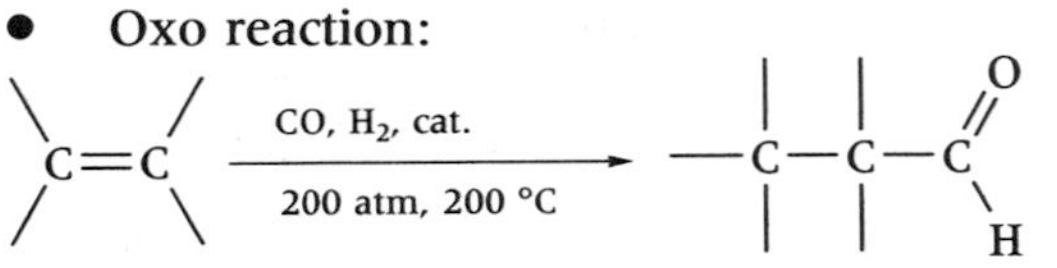

$$\overset{\diagdown}{\underset{\diagup}{C}}=\overset{\diagup}{\underset{\diagdown}{C}} \xrightarrow[\text{200 atm, 200 °C}]{CO,\ H_2,\ cat.} -\underset{\mid}{C}-\underset{\mid}{C}-\overset{O}{\underset{H}{C^{/\!/}}}$$

increase number of carbon atoms

- Markovnikoff's rule:
  When hydrogen chloride adds to a double bond the hydrogen joins to the carbon that already has the most hydrogen atoms bonded to it.

- Carbocations:
  (i) Are groups in which a carbon atom carries a positive charge.
  (ii) Order of their energetic stability is primary < secondary < tertiary.

Alkynes

Have the general formula C_nH_{2n-2}; e.g. ethyne, C_2H_2; propyne, C_3H_4.

- Preparation:
  Ethyne is made from water and a carbide;

  $$e.g.\ CaC_2(s) + H_2O(l) \rightarrow Ca(OH)_2(s) + C_2H_2(g)$$

Properties

- Oxidation:

  $$-C\equiv C- \xrightarrow{burn} CO_2 + H_2O \quad \text{very sooty flame}$$

- Addition:

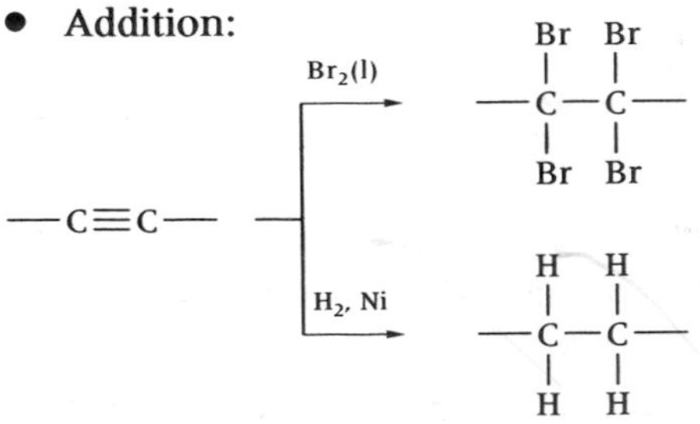

$$-C\equiv C-$$

reagent	product	note
$Br_2(l)$	$-\overset{Br}{\underset{Br}{C}}-\overset{Br}{\underset{Br}{C}}-$	bromine decolourised
H_2, Ni	$-\overset{H}{\underset{H}{C}}-\overset{H}{\underset{H}{C}}-$	hydrogenation reaction

- Polymerisation:

  $$3C_2H_2 \xrightarrow{400\,°C} C_6H_6$$
  benzene

113

Aromatic hydrocarbons

113.1   What are aromatic hydrocarbons?

Aromatic hydrocarbons contain delocalised electrons around a ring of carbon atoms. As a group, they are also called *arenes*. The most famous arene is benzene, C_6H_6. It is a colourless liquid at room temperature. (Density, $0.87\,g\,cm^{-3}$; melting point, $5.7\,°C$; boiling point, $80.3\,°C$.) The structure of benzene is unusual because it appears to have three single bonds and three double bonds between its six carbon atoms. However, you should know that there are no individual π bonds in benzene (see Unit 16). The π electrons are delocalised around the ring (Figure 113.1). This delocalisation brings a degree of energetic stability to benzene that other unsaturated compounds may not have.

If we remove one of the hydrogen atoms from the ring and substitute it by a methyl group, we have a new aromatic hydrocarbon, methylbenzene. Similarly we can exchange a hydrogen atom for an ethyl group, which gives ethylbenzene. These hydrocarbon groups attached to the ring are called the *side chains*. Some arenes are shown in Table 113.1.

If you build a model of benzene and replace two hydrogen atoms by methyl groups, you will find that you can do this in three different ways, i.e. there are three *isomers* (Figure 113.2). Notice that we base the name of the isomers on numbers given to each of the carbon atoms in the ring.

More complicated arenes are made when two or more rings are joined together. This happens, for example, in naphthalene and anthracene (Figure 113.3). Molecules like these have their π clouds stretching over two or more rings. The delocalisation is even more extensive than in benzene. Needless to say, in these compounds many isomers are possible when different groups are substituted for the hydrogen atoms.

113.1 (i) Write the equation for the burning of benzene in oxygen.

(ii) Why is it that if benzene is burned in air it produces a very smoky flame and large amounts of soot are sent into the air?

(a)

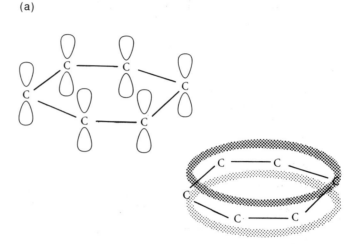

Delocalised π cloud of electrons

(b)

(c)

All represent the same molecule

Figure 113.1 *(a) The bonding in benzene, C_6H_6. The overlap of a p orbital on each carbon atom leads to π molecular orbitals. (b) Three ways of representing benzene. The third one is the neatest, and the one we shall normally use. (c) Also, notice that the symmetry of the benzene molecule means that diagrams like those shown represent the same molecule*

Table 113.1. Some aromatic hydrocarbons (arenes)

Compound	Structure	Melting point/°C	Boiling point/°C
Benzene		5.7	80.3
Methylbenzene	CH₃	−94.5	110.8
Ethylbenzene	C₂H₅	−94.5	136.3
1,2-Dimethylbenzene	CH₃ CH₃	−25.0	144.6
1,3-Dimethylbenzene	CH₃ CH₃	−47.7	139.3
1,4-Dimethylbenzene	CH₃ CH₃	13.4	138.5
Naphthalene		80.4	218.1
Anthracene		217	354

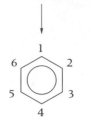

Position of first side group

The numbering system of benzene

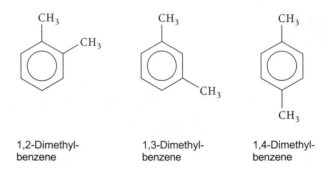

| 1,2-Dimethyl-benzene | 1,3-Dimethyl-benzene | 1,4-Dimethyl-benzene |

Figure 113.2 *There are three isomers of dimethylbenzene. Where there is a choice, we use the lowest possible sequence of numbers*

113.2 Ways of making arenes

(a) Making benzene

Benzene and other arenes can be obtained by distilling coal. This is a particularly messy process in the laboratory, and requires a lengthy business of separating the products from one another. However, in industry it is an economic way of isolating benzene. There is strong demand for coke, which is produced by heating coal in the absence of air. For every tonne of coal turned into coke, about 70 dm³ of coal tar is made. This is an oily liquid, which contains a variety of products. If the coal tar is separated by fractional distillation, around 30 dm³ of benzene can be collected. Methylbenzene, naphthalene and anthracene are also obtained in smaller quantities.

In the laboratory a quicker way to make benzene is to heat the calcium salt of benzoic acid, $(C_6H_5COO)_2Ca$, with soda lime (soda lime contains calcium hydroxide together with sodium hydroxide):

$$(C_6H_5COO)_2Ca(s) + Ca(OH)_2(s) \rightarrow 2C_6H_6(l) + 2CaCO_3(s)$$

Note: arenes are often carcinogenic – you should use them only with great caution.

(b) Making arenes

There is one method that is extremely useful for making a wide variety of arenes. It is the *Friedel–Crafts alkylation*

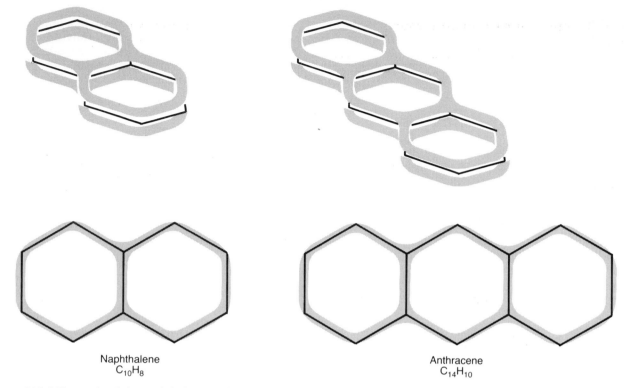

Naphthalene
$C_{10}H_8$

Anthracene
$C_{14}H_{10}$

Figure 113.3 *The π clouds in naphthalene and anthracene stretch over all the carbon atoms*

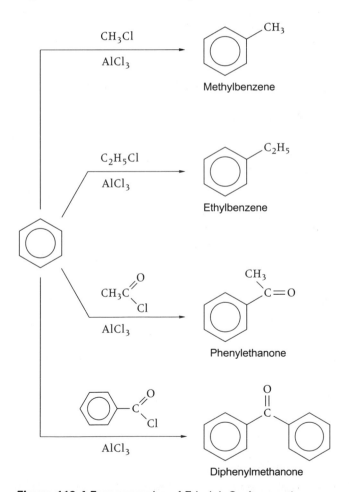

CH_3Cl

$AlCl_3$

Methylbenzene

C_2H_5Cl

$AlCl_3$

Ethylbenzene

$AlCl_3$

Phenylethanone

$AlCl_3$

Diphenylmethanone

Figure 113.4 *Four examples of Friedel–Crafts reactions*

reaction. The essence of the reaction is to take a chloroalkane and react it with benzene in the presence of aluminium trichloride. Some of the possibilities are illustrated in Figure 113.4.

The reason why the reaction takes place is that the benzene ring is attacked by positively charged alkyl groups. For example, if we use chloromethane, CH_3Cl, positively charged methyl groups, CH_3^+, are generated. (The aluminium trichloride plays an active part in generating them.) Such groups are *electrophiles*; that is, they seek out centres of negative charge. The benzene ring with its rich cloud of delocalised π electrons is such a centre. We could write the equation in this way:

In panel 113.1 you will find details of the mechanism of the Friedel–Crafts reaction.

> **113.2** A second method of making arenes is similar to the Wurtz synthesis of alkanes. It is called the Wurtz–Fittig reaction. The method involves heating sodium with bromobenzene and a bromoalkane. Predict the equation for the reaction that would take place if bromoethane were used.

113.3 What reagents would you mix with benzene if you wanted to make (i) ethylbenzene, (ii) 2-phenylpropane?

113.3 The reactions of arenes

The reactions of arenes are sometimes similar, but often different, to those of non-aromatic hydrocarbons. We can split the reactions into two groups:

(i) those that affect the benzene ring,
(ii) those that affect the side chain.

With the *benzene ring*, we can have two types of reaction. First, there are the *addition* reactions when the π cloud around the ring is broken and atoms add on to the ring. These are different to *substitution* reactions. Here, hydrogen atoms are replaced by other groups, but the π cloud is left intact.

If the benzene ring in an aromatic compound is attacked, we sometimes say that this is an attack on the nucleus, as opposed to the side chain. This is an unfortunate use of the word because it does *not* mean that benzene rings are attacked by nucleophiles. Indeed, the reverse is true. The typical reagent to bring about substitution in a benzene ring is an electrophile.

There are several differences between the reactions of the benzene ring and other unsaturated compounds, such as alkenes. The reason is that the increased delocalisation of the π electrons has an influence on both the energetics of reactions and their kinetics.

With the *side chain*, several things can happen, the most important of which are substitution and oxidation reactions.

In the next three sections we shall look at examples of each of these reactions.

113.4 Reactions of the side chain

(a) *Oxidation*

The side chain, but not the benzene ring, is easily oxidised by reagents such as hot potassium manganate(VII), potassium dichromate(VI), or even nitric acid. The main product is benzoic acid, C_6H_5COOH, but carbon dioxide will also be given off. This acid is made no matter the length of the hydrocarbon side chain. For example,

benzoic acid

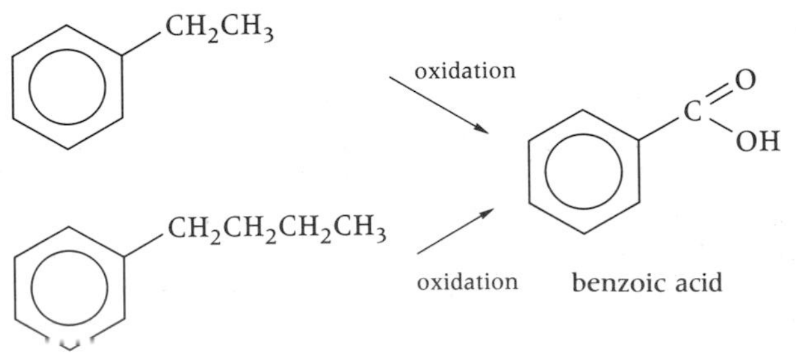

It is possible to oxidise a side chain in a milder way by using chromium(VI) dichloride dioxide (chromyl chloride), CrO_2Cl_2. This reagent can be made by mixing potassium dichromate(VI) with hydrochloric acid. It will oxidise side chains to the aldehyde group, CHO, rather than to an acid. For example,

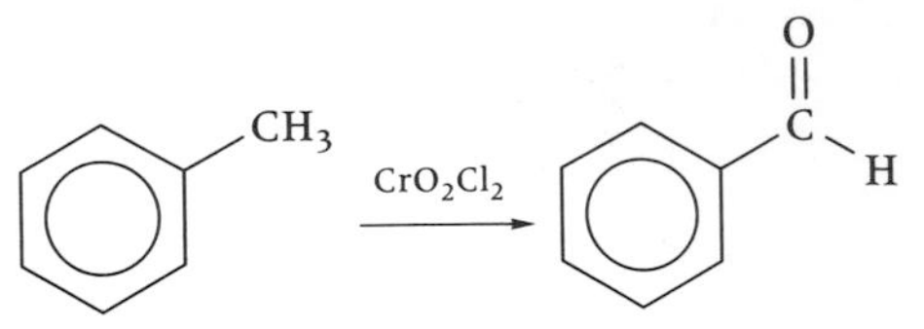

methylbenzene benzaldehyde

(b) *With halogens*

We know that alkanes undergo substitution reactions with halogens at high temperatures or in ultraviolet light. The reaction is caused by the presence of free radicals. Similar reactions take place with the side chain of arenes. For example, if ultraviolet light initiates the reaction between methylbenzene vapour and chlorine, the products may include:

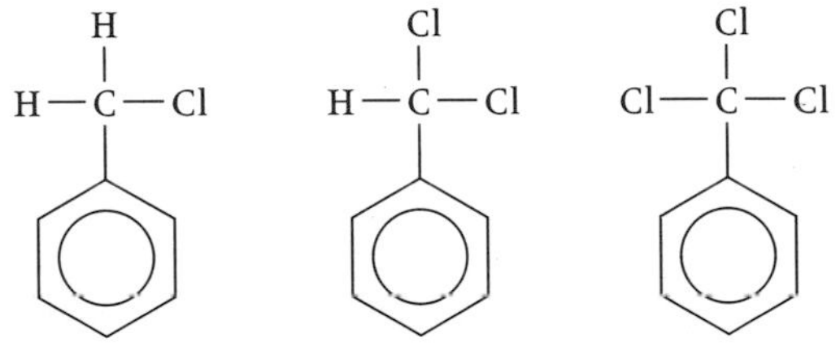

The thing to notice is that the benzene ring is not attacked. Indeed the rule is that:

Free radicals do not attack a benzene ring if there is a hydrocarbon side chain available.

If the side chain is more complicated than a methyl group, then many more possible substitution products are possible. For example, with ethylbenzene, there are five hydrogen atoms that can be replaced by halogens. However, some products are much more favoured than others. Especially, if a little chlorine reacts with ethylbenzene, the major product is 1-chloro-1-phenylethane:

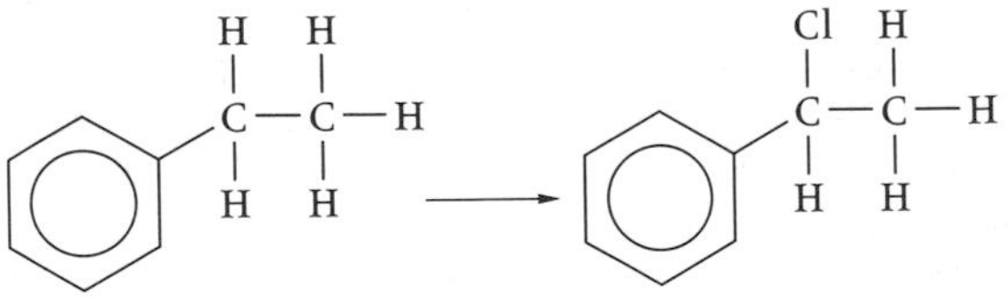

1-chloro-1-phenylethane

Indeed it is always the case that the hydrogen atoms on the carbon atom closest to the benzene ring are the most easily replaced by halogens. The reason for this is complicated and too involved for us to deal with

113.4 Figure 113.5 shows an apparatus set up by a student intending to oxidise methylbenzene using potassium manganate(VII) solution. What is wrong with it?

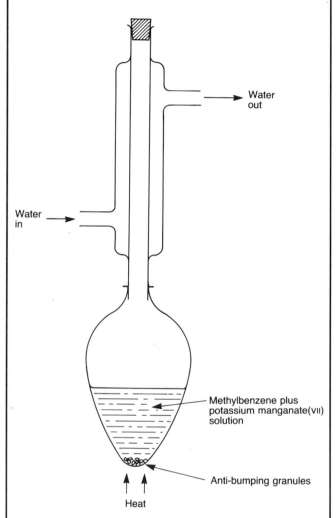

Water out

Water in

Methylbenzene plus potassium manganate(VII) solution

Anti-bumping granules

Heat

Figure 113.5 Proposed arrangement of apparatus for question 113.4

113.5 Benzoic acid is a white solid at room temperature. It is only very slightly soluble in cold water. If the oxidation of methylbenzene is carried out in acid conditions, a white precipitate appears in the flask. In alkaline conditions no precipitate appears. However, if acid is added at the end of the experiment, the precipitate does form. Explain these observations. (Hint: benzoic acid is a weak acid.)

113.6 1,3-Dimethylbenzene is oxidised by potassium dichromate(VI) solution. Draw the product of the reaction.

113.5 Addition reactions of the benzene ring

Hydrogen will add to benzene if it is reacted with hydrogen in the presence of a nickel catalyst at around 150 °C:

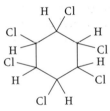

$$C_6H_6(g) + 3H_2(g) \rightarrow C_6H_{12}(g)$$
cyclohexane

This is much the same reaction as the hydrogenation of alkenes. It is possible to get halogens to add to a benzene molecule. A mixture of benzene, halogen and ultraviolet light will do the trick. For example, 1 mol of benzene, C_6H_6, takes up 3 mol of chlorine to give $C_6H_6Cl_6$. Notice the difference between this and the normal free radical reaction: the six hydrogen atoms on the ring are *not* substituted. The chlorine atoms add on to the ring. As you might expect, the structure of the product is based on cyclohexane:

1,2,3,4,5,6-hexachlorocyclohexane

This substance is also known as benzene hexachloride (even though it does not contain a benzene ring). It is used as the weedkiller Lindane.

The direct reaction of benzene with bromine, or bromine water, is extremely slow at room temperature. For this reason bromine is not decolourised by benzene. This is a distinct difference to non-aromatic alkenes or alkynes.

113.6 Electrophilic substitution of benzene rings

In the Friedel–Crafts alkylation reaction we have already seen one type of electrophilic substitution of a benzene ring. Other reactions of this type, which are also called *aromatic substitutions*, are listed in Table 113.2. We shall take them one by one. You will find brief details of the mechanisms of the reactions in panel 113.1.

Table 113.2. Electrophilic substitution reactions of a benzene ring

Type of reaction	Example	Reagent	Electrophile
Alkylation	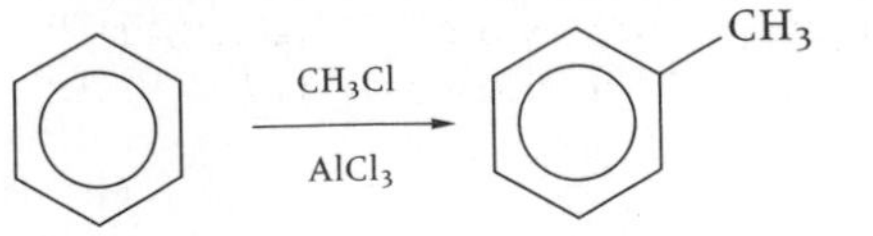 benzene + CH_3Cl, $AlCl_3$ → methylbenzene (CH_3)	Halogenoalkane and aluminium trichloride	$CH_3^{\oplus}$
Acylation	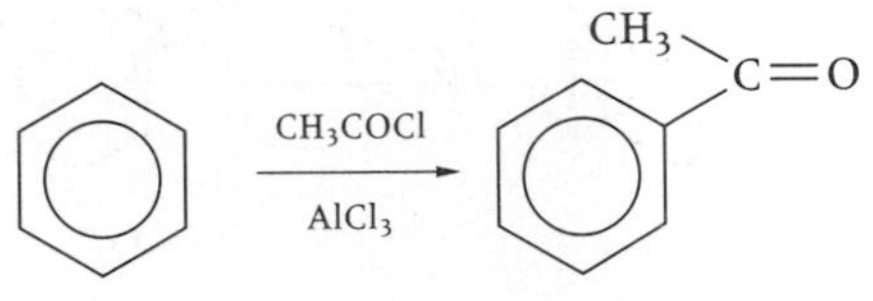 benzene + CH_3COCl, $AlCl_3$ → CH_3–C=O	Acyl chloride and aluminium trichloride	$CH_3CO^{\oplus}$
Halogenation	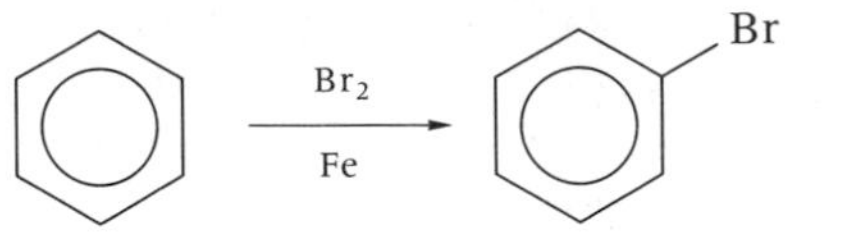 benzene + Br_2, Fe → Br	Chlorine or bromine and iron filings	$Cl^{\oplus}$, $Br^{\oplus}$
Nitration	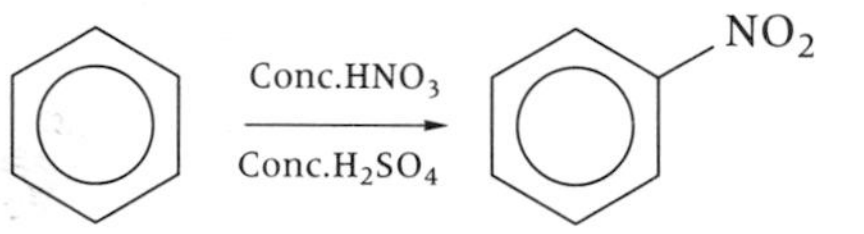 benzene + Conc. HNO_3, Conc. H_2SO_4 → NO_2	Conc. nitric acid and sulphuric acid	$NO_2^{\oplus}$ (nitryl cation)
Sulphonation	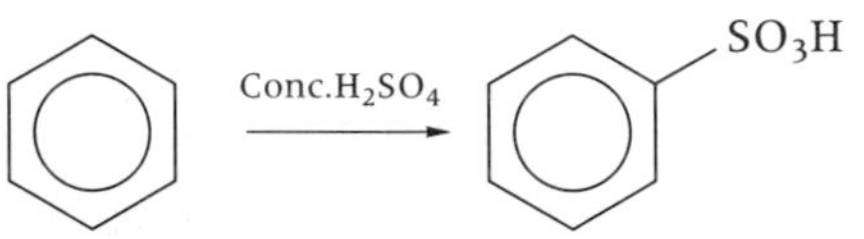 benzene + Conc. H_2SO_4 → SO_3H	Conc. sulphuric acid	$^{\delta+}SO_3^{\delta-}$
Diazotisation	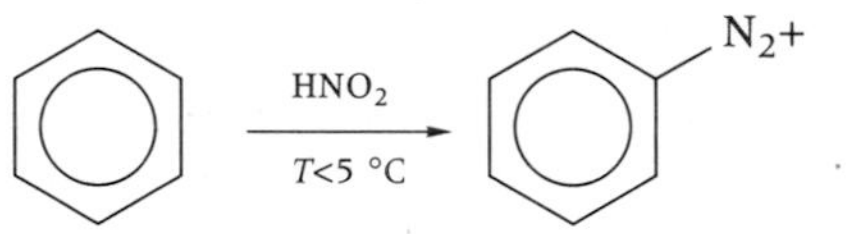 benzene + HNO_2, $T<5\,°C$ → N_2^+	Phenylamine, acid and sodium nitrite	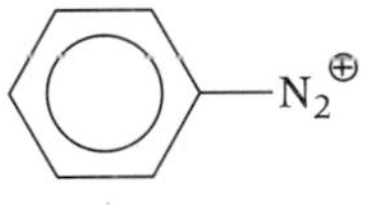 —$N_2^{\oplus}$ (benzene-diazonium cation)

(a) Alkylation

We need say no more about this than you will find in section 113.2 and panel 113.1.

(b) Acylation

The acyl group is a combination of an alkyl group and a carbonyl group (>C=O). Among the most important molecules that contain an acyl group are the acid chlorides. Typical ones are:

$$CH_3-C\overset{O}{\underset{Cl}{<}} \qquad CH_3CH_2-C\overset{O}{\underset{Cl}{<}}$$

ethanoyl chloride   propanoyl chloride

The mechanism of their attack on a benzene ring is very similar to the way alkylation takes place. For example, the carbocation produced when ethanoyl chloride is mixed with aluminium trichloride is:

$$CH_3-C\overset{O}{\underset{\oplus}{<}}$$

This is the electrophile that attacks the ring.

We can show the result of the reactions in the following examples:

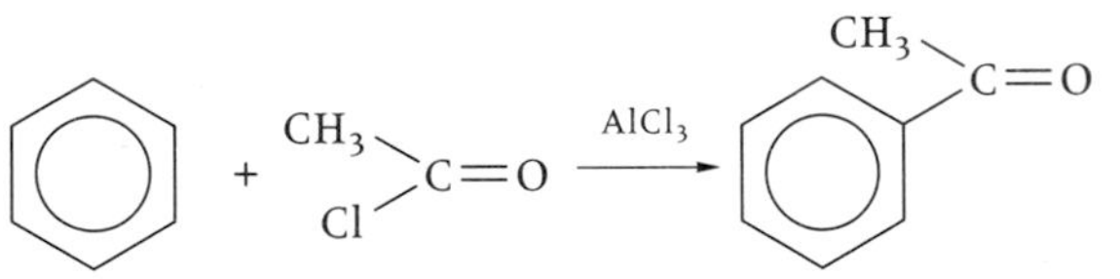

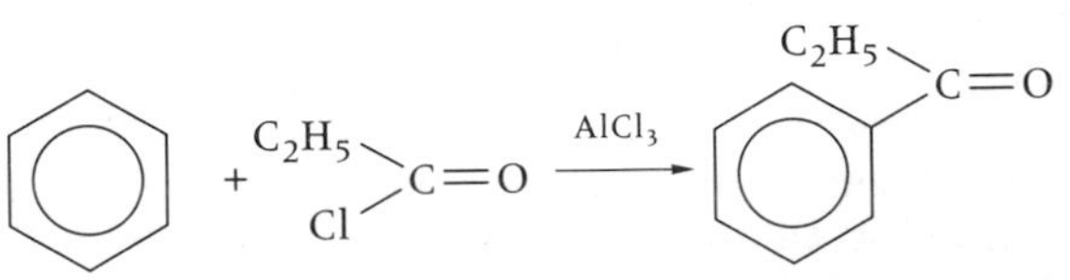

benzene + CH_3\C=O (Cl) $\xrightarrow{AlCl_3}$ CH_3\C=O

benzene + C_2H_5\C=O (Cl) $\xrightarrow{AlCl_3}$ C_2H_5\C=O

(c) Halogenation

We have made the point that benzene will not decolourise bromine on its own. However, if some iron filings are present, a reaction starts and the bromine *is* decolourised. (Hydrogen bromide is given off as well.) The iron is said to act as a *halogen carrier*. It promotes the production of Br^+ ions, which attack the ring:

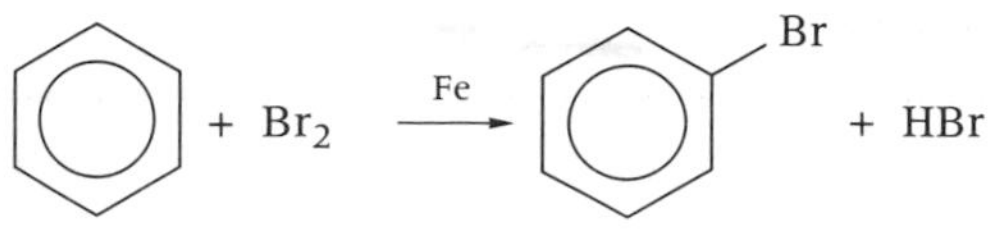

(d) Nitration

In nitration, nitryl cations, NO_2^+, made by the reaction between concentrated nitric and sulphuric acids, are the electrophiles:

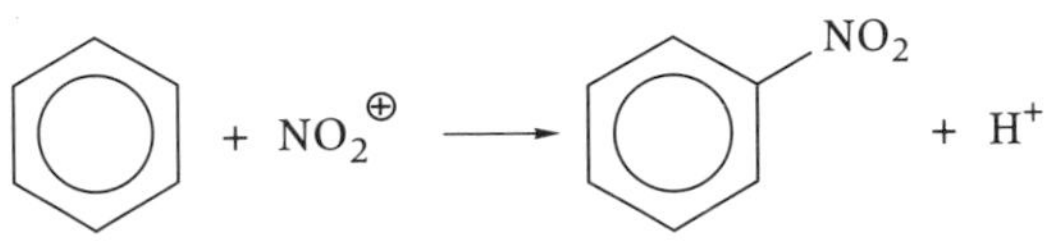

For obvious reasons you may hear of a mixture of concentrated nitric and sulphuric acids being called *nitrating mixture*.

(e) Sulphonation

The overall result of reacting an arene with concentrated sulphuric acid is that the benzene ring exchanges a hydrogen atom for a hydrogensulphite group:

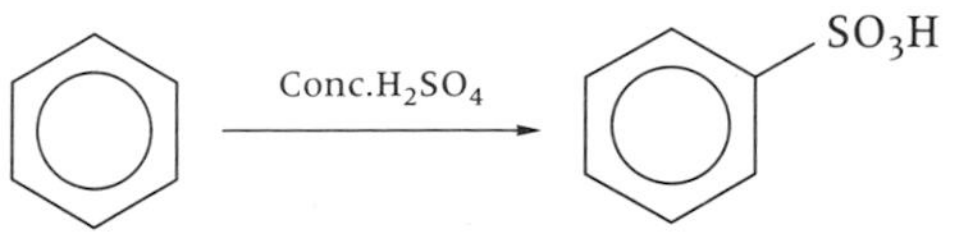

(Actually the reaction involves sulphur trioxide as an intermediate.)

(f) Diazotisation

If you mix phenylamine together with an ice-cold solution of sodium nitrite in hydrochloric acid, the following reaction takes place:

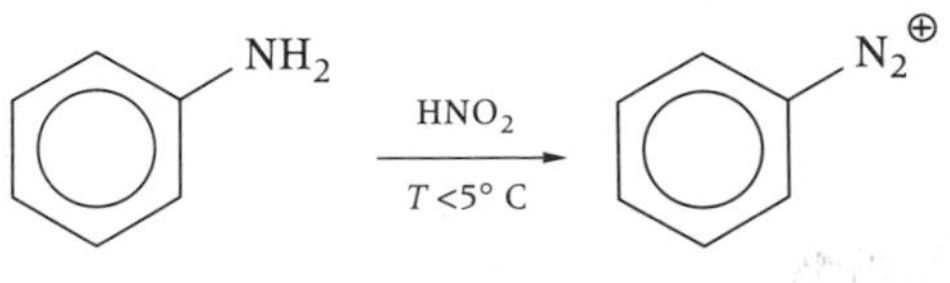

benzenediazonium ion

The positively charged benzenediazonium ion is not a very good electrophile. It will attack a benzene ring that has an activating group attached, especially phenol and phenylamine. The reaction is known as a *coupling* reaction. For example,

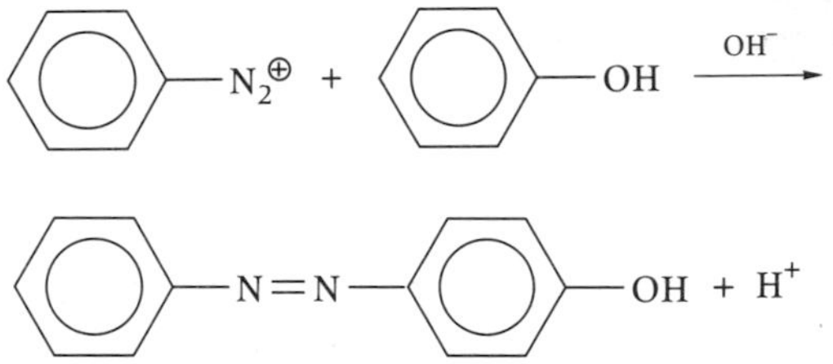

Coupling reactions like this have been widely used in making dyes. You will find details in section 122.8.

Panel 113.1

The mechanisms of electrophilic substitution reactions

Friedel–Crafts alkylation

Let us suppose that we are using chloromethane to convert benzene into methylbenzene. Chloromethane is a polar molecule. The chlorine atom is highly electronegative: it carries a slightly negative charge and has three lone pairs around it. Also, aluminium trichloride is a Lewis acid (see section 74.4). This means that the molecule can accept a lone pair of electrons in making a fourth bond to the aluminium atom. If we bring a chloromethane molecule near to an aluminium trichloride molecule, a lone pair on the chlorine atom can form a coordinate bond. In the process the bond between the methyl group and the chlorine atom weakens and finally breaks. This frees a methyl carbocation, CH_3^+. It is this electrophile that attacks the ring:

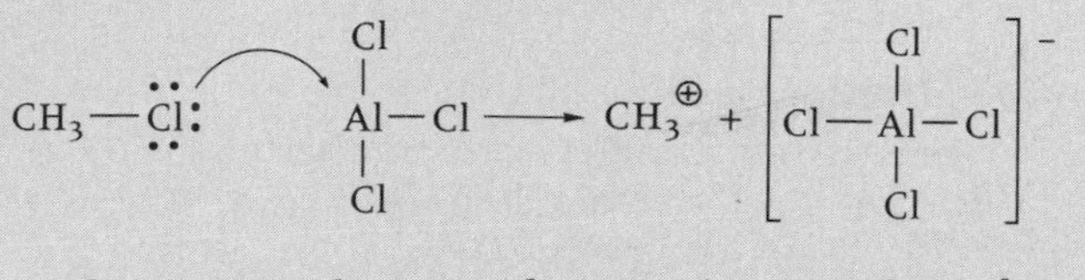

The intermediate in the reaction consists of a benzene ring that has both a methyl group and a hydrogen atom attached by σ bonds to one of the carbon atoms. It is known as a *sigma complex*, or sometimes as a *Wheland intermediate*. The intermediate breaks up by the loss of a hydrogen ion, which carries the extra positive charge away with it:

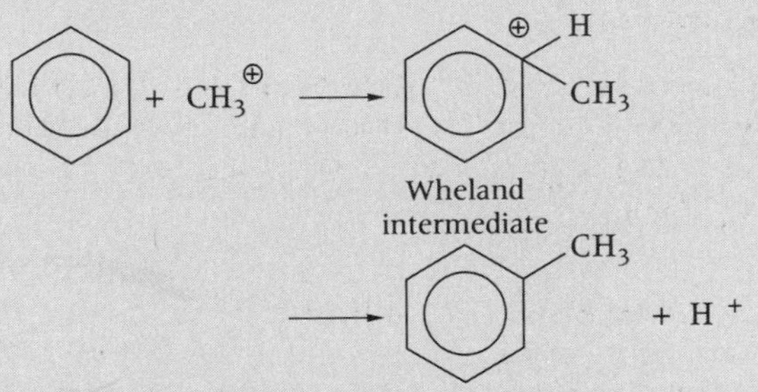

Incidentally, it is possible for molecules to be held by a benzene ring by attraction with the π cloud. Bromine and hydrogen chloride are known to form *pi complexes* in this way.

Friedel–Crafts acylation

Here the aluminium trichloride brings about the production of a different type of carbocation than in the alkylation reaction; but the basic mechanism is the same. It is summed up in the next few equations:

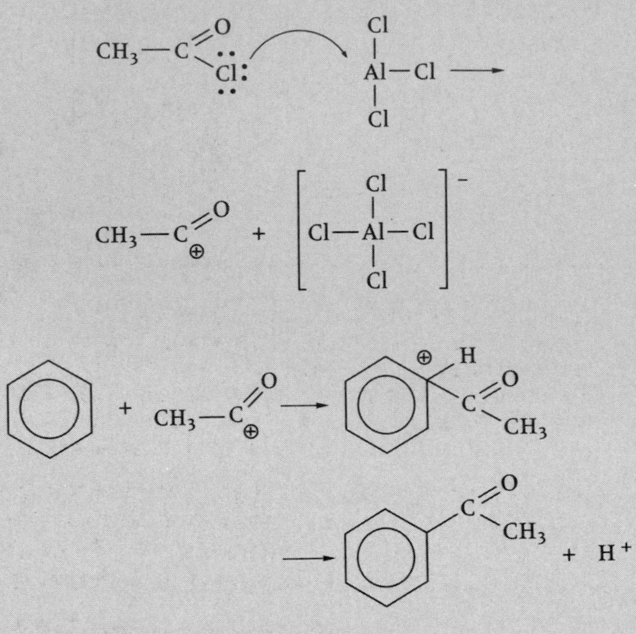

$$CH_3-C{\overset{O}{\underset{Cl}{}}} \quad \begin{array}{c} Cl \\ | \\ Al-Cl \\ | \\ Cl \end{array} \longrightarrow$$

$$CH_3-C{\overset{O}{\underset{\oplus}{}}} \quad + \quad \left[\begin{array}{c} Cl \\ | \\ Cl-Al-Cl \\ | \\ Cl \end{array} \right]^-$$

Halogenation

If we assume that bromine and iron filings are mixed, the iron(III) bromide produced acts as a Lewis acid, accepting a pair of electrons from a bromine atom at one end of a bromine molecule. The molecule is polarised by the transfer of some electron density to the iron atom, and the second bromine atom in the molecule takes on a positive character. If the process continues successfully, a positively charged bromine atom is made. This is the electrophile that attacks the ring:

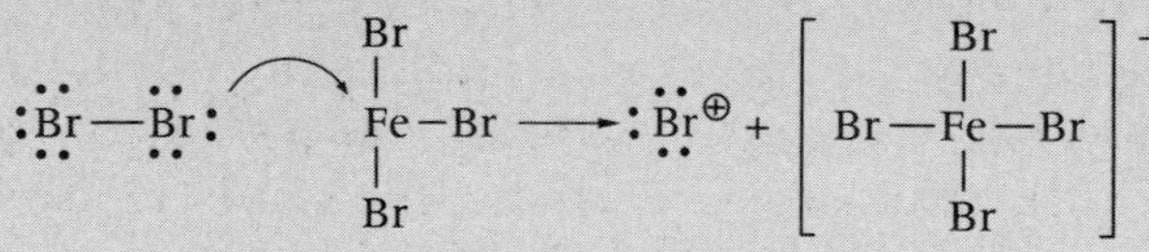

Nitration

The mechanism of nitration has quite a history. It caused a good deal of controversy for a number of years. Given that the nitro group, NO_2, finds its way onto the benzene ring, it is natural to assume that the attacking electrophile is the nitryl cation, NO_2^+. The problem is to explain how this ion is made, and to find experimental evidence for its existence.

We now believe it is made in the equilibrium

$$HNO_3 + 2H_2SO_4 \rightleftharpoons NO_2^+ + 2HSO_4^- + H_3O^+$$

which itself is the summary of several intermediate equilibria. Direct evidence for the presence of the nitryl cation in the mixture comes from spectroscopy, and from the fact that it is possible to isolate salts that contain the ion.

The reaction between the ion and the benzene ring follows similar stages to the other reactions:

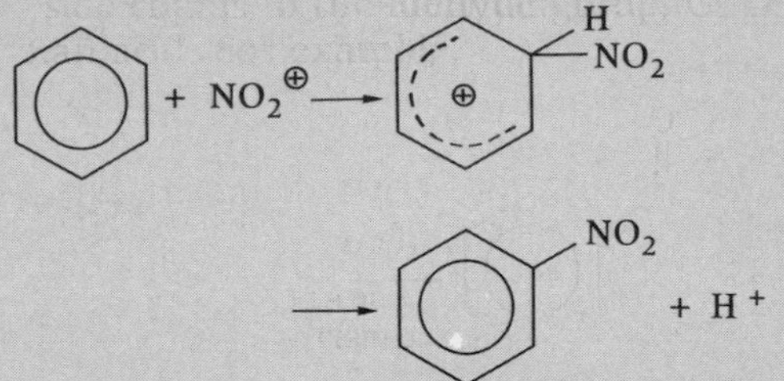

Sulphonation

In the sulphonation reaction, the electrophile is sulphur trioxide, SO_3. It is made by the autoionisation of sulphuric acid:

$$2H_2SO_4 \rightleftharpoons HSO_4^- + H_3O^+ + SO_3$$

In SO_3, sulphur has only three pairs of electrons involved in bonding to the oxygen atoms. Therefore it has space for a fourth pair before it completes its octet. Also the sulphur atom will carry a partial positive charge owing to the electron withdrawing nature of the oxygen atoms bonded to it. The reaction is:

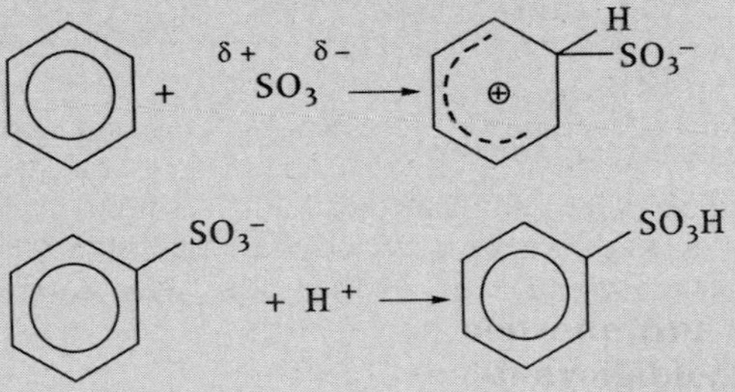

Diazotisation

We shall ignore the details of how the benzenediazonium ion comes to be made. The key thing is that the nitrogen atoms on the end of the ion carry the positive charge that we expect of many electrophiles. It is most important that the reaction mixture is kept cold (less than 5 °C), otherwise the ion decomposes. The reaction is:

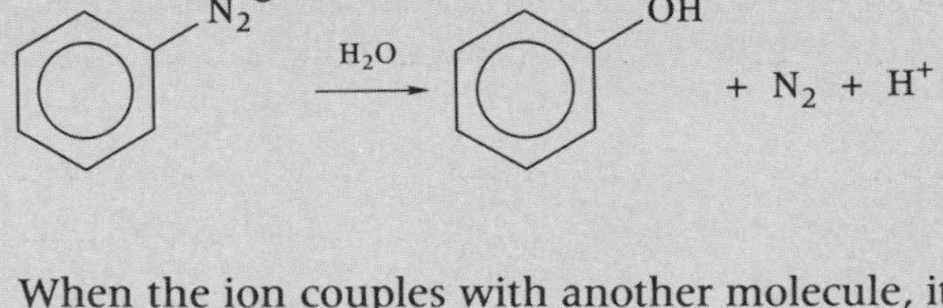

When the ion couples with another molecule, it is the *para* position that is attacked. (See the next unit for an explanation of the term '*para*'.) For example, with phenol and with phenylamine:

Panel 113.1 – cont.

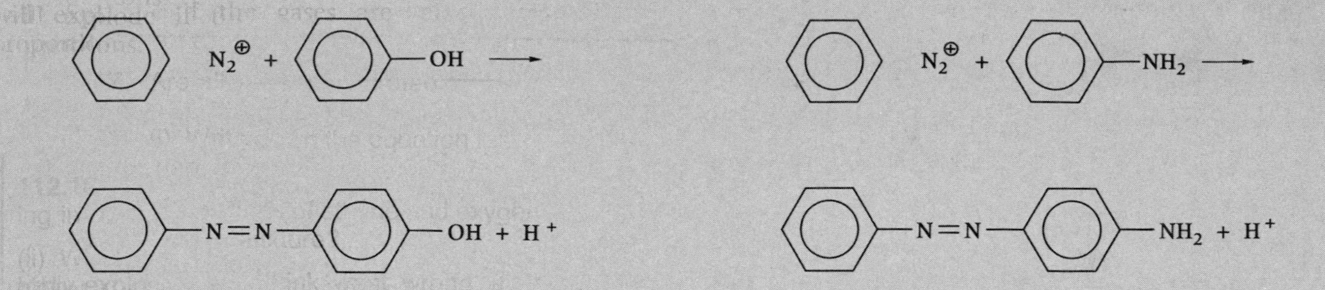

$$\text{C}_6\text{H}_5\text{-N}_2^{\oplus} + \text{C}_6\text{H}_5\text{-OH} \longrightarrow \text{C}_6\text{H}_5\text{-N=N-C}_6\text{H}_4\text{-OH} + \text{H}^+$$

$$\text{C}_6\text{H}_5\text{-N}_2^{\oplus} + \text{C}_6\text{H}_5\text{-NH}_2 \longrightarrow \text{C}_6\text{H}_5\text{-N=N-C}_6\text{H}_4\text{-NH}_2 + \text{H}^+$$

113.7 If the key thing about the use of aluminium trichloride in the Friedel–Crafts reaction is that it is a Lewis acid (or, alternatively, that it is electron deficient), which other molecule could be used in its place? (Hint: look at section 74.4.)

113.8 In 1825 Michael Faraday wrote the following in the *Philosophical Magazine*:

The object of the paper which I have the honour of submitting to the attention of the Royal Society, is to describe particularly two new compounds of carbon and hydrogen, and generally, other products obtained during the decomposition of oil by heat · . .. Bi-carburet of hydrogen appears in common circumstances as a colourless transparent liquid, having an odour resembling that of oil gas, and partaking also of that of almonds. Its specific gravity is nearly 0.85 at 60°. When cooled to about 32° it crystallizes, becoming solid . . .. Its boiling point in contact with glass is 186° . . .. Chlorine introduced to the substance in a retort exerted but little action until placed in sunlight, when dense fumes were formed, without the evolution of much heat; and ultimately muriatic acid was produced . . .. The following is a result of what obtained when it was passed over heated oxide of copper: 0.776 grain of the substance produced 5.6 cubic inches of carbonic gas, at a temperature of 60°, and pressure 29.98 inches; and 0.58 grain of water were collected . . .. [His calculation showed that] making the hydrogen 1, the carbon is not far removed from 12, or two proportionals . . .. This result is confirmed by such data as I have been able to obtain by detonating the vapour of the substance with oxygen. Thus in one experiment 8092 mercury grain measures of oxygen at 62° had such quantity of the substance introduced into it as would entirely rise in the vapour; the volume increased to 8505: hence the vapour amounted to 413 parts, or 1/206 of the mixture nearly. Seven volumes of this mixture were detonated in an eudiometer tube by an electric spark, and diminished in consequence nearly to 6.1: these, acted upon by potash, were further diminished to 4, which were pure oxygen. Hence 3 volumes of mixture had been detonated, of which nearly 0.34 was vapour of the substance, and 2.65 oxygen. The carbonic acid amounted to 2.1 volumes . . ..

(i) Before we go into the chemistry of Faraday's observations, you need to know that he measured temperatures in degrees Fahrenheit. Find out the Celsius equivalent of 32°, 60° and 186° Fahrenheit.

(ii) Do Faraday's measurements of the specific gravity and boiling point of bi-carburet of hydrogen give you a guide to the identity of the compound? (You can assume that specific gravity is equivalent to density.)

(iii) Suggest the modern name of 'muriatic acid', and explain the reaction that made it.

(iv) 1 grain is equivalent to 0.065 g, and 1 cubic inch is approximately 16.4 cm^3. What substance do you think Faraday used to measure pressure? Work out the pressure of 29.98 inches in more common units.

(v) What is the modern name for carbonic gas?

(vi) Calculate the mass of carbon and of hydrogen that the 0.776 grain of the substance contained; hence calculate the empirical formula of the compound. Suggest the likely molecular formula.

(vii) Faraday's comment about the proportionality of carbon and hydrogen implies that the empirical formula was C_2H. What might be the reason for him suggesting this formula?

(viii) Check to see if his measurements on the reaction of the compound with oxygen fit the molecular formula that you have suggested for the compound.

Answers

113.1 (i) $2C_6H_6(l) + 15O_2(g) \rightarrow 12CO_2(g) + 6H_2O(g)$
(ii) Owing to the very large proportion of carbon in benzene, a great deal of oxygen is needed to convert all of it to carbon dioxide. In air there is not enough oxygen available close to the flame, and some of the carbon remains as soot.

113.2

$$C_6H_5Br + C_2H_5Br + 2Na \rightarrow C_6H_5C_2H_5 + 2NaBr$$
ethylbenzene

113.3 (i) Chloroethane, (ii) 2-chloropropane, both together with aluminium trichloride

113.4 The stopper should not be in the top of the condenser. Carbon dioxide is given off in the reaction, so with the stopper in, pressure would build up. At the least, the stopper would be blown out; but if it were stuck, a much more serious explosion could take place.

113.5 In water, the equilibrium

$$C_6H_5COOH(s) + H_2O(l) \rightleftharpoons C_6H_5COO^-(aq) + H_3O^+(aq)$$
white solid   colourless solution

lies far to the left-hand side. If alkali is present, oxonium ions are removed. Le Chatelier's principle confirms that the equilibrium should move to the right, i.e. in the direction that tends to replace the lost oxonium ions. Thus we should see a colourless solution. If acid is added, the equilibrium will shift to the left. This is the direction that produces insoluble benzoic acid crystals.

113.6 Both methyl groups are oxidised to acid groups. Benzene-1,3-dicarboxylic acid is made:

113.7 Boron trichloride is the one you were expected to think of. There are others, e.g. iron(III) chloride.

113.8 (i) The temperatures are, respectively, 0 °C, 15.6 °C and 85.6 °C.

(ii) The specific gravity is very close to that of benzene (see section 113.1), although the melting and boiling points are rather different. We might expect all his results to be inaccurate by today's standards: he obtained his sample from an oily by-product of a process used by the Portable Gas Company, not a chemical supplier.

(iii) We know it as hydrochloric acid. This is the free radical chlorination of benzene.

(iv) His pressure 29.98 inches refers to the height of a column of mercury supported by the gas. This height is nearly 76 cm of mercury, or a pressure of 1 atm, or 101.325 kPa.

(v) Carbon dioxide.

(vi) 0.776 grain is 0.052 g; 5.6 cubic inches is 91.84 cm^3; 0.58 grain is 0.038 g. Thus we have: 0.052 g of substance reacted with CuO to give 91.84 cm^3 of carbon dioxide (at 15.6 °C) and 0.038 g water. If we convert to s.t.p., the volume of CO_2 would be 91.84 cm^3 × (273 K/288.6 K), i.e. 86.88 cm^3. Therefore,

number of moles of CO_2
= 86.88 cm^3/22 400 cm^3 mol^{-1} = 0.0039 mol

This is also the number of moles of carbon that the original compound contained. (The carbon cannot come from the CuO.) Also

number of moles of water
= 0.038 g/18 g mol^{-1} = 0.0021 mol

This contains 2 × 0.0021 mol = 0.0042 mol of hydrogen. (All the hydrogen must also have been in the original compound.) Thus the ratio of moles of carbon to moles of hydrogen is 0.0039 mol to 0.0042 mol, or approximately 1:1. The empirical formula is CH.

We have reason to think it was benzene, C_6H_6. We know that 1 mol of benzene would contain 6 mol of carbon; so 0.0039 mol of carbon should come from 0.0039 mol/6 = 65 × 10^{-5} mol of benzene. The mass of this amount of benzene is 78 g mol^{-1} × 65 × 10^{-5} mol = 0.0507 g. This is fairly close to the mass of the compound that Faraday used.

(vii) In Faraday's time, the relative atomic mass of carbon was thought to be 6 (not 12), so he thought it contained twice as much carbon as we calculate.

(viii) When benzene vapour completely burns in oxygen the equation is

$$C_6H_6(g) + (15/2)O_2(g) \rightarrow 6CO_2(g) + 3H_2O(l)$$

where we assume that the water condenses to a liquid at 62 °F, or 16.7 °C. The equation tells us that (under the same conditions of temperature and pressure) 1 vol of benzene should react completely with 15/2 vol of oxygen to give 6 vol of carbon dioxide. Alternatively,

1/6 vol C_6H_6 reacts with 15/12 vol O_2 to give 1 vol CO_2

i.e. 0.167 vol C_6H_6 reacts with 1.25 vol O_2 to give 1 vol CO_2, and 0.351 vol C_6H_6 reacts with 2.625 vol O_2 to give 2.1 vol CO_2. (Here we have scaled the figures to give the same volume of CO_2 that Faraday used.) You can see that the results agree with his figures to reasonable accuracy. In brief, Faraday had discovered benzene.

- Aromatic compounds contain one or more benzene rings.

Benzene, C_6H_6

    (i) Has a delocalised ring of electrons.

    (ii) Is more energetically stable than a similar molecule with separate single and double bonds.

    (iii) Does not easily undergo addition reactions.

    (iv) Is attacked by electrophiles, e.g. NO_2^+, CH_3^+.

- Preparation:
  Decarboxylation of calcium benzoate by heating with soda lime;

$$(C_6H_5COO)_2Ca(s) + Ca(OH)_2(s) \rightarrow 2C_6H_6(l) + 2CaCO_3(s)$$

Properties

- Oxidation:
  Burns in air with very sooty flame. But in oxygen

$$2C_6H_6(l) + 15O_2(g) \rightarrow 12CO_2(g) + 6H_2O(g)$$

- Hydrogenation:

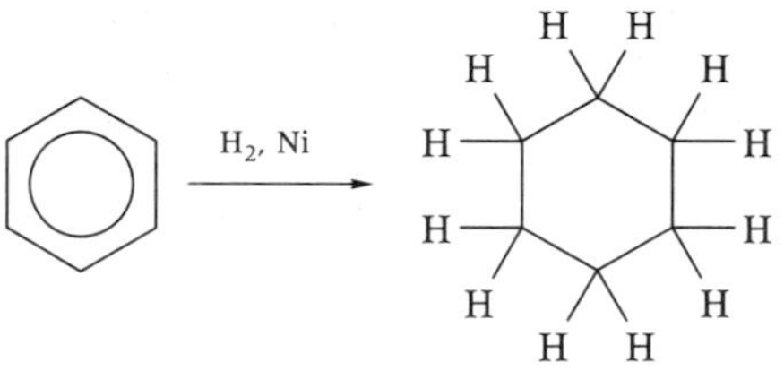

cyclohexane

- Chlorination:
  If there is no side chain,

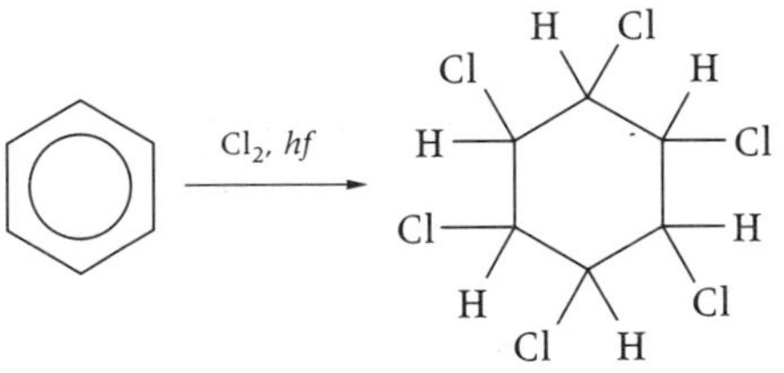

free radical reaction

- Electrophilic attack:
  This is summarised in Table 113.2.

Arenes

- Preparation:
  Friedel–Crafts reaction: benzene reacted with a halogenoalkane and aluminium trichloride; e.g.

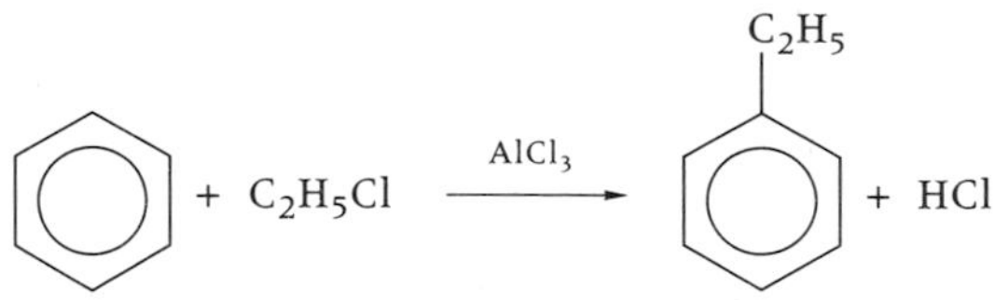

Reactions

- Of the side chain:
  (i) Oxidation

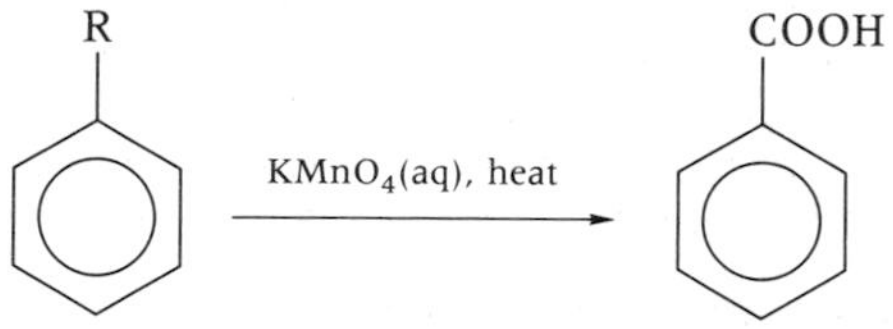

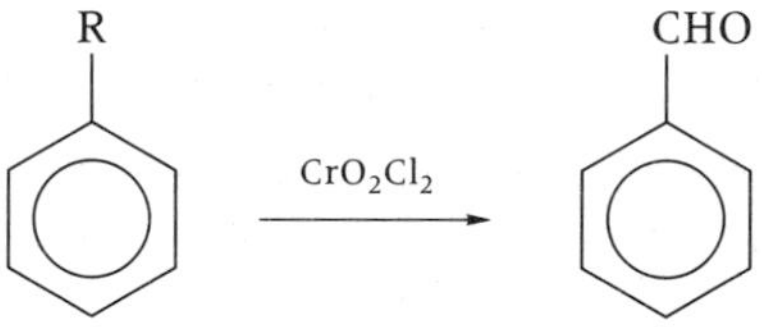

    (ii) Halogenation: ultraviolet light induces attack on the side chain, not the ring; e.g.

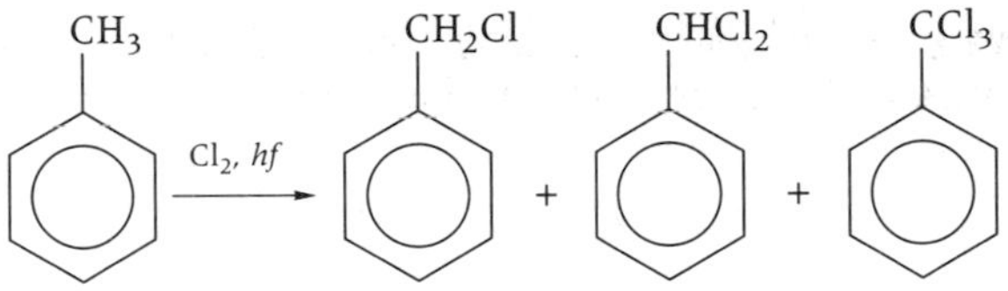

free radicals involved.

- Of the ring:
  (i) Hydrogenation by heating with hydrogen and a nickel catalyst (see above).
  (ii) Electrophilic attack takes place; e.g. ring attacked in Friedel–Crafts reactions (see above).

- Diazotisation:
  Conversion of an aromatic amine into a diazonium ion. The latter can couple with, for example, phenol;

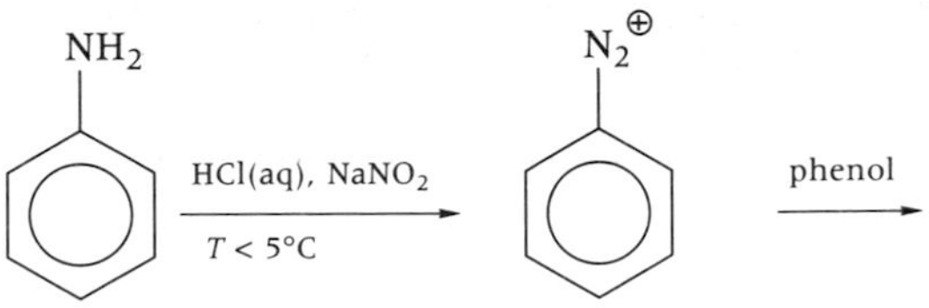

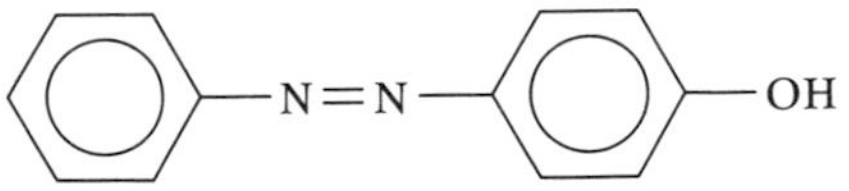

114

More about electrophilic substitution

114.1 How a group on a benzene ring influences electrophilic substitution

We know how to substitute different groups in a benzene ring. Now we shall see how these groups affect the way the ring reacts if we try to substitute a second group. It turns out that the presence of the first group has two main effects.

First, it can change *how rapidly* the ring is attacked. Some groups make the ring react faster, some slower. Thus some groups *activate* and some *deactivate* the ring.

Secondly, it influences where on the ring the second group goes. This is the *orientation* effect.

(a) Activating and deactivating effects

You can see the activating influence of some groups very easily. Phenol, C_6H_5OH, consists of a benzene ring with an OH group bonded to it. At room temperature phenol is a pale pink solid, and is partially soluble in water. If you dissolve a little phenol in water and add bromine water, you will see an immediate white precipitate appear. The precipitate is 2,4,6-tribromophenol. Clearly the OH group activates the ring. In Table 114.1 you will find a list of activating and deactivating groups.

Table 114.1. Activating and deactivating groups in electrophilic substitution

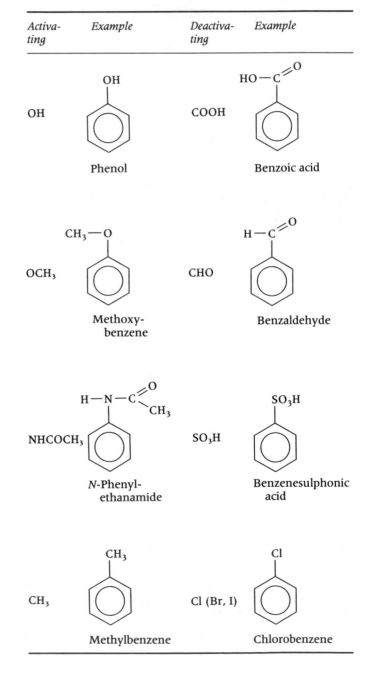

Activating	Example	Deactivating	Example
NH$_2$	Phenylamine	NO$_2$	Nitrobenzene
OH	Phenol	COOH	Benzoic acid
OCH$_3$	Methoxy-benzene	CHO	Benzaldehyde
NHCOCH$_3$	N-Phenyl-ethanamide	SO$_3$H	Benzenesulphonic acid
CH$_3$	Methylbenzene	Cl (Br, I)	Chlorobenzene

(b) *Orientation effects*

Suppose we have a benzene ring with a methyl group already bonded to it. If methylbenzene is attacked by a nitro group, then in principle we might get any one (or a mixture) of the following molecules:

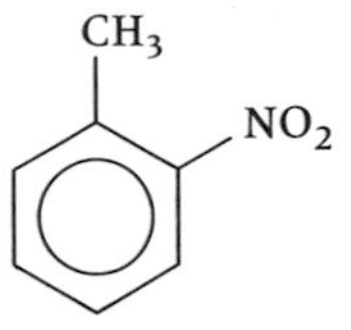

methyl-2-nitrobenzene
(*ortho*-nitrotoluene)

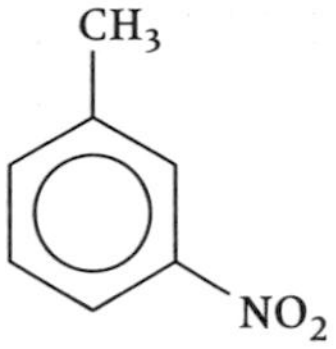

methyl-3-nitrobenzene
(*meta*-nitrotoluene)

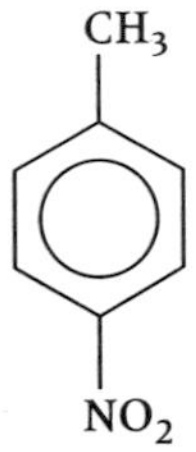

methyl-4-nitrobenzene
(*para*-nitrotoluene)

Here we have given the old names of the products as well as their modern names. The positions that we label by numbers were, and often still are, called the *ortho*, *meta* and *para* positions (Figure 114.1).

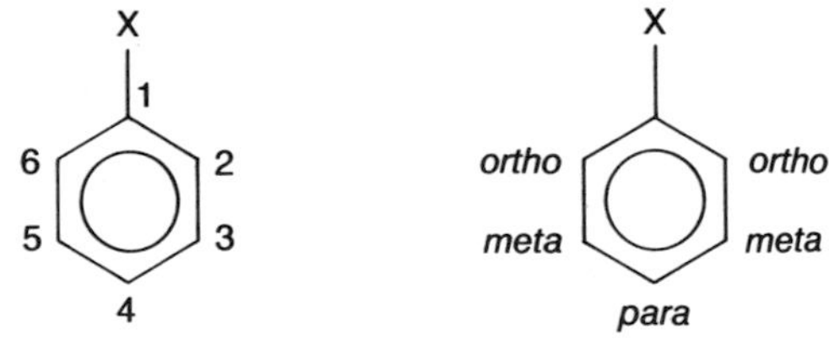

Figure 114.1 *The new and old ways of naming the positions around a benzene ring*

If the experiment is carried out and the products analysed, we find that there is less than 5% of methyl-3-nitrobenzene produced. This compares with about 58% methyl-2-nitrobenzene and 38% methyl-4-nitrobenzene. Using the older names, we say that:

> **The methyl group directs *ortho* and *para*.**

However, if we nitrate nitrobenzene itself, over 90% of the product is 1,3-dinitrobenzene (the *meta* product). In this case:

> **The nitro group directs *meta*.**

Table 114.2 summarises the results of electrophilic substitution for a number of different compounds.

Table 114.2. Orientation effects in electrophilic substitution

Groups that direct *ortho* and *para*	NH$_2$, OH, OCH$_3$, NHCOCH$_3$, CH$_3$, Cl, Br, I
Groups that direct *meta*	NO$_2$, COOH, CHO, SO$_3$H

If you compare Tables 114.1 and 114.2 you will see that, apart from the halogens (see section 114.4), groups that direct *ortho* and *para* also activate the ring, and groups that direct *meta* deactivate the ring. Now it is time for us to explain these observations.

114.2 Why do some groups activate and some deactivate the ring?

Electrophiles seek out centres of negative charge, so we should expect to find that activating groups have a way of increasing, and deactivating groups decreasing, the electron density around a benzene ring.

We can understand the deactivating effect of the nitro group by noticing that the two oxygen atoms will draw electron density towards themselves, leaving the nitrogen atom slightly positively charged. This, in turn, will draw electrons away from the ring.

We have evidence that this happens from the dipole moment of nitrobenzene. The ring is the positive end of the dipole, the nitro group the negative end. The orientation of the dipole with the other deactivating groups also shows that the benzene ring suffers a withdrawal of charge. Table 114.3 lists the dipole moments of some substituted arenes.

Table 114.3. The dipole moment of some substituted benzene molecules

Activating groups		Deactivating groups	
Diagram	Dipole moment/D	Diagram	Dipole moment/D
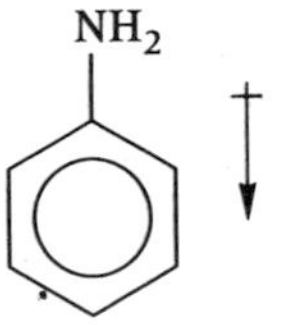 NH$_2$	1.53	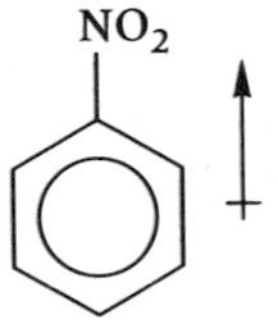 NO$_2$	4.22
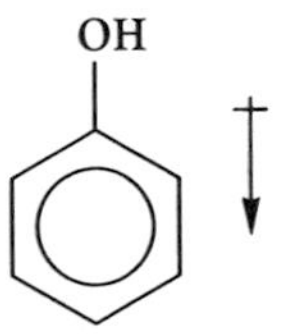 OH	1.45	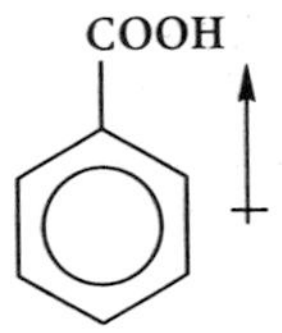 COOH	1.71
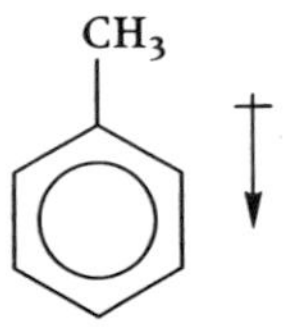 CH$_3$	0.36	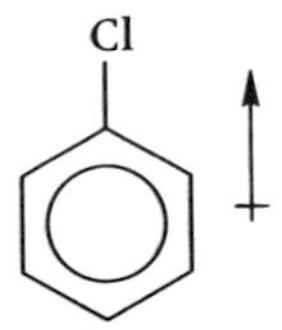 Cl	1.67

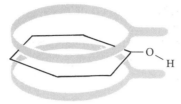

Phenol, C_6H_5OH

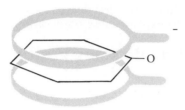

Phenoxide ion, $C_6H_5O^-$

Figure 114.2 *The π electrons in phenol and the phenoxide ion are delocalised around the ring and over the oxygen. However, the delocalisation is greater in the phenoxide ion. This is responsible for the acidity of phenol in water*

Now we come to the problem of why compounds like phenol and methoxybenzene activate the ring even though they have an electronegative oxygen atom directly attached to the ring. If anything, we might expect these molecules to be severely deactivated. However, the dipole moments of the molecules tell us that the ring is definitely richer in electrons than the OH or CH_3O group bonded to it. The reason is that a lone pair on the oxygen atom can interact with the π cloud on the ring (Figure 114.2). This interaction means that, although electron density may be drawn to the oxygen atom through the σ bond joining it to the ring, electron density is fed into the ring through the π system. Indeed, when a solution of phenol in water reacts with bromine, a small but significant number of phenol molecules exist as phenoxide ions. Phenol is a weak acid in water:

$$C_6H_5OH(aq) + H_2O(l) \rightleftharpoons C_6H_5O^-(aq) + H_3O^+(aq)$$
$$\text{phenoxide ion}$$

Given that the oxygen now carries a complete unit of negative charge, the tendency for electron density to build up on the ring is even greater than in the phenol molecule.

114.3 Why do some groups direct *ortho/para* and some *meta*?

(a) Ortho *and* para *direction*

We shall take the nitration of phenol as an example of a reaction in which the *ortho* and *para* products predominate. Phenol can be nitrated with dilute nitric acid, with the mechanism involving attack by nitrosonium ions, $^{\oplus}NO$, rather than nitryl cations, which are present in a mixture of concentrated nitric and sulphuric acids. Let us assume that the nitrosonium ion makes a σ complex to the ring at the *ortho* position. It is helpful to start by showing the benzene ring as if it had three separate double bonds (Figure 114.3). The bond to the nitrosonium ion is made by the two π electrons originally joining the carbon atoms. This means that the carbon atom on the other end of the π bond loses electron density and takes up the positive charge.

The π electrons further round the ring will feel the presence of this charge, and we can imagine that they would move towards it. We show this shift of the π electrons as the movement of the π bond. Another movement like this shifts the π cloud further round the ring, and the positive charge is left on the carbon atom bonded to the OH group (Figure 114.3). This is where the lone pairs on the oxygen have an important part to play. As we have seen, one of the lone pairs can bond to this carbon atom. In this way the positive charge that the nitrosonium ion brought with it is spread over four atoms. We know that spreading charge around a molecule (or ion) rather than allowing it to congregate in one place leads to an increase in energetic stability. You might like to show that a similar set of structures can be drawn for *para* attack by $^{\oplus}NO$.

The diagrams that we have drawn in Figure 114.3 are called *canonical structures*. The more canonical structures that you can draw for a given ion or molecule, the greater the energetic stability.

Now see what happens if a nitrosonium ion attacks the *meta* position. We can draw the canonical structures in Figure 114.4. Here there are only three structures possible. A lone pair on the oxygen atom cannot bond to the ring in the same way as before because the positive charge does not arrive on the carbon atom holding the OH group. In this way we see that the positive charge on the σ complex during *meta* attack is not delocalised as much as in *ortho* attack. Thus the *meta* complex is less energetically stable than the *ortho* and *para* complexes.

We seem to have explained the relative energetic stabilities of the three different σ complexes fairly well. We now have to explain why this has an effect on the relative quantities of the three final products. To do this you will need to have read the units on kinetics (Units 77–79). There we said that the higher the activation energy for a reaction, the less chance there is for reactants to combine to give products. We shall assume that the σ complexes are the transition states in the reactions. (In practice, this assumption is not strictly correct. However, we shall make no great error by ignoring the

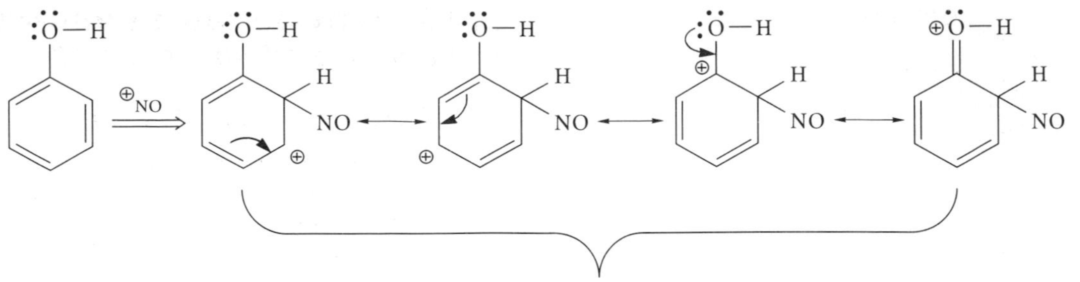

Canonical structures

The final step in the reaction is

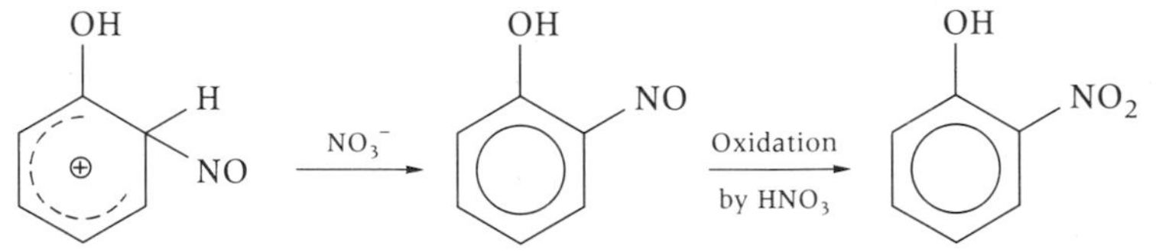

Figure 114.3 Ortho *attack by a nitrosonium ion,* ^{+}NO, *gives a Wheland intermediate, for which we can draw four different structures (the canonical structures). The curly arrows represent the movement of a pair of electrons. The structures imply that the positive charge is spread around the ring, and over the oxygen atom. The greater the spread of charge, the greater the energetic stability*

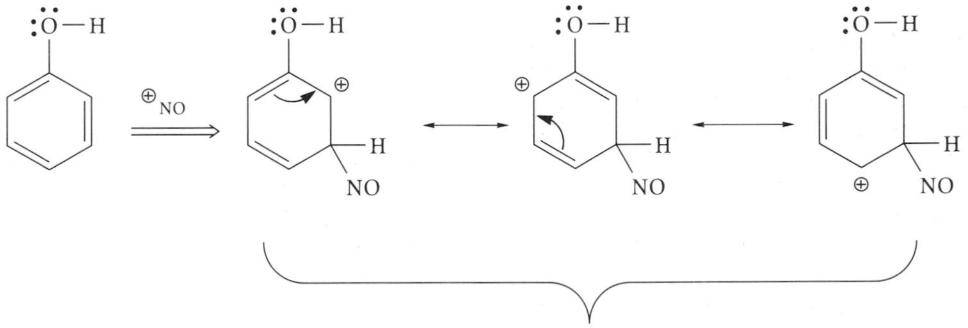

Three canonical structures

Figure 114.4 *Only three canonical structures can be drawn for* meta attack on phenol. The Wheland intermediate is not stabilised to the same extent as that for ortho or para attacks. Notice that a lone pair on the oxygen cannot bond to the ring here, as that would necessitate five bonds to a carbon atom (which is impossible)*

complications.) This being so, we would expect the activation energy for *ortho* and *para* attack to be lower than for *meta* attack. (The *ortho* and *para* σ complexes are energetically more stable, i.e. at a *lower* energy, than the *meta* σ complex.) As a result, the *ortho* and *para* complexes should be made faster than the *meta* complex. Indeed, this fits with the observation that the major product of the reaction has an NO_2 group at the *ortho* and *para* positions. (About 40% 2-nitrophenol and 13% 4-nitrophenol; there are many side reactions as well.) The final step in the reaction is the very rapid oxidation of the nitrosophenols by nitric acid.

(b) Meta *direction*

The nitration of benzoic acid by a mixture of concentrated sulphuric and nitric acids gives about 80% of the *meta* product, 3-nitrobenzoic acid. To see why the acid group directs *meta*, look at Figure 114.5, where the canonical structures for *ortho* and *meta* attack are drawn.

First, notice that the electronegative oxygen atom double bonded to the carbon atom in the acid group draws electrons towards it. This leaves the carbon atom with a slight positive charge. There are three canonical structures in each case, so at first sight it seems that the charge is delocalised to the same extent. However, if you look at the third structure for *ortho* attack, you will see that the positive charge on the ring is brought adjacent to the slight positive charge already present on the carbon atom of the acid group. In fact, this is energetically *unfavourable*. Therefore, our count of canonical structures gives three in favour of the *meta* σ complex and two in favour of the *ortho* σ complex. This is a typical result for groups that direct *meta*.

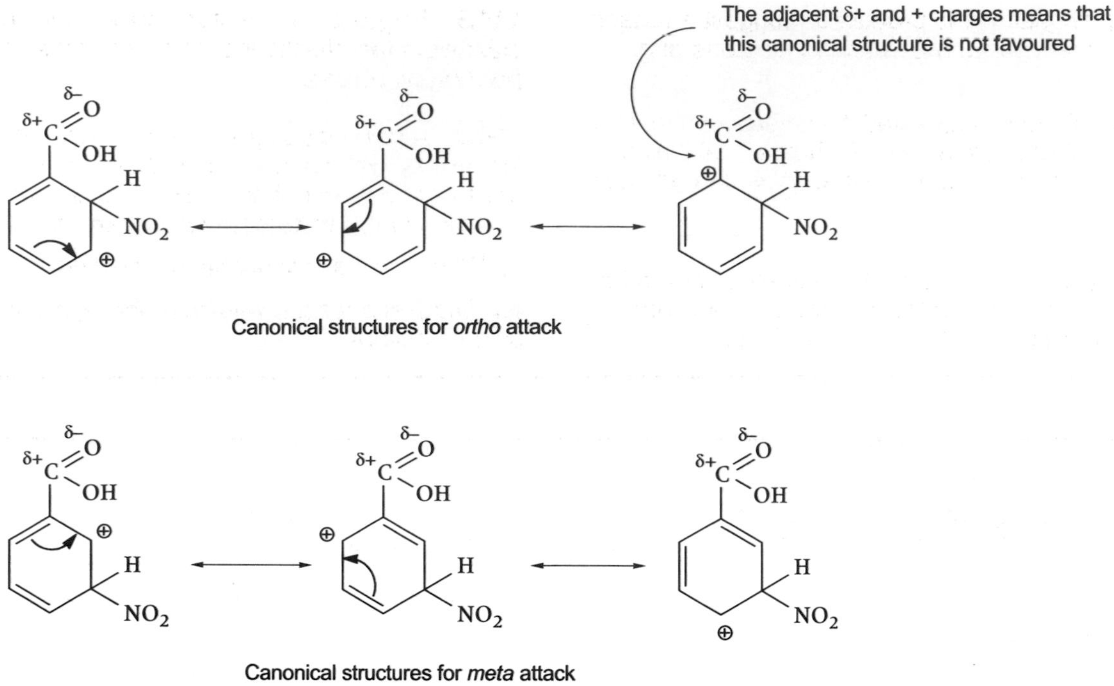

Canonical structures for *ortho* attack

Canonical structures for *meta* attack

Figure 114.5 *Meta attack on benzoic acid is favoured. One of the canonical structures in ortho (and para) attack leads to δ+ and + charges coming close together. This is energetically unfavourable. For meta attack, this repulsion does not occur*

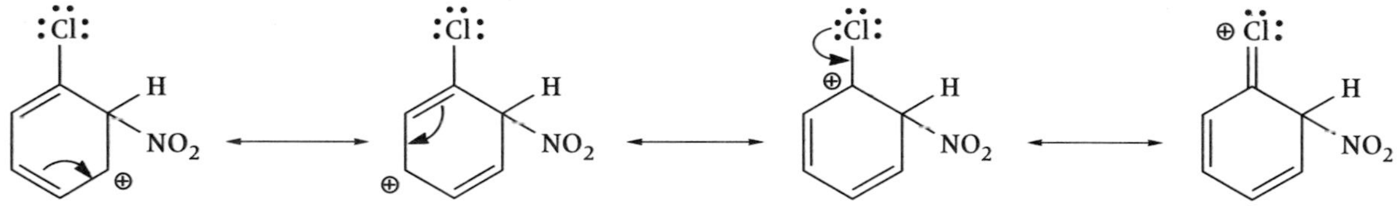

Figure 114.6 *The four canonical structures for ortho attack on chlorobenzene*

114.4 Why do the halogens behave differently to other groups?

The usual pattern of events is for groups that direct *ortho* and *para* to activate the ring, and groups that direct *meta* to deactivate the ring. However, the halogens are exceptions. They direct *ortho* and *para* but deactivate the ring. The electron withdrawing nature of the halogens means that they withdraw electron density from the ring. This is shown, for example, by the orientation of the dipole moment of chlorobenzene. As a result, an incoming electrophile will find the benzene ring in chlorobenzene less attractive than in benzene itself.

Here we have an explanation of why the rate of the reaction is slower.

Now let us look at the electron structure of a halogen atom when it is joined to a benzene ring. There are three lone pairs around the halogen atom (compare the bonding in hydrogen chloride; section 14.2). In Figure 114.6 the canonical structures for *ortho* attack of a nitryl cation on chlorobenzene are drawn.

The key thing to spot is that, like the oxygen in phenol or the nitrogen in phenylamine, a lone pair on the chlorine can link with the π cloud on the ring. This tends to stabilise the *ortho* and *para* σ complexes at the expense of the *meta* σ complex. Thus we can thank the lone pairs on the halogen atoms for the *ortho*/*para* directing ability.

114.1 We can write a substituted benzene as C_6H_5X. Similarly we can use the letter E to stand for an electrophilic reagent. Assume two E groups successfully attack C_6H_5X.

(i) Is the formula of the product $C_6H_5XE_2$, $C_6H_4XE_2$, or $C_6H_3XE_2$?

(ii) How many different products are possible?

(iii) Draw their structures.

114.2 If bromine water is added to a solution of phenylamine, a white precipitate of 2,4,6-

tribromophenylamine is produced. Suggest a reason why the amine group activates the benzene ring.

114.3 A student suggested that each of the four canonical structures drawn in Figure 114.3 would exist for a quarter of the time. Was the student correct?

114.4 If you have not already done so, draw out the canonical structures for the σ complex made when $\oplus$NO attacks phenol at the *para* position.

114.5 Suggest a reason why the benzene-diazonium ion attacks the *para* rather than the *ortho* position on phenol.

114.6 A student suggested that, ignoring complicating factors, when a group substitutes at the *ortho* and *para* positions, we should expect to get about twice as much of the *ortho* product as *para* product.

(i) Why is this a reasonable suggestion?

(ii) Suggest a reason why, in practice, this ideal ratio seldom occurs.

Answers

114.1 (i) $C_6H_3XE_2$. This is a substitution reaction, so for each E group that joins, one hydrogen atom must be lost.

(ii) Six.

(iii) The structures are shown in Figure 114.7.

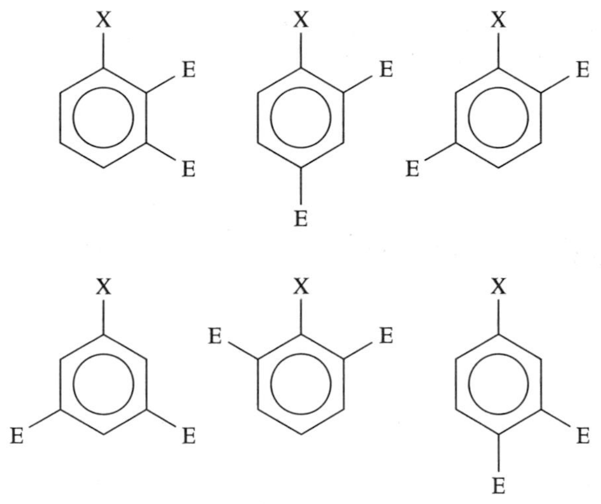

Figure 114.7 *Diagrams for answer to question 114.1*

114.2 Like the oxygen atom in phenol, the nitrogen has a lone pair of electrons that can interact with the π cloud on the benzene ring. You should be able to draw the same set of canonical structures for an electrophile attacking phenylamine as in the case of phenol.

114.3 No. If you look back at section 14.7 you will find that we said that it is a mistake to think that an ion or molecule constantly changes from one structure to another. Canonical structures are *our* way of explaining

the way chemicals behave. Presumably the ion or molecule knows nothing about them!

114.4 The structures are drawn in Figure 114.8.

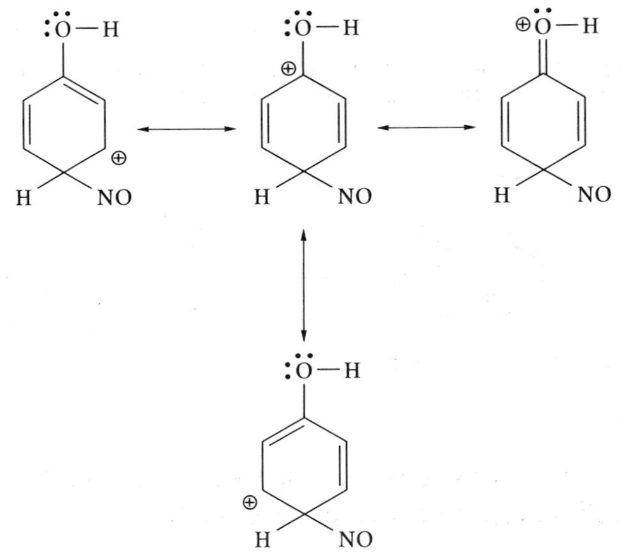

Figure 114.8 *Diagrams for answer to question 114.4*

114.5 There is steric hindrance at work. The large benzenediazonium ion cannot easily get past the group next to the *ortho* positions.

114.6 (i) There are two *ortho* for each *para* position. Hence the ratio should be 2:1.

(ii) There are many factors to take into account. The one you should suggest is that steric hindrance will decrease the likelihood of groups bonding at the *ortho* position.

UNIT 114 SUMMARY

- Groups that direct *ortho* and *para*, i.e. 2 and 4 positions on a benzene ring, tend to feed electrons into the ring.

- Groups that direct *meta*, i.e. 3 position on a benzene ring, tend to withdraw electrons from the ring.
- Tables 114.1 and 114.2 provide information about activating and deactivating groups.

115

Organic halogen compounds

115.1 What are halogenoalkanes?

The halogenoalkanes are hydrocarbon chains that have one or more hydrogen atoms exchanged for halogen atoms. Typical examples are listed in Table 115.1. (We shall consider aromatic halogen compounds in later sections.)

Halogenoalkanes have many uses. A number of them, like tetrachloromethane, have been used as dry cleaning fluids. Others, especially the brominated hydrocarbons such as 1,2-dibromoethane, are used in fire extinguishers. This compound is also used as a petrol additive. Tetraethyl-lead(IV) added to petrol produces free radicals, which help petrol burn evenly. It also liberates lead atoms; these are mopped up by 1,2-dibromoethane. This is a sensible way to protect car engines from seizing up, but it is not good for those of us who have to breathe the air polluted by lead compounds. They are known to be poisonous, and cause brain damage in children.

In section 81.4 we mentioned the extremely serious effects on the ozone layer around the Earth of chlorofluorocarbons (CFCs), which are used in aerosols and refrigerators.

115.2 Methods of making halogenoalkanes

The main way of making a halogenoalkane is to replace the OH group of an alcohol by a halogen. The laboratory preparation of iodoethane is described in panel 115.1 and Figures 115.1 and 115.2. This shows you the principal stages found in many organic preparations.

In one method of making bromoethane, hydrogen bromide is made by the reaction between concentrated sulphuric acid and sodium bromide, and then reacts with ethanol:

$$H_2SO_4 + NaBr \rightarrow NaHSO_4 + HBr$$
$$C_2H_5OH + HBr \rightarrow C_2H_5Br + H_2O$$

This method can be used to make many halogenoalkanes. The essential thing is to choose the right alcohol for the halogenoalkane you need. In the case of iodoethane, a mixture of red phosphorus and iodine is reacted with ethanol:

$$2P + 3I_2 \rightarrow 2PI_3$$
$$3C_2H_5OH + PI_3 \rightarrow 3C_2H_5I + H_3PO_3$$

There are two other ways of making halogenoalkanes from alcohols. You could add a phosphorus halide to the alcohol. For example,

$$\underset{\substack{\text{phosphorus} \\ \text{pentachloride}}}{PCl_5} + C_2H_5OH \rightarrow \underset{\substack{\text{phosphorus} \\ \text{trichloride} \\ \text{oxide}}}{POCl_3} + HCl + \underset{\text{chloroethane}}{C_2H_5Cl}$$

This reaction has the virtue of taking place in the cold. However, it is more often regarded as a test for the presence of an OH group. The white fumes of hydrogen chloride given off in the reaction give a tell-tale sign of the reaction taking place.

A similar reaction occurs between an alcohol and sulphur dichloride oxide:

$$SOCl_2 + C_2H_5OH \rightarrow SO_2 + HCl + C_2H_5Cl$$

The advantage of this reaction is that the two side products are gases and are easily removed from the reaction mixture.

115.1 Using one of these methods, which bromoalkanes would be made from the following alcohols: (i) propan-1-ol; (ii) butan-1-ol; (iii) butan-2-ol?

115.3 Halogenoalkanes are attacked by nucleophiles

The majority of halogenoalkanes have a dipole moment (Table 115.2). This is a result of the electron withdrawing ability of halogen atoms. The halogen atom carries a slight negative charge, with the remainder of the

Table 115.1. Halogenoalkanes

Name	Formula	Structure	Melting point/°C	Boiling point/°C
Chloromethane	CH_3Cl	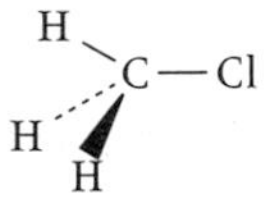	−97.6	−24.1
Dichloromethane	CH_2Cl_2	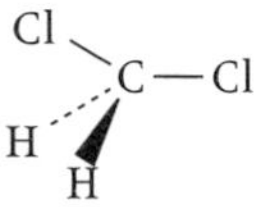	−95.0	139.9
Trichloromethane (chloroform)	$CHCl_3$	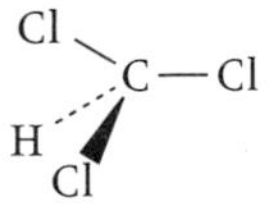	−63.3	61.9
Tetrachloromethane (carbon tetrachloride)	CCl_4	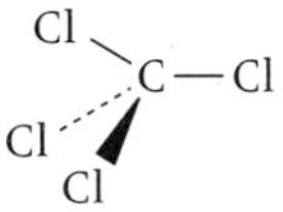	−22.8	76.7
Chloroethane	CH_3CH_2Cl	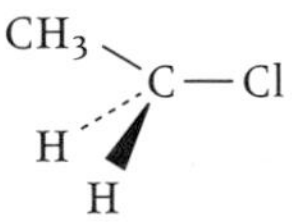	−136.3	12.4
Bromoethane	CH_3CH_2Br	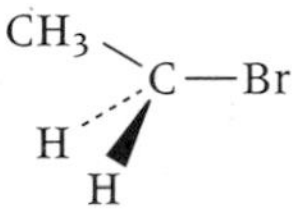	−118.5	38.5
1,2-Dibromoethane	CH_2BrCH_2Br	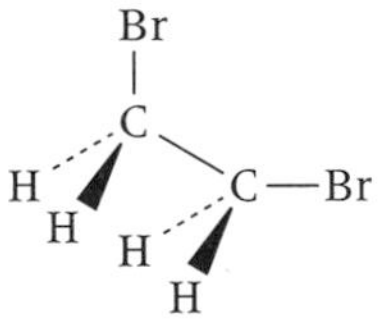	9.9	13.5
Iodoethane	CH_3CH_2I	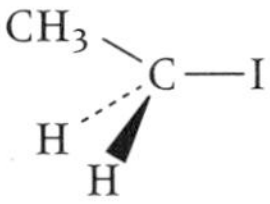	−111.0	72.5
Triiodomethane (iodoform)	CHI_3	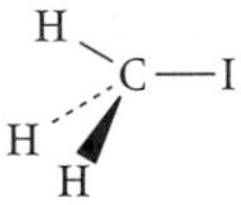	119.0	218 (approx.)
Difluorochloromethane	$CHClF_2$	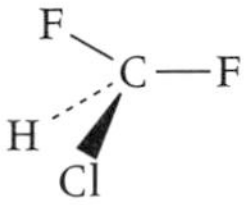	−146.0	−40.8
Trifluorochloromethane	$CClF_3$	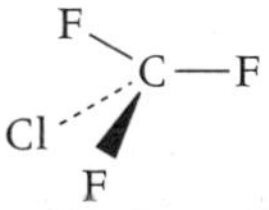	−181.0	−81.4

Panel 115.1

The preparation of iodoethane

Organic preparations tend to follow a similar pattern. The order of events is often like this:

(i) *Stage 1*. Heat the reaction mixture, probably under reflux, for up to an hour.
(ii) *Stage 2*. Distil the final mixture to separate the bulk of the product.
(iii) *Stage 3*. Purify the product, e.g. by drying and redistillation.

The details below will show you these stages in the particular case of preparing iodoethane.

Warning: This reaction sometimes has a sting in its tail; there have been occasions when the reaction mixture has exploded. Do *not* attempt the reaction except under expert guidance.

Stage 1

Ethanol, red phosphorus and iodine are mixed in a flask (see question 115.9 for quantities). The best type is a Quickfit flask, or a similar type, which has a ground glass joint. The flask fits precisely to a Liebig condenser as shown in Figure 115.1. We say that the apparatus has been set up for heating the reactants under *reflux*. Owing to their covalent nature, many organic liquids vaporise easily. If, for example, the reaction flask were heated without the condenser, the ethanol would soon evaporate. With the condenser present, the vapour condenses and drops back into the flask. The condenser also prevents the product (iodoethane) escaping.

The mixture is heated over a water bath for about half an hour. This, too, is a common approach to organic reactions. It is unwise to heat the reactants directly with a bunsen burner. The bunsen flame can cause the temperature to rise too quickly, and tends to heat one part of the flask more than another. (Alternatives to water baths are sand trays or electric mantle heaters.)

Stage 2

Once the reaction is complete, the products have to be separated from the unused reactants, and side products, in the flask. The usual way of doing this is to distil the mixture. The apparatus is dismantled and set up for distillation as shown in Figure 115.2. Distillation will only be effective if the product has a significantly different boiling point from the other chemicals in the flask. Iodoethane has a boiling point of 72.5°C, whereas ethanol boils at 78°C. Thus iodoethane should evaporate in greater quantity than ethanol, and reach the condenser first. The thermometer at the neck of the condenser should read 72.5°C when iodoethane distils over. If the temperature rises above this, it is a sign that ethanol is beginning to distil and the distillation is stopped.

Stage 3

The distillate (the liquid collected from the distillation) should be mainly iodoethane, but it will be contaminated by a mixture of ethanol and phosphonic acid (H_3PO_3). The distillate is placed in a separating funnel and washed with a little distilled water. The separating funnel allows the lower, oily layer of iodoethane to be run off. The ethanol and phosphonic acid will remain in the water. Finally the iodoethane is left in contact with a drying agent to remove water. Anhydrous calcium chloride and anhydrous sodium sulphate are common drying agents. The final part of the experiment is to redistil the dried product (in a clean apparatus, of course).

It is not uncommon for the yield in an organic preparation to be around 60%. This is due to the reactants giving side products, and to the losses that are involved in the separation.

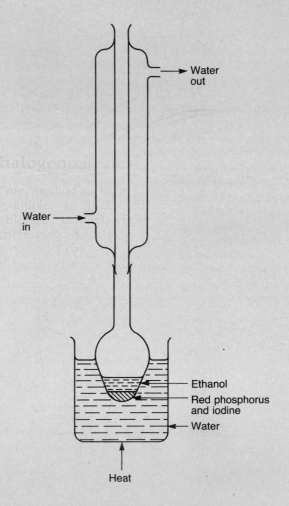

Figure 115.1 *An apparatus for making iodoethane*

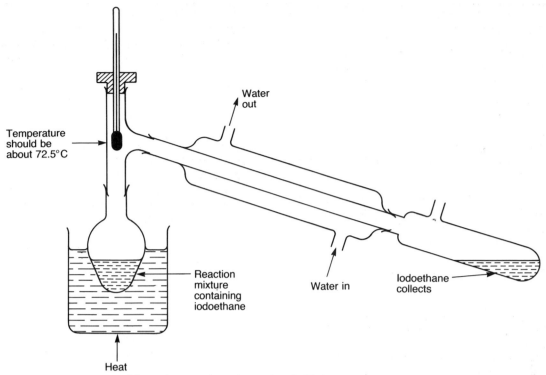

Figure 115.2 *Iodoethane is separated from the reaction mixture by distillation*

Table 115.2. The dipole moments of some halogenoalkanes

Halogenoalkane	Dipole moment/D
CH_3Cl	1.86
CH_3Br	1.79
CH_3I	1.64
C_2H_5Cl	1.98
C_2H_5Br	2.02
C_2H_5I	1.90

molecule having a slight positive charge. The carbon atom bonded to the halogen can act as a centre of attraction for *nucleophiles* (Figure 115.3). Atoms or molecules with lone pairs of electrons and/or a negative charge make good nucleophiles. However, equally important as far as a successful reaction is concerned is that the halogen atom makes a *good leaving group*. For example, iodine atoms can readily change into iodide ions, I⁻. This encourages the carbon–iodine bond to break, thus disrupting the original molecule and allowing the incoming nucleophile to bond to the carbon atom. Please read Unit 81 for more information about this type of reaction.

The majority of halogenoalkanes fit the same pattern in their reactions:

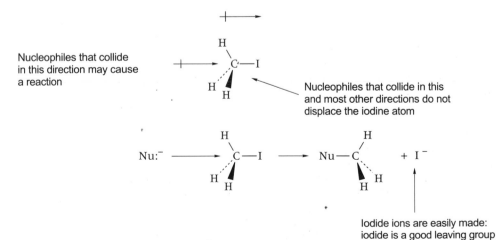

Figure 115.3 *Key factors influencing the reaction of a halogenoalkane with a nucleophilea*

Table 115.3. The substitution reactions of halogenoalkanes

Nucleophile	Type of product	Example	Conditions
		C_2H_5I changes to:	Starting material: iodoethane, C_2H_5I
OH^-, H_2O	Alcohol	C_2H_5OH, ethanol	Hydrolysis: heat with potassium hydroxide solution, or damp silver oxide
CH_3O^-	Ether	$C_2H_5OCH_3$, methoxyethane	Williamson ether synthesis: warm with sodium methoxide
CN^-	Nitrile	C_2H_5CN, propanenitrile	Heat with an alcoholic solution of potassium cyanide
NH_3	Amine	$C_2H_5NH_2$, ethylamine	Heat in a sealed container with ammonia in alcohol
$C_2H_5^-$	Alkane	C_4H_{10}, butane	Wurtz synthesis of alkanes: heat with sodium

---

Halogenoalkanes undergo nucleophilic substitution reactions.

The exceptions are molecules like tetrachloromethane, which have no hydrogen atoms remaining, and no overall dipole moment. Table 115.3 summarises the different possibilities. Here are brief details about each one.

(a) Hydrolysis

Hydrolysis using water on its own has little effect on a halogenoalkane. However, the rate of reaction is increased by boiling with alkali such as potassium hydroxide. The reaction is then called an alkaline hydrolysis. The equations are simple to write down because the overall change is the replacement of the halogen atom by an OH group. For example,

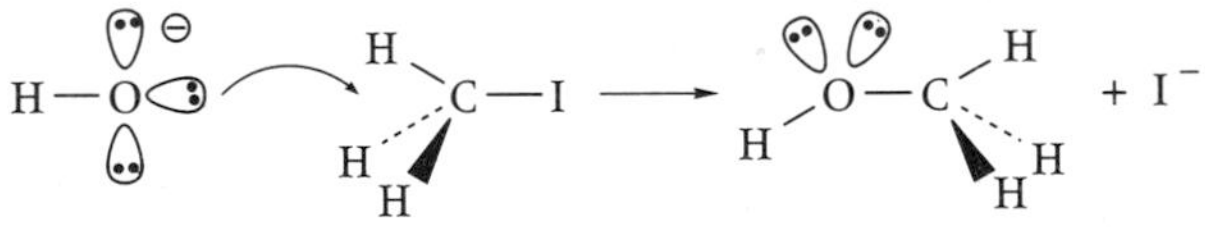

$$OH^- + CH_3I \longrightarrow CH_3OH + I^-$$
iodomethane   methanol

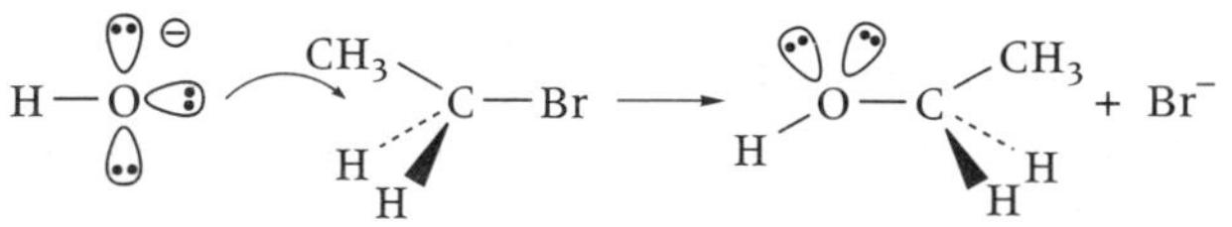

$$OH^- + C_2H_5Br \longrightarrow C_2H_5OH + Br^-$$
bromoethane   ethanol

Both these examples are with primary halogenoalkanes; secondary and tertiary halogenoalkanes react in similar fashion, although there are some subtleties. The mechanism of the nucleophilic attack on halogenoalkanes by hydroxide ions has been studied in detail. You will find information about it in Unit 81.

(b) Ether formation

This reaction of halogenoalkanes is more often thought of as a method of preparation of ethers. It is known as the *Williamson ether synthesis*. The first step in the method is to make a solution of an *alkoxide* ion. The simplest way of doing this is to dissolve sodium in an alcohol. For example,

$$2CH_3OH + 2Na \rightarrow 2CH_3O^- + 2Na^+ + H_2$$
methanol   methoxide ion

$$2C_2H_5OH + 2Na \rightarrow 2C_2H_5O^- + 2Na^+ + H_2$$
ethanol   ethoxide ion

Methoxide and ethoxide ions are particular examples of alkoxide ions. The alkoxide ion is the nucleophile, which then goes on to react with a halogenoalkane:

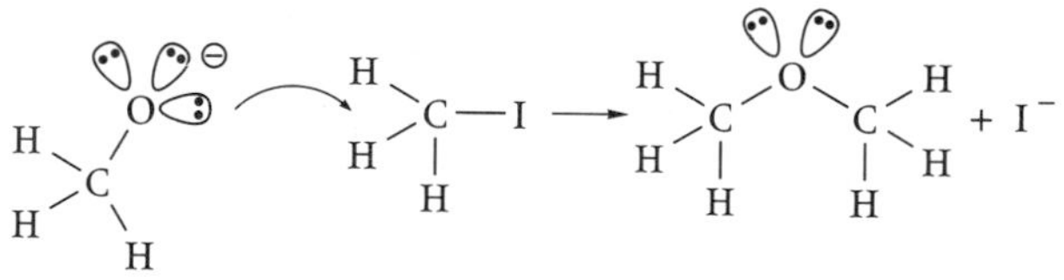

$$CH_3O^- + CH_3I \rightarrow CH_3OCH_3 + I^-$$
methoxide ion   iodomethane   methoxymethane (dimethyl ether)

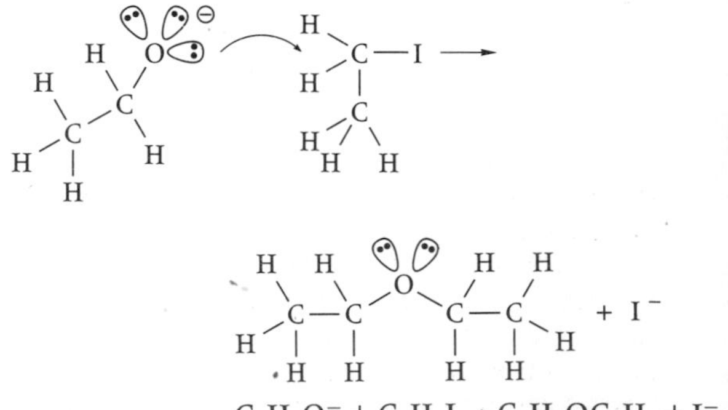

$$C_2H_5O^- + C_2H_5I \rightarrow C_2H_5OC_2H_5 + I^-$$
ethoxide ion   iodoethane   ethoxyethane (diethyl ether)

In practice the reactions are fairly easy to carry out, provided you take care over the risk of fire; ethers are very flammable.

(c) Nitrile formation

Nitriles used to be called cyanides; they contain the —CN group. Being able to introduce this group into an organic molecule can be extremely useful because it increases the number of carbon atoms in the molecule. The method of converting a halogenoalkane into a nitrile is straightforward. The halogenoalkane is warmed with an inorganic cyanide, such as potassium cyanide, in ethanol. The ethanol is a convenient solvent for both reactants, and thereby allows them to react together:

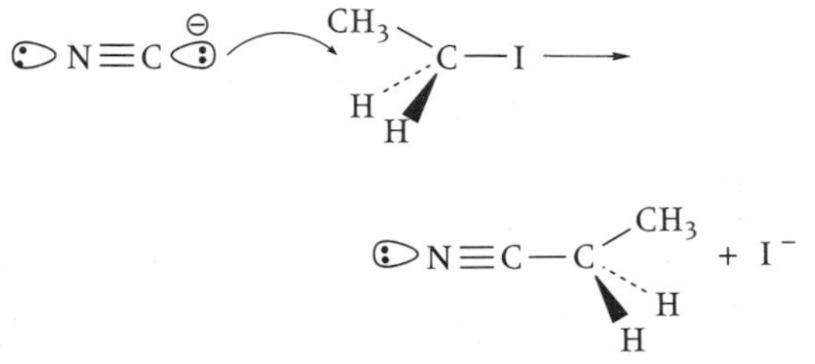

$$CN^- + C_2H_5I \;\rightarrow\; C_2H_5CN \;\; + I^-$$
iodoethane   propanenitrile

(d) Amine formation

Simple amines contain the —NH$_2$ group. This group will replace a halogen atom if the halogenoalkane is heated with ammonia in alcohol under pressure. An ammonia molecule will act as a nucleophile owing to its lone pair of electrons. In the reaction with a halogenoalkane, one of the hydrogen atoms from the ammonia is lost and combines with the halogen atom displaced by the action of the lone pair. For example,

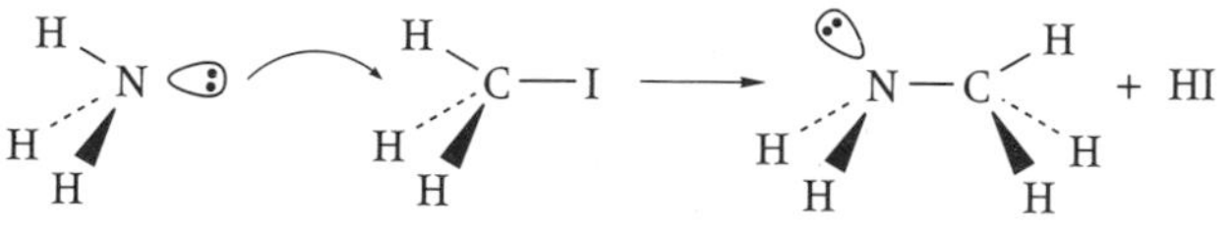

$$NH_3 + CH_3I \;\rightarrow CH_3NH_2 \;\; + HI$$
iodo-   methylamine
methane

This reaction illustrates the fact that nucleophiles do not have to have a negative charge to react with halogenoalkanes. However, it is not a good way of preparing amines because a number of side products always appear. Let us assume that methylamine molecules have been made in the above reaction. The nitrogen atom in methylamine still has its lone pair of electrons. Therefore methylamine can also be a nucleophile and attack iodomethane molecules. This reaction yields a more complicated amine called a secondary amine:

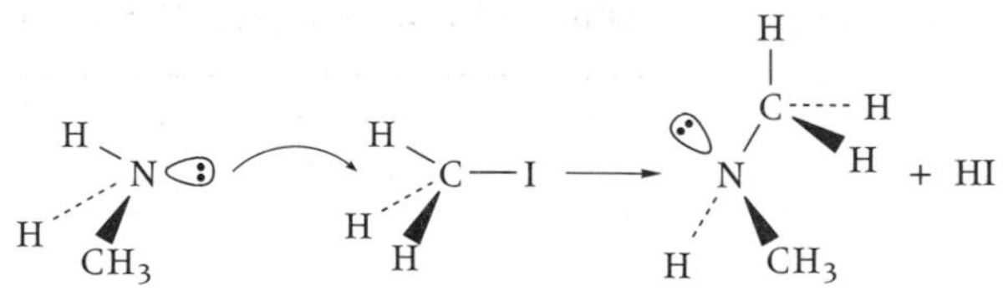

$$CH_3NH_2 + CH_3I \rightarrow (CH_3)_2NH \;\;\; + HI$$
dimethylamine

There are even more complicated products to be found in the final mixture (see question 115.8).

(e) Wurtz synthesis of alkanes

We discussed this method in section 111.4. The only extra piece of information we might mention here is that the effect of the sodium is to convert the halogenoalkane into a *carbanion*. A carbanion is an organic group with a negative charge. For example,

$$C_2H_5I + 2Na \rightarrow C_2H_5^- \;\;\; + I^- + 2Na^+$$
ethyl
carbanion

The carbanion is the nucleophile that attacks another halogenoalkane molecule:

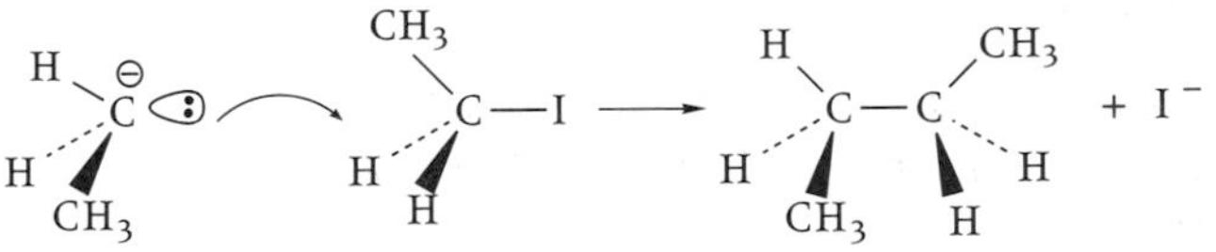

$$C_2H_5^- + C_2H_5I \rightarrow C_4H_{10} + I^-$$

115.2   In a previous unit we found that alkanes will react with halogens to give halogenoalkanes.

(i) What conditions are needed for the reactions to take place?

(ii) Why is the method not a good way of making halogenoalkanes in the laboratory?

115.3   Instead of phosphorus pentachloride, sulphur dichloride oxide, SOCl$_2$, can be used to convert alcohols to chloroalkanes. For example, if this reagent is added to ethanol, C$_2$H$_5$OH, the major product is chloroethane:

$$C_2H_5OH + SOCl_2 \rightarrow C_2H_5Cl + ? + ?$$

(i) What might be the two molecules missing from the equation?

(ii) Why is the separation of the chloroethane easier in this reaction than with phosphorus pentachloride?

115.4   Which of these molecules have a dipole moment: (i) dichloromethane, (ii) tetrachloromethane?

115.5   Suggest a method for making 1,2-dibromoethane, without using an alcohol.

115.6 Why is it easier to carry out experiments with iodoethane rather than chloroethane?

115.7 Draw a diagram of the apparatus you would use to perform the alkaline hydrolysis of iodoethane. Suggest a method of separating the ethanol from the reaction mixture.

115.8 Explain why (i) the reaction of iodomethane with ammonia gives trimethylamine, $(CH_3)_3N$, as one of the products; (ii) an ionic solid of formula $(CH_3)_4NI$ is also found.

115.4   An elimination reaction of halogenoalkanes

You may remember that one method of making an alkene is to heat a halogenoalkane with a concentrated alcoholic solution of potassium hydroxide. (This is the dehydrohalogenation reaction.) A hydrogen atom and the halogen atom are lost, or *eliminated*, from the halogenoalkane, e.g.

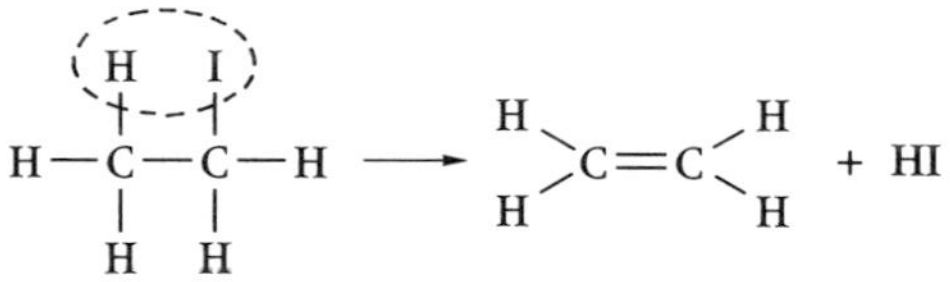

$$C_2H_5I \rightarrow C_2H_4 + HI$$
iodo-   ethene
ethane

115.9 In making iodoethane, a mixture of $5\,cm^3$ of ethanol, $5\,g$ of iodine and $0.5\,g$ of red phosphorus were used. The final mass of iodoethane was $5.1\,g$.

(i) The density of ethanol is approximately $0.8\,g$ cm^{-3}. What mass of ethanol was taken? How many moles is this? $M(C_2H_5OH) = 46\,g\,mol^{-1}$.

(ii) How many moles of phosphorus and of iodine were used? $M(P) = 31\,g\,mol^{-1}$, $M(I_2) = 254\,g\,mol^{-1}$.

(iii) What is the ratio (in whole numbers) of the moles of C_2H_5OH, P and I_2?

(iv) Using your answers so far, and by consulting the equation for the reaction, which of the chemicals was in excess? Which substance governs the quantity of iodoethane that can be made?

(v) Calculate the maximum mass of iodoethane that could be made.

(vi) What was the yield of the reaction?

115.5   There are two types of aromatic halogen compound

The two types are:

> **Those with the halogen atom(s) on a benzene ring, sometimes called *halogenoarenes*.**
>
> **Those with the halogen atom on a hydro-carbon side chain joined to a benzene ring, sometimes called *aralkyl halides*.**

Table 115.4 lists examples of both types. There is one important pattern that emerges in the properties of these two types of compound. It is this:

> ***Aralkyl halides* behave like halogenoalkanes. *Halogenoarenes* do not behave like halogenoalkanes.**

Similarly, the methods we use to prepare them are different. If you know the preparation and properties of halogenoalkanes, e.g. iodoethane, then you should be able to predict those of aralkyl halides, e.g. (iodomethyl)benzene.

115.6   How do we prepare halogenoarenes?

One method is to use the *Friedel–Crafts reaction*. For example, to make chlorobenzene we could pass chlorine through benzene containing aluminium trichloride. An electrophilic attack on the ring takes place and a chlorine atom takes the place of one of the hydrogen atoms:

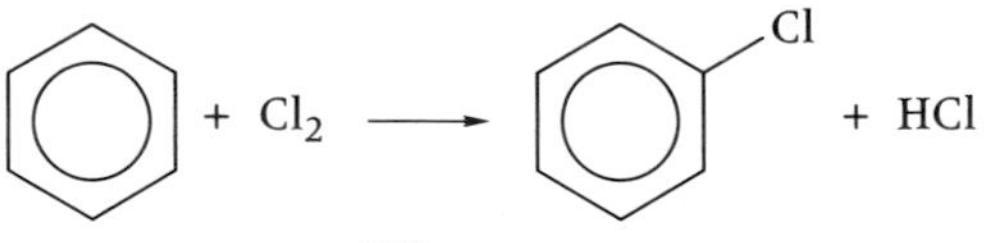

$$C_6H_6 + Cl_2 \xrightarrow{AlCl_3} C_6H_5Cl + HCl$$

A second and simpler method is to use phenylamine, $C_6H_5NH_2$, as a starting material. Phenylamine can be easily converted into a *benzenediazonium* ion by mixing it with an ice-cold solution of sodium nitrite. This solution contains nitrous acid, HNO_2:

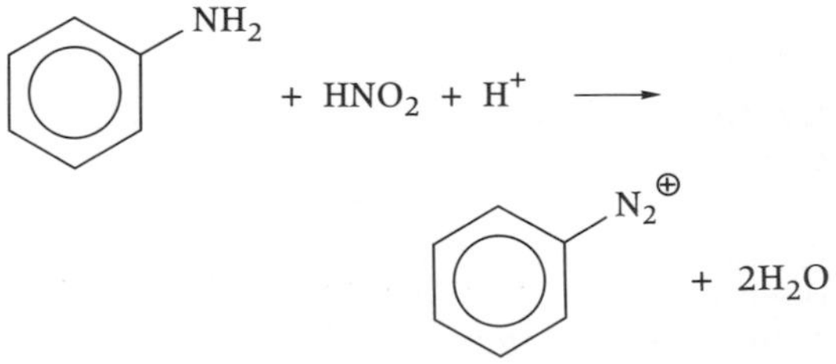

$$C_6H_5NH_2 + HNO_2 + H^+ \rightarrow C_6H_5N_2^\oplus + 2H_2O$$
phenylamine   benzene-
                               diazonium ion

Table 115.4. Some aromatic halogen compounds

Name	Formula	Structure	Melting point/°C	Boiling point/°C
Halogenoarenes				
Chlorobenzene	C_6H_5Cl	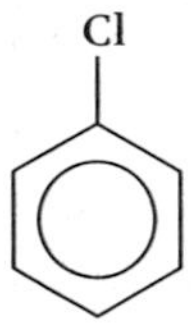	−45.5	132
1,2-Dichlorobenzene (*ortho*-dichlorobenzene)	$C_6H_4Cl_2$	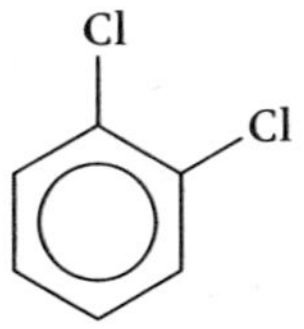	−17	180.5
1,3-Dibromobenzene (*meta*-dibromobenzene)	$C_6H_4Br_2$	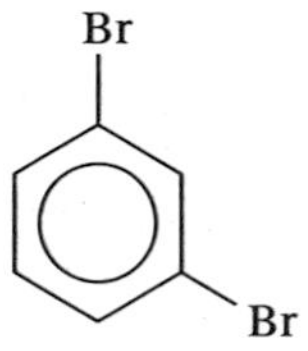	−7	218
1,4-Dichlorobenzene (*para*-dichlorobenzene)	$C_6H_4Cl_2$	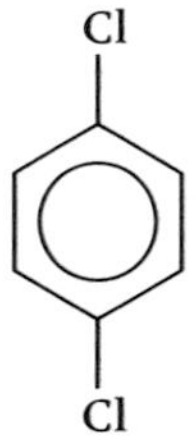	53.1	174
Chloro-2-methylbenzene (*ortho*-chlorotoluene)	$C_6H_4CH_3Cl$	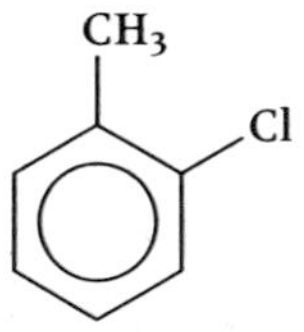	−34	159
Aralkyl halides				
(Chloromethyl)benzene	$C_6H_5CH_2Cl$	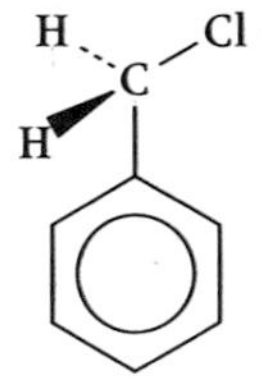	−39	179.3
(Dichloromethyl)benzene	$C_6H_5CHCl_2$	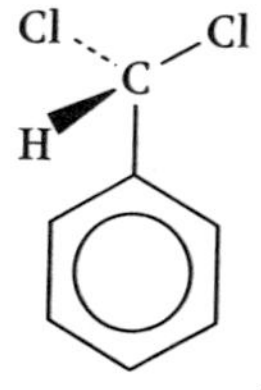	−16.4	205.2
1-Chloro-2-phenylethane	$C_6H_5CH_2CH_2Cl$	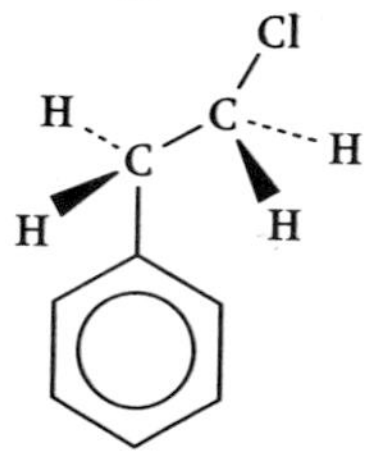	*	197

*Data not available

One of the properties of the benzenediazonium ion is that it is very reactive. You will find in section 122.7 that it reacts with a wide variety of substances, one of which is a halide ion. For example, if we add potassium iodide to a solution containing benzenediazonium ions, and warm the mixture, iodobenzene is made:

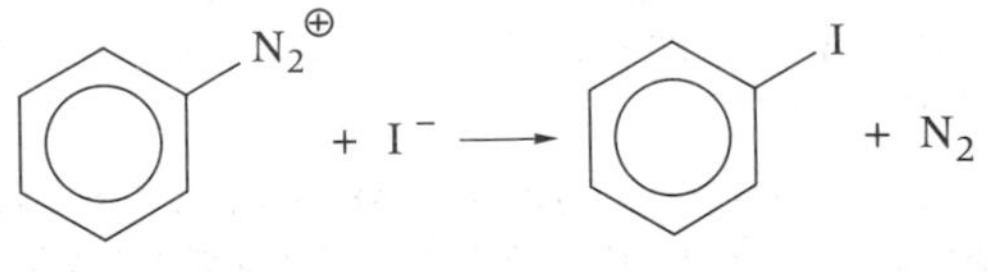

$$C_6H_5N_2^{\oplus} + I^- \rightarrow C_6H_5I + N_2$$
iodo-
benzene

This is one of the few ways of introducing an iodine atom into a benzene ring.

Chlorobenzene (or bromobenzene) can be made by *Sandmeyer's reaction*. Here, copper(I) chloride and hydrochloric acid (or copper(I) bromide and hydrobromic acid) are added to the benzenediazonium ion solution.

115.10 Sketch the apparatus you would use to make chlorobenzene from benzene by the Friedel–Crafts reaction. Make sure you take account of the states of the reactants and products. One problem you have to overcome is the fact that once all the benzene is converted into chlorobenzene, a second chlorine would be substituted into the ring. How will you know when to stop the reaction before this happens?

115.11 The separation of the chlorobenzene takes place in the following steps:

(a) After cooling, the contents of the reaction flask are poured into an approximately equal volume of water.

(b) The chlorobenzene layer is run off.

(c) It is then washed with sodium hydroxide solution, followed by water.

(d) Anhydrous calcium chloride is added, and the mixture left until it becomes clear.

(e) The liquid is distilled.

Here are some questions about this process.

(i) What is removed by the water in (a)?

(ii) What piece of apparatus would you use in step (b)? The density of chlorobenzene is about 1.1 g cm^{-3}. Will it be the top or the bottom layer?

(iii) What *two* effects will the sodium hydroxide have?

(iv) Describe what you would do to carry out step (c).

(v) What is the purpose of the anhydrous calcium chloride? Why might the chlorobenzene look cloudy?

(vi) What else might be present apart from chlorobenzene? At what temperature would you expect the chlorobenzene to distil over?

115.12 What changes to your apparatus, and to the method, would you make if you were to make bromobenzene instead of chlorobenzene?

115.13 What will be made if chlorobenzene is itself chlorinated by the Friedel–Crafts reaction?

115.7 What are the reactions of halogenoarenes?

The main difference between a halogenoarene, e.g. chlorobenzene, and a halogenoalkane, e.g. chloromethane, is that the halogenoarene is much more resistant to nucleophilic attack. For example, the chlorine in chlorobenzene can be replaced by an OH group but only at high pressure and temperature (around 300 atm and 300 °C):

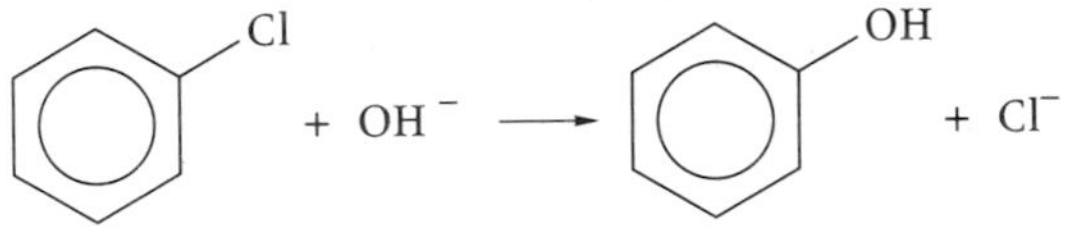

$$C_6H_5Cl + OH^- \xrightarrow[300\ \text{atm}]{300°\text{C}} C_6H_5OH + Cl^-$$
chlorobenzene   phenol

For obvious reasons this reaction is hopeless as a laboratory method of preparing phenol; but on a much larger scale it is used as an industrial method of manufacturing phenol.

Another industrial use of chlorobenzene is in the manufacture of phenylamine. The chlorobenzene is reacted with ammonia in the presence of a copper(I) oxide catalyst. Again, a high pressure and temperature are needed:

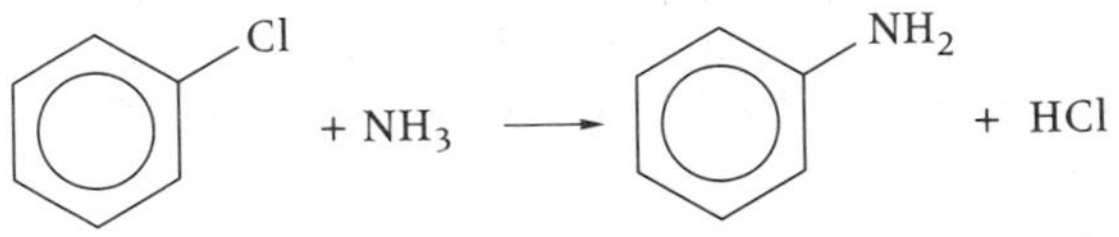

$$C_6H_5Cl + NH_3 \xrightarrow[\text{catalyst}]{Cu_2O} C_6H_5NH_2 + HCl$$
chlorobenzene   phenylamine

115.14 What will be the product of the reaction between 1-chloro-1-phenylethane, $C_6H_5CHClCH_3$,

and alcoholic potassium hydroxide?

115.15 Two bottles contained iodobenzene and (iodomethyl)benzene. However, the bottles were only labelled as A and B. It was not known which chemical was in which bottle. A sample from each bottle was put in two test tubes and warmed with sodium hydroxide solution. After cooling, dilute nitric acid was added followed by dilute silver nitrate solution. The results were:

Sample from bottle A gave a yellow precipitate.
Sample from bottle B gave no change.

(i) What was the yellow precipitate?

(ii) Why was dilute nitric acid added?

(iii) Which chemical was in which bottle?

(iv) Briefly explain the reaction.

(v) What would have been the result of using iodoethane in the experiment?

115.16 You have to make 1-chloro-2-phenylethane starting from an alcohol. Draw the structure of the alcohol you would use and say how you would perform the conversion.

115.17 A chemist wanted to make chloro-2-methylbenzene. The method the chemist chose was to react methylbenzene with chlorine in the presence of ultraviolet light.

(i) Why was this the wrong method?

(ii) What method should be used?

115.18 The length of the carbon–chlorine bond in chlorobenzene is 169 pm, while in most chloroalkanes it is 177 pm.

(i) What does this indicate about the carbon–chlorine bond in chlorobenzene?

(ii) How does this information fit with the relative unreactivity of chlorobenzene to nucleophilic attack, e.g. by hydroxide ions?

Answers

115.1 (i) 1-Bromopropane, (ii) 1-bromobutane, (iii) 2-bromobutane.

115.2 (i) High temperature or ultraviolet light.

(ii) This is a free radical reaction. It is difficult to control the proportions of the various products that can be made. This leads to problems over separation at the end of the reaction.

115.3 (i) Sulphur dioxide, SO_2, and hydrogen chloride, HCl.

(ii) The products are gases, the majority of which will bubble out of the reaction mixture, thus leaving a reasonably pure product.

115.4 Dichloromethane. Tetrachloromethane has polar bonds, but is symmetrical overall, so does not have a dipole moment.

115.5 One method is to mix bromine and ethene:

$$C_2H_4 + Br_2 \rightarrow CH_2BrCH_2Br$$

115.6 From Table 115.1 you will find that chloroethane is a gas while iodoethane is a liquid at room temperature and pressure. It is much easier to handle liquids rather than gases in a reaction. For other reasons, see the next unit.

115.7 Reflux is necessary: see Figure 113.5, but *not* with the stopper in. Ethanol is the lowest boiling point component of the mixture, so should distil over first. However, it will be contaminated by water vapour. The

ethanol can be dried using calcium oxide. Alternatively, but less pleasant, the azeotropic mixture (see section 63.4) made by ethanol and water can be separated by adding benzene and redistilling.

115.8 (i) The first two products of the reaction are methylamine, CH_3NH_2, and dimethylamine, $(CH_3)_2NH$. Both these molecules, like ammonia, have a lone pair on the nitrogen. Trimethylamine is produced by the nucleophilic attack of the lone pair on the nitrogen of dimethylamine on the carbon atom of iodomethane:

(ii) In Unit 15 we saw that ammonia will react with hydrogen chloride to give the ionic solid ammonium chloride, NH_4Cl. We said that this can be explained by assuming that the lone pair on the nitrogen forms a coordinate bond with a positively charged hydrogen atom. In our organic reaction, instead of hydrogen chloride, we have iodomethane. The carbocation CH_3^+ takes the place of the hydrogen ion, and iodide takes the place of the chloride ion:

Answers – cont.

115.9 (i) Mass of $C_2H_5OH = 4.2\,g$,
i.e. $4.2\,g/46\,g\,mol^{-1} = 0.09\,mol$.

(ii) $0.02\,mol$ of P; $0.02\,mol$ of I_2.

(iii) The ratio is approximately $5\,mol$ $C_2H_5OH:1\,mol$ P:1 mol I_2.

(iv) The equation tells us that $3\,mol$ of C_2H_5OH combine with $1\,mol$ of PI_3 to give $3\,mol$ of C_2H_5I. Also, $1\,mol$ of PI_3 is obtained from $2\,mol$ of P and $3\,mol$ of I_2. The reactants need to be in the ratio $3\,mol$ C_2H_5OH to $1\,mol$ P to $1.5\,mol$ I_2. The ethanol and phosphorus are in excess. This tells us that the iodine would be consumed before either of these chemicals was used up. The iodine governs the amount of C_2H_5I that can be made.

(v) We can obtain $0.04\,mol$ of C_2H_5I, i.e. $6.24\,g$.

(vi) The yield was $(5.1\,g/6.24\,g) \times 100\% = 81.7\%$.

115.10 The apparatus is sketched in Figure 115.4. You would have to weigh the flask and benzene before starting the reaction. From time to time during the reaction, the supply of chlorine would have to be stopped, the flask disconnected from the apparatus and reweighed. This process would be continued until the benzene had increased in weight by the appropriate amount. (1 mol of benzene would increase from $78\,g$ to $112.5\,g$.)

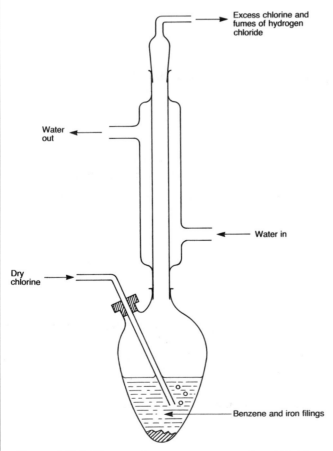

Figure 115.4 *Diagram for answer to question 115.10: a method of making chlorobenzene. The flask can be warmed using a water bath. The apparatus must be set up in a fume cupboard*

115.11 (i) The majority of the hydrogen chloride dissolved in the mixture will be transferred to the water.

(ii) A separating (or tap) funnel. Chlorobenzene is more dense than water, so it will be the bottom layer.

(iii) It will neutralise any remaining hydrogen chloride, and it will convert any chlorine dissolved in the organic layer into water soluble chlorate(I) and chloride ions (see section 101.6). Washing with water helps to remove the inorganic impurities.

(iv) The chlorobenzene and sodium hydroxide would be shaken together in a separating funnel (with the tap closed and its stopper in!). The separating funnel should be held upside down from time to time and the tap opened carefully to prevent the build up of pressure. The lower chlorobenzene layer should be run off and the sodium hydroxide solution discarded. This procedure should be repeated with the chlorobenzene and water.

(v) It dries the chlorobenzene. The cloudiness is due to tiny droplets of water that remain in the chlorobenzene after shaking with water.

(vi) Unreacted benzene and products like 1,2-dichlorobenzene would be impurities. The chlorobenzene should distil over in a range a few degrees either side of its boiling point, say 128 to 136°C.

115.12 You could use iron filings as the halogen carrier instead of aluminium trichloride. The apparatus of Figure 115.5 could be used. It would not be necessary to disconnect the flask and weigh it during the reaction. Once you knew the weight of benzene you were to use, you would put the required weight of bromine into the dropping funnel.

115.13 A chlorine atom directs *ortho* and *para* (even though it deactivates the ring). Therefore you would expect to get 1,2-dichlorobenzene and 1,4-dichlorobenzene.

115.14 This is the dehydrohalogenation reaction. The product is phenylethene (also known as styrene):

115.15 (i) Silver iodide.

(ii) The acid removes hydroxide ions. If they are present, a dirty grey deposit of silver(I) oxide is made, which prevents the silver ions reacting with iodide ions.

(iii), (iv) You should know that a halogen atom attached directly to the ring will not react easily with nucleophiles, so the iodobenzene was in bottle B.

(v) Iodoethane is relatively easily attacked by nucleophiles. The iodide ion is released into the solution as the iodoethane is converted into ethanol.

115.16 The alcohol is 2-phenylethanol:

Answers – cont.

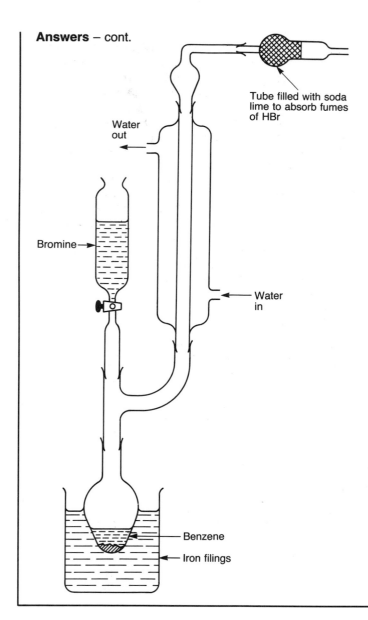

Tube filled with soda lime to absorb fumes of HBr

Water out

Bromine

Water in

Benzene

Iron filings

The conversion could be achieved by reacting the alcohol with phosphorus pentachloride or sulphur dichloride oxide, $SOCl_2$.

115.17 (i) The free radical reaction started by the ultraviolet light will mean that the methyl side chain is attacked, not the ring (see section 113.4).

(ii) A Friedel–Crafts reaction: chlorine and aluminium trichloride.

115.18 (i) Two things: first, the shorter bond is the stronger of the two; secondly, it fits with the picture we drew in the last unit, in which we said that the chlorine atom in chlorobenzene could use one of its lone pairs in bonding with the π system on the ring. This gives a measure of double bond character to the bond. (Compare the difference in bond lengths in ethane and ethene; section 33.5.)

(ii) It fits nicely. The stronger the bond, the more difficult it should be to break. However, there is another reason for the difference in reactivity. The geometry around the carbon bonded to the chlorine atom in chloroalkanes is tetrahedral, while in chlorobenzene it is planar. There is not such a clear pathway for the nucleophile to attack the carbon atom in chlorobenzene.

Figure 115.5 *Diagram for answer to question 115.12. The apparatus should be used in a fume cupboard*

UNIT 115 SUMMARY

Halogenoalkanes

Are hydrocarbons in which one or more hydrogen atoms are replaced by halogen atoms;
e.g. bromomethane, CH_3Br;
1,2-dichloroethane, CH_2ClCH_2Cl.

- Preparation:
  (i) React a halogenoalkane with a sodium halide and concentrated sulphuric acid;

  e.g. $C_2H_5OH + HBr \rightarrow C_2H_5Br + H_2O$

  (ii) Iodoethane can be made from ethanol, iodine and red phosphorus;

  $3C_2H_5OH + PI_3 \rightarrow 3C_2H_5I + H_3PO_3$

Reactions

- Undergo nucleophilic attack. The possibilities are summarised in Table 115.3.
- Take part in elimination reactions;

  e.g. $C_2H_5I \xrightarrow[\text{KOH}]{\text{alcoholic}} C_2H_4 + HI$

  dehydrohalogenation; makes alkenes.

Aralkyl halides

Have halogen atoms attached to the side chain of an aromatic molecule, e.g. $C_6H_5CH_2Cl$. They behave like halogenoalkanes.

Organic halogen compounds 779

Summary – cont.

Halogenoarenes

(i) Have one or more halogen atoms attached to a benzene ring.

(ii) Do not undergo nucleophilic attack as easily as halogenoalkanes; halogen atoms are extremely hard to remove from a benzene ring.

- Preparation:

  (i) Friedel–Crafts reaction; e.g.

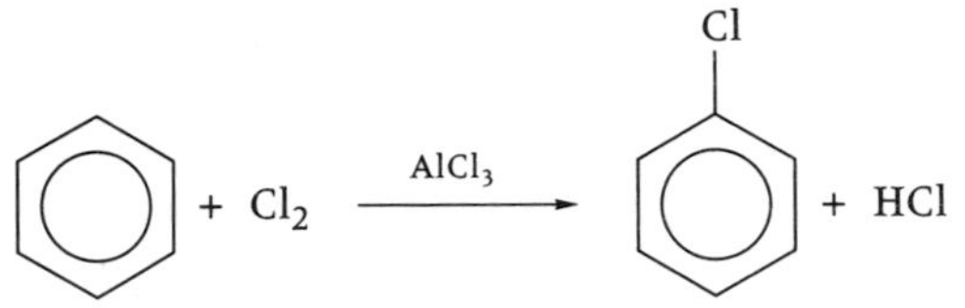

  (ii) From diazonium compounds; e.g.

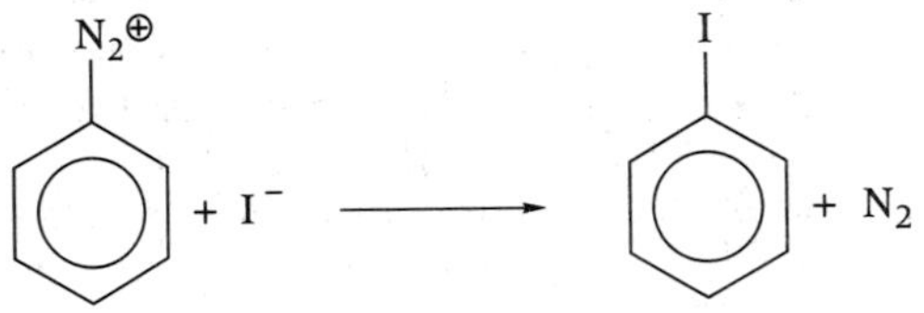

Reactions

- Chlorobenzene can be converted into phenol and phenylamine;

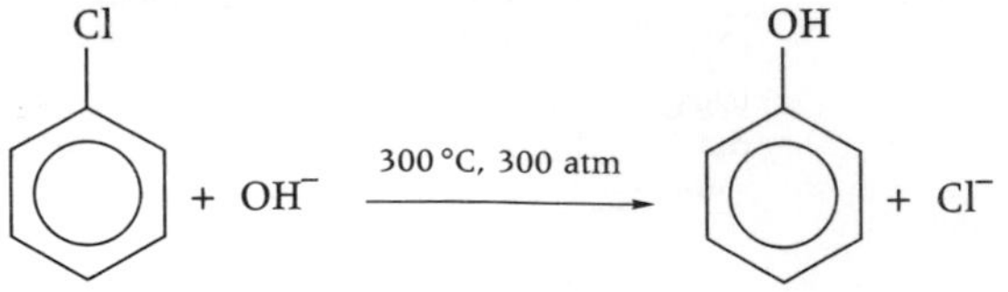

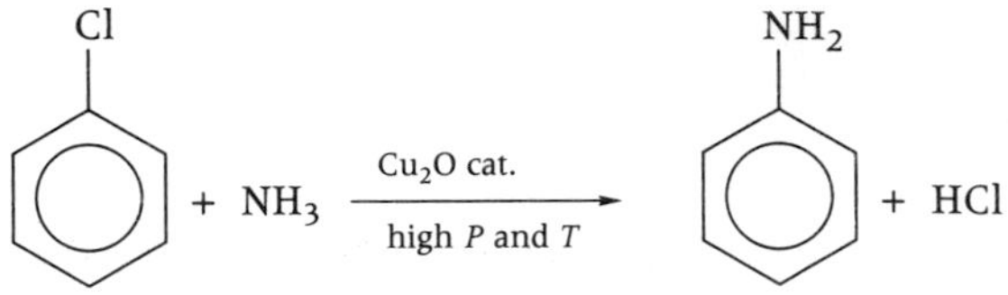

116

Alcohols

116.1 The structures and uses of alcohols

Alcohols contain one or more OH groups. The simpler ones, the *monohydric alcohols*, contain only one OH group per molecule and have the general formula $C_nH_{2n+1}OH$. Ethanol, C_2H_5OH, is especially well known (and loved by some) because it is the ingredient in alcoholic drinks that causes people to become drunk. If there is more than one OH group in the molecule, it is called a *polyhydric alcohol*. Ethane-1,2-diol (glycol), CH_2OHCH_2OH, with two OH groups is a dihydric alcohol familiar to car and lorry owners when they add it to water in radiators; it is commonly known as antifreeze. Propane-1,2,3-triol (glycerol), $CH_2OHCHOHCH_2OH$, is a trihydric alcohol that is often used in medicine and cosmetics as a basis for the preparation of creams and ointments.

Alcohols have many uses in the chemical industry, especially as solvents and as intermediates in the manufacture of chemicals such as esters. The use of tetraethyl-lead(IV) as a petrol additive is declining. It is being replaced by adding small amounts of alcohol to petrol. The alcohol also has the ability to encourage the smooth burning of petrol. In some countries alcohol rather than petrol is used as a fuel for motor vehicles.

It is the OH group that is responsible for the miscibility of many alcohols with water. Just as a water molecule can hydrogen bond with another water molecule, so it can with an alcohol (see Figure 116.1).

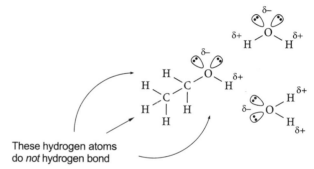

These hydrogen atoms do *not* hydrogen bond

Figure 116.1 *Ethanol can be hydrogen bonded to water molecules*

Two models showing the shapes of methanol and ethanol.

Table 116.1. Common alcohols

Name	Formula	Structure	Melting point/°C	Boiling point/°C
Methanol	CH_3OH	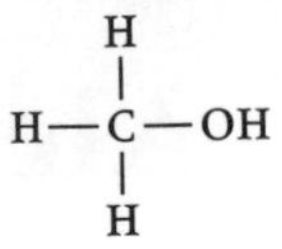	−97.5	64.7
Ethanol	C_2H_5OH	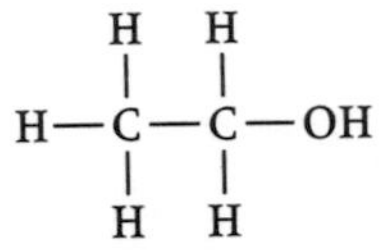	−114	78.5
Propan-1-ol	C_3H_7OH	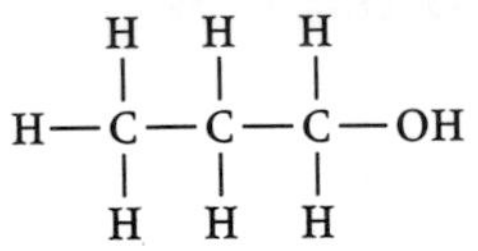	−126	97.4
Propan-2-ol	$(CH_3)_2CHOH$	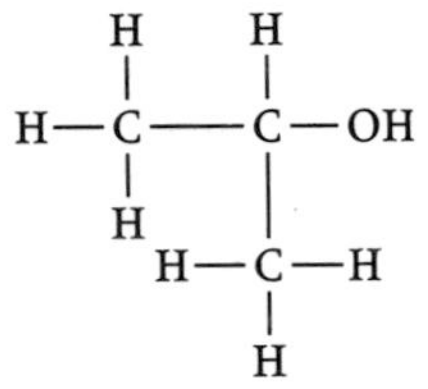	−88.4	82.7
2-Methylpropan-2-ol	$(CH_3)_3COH$	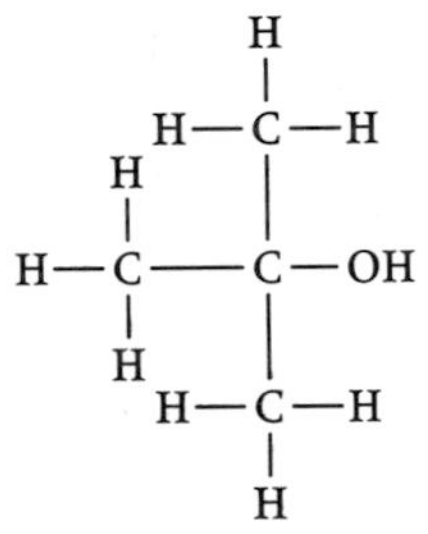	25.5	83
Butan-1-ol	C_4H_9OH	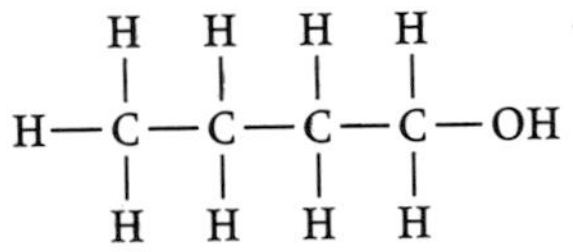	−89.2	117.9
Ethane-1,2-diol (glycol)	CH_2OHCH_2OH	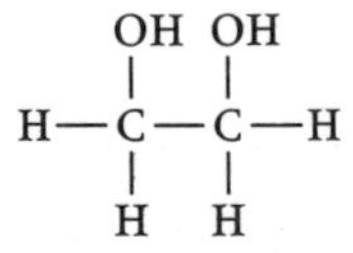	−13.4	197
Propane-1,2,3-triol (glycerol)	$CH_2OHCHOHCH_2OH$	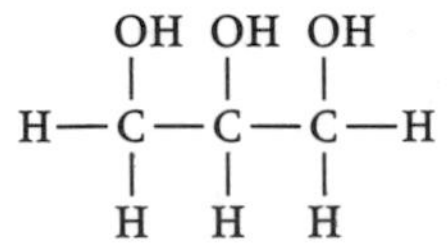	18	290

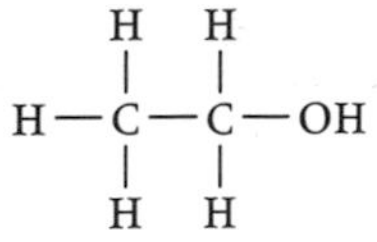

H H
H—C—C—OH
H H

CH₃CH₂OH

Ethanol

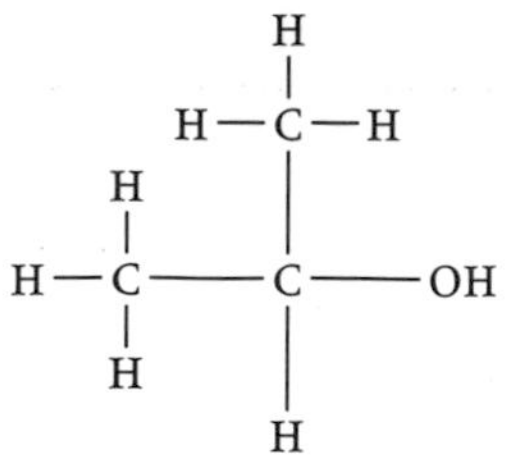

CH_3CH_2OH

Ethanol

H
H—C—H
H
H—C—C—OH
H
H

$(CH_3)_2CHOH$

Propan-2-ol

H
H—C—H
H
H—C—C—OH
H
H—C—H
H

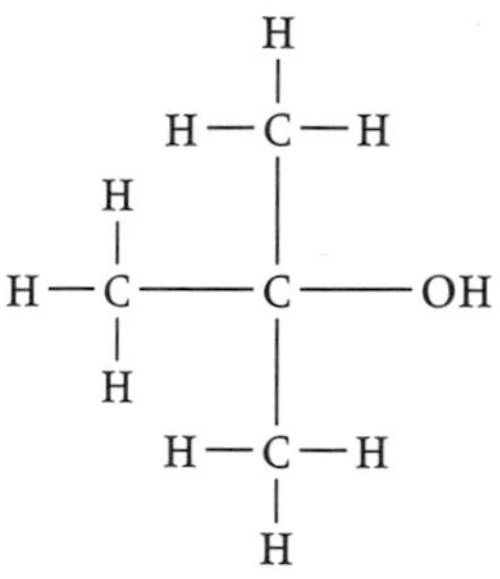

$(CH_3)_3COH$

2-Methylpropan-2-ol

Figure 116.2 *Typical primary, secondary and tertiary alcohols*

Each of the alcohols in Table 116.1 fits into one of three categories. They are either *primary*, *secondary*, or *tertiary* alcohols. Typical structures are shown in Figure 116.2. You will see that, respectively, they have one methyl, two methyl and three methyl groups attached to the carbon bonded to the OH group. The general formulae of the three varieties are:

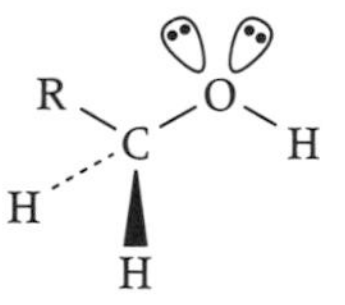

RCH_2OH
primary

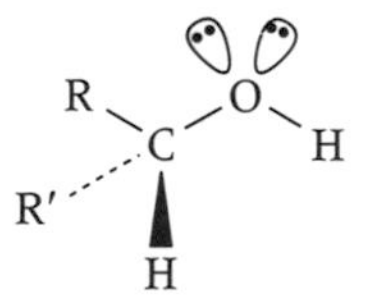

$RR'CHOH$
secondary

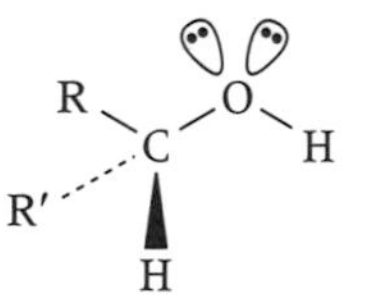

$RR'R''COH$
tertiary

In some reactions, primary, secondary and tertiary alcohols behave in different ways. Similarly, aromatic alcohols often behave differently to the non-aromatics, and we shall study the former in the next unit.

116.1 Explain why it is sometimes said that water is the simplest alcohol.

116.2 Why do alcohols with a long carbon backbone become immiscible with water?

116.3 Explain why you would expect alcohols to have a dipole moment.

116.4 An alcohol has the formula $C_6H_{11}OH$ and it does not react with bromine or bromine water.

(i) What does the information about the bromine tell you?

(ii) What is the structure of the alcohol?

(iii) Suggest a name for it.

116.5 What are the names of these alcohols?

(i)
$$CH_3—CH_2—CH_2—\overset{\overset{\displaystyle OH}{|}}{\underset{\underset{\displaystyle H}{|}}{C}}—CH_3$$

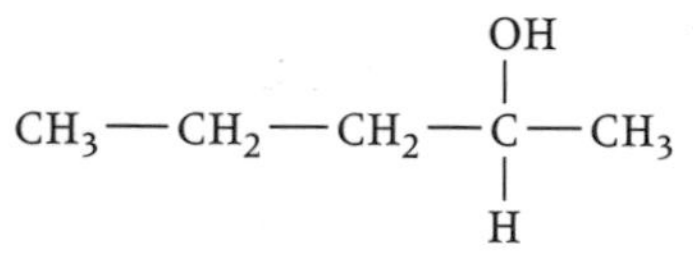

(ii)
$$CH_3—CH_2—\overset{\overset{\displaystyle OH}{|}}{\underset{\underset{\displaystyle CH_3}{|}}{C}}—CH_3$$

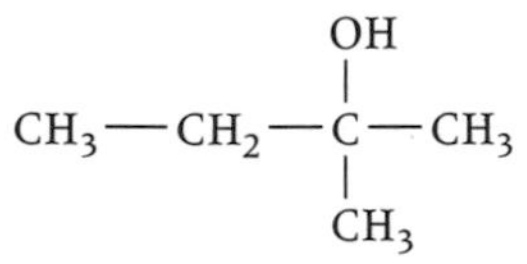

116.2 How we can make alcohols

There are many ways of making alcohols. We shall look at two large-scale methods of making them, and four methods that are more often used in the laboratory.

(a) Fermentation

Fermentation is the traditional way of making ethanol. Ethanol is made by the action of enzymes on sugars, particularly glucose and fructose, both of which have the formula $C_6H_{12}O_6$. One enzyme, *zymase*, which is extremely effective in causing fermentation, is found in yeasts. In wine making, grapes are the source of sugars and yeast. As grapes ripen, the amount of sugar inside them increases, and yeasts grow on the outer skin. By crushing the grapes the sugary juices and yeast are brought into contact and fermentation starts. The sugar sucrose, $C_{12}H_{22}O_{11}$, which is itself resistant to attack by zymase, can be broken into glucose and fructose by another enzyme, invertase:

$$C_{12}H_{22}O_{11} + H_2O \rightarrow C_6H_{12}O_6 + C_6H_{12}O_6 \quad \text{catalysed by}$$
sucrose $\qquad\qquad$ glucose $\quad$ fructose $\qquad$ invertase

If you have made wine, beer, or lager at home you will know that the fermentation has to be done in a bottle fitted with an air lock. During fermentation carbon dioxide is released. The gas quickly displaces air from the bottle, and the fermentation takes place in *anaerobic* conditions. You can tell how well the fermentation is proceeding by studying the rate at which bubbles of carbon dioxide are released.

We can write the reaction taking place in this way:

$$C_2H_{12}O_6 \rightarrow 2C_2H_5OH + 2CO_2 \quad \text{catalysed by zymase}$$
sugar $\qquad\qquad$ ethanol

The conversion of sugar into ethanol will not continue

Scotch whisky is distilled in copper stills behind the control panel.

Table 116.2. Table of alcohol consumption statistics*†

Country	Year	Spirits	Wine	Beer	Total alcohol
France	1968	7.96	146.72	78.26	22.62
Italy	1969	4.23	152.51	14.95	14.99
USA	1971	9.92	7.87	98.03	10.15
UK	1970	2.14	4.91	114.71	7.21
Ireland	1970	3.71	4.81	98.71	7.06

*The figures give the number of litres drunk per person aged 15 years or over in the year shown in the table. The spirits, wine and beer columns give the volume of each liquid consumed (alcohol plus water). The final column gives the total equivalent volume of pure alcohol (C_2H_5OH) for all three types of alcoholic drink. Notice the differences in the drinking habits of the people of the different countries: e.g. French and Italian people tend to drink significantly more wine than beer, while in the USA, UK and Ireland the opposite is true
†Data taken from the *Encyclopaedia Britannica*, 15th edn, 1985, vol. 13, p. 223

indefinitely because, once the percentage of alcohol exceeds 14%, the action of zymase is inhibited.

It is important that air does not get into the mixture. Oxygen will oxidise the alcohol and convert it into ethanoic acid (old name, acetic acid). This acid gives vinegar its sharp taste, and its presence ruins the taste of an alcoholic drink. Many years ago Louis Pasteur proved that bacteria present in air will also cause wine to go off.

Given that many people find the effects of drinking alcohol pleasant, it is not surprising that ways have been found to increase the proportion of alcohol above the 14% limit. Typically this is done by distillation. The method of distillation, and the number of times it is carried out, give different qualities of various spirits such as vodka, brandy and whisky. Table 116.2 shows the consumption of alcoholic drinks in various countries.

The taste of different wines and spirits depends on many things, especially the particular mix of more complicated alcohols and esters that are made during the fermentation and while the alcoholic liquid is left to mature. If, for example, wine or spirits are kept in wooden barrels, then the wood itself gives a particular taste.

Methanol, CH_3OH, was once obtained by heating wood and collecting the tarry liquid given off. The liquid contains a few per cent of methanol. Hence the old name for methanol was wood alcohol. Like ethanol, methanol causes intoxication; but in addition it causes blindness and brain damage to those who drink it often. Together with a coloured dye, methanol can be added to ethanol and sold as 'methylated spirits'. Industrial methylated spirits (IMS) may also contain another poisonous organic reagent, pyridine.

(b) Hydration of alkenes

In Unit 112 we discussed the way in which alkenes take part in addition reactions. One of these reactions involved absorbing an alkene in concentrated sulphuric acid, and then diluting the acid with water. The result is an alcohol. For example,

ethene $\longrightarrow$ ethanol

propene $\longrightarrow$ propan-2-ol

Notice that, in effect, this is the addition of H—OH across the double bond in the alkene. The addition is in accordance with Markovnikoff's rule.

On an industrial scale this reaction uses alkenes made by the cracking of hydrocarbons in oil. The alkene is passed into sulphuric acid at around 80 °C and 30 atm pressure. The acid is diluted and treated with steam to release the alcohol. The final stage is the distillation of the solution. A problem that arises at this stage is that ethanol, and many other alcohols, give azeotropic (constant boiling) mixtures with water (see section 63.4). However, ethanol can be completely separated by distilling the azeotropic mixture with benzene.

(c) Grignard reagents

We have met Grignard reagents before. Their general formula is RMgX, where R is an organic radical and X is a halogen, often bromine. To make an alcohol we would normally react the Grignard reagent with an aldehyde or ketone. The functional groups of aldehydes and ketones are very similar:

$$R-C\underset{H}{\overset{O}{\lessgtr}} \qquad R-C\underset{R'}{\overset{O}{\lessgtr}}$$

an aldehyde   a ketone

The carbon atoms are ideal sites for nucleophilic attack. We know that the organic radical bonded to the magnesium atom carries a negative charge, and it acts as a nucleophile. Now let us see what happens when a suitable Grignard reagent reacts with an aldehyde or ketone. Let us assume that we are using CH_3MgBr:

$$H-C\underset{H}{\overset{O}{\lessgtr}} + CH_3MgBr \longrightarrow CH_3-\underset{H}{\overset{H}{\underset{|}{\overset{|}{C}}}}-OMgBr$$

methanal

$$CH_3-\underset{H}{\overset{H}{\underset{|}{\overset{|}{C}}}}-OMgBr + H_2O \longrightarrow$$

$$CH_3-\underset{H}{\overset{H}{\underset{|}{\overset{|}{C}}}}-OH \qquad \text{primary alcohol}$$

ethanol

The overall result is for the methyl group to bond to the carbon atom and a hydrogen atom to the oxygen. Thus, we can summarise the reactions in this way:

$$CH_3-C\underset{H}{\overset{O}{\lessgtr}} \xrightarrow[\text{(ii) } H_2O]{\text{(i) } CH_3MgBr} CH_3-\underset{H}{\overset{OH}{\underset{|}{\overset{|}{C}}}}-CH_3 \quad \begin{array}{l}\text{secondary}\\\text{alcohol}\end{array}$$

ethanal   propan-2-ol

$$CH_3-C\underset{CH_3}{\overset{O}{\lessgtr}} \xrightarrow[\text{(ii) } H_2O]{\text{(i) } CH_3MgBr} CH_3-\underset{CH_3}{\overset{OH}{\underset{|}{\overset{|}{C}}}}-CH_3 \quad \begin{array}{l}\text{tertiary}\\\text{alcohol}\end{array}$$

propanone   2-methylpropan-2-ol

Of course, if we were to use a C_2H_5 group instead of CH_3 in the Grignard reagent, more complex alcohols would be made. Similarly, by changing the aldehyde or ketone we can make a variety of different alcohols.

(d) Reduction of carbonyl compounds

Aldehydes and ketones can be reduced with hydrogen in the presence of a catalyst, or by reacting them with lithium tetrahydridoaluminate(III), $LiAlH_4$ (or sodium tetrahydridoborate(III), $NaBH_4$). The pattern is:

| Aldehydes are reduced to primary alcohols. |
| Ketones are reduced to secondary alcohols. |

Here are three examples:

$$H-C\underset{H}{\overset{O}{\lessgtr}} \xrightarrow{H_2, Ni} H-\underset{H}{\overset{H}{\underset{|}{\overset{|}{C}}}}-OH \qquad CH_3OH$$

methanal   methanol

$$CH_3-C\underset{H}{\overset{O}{\lessgtr}} \xrightarrow[\text{(ii) acid}]{\text{(i) } LiAlH_4} CH_3-\underset{H}{\overset{H}{\underset{|}{\overset{|}{C}}}}-OH \qquad CH_3CH_2OH$$

ethanal   ethanol

$$CH_3-C\underset{CH_3}{\overset{O}{\lessgtr}} \xrightarrow[\text{(ii) acid}]{\text{(i) } NaBH_4} CH_3-\underset{H}{\overset{OH}{\underset{|}{\overset{|}{C}}}}-CH_3 \qquad \begin{array}{l}(CH_3)_2\\CHOH\end{array}$$

propanone   propan-2-ol

(e) Hydrolysis of halogenoalkanes

We have covered this type of reaction in some detail as a property of halogenoalkanes. We need only mention that halogenoalkanes will have their halogen replaced by an OH group if we boil them with sodium hydroxide solution. For example,

$$C_2H_5I + OH^- \rightarrow C_2H_5OH + I^-$$

(f) Hydroboration

Diborane, B_2H_6, is an interesting molecule in its own right. It is one of the electron deficient molecules that we discussed in section 90.7. In organic chemistry it has the welcome property of combining with alkenes to give a product that, on oxidation with hydrogen peroxide, turns into an alcohol:

$$\underset{H}{\overset{H}{>}}C=C\underset{H}{\overset{H}{<}} \xrightarrow[\text{(ii) } H_2O_2]{\text{(i) } B_2H_6} H-\underset{H}{\overset{H}{\underset{|}{\overset{|}{C}}}}-\underset{H}{\overset{H}{\underset{|}{\overset{|}{C}}}}-OH$$

ethene   ethanol

$$\underset{H}{\overset{CH_3}{>}}C=C\underset{H}{\overset{H}{<}} \xrightarrow[\text{(ii) } H_2O_2]{\text{(i) } B_2H_6} CH_3-\underset{H}{\overset{H}{\underset{|}{\overset{|}{C}}}}-\underset{H}{\overset{H}{\underset{|}{\overset{|}{C}}}}-OH$$

propene   propan-1-ol

$$\underset{H}{\overset{C_2H_5}{>}}C=C\underset{H}{\overset{H}{<}} \xrightarrow[\text{(ii) } H_2O_2]{\text{(i) } B_2H_6} C_2H_5-\underset{H}{\overset{H}{\underset{|}{\overset{|}{C}}}}-\underset{H}{\overset{H}{\underset{|}{\overset{|}{C}}}}-OH$$

but-1-ene   butan-1-ol

Like the preparation of alcohols by absorption in concentrated sulphuric acid, the net result is the addition of H—OH across the double bond. However, notice that in hydroboration the addition does *not* obey Markovnikoff's rule; we say it is an *anti-Markovnikoff* addition.

116.6 The modern method of making methanol is by the use of synthesis gas, a mixture of carbon monoxide and hydrogen, which is itself made from natural gas and steam (see section 90.2):

$$CO(g) + 2H_2(g) \rightarrow CH_3OH(g)$$

The reaction takes place above 350°C, at around 20 atm and with a catalyst of zinc oxide and chromium(III) oxide. Use the data in Table 44.1 to calculate the enthalpy change of this reaction under standard conditions (25°C and 1 atm).

116.7 You have been given the task of studying the rate of fermentation of grape juice. What method would you use?

116.8 What would you expect to be made if the Grignard reagent C_2H_5MgBr reacted with butanone,

and the product hydrolysed?

116.9 Which alcohol will be made if pentan-3-one,

is reacted with lithium tetrahydridoaluminate(III)?

116.10 What will be the result of using pent-2-ene, (i) by absorbing it in concentrated sulphuric acid, followed by dilution with water, (ii) by reacting it with diborane, followed by hydrogen peroxide?

116.3 The oxidation reactions of alcohols

The easiest way to oxidise an alcohol is to burn it. You may have seen the characteristic pale blue flame of ethanol burning if you have ever set fire to brandy on a Christmas pudding:

$$2C_2H_5OH(l) + 7O_2(g) \rightarrow 4CO_2(g) + 6H_2O(g)$$

However, in the laboratory we can use less drastic, and more useful, methods of oxidation. A common oxidising agent to use is acidified sodium dichromate(VI). Owing to the production of chromium(III)

Table 116.3. The oxidation of primary, secondary and tertiary alcohols

With acidified sodium dichromate(VI) solution

Primary alcohols →aldehydes →acids

Secondary alcohols→ketones

Tertiary alcohols→no reaction

With copper at 250°C

Primary alcohols give aldehydes

Secondary alcohols give ketones

Tertiary alcohols do not react

---

ions, this orange coloured solution changes to green if it meets a reducing agent. Primary, secondary and tertiary alcohols react in different ways, indicated in Table 116.3. Alcohols can also be oxidised to aldehydes or ketones by passing their vapour over copper powder kept at around 250°C. Similarly, we can use the reaction as a way of distinguishing the three types of alcohol. Especially, if you warm ethanol with acidified sodium dichromate(VI) solution, you will soon notice the characteristic smell of ethanal. Be careful if you try this. The reaction mixture will oxidise you or your clothes if it gets on them. Also, ethanol and ethanal vapours are highly flammable. Do *not* heat the mixture over a naked flame. Put it in a beaker of hot water.

116.11 The oxidation of primary and secondary alcohols can be the basis of preparing aldehydes and ketones. Which alcohol would you oxidise if you wanted to make the following?

(i) butanal,

$$C_3H_7 - C {\overset{\displaystyle O}{\underset{\displaystyle H}{<}}}$$

(ii) butanone,

$$C_2H_5 - C {\overset{\displaystyle O}{\underset{\displaystyle CH_3}{<}}}$$

116.12 Using a simple definition of oxidation, explain why we have called the conversion of an alcohol into an aldehyde or ketone an oxidation reaction.

116.4 Two reactions of the OH group

(a) With sodium

There are two ways of showing that alcohols have an OH group. If you drop a small piece of sodium into ethanol in a test tube, you will see it sink to the bottom of the tube. You will also see bubbles of gas given off. The gas is hydrogen:

$$2C_2H_5OH + 2Na \rightarrow 2C_2H_5O^-Na^+ + H_2$$
ethanol   sodium
ethoxide

This reaction shows that alcohols can lose their hydrogen from the OH group. This is a reaction that we would normally associate with acids. (Remember, hydrogen is given off by the reaction of most metals with an inorganic acid.) However, water itself is a much stronger acid than an alcohol. Only OH groups on a benzene ring are appreciably acidic in water, e.g. phenol C_6H_5OH.

(b) With phosphorus pentachloride or sulphur dichloride oxide

The second reaction that shows the presence of the OH group can also be used as a method of preparing halogenoalkanes. With either of these two reagents, we see white fumes of hydrogen chloride given off when they are added to an alcohol, e.g.

$$C_2H_5OH + PCl_5 \rightarrow C_2H_5Cl + POCl_3 + HCl$$
white
fumes

$$C_2H_5OH + SOCl_2 \rightarrow C_2H_5Cl + SO_2 + HCl$$
white
fumes

You should be careful not to read too much into these two tests. They do *not* prove that the chemical being

tested is an alcohol. An organic acid like ethanoic acid, CH_3COOH, will also give off hydrogen with sodium, and white fumes of hydrogen chloride with PCl_5 or $SOCl_2$. The tests show that a molecule *has an OH group*.

116.13 If you were using PCl_5 or $SOCl_2$ to test for the presence of OH groups, why must the chemicals and apparatus be dry?

116.5 Reactions in which the OH group is replaced by another group

We have just discussed examples of these reactions, those with PCl_5 or $SOCl_2$. Here the OH group is replaced by a halogen. If you look back at the methods of making halogenoalkanes, you will find several possibilities: in particular, reacting an alcohol with a sodium halide and concentrated sulphuric acid, or with red phosphorus and bromine or iodine. It is interesting to compare the reactivities of primary, secondary and tertiary alcohols in these reactions. The order is

tertiary   >   secondary   >   primary
most reactive   least reactive

The reason for this difference in reactivity lies in the nature of the intermediates formed in the reactions. Often these involve carbocations. As we saw in Unit 112, a tertiary carbocation is more favoured than a secondary, and a primary carbocation least favoured. For example, if an alcohol reacts with a hydrogen halide, the first thing that happens is that the oxygen of the OH group becomes protonated.

(i) *Stage 1*: Protonation of the oxygen atom.

(ii) *Stage 2*: Water can be lost from the molecule, leaving a carbocation behind.

(iii) *Stage 3*: It is the carbocation that reacts with the halide ion, X^-.

It is stage 2 that is encouraged if the carbocation made is energetically favoured. For this reason, we would expect 2-methylpropan-2-ol to be the most reactive because it gives a tertiary carbocation in stage 2.

116.14 Can alcohols act as nucleophiles? Explain your answer.

116.15 To convert butan-1-ol to 1-chlorobutane using hydrogen chloride requires heat and a zinc chloride catalyst. The isomeric alcohol 2-methyl-propan-2-ol is converted to 2-chloro-2-methylpropane by concentrated hydrochloric acid at room temperature. Why is there such a difference in reactivity?

116.6 Dehydration reactions

Alcohols can be dehydrated by passing them over hot aluminium oxide, or by heating them at about $180\,^{\circ}C$ with concentrated sulphuric acid. We looked at both types of reaction in section 112.2 because they are used to prepare alkenes. For example,

$$CH_3CH_2OH - H_2O \rightarrow C_2H_4$$
ethanol ethene

$$CH_3CH_2CHOHCH_3 - H_2O \rightarrow CH_3CH{=}CHCH_3$$
butan-2-ol but-2-ene

Notice that, in the second example, but-2-ene rather than but-1-ene is the major product.

116.7 When alcohols react with acids, esters are made

If you mix a little ethanol with an approximately equal amount of ethanoic acid, and warm them together (using a water bath) you will eventually detect the very sweet smell of ethyl ethanoate. (It smells of pear drops.) However, if you were first to add a few drops of concentrated sulphuric acid to the reaction mixture, you would detect the smell much more quickly. The acid catalyses the reaction, which is in fact reversible:

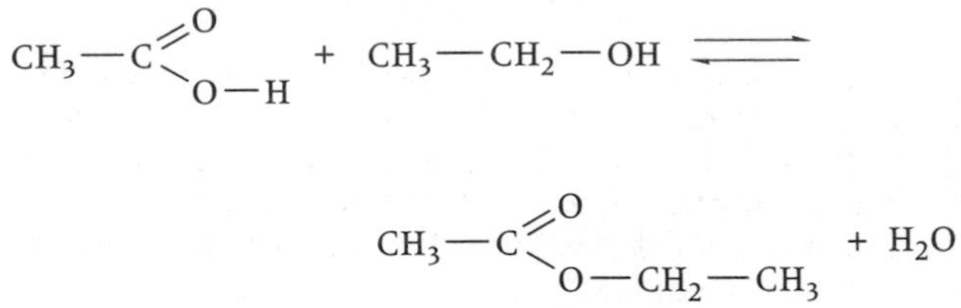

$$CH_3COOH + CH_3CH_2OH \rightleftharpoons CH_3COOCH_2CH_3 + H_2O$$
ethanoic acid   ethanol   ethyl ethanoate

Other alcohols and acids react in similar ways. We shall return to this reaction in section 120.6. Esters can also be made by reacting an alcohol with an acid chloride. For example, ethyl ethanoate can be made from ethanol and ethanoyl chloride. This reaction is very much faster than the previous method:

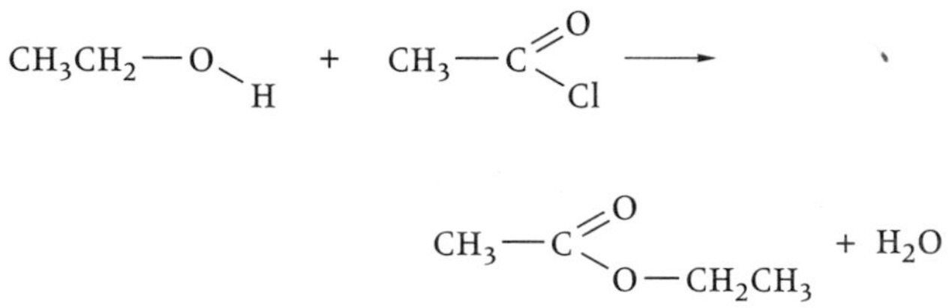

$$CH_3CH_2OH + CH_3COCl \rightarrow CH_3COOCH_2CH_3 + H_2O$$
ethanol   ethanoyl   ethyl
              chloride   ethanoate

116.16 Draw the structure of the ester made when propan-1-ol reacts with ethanoic acid. Suggest a name for the molecule.

Answers

116.1 We can write the formula of water as H—OH, and an alcohol as R—OH. The similarity is clear; both have an OH group.

116.2 The miscibility of alcohols with water depends on (i) the attractions between the molecules owing to the hydrogen bonds between them and (ii) the entropy of mixing. If the carbon chain becomes too long, its presence in water disrupts many hydrogen bonds between the water molecules. This is energetically unfavourable. Eventually the enthalpy change for mixing becomes positive and overcomes the favourable entropy of mixing. (Remember, $\Delta G = \Delta H - T\Delta S$, and we must have $\Delta G < 0$ for the process to occur.)

116.3 The electronegative oxygen attracts electron density towards itself. This leaves the hydrogen and car-bon atoms attached to it with a slight positive charge. Owing to the unsymmetrical distribution of charge, a dipole moment results.

116.4 (i) The molecule has no double or triple bonds.

(ii) Its structure is:

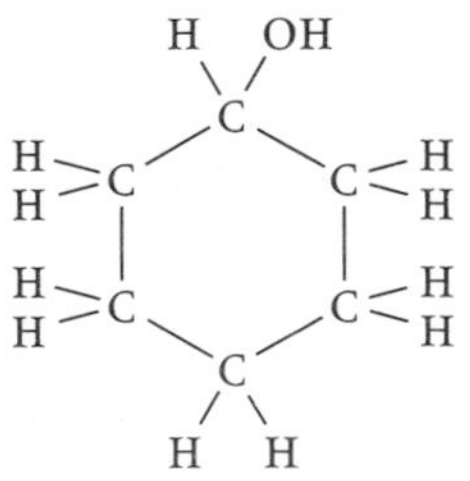

(iii) Cyclohexanol.

116.5 (i) Pentan-2-ol.

(ii) 2-Methylbutan-2-ol.

116.6 We have,

$$\Delta H^{\circ} \text{(reaction)} = \Delta H_f^{\circ}(CH_3OH) - \Delta H_f^{\circ}(CO)$$
$$= -238.9 \text{ kJ mol}^{-1} - (-110.5 \text{ kJ mol}^{-1})$$
$$= -128.4 \text{ kJ mol}^{-1}$$

116.7 A simple method would be to measure the volume of carbon dioxide given off over a given time interval, say every 10 minutes. If you wanted to use a home brewing kit, and you had the patience, you could simply count the number of bubbles of gas escaping in a given time interval.

116.8 2-Methylbutan-2-ol, a tertiary alcohol:

$$CH_3 - CH_2 - \overset{\overset{\displaystyle OH}{|}}{\underset{\underset{\displaystyle CH_3}{|}}{C}} - CH_3$$

116.9 Pentan-3-ol:

$$CH_3 - CH_2 - \overset{\overset{\displaystyle OH}{|}}{\underset{\underset{\displaystyle H}{|}}{C}} - CH_2 - CH_3$$

116.10 (i) Markovnikoff addition: the product is pentan-2-ol:

$$CH_3 - CH_2 - CH_2 - \overset{\overset{\displaystyle OH}{|}}{\underset{\underset{\displaystyle H}{|}}{C}} - CH_3$$

(ii) Anti-Markovnikoff addition: the product is pentan-1-ol:

$$CH_3 - CH_2 - CH_2 - CH_2 - CH_2OH$$

116.11 (i) Butan-1-ol:

$$CH_3 - CH_2 - CH_2 - CH_2 - OH$$

(ii) Butan-2-ol:

$$CH_3 - CH_2 - \overset{\overset{\displaystyle OH}{|}}{\underset{\underset{\displaystyle H}{|}}{C}} - CH_3$$

116.12 Oxidation is the loss of hydrogen. An aldehyde or ketone contains fewer hydrogen atoms than its parent alcohol.

116.13 Water contains OH groups, so it too reacts with the reagents. If water is present, the test will appear to work even if there are no OH groups in the organic compound.

116.14 Yes. The reaction of an alcohol with, say, hydrogen chloride can be thought of as a nucleophilic attack by one of the lone pairs on the oxygen on the hydrogen bonded to the chlorine atom. Normally, though, we talk about the lone pair being protonated.

116.15 The intermediate formed in the reaction with 2-methylpropan-2-ol is a tertiary carbocation, so it is more energetically favoured than the primary carbocation obtained from butan-1-ol. This difference in reactivity is the basis of the *Lucas test* for distinguishing primary, secondary and tertiary alcohols. The alcohol is mixed with concentrated hydrochloric acid and zinc chloride. The solution goes cloudy when chloroalkanes are made. You should now know that tertiary alcohols give the cloudiness most quickly, and primary alcohols the least quickly (if at all).

116.16 The ester is propyl ethanoate:

$$CH_3 - C\overset{\displaystyle O}{\underset{\displaystyle O - CH_2 - CH_2 - CH_3}{\diagup}}$$

UNIT 116 SUMMARY

- Alcohols:
  - (i) Have one or more OH groups as their functional groups.
  - (ii) The non-aromatic monohydric alcohols have the general formula $C_nH_{2n+1}OH$; e.g. methanol, CH_3OH; ethanol, C_2H_5OH.
  - (iii) Primary alcohols have formula RCH_2OH, secondary $RR'CHOH$, tertiary $RR'R''COH$.
- Preparation:
  - (i) By fermentation of sugars;

    e.g. $C_6H_{12}O_6 \rightarrow 2C_2H_5OH + CO_2$

    catalysed by zymase
  - (ii) Hydration of alkenes (an addition reaction);

    e.g. $C_2H_4 \xrightarrow[\text{(ii) H}_2\text{O}]{\text{(i) conc. H}_2\text{SO}_4} C_2H_5OH$

  - (iii) Using a Grignard reagent. Aldehydes give secondary alcohols; ketones give tertiary alcohols; e.g.

    $$R - C\overset{\displaystyle O}{\underset{\displaystyle H}{\diagdown}} \xrightarrow[\text{(ii) H}_2\text{O}]{\text{(i) CH}_3\text{MgBr}} R - \overset{\overset{\displaystyle OH}{|}}{\underset{\underset{\displaystyle H}{|}}{C}} - CH_3$$

  - (iv) Reduction of aldehydes or ketones by lithium tetrahydridoaluminate(III), or using hydrogen and a nickel catalyst; e.g.

    $$R - C\overset{\displaystyle O}{\underset{\displaystyle R'}{\diagdown}} \xrightarrow[\text{(ii) H}_2\text{O}]{\text{(i) LiAlH}_4} R - \overset{\overset{\displaystyle OH}{|}}{\underset{\underset{\displaystyle H}{|}}{C}} - R'$$

Summary – cont.

(v) Hydrolysis of halogenoalkanes by heating with alkali;

e.g. $C_2H_5I + OH^- \rightarrow C_2H_5OH + I^-$

(vi) Hydroboration of alkenes using diborane and hydrogen peroxide; e.g.

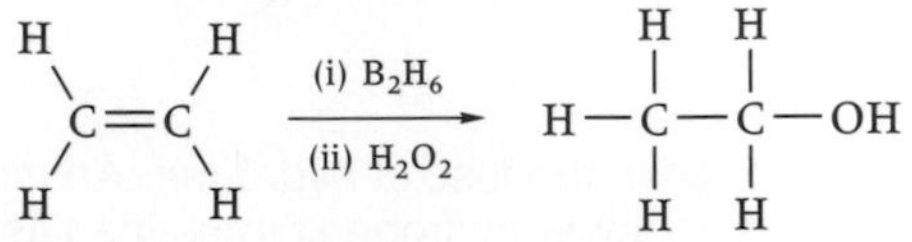

$$\underset{H}{\overset{H}{\diagdown}}C=C\underset{H}{\overset{H}{\diagup}} \xrightarrow[\text{(ii) } H_2O_2]{\text{(i) } B_2H_6} \quad H-\underset{\underset{H}{|}}{\overset{\overset{H}{|}}{C}}-\underset{\underset{H}{|}}{\overset{\overset{H}{|}}{C}}-OH$$

Reactions

- Oxidation:

  (i) Alcohols burn;

  e.g. $2C_2H_5OH(l) + 7O_2(g) \rightarrow$
  $$4CO_2(g) + 6H_2O(g)$$

  (ii) With acidified sodium dichromate(VI) solution;

  primary alcohols $\rightarrow$ aldehydes $\rightarrow$ acids
  secondary alcohols $\rightarrow$ ketones

  (iii) React with copper at $250\,^\circ C$;

  primary alcohols $\rightarrow$ aldehydes
  secondary alcohols $\rightarrow$ ketones

- With sodium:
  Hydrogen released;

  e.g. $2C_2H_5OH + 2Na \rightarrow 2C_2H_5O^-Na^+ + H_2$

- Dehydration:
  With concentrated sulphuric acid;

  e.g. $CH_3CH_2CHOHCH_3 - H_2O \rightarrow CH_3CH = CHCH_3$

- Esterification:

  (i) Warm an alcohol with an organic acid plus a little concentrated sulphuric acid;

  e.g. $CH_3COOH + CH_3CH_2OH \rightleftharpoons$
  $$CH_3COOCH_2CH_3 + H_2O$$
  ethyl ethanoate

  (ii) React with an acid chloride;

  e.g. $CH_3CH_2OH + CH_3COCl \rightarrow$
  ethanoyl chloride
  $$CH_3COOCH_2CH_3 + HCl$$

- Test for OH group:
  White fumes of HCl with phosphorus pentachloride or sulphur dichloride oxide;

  e.g. $C_2H_5OH + SOCl_2 \rightarrow C_2H_5Cl + SO_2 + HCl$

117

Aromatic alcohols

117.1 There are two kinds of aromatic alcohol

Aromatic alcohols contain an OH group, but it can be attached either directly to the benzene ring or to a hydrocarbon side chain. Table 117.1 shows you some examples. As a general rule, if the OH group is on the side chain, the alcohol can be made by the methods we met in the last unit. Similarly it will have chemical properties in common with them. Questions 117.1 to 117.4 will check your understanding of these points.

When the OH group is bonded to a benzene ring there are two major effects: first, the properties of the OH group are changed; and secondly, the properties of the benzene ring are changed as well. We shall see these effects at work in the following sections.

The most common aromatic alcohol is phenol, C_6H_5OH. At room temperature pure phenol is a clear, colourless crystalline solid. However, it is often pale pink in colour owing to impurities being present. It has a distinctive smell of antiseptic, which is to be expected because phenol *is* an antiseptic. If you use phenol, take care with it. The crystals or a solution in water will irritate your skin, and may cause it to blister.

117.1 Suggest a way of making 2-phenylethanol starting with 1-iodo-2-phenylethane.

117.2 Which two chemicals would you mix if you wanted to make the ester, phenylethyl ethanoate:

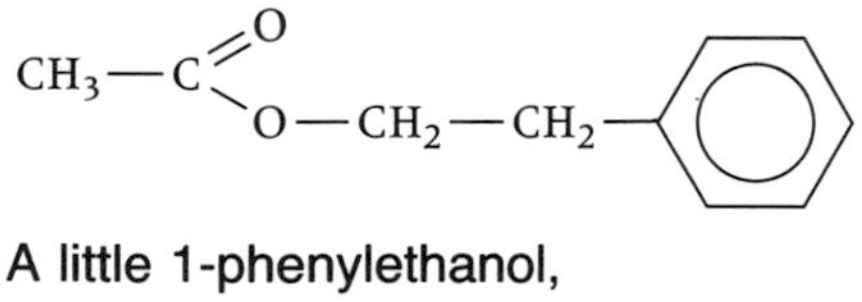

117.3 A little 1-phenylethanol,

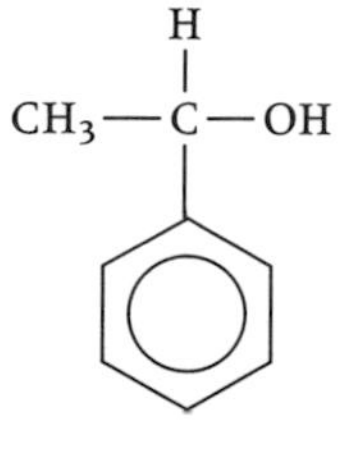

was oxidised by acidified potassium dichromate(VI) solution.

(i) What type of alcohol is 1-phenylethanol?

(ii) What type of product do these alcohols give on oxidation?

(iii) Draw the structure of the molecule made in this reaction.

117.4 2-Phenylethanol is subjected to dehydration by reacting it with concentrated sulphuric acid. Draw the structure of the product of the reaction.

117.2 How to make phenol

It is an unrewarding task to persuade benzene to swap one of its hydrogen atoms for an OH group directly. Similarly, trying to replace a halogen on a benzene by an OH group is far from easy (see section 115.7). In the laboratory the best method is to convert phenylamine into phenol. If we put phenylamine into an ice cold solution of sodium nitrite in hydrochloric acid, it is converted into a benzenediazonium ion by the nitrous acid present:

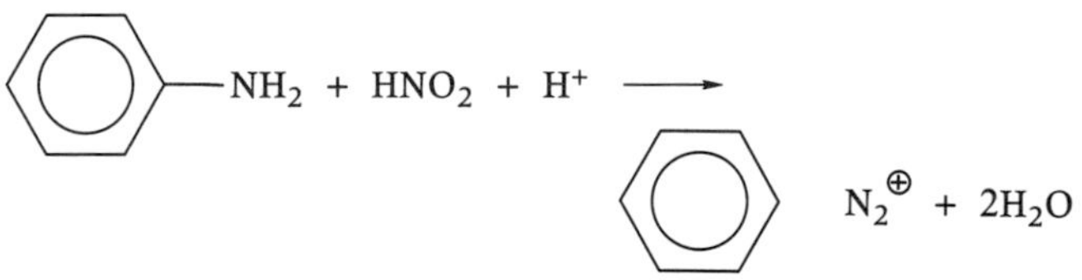

$$C_6H_5NH_2 + HNO_2 + H^+ \rightarrow C_6H_5N_2^{\oplus} + 2H_2O$$
phenylamine   benzene diazonium ion

We can convert the benzenediazonium ion into phenol by warming the solution. As the temperature increases, the benzenediazonium ion is rapidly hydrolysed:

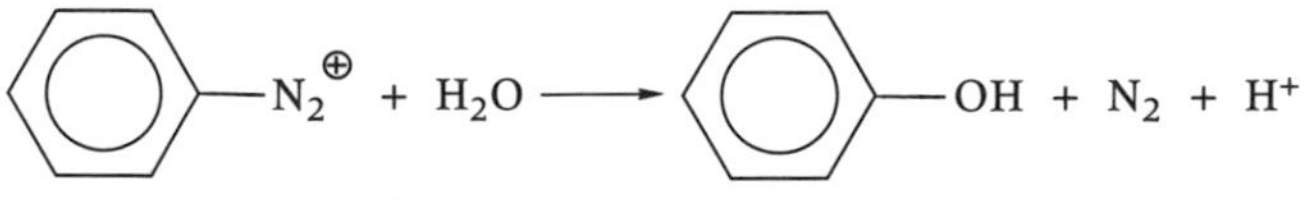

$$C_6H_5N_2^{\oplus} + H_2O \rightarrow C_6H_5OH + N_2 + H^+$$
phenol

Table 117.1. Some aromatic alcohols

Name	Formula	Structure	Melting point/°C	Boiling point/°C
Phenol	C_6H_5OH	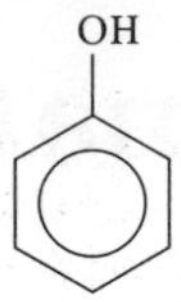	41	182
Benzene-1,2-diol (catechol)	$C_6H_4(OH)_2$	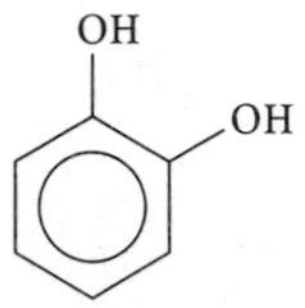	104	246
Benzene-1,3-diol (resorcinol)	$C_6H_4(OH)_2$	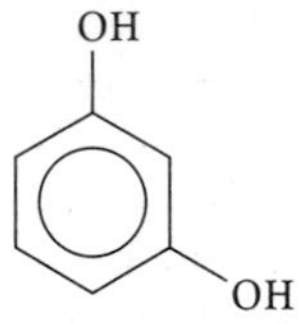	110	281
Benzene-1,4-diol (hydroquinone)	$C_6H_4(OH)_2$	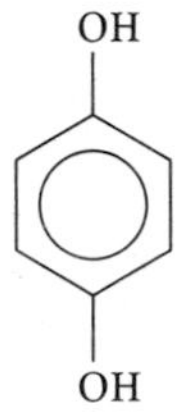	173	286
2-Methylphenol (*ortho*-cresol)	$CH_3C_6H_4OH$	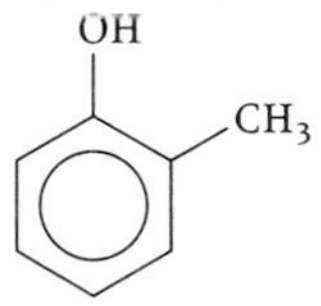	31	191
3-Methylphenol (*meta*-cresol)	$CH_3C_6H_4OH$	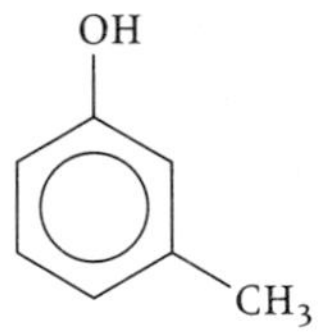	11	201
4-Methylphenol (*para*-cresol)	$CH_3C_6H_4OH$	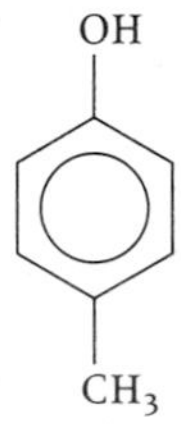	35	202
Phenylmethanol (benzyl alcohol)	$C_6H_5CH_2OH$	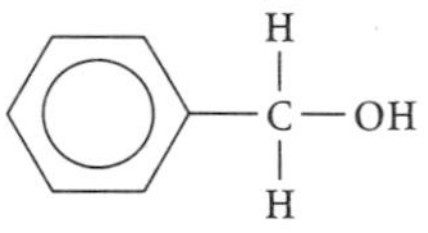	−15	205
2-Phenylethanol	$C_6H_5CH_2CH_2OH$	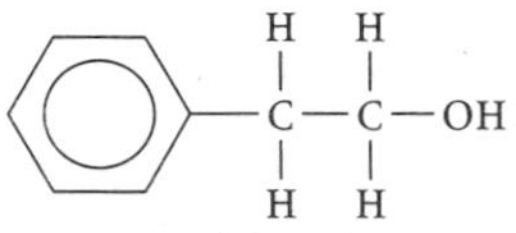	−27	221

In industry. phenol is made by the direct reaction, at around 300 atm and 300 °C, of chlorobenzene and an 8% solution of sodium hydroxide:

$$C_6H_5Cl + OH^- \rightarrow C_6H_5OH + Cl^-$$
$$\text{phenol}$$

Another method is to heat the sodium salt of benzenesulphonic acid, $C_6H_5SO_3^-Na^+$, with sodium hydroxide. The first product is sodium phenoxide, $C_6H_5O^-Na^+$. Phenol is released by adding hydrochloric acid:

SO₃⁻Na⁺ →(NaOH) O⁻Na⁺ →(HCl) OH

Given the rather drastic conditions used in the industrial process, with all the expense that it involves, you should guess that there is a considerable demand for phenol. Indeed, phenol has many uses; for example, as an antiseptic, but more importantly in the manufacture of dyes and polymers.

> **117.5** If you were carrying out the preparation of phenol from phenylamine, how would you know if the reaction between the benzenediazonium ion and water was taking place?

117.3 Why is phenol acidic?

Phenol is partially soluble in water, and its solution has a pH of around 5 or 6, showing that it is a weak acid. This makes phenol clearly different to non-aromatic alcohols. We can write the reaction with water as

$$C_6H_5OH(aq) + H_2O(l) \rightleftharpoons C_6H_5O^-(aq) + H_3O^+(aq)$$
$$\text{phenoxide ion}$$

The reason why phenol is acidic lies in the nature of the phenoxide ion. One of the lone pairs on the oxygen atom can become involved with the π cloud on the benzene ring. When this happens, charge is delocalised, and the phenoxide ion is energetically favoured in a way that is impossible with alcohols like ethanol (Figure 117.1).

That phenol is only a weak acid is shown not only by its pH but by its inability to give carbon dioxide with a carbonate or hydrogencarbonate. However, it will dissolve in sodium hydroxide solution:

OH (s) + OH⁻(aq) ⟶ O⁻ (aq) + H₂O(l)

Its pK_a is 9.96. This compares with pK_a values of 4.8

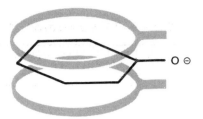

Phenoxide ion, $C_6H_5O^-$

Figure 117.1 *The spread of negative charge over the benzene ring gives energetic stability to the phenoxide ion. The charge on ions such as $C_2H_5O^-$ cannot be delocalised in the same way, so there is no gain in energetic stability for them*

for ethanoic acid, 4.2 for benzoic acid and 16 for ethanol. Thus, phenol is markedly more acidic than non-aromatic alcohols, but not as acidic as typical organic weak acids.

> **117.6** Phenol is only partially soluble, but the phenoxide ion is very soluble in water. If you place a few crystals of phenol in a test tube and add water to them you will see two layers: one is mainly phenol with a little water in it, the other is mainly water containing a little phenol. (Look back at Unit 61 for more details about this.)
> What would you expect to see if (i) you added sodium hydroxide solution to the test tube, and then (ii) added hydrochloric acid?

117.4 Phenol is more reactive than benzene

We shall split the reactions of phenol into those which primarily involve the benzene ring, and those which affect the OH group. In this section we shall concentrate on the ring. The key thing to realise is that:

> **The OH group activates the ring towards electrophilic substitution.**

One result of this is that the conditions used in reactions with phenol are often much less drastic than they are with benzene alone.

(a) Bromination

If you add bromine or bromine water to a solution of phenol, there is an immediate white precipitate of 2,4,6-tribromophenol:

OH + 3Br₂ ⟶ (2,4,6-tribromophenol with Br at 2,4,6 positions) + 3HBr

(b) Nitration

Similarly, phenol is nitrated very easily. Dilute nitric acid will give a mixture of products:

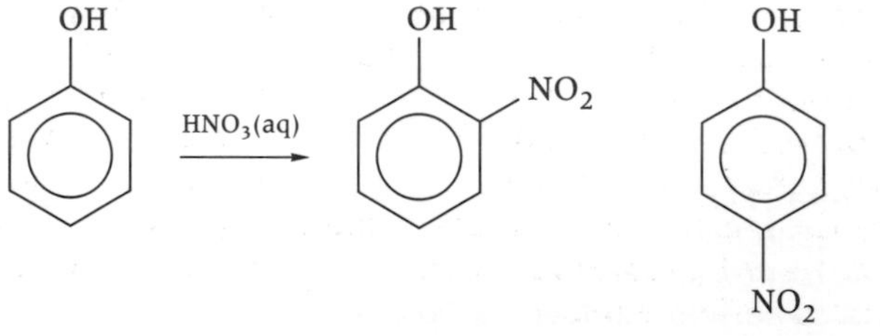

2-nitrophenol   4-nitrophenol

Notice the difference with the nitration of benzene, which requires a mixture of concentrated nitric and sulphuric acids. As we saw in the last unit, the reaction with phenol involves nitrosonium ions, $^{\oplus}NO$, rather than nitryl cations, NO_2^+.

(c) Sulphonation

Warm concentrated sulphuric acid converts phenol into 2-hydroxybenzenesulphonic acid:

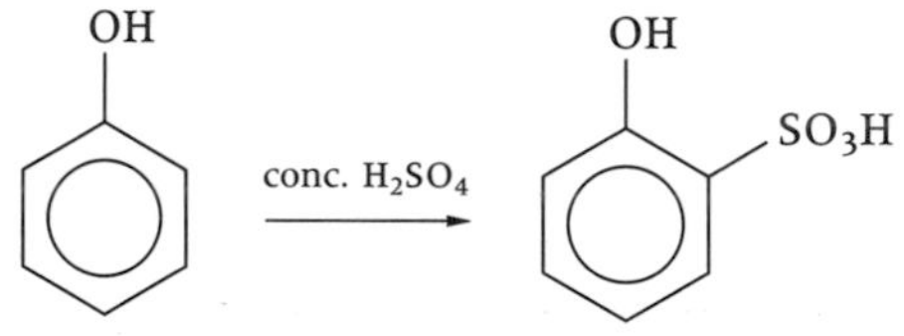

117.5   Reactions of the OH group on phenol

Here we shall look at some reactions that phenol has in common with aliphatic alcohols like methanol and ethanol. However, sometimes these are reactions of the phenoxide ion rather than phenol itself.

(a) Ether formation

If phenol is warmed with iodoethane and sodium hydroxide solution, an ether is produced. The sodium hydroxide has the effect of converting phenol into the phenoxide ion; this is the nucleophile that attacks the iodoethane.

(i) $C_6H_5OH + OH^- \rightarrow C_6H_5O^- + H_2O$

(ii)

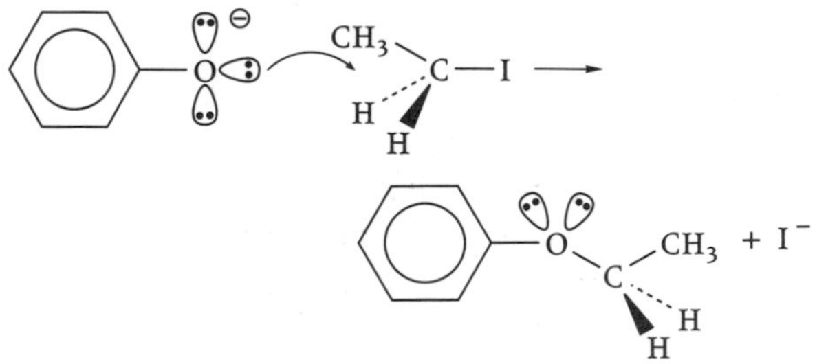

$C_6H_5O^- + C_2H_5I \rightarrow C_6H_5OC_2H_5 + I^-$
ethoxybenzene

This reaction is similar to the Williamson ether synthesis you will find in section 121.2.

(b) Esterification

Phenol cannot be converted into an ester in the same way as aliphatic alcohols. However, it will react with acid chlorides like ethanoyl chloride:

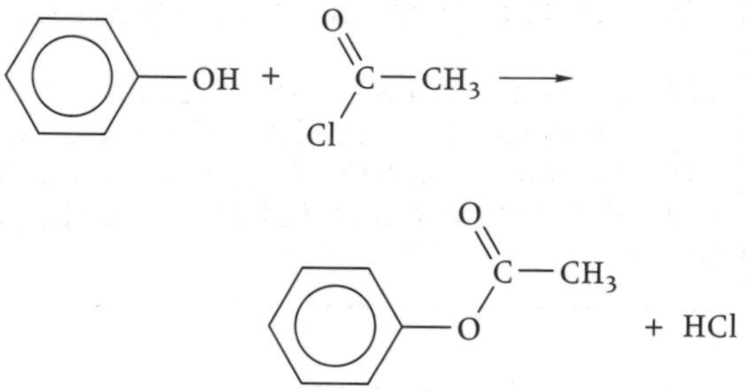

$C_6H_5OH + CH_3COCl \rightarrow C_6H_5OCOCH_3 + HCl$
phenyl
ethanoate

(c) Displacement by halogens

If phosphorus pentachloride or sulphur dichloride oxide is added to an aliphatic alcohol, the OH group is immediately displaced, and a chlorine atom takes its place. With phenol, this type of reaction will take place, but very much less easily. For this reason, the use of these reagents is *not* a good test for an OH group directly attached to a benzene ring.

117.7   Draw out the structure of ethoxybenzene showing all the bonds and the shape of the molecule.

117.8   Phenol can be esterified by adding a few drops of ethanoyl chloride to phenol crystals. The reaction is vigorous, so be careful if you attempt it. A student suggested that, because the reaction is violent, it would be better to dilute the phenol by dissolving a little of it in water and then adding the ethanoyl chloride. What do you think of this suggestion?

117.9   Benzoyl chloride, C_6H_5COCl, has a structure very much like ethanoyl chloride.

(i) Draw out the structure of the molecule.

(ii) Predict what, if anything, would happen if this liquid is added to phenol.

117.10   Phenol can be reduced in two ways. (a) After phenol is vaporised and passed over hot powdered zinc, a liquid can be condensed and the zinc is converted to zinc oxide. (b) If phenol is hydrogenated by passing phenol vapour and hydrogen over a nickel catalyst, another liquid is made. It has the formula $C_6H_{11}OH$.

(i) What are the two liquids?

(ii) Write the equations for the reactions.

117.11 Suggest a way of making benzene-1,3-diol from 1,3-dinitrobenzene.

117.12 Why is the boiling point of 2-nitrophenol very much lower than that of either 3-nitrophenol or 4-nitrophenol? Look at section 21.3 if you are stuck.

117.13 The acid dissociation constant of phenol is approximately 10^{-10} mol dm^{-3}. Calculate the pH of a 0.01 mol dm^{-3} solution of phenol. You can remind yourself about the method of calculation by looking at Unit 75.

117.6 A test for phenol

There is a very simple test for phenol. It is to add a few drops of a neutral solution of iron(III) chloride. A blue or blue-violet colour is produced owing to a complex made between the phenol and the iron(III) ion. Similar compounds with an OH group on a benzene ring give coloured complexes. However, they are not always blue. You should only use this test as a guide; if you see no colour change, it is almost certain that the compound you are using is not a phenol. If you do get a colour change, then it is likely (but not certain) that it is a phenol. The reason why it is not certain is that other classes of compound can also give colour changes with iron(III) chloride solution. Among the most important of these are the salts of many organic acids, and compounds that contain the enol group,

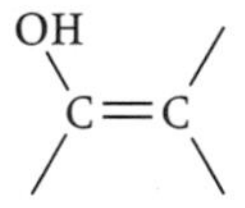

117.7 A polymerisation reaction of phenol

One of the first plastics that came into widespread use just after the Second World War was called Bakelite. It is a hard, but brittle, material and has found a large number of uses, e.g. in electrical fittings such as plugs and sockets. It is made through the reaction of phenol with methanal, catalysed by acid or alkali. The product of the reaction is a structure consisting of phenol molecules linked together by bonds to CH_2 groups. A typical section of polymer looks like this:

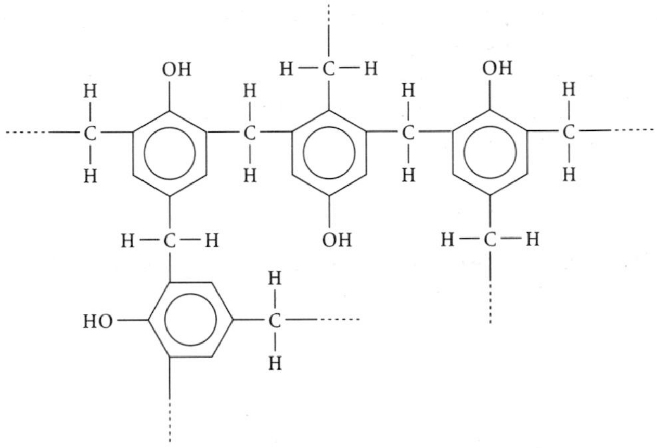

Notice the way the molecules are linked in two directions, both along a chain, and across from one chain to another. This represents a polymer that is *crosslinked*. Crosslinking leads to a very rigid structure because the various groups are not free to twist round and move their positions.

Answers

117.1 Reflux with sodium hydroxide solution.

117.2 2-Phenylethanol and ethanoyl chloride.

117.3 (i) A secondary alcohol.

(ii) Ketones.

(iii) Phenylethanone:

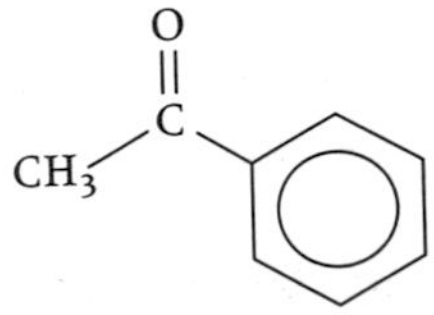

117.4 Phenylethene:

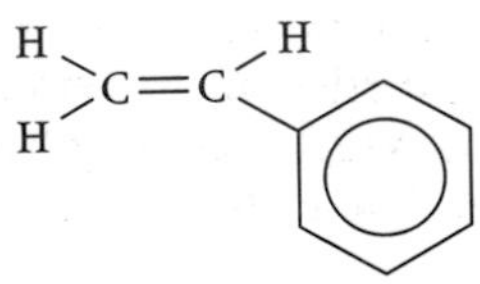

117.5 You would see bubbles of nitrogen given off.

117.6 (i) The solution will clear because all the phenol molecules will be converted into soluble phenoxide ions.

(ii) The solution will become cloudy owing to phenol being liberated:

$$C_6H_5O^-(aq) + H^+(aq) \rightarrow C_6H_5OH(s)$$

117.7

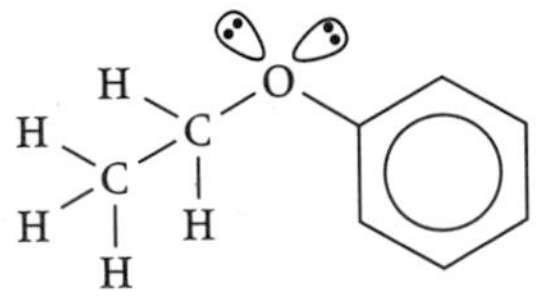

117.8 Rather silly. The ethanoyl chloride would react with the water and not the phenol.

Answers – cont.

117.9 (i)

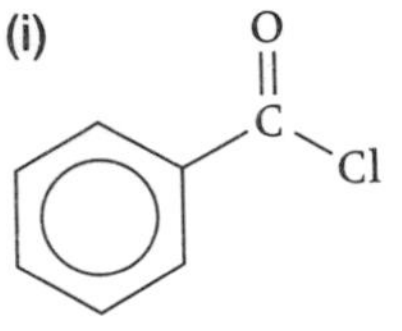

(ii) It will give the ester phenyl benzoate:

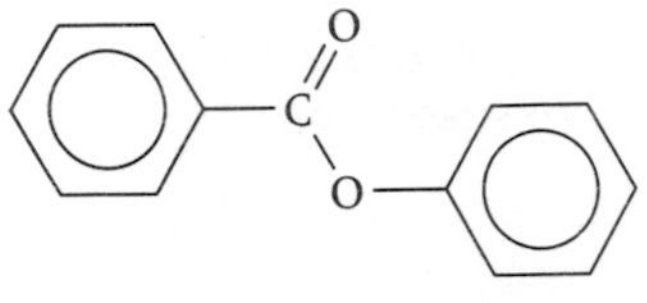

117.10 (i) Benzene and cyclohexanol.

(ii) $C_6H_5OH + Zn \rightarrow C_6H_6 + ZnO$

$C_6H_5OH + 3H_2 \rightarrow C_6H_{11}OH$

117.11 React 1,3-dinitrobenzene with sodium nitrite and hydrochloric acid.

117.12 There is *intra*molecular hydrogen bonding in 2-nitrophenol, and *inter*molecular hydrogen bonding in 4-nitrophenol.

117.13 We start with the equation

$$C_6H_5OH(aq) \rightleftharpoons C_6H_5O^-(aq) + H^+(aq)$$

for which

$$K_a = \frac{[C_6H_5O^-(aq)][H^+(aq)]}{[C_6H_5OH(aq)]}$$

(As usual, the concentrations refer to equilibrium conditions.) Given that a weak acid is only very slightly dissociated into ions, we can put $[C_6H_5OH(aq)] = 0.01$ mol dm^{-3}. Hence,

$[C_6H_5O^-(aq)][H^+(aq)] = 10^{-10}$ mol $dm^{-3} \times 0.01$ mol dm^{-3}

But also, from the equation we know that

$$[C_6H_5O^-(aq)] = [H^+(aq)]$$

Therefore,

$[H^+(aq)]^2 = 10^{-12}$ mol^2 dm^{-6}

$[H^+(aq)] = 10^{-6}$ mol dm^{-3}

Finally, because $pH = -\lg[H^+(aq)]$, we have $pH = 6$.

UNIT 117 SUMMARY

- Aromatic alcohols:
  (i) Aromatic alcohols have OH group(s) on hydrocarbon side chains.
  (ii) Phenols have OH groups directly attached to a benzene ring.
  (iii) The properties of phenols are markedly different to those of non-aromatic alcohols. It is best to think of phenols as a class of compound separate from alcohols.
  (iv) Phenol is more reactive than benzene.

Phenol, C₆H₅OH

- Preparation:
  By hydrolysing a benzenediazonium ion solution;

  (i) $C_6H_5NH_2 + HNO_2 + H^+ \rightarrow C_6H_5N_2^{\oplus} + 2H_2O$
  phenylamine   benzene diazonium ion

  (ii) $C_6H_5N_2^{\oplus} + H_2O \rightarrow C_5H_5OH + N_2 + H^+$

Properties

- Acidity:
  Phenol is acidic in water owing to stability of the phenate ion, $C_6H_5O^-$;

  $$C_6H_5OH(aq) \rightleftharpoons C_6H_5O^-(aq) + H^+(aq)$$

- Bromination:
  Immediate white precipitate of 2,4,6-tribromophenol with bromine.

- Nitration:
  Easily nitrated with dilute nitric acid.
- Sulphonation:
  2-Hydroxybenzenesulphonic acid made with warm concentrated sulphuric acid.

Reactions

- Ether formation:
  By reaction with iodoethane and sodium hydroxide solution, ethoxybenzene, $C_6H_5OC_2H_5$, is made.
- Ester formation:
  An acid chloride must be used; e.g.

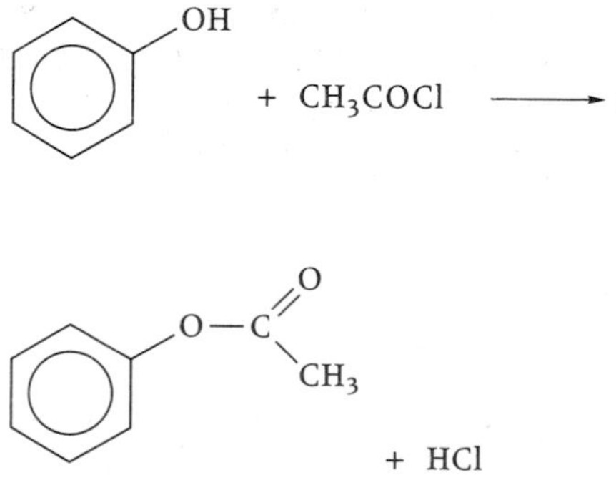

- Polymerisation:
  With methanal and acid, Bakelite is made.
- Test for phenol, and enols in general:
  A blue or blue-violet colour is produced with neutral solution of iron(III) chloride.

118

Aldehydes and ketones

118.1 Aldehydes and ketones contain a carbonyl group

All aldehydes and ketones contain the carbonyl group

$$\text{>C=O}$$

(see Figure 118.1). Aldehydes have at least one hydrogen atom bonded to the carbon atom as well as the oxygen. Ketones have two organic radicals attached to the carbon atom. You can tell whether a substance is an aldehyde or a ketone by looking at its formula: aldehydes have a common formula RCHO

$$R-C\overset{\displaystyle O}{\underset{\displaystyle H}{\diagup}}$$

and ketones RR'CO

$$R-C\overset{\displaystyle O}{\underset{\displaystyle R'}{\diagup}}$$

where R and R' are two organic radicals. The names of aldehydes always end in *-al* and ketones in *-one*. Table 118.1 lists examples of each type, and Table 118.2 their uses.

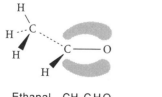

Ethanal, CH_3CHO

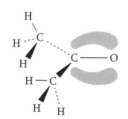

Propanone, $(CH_3)_2CO$

Figure 118.1 *Both aldehydes and ketones are planar in the neighbourhood of the carbonyl (>C=O) group. Notice that the π bond is distorted towards the electronegative oxygen atom*

Aldehydes and ketones are often sweet smelling, volatile, liquids (Figure 118.2). Ethanal has a characteristic smell, which may be detected near rotting fruit. (You might try leaving an apple to rot; eventually you will smell ethanal coming from it.)

118.1 Why do we not write the name of butanone as butan-2-one?

118.2 One reason why ketones are good solvents is that, like water, they are polar molecules.

(i) Draw a diagram of propanone and show by an arrow the direction of the dipole moment.

(ii) Briefly explain why propanone is a liquid at room temperature and pressure.

118.3 Draw a diagram showing the structure of cyclohexanone, $C_5H_{10}CO$.

118.2 The manufacture of simple aldehydes and ketones

(a) *Methanal*

Methanal can be manufactured by the controlled oxidation of methanol. A temperature between 600 and 650° C together with a silver catalyst are used:

$$2CH_3OH + O_2 \xrightarrow{600\,°C,\ Ag} 2HCHO + 2H_2O$$

Methanal is completely miscible with water, and is sold as a solution called *formalin*. Formalin solution is a good disinfectant, with a characteristic smell. If you see biological samples preserved in bottles, it is very likely that formalin is the solution in which they are kept. Although it is widely available, take care if you use the solution. With some acids it can produce powerful carcinogens.

Table 118.1. Common aldehydes and ketones

Name	Formula	Structure	Melting point/°C	Boiling point/°C
Aldehydes				
Methanal	HCHO	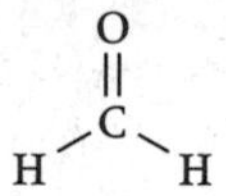	−91.9	−19
Ethanal	CH_3CHO	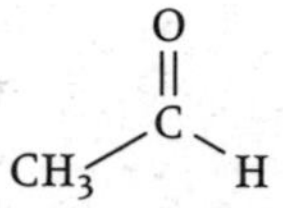	−122.9	20.6
Propanal	C_2H_5CHO	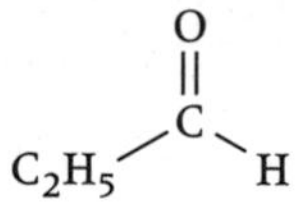	−79.9	48.2
Benzaldehyde	C_6H_5CHO	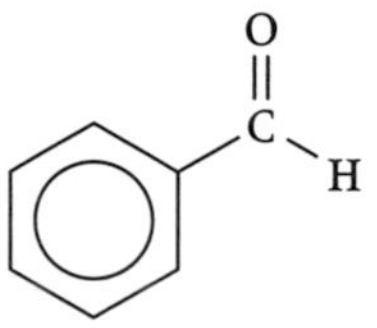	22.3	178.2
Ketones				
Propanone (acetone)	CH_3COCH_3	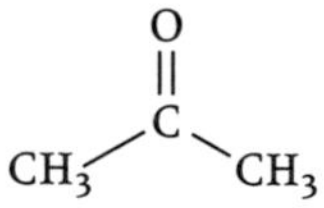	−94.6	56.4
Butanone	$C_2H_5COCH_3$	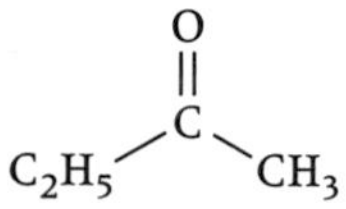	−86.5	79.8
Pentan-2-one	$C_3H_7COCH_3$	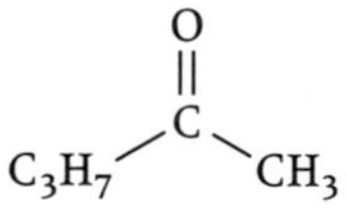	−78	102
Pentan-3-one	$C_2H_5COC_2H_5$	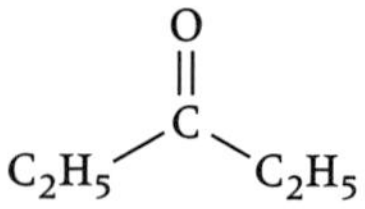	−38.8	102.1
Phenylethanone (acetophenone)	$C_6H_5COCH_3$	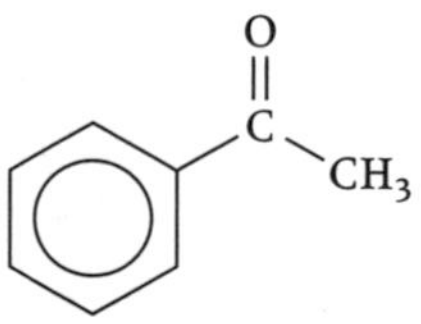	19.8	202.2
Diphenylmethanone (benzophenone)	$C_6H_5COC_6H_5$	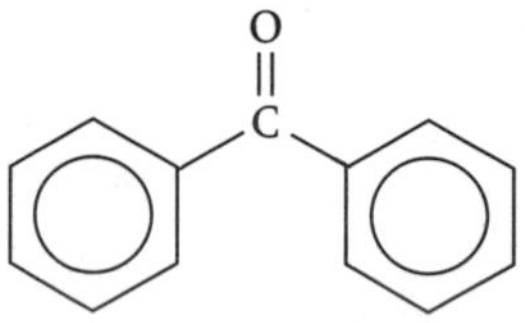	48	306

Two models showing the shapes of methanal and ethanal.

Table 118.2. The uses of aldehydes and ketones

	Uses
Aldehydes	
Methanal	In solution with water, sold as formalin: a disinfectant (formaldehyde) and preservative
Ethanal	Used in the manufacture of ethanoic acid and its derivatives, e.g. ethanoic anhydride
General	More complex aldehydes are found in perfumes and flavourings
Ketones	
Propanone (acetone)	As a solvent. This is a more important use than you might think. For example, many artificial fibres are manufactured from cellulose acetate. Propanone dissolves the acetate. The solution is forced through tiny holes into a warm atmosphere. The propanone evaporates easily, leaving fine fibres, which can be woven into clothes and furniture coverings
General	As intermediates in the manufacture of other chemicals, e.g. cyclohexanone is used in one stage of Nylon manufacture

(b) Ethanal

Ethanal is made on an industrial scale by the *Wacker process* (of which there are several variations). The essence of the process is to use palladium(II) chloride to convert a mixture of ethene and water to ethanal:

$$C_2H_4 + H_2O + PdCl_2 \rightarrow CH_3CHO + Pd + 2HCl$$

However, this alone would be an extremely expensive method of making ethanal. The beauty of the Wacker

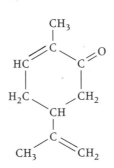

Cinnamon flavour

3-Phenyl-2-propenal, otherwise known as cinnamaldehyde

Vanilla flavour

3-Methoxy-4-hydroxy-benzaldehyde, also known as vanillin

Spearmint flavour

Carvone; this is a ketone built from isoprene

Peppermint flavour

Menthone; this is also derived from isoprene

Figure 118.2 *Two aldehydes and two ketones used for flavourings. (The isoprene structure is shown in Unit 112)*

process is that the palladium(II) chloride is regenerated by reacting it with copper(II) chloride:

$$Pd + 2CuCl_2 \rightarrow PdCl_2 + 2CuCl$$
copper(II) chloride   copper(I) chloride

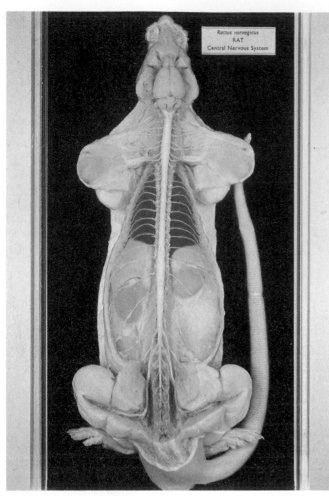

Specimens like this, of a rat dissection showing brain and spinal nerves, are preserved in formalin.

The final trick is to use oxygen (from the air) to oxidise the copper(I) chloride back to copper(II) chloride:

$$4CuCl + O_2 + 4HCl \rightarrow 4CuCl_2 + 2H_2O$$

Thus both the palladium(II) chloride and copper(II) chloride can be continuously recycled.

(c) *Propanone*

Propanone has been manufactured by many processes. One of the first methods was to isolate calcium ethanoate from wood ash and heat it:

$$(CH_3COO)_2Ca \rightarrow CaCO_3 + CH_3COCH_3$$

This rather old-fashioned method has become redundant. During the First World War a biochemical method was developed, which uses specially adapted bacteria to convert carbohydrates into propanone (and other substances). While feasible, the economics of the process requires a very cheap source of carbohydrate. At present propanone is mainly manufactured from cumene or from propan-2-ol.

Cumene is the aromatic hydrocarbon $C_6H_5CH(CH_3)_2$. Its systematic name is (1-methylethyl)benzene. By reaction with oxygen followed by treatment with acid, it is converted into phenol and propanone:

phenol　　　propanone

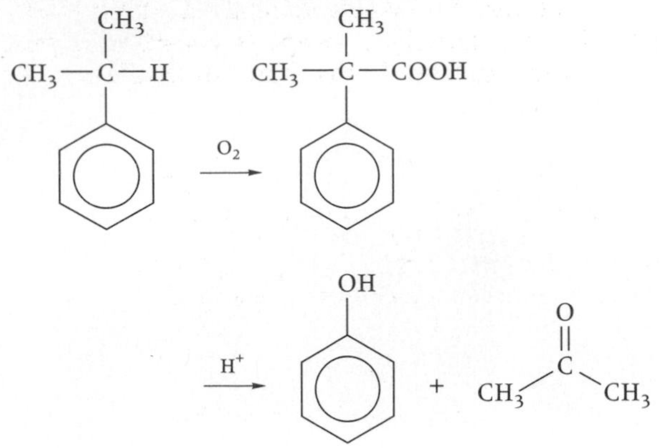

A second process that converts propan-2-ol into propanone is more straightforward, but less widely used. The alcohol vapour is passed over a zinc oxide catalyst at between 400 and 600 °C:

$$CH_3CHOHCH_3 \xrightarrow[500\,°C]{ZnO\ cat.} CH_3COCH_3 + H_2$$

118.3 Two methods of preparing aldehydes or ketones

(a) *Oxidation of an alcohol*

This is the simplest method.

> **Primary alcohols are oxidised to aldehydes.**
>
> **Secondary alcohols are oxidised to ketones.**

The oxidation can be achieved by using an acidified solution of sodium dichromate(VI), or by passing the alcohol vapour over hot copper powder. For example,

CH_3CH_2OH ethanol　　　CH_3CHO ethanal

$(CH_3)_2CHOH$ propan-2-ol　　　$(CH_3)_2CO$ propanone

$C_6H_5CH_2OH$ phenylmethanol　　　C_6H_5CHO benzaldehyde

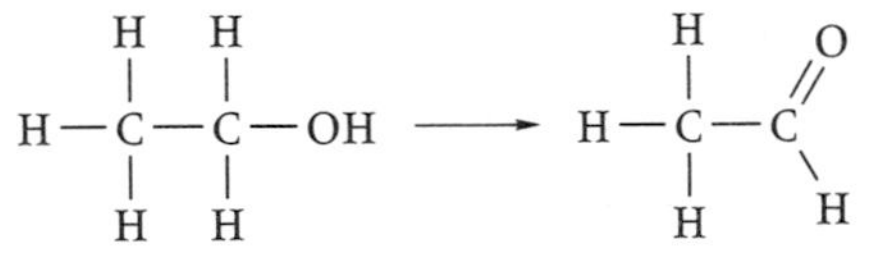

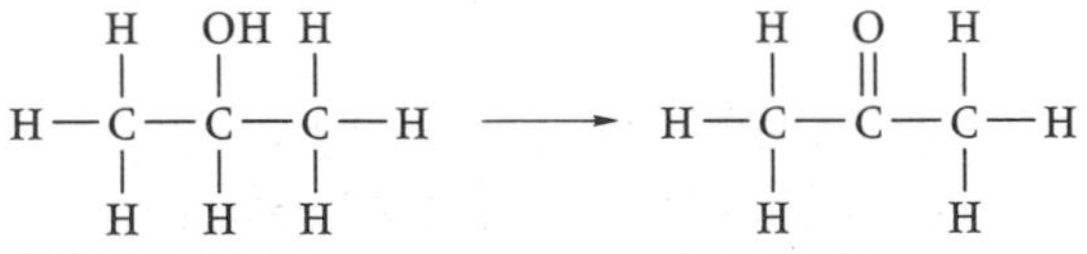

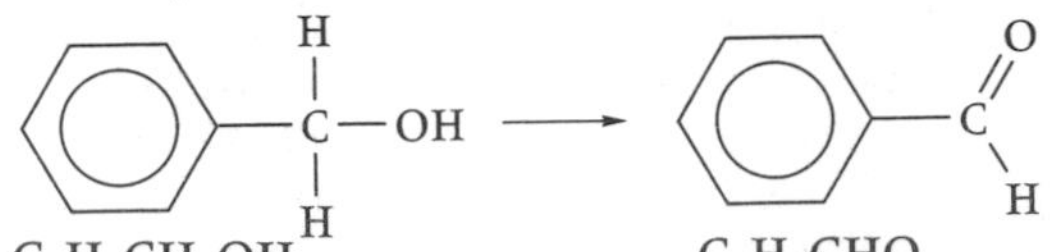

(b) *Ozonolysis*

Ozonolysis can be used to convert an alkene into aldehydes or ketones. You will find details of the method in section 112.5. Ozonolysis swaps two carbonyl groups for the carbon–carbon double bond in an alkene. Here are three examples:

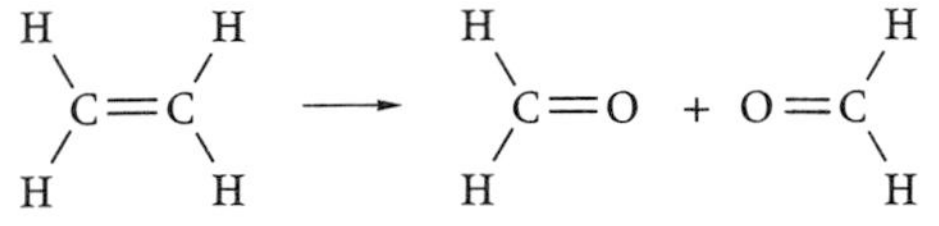

ethene → methanal + methanal

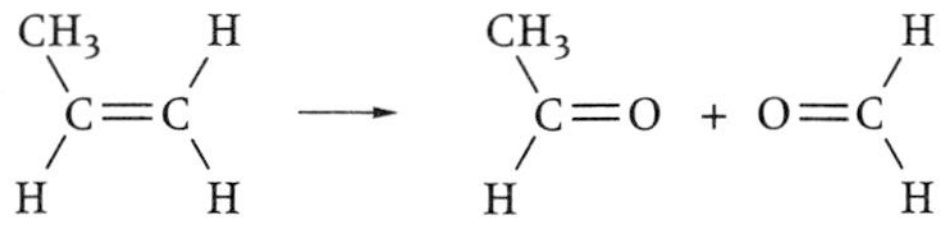

but-1-ene → propanal + methanal

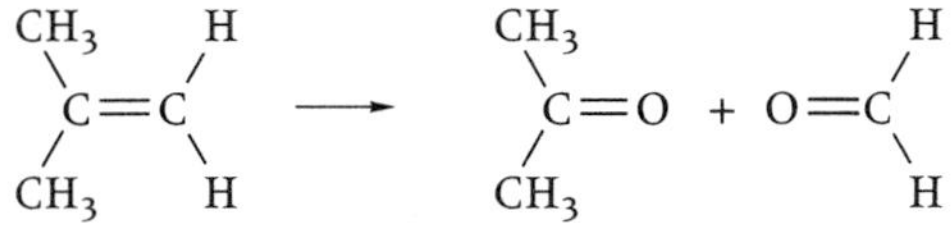

2-methylpropene → propanone + methanal

Ozonolysis can also be used as a method of discovering where a double bond is to be found in a molecule; see question 118.6.

118.4 Which alcohol would you choose to oxidise in order to make (i) 2-methylpropanal, (ii) pentan-2-one?

118.5 Say which chemicals, and sketch the apparatus, you would use to make propanal by oxidising an alcohol.

118.6 After ozonolysis of an alkene, the products were butanone and pentan-2-one.

(i) What might be the original alkene?

(ii) Is there more than one possibility?

118.4 Two methods of preparing aromatic aldehydes and ketones

Benzaldehyde can be made by several methods. The first is to react methylbenzene with chlorine in the presence of ultraviolet light in such quantities to optimise the amount of dichloromethylbenzene, $C_6H_5CHCl_2$, made in the reaction. The final step is to reflux with water. The hydrolysis produces benzaldehyde.

(i)

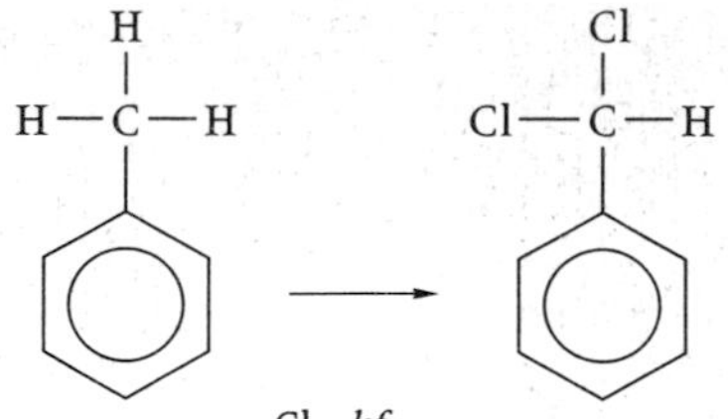

$$C_6H_5CH_3 \xrightarrow{Cl_2,\ hf} C_6H_5CHCl_2$$

(ii)

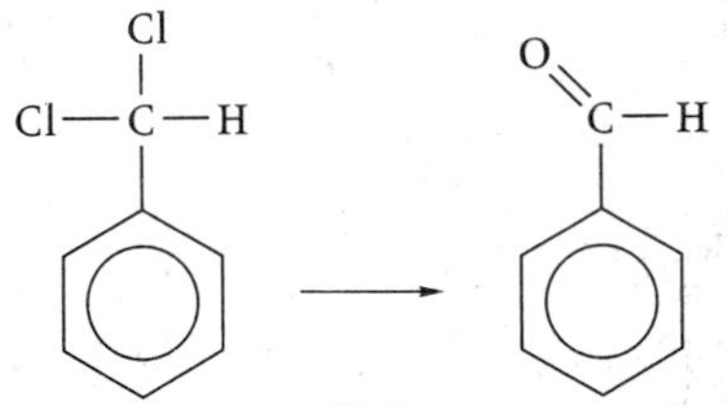

$$C_6H_5CHCl_2 \xrightarrow{H_2O} C_6H_5CHO$$
benzaldehyde

A second approach is to reduce benzoyl chloride, C_6H_5COCl, by hydrogen using a palladium catalyst. However, the catalyst must not be too efficient, otherwise the benzoyl chloride is converted into an alcohol. For this reason the catalyst is mixed with barium sulphate, or another substance, to inhibit its action:

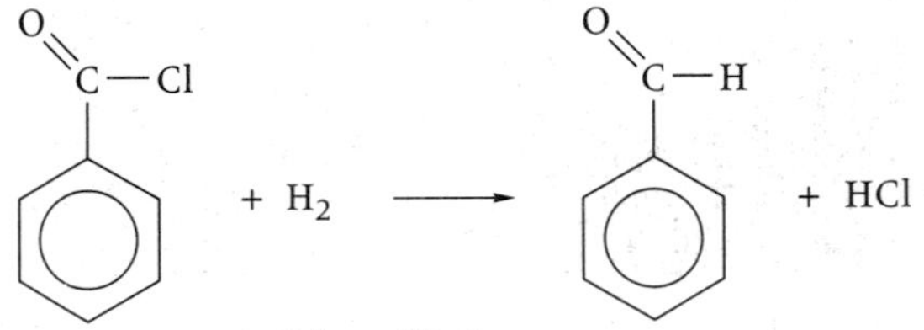

$$C_6H_5COCl + H_2 \xrightarrow{Pd,\ modified} C_6H_5CHO + HCl$$

This method of reduction is known as the *Rosenmund reaction*.

Friedel–Crafts reactions are convenient ways of making aromatic ketones. The general method is shown by the reaction of benzene with ethanoyl chloride using aluminium trichloride as a catalyst. We have seen this reaction before (panel 113.1) and shown that it involves an electrophilic attack on the benzene ring. The ethanoyl chloride is converted into a carbocation through its interaction with the aluminium trichloride:

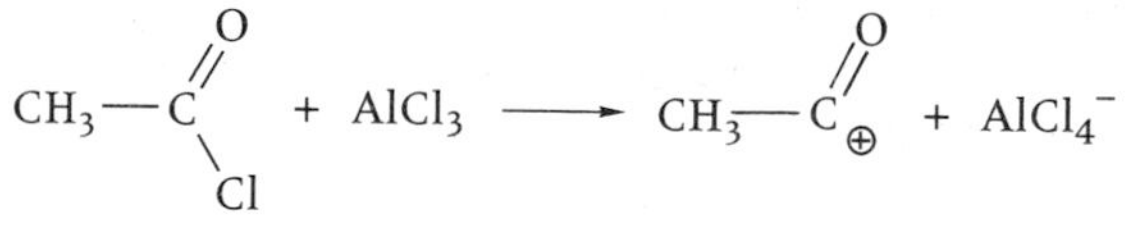

$$CH_3COCl + AlCl_3 \rightarrow CH_3CO^+ + AlCl_4^-$$

This carbocation attacks the benzene ring:

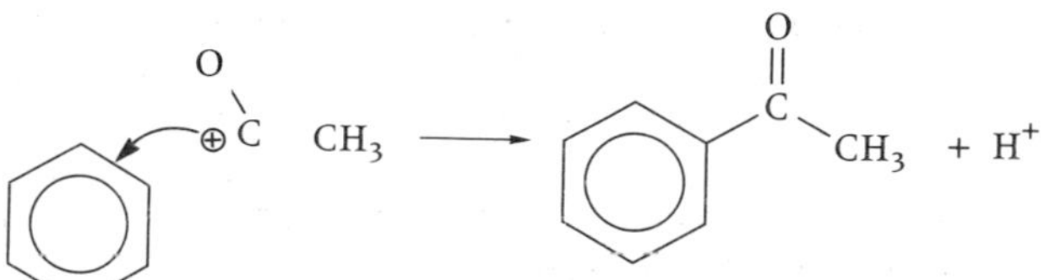

$$C_6H_6 + CH_3CO^+ \rightarrow C_6H_5COCH_3 + H^+$$
phenylethanone

Aldehydes and ketones   801

118.5 Aldehydes and ketones undergo addition reactions

We have already said that the carbonyl group is inherently polar owing to the electronegative oxygen drawing electron density towards itself. Also, the atoms attached directly to the carbon of the carbonyl group all lie in the same plane (see Figure 118.1). The carbon atom bonded to the oxygen not only carries a slight positive charge, but is also open to attack by incoming nucleophiles from above or below the plane of the molecule; there is little chance of steric hindrance stopping a reaction taking place.

The main feature of nucleophilic attack on the carbonyl group is that the attacking group bonds to the carbon atom, and the π bond to the oxygen atom breaks. This leaves the oxygen atom with the ability to make a σ bond to another species, usually a hydrogen ion. Rather than displace another group (which happens in nucleophilic attack on halogenoalkanes), the attacking nucleophile adds on to the aldehyde or ketone; hence this type of reaction is called an *addition reaction* (Table 118.3). The reaction with cyanide ions is typical of addition.

(a) Cyanide addition

This particular reaction is useful because it provides us with a pathway for making mixed alcohols and acids (Figure 118.3). If a nitrile is hydrolysed it is converted into a carboxylic acid. Thus, we can carry out the changes

Table 118.3. Some addition reactions of aldehydes and ketones

Nucleophile	Product of reaction with:		
	Ethanal	*Benzaldehyde*	*Propanone*
Cyanide ions, CN^-			
Hydrogensulphite, HSO_3^-			
Alcohols, e.g. C_2H_5OH		No reaction	

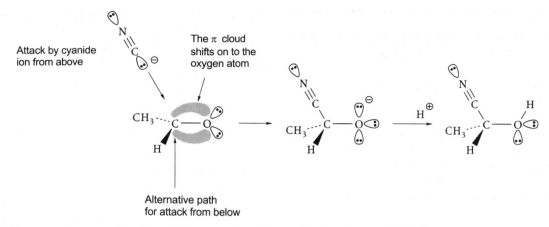

Attack by cyanide ion from above

The π cloud shifts on to the oxygen atom

Alternative path for attack from below

Figure 118.3 *One way of representing the addition reaction of a cyanide ion with an aldehyde*

$$CH_3CHO \xrightarrow{CN^-,\ H^+} CH_3CH(OH)CN \xrightarrow{H^+} CH_3CH(OH)COOH$$

ethanal   2-hydroxypropane-nitrile   2-hydroxypropanoic acid

2-Hydroxypropanoic acid is also known as lactic acid. It has been widely studied because it is one of the substances made when milk goes sour. The acid is formed by the action of a microorganism, *Bacillus acidi lactici*, on sugars. In the early 1900s large amounts of the bacteria were isolated and used to prepare lactic acid on an industrial scale. The acid was used in the dyeing and leather industries.

(b) Hydrogensulphite addition

Hydrogensulphite ions, HSO_3^-, will add to both aldehydes and ketones. The net result of the addition is shown in this reaction:

$$CH_3COCH_3 + HSO_3^- \rightarrow CH_3C(OH)(SO_3^-)CH_3$$

(c) Alcohol addition

Aldehydes, but not ketones, will give addition reactions with alcohols provided all the reagents are dry, and that hydrogen chloride is used to catalyse the reaction. The most common example of this type of addition is ethanol adding to ethanal:

$$CH_3CHO + 2C_2H_5OH \xrightarrow{dry\ HCl} CH_3CH(OC_2H_5)_2 + H_2O$$

ethanal   1,1-diethoxyethane

The product, 1,1-diethoxyethane, was once known as acetal. It is a type of ether.

118.9 It is possible to detect a compound called a hemiacetal in a mixture of an aldehyde and an alcohol. In a mixture of ethanol and ethanal, the hemiacetal has the formula $C_4H_{10}O_2$. It has a structure very similar to 1,1-diethoxyethane (acetal). What is the structure of the hemiacetal?

118.10 (i) Explain why 2-hydroxypropanoic acid should be optically active.

(ii) Why does a solution of 2-hydroxypropanoic acid made by the method

$$CH_3CHO \rightarrow CH_3CH(OH)CN \rightarrow CH_3CH(OH)COOH$$

have no optical activity. (Hint: think about the way the cyanide ions can attack the aldehyde.)

118.6   Condensation reactions

Owing to the lone pair on its nitrogen atom, ammonia can add to an aldehyde or ketone. For example, with ethanal the reaction follows the pattern of Table 118.4:

$$CH_3CHO + NH_3 \rightarrow CH_3CH(OH)NH_2$$

However, this type of compound is usually unstable, and in any case experience shows that other nitrogen containing compounds are much more useful in their reactions. As an illustration of the type of reaction that can take place, let us look at how hydrazine, N_2H_4, reacts with ethanal. The first step is similar to the reaction with ammonia:

Table 118.4. Condensation reactions of aldehydes and ketones

Reagent	Type of product	Typical reaction
NH_2-NH_2 Hydrazine	$\underset{R'}{\overset{R}{>}}C=N-NH_2$ Hydrazone	$\underset{CH_3}{\overset{O}{\parallel}}{\overset{}{C}}-H \longrightarrow \underset{H}{\overset{CH_3}{>}}C=N-NH_2$
$NH_2-N\overset{H}{-}C_6H_5$ Phenylhydrazine	$\underset{R'}{\overset{R}{>}}C=N-N\overset{H}{-}C_6H_5$ Phenylhydrazone	$CH_3-\overset{O}{\overset{\parallel}{C}}-CH_3 \longrightarrow \underset{CH_3}{\overset{CH_3}{>}}C=N-N\overset{H}{-}C_6H_5$ or $\underset{CH_3}{\overset{CH_3}{>}}C=N-NHC_6H_5$
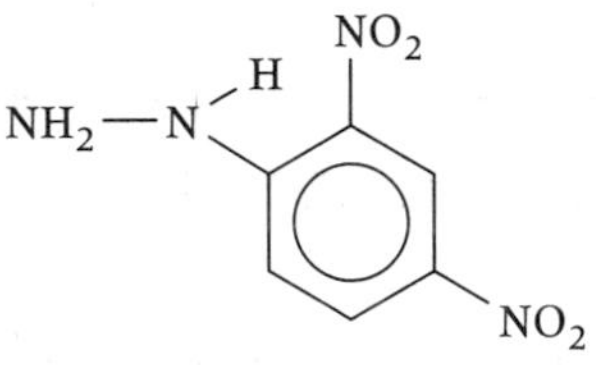 $NH_2-N\overset{H}{-}C_6H_3(NO_2)_2$ 2,4-Dinitrophenylhydrazine	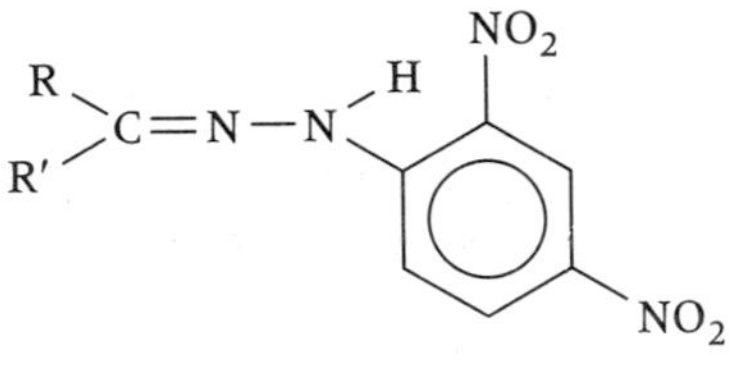 $\underset{R'}{\overset{R}{>}}C=N-N\overset{H}{-}C_6H_3(NO_2)_2$ 2,4-Dinitrophenylhydrazone	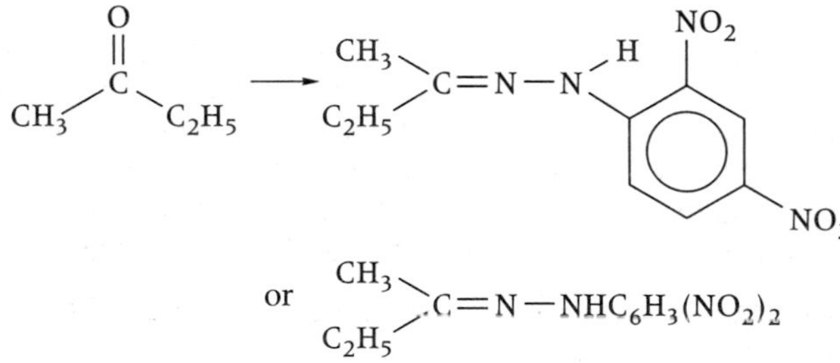 $CH_3-\overset{O}{\overset{\parallel}{C}}-C_2H_5 \longrightarrow \underset{C_2H_5}{\overset{CH_3}{>}}C=N-N\overset{H}{-}C_6H_3(NO_2)_2$ or $\underset{C_2H_5}{\overset{CH_3}{>}}C=N-NHC_6H_3(NO_2)_2$

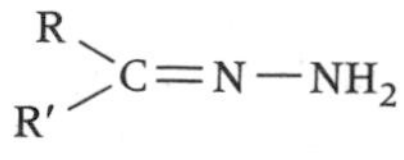

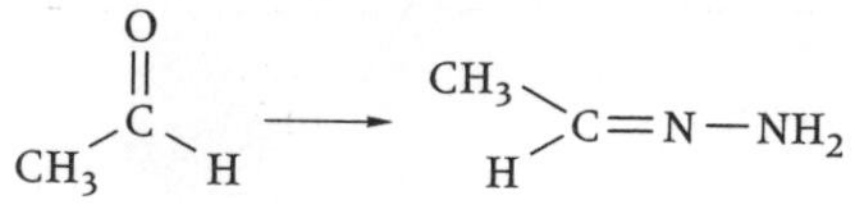

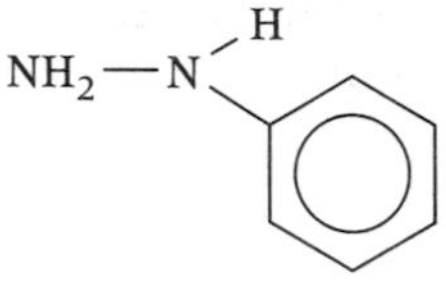

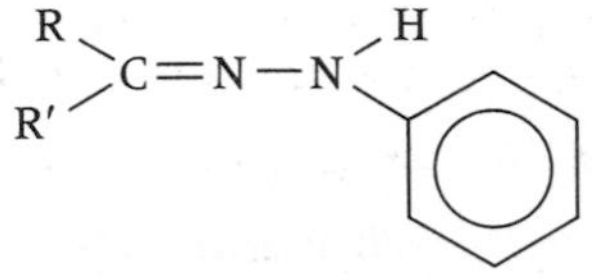

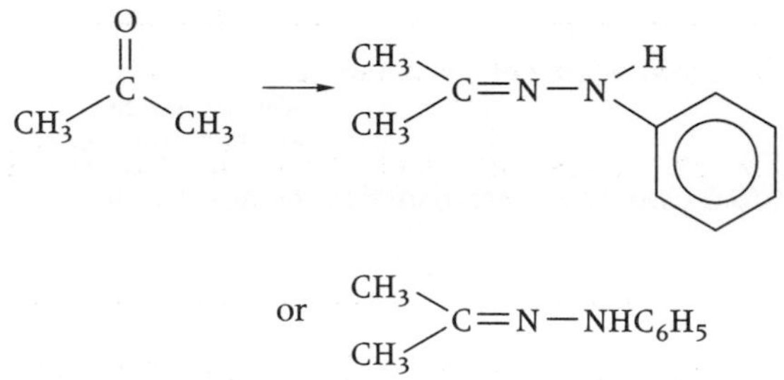

is also used to describe organic reactions in which relatively small molecules other than water are released.) Notice, however, that it is only a special type of nucleophilic attack. If it were not for the lone pair on the nitrogen atoms of hydrazine, the reaction would not take place.

There are several variations on this theme, collected together in Table 118.4. We have used ethanal in the example above, but all aldehydes and ketones give similar reactions.

The reason why these addition reactions are important is that they can be used as a means of identifying aldehydes and ketones. The reagent 2,4-dinitrophenylhydrazine is especially important in this respect. It is usually made up in a solution with methanol together with a little concentrated sulphuric acid, and is orange in colour. When it reacts with an aldehyde or ketone, the product is invariably a solid, usually orange-yellow in colour. After the precipitate is washed and dried it can be tested in a melting point apparatus. The melting points of a large number of 2,4-dinitrophenyl-hydrazones have been tabulated (Table 118.5), so if the measured melting point is compared with those in a

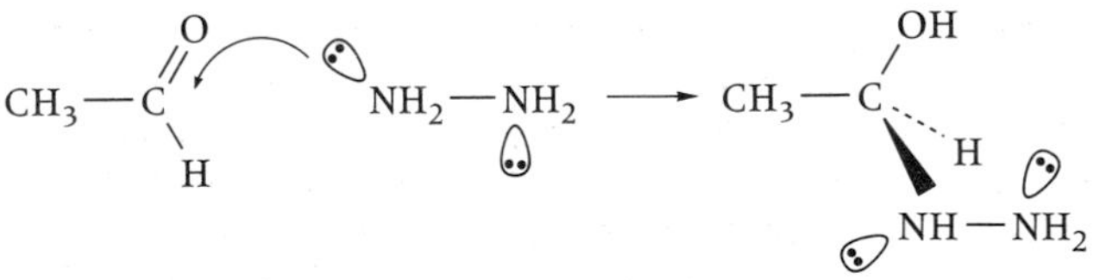

$$CH_3CHO + NH_2-NH_2 \rightarrow CH_3CH(OH)NH-NH_2$$

This molecule has only a fleeting existence. It breaks down to give a molecule called a *hydrazone*:

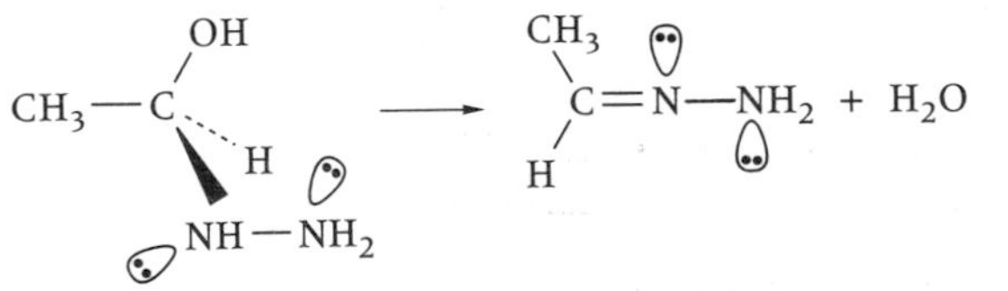

$$CH_3CH(OH)NH-NH_2 \rightarrow CH_3CH=N-NH_2 + H_2O$$

Overall we can summarise the reaction as

ethanal + hydrazine → a hydrazone + water

It is because water is released in the reaction that it is classified as a *condensation* reaction. (However, this term

Table 118.5. The melting points of some 2,4-dinitrophenylhydrazones*

Hydrazone formed with	Melting point/° C
Ethanal	168
Propanone	128
Butanal	123
Butanone	115
Benzaldehyde	237
Diphenylmethanone	238

*The melting points are not always sharp

data book the particular aldehyde or ketone can be identified.

118.11 Hydroxylamine, NH_2OH, gives condensation reactions with aldehydes and ketones. Predict the result of reacting this substance with propanone. Draw the structure of the product.

118.12 If you do an experiment to identify an aldehyde or ketone using 2,4-dinitrophenylhydrazine solution, be careful over three points. First, use the solution sparingly; it is quite common for solid 2,4-dinitrophenylhydrazine to precipitate out. You will not get too far if you mistake this for the hydrazone.

(i) Secondly, why is it important that the precipitate is washed and dried properly?

(ii) Thirdly, do not increase the temperature of the melting point apparatus too rapidly. Why?

118.7 Aldehydes and ketones can be reduced

Aldehydes and ketones can both be reduced. With lithium tetrahydridoaluminate(III), or hydrogen and nickel catalyst, the carbonyl group is converted into an alcohol:

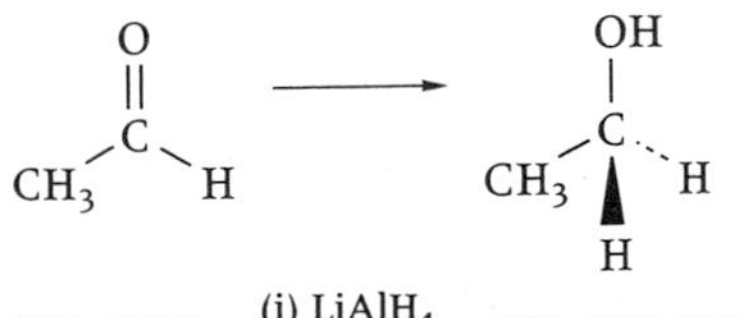

$$CH_3CHO \xrightarrow[\text{(ii) } H^+]{\text{(i) LiAlH}_4} CH_3CH_2OH$$
ethanal → ethanol

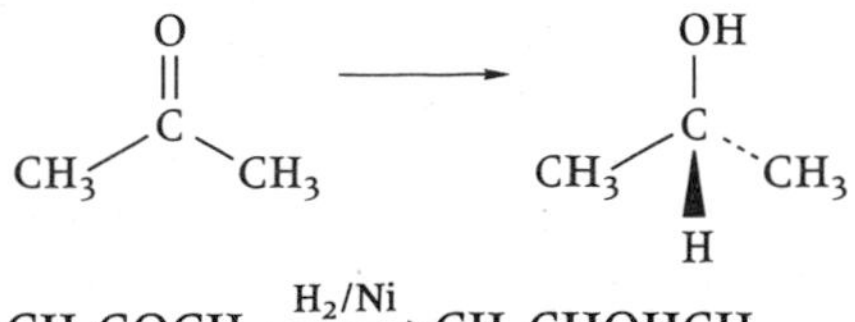

$$CH_3COCH_3 \xrightarrow{H_2/Ni} CH_3CHOHCH_3$$
propanone → propan-2-ol

You should be able to see that the rule is that aldehydes are reduced to primary, and ketones to secondary alcohols.

In other circumstances the reduction can be sufficient to replace the oxygen completely by hydrogen. There are two methods for achieving this change. Both use the condensation reaction with hydrazine as their starting point. Where they differ is in the treatment of the resulting hydrazone.

In *Wolff–Kishner reduction* the hydrazone is heated with an alkali. In the *Clemmensen reduction* the same result is achieved using zinc amalgam and hydrochloric acid. Here is an example of each method:

(i) Wolff–Kishner reduction

$$CH_3COCH_3 \xrightarrow{N_2H_4} (CH_3)_2C{=}N{-}NH_2 \xrightarrow{OH^-} (CH_3)_2CH_2$$
propanone → propane

(ii) Clemmensen reduction

$$C_6H_5CHO \xrightarrow{N_2H_4} C_6H_5CH{=}N{-}NH_2 \xrightarrow{H^+, \text{ Zn/Hg}} C_6H_5CH_3$$
benzaldehyde → methyl-benzene

118.13 What will be made if butanone is reduced using $LiAlH_4$?

118.14 How would you convert butanone into butane?

118.8 Aldehydes are good reducing agents

Aldehydes are good reducing agents, i.e. they can be oxidised easily. We have seen this when we discussed the oxidation of primary alcohols. If the alcohol is *not* refluxed with the oxidising agent, an aldehyde can be collected. However, with reflux, the aldehyde is itself oxidised to a carboxylic acid. Thus, when they act as reducing agents, *aldehydes are oxidised to acids*. For example,

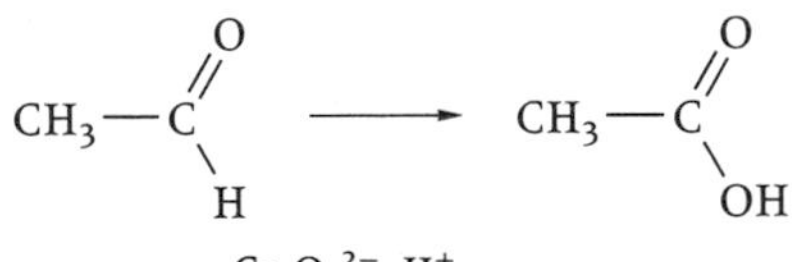

$$CH_3CHO \xrightarrow{Cr_2O_7^{2-}, \ H^+} CH_3COOH$$
ethanal → ethanoic acid

(a) Silver mirror test

When aqueous ammonia is added to silver nitrate solution, the silver ions are converted to diamminesilver(I) ions, $Ag(NH_3)_2^+$. This mixture is sometimes known as ammoniacal silver nitrate. (A similar reagent, 'Tollens' reagent, can also be used.) If an aldehyde is mixed with the solution in a test tube, silver is produced. This can be seen as a mirror on the inside of the test tube; hence this is called the silver mirror test for aldehydes:

$$CH_3CHO(aq) + 2Ag(NH_3)_2^+(aq) + 3OH^-(aq) \rightarrow$$
$$CH_3COO^-(aq) + 2Ag(s) + 4NH_3(aq) + 2H_2O(l)$$

(b) *Fehling's test*

Aldehydes also have sufficient reducing ability to convert copper(II) to copper(I). Usually the copper(II) is held as a complex ion in an alkaline solution of copper(II) tartrate known as Fehling's solution. The solution is a clear royal blue, but when it is warmed with an aldehyde an orange-yellow precipitate of copper(I) oxide is formed.

Ethanal also gives the iodoform test – see the next section.

118.15 In the reaction of ethanal with Fehling's solution, assume that the reaction involves copper(II) ions and hydroxide ions. Write down the equation for the reaction. (The formula of copper(I) oxide is Cu_2O).

118.16 In the equation for the reaction of Tollens' reagent with aldehydes, how do we know that a redox reaction has taken place?

118.9 Ketones are hard to oxidise

Most ketones are difficult to oxidise. If they are refluxed for long enough with vigorous oxidising agents like alkaline potassium manganate(VII) or concentrated nitric acid, they will split apart near the carbonyl group. Carboxylic acids are the products. For example, the ketone $CH_3CH_2COCH_2CH_2CH_3$ gives CH_3CH_2COOH and $CH_3CH_2CH_2COOH$, but other acids are also produced; so no one would use this reaction unless they were desperate.

One oxidation reaction of ketones that is useful is the *iodoform reaction*. This is given by methylketones such as propanone and butanone.

You could try out the reaction by warming propanone with a solution of iodine in sodium hydroxide. This solution contains the iodate(I) ion, IO^-, an oxidising agent. You should see a yellow precipitate of triiodomethane (iodoform) appear:

$$CH_3COCH_3(aq) + 3IO^-(aq) \rightarrow$$
$$CH_3COO^-(aq) + CHI_3(s) + 2OH^-(aq)$$
yellow

Be careful that you realise that this is *not* a test for ketones. The test will work with any molecule that has the group

$$CH_3 - \underset{X}{\overset{O}{\underset{\|}{C}}}$$

or can be oxidised to give this structure. Ethanal will also give the test, and so will ethanol because it is oxidised to ethanal by the iodate(I) ion.

118.17 Which of the following aldehydes or ketones would react positively in the iodoform reaction: (i) butanone; (ii) butanal; (iii) pentan-3-one; (iv) pentan-2-one; (v) benzaldehyde?

118.10 Reactions with halogens

If iodine solution is added to an aldehyde or ketone, the colour of the solution fades. A hydrogen atom on the carbon atom next door to the carbonyl group is replaced by iodine. A typical reaction is

$$CH_3 - \overset{O}{\overset{\|}{C}} \diagdown CH_3 + I_2 \longrightarrow CH_3 - \overset{O}{\overset{\|}{C}} \diagdown CH_2I + HI$$

$$CH_3COCH_3 + I_2 \rightarrow CH_3COCH_2I + HI$$

However, this equation hides the way that the reaction takes place. In the units on kinetics we found that the iodine does not react directly with the propanone. Rather it reacts with the *enol* form of the ketone, which exists in equilibrium with the *keto* form. This equilibrium is known as the *keto–enol tautomerism*:

$$\underset{CH_3}{\overset{O}{\overset{\|}{C}}} \diagdown CH_3 \rightleftharpoons \overset{OH}{\underset{CH_3}{C}} = \overset{H}{\underset{H}{C}}$$

$$CH_3COCH_3 \rightleftharpoons CH_3COH=CH_2$$
keto form   enol form

The iodine takes part in a reaction with the double bond. You will find further details about the reaction in Unit 81. The fact that hydrogen atoms on the carbon atom adjacent to the carbonyl group (the alpha, α, hydrogen atoms) can be replaced by halogens is also shown by the reaction of ethanal with chlorine. Substitution takes place in the presence of ultraviolet light:

$$\underset{CH_3}{\overset{O}{\overset{\|}{C}}} \diagup H + 3Cl_2 \longrightarrow \underset{CCl_3}{\overset{O}{\overset{\|}{C}}} \diagup H + 3HCl$$

$$CH_3CHO + 3Cl_2 \rightarrow CCl_3CHO + 3HCl$$
trichloroethanal

The product, trichloroethanal, is also known as chloral. When it is heated with chlorobenzene and concentrated sulphuric acid, dichlorodiphenyltrichloroethane, which is better known as DDT, is made:

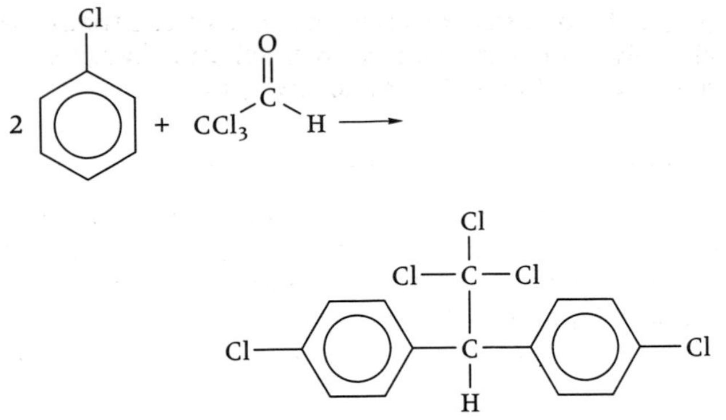

$$2C_6H_5Cl + CCl_3CHO \rightarrow DDT$$

DDT is one of the most effective insecticides that has ever been invented. It was first made in 1873, but its use only became widespread during the Second World War. Since then it must have prevented the premature deaths of millions of people owing to its ability to kill fleas, lice and mosquitoes, which are carriers of diseases such as typhus and malaria. DDT is a remarkably stable molecule (both energetically and kinetically). Once, this was thought to be one of its greatest virtues; but since the 1960s it has become clear that its longevity makes DDT a great environmental hazard. Essentially, the reason is that DDT finds its way into the food chain of many animals, including humans (Table 118.6). For example, it weakens the shells of birds' eggs and prevents successful breeding. In many parts of the world its use has been banned, although it is still used in some countries because it is one of the few cheap insecticides available.

Table 118.6. How DDT is passed along a food chain to peregrine falcons around Lake Michigan, Canada*

	DDT concentration /parts per million (of body weight)
Lake mud	0.014
Lake water	0.000 02
Amphipods eaten by fish	0.410
Trout	6
Herring gulls	99
Peregrine falcons	5 000

*Table adapted from: Heaton, C. A. (ed.) (1986). *The Chemical Industry*, Blackie, Glasgow, p. 270

118.11 Reaction with phosphorus pentachloride

You should know that phosphorus pentachloride can be used to test for the presence of OH groups. Although aldehydes and ketones do not contain this group, they still react but in a different way. The oxygen atom is replaced by two chlorine atoms:

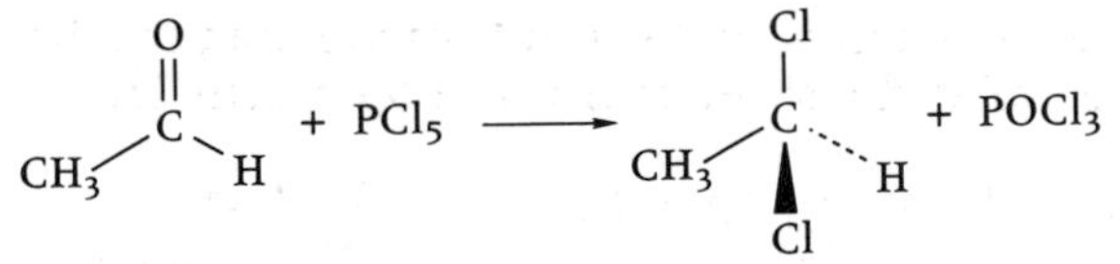

$$CH_3CHO + PCl_5 \rightarrow CH_3CHCl_2 + POCl_3$$

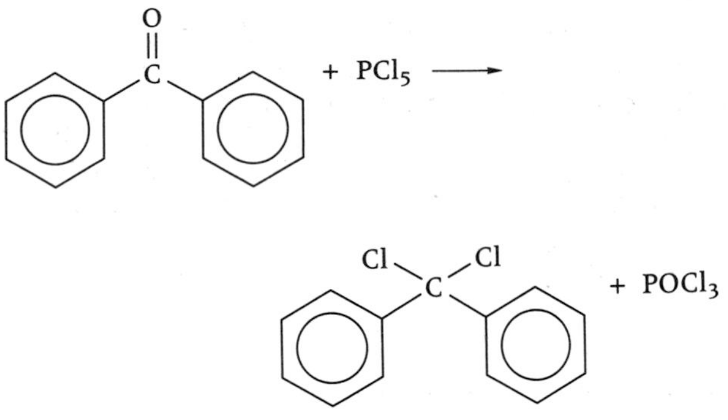

$$C_6H_5COC_6H_5 + PCl_5 \rightarrow C_6H_5CCl_2C_6H_5 + POCl_3$$

Notice that there are no fumes of hydrogen chloride given off.

> **118.18** A student wrote the following: 'The organic liquid became warm when it reacted with phosphorus pentachloride. The reaction clearly showed that the liquid was an alcohol.' What do you think of this statement?

118.12 Polymerisation reactions of aldehydes

Methanal and ethanal can be persuaded to polymerise. Poly(methanal) (sometimes called paraformaldehyde) is a polymer of methanal. It can be made by evaporating an aqueous solution of methanal, which leaves it as a white powder. It has long chains of repeating CH_2O units, like this: $-CH_2-O-CH_2-O-CH_2-O-$.

In the presence of acid and at low temperatures, methanal molecules will combine to make a trimer, which is simply known as the methanal trimer (or metaformaldehyde):

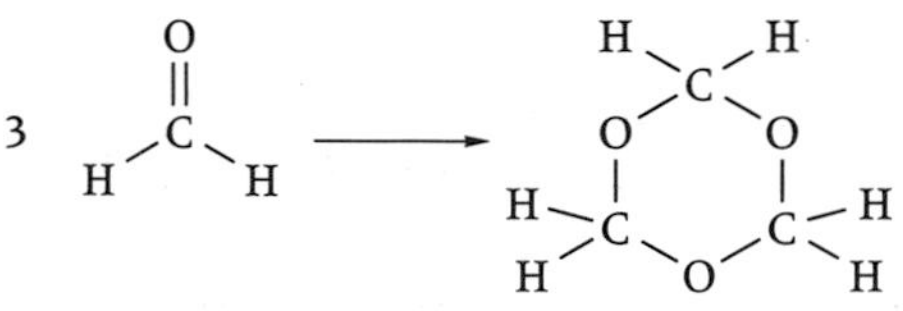

methanal trimer

A very little concentrated sulphuric acid added to ethanal will convert it to a liquid trimer consisting of three ethanal molecules combined in a ring-like structure:

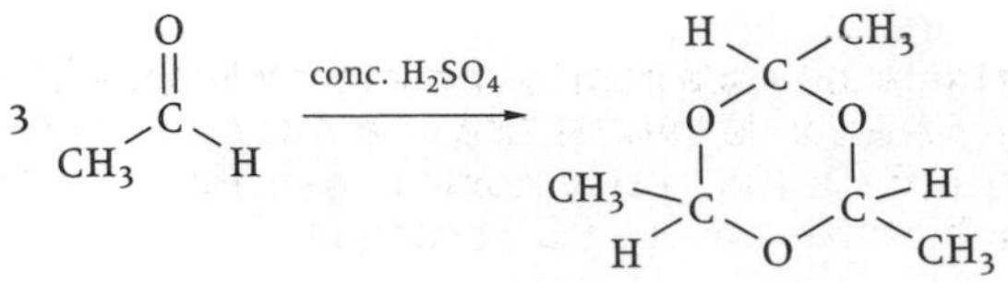

ethanal trimer (paraldehyde)

It is possible that you have heard of a substance called 'meta'. This is the shortened name of metaldehyde, which should really be called the ethanal tetramer. It is a white solid made by passing hydrogen chloride into cold ethanal. It is used as a fuel, usually for steam powered models, and as the active ingredient of slug pellets.

118.19 Predict and draw the structure of the ethanal tetramer.

118.13 The aldol condensation

The aldol condensation is a reaction that aldehydes undergo. The simplest example, from which the reaction gets its name, is between two ethanal molecules in the presence of hydroxide ions. Overall we can write the reaction like this:

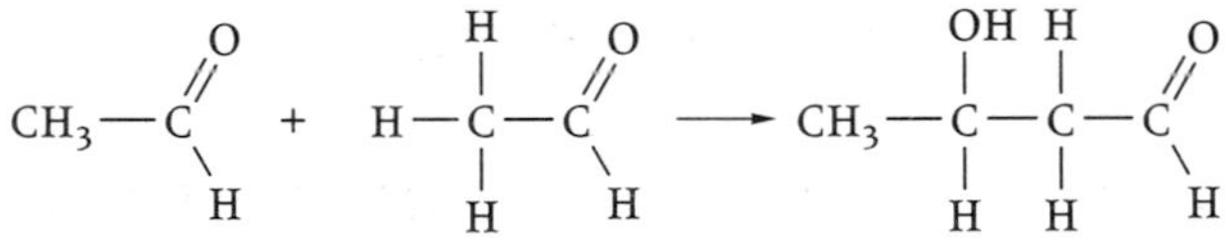

$$2CH_3CHO \xrightarrow{OH^-} CH_3CH(OH)CH_2CHO$$

ethanal   aldol, i.e. 3-hydroxybutanal

However, this hides from us the mechanism of the reaction. In section 118.9 we noticed that the α hydrogen atoms (those on the carbon atom next to the carbonyl group) are more liable to be lost from the molecule than the hydrogen atoms in, say, an alkane. We can say that they are *more acidic* than normal. With hydroxide ions, an equilibrium is set up:

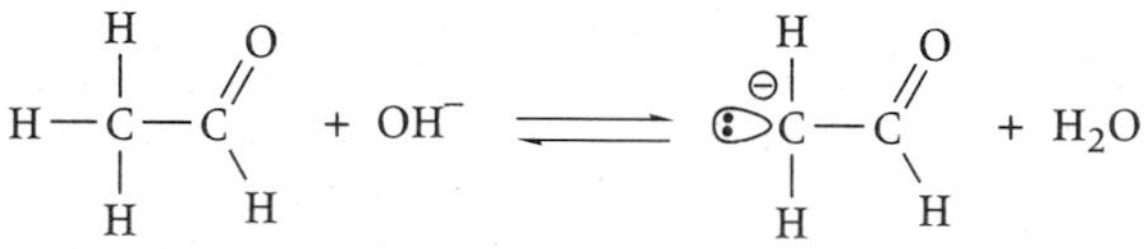

$$CH_3CHO + OH^- \rightleftharpoons CH_2CHO^- + H_2O$$

The negative ion (a carbanion) is a nucleophile, and, like the other nucleophiles we met in section 118.4, it will attack the carbonyl group on another ethanal molecule:

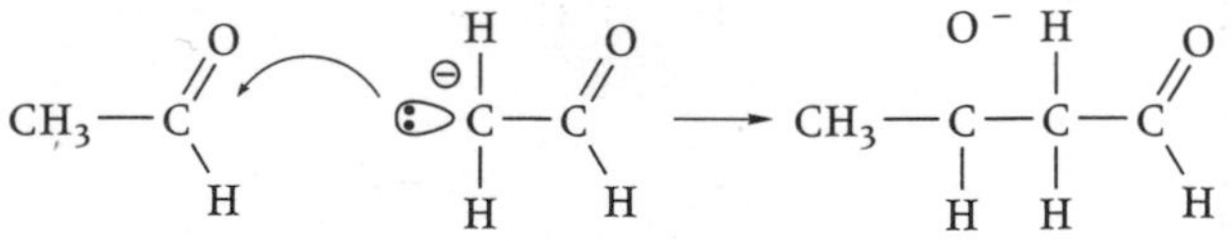

$$CH_3CHO + {}^-CH_2CHO \rightarrow CH_3CH(O^-)CH_2CHO$$

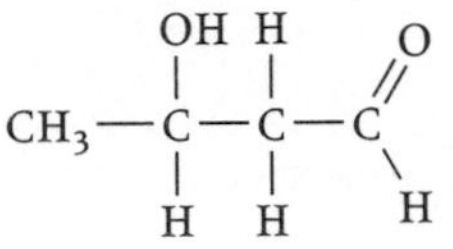

$$CH_3CH(O^-)CH_2CHO + H_2O \rightarrow$$
$$CH_3CH(OH)CH_2CHO + OH^-$$

This reaction happens with other aldehydes. It also takes place with ketones provided that they have α hydrogen atoms. Thus, propanone will give a similar reaction (but with difficulty):

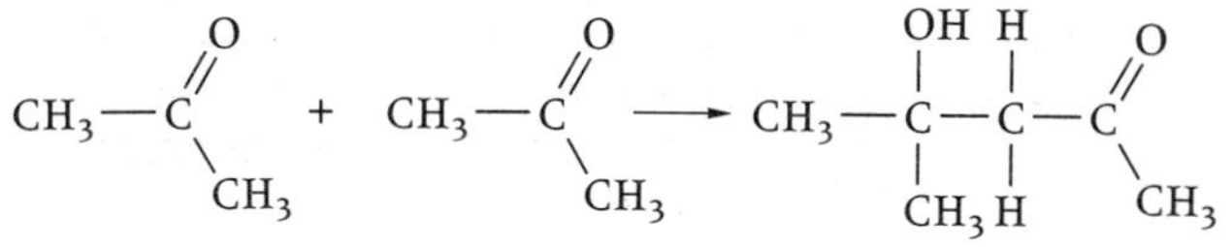

$$2CH_3COCH_3 \rightarrow CH_3CCH_3OHCH_2COCH_3$$

118.14 Cannizzaro's reaction

This is the reaction in which aldehydes without α hydrogen atoms can regain their reputation. Methanal shows the pattern. It is converted by alkali into a mixture of methanol and the methanoate ion:

$$2HCHO + OH^- \rightarrow CH_3OH + HCOO^-$$

The mechanism of this reaction would take us too far afield, but the pattern should be clear to you. For example, benzaldehyde gives a mixture of phenylmethanol and the benzoate ion:

$$2C_6H_5CHO + OH^- \rightarrow C_6H_5CH_2OH + C_6H_5COO^-$$

Answers

118.1 If the carbonyl group were on one of the two end carbon atoms, the molecule would be an aldehyde, so it must be on one of the two carbon atoms in the middle of the molecule. It does not matter at which end you start numbering, the carbonyl group *must* be on the second carbon atom. As there is no other possible place, there is no need to put in the number.

118.2 (i)

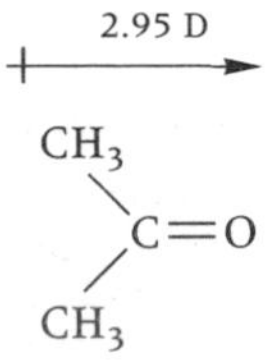

Answers – cont.

(ii) The attractions between dipole moments on different molecules are sufficiently strong to keep the molecules together as a liquid (see Unit 20).

118.3

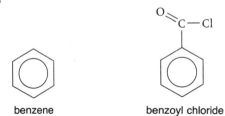

118.4 (i) 2-Methylpropan-1-ol:

$$CH_3 - \overset{\overset{\displaystyle CH_3}{|}}{\underset{\underset{\displaystyle H}{|}}{C}} - CH_2 - OH$$

(ii) Pentan-2-ol:

$$CH_3 - CH_2 - CH_2 - \overset{\overset{\displaystyle OH}{|}}{\underset{\underset{\displaystyle H}{|}}{C}} - CH_3$$

118.5 The reaction would be between propan-1-ol and sodium dichromate(VI) acidified with concentrated sulphuric acid. The mixture should be heated, with the aldehyde vapour being condensed using an apparatus like that of Figure 115.2.

118.6 (i), (ii) *Cis* and *trans* isomers of the following alkene are possible:

$$\underset{\displaystyle CH_3}{\overset{\displaystyle C_3H_7}{}}C=C\underset{\displaystyle CH_3}{\overset{\displaystyle C_2H_5}{}}$$

$$\underset{\displaystyle C_3H_7}{\overset{\displaystyle CH_3}{}}C=C\underset{\displaystyle CH_3}{\overset{\displaystyle C_2H_5}{}}$$

cis-3,4-dimethylhept-3-ene   *trans*-3,4-dimethylhept-3-ene

118.7 (i) Phenylmethanol, $C_6H_5CH_2OH$.

(ii) The catalyst has been *poisoned*.

118.8

benzene   benzoyl chloride

118.9

$$C_2H_5 - \overset{\overset{\displaystyle OH}{|}}{\underset{\underset{\displaystyle CH_3}{}}{C}} \cdots H$$

118.10 (i) It has an asymmetric carbon atom.

(ii) The cyanide ions can attack the aldehyde from above or below the plane of the molecule. The products of the two reactions are mirror images of one another.

Hence the acids made from them are mirror images. The resulting solution is inactive because it contains equal amounts of the two optical isomers. See Unit 110 for more information about optical activity.

$$\underset{\displaystyle H}{\overset{\displaystyle COOH}{\underset{\displaystyle CH_3}{}}C}\,OH \qquad HO\,\underset{\displaystyle H}{\overset{\displaystyle COOH}{}}C\,CH_3$$

118.11 The product of the reaction of hydroxylamine and an aldehyde or ketone is called an oxime:

$$\rangle C = N - OH$$

118.12 (i) If there is an impurity present, the melting point of the hydrazone will be lowered, so a faulty identification can occur.

(ii) The temperature of the apparatus may increase faster than the thermometer can respond. For example, the thermometer may read 110°C when the sample is really at 120°C. You would report a melting point 10°C lower than it should be.

118.13 Butan-2-ol:

$$CH_3 - CH_2 - \overset{\overset{\displaystyle OH}{|}}{\underset{\underset{\displaystyle H}{|}}{C}} - CH_3$$

118.14 Reduce it by passing the vapour together with hydrogen over a pure palladium catalyst.

118.15 $CH_3CHO + 2Cu^{2+} + 5OH \rightarrow$
$$CH_3COO^- + Cu_2O + 3H_2O$$

118.16 The silver ion, Ag^+, gains an electron when it changes to a silver atom. Hence it must have been reduced. Likewise, the aldehyde has been oxidised.

118.17 Only butanone and pentan-2-one, because they are the only ones that have a CH_3 group next to the carbonyl group.

118.18 The student's report is not clear. Alcohols, acids, aldehydes and ketones react with phosphorus pentachloride. Only alcohols and acids give white fumes of hydrogen chloride (because they have an OH group). If the student saw white fumes, the liquid *might* have been an alcohol; but he or she could not be sure it was not an acid without doing further tests. If there were no fumes, the liquid was an aldehyde or ketone.

118.19

- Aldehydes and ketones:
    (i) Contain the carbonyl group;
    aldehydes

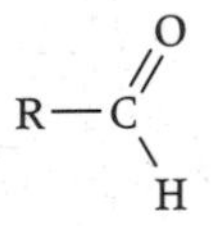

$$R-\overset{\displaystyle O}{\underset{\displaystyle H}{C}}$$

    ketones

$$R-\overset{\displaystyle O}{\underset{\displaystyle R'}{C}}$$

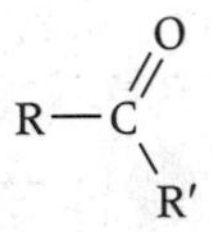

    (ii) Undergo addition reactions.
- Manufacture of ethanal:
    The Wacker process;

$$C_2H_4 + H_2O + PdCl_2 \rightarrow CH_3CHO + Pd + 2HCl$$

    with the palladium being recycled using copper(II) chloride and oxygen.
- Preparation:
    (i) Oxidation of alcohols
    By acidified solution of sodium dichromate(VI) (no reflux). By passing over hot copper powder. Primary alcohols are oxidised to aldehydes. Secondary alcohols are oxidised to ketones.

    e.g. $C_6H_5CH_2OH \rightarrow C_6H_5CHO$
    phenylmethanol   benzaldehyde
    $(CH_3)_2CHOH \rightarrow (CH_3)_2CO$
    propan-2-ol   propanone

    (ii) Benzaldehyde also made by hydrolysis of dichloromethylbenzene;

$$C_6H_5CHCl_2 \xrightarrow{H_2O} C_6H_5CHO$$

    (iii) Aromatic ketones made by Friedel–Crafts acylation;
    e.g.

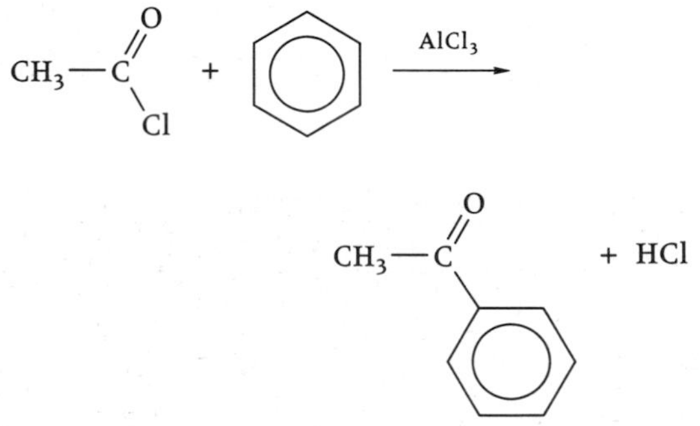

Reactions

- Addition and condensation:
    See Tables 118.3 and 118.4. Especially note
    (i) Addition of cyanide ions;
    e.g. $CH_3CHO \xrightarrow{CN^-, H^+} CH_3CH(OH)CN \xrightarrow{H^+} CH_3CH(OH)COOH$

    (ii) Reaction with 2,4-dinitrophenylhydrazine used to characterise aldehydes and ketones.
- Reduction:
    Using lithium tetrahydridoaluminate(III);

    aldehydes → primary alcohols
    ketones → secondary alcohols
- Oxidation:
    Using warm acidified sodium dichromate(VI) under reflux;

    aldehydes → acids
    e.g. $CH_3CHO \rightarrow CH_3COOH$
- Phosphorus pentachloride:
    Carbonyl oxygen replaced by chlorine atoms (no fumes of HCl);

    e.g. $CH_3CHO + PCl_5 \rightarrow CH_3CHCl_2 + POCl_3$
- Polymerisation of ethanal:
    With concentrated sulphuric acid: ethanal trimer.
    With hydrogen chloride: ethanal tetramer.
- Propanone and iodine:
    Iodine decolourised. Reaction involves the keto–enol tautomerism;

$$\overset{\displaystyle O}{\underset{\displaystyle \parallel}{CH_3C}}-CH_3 \rightleftharpoons CH_3\overset{\displaystyle OH}{\underset{\displaystyle \vert}{C}}=CH_2$$

- Iodoform reaction:
    A molecule with the group

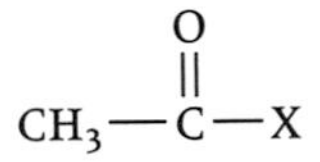

$$CH_3-\overset{\displaystyle O}{\overset{\displaystyle \parallel}{C}}-X$$

    or one that can be oxidised to this group gives a yellow precipitate of iodoform, CHI_3, when warmed with iodine in sodium hydroxide solution.
- Tests for aldehydes:
    (i) Ammoniacal silver nitrate solution → silver mirror.
    (ii) Orange precipitate with Fehling's solution.

Right. Clearing of tropical rain forest may have a marked effect on the world's climate. See Unit 91. The three photographs show parts of the rain forest in Malaysia before, during and after clearing.

The solutions below show the attractive colours of many transition metal compounds. From the left the colours are due to chromium(III) (in $K_2Cr_2O_7(aq)$), manganese(VII) (in $KMnO_4(aq)$, cobalt(III) (in $CoCl_3(aq)$), nickel(II) (in $NiSO_4(aq)$), and copper(II) (in $CuSO_4(aq)$). See Unit 105 (and Unit 25 in Volume 1).

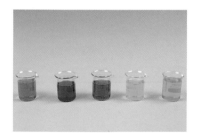

Top left. Samples of ruby (red) and sapphire (blue) crystals. Ruby contains small amounts of chromium ions, sapphire contains iron, cobalt or titanium ions. Large ruby crystals are artificially grown in order to make lasers. Both minerals are based on a corundum structure. See Unit 94.

Middle right. A sample of the mineral Blue John, an elegant variety of the mineral fluorspar, CaF_2. See Unit 93.

Middle left and bottom left and right. Quartz, opal and amethyst are based on the same formula SiO_2 and are framework silicates. The colours of opal and amethyst are due to small quantities of impurities such as manganese. See Unit 96 for more about silicates.

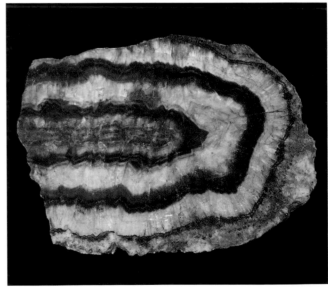

Middle left. Carbon dioxide fire extinguishers produce clouds of cold, dense vapour in use. See Unit 95.

Top right. The characteristic haze of photochemical smog covering Los Angeles, USA. See Unit 97.

Bottom right. The excess growth of algae (an algal bloom) like that shown here on the Basingstoke canal in the UK has been encouraged by excess nitrate in water. See Unit 97.

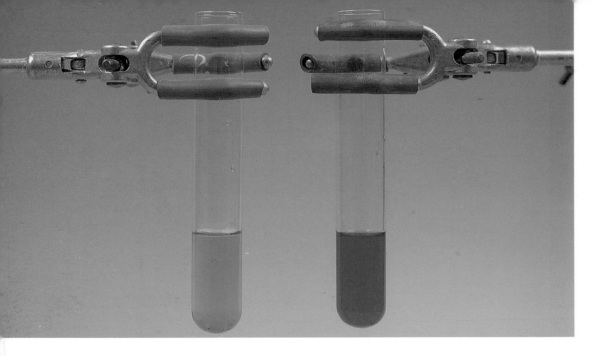

Top. These tubes show the result of adding 1,1,1-trichloroethane to a solution of bromine in water (left) and iodine in water (right). This is a good way of testing for bromine and iodine. See Unit 101.

Middle and bottom left. The models in the upper photograph show the shape of the mirror image forms of lactic acid. The lower photograph is a computer image of the same pair of isomers. The computer has drawn the molecules in a ball-and-stick fashion. The hydrogen atoms are not shown. See Unit 110 for information about mirror images.

Bottom right. Sulphur at the bottom of the tube has been warmed and a few drops of fuming nitric acid have been added. The brown fumes of nitrogen dioxide above the sulphur are a sure sign that the sulphur has been oxidised by the acid. See Unit 97.

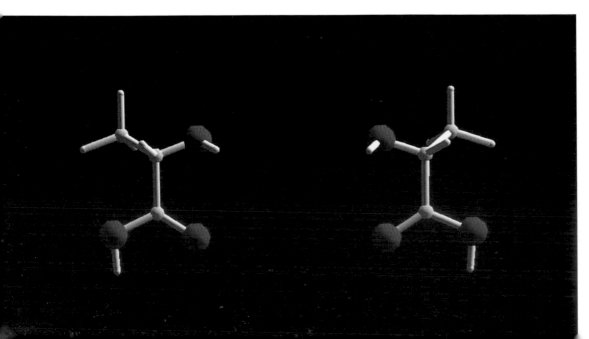

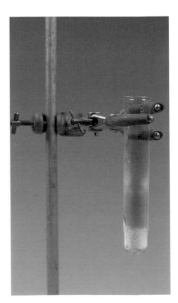

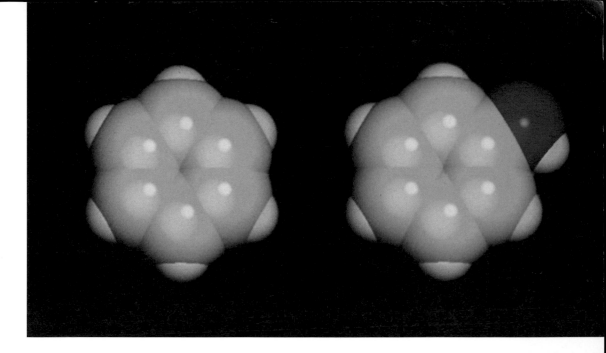

Top. These images drawn by computer represent space-filling models of benzene C_6H_6 (left) and phenol C_6H_5OH (right). They give a good idea of the relative amounts of space taken up by the carbon and hydrogen atoms. See Unit 117 for information about phenol.

Middle. Neon lights in Piccadilly circus.

Bottom left. Brightly coloured clothes are partly the result of research in organic chemistry. See Unit 122.

Bottom right. The results of a successful silver mirror test (left) and a Fehling's test (right) for an aldehyde. See Unit 118.

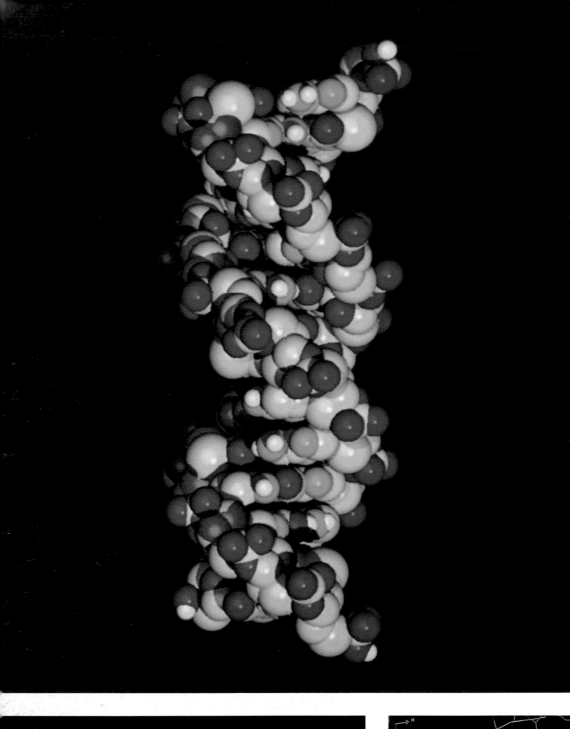

Left. A computer-drawn representation of a fragment of DNA. You should be able to see the two strands of the helix winding around each other. See Unit 124 for information about DNA.

Bottom left. This computer-drawn image represents the structure of an enzyme which happens to have two active sites. The blue, yellow, and red regions are the two substrate molecules which fit in the active sites. The bonds between atoms of the enzyme are shown by short lines. Red lines indicate carbon–oxygen bonds; blue lines represent carbon–nitrogen bonds.

Bottom right. The third image is a magnified view of the region around the substrate. The substrate is held in the active site by hydrogen bonds. If you look carefully at the arrangement of the bonds you should be able to spot several peptide links (—CONH—). See Unit 123 for information about peptide links and enzymes.

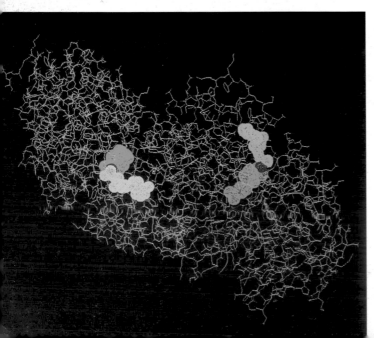

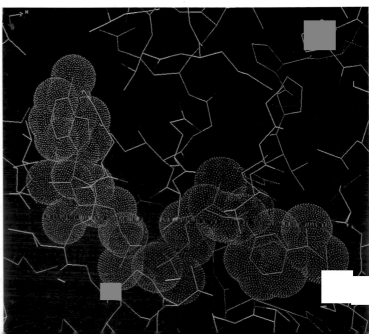

Top. You might like to count the number of ways in which these climbers rely on nylon for their survival. See Unit 127.
Bottom. Although most detergents are biodegradable, some can give precipitates with metal ions, producing a scum like this on a river in the Peak District, UK. See Unit 129.

Foam fire fighting, emulsion paints, and mayonnaise make use of surfactants to preserve their colloidal nature. See Unit 129 (and Unit 23 in Volume 1).

119

Carboxylic acids

119.1 Most carboxylic acids are weak acids

The majority of non-aromatic carboxylic acids have the general formula $C_nH_{2n+1}COOH$. Usually they are weak acids. The most common one is ethanoic acid (acetic acid), CH_3COOH, which is the key ingredient of vinegar. It is the COOH group in the molecule that is responsible for its acidity. In water the molecule will dissociate into ions:

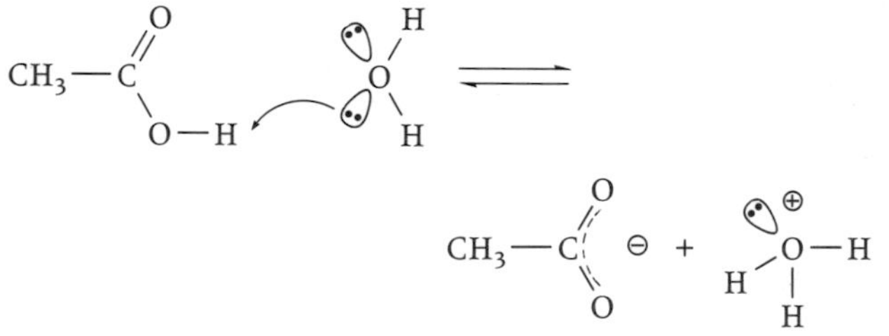

$$CH_3COOH(l) + H_2O(l) \rightleftharpoons CH_3COO^-(aq) + H_3O^+(aq)$$
$$\text{ethanoate ion}$$

The ethanoate ion is energetically stabilised by the spreading of the negative charge over a carbon and two oxygen atoms:

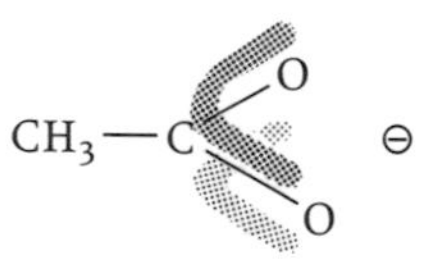

In Unit 75 we said that ethanoic acid is only slightly dissociated into ions; that is, the equilibrium lies far over to the left-hand side of the equation. This is indicated by the value of its acid dissociation constant, K_a, or its pK_a value. In Table 119.1 you will find information about some carboxylic acids and their pK_a values. The pK_a of an acid is defined as $-\lg K_a$. This means that the smaller the size of K_a, the larger the value of pK_a, and vice versa. Essentially, the smaller the value of pK_a, the stronger is the acid.

Two models showing the shapes of methanoic acid and ethanoic acid.

Table 119.1. Some carboxylic acids

Name	Formula	Structure	Melting point/°C	Boiling point/°C	pK$_a$
Methanoic acid	HCOOH	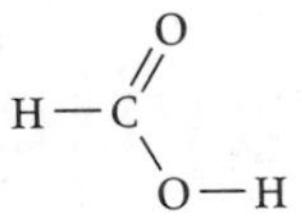	8.6	100.7	3.8
Ethanoic acid	CH$_3$COOH	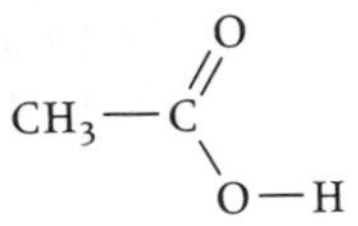	16.8	118.0	4.8
Propanoic acid	C$_2$H$_5$COOH	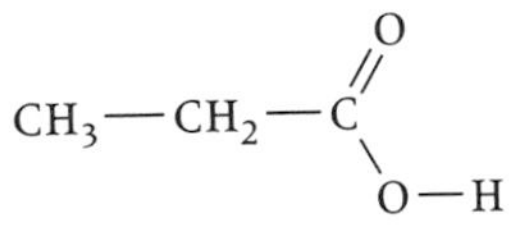	−20.7	141.1	4.9
Chloroethanoic acid	CH$_2$ClCOOH	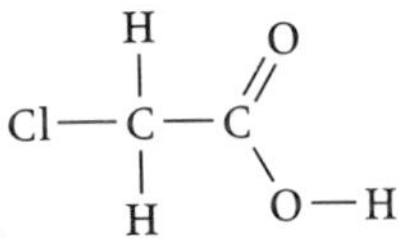	63.2	189.7	2.9
Dichloroethanoic acid	CHCl$_2$COOH	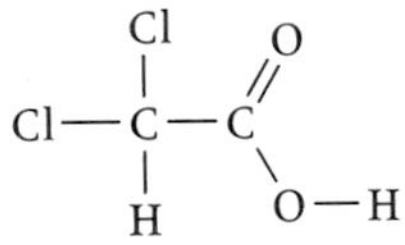	11.0	192.7	1.3
Trichloroethanoic acid	CCl$_3$COOH	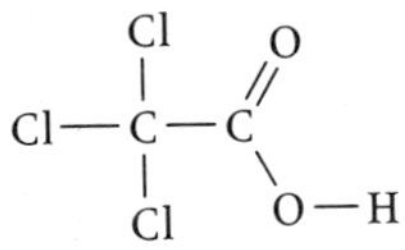	56.5	197.7	0.7
Benzoic acid (benzenecarboxylic acid)	C$_6$H$_5$COOH	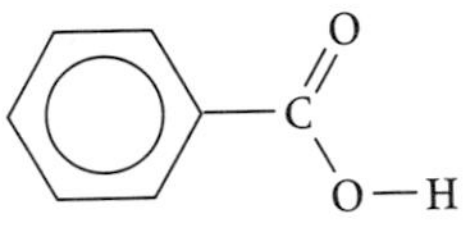	121.9	249.2	4.2
Phenylethanoic acid	C$_6$H$_5$CH$_2$COOH	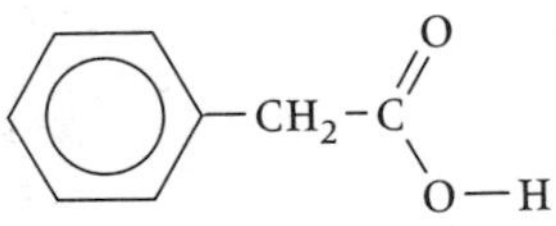	77	266	4.3

119.2 Ways of making carboxylic acids

Some of the methods of making carboxylic acids in the laboratory we have met before; but one that is a new reaction uses *nitriles*. This is an important process, which you should do your best to remember.

(a) *Hydrolysis of nitriles*

A nitrile contains a CN group. Examples are ethanenitrile, CH$_3$CN, and propanenitrile, C$_2$H$_5$CN. Often they are liquids with a sweet smell. If ethanenitrile is refluxed with an alkali, the following reaction takes place:

$$CH_3-C\equiv N \ + \ H_2O \ + \ OH^- \longrightarrow CH_3-C\overset{O}{\underset{O}{\overset{\|}{|}}}{}^\ominus + NH_3$$

$$CH_3CN(aq) + H_2O(l) + OH^-(aq) \rightarrow$$

ethanenitrile

$$CH_3COO^-(aq) + NH_3(g)$$

ethanoate ion

Once the reaction ceases, acid can be added. This protonates the ethanoate ions and ethanoic acid molecules are produced:

$$CH_3-C\overset{O}{\underset{O}{\overset{\|}{|}}}{}^\ominus + H^+ \longrightarrow CH_3-C\overset{O}{\underset{\ddot{O}-H}{\overset{\|}{}}}$$

$$CH_3COO^-(aq) + H^+(aq) \rightarrow CH_3COOH(aq)$$

You will discover why nitriles are so important in Unit 120.

(b) Oxidation of primary alcohols

You should find this a familiar reaction by now. The usual oxidising agent is sodium dichromate(VI) solution. This is refluxed with a primary alcohol that has the same number of carbon atoms as the acid you want to make. For example,

$$CH_3CH_2OH \rightarrow CH_3COOH$$

ethanol   ethanoic acid

(c) Oxidation of an arene

We have seen that one of the reactions that all arenes have in common is that, irrespective of the length of the side chain, it will be oxidised down to an acid group. In this case the oxidising agent is usually an acid or alkaline solution of potassium manganate(VII). For example,

ethylbenzene   benzoic acid

(d) Grignard reagents

We can write a general carboxylic acid as RCOOH, where R might be CH_3, C_2H_5, etc. To make the acid, a Grignard reagent of formula RMgI would be needed. It can be made by mixing magnesium powder with the iodoalkane, RI, dissolved in dry ether. Finally, with RMgI dissolved in ether, dry carbon dioxide should be bubbled through the solution. (Alternatively, solid carbon dioxide could be used.) Once the initial reaction

subsides, the addition of acid liberates the carboxylic acid. We have,

$$RMgI + CO_2 \rightarrow RCOOMgI \xrightarrow{H^+} RCOOH + Mg^{2+} + I^-$$

For example, C_3H_7MgI would give us propanoic acid, C_3H_7COOH.

119.1 When it is completely free of moisture, ethanoic acid is known as *glacial* ethanoic acid. At different times of the year, students have been known to say that glacial ethanoic acid is a solid; at others, that it is a liquid. Explain.

119.2 (i) In the preparation of a carboxylic acid from a nitrile using alkaline hydrolysis, how would you know that the reaction is complete?

(ii) It is possible to use acid in the hydrolysis reaction rather than alkali. A similar reaction takes place, but very little ammonia is given off. Why?

119.3 If you were to attempt to make methanoic acid, HCOOH, which nitrile would you have to use. Why would you be wise to avoid trying this reaction?

119.4 Sketch the apparatus you would use to oxidise ethanol to ethanoic acid.

119.5 Use the values of standard redox potentials in Table 70.1 to compare the oxidising ability of dichromate(VI) ions and manganate(VII) ions in acid solution, and manganate(VII) ions in alkaline solution.

119.6 Suggest ways of carrying out these changes:

(i) $C_6H_5CH_3$ to C_6H_5COOH;

(ii) C_2H_5Br to C_2H_5COOH;

(iii) $CH_3C_6H_4I$ to $CH_3C_6H_4COOH$.

(There are two ways of doing (ii) and (iii).)

119.3 The reactions of carboxylic acids

The carboxylic acid group is a combination of two smaller groups, one of which, the OH group, we already know something about. The other is the carbonyl group

$$\overset{}{\underset{}{>}}C=O$$

The reactions of the OH group in a carboxylic acid are often similar to those of the OH group in an alcohol. Even the acidity of a carboxylic acid is really an extension of a property of alcohols. They too can be classified as acids, but they are so weak that we often choose to ignore this feature of their chemistry.

However, carboxylic acids show few of the reactions that we met in the previous unit on aldehydes and ketones. Especially, carboxylic acids do *not* undergo addition reactions. Essentially, the reason for this is that nucleophiles that will attack aldehydes or ketones tend to act as bases in the presence of carboxylic acids, i.e. they are protonated by the hydrogen of the acid OH group. At the same time, owing to its negative charge, the $RCOO^-$ ion left is no longer attractive to nucleophiles. Alcohols represent a class of nucleophile that are not deactivated by acids; see Unit 120.

In the next few paragraphs we shall split the reactions of carboxylic acids into those which mainly affect the OH group and those which affect the carbonyl group.

(a)   Reactions of the OH group

Carboxylic acids are neutralised by alkalis

Owing to their weakness, titrations of carboxylic acids and alkalis like sodium hydroxide are not easy to perform accurately. However, with the right choice of indicator, the endpoint of a reaction like

$$CH_3COOH(aq) + OH^-(aq) \rightarrow CH_3COO^-(aq) + H_2O(l)$$

can be found.

With carbonates, or hydrogencarbonates, aqueous solutions of the acids generate sufficient hydrogen ions to give off carbon dioxide, e.g.

$$CO_3{}^{2-}(aq) + CH_3COOH(aq) \rightarrow$$
$$CH_3COO^-(aq) + H_2O(l) + CO_2(g)$$

Benzoic acid, C_6H_5COOH, is also a weak acid, but it is much less soluble in water than ethanoic acid. If it is made in an aqueous solution, it will appear as a white precipitate. However, it will dissolve in alkali because the negative benzoate ion is much more soluble in water. Likewise, a solution of a benzoate will precipitate benzoic acid crystals if a strong acid, e.g. sulphuric acid, is added:

$$C_6H_5COOH \underset{acid}{\overset{alkali}{\rightleftharpoons}} C_6H_5COO^-$$
white solid   colourless in solution

With sodium

Carboxylic acids give off hydrogen when reacted with sodium. For example,

$$2CH_3COOH(l) + 2Na(s) \rightarrow 2CH_3COO^-Na^+(s) + H_2(g)$$
sodium ethanoate

Sodium ethanoate is an ionic solid, like the sodium salts of inorganic acids.

With phosphorus pentachloride

The OH group is replaced by a chlorine atom. For example,

$$CH_3COOH(l) + PCl_5(s) \rightarrow$$
$$CH_3COCl(l) + POCl_3(s) + HCl(g)$$

The tell-tale signs of hydrogen chloride are also produced with sulphur dichloride oxide, $SOCl_2$.

Esters are made with alcohols and acid chlorides

Esterification reactions between alcohols and carboxylic acids have been studied in great detail. Especially, the reaction between ethanol and ethanoic acid has been investigated as an example of an equilibrium reaction (see section 54.2):

$$CH_3COOH(l) + C_2H_5OH(l) \rightleftharpoons CH_3COOC_2H_5(l) + H_2O(l)$$
ethyl ethanoate

The combination of ethanol with ethanoic acid is slow compared to its reaction with ethanoyl chloride, CH_3COCl. This substance is a liquid that fumes in air, but it can be carefully dripped onto ethanol. It reacts with a 'fizz' and white fumes of hydrogen chloride are given off. The ester is made immediately:

$$C_2H_5OH(l) + CH_3COCl(l) \rightarrow CH_3COOC_2H_5(l) + HCl(g)$$

(b)   Reactions of the carbonyl group

The only reaction of note that we need to consider is the reduction of the carbonyl group by lithium tetrahydridoaluminate(III), $LiAlH_4$. It is converted into a CH_2 group. This means that the acid $RCOOH$ is changed to the alcohol RCH_2OH. For example,

$$CH_3COOH \xrightarrow{LiAlH_4} CH_3CH_2OH$$
ethanoic acid   ethanol

One feature of the chemistry of lithium tetrahydridoaluminate(III) is that it will perform this type of reduction but leave a carbon–carbon double bond unaffected.

119.7 Ethane-1,2-dioic acid (oxalic acid) is a dicarboxylic acid; it has two acid groups in its molecule. This is shown by its formula, which is sometimes written $(CH_2COOH)_2$.

(i) Draw out the structure of the acid.

(ii) Given $50\ cm^3$ of a $0.1\ mol\ dm^{-3}$ solution of the acid, what volume of $0.2\ mol\ dm^{-3}$ sodium hydroxide would be needed to neutralise the acid completely?

119.8 Which indicator would you use in a titration of ethanoic acid with sodium hydroxide solution?

119.9 If you use ethanoyl chloride to esterify ethanol, you may detect the smell of vinegar near the bottle containing the chloride. Why?

119.10 How could you make ethanoyl chloride from ethanoic acid?

119.11 What would be the outcome of reacting methanol, CH_3OH, with ethanoyl chloride?

119.12 Benzoyl chloride, C_6H_5COCl, is similar in structure to ethanoyl chloride.

(i) Draw the structure of the molecule.

(ii) What would be made if benzoyl chloride was added to ethanol? Draw the structure of the molecule.

(iii) What would be made if you added benzoyl chloride to water? What would you see?

119.4   The chloroethanoic acids

Table 119.1 includes data on three chloroethanoic acids. You can tell from their pK_a values that the strength of

the acid increases as the number of chlorine atoms in the molecule increases. Consult Unit 75 if you wish to understand the reason for this.

This feature of the acids is interesting, but chloroethanoic acid is an important chemical for another reason. It can be used to prepare amino acids. To make chloroethanoic acid, chlorine can be bubbled through the acid with iodine or red phosphorus present as a catalyst. The reaction has to be stopped when the theoretical increase in weight has been achieved:

$$CH_3COOH(l) + Cl_2(g) \rightarrow CH_2ClCOOH(l) + HCl(g)$$

Chloroethanoic acid can be easily converted into the amino acid glycine by mixing it with concentrated aqueous ammonia, although the separation of the amino acid is time consuming:

$$CH_2ClCOOH(l) + NH_3(aq) \rightarrow CH_2NH_2COOH(l) + HCl(aq)$$
glycine

119.13 How would you expect the pK_a value of fluoroethanoic acid to compare with that of chloroethanoic acid?

119.14 Chloroethanoic acid undergoes hydrolysis easily if it is warmed with water or alkali. What might be the structure of the molecule made in the reaction? (Hint: as well as an acid, what other type of homologous series does chloroethanoic acid resemble?)

Answers

119.1 The melting point of pure ethanoic acid is $16.8°C$. Thus, in winter, it is quite possible for the acid to solidify; in summer, it is more likely to be a liquid.

119.2 (i) No more ammonia would be given off.

(ii) The ammonia made in the reaction is converted into ammonium ions, NH_4^+, by reaction with hydrogen ions from the acid.

119.3 The nitrile would have to be HCN, i.e. hydrogen cyanide. This gas is extremely poisonous, and in ordinary laboratories should be avoided at all costs.

119.4 Use a reflux apparatus like that shown in Figure 115.4.

119.5 The manganate(VII) solution in alkali is the strongest oxidising agent. Its E° is $+1.70\ V$ compared to $+1.51\ V$ for acidified manganate(VII) and $+1.33\ V$ for acidified dichromate(VI).

119.6 (i) Heat with alkaline potassium manganate(VII) solution.

(ii) Either (a) convert the C_2H_5Br into a Grignard reagent and treat with carbon dioxide followed by acid; or (b) react it with an alcoholic solution of potassium cyanide to make propanenitrile, C_2H_5CN, and then reflux with alkali. Acid would be added at the end of the reaction to liberate C_2H_5COOH molecules, otherwise they would remain as ethanoate ions, $C_2H_5COO^-$.

(iii) The same two methods as in the answer to (ii) could be used.

119.7 (i)

Answers – cont.

(ii) This is a dibasic acid, so $50\,cm^3$ of a $0.1\,mol\,dm^{-3}$ solution would need $100\,cm^3$ of $0.1\,mol\,dm^{-3}$ sodium hydroxide, or $50\,cm^3$ of $0.2\,mol\,dm^{-3}$ alkali.

119.8 The common one to use for the titration of a weak acid with a strong base is phenolphthalein.

119.9 When ethanoyl chloride fumes in air it is because of its reaction with water vapour. The reaction gives ethanoic acid (hence the smell) as well as hydrogen chloride:

$$CH_3COCl + H_2O \rightarrow CH_3COOH + HCl$$

119.10 Add phosphorus pentachloride or, better, sulphur dichloride oxide to the acid (see section 119.3).

119.11 Methyl ethanoate is made:

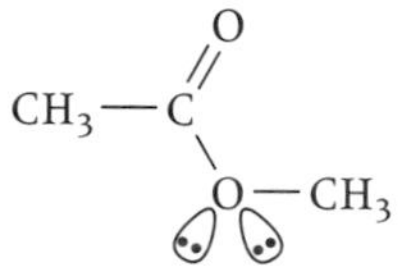

119.12 (i) Benzoyl chloride:

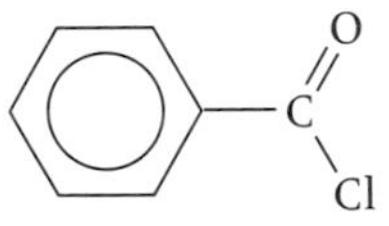

(ii) Ethyl benzoate is made:

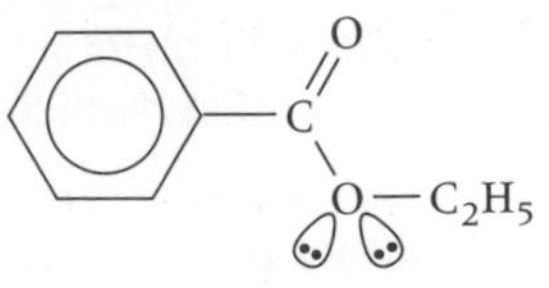

(iii) Benzoic acid, C_6H_5COOH, is formed through the reaction

$$C_6H_5COCl + H_2O \rightarrow C_6H_5COOH + HCl$$

This acid is insoluble in water, and you would see a white precipitate or cloudiness appear.

119.13 We might expect that a highly electronegative fluorine atom would pull electron density towards itself even more strongly than a chlorine atom. This would lead to the charge of the FCH_2COO^- ion being spread widely over the ion, and thereby increased energetic stability. This explanation fits the facts: the pK_a of fluoroethanoic acid is about 2.6. However, the true reason is more likely to be found in differences in entropies of solvation of the ions involved.

119.14 If we forget about the acid group, the molecule looks like a halogenoalkane. When a halogenoalkane is hydrolysed, the halogen is replaced by an OH group. The same thing happens here. The product is hydroxyethanoic acid, $OHCH_2COOH$:

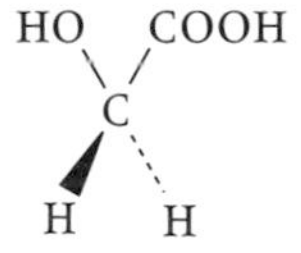

UNIT 119 SUMMARY

- Carboxylic acids:
    - (i) Contain the group

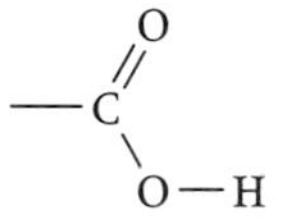

    - (ii) Non-aromatic acids have the general formula $C_nH_{2n+1}COOH$.
    - (iii) Are weak acids.
- Preparation:
    - (i) Hydrolysis of nitriles;

    e.g. $CH_3CN(aq) + H_2O(l) + OH^-(aq) \rightarrow$
    $$CH_3COO^-(aq) + NH_3(g)$$
    then

    $$CH_3COO^-(aq) + H^+(aq) \rightarrow CH_3COOH(aq)$$

    - (ii) Oxidation of primary alcohols under reflux with acidified sodium dichromate(VI) solution;

    e.g. $CH_3CH_2OH \rightarrow CH_3COOH$

    - (iii) From a Grignard reagent;

    $$RMgI + CO_2 \rightarrow RCOOMgI \xrightarrow{H^+} RCOOH + Mg^{2+}$$

    - (iv) Benzoic acid made by oxidation of a hydrocarbon side chain on a benzene ring using an acid or alkaline solution of potassium manganate(VII);

    e.g. $C_6H_5CH_3 \rightarrow C_6H_5COOH$

Reactions

- Weak acids:
    Carboxylic acids are only partially dissociated into ions.
- With sodium:
    Hydrogen released.
- Phosphorus pentachloride:
    White fumes of hydrogen chloride released;

    e.g. $CH_3COOH(l) + PCl_5(s) \rightarrow$
    $$CH_3COCl(l) + POCl_3(s) + HCl(g)$$

Summary – cont.
- Esterification:
  Esters are made with alcohols and acid chlorides;

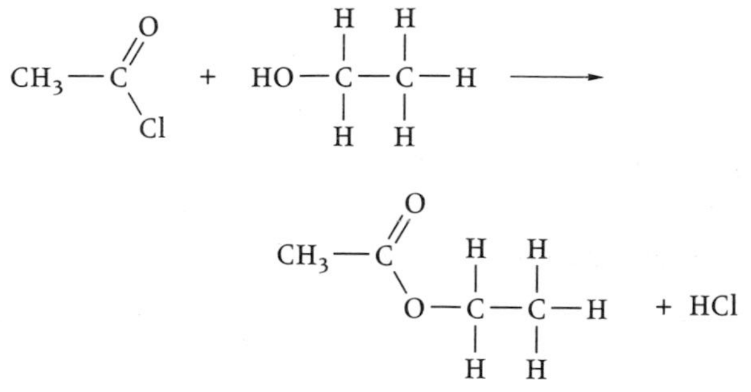

$$CH_3-C\underset{Cl}{\overset{O}{<}} \quad + \quad HO-\underset{\underset{H}{|}}{\overset{\overset{H}{|}}{C}}-\underset{\underset{H}{|}}{\overset{\overset{H}{|}}{C}}-H \quad \longrightarrow$$

$$CH_3-C\underset{O-\underset{\underset{H}{|}}{\overset{\overset{H}{|}}{C}}-\underset{\underset{H}{|}}{\overset{\overset{H}{|}}{C}}-H}{\overset{O}{<}} \quad + \quad HCl$$

- Reduction:
  Lithium tetrahydridoaluminate(III) reduces acids to alcohols;

  e.g. $CH_3COOH \rightarrow CH_3CH_2OH$

- Chloroethanoic acid:
  (i) Is a stronger acid than ethanoic acid.
  (ii) Can be used to prepare the amino acid glycine;

  $CH_2ClCOOH(l) + NH_3(aq) \rightarrow$
  $\qquad\qquad CH_2NH_2COOH(l) + HCl(aq)$

  α-Chloro acids can be used to make other amino acids.

120

Carboxylic acid derivatives

120.1 What are the derivatives of carboxylic acids?

The functional group of carboxylic acids is the COOH group

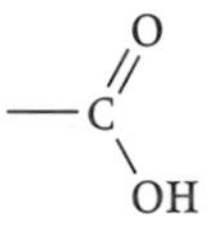

Derivatives of the acids are normally made by swapping the OH group for another atom or group. The carbonyl part of the acid is usually left intact. As a consequence we can represent many of the derivatives by the common formula RCOX

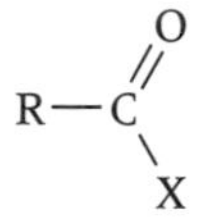

The main types are listed in Table 120.1.

120.2 Acid chlorides

Acid chlorides are easy to make. We have seen one way before: add phosphorus pentachloride to a carboxylic acid. However, this is not a good method to use in the laboratory. It is much easier to use either phosphorus trichloride, PCl_3, or sulphur dichloride oxide, $SOCl_2$. These two substances are liquids, so their addition to the acid can be more easily controlled. Also, as we have noted before, the side products of the reaction with $SOCl_2$ are gases. These are more easily removed from the reaction mixture, thus making it simpler to purify the acid chloride. The reactions are:

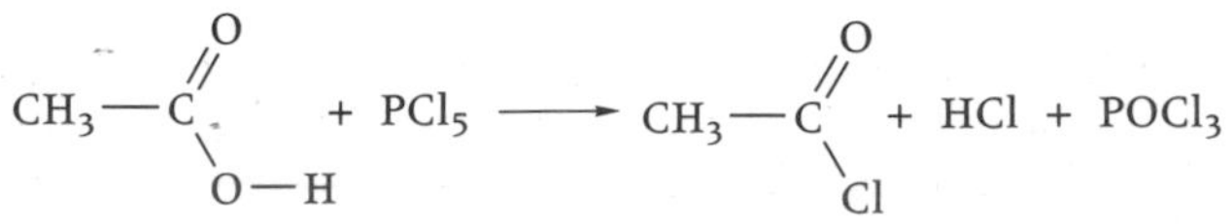

$$CH_3COOH + PCl_5 \rightarrow CH_3COCl + HCl + POCl_3$$
ethanoic acid → ethanoyl chloride

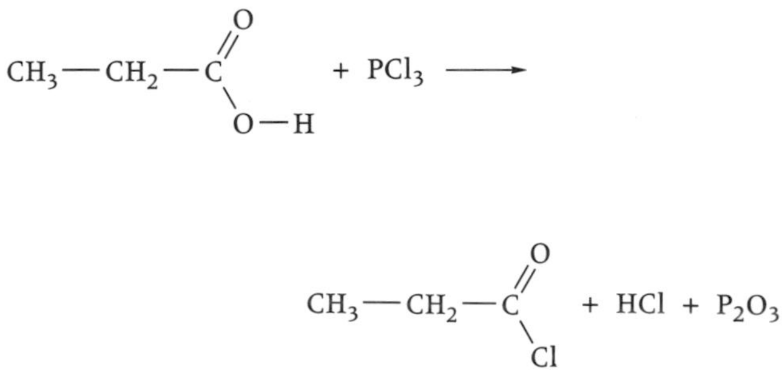

$$CH_3CH_2COOH + PCl_3 \rightarrow CH_3CH_2COCl + HCl + P_2O_3$$
propanoic acid → propanoyl chloride

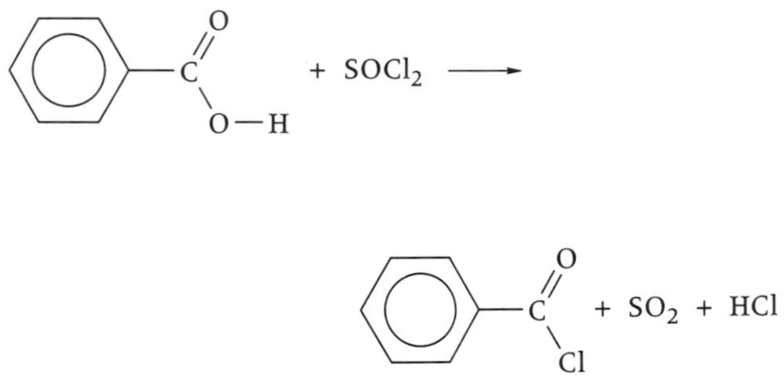

$$C_6H_5COOH + SOCl_2 \rightarrow C_6H_5COCl + SO_2 + HCl$$
benzoic acid → benzoyl chloride

(a) Reactions of acid chlorides

Nucleophilic attack

Acid chlorides are very reactive. The carbon atom bonded to the oxygen and chlorine atoms loses a lot of electron density to these electronegative atoms. This leaves the carbon atom with a significant amount of positive charge and it becomes an attractive site for attack by nucleophiles.

With its lone pairs of electrons, water is able to decompose acid chlorides; it converts them back into acids, e.g.

Table 120.1. The derivatives of carboxylic acids

Type	Functional group	Typical examples	
Acid chloride	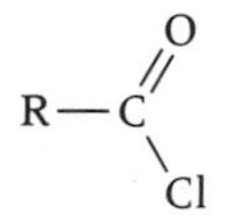 $R-\overset{\displaystyle O}{\underset{\displaystyle Cl}{C}}$	Ethanoyl chloride	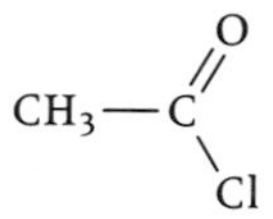 $CH_3-\overset{\displaystyle O}{\underset{\displaystyle Cl}{C}}$
		Benzoyl chloride	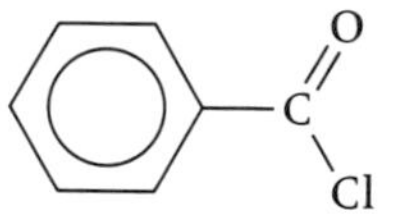 (benzoyl chloride structure)
Acid anhydride	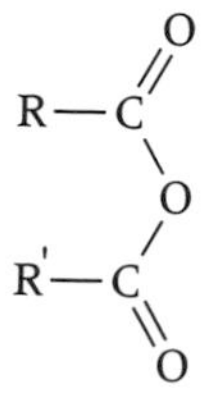 $R-\overset{\displaystyle O}{C}$ ... $R'-\overset{}{\underset{\displaystyle O}{C}}$ with bridging O	Ethanoic anhydride	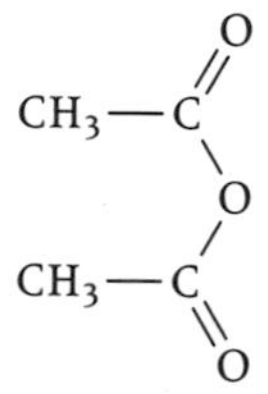 CH_3-C ... CH_3-C with bridging O
Amide	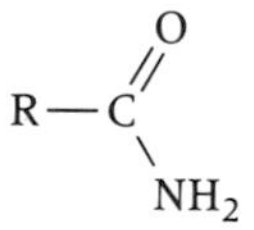 $R-\overset{\displaystyle O}{\underset{\displaystyle NH_2}{C}}$	Ethanamide	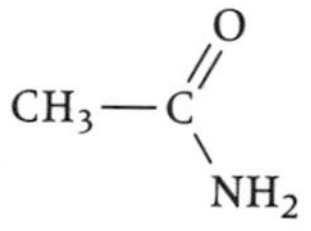 $CH_3-\overset{\displaystyle O}{\underset{\displaystyle NH_2}{C}}$
		Benzamide	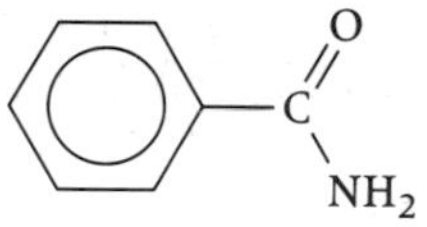 (benzamide structure)
Ester	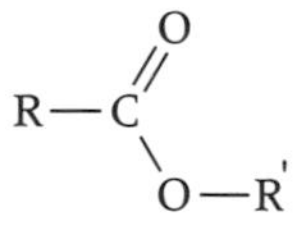 $R-\overset{\displaystyle O}{\underset{\displaystyle O-R'}{C}}$	Ethyl ethanoate	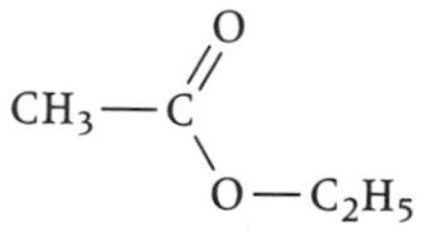 $CH_3-\overset{\displaystyle O}{\underset{\displaystyle O-C_2H_5}{C}}$
		Methyl benzoate	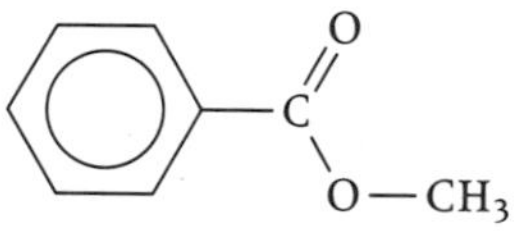 (methyl benzoate structure $O-CH_3$)

as Z—H, where Z is the atom or group that has a lone pair. The result of the reaction is

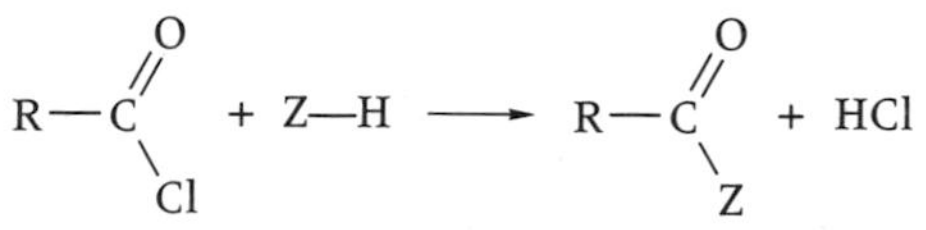

$$R-\overset{\displaystyle O}{\underset{\displaystyle Cl}{C}} \ + \ Z-H \ \longrightarrow \ R-\overset{\displaystyle O}{\underset{\displaystyle Z}{C}} \ + \ HCl$$

$$RCOCl + Z—H \rightarrow RCOZ + HCl$$

We can fit ammonia, NH_3, to this pattern by writing it as NH_2—H. Similarly, amines like methylamine, CH_3NH_2, become CH_3NH—H; and alcohols like ethanol, C_2H_5OH, become C_2H_5O—H.

Table 120.2 shows you the pattern of the reactions that take place with ammonia, amines and alcohols.

(reaction mechanism scheme: benzoyl chloride + H₂O → benzoic acid + HCl)

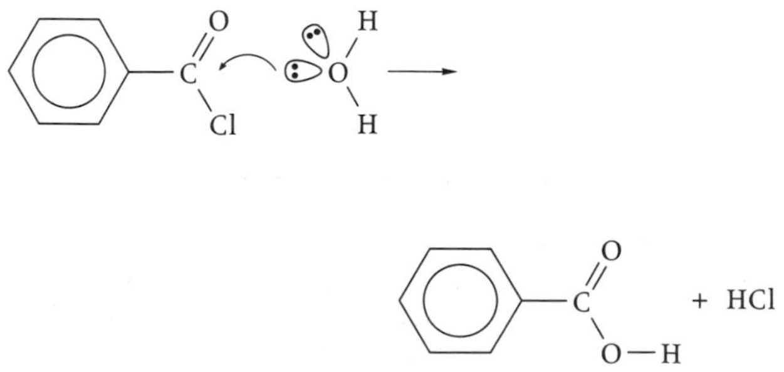

$$C_6H_5COCl + H_2O \rightarrow C_6H_5COOH + HCl$$
benzoyl   benzoic
chloride   acid

This is just one example of a general pattern. We can write the typical nucleophiles that attack acid chlorides

Table 120.2. The reactions of acid chlorides

Nucleophile	Type of product	After reaction with*	
		Ethanoyl chloride	Benzoyl chloride
Ammonia, NH₂—H	Amide	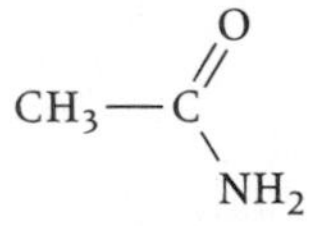 Ethanamide	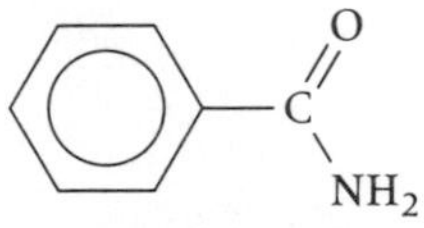 Benzamide
Amines, RNH—H e.g. CH₃NH—H (methylamine	Substituted amide	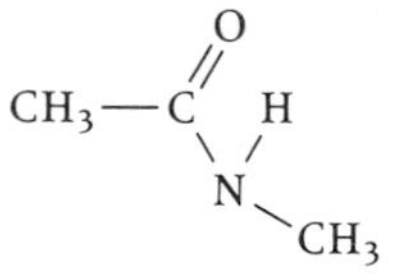 N-methylethanamide	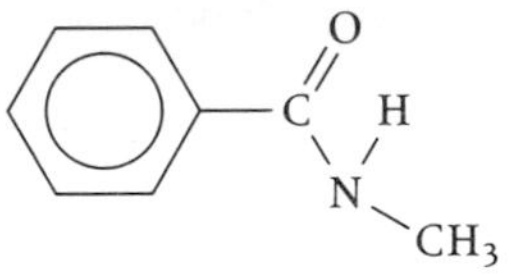 N-methylbenzamide
Alcohols, e.g. C₂H₅O—H (ethanol)	Ester	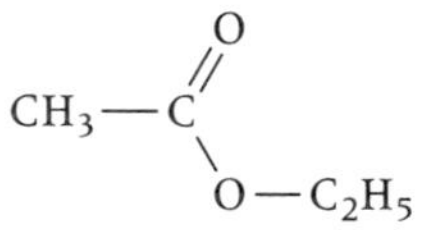 Ethyl ethanoate	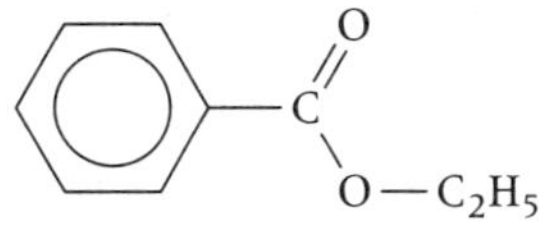 Methyl benzoate

*The structures of ethanoyl chloride and benzoyl chloride are

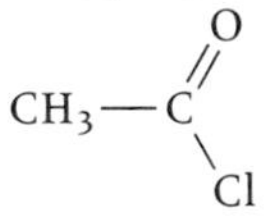

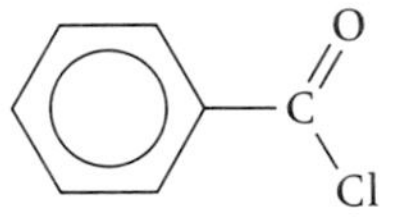

Acetylation reactions

We can look at these reactions in a different way. For instance, if we concentrate on the change to the methylamine molecule when it reacts with ethanoyl chloride, it gains the CH_3CO group:

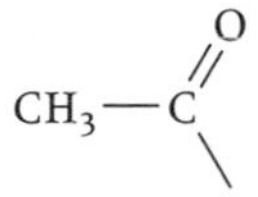

This is the ethanoyl or *acetyl* group. We say that methylamine has been *acetylated*. Likewise, ethanoyl chloride will acetylate water, ammonia and alcohols. Similarly, benzoyl chloride will *benzoylate* these molecules.

It is possible to acetylate molecules using a different reagent, ethanoic anhydride, but we shall return to this later.

If you wanted to acetylate (or benzoylate) an aromatic molecule, you would use a Friedel–Crafts reaction in which the molecule is reacted with the acetylating agent and aluminium trichloride. For example (see also Figure 120.1),

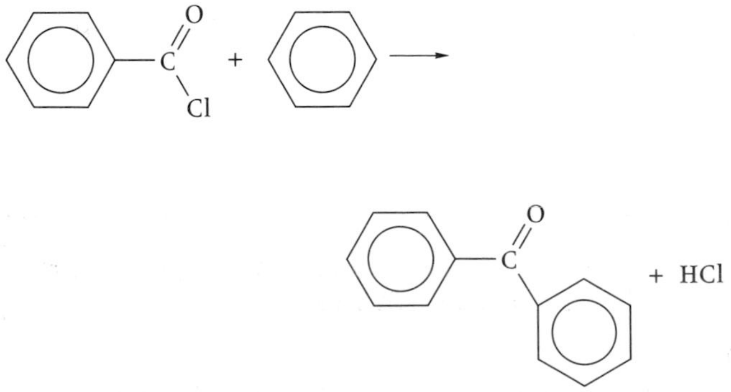

$$C_6H_5COCl + C_6H_6 \xrightarrow{AlCl_3} C_6H_5COC_6H_5 + HCl$$
diphenyl-methanone

Note: Acetylation and benzoylation are two examples of the general reaction called *acylation*. Acylation refers to the introduction of an $R—C{=}O$ group into a molecule.

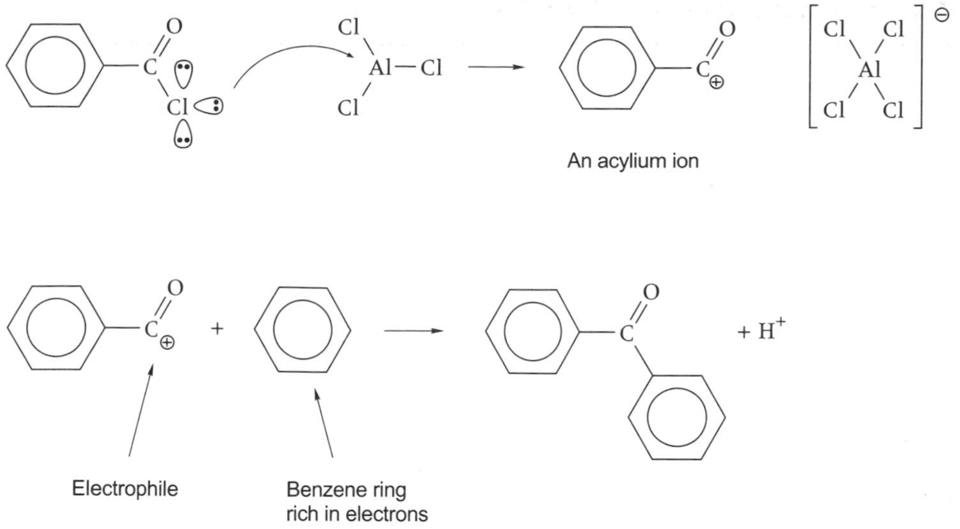

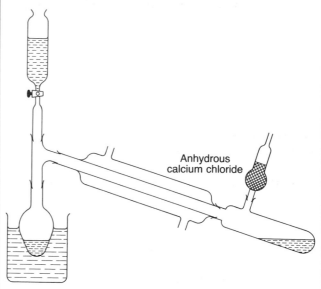

Figure 120.1 *The reaction between benzoyl chloride and benzene involves the production of an electrophile. This is the basis of all Friedel–Crafts reactions. A similar acylium ion (CH$_3$—C$^+$=O) is made with ethanoyl chloride and ethanoic anhydride*

Reduction of acid chlorides

The main ways of reducing acid chlorides are as follows.

(i) With a modified palladium catalyst and hydrogen, they are converted into aldehydes (section 118.4):

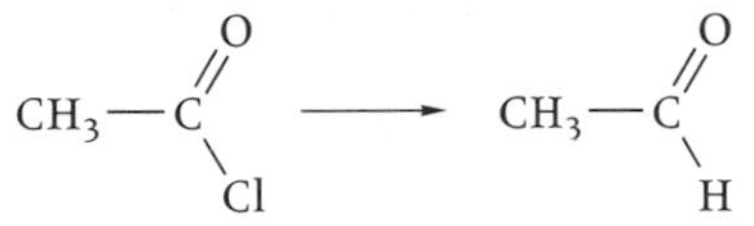

$$CH_3COCl \xrightarrow{H_2, \ Pd/BaSO_4} CH_3CHO$$

(ii) With lithium tetrahydridoaluminate(III) they are converted into alcohols:

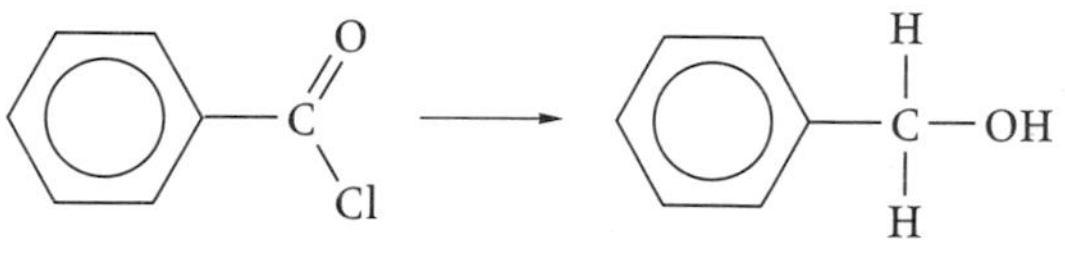

$$C_6H_5COCl \xrightarrow{LiAlH_4} C_6H_5CH_2OH$$

120.1 With which other types of compound does phosphorus pentachloride react?

120.2 Look at Figure 120.2, which shows an apparatus that can be used to prepare ethanoyl chloride from ethanoic acid and phosphorus trichloride. The reaction is exothermic. The water bath is heated only after the reaction has finished.

(i) Copy and label the diagram.

(ii) Why should the reagents be mixed slowly?

(iii) Why is the water bath present while the reaction takes place?

(iv) What is the purpose of the anhydrous calcium chloride? (Hint: look at the properties of ethanoyl chloride.)

Figure 120.2 *The apparatus for question 120.2*

120.3 Acid anhydrides

One of the most important anhydrides is ethanoic anhydride. As the name suggests, it is made by removing the elements of water from ethanoic acid. (See section 99.9 for information about inorganic anhydrides.) The reaction is:

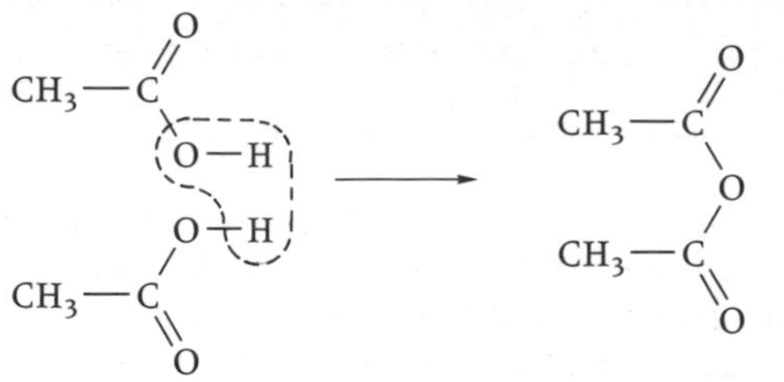

$$2CH_3COOH - H_2O \rightarrow (CH_3CO)_2O$$
ethanoic   ethanoic
acid   anhydride

In practice we can make an anhydride in several ways. In the first, the acid vapour is passed over hot zinc oxide. However, the most convenient method is to start with the sodium salt of the acid, and use ethanoyl chloride as an acetylating agent:

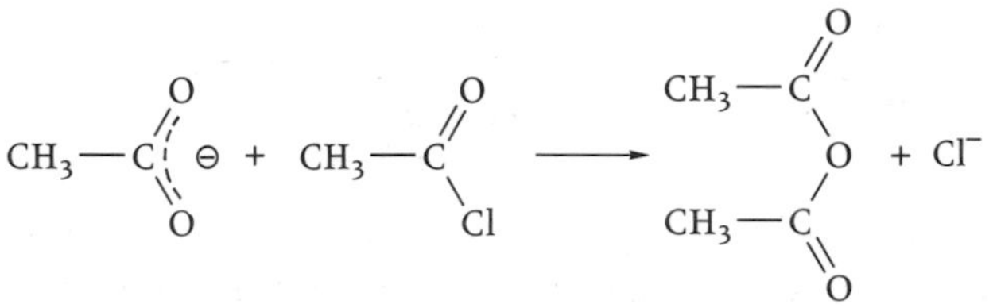

$$CH_3COO^-Na^+ + CH_3COCl \rightarrow (CH_3CO)_2O + NaCl$$
sodium   ethanoyl   ethanoic
ethanoate   chloride   anhydride

(a) *Reactions of anhydrides*

Ethanoic anhydride can be used as an acetylating agent. It behaves in much the same way as ethanoyl chloride, except that whereas ethanoyl chloride gives off hydrogen chloride, with ethanoic anhydride we get ethanoic acid:

(i) With ammonia

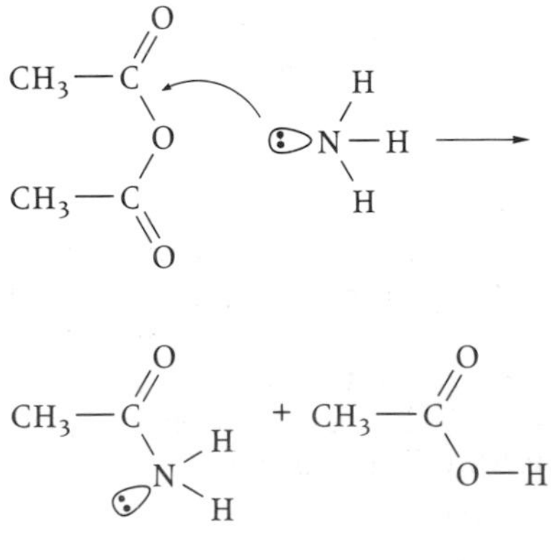

$$(CH_3CO)_2O + NH_3 \rightarrow CH_3CONH_2 + CH_3COOH$$
ethanamide   ethanoic acid

(ii) With amines

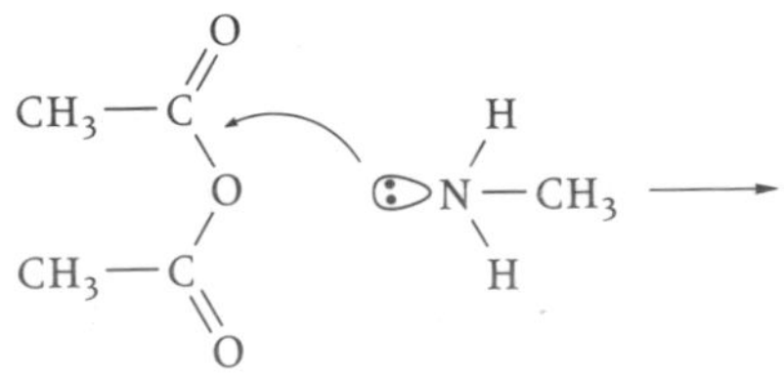

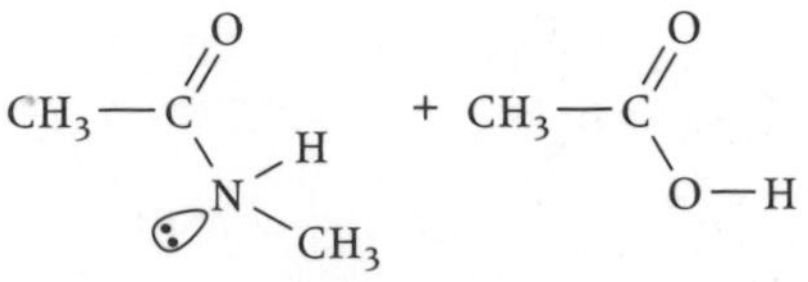

$$(CH_3CO)_2O + CH_3NH_2 \rightarrow CH_3CONHCH_3 + CH_3COOH$$
methyl-   N-methyl-   ethanoic acid
amine   ethanamide

(iii) With ethanol

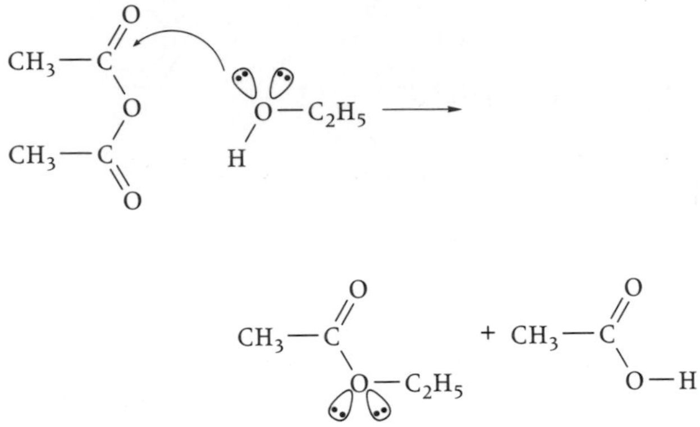

$$(CH_3CO)_2O + C_2H_5OH \rightarrow CH_3COOC_2H_5 + CH_3COOH$$
ethanol   ethyl ethanoate   ethanoic acid

Like the acid chlorides, anhydrides will take part in Friedel–Crafts reactions. We could, for example, use ethanoic anhydride together with aluminium trichloride to convert benzene into phenylethanone:

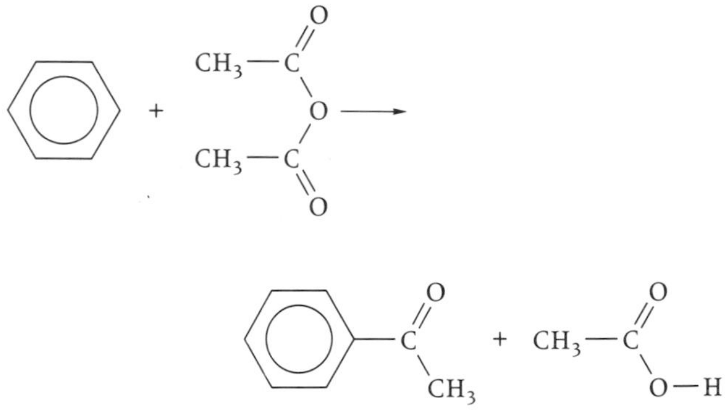

$$C_6H_6 + (CH_3CO)_2O \xrightarrow{AlCl_3} C_6H_5COCH_3 + CH_3COOH$$

120.3 (i) How would you make benzoic anhydride, $(C_6H_5CO)_2O$?

(ii) Draw the structures of the molecules made if benzoic anhydride were to react with ammonia, ethylamine and methanol.

120.4 Benzene-1,2-dicarboxylic anhydride (otherwise known as phthalic anhydride) is a rather unusual anhydride:

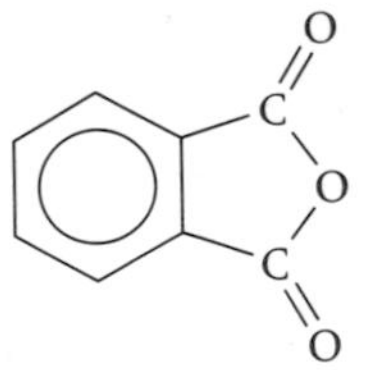

Draw the structure of the acid from which it might be made.

120.4 Amides

Amides contain the functional group $CONH_2$

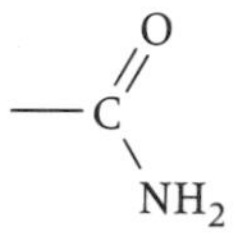

They cannot be made directly from carboxylic acids. First, we must convert the acid into its ammonium salt. This is done by reacting the acid with ammonium carbonate; e.g.

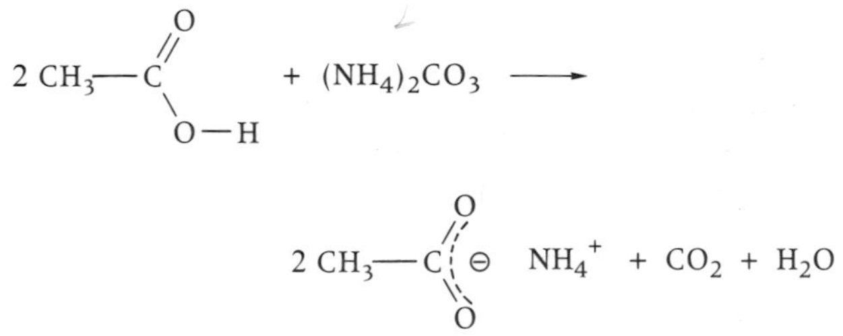

$$2CH_3COOH + (NH_4)_2CO_3 \rightarrow 2CH_3COO^-NH_4^+ + CO_2 + H_2O$$

Then the ammonium salt is decomposed by heating:

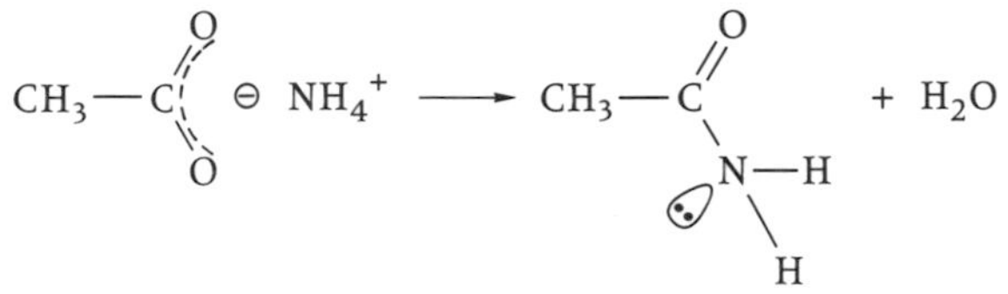

$$CH_3COO^-NH_4^+ \xrightarrow{\text{heat}} CH_3CONH_2 + H_2O$$
$$\text{ethanamide}$$

Ethanamide was once known as acetamide. Unless it is completely pure, it has a very distinctive smell – of mice. It is used in the plastics industry as well as in such widely different activities as the manufacture of cosmetics and explosives.

(a) Reactions of amides

The Hofmann degradation

One of the most important reactions of amides is the *Hofmann degradation*. It converts amides into amines; but of crucial importance is the fact that:

> **The amine has one less carbon atom than the amide from whence it came.**

To carry out this reaction you combine an amide with a mixture of bromine and sodium hydroxide solution. The latter two reagents make the bromate(I) ion. This is an oxidising agent and, together with hydroxide ions, is the active ingredient in the reaction. One example is:

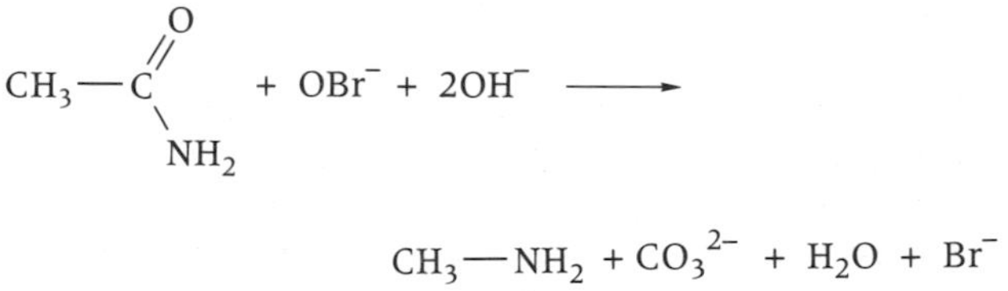

$$CH_3{-}NH_2 + CO_3^{2-} + H_2O + Br^-$$

$$CH_3CONH_2 + OBr^- + 2OH^- \rightarrow$$
ethanamide
(two carbons)
$$\quad CH_3NH_2 + CO_3^{2-} + H_2O + Br^-$$
$$\quad \text{methylamine}$$
$$\quad \text{(one carbon)}$$

The Hofmann degradation is a reaction of which you should take note. It is one of the few ways of reducing the number of carbon atoms in a molecule. With its use we can solve problems like: how could we change propanol into ethanol? This is an example of *descending a homologous series*. The scheme of the conversion would be

$$C_3H_7OH \rightarrow C_2H_5COOH \rightarrow$$
$$\quad C_2H_5CONH_2 \rightarrow C_2H_5NH_2 \rightarrow C_2H_5OH$$

The steps in the changes are achieved in this way:

(i) $C_3H_7OH \rightarrow C_2H_5COOH$, oxidise with acidified sodium dichromate(VI);
(ii) $C_2H_5COOH \rightarrow C_2H_5CONH_2$, heat with ammonium carbonate;
(iii) $C_2H_5CONH_2 \rightarrow C_2H_5NH_2$, Hofmann degradation;
(iv) $C_2H_5NH_2 \rightarrow C_2H_5OH$, warm with acidified sodium nitrite.

Reaction with nitrous acid, HNO_2

In section 117.2 we found that a warm mixture of sodium nitrite, $NaNO_2$, and dilute hydrochloric acid would convert phenylamine into phenol. We also saw that the active ingredient in this mixture is nitrous acid. This conversion of an NH_2 group into an OH group will also work with amides. For example,

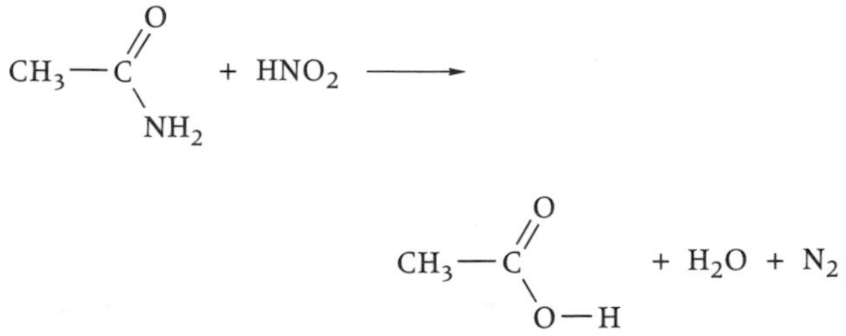

$$CH_3CONH_2 + HNO_2 \rightarrow CH_3COOH + H_2O + N_2$$
ethanamide   ethanoic acid

You can see that the pattern is to convert an amide into an acid. Notice that if you were to perform this reaction, you would see bubbles of nitrogen given off.

Making nitriles

Amides can be converted into nitriles by heating them with phosphorus(v) oxide, P_4O_{10}. The effect of this oxide is to *dehydrate* the amide. We can best show the reaction by the loss of water:

$$CH_3-\overset{\displaystyle O}{\underset{\displaystyle NH_2}{C}} \ - \ H_2O \longrightarrow CH_3-CN$$

$$CH_3CONH_2 - H_2O \xrightarrow{P_4O_{10}} CH_3CN$$
ethanamide ethanenitrile

Hydrolysis

Ammonia is given off when amides are heated with alkali. For example,

$$C_2H_5-\overset{\displaystyle O}{\underset{\displaystyle NH_2}{C}} \ + \ OH^- \longrightarrow C_2H_5-\overset{\displaystyle O}{\underset{\displaystyle O^-}{C}} \ + \ NH_3$$

$$C_2H_5CONH_2 + OH^- \rightarrow C_2H_5COO^- + NH_3$$
propanamide propanoate ion

If acid is used instead of alkali, the ammonia stays in solution as ammonium ions, NH_4^+.

120.5   Here is a description of a method for making ethanamide adapted from Waddington, D. J. and Finlay, H. S. (1967). *Organic Chemistry Through Experiment*, 3rd edn, Mills and Boon, London, p. 60:
  Place 3 g of ammonium carbonate in a flask. Add 6 cm³ of glacial ethanoic acid carefully. Boil gently under reflux for half an hour. Allow the apparatus to cool and arrange the apparatus for distillation. Heat the flask but do not allow the thermometer to exceed 180° C. Reject the distillate. Clean and dry the receiver and attach it to the condenser again. Continue the distillation, using the condenser as an air condenser. Collect the fraction in the range 210–225° C. The distillate solidifies on cooling and may be purified by recrystallisation from propanone. Yield about 2 g.

(i) How many moles of ammonium carbonate and ethanoic acid were used? (The density of ethanoic acid is 1.05 g cm⁻³.)

(ii) How do these figures compare with those in the equation for the reaction? Comment on your answer.

(iii) If the yield is 2 g of ethanamide, how many moles of product are made?

(iv) What is the percentage yield?

(v) What is happening during the reflux part of the reaction?

(vi) What are the most likely substances to be collected after distillation at 180° C?

(vii) Why is an air condenser essential for the second distillation?

(viii) How would you perform the recrystallisation process?

(ix) If you were of an enquiring nature and wanted to discover whether the product really was ethanamide, what would you do?

120.6 The Hofmann degradation decreases the number of carbon atoms in a chain by one. Give a reaction that does the opposite (i.e. increases the number of carbon atoms by one).

120.5   The reactions of urea, $(NH_2)_2CO$

Urea is an amide, but it has two NH_2 groups attached to a carbonyl group.

$$\overset{\displaystyle H_2N}{\underset{\displaystyle H_2N}{>}}C=O$$

Both of these groups are destroyed when it reacts with nitrous acid:

$$(NH_2)_2CO + 2HNO_2 \rightarrow CO_2 + N_2 + 3H_2O$$

Urea can be hydrolysed in three ways. Acid or alkaline hydrolysis can be used:

$$(NH_2)_2CO + 2H^+ + H_2O \rightarrow CO_2 + 2NH_4^+$$
$$(NH_2)_2CO + 2OH^- \rightarrow CO_3^{2-} + 2NH_3$$

or, more efficiently, it is decomposed by the enzyme *urease*:

$$(NH_2)_2CO + H_2O \xrightarrow{urease} 2NH_3 + CO_2$$

If you heat urea in a test tube you will smell ammonia being given off. The solid remaining is called *biuret*. Its structure is important because it contains a *peptide link*

$$\overset{\displaystyle O}{\underset{\displaystyle H}{>}}C-\overset{\displaystyle}{\underset{\displaystyle}{N}}<$$

The peptide link is a feature of proteins, which we shall examine in Unit 123. The reaction is:

$$2 \ \overset{\displaystyle H_2N}{\underset{\displaystyle H_2N}{>}}C=O \longrightarrow H_2N-\overset{\displaystyle O}{C}-\overset{\displaystyle}{\underset{\displaystyle H}{N}}-\overset{\displaystyle O}{C}-NH_2 \ + \ NH_3$$

$$2(NH_2)_2CO \rightarrow NH_2-CO-NH-CO-NH_2 + NH_3$$
biuret

There is a test that shows the presence of a peptide link. A few drops of copper(II) sulphate solution are added to a solution of the sample followed by further drops of sodium hydroxide solution. If a peptide link is present, a violet colour should appear. This test is known as the *biuret test*.

Urea is also able to take part in polymerisation reactions; but we shall leave these until later (Units 127 and 128).

Being an amide, we should expect urea to undergo a Hofmann degradation, but given that there is only one carbon atom present, we can hardly expect to get a useful product out of the reaction. In fact, urea is oxidised in the following manner:

$$(NH_2)_2CO + 3OBr^- + 2OH^- \rightarrow CO_3^{2-} + N_2 + 3H_2O + 3Br^-$$

120.6 Esters

Life would not be the same without esters! They are to be found in an immense range of materials, both naturally occurring and man-made. They are responsible for the odour of many foodstuffs, and therefore for much of our sense of taste, and for the perfumes of alcoholic drinks, oils, waxes, flowers and cosmetics. Aspirin, one of the most remarkable successes of the pharmaceutical industry, is also an ester. Two esters are shown in Figure 120.3.

The classical method of making an ester is to warm an alcohol with a carboxylic acid, together with a little concentrated sulphuric acid as a catalyst. You can do this for yourself with ethanol and ethanoic acid:

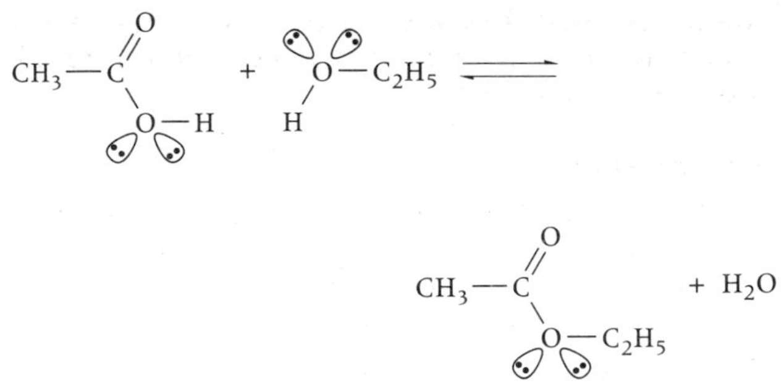

$$CH_3COOH + C_2H_5OH \rightleftharpoons CH_3COOC_2H_5 + H_2O$$

If you would like to know the mechanism for the reaction, consult panel 120.1.

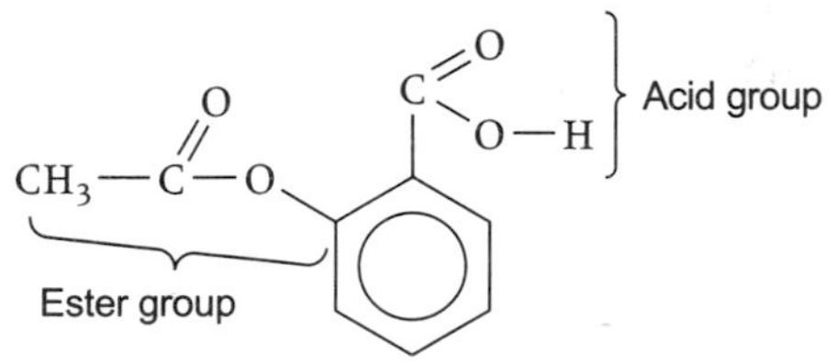

Aspirin, perhaps the most commonly available pharmaceutical

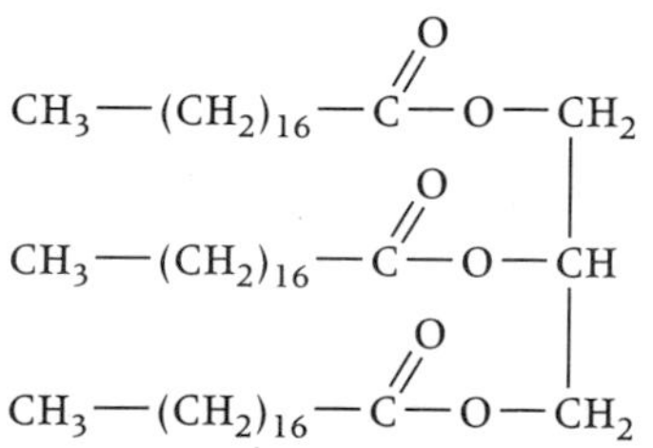

Tristearin, an example of a fat (fats are complicated esters)

Figure 120.3 *Two esters with markedly different properties*

Panel 120.1

The mechanism of an esterification reaction

The basic information that we have about the esterification reaction between ethanoic acid and ethanol is that the equation for the reaction is

$$CH_3COOH + C_2H_5OH \rightleftharpoons CH_3COOC_2H_5 + H_2O$$

and that it is catalysed by acid. From what we have said about the slight positive charge carried by the carbon atom of a carbonyl group, it seems possible that this carbon is the site for nucleophilic attack. The nucleophile is the alcohol molecule, or, more precisely, one of the lone pairs on the oxygen atom. However, nuclear magnetic resonance spectroscopy (which tells us where the protons are in a molecule) gives us evidence that a significant number of ethanoic acid molecules exist as carbocations, owing to the carbonyl oxygen being protonated:

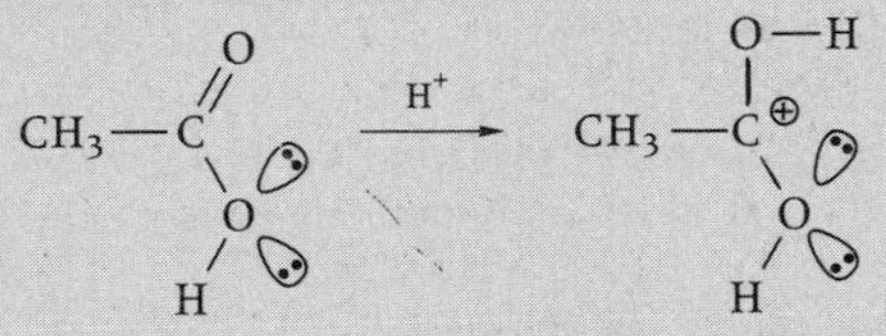

Ethanol molecules attack the carbocations, rather than ethanoic acid molecules:

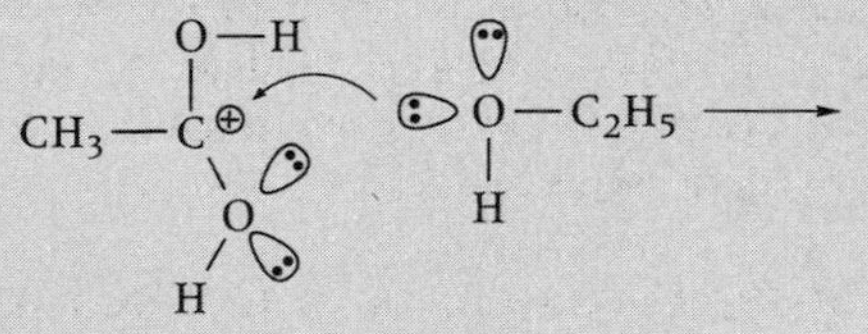

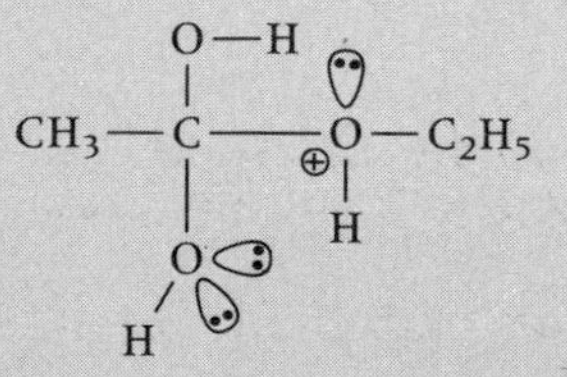

It is at this stage that a proton is transferred from the incoming alcohol molecule to the OH group of the acid. A water molecule is then lost, leaving another carbocation, but one that suffers a rearrangement by the loss of the proton on the original carbonyl oxygen. This regenerates the carbonyl group, and releases a hydrogen ion back into the solution:

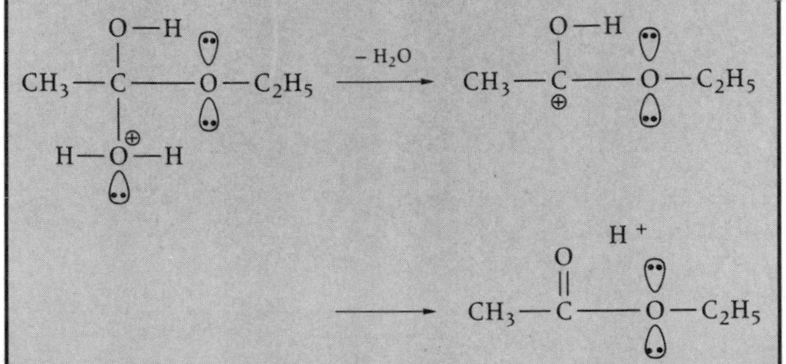

It is often the case that a water molecule is lost from an intermediate in a reaction scheme. A water molecule is said to be a *good leaving group*.

Alternative methods are to react an alcohol with an acid chloride or an acid anhydride:

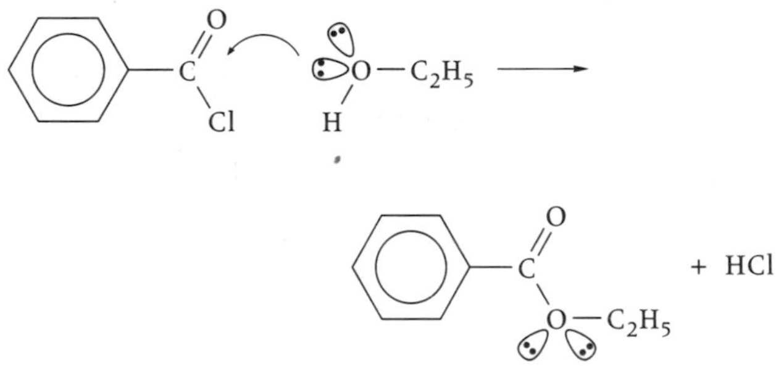

$C_6H_5COCl + C_2H_5OH \rightarrow C_6H_5COOC_2H_5 + HCl$
benzoyl chloride   ethyl benzoate

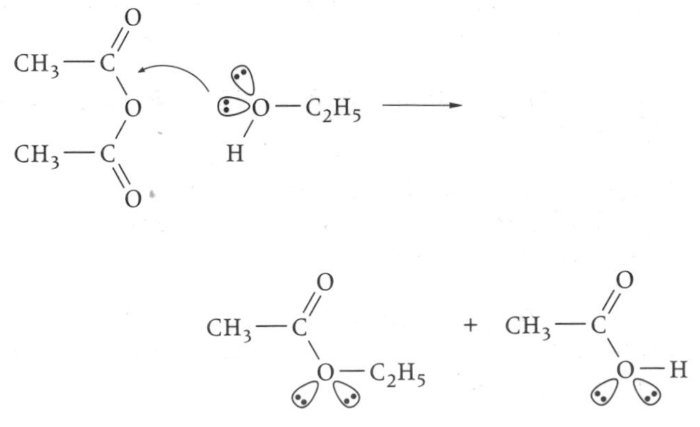

$(CH_3CO)_2O + C_2H_5OH \rightarrow CH_3COOC_2H_5 + CH_3COOH$
ethanoic anhydride   ethyl ethanoate

(a) *Reactions of esters*

Hydrolysis

Esters can be hydrolysed by warming them with water, acid or, with greater efficiency, alkali. The reaction that takes place is essentially the reverse of esterification:

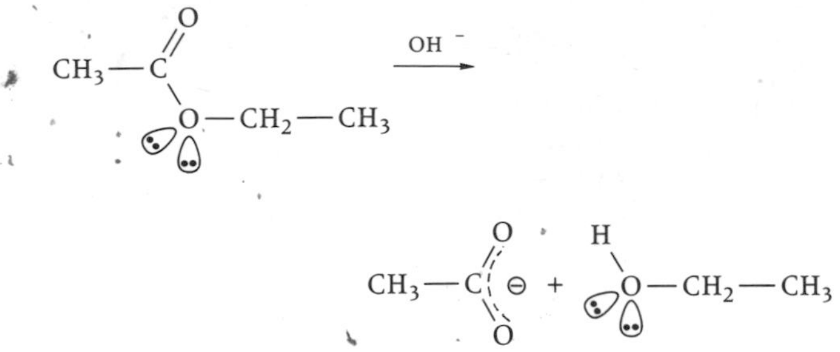

$CH_3COOC_2H_5 \xrightarrow{OH^-} CH_3COO^- + C_2H_5OH$

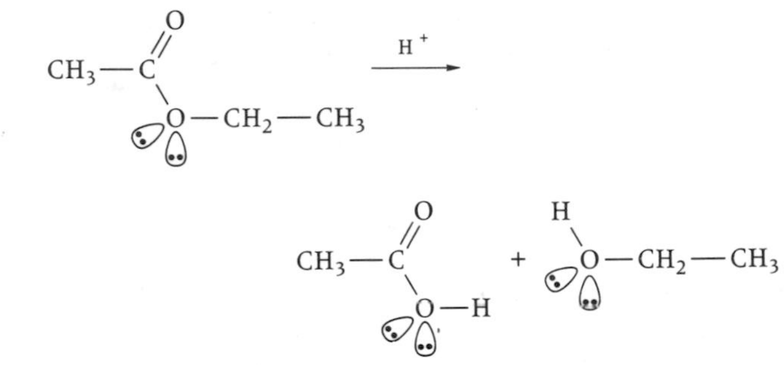

$CH_3COOC_2H_5 \xrightarrow{H^+} CH_3COOH + C_2H_5OH$

Reduction

Esters can be reduced by lithium tetrahydridoaluminate(III), or by sodium in ethanol. In either case the ester is converted into a mixture of alcohols:

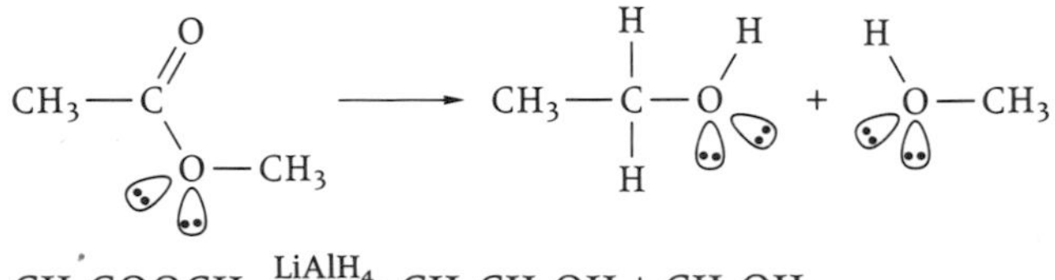

$CH_3COOCH_3 \xrightarrow{LiAlH_4} CH_3CH_2OH + CH_3OH$
ethyl ethanoate   ethanol   methanol

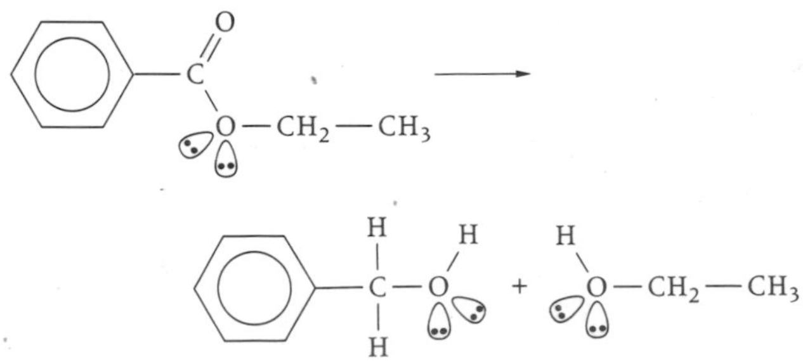

$C_6H_5COOC_2H_5 \xrightarrow{Na/C_2H_5OH} C_6H_5CH_2OH + C_2H_5OH$
ethyl benzoate   phenylmethanol   ethanol

With ammonia

Esters will react with ammonia. The product is an amide. For example,

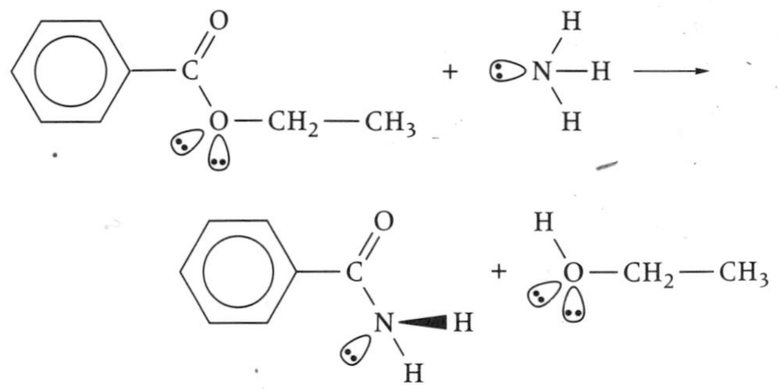

$$C_6H_5COOC_2H_5 + NH_3 \rightarrow C_6H_5CONH_2 + C_2H_5OH$$

ethyl benzoate   benzamide

Ammonia is not such a strong nucleophile as, say, hydroxide ion. For an effective reaction with ammonia, the ester and ammonia have to be heated together in a sealed tube.

120.7 Nitriles

Nitriles contain the CN group. Sometimes they are thought of as derivatives of carboxylic acids; but rather the connection is the other way round. We make carboxylic acids from nitriles. The method is fairly easy. Hydrolyse the nitrile using acid or alkali. With acid, carboxylic acid molecules are made; in alkali, the reaction leaves us with the carboxylic acid anions. Two examples are:

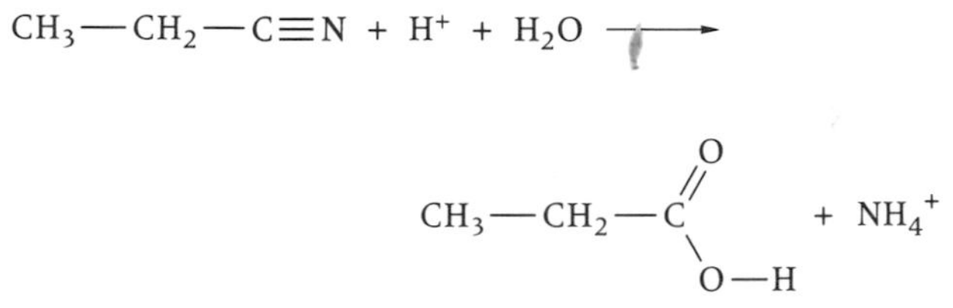

$$C_2H_5CN + H^+ + H_2O \rightarrow C_2H_5COOH + NH_4^+$$

propanenitrile   propanoic acid

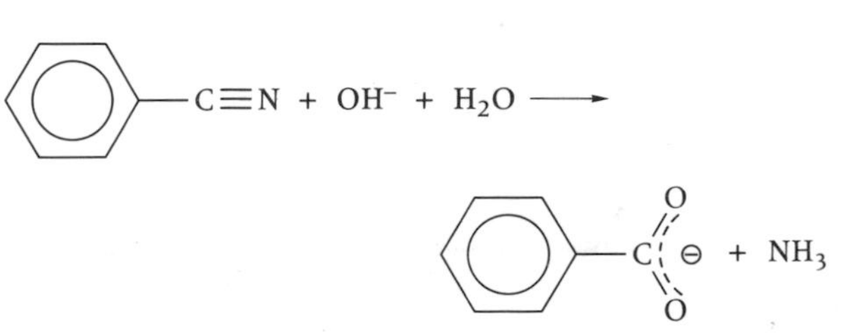

$$C_6H_5CN + OH^- + H_2O \rightarrow C_6H_5COO^- + NH_3$$

benzonitrile   benzoate ion

(a) The importance of nitriles

Nitriles provide a way of increasing the number of carbon atoms in a chain. This is the process of *ascending a homologous series*. A standard route is as follows, illustrated by taking iodoethane as our starting point:

$$C_2H_5I \rightarrow C_2H_5CN \rightarrow C_2H_5COOH \rightarrow$$
iodo
ethane
$$C_2H_5CH_2OH \rightarrow C_2H_5CH_2I$$
iodopropane

Once the new halogenoalkane is made, the entire range of homologous series can be reached. Questions 120.7 to 120.11 will lead you through a few examples.

Nitriles can also be converted into amines. This is achieved by reduction with hydrogen and a nickel catalyst at 140 °C:

$$CH_3CN + 2H_2 \xrightarrow{\text{Ni cat.}} CH_3CH_2NH_2$$

The same result can be achieved using sodium in ethanol, or lithium tetrahydridoaluminate(III).

(b) Preparation of nitriles

We can make nitriles by dehydrating amides with phosphorus(V) oxide (see section 120.4), or by heating halogenoalkanes with an alcoholic solution of potassium cyanide (see section 115.3); e.g.

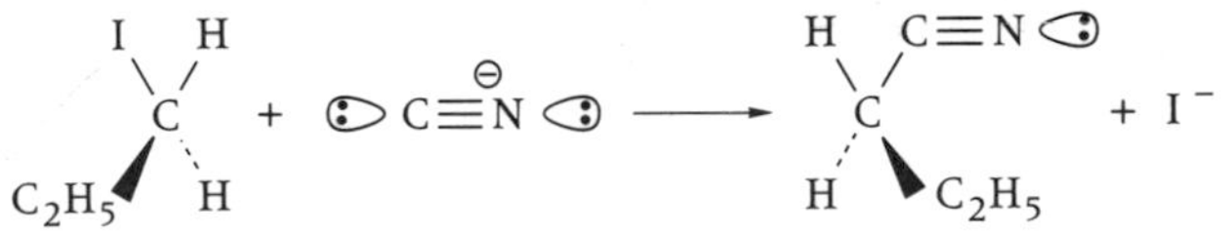

$$C_3H_7I + CN^- \rightarrow C_3H_7CN + I^-$$
iodopropane   butanenitrile

120.7 In the scheme for converting iodoethane into iodopropane, write down the reaction conditions for each stage.

120.8 How would you make propanoic acid starting with iodoethane?

120.9 How would you make ethanol starting from iodomethane?

120.10 How would you make ethyl propanoate, $C_2H_5COOC_2H_5$, starting with methyl iodide as the major organic reagent?

120.11 If you have answered question 120.10, why would you be wise to start with a substance other than methyl iodide?

120.12 Devise a reaction scheme that would convert propanoic acid into iodoethane.

Answers

120.1 It is used as a test for compounds that contain OH groups such as alcohols, giving off fumes of hydrogen chloride. It will also react with carbonyl groups (see section 118.11).

120.2 (i) The ethanoic acid goes in the flask, and phosphorus trichloride in the dropping funnel.

(ii) The reaction is exothermic, so there is a danger of it being very violent if the reactants are mixed too quickly.

(iii) It cools the mixture.

(iv) It prevents water vapour from the atmosphere entering the apparatus and reacting with the ethanoyl chloride.

120.3 (i) Either pass benzoic acid vapour over hot zinc oxide, or (better) add ethanoyl chloride to sodium benzoate.

(ii)

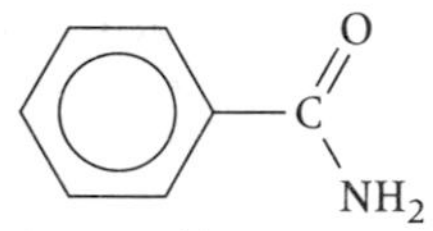

benzamide

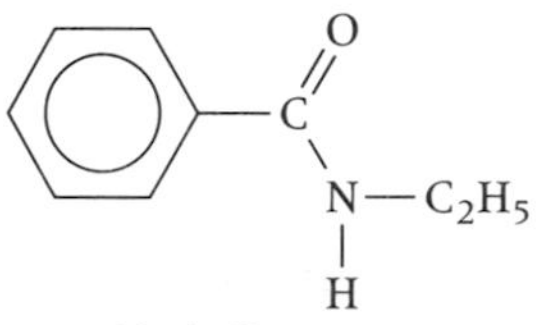

N-ethylbenzamide

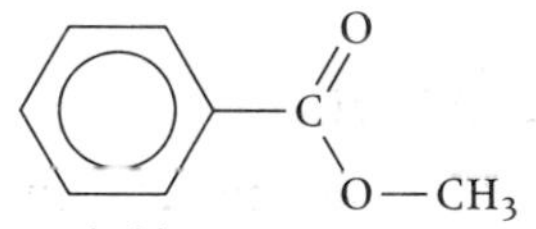

methyl benzoate

120.4 1,2-Benzenedioic acid:

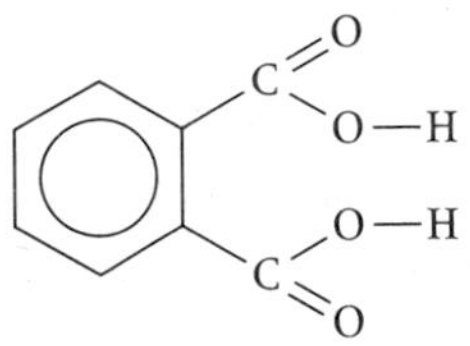

120.5 (i) One mole of ammonium carbonate has a mass of 96 g, so 3 g represents 0.03 mol. 6 cm^3 of ethanoic acid represents 6.3 g; 1 mol of ethanoic acid has a mass of 60 g, so the number of moles used is about 0.1 mol.

(ii) From the equation, 2 mol of CH_3COOH react with 1 mol of $(NH_4)_2CO_3$ to give 2 mol of CH_3COONH_4. In the recipe, we have a ratio of 0.1 mol of CH_3COOH to 0.03 mol of $(NH_4)_2CO_3$, i.e. about 3 mol to 1 mol. This ratio ensures that all the $(NH_4)_2CO_3$ is used up.

(iii) One mole of ethanamide has a mass 59 g, so the yield is about 0.03 mol.

(iv) From the two equations, 1 mol of $(NH_4)_2CO_3$ should give 2 mol of ethanamide, CH_3CONH_2. Therefore, 0.03 mol of $(NH_4)_2CO_3$ should give 0.06 mol of ethanamide. Notice that we must not choose ethanoic acid with which to calculate the yield because it is in excess (not all of it is used up). The yield is

$$\frac{0.03 \text{ mol}}{0.06 \text{ mol}} \times 100\% = 50\%$$

(v) CO_2 and H_2O are released, while the salt $CH_3COO^-NH_4^+$ is made.

(vi) Ethanoic acid that is left over, and water that is made in the reaction.

(vii) If it were water cooled, the ethanamide would crystallise inside the condenser and be particularly hard to collect.

(viii) Dissolve the crystals in the minimum of warm propanone, and allow to cool. Filter off the crystals and allow to dry. (Keep the crystals in a desiccator to keep them dry.)

(ix) This would depend on your resources. A simple test would be to use a melting point apparatus to determine its melting point and compare it with the value in data tables. More sophisticated methods might include using a mass spectrometer and infrared (vibrational) spectroscopy.

120.6 Making a nitrile, e.g.

$$CH_3I + CN^- \rightarrow CH_3CN + I^-$$

120.7 $C_2H_5I \rightarrow C_2H_5CN$; warm with alcoholic potassium cyanide.
$C_2H_5CN \rightarrow C_2H_5COOH$; warm with acid or alkali.
$C_2H_5COOH \rightarrow C_2H_5CH_2OH$; reduce with $LiAlH_4$.
$C_2H_5CH_2OH \rightarrow C_2H_5CH_2I$; warm with iodine and red phosphorus.

120.8 Convert iodoethane into propanenitrile (see question 120.7) and then hydrolyse with acid or alkali. After alkaline hydrolysis, you would have to add acid to convert the $C_2H_5COO^-$ ions into C_2H_5COOH molecules.

120.9 Convert the iodomethane into ethanenitrile, CH_3CN, then the nitrile into ethanoic acid, CH_3COOH. (See questions 120.7 and 120.8 for the method.) Finally, reduce the acid with lithium tetrahydridoaluminate(III). The result is ethanol, CH_3CH_2OH.

120.10 If you look at the structure of the molecule you should see that it is an ester. We can make an ester by one of the methods of section 120.5. One way would be to react propanoic acid with ethanol in the presence of some concentrated sulphuric acid; another would be to react propanoyl chloride with ethanol. If we choose the first method, the reaction scheme would look like this:

$$CH_3I \rightarrow CH_3CN \rightarrow CH_3COOH \rightarrow CH_3CH_2OH \rightarrow$$
ethanol

$$CH_3CH_2I \xrightarrow{CN^-} CH_3CH_2CN \rightarrow CH_3CH_2COOH$$
propanoic acid

Thus the route to the acid also provides the alcohol we need. You will find the reagents needed listed in question 120.7, or elsewhere in the previous organic units.

120.11 If we assume a 60% yield at each stage, then starting with 1 mol of methyl iodide, we would end up with only 0.05 mol (i.e. $(0.6)^6$ mol) of propanoic acid, and even less of the final ester. The preparation of an organic compound should be done with as few stages as possible.

UNIT 120 SUMMARY

- Derivatives of carboxylic acids:

  Include acid chlorides

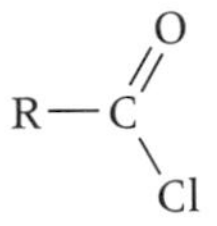

  $$R-\overset{\displaystyle O}{\underset{\displaystyle Cl}{C}}$$

  acid anhydrides

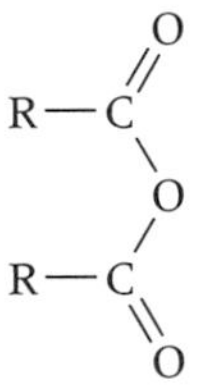

  $$R-\overset{\displaystyle O}{C}\diagdown_{O}$$
  $$R-\overset{\displaystyle }{\underset{\displaystyle O}{C}}$$

  amides

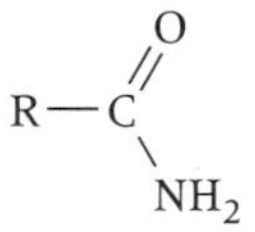

  $$R-\overset{\displaystyle O}{\underset{\displaystyle NH_2}{C}}$$

  esters

  $$R-\overset{\displaystyle O}{\underset{\displaystyle O-R'}{C}}$$

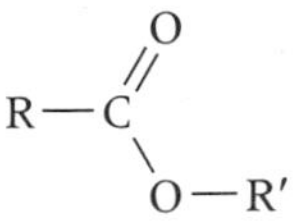

- Preparation:
  (i) Acid chlorides: acid plus phosphorus pentachloride, phosphorus trichloride, or sulphur dichloride oxide;

  e.g. $CH_3COOH + SOCl_2 \rightarrow$
  $\qquad\qquad CH_3COCl + SO_2 + HCl$

  (ii) Acid anhydrides: using an acid chloride;

  e.g. $CH_3COO^-Na^+ + CH_3COCl \rightarrow$
  $\qquad\qquad (CH_3CO)_2O + NaCl$
  $\qquad\qquad$ ethanoic
  $\qquad\qquad$ anhydride

  (iii) Amides: heat the ammonium salt of a carboxylic acid;

  e.g. $CH_3COO^-NH_4^+ \rightarrow CH_3CONH_2 + H_2O$

  (iv) Esters: made from an alcohol plus an acid chloride, or a carboxylic acid with an acid catalyst;

  e.g. $C_3H_7OH + CH_3COCl \rightarrow$
  $\qquad\qquad CH_3COOC_3H_7 + HCl$
  $\qquad\qquad$ propyl ethanoate

Properties

- Acid chlorides:
  (i) Undergo nucleophilic attack. The general pattern is

  $RCOCl + ZH \rightarrow RCOZ + HCl$

  e.g. water $Z = OH$, acids made; ammonia $Z = NH_2$, amides made; alcohols $Z = RO$, esters made.

  (ii) Used as acetylating agents:
  e.g.
  $C_6H_5COCl + C_6H_6 \xrightarrow{AlCl_3} C_6H_5COC_6H_5 + HCl$
  $\qquad\qquad\qquad$ diphenylmethanone

  (iii) Reduction by hydrogen with a catalyst;

  $CH_3COCl \xrightarrow{H_2,\ Pd/BaSO_4} CH_3CHO$

  (iv) Reduction by lithium tetrahydridoaluminate(III) gives alcohols;

  $C_6H_5COCl \rightarrow C_6H_5CH_2OH$

- Acid anhydrides:
  Acetylating agents, similar to acid chlorides but ethanoic acid made as a side product rather than hydrogen chloride.

- Amides:
  (i) Hofmann degradation: heat an amide with bromine in sodium hydroxide solution. The product is an amine with one less carbon atom than the amide;

  e.g. $CH_3CONH_2 \rightarrow CH_3NH_2$

  (ii) With nitrous acid, amide is converted into a carboxylic acid;

  e.g. $CH_3CONH_2 \rightarrow CH_3COOH$

  (iii) Conversion into nitriles by heating with phosphorus(V) oxide;

  e.g. $CH_3CONH_2 - H_2O \rightarrow CH_3CN$

Summary – cont.

    (iv) Hydrolysis; by heating with alkali;

        e.g. $C_2H_5CONH_2 + OH^- \rightarrow C_2H_5COO^- + NH_3$

- Esters:

    (i) Undergo hydrolysis by warming with acid or alkali; conversion to an alcohol and carboxylic acid.

    (ii) Reduced by lithium tetrahydridoaluminate(III) or sodium in ethanol to a mixture of alcohols;

        e.g. $CH_3COOCH_3 \xrightarrow{LiAlH_4} CH_3CH_2OH + CH_3OH$

    (iii) With ammonia, an amide and alcohol are produced;

        e.g. $C_6H_5COOC_2H_5 + NH_3 \rightarrow$
                        $C_6H_5CONH_2 + C_2H_5OH$

- Biuret test:

    A violet colour with a mixture of copper(II) sulphate solution and sodium hydroxide solution shows presence of a peptide link,

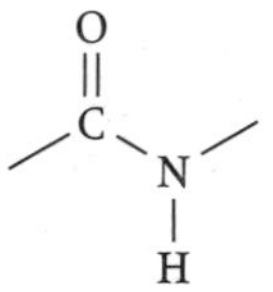

$$
\begin{array}{c}
\quad\quad O \\
\quad\quad \| \\
\diagdown \text{C} \diagdown \\
\quad\quad \text{N} \diagup \\
\quad\quad | \\
\quad\quad \text{H}
\end{array}
$$

Nitriles

- Preparation:

    (i) Dehydrating of amides with phosphorus(V) oxide;

        e.g. $C_2H_5CONH_2 - H_2O \rightarrow C_2H_5CN$

    (ii) Heating halogenoalkanes with an alcoholic solution of potassium cyanide;

        e.g. $C_3H_7I + CN^- \rightarrow C_3H_7CN + I^-$

Reactions

- Undergo hydrolysis to carboxylic acids;

    e.g. $\underset{\text{benzonitrile}}{C_6H_5CN} + OH^- + H_2O \rightarrow \underset{\text{benzoate ion}}{C_6H_5COO^-} + NH_3$

- Reduced to amines;

    $CH_3CN + 2H_2 \xrightarrow{Ni\ cat.} CH_3CH_2NH_2$

- Importance:

    Used to increase the number of carbon atoms in a chain, i.e. ascend a homologous series;

    e.g. $C_2H_5I \rightarrow C_2H_5CN \rightarrow C_2H_5COOH \rightarrow$
                  $C_2H_5CH_2OH \rightarrow C_2H_5CH_2I$

121

Ethers

121.1 **What are ethers?**

Ethers contain two organic radicals bonded to the same oxygen atom. The most common ether is ethoxyethane, $C_2H_5OC_2H_5$. Sometimes it is simply called 'ether' or 'diethyl ether'. In common with other members of the homologous series, it is a volatile liquid. It once found use as an anaesthetic. Some ethers are shown in Table 121.1.

If, like ethoxyethane, the ether contains two identical organic radicals, then it is a symmetrical ether. If the radicals are different, we have an unsymmetrical or mixed ether. Methoxyethane is an unsymmetrical ether:

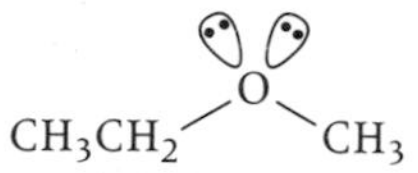

$$CH_3CH_2 \overset{\displaystyle \cdot\cdot}{\underset{\displaystyle O}{}} CH_3$$

Like water, ethers have a triangular shape and they are slightly polar. They are good solvents for a wide range of molecules. Especially, ethoxyethane is widely used on an industrial scale as a solvent. Owing to their volatility and to their sweet smell, some aromatic ethers are used in cosmetics and perfumes.

In the laboratory one of the important things to know about ethers is that they catch fire very easily (their *flash points* are very low). Ethoxyethane has been involved in many fires. Its vapour is dense and if it escapes from an

Table 121.1. Examples of ethers

Name	Formula	Structure	Melting point/°C	Boiling point/°C
Methoxymethane	CH_3OCH_3	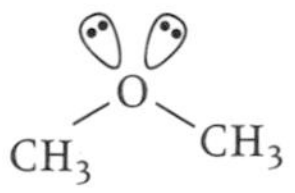	−140	−24
Ethoxyethane	$C_2H_5OC_2H_5$	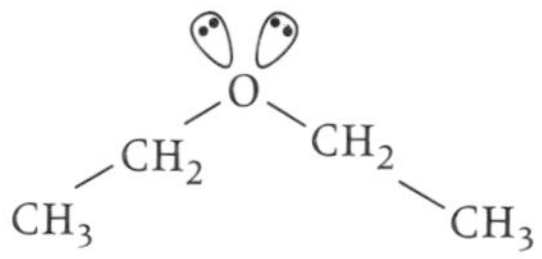	−116	34.6
Methoxybenzene	$C_6H_5OCH_3$	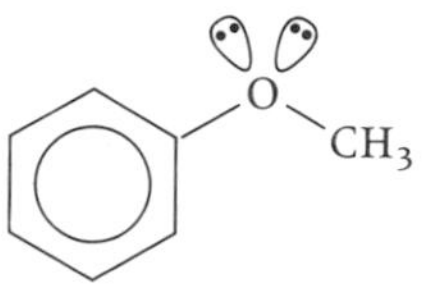	−37.2	154.2
Tetrahydrofuran	$CH_2CH_2CH_2CH_2O$	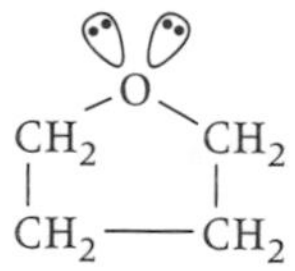	−65	67

Two models showing the shapes of methoxyethane (bottom) and ethoxyethane (top).

apparatus, it can travel over benches and floors. If it reaches the site of a flame some distance away, then an explosion can occur. You should never heat a flask or test tube containing an ether using a naked flame; a preheated water bath is best. Ether has also been known to explode owing to the formation of highly reactive peroxides in the liquid.

There is no obvious site for electrophilic or nucleophilic attack on an ether; ethers are relatively unreactive molecules.

121.2 The preparation of ethers

There are two ways of preparing ethers.

(a) *Dehydration of an alcohol*

We have seen this reaction before as a property of alcohols. The method is to heat an alcohol with concentrated sulphuric acid at about 140 °C. The overall effect of the acid is to take one molecule of water from two molecules of the alcohol. If we wanted to make ethoxyethane, we would use ethanol:

$$CH_3-CH_2-OO-CH_2-CH_3 \xrightarrow{\;-H_2O\;}$$
$$HH$$

$$CH_3-CH_2-O$$
$$CH_2$$
$$CH_3$$

$$2C_2H_5OH - H_2O \rightarrow C_2H_5OC_2H_5$$

By writing the equation in this way, it appears that the acid is not destroyed. Provided we keep the ethanol in excess, we should be able to keep the reaction going indefinitely. For a limited period of time this is so, and we can drip ethanol into the reaction flask to replace that which is changed into ether. The reaction is called *Williamson's continuous ether process*. However, owing to side reactions, in time the acid is reduced to sulphurous acid, and no more ether is made.

(b) Williamson's ether synthesis

The first step in the method is to make a solution of an *alkoxide* ion. The simplest way of doing this is to dissolve sodium in an alcohol. For example, to make ethoxyethane we would use ethanol:

$$2C_2H_5OH + 2Na \rightarrow 2C_2H_5O^- + 2Na^+ + H_2$$

ethanol   ethoxide ion

In section 115.3 we found that the ethoxide ion is a nucleophile, which will react with a halogenoalkane. We would add iodoethane to the flask that contains the ethoxide ion, and reflux them for 10 minutes:

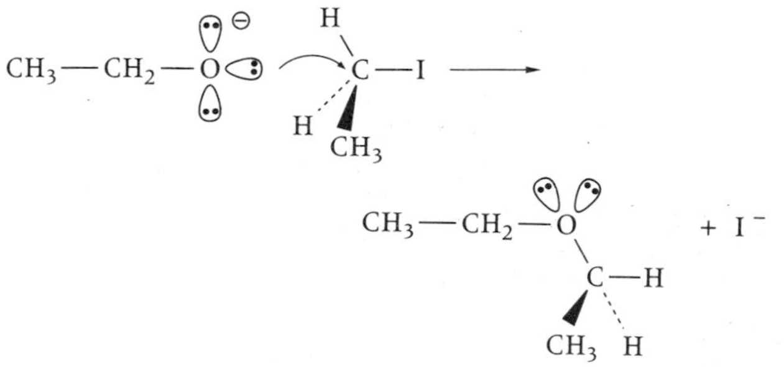

$$C_2H_5O^- + C_2H_5I \rightarrow C_2H_5OC_2H_5 + I^-$$

ethoxide   iodo-   ethoxyethane
ion   ethane

The ether is removed from the reaction flask by distillation, but the distillate is contaminated by ethanol. If the distillate is washed with brine, the ethanol dissolves in the brine, but the ether remains as a separate layer. After separation, the ether can be dried with calcium chloride and redistilled if necessary.

121.3 The reactions of ethers

We have said that ethers are not very reactive. However, the one major reaction they undergo is *cleavage*. If an ether is refluxed with a concentrated solution of hydrogen iodide, the organic radicals are converted into halogenoalkanes:

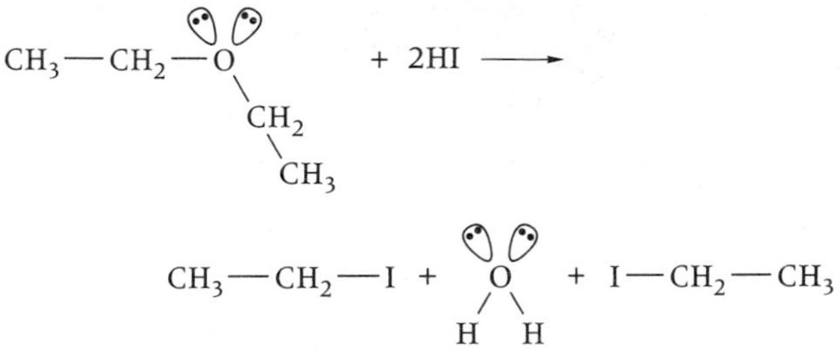

$$C_2H_5OC_2H_5 + 2HI \rightarrow 2C_2H_5I + H_2O$$

Methoxybenzene and other similar aromatic ethers are a little different. Phenol is one of the products:

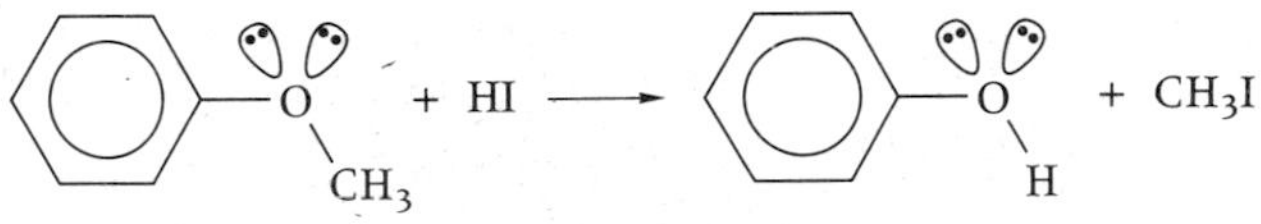

$$C_6H_5OCH_3 + HI \rightarrow C_6H_5OH + CH_3I$$
phenol

A cleavage reaction also takes place on heating with phosphorus pentachloride. The ether is converted into chloroalkanes, but no hydrogen chloride is given off. For example,

$$C_2H_5OCH_3 + PCl_5 \rightarrow C_2H_5Cl + CH_3Cl + POCl_3$$

121.1 The name of $(CH_3CH_2)(CH_3)CHOCH_3$

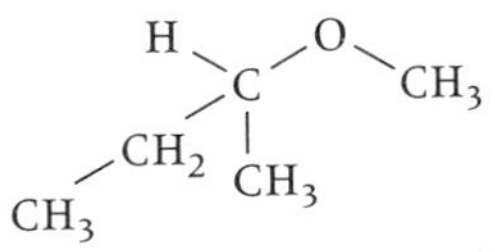

is 2-methoxybutane. What are the names of the following ethers?

(i) $(CH_3CH_2)(CH_3)CHOC_2H_5$

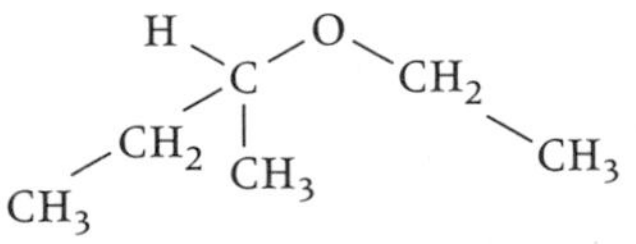

(ii) $(CH_3CH_2CH_2)(CH_3)CHOCH_3$

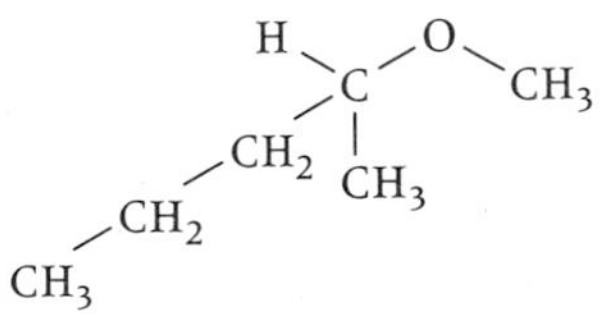

121.2 Why do the melting and boiling points of the ethers suggest that they are only *slightly* polar?

121.3 One mole of water will dissolve about 0.02 mol of ethoxyethane. This solubility is partly due to a small amount of hydrogen bonding that can take place between water and ethoxyethane. Explain, using a diagram, how this hydrogen bonding occurs.

121.4 However, ethoxyethane is much more soluble in an acid like hydrochloric acid. Suggest a reason for this. (Hint: first, take notice of the oxygen in an ether, then think about what happens to hydrogen ions in water.)

121.5 What is the equation for the burning of ethoxyethane in oxygen?

121.6 An organic compound had the formula C_2H_6O. Can you be sure of the name and formula of the compound?

121.7 In the Williamson continuous ether process:

(i) What will be made if the alcohol is *not* kept in

excess, or if the temperature goes up to around $180°C$?

(ii) Why is alkali used to wash the ether before it is dried with calcium chloride and redistilled?

121.8 Williamson's ether synthesis can be used to make mixed ethers. Suggest a way of making methoxyethane, $C_2H_5OCH_3$. Sketch the apparatus that you would use.

121.9 At the end of a Williamson ether synthesis, a white crystalline solid is seen in the reaction flask. What is it?

Answers

121.1 (i) 2-Ethoxybutane; (ii) 2-methoxypentane.

121.2 The melting and boiling points are low. This tells us that the intermolecular forces are weak. Therefore the attractive forces between the dipoles must be weak. Hydrogen bonding does not occur in ethers because none of the hydrogen atoms are directly attached to the electronegative oxygen atom.

121.3

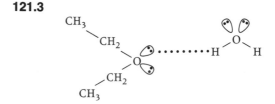

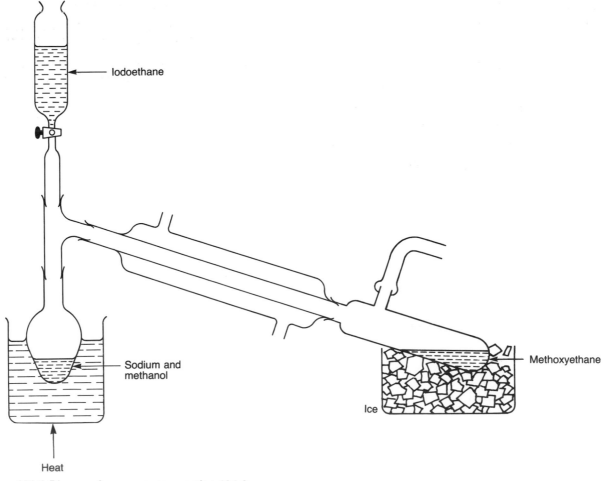

Figure 121.1 *Diagram for answer to question 121.8*

Answers – cont.

Hydrogen bonding can occur between the oxygen atom on an ether and hydrogen atoms on surrounding water molecules. Hydrogen bonding *cannot* occur between oxygen atoms in water molecules and hydrogen atoms bonded to carbon atoms in the ether. These bonds are not sufficiently polar to give the hydrogen atoms an appreciable positive charge.

121.4 Just as the oxygen atom in water has one of its lone pairs protonated to make the oxonium ion, so too a lone pair on the oxygen atom in an ether can be protonated.

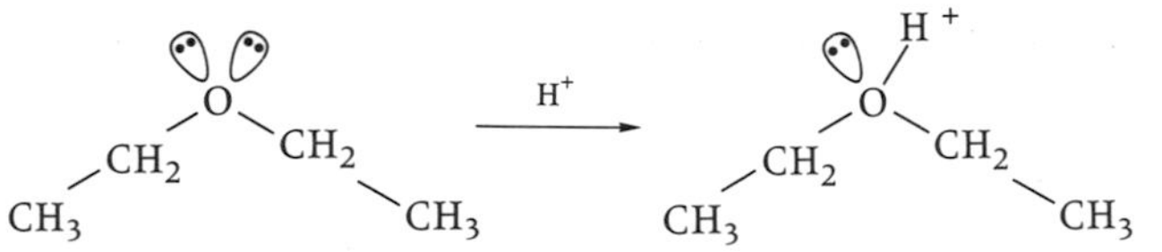

121.5 $C_2H_5OC_2H_5 + 6O_2 \rightarrow 4CO_2 + 5H_2O$

121.6 No; isomers are possible. It could be ethanol, C_2H_5OH, or methoxymethane, CH_3OCH_3.

121.7 (i) Ethene, C_2H_4 (see section 112.2).

(ii) Acidic fumes contaminate the ether. There are fumes from the sulphuric acid and sulphur dioxide is produced when the sulphuric acid oxidises some of the alcohol molecules.

121.8 You could use (i) sodium dissolved in methanol to make methoxide ions, which is then reacted with iodoethane; or (ii) sodium dissolved in ethanol to make ethoxide ions, which is then reacted with iodomethane. The first method is best because iodoethane is a liquid. (Iodomethane is a gas and more difficult to handle.) The apparatus is shown in Figure 121.1.

121.9 Sodium iodide, NaI.

UNIT 121 SUMMARY

- Ethers:
  - (i) Are of the general form R—O—R′.
  - (ii) Show few reactions.
  - (iii) Are highly volatile.
- Preparation:
  - (i) Dehydration of an alcohol by heating with concentrated sulphuric acid at about 140°C;

    e.g. $2C_2H_5OH - H_2O \rightarrow C_2H_5OC_2H_5$

  - (ii) Williamson's ether synthesis. Dissolve sodium in an alcohol, then reflux with an iodoalkane;

    e.g. $2C_2H_5OH + 2Na \rightarrow 2C_2H_5O^- + 2Na^+ + H_2$

    $C_2H_5O^- + C_2H_5I \rightarrow C_2H_5OC_2H_5 + I^-$

- Reactions:
  - (i) Cleaved by refluxing with a concentrated solution of hydrogen iodide;

    e.g. $C_2H_5OC_2H_5 + 2HI \rightarrow 2C_2H_5I + H_2O$
    $C_6H_5OCH_3 + 2HI \rightarrow C_6H_5OH + CH_3I$

  - (ii) Cleaved by heating with phosphorus penta-chloride; no hydrogen chloride given off;

    e.g. $C_2H_5OCH_3 + PCl_5 \rightarrow$
    $\qquad C_2H_5Cl + CH_3Cl + POCl_3$

122

Amines

122.1 There are several types of amine

The key thing about amines is that there is a nitrogen atom bonded to an organic radical. The simplest aliphatic amine is methylamine, CH_3NH_2

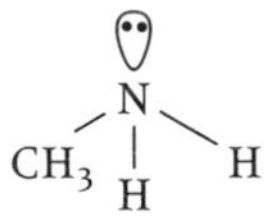

which is similar to ammonia, with one of the hydrogen atoms being replaced by a methyl group. This type of amine is a *primary* amine. If two hydrogen atoms are replaced, then we have a *secondary* amine. With all three hydrogen atoms replaced, a *tertiary* amine is the result. Phenylamine, $C_6H_5NH_2$, is the simplest aromatic primary amine. Some amines are listed in Table 122.1.

Some of the most interesting amines have two NH_2 groups in the same molecule. For example, ethane-1,2-diamine is used as a chelating agent in transition metal chemistry; and hexane-1,6-diamine is used in making Nylon.

Amines are poisonous, many being gases with an unpleasant fishy smell (Figure 122.1). They are widely used as intermediates in the preparation and manufacture of pharmaceuticals, dyes and fabrics. Two common substances related to amines are shown in Figure 122.2.

A common industrial method of manufacturing aliphatic amines is to pass a mixture of ammonia and the vapour of an alcohol over a hot silica or aluminium oxide catalyst. The effect of the catalyst is to remove the elements of water:

$$CH_3OH + NH_3 \rightarrow CH_3NH_2 + H_2O$$

methanol   methyl-   removed by
              amine   the catalyst

Often a mixture of amines results, but sometimes the proportions of each type can be controlled by using a specific set of conditions (temperature and catalyst).

To make phenylamine, a different type of reaction is needed. This time the starting materials are chlorobenzene and ammonia. At a pressure of 50 atm

their vapours are passed over a copper catalyst kept at 200 °C:

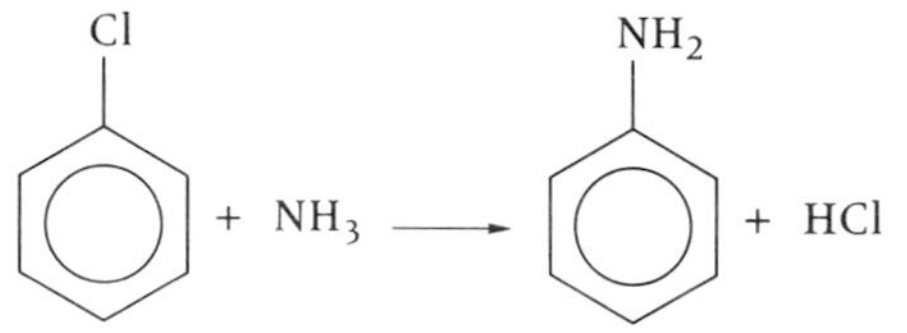

$$C_6H_5Cl + NH_3 \xrightarrow{Cu,\ 200\,°C} C_6H_5NH_2 + HCl$$

122.2 The preparation of amines

We have discussed all but one of the methods of making amines in previous chapters. Here is a summary of the methods that you should have already met.

(a) From halogenoalkanes

The halogenoalkane is heated under pressure with ammonia in alcohol. A mixture of primary, secondary and tertiary amines may be recovered. For example,

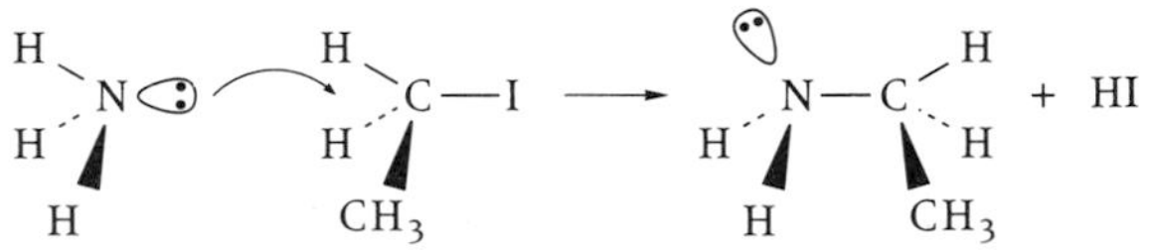

$$NH_3 + C_2H_5I \rightarrow C_2H_5NH_2 + HI$$
iodoethane   ethylamine

The primary amine can go on to make secondary and tertiary amines. You will find details about this complication in section 122.4.

(b) Reduction of nitriles

The vapour of a nitrile is mixed with hydrogen and passed over a nickel catalyst at a temperature of 140 °C. For example,

$$CH_3 - CH_2 - C \equiv N + 2H_2 \longrightarrow$$

$$CH_3 - CH_2 - CH_2 - NH_2$$

$$C_2H_5CN + 2H_2 \rightarrow C_2H_5CH_2NH_2$$
propanenitrile   propylamine

Table 122.1. Examples of amines

Name	Formula	Structure	Melting point/°C	Boiling point/°C
Primary amines				
Methylamine	CH_3NH_2	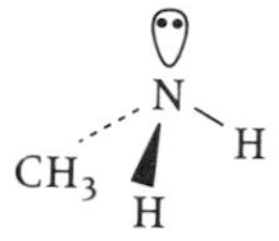	−93.3	−6.2
Ethylamine	$CH_3CH_2NH_2$	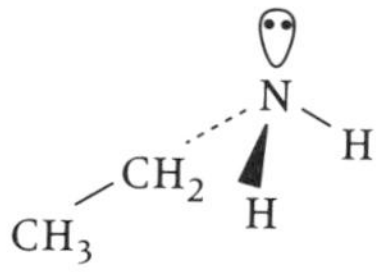	−80.9	16.7
2-Aminopropane	$CH_3CHNH_2CH_3$	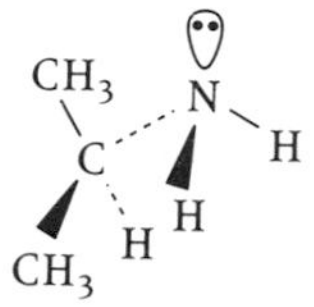	−95.1	32.6
Phenylamine (aniline)	$C_6H_5NH_2$	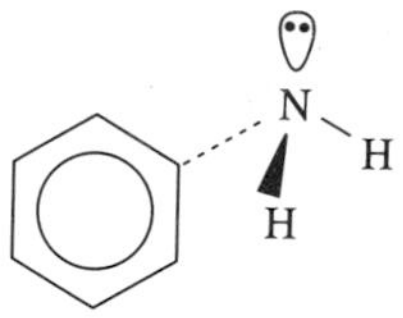	−6.2	184.3
Ethane-1,2-diamine	$NH_2CH_2CH_2NH_2$	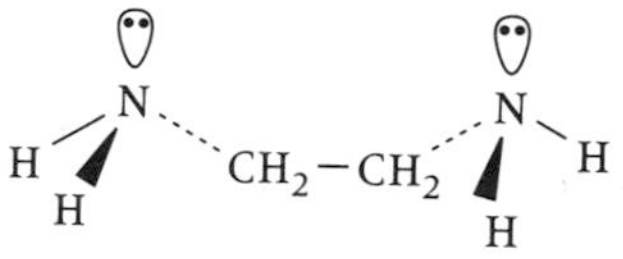	8	117
Secondary amines				
Dimethylamine	$(CH_3)_2NH$	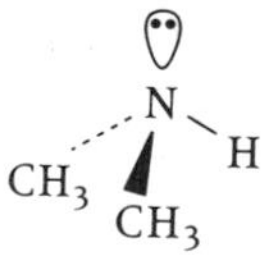	−92	7
Diethylamine	$(CH_3CH_2)_2NH$	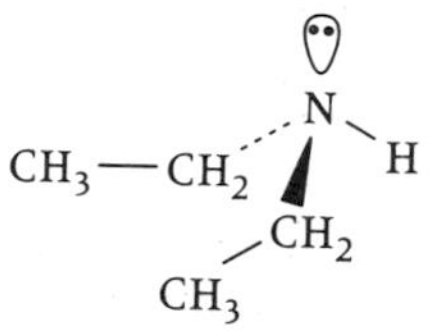	−39	55
N-Methylphenylamine	$C_6H_5NHCH_3$	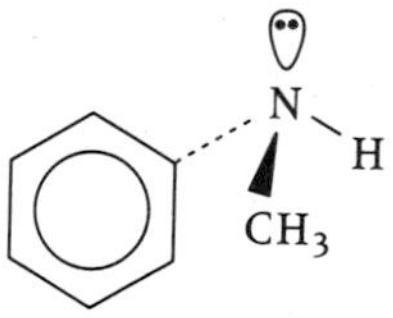	−57	196
Tertiary amines				
Trimethylamine	$(CH_3)_3N$	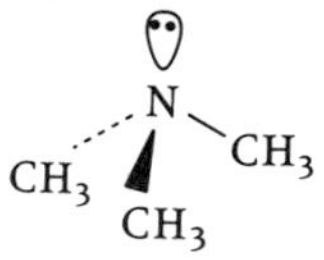	−117.2	3
Triethylamine	$(CH_3CH_2)_3N$	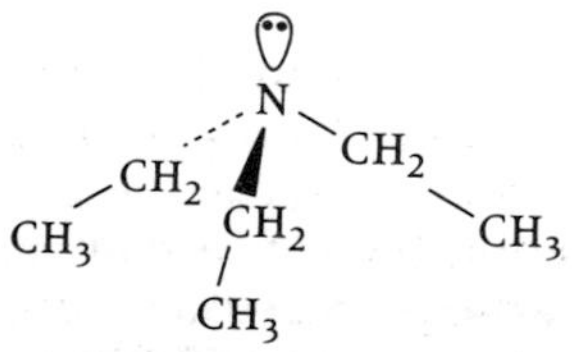	−114.6	89.7

(a) Muscarine

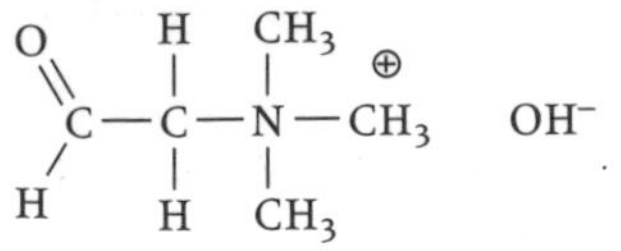

(b) Putrescine

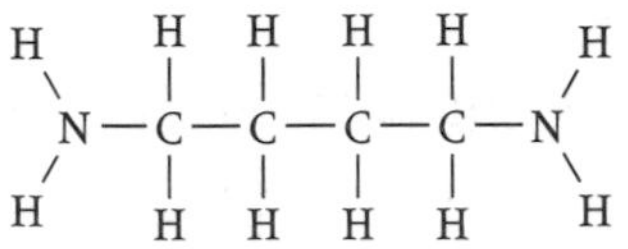

(c) Cadaverine

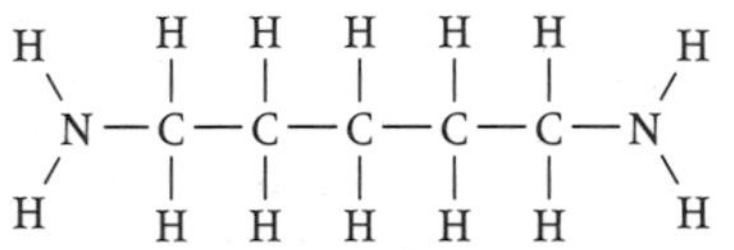

(d) Cocaine

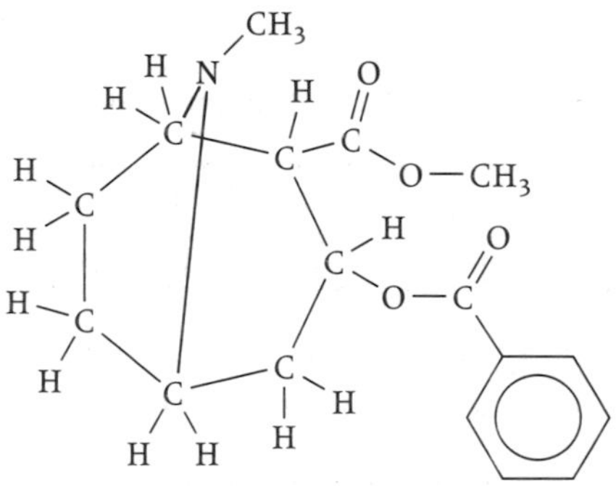

Figure 122.1 *Four unpleasant amines. (a) Muscarine is a tertiary amine found in some highly poisonous mushrooms. (b) Putrescine (butane-1,4-diamine) has a particularly foul smell – of dead bodies and other decaying animal matter. (c) Cadaverine (pentane-1,5-diamine) competes with putrescine for foulness. It too is associated with corpses (cadavers). (d) Cocaine is an infamous amine that causes immense harm to those who use it*

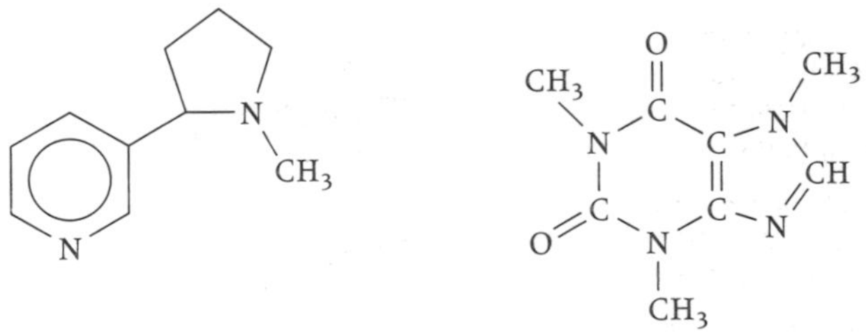

Nicotine

Caffeine

Figure 122.2 *These molecules are related to amines, but have their nitrogen atoms joined in rings with carbon atoms. They are examples of heterocyclic compounds. Both have effects on the nervous system. Nicotine is a dangerous component of tobacco. Caffeine is less dangerous and is found in coffee*

Reduction can also be achieved using sodium in ethanol, or lithium tetrahydridoaluminate(III).

(c)   *The Hofmann degradation of amides*

An amide is refluxed with a mixture of bromine and sodium hydroxide solution:

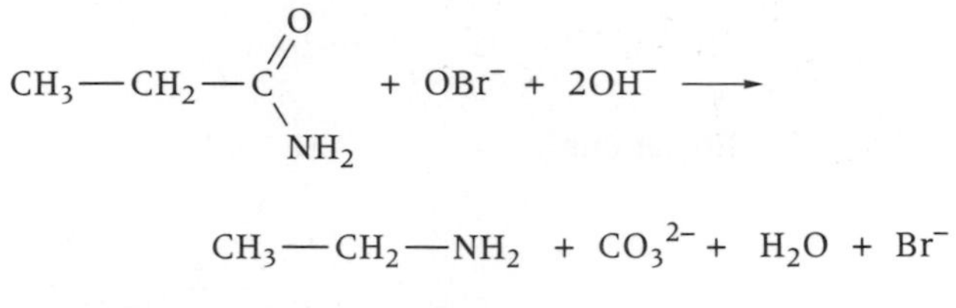

$$CH_3-CH_2-NH_2 + CO_3^{2-} + H_2O + Br^-$$

$C_2H_5CONH_2 + OBr^- + 2OH^- \rightarrow$
propanamide
$\qquad\qquad\qquad C_2H_5NH_2 + CO_3^{2-} + H_2O + Br^-$
$\qquad\qquad\qquad$ ethylamine

(d)   *Reduction of a nitro group*

This is a reaction that we have not met before. It is especially useful for making phenylamine, or other aromatic amines, where the amine group is directly bonded to the benzene ring. The aim is to convert a nitro group, $-NO_2$, into the amine group, $-NH_2$. We shall use the change of nitrobenzene into phenylamine as our example. The essence of the method is to reflux nitrobenzene with a few pieces of tin and concentrated hydrochloric acid. The metal and acid give off hydrogen, which performs the reduction. We can write the reactions in a simplified way as follows.

First, the tin reduces the hydrogen ions:

$$Sn + 4H^+ \rightarrow Sn^{4+} + 2H_2$$

While at the same time the hexachlorostannate(IV) complex ion is formed:

$$Sn^{4+} + 6Cl^- \rightarrow SnCl_6^{2-}$$

The nitrobenzene is reduced by the hydrogen:

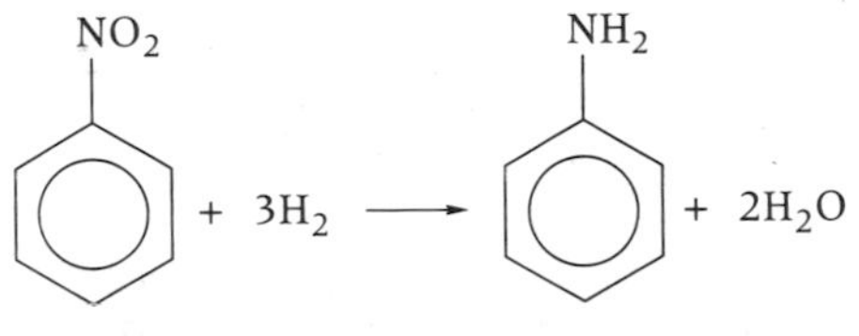

$$C_6H_5NO_2 + 3H_2 \rightarrow C_6H_5NH_2 + 2H_2O$$

Given the melting and boiling points of phenylamine, you might expect to see the product as a liquid. In fact, when the flask cools you would discover that the contents become completely solid. The reason for this is the ease with which amines can act as bases. The nitrogen atom of an amine group has a lone pair of electrons, which can be protonated:

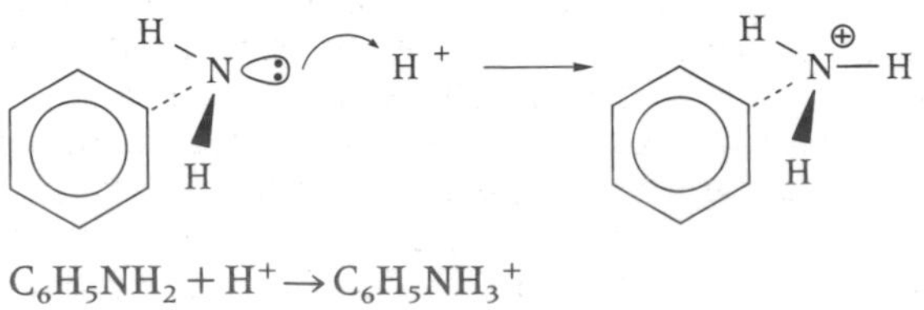

$$C_6H_5NH_2 + H^+ \rightarrow C_6H_5NH_3^+$$

The solid in the flask is an ionic organic salt having the composition $(C_6H_5NH_3^+)_2SnCl_6^{2-}$.

If we want to liberate the phenylamine we need to destroy the solid. The most convenient way is to rip the extra proton from the nitrogen lone pair. We use a concentrated solution of sodium hydroxide to do this:

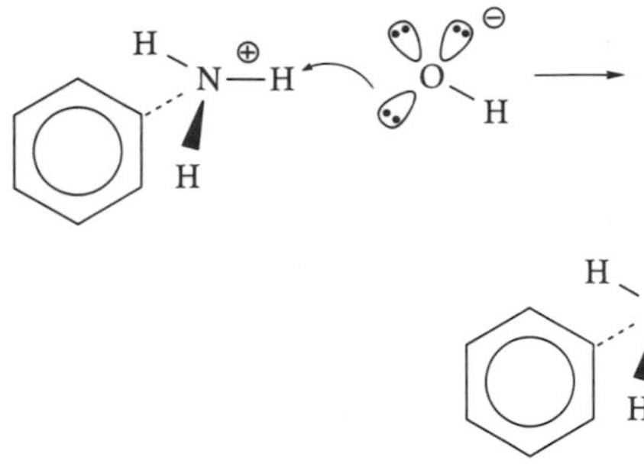

$$C_6H_5NH_3^+ + OH^- \rightarrow C_6H_5NH_2 + H_2O$$

The final stage is to remove the phenylamine from the solution. Owing to the low volatility of phenylamine, steam distillation is used. The presence of the steam has two effects. First, it lowers the temperature at which the phenylamine will distil over; and secondly, it helps to heat the mixture. (You might like to look at section 63.5 to remind yourself about the principles involved in steam distillation.) After some purification, the distillate yields pure phenylamine. *Warning*: phenylamine is extremely poisonous.

122.1 (i) Apart from its use in making amines, what is the importance of the Hofmann degradation reaction?

(ii) If you were to make phenylamine using a Hofmann degradation, what would be your organic starting material?

122.2 Imagine that you could not obtain a supply of nitrobenzene in order to make phenylamine. How would you convert benzene into nitrobenzene?

122.3 Suggest a way of making (phenylmethyl)amine, $C_6H_5CH_2NH_2$, starting from methylbenzene, $C_6H_5CH_3$.

122.3 Amines are bases

You might remember that the Lewis theory claims that acids are electron pair acceptors, and bases are electron pair donors. With its lone pair of electrons on the nitrogen atom, an amine can behave as a Lewis base. The simplest reaction that shows us this property is that, like ammonia, some amines are very soluble in water. For example, methylamine, ethylamine and propylamine are all very soluble. When they dissolve, an equilibrium

is set up for which we can define an equilibrium constant.

(i) Ammonia

$$NH_3(aq) + H_2O(l) \rightleftharpoons NH_4^+(aq) + OH^-(aq)$$

$$K_b = \frac{[NH_4^+(aq)][OH^-(aq)]}{[NH_3(aq)]}$$
$$= 1.75 \times 10^{-5} \text{ mol dm}^{-3}$$

(ii) Methylamine

$$CH_3NH_2(aq) + H_2O(l) \rightleftharpoons CH_3NH_3^+(aq) + OH^-(aq)$$

$$K_b = \frac{[CH_3NH_3^+(aq)][OH^-(aq)]}{[CH_3NH_2(aq)]}$$
$$= 4.4 \times 10^{-4} \text{ mol dm}^{-3}$$

The larger the value of K_b, the further the equilibrium lies to the right, and the stronger is the basic nature of the amine. Be careful to take account of the powers of 10 when you look at K_b values. For example, 10^{-4} is 10 times larger than 10^{-5}, so the K_b of methylamine shows it to be a *stronger* base than ammonia.

A sensible measure that we use to compare bases is their pK_b values. We define the pK_b by

$$pK_b = -\lg K_b$$

For example, the pK_b of ammonia is 4.8, and of methylamine is 3.4. From this we can tell that:

The smaller the value of pK_b, the stronger the base.

Table 122.2 gives you pK_b values of a number of amines. We shall now try to explain some of the trends in the values.

Perhaps the most obvious thing about these values is that the aromatic amines are much weaker bases than the others. Also, they show us that secondary amines

Table 122.2. Amines and their pK_b values

Amine	Formula	pK_b
(Ammonia)	(NH_3)	(4.8)
Primary amines		
Methylamine	CH_3NH_2	3.4
Ethylamine	$CH_3CH_2NH_2$	3.3
Propylamine	$CH_3CH_2CH_2NH_2$	3.4
Secondary amines		
Dimethylamine	$(CH_3)_2NH$	3.3
Diethylamine	$(CH_3CH_2)_2NH$	3.0
Tertiary amines		
Trimethylamine	$(CH_3)_3N$	4.2
Triethylamine	$(CH_3CH_2)_3N$	3.1
Aromatic amines		
Phenylamine	$C_6H_5NH_2$	9.4
N-Methylphenylamine	$C_6H_5NHCH_3$	9.2
N-Phenylphenylamine	$(C_6H_5)_2NH$	13.2

tend to be stronger bases than primary amines. However, this trend does not continue with the tertiary amines. They are weaker than, or little different to, secondary amines.

There are three factors that we must take into account if we are to explain these observations. We would expect the base strength to increase:

(i) if the lone pair is more available for protonation;

(ii) if the protonated positive ion is energetically favoured by delocalisation of charge;

(iii) if the positive ion is readily solvated by the surrounding water molecules.

Likewise, the opposites of these three factors should reduce the base strength. Let us look at the differences between the non-aromatic amines first.

(a) Why are secondary amines more basic than primary amines?

We know that methyl groups give a positive inductive effect. That is, they are able to act as sources of electron density. How we use this idea to explain why, for example, dimethylamine is a stronger base than methylamine, depends on your point of view. We can say that the extra methyl group feeds electron density towards the nitrogen atom, and this in turn results in a lone pair that is a richer source of electron density. Alternatively (or as well), we can concentrate on the nature of the protonated ion and say that, by feeding electron density to the nitrogen atom, the extra positive charge is spread further over the entire molecule (Figure 122.3). By now you will recognise that this would lead to a more energetically favourable situation.

On this basis we would predict the order of base strength to be

$$(CH_3)_3N > (CH_3)_2NH > CH_3NH_2$$
most basic   least basic

Of course, if this were all there were to explaining relative base strengths, we would predict that trimethylamine would be a stronger base than dimethylamine; but it is not. Here we shall turn to the business of solvation that we mentioned under (iii) above. As a general rule, the greater the amount of

hydrocarbon in a molecule, the less soluble it is in water; which is another way of saying that hydrocarbon groups cannot be solvated efficiently by water. Especially, hydrogen atoms directly bonded to carbon atoms are insufficiently polar to take part in hydrogen bonding. However, the hydrogen atoms attached to a nitrogen atom can and do take part in hydrogen bonding. If these hydrogen atoms are replaced by methyl (or other hydrocarbon) groups, then we would expect the amount of hydrogen bonding to decrease. In this way we would expect the degree of hydrogen bonding, and solvation, of the methylamines to follow the order

$$CH_3NH_2 > (CH_3)_2NH > (CH_3)_3N$$
most solvated,   least solvated,
most favoured   least favoured
in solution   in solution

As a result, the base strength should follow the order

$$CH_3NH_2 > (CH_3)_2NH > (CH_3)_3N$$
most basic   least basic

Clearly we have a problem. Our two predictions of the order of base strength are opposite to one another. Faced with this situation, we turn to the experimental results, which tell us that

$$(CH_3)_2NH > CH_3NH_2 > (CH_3)_3N$$
most basic   least basic

The most we can say is that it looks as if the positive inductive effect is more important than solvation until there is too large a proportion of hydrocarbon in the molecule. The change-over comes with tertiary amines.

Note that the order of the base strengths depends on the solvent: in non-aqueous solvents, trimethylamine can be a stronger base than the other two.

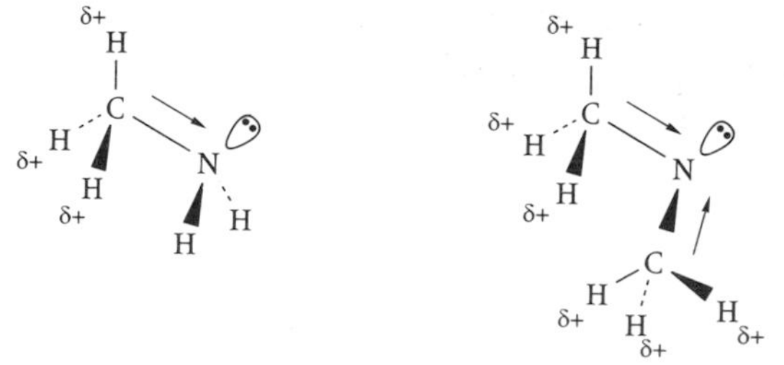

(b) Amines dissolve in acids

Normally amines dissolve easily in strong acids like hydrochloric acid. This is because the lone pair becomes protonated, turning the amine into a positively charged ion. The ion can be solvated by water molecules in the same way as many other similar ions:

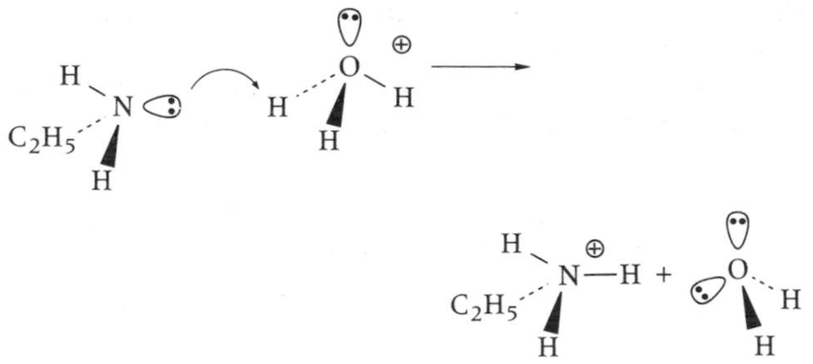

$$C_2H_5NH_2 + H_3O^+ \rightarrow C_2H_5NH_3^+ + H_2O$$

or

$$C_2H_5NH_2 + H^+ \rightarrow C_2H_5NH_3^+$$

Figure 122.3 Methyl groups can feed in electron density to the nitrogen atom in an amine. (Another way of saying this is that methyl groups are polarisable). The hydrogen atoms in the methyl groups take on a slight positive charge, δ+

122.4 Why are hydrogen atoms bonded to a nitrogen atom able to take part in hydrogen bonding with water molecules?

122.5 Imagine that you have a flask containing ethylamine dissolved in hydrochloric acid. If you carefully evaporated the solution, you would be left with a white solid.

(i) What would be its formula?

(ii) What would happen if you heated a few crystals of the solid with sodium hydroxide solution?

(iii) Write an equation for the reaction.

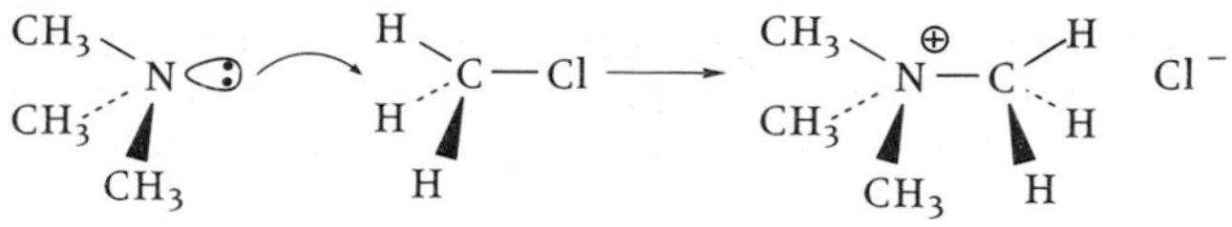

$$(CH_3)_3N + CH_3Cl \rightarrow (CH_3)_4N^+Cl^-$$
tetramethylammonium
chloride

If an amine has a methyl group added to it, we say that the amine has been *methylated*. Methylation reactions have been used to give information about the structures of organic compounds. Try question 122.16 later if you would like to see how this is done.

122.4 Amines react with halogenoalkanes

In section 122.2 we saw that, under pressure, halogenoalkanes will react with ammonia in a solution with alcohol. Owing to its lone pair of electrons, an ammonia molecule can act as a nucleophile. It will attack the carbon atom to which the halogen atom is attached:

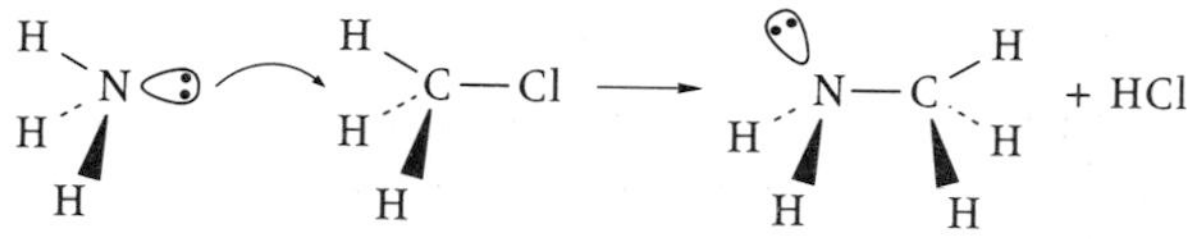

$$NH_3 + CH_3Cl \rightarrow CH_3NH_2 + HCl$$

Given that amines are similar to ammonia, we should expect them to behave as nucleophiles as well. They do:

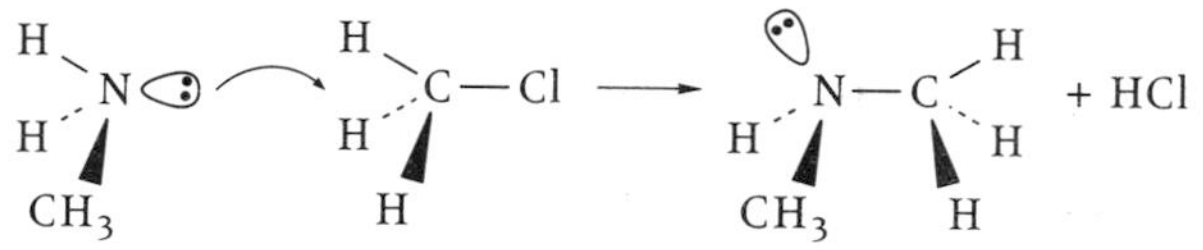

$$CH_3NH_2 + CH_3Cl \rightarrow (CH_3)_2NH + HCl$$
methylamine   dimethylamine

Here we have an example of a primary amine (methylamine) reacting with a halogenoalkane to give a secondary amine (dimethylamine).

However, with the lone pair intact, the secondary amine can react further to give a tertiary amine:

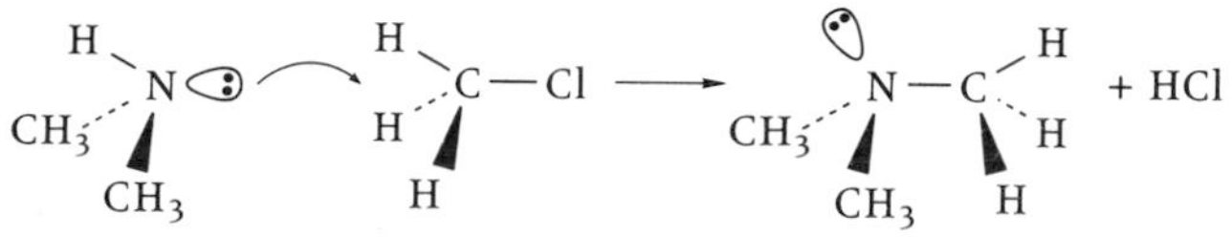

$$(CH_3)_2NH + CH_3Cl \rightarrow (CH_3)_3N + HCl$$
trimethylamine

Likewise, a *quaternary* amine can be produced, although it is to be found as an ionic salt:

122.6 What would you expect to see if you bubbled methylamine through a solution of copper(II) sulphate? (Hint: see section 105.3.)

122.7 What is the connection between the reaction of amines with halogenoalkanes and the Lewis theory of acids and bases?

122.8 Suppose tetramethylammonium chloride, or another quaternary ammonium salt, is dissolved in water. What, if anything, would happen if you were to add silver nitrate solution?

122.5 The acetylation and benzoylation of amines

In Unit 120 we found that ethanoyl and benzoyl chloride are very reactive. They have a carbon atom that is ripe for nucleophilic attack and, as we have seen, amines can act as nucleophiles. The effect of the reaction is that the amine group is acetylated, or benzoylated, as the case may be:

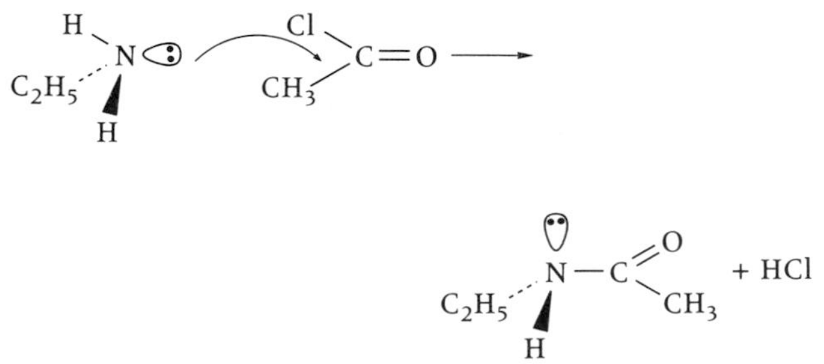

$$C_2H_5NH_2 + CH_3COCl \rightarrow C_2H_5NHCOCH_3 + HCl$$
ethylamine   ethanoyl   N-ethylethanamide
             chloride

$$CH_3NH_2 + C_6H_5COCl \rightarrow CH_3NHCOC_6H_5 + HCl$$
methyl-   benzoyl   N-methylbenzamide
amine   chloride

A variation on this type of reaction is known as the *Schotten–Baumann* reaction. Between 1884 and 1886 Carl Schotten and Eugen Baumann investigated the reaction that has made their names famous. It involves the reaction between benzoyl chloride and phenylamine in the presence of sodium hydroxide. The alkali enhances the reaction, perhaps by removing hydrogen ions that would otherwise protonate the nitrogen on the phenylamine and prevent the reaction taking place:

$$C_6H_5NH_2 + C_6H_5COCl + OH^- \rightarrow$$
$$C_6H_5NHCOC_6H_5 + H_2O + Cl^-$$

In fact the Schotten–Baumann reaction can be done with other aromatic amines and aromatic acid chlorides. The same name is given to the reaction of benzoyl chloride with phenol in the presence of alkali.

122.9 Ethanoyl chloride is a good acetylating agent; so too is ethanoic anhydride. Look back at section 120.3 if you have forgotten the structure of ethanoic anhydride, and predict the result of reacting it with phenylamine.

122.10 What will be made in the Schotten–Baumann reaction if phenol and benzoyl chloride are mixed in a solution of sodium hydroxide?

122.6 Reactions with nitrous acid at ordinary temperatures

Nitrous acid, HNO_2, is easy to make; but it decomposes easily as well. You can make the acid by dissolving sodium nitrite in iced water, and adding ice cold dilute hydrochloric acid to the solution. You should see a rather fine pale blue colour appear. However, be careful about this reaction. Nitrites are known carcinogens, so you should avoid contact with sodium nitrite or its solutions. Likewise, without permission from your teacher or lecturer, you should not attempt the reactions that we shall meet in this section. In any event it is best to wear protective gloves, and eye protection of course.

Essentially there are two things that can happen when an amine reacts with nitrous acid; either the amine changes into an alcohol (which is not of much use), or it does something far more interesting. It is the temperature that determines which of the two alternatives takes place. In this section we shall deal with the first possibility.

(a) At temperatures above 5°C

> **Primary aliphatic amines and phenylamine are converted into alcohols.**

The sign that this is happening is that bubbles of gas appear in the solution. The bubbles are nitrogen gas. For example,

$$C_2H_5NH_2 + HNO_2 \xrightarrow{T>5\,^\circ C} C_2H_5OH + N_2 + H_2O$$
ethylamine   ethanol

$$C_6H_5NH_2 + HNO_2 \xrightarrow{T>5\,^\circ C} C_6H_5OH + N_2 + H_2O$$
phenylamine   phenol

The intermediate in these reactions is a *diazonium ion*. You can spot a diazonium ion in an equation owing to it having an $-N_2^+$ group in place of the $-NH_2$ group. The formula N_2^+ is a shorthand for $-N^+\equiv N$. It is this ion which is destroyed as the temperature increases much above 5 °C. First we have

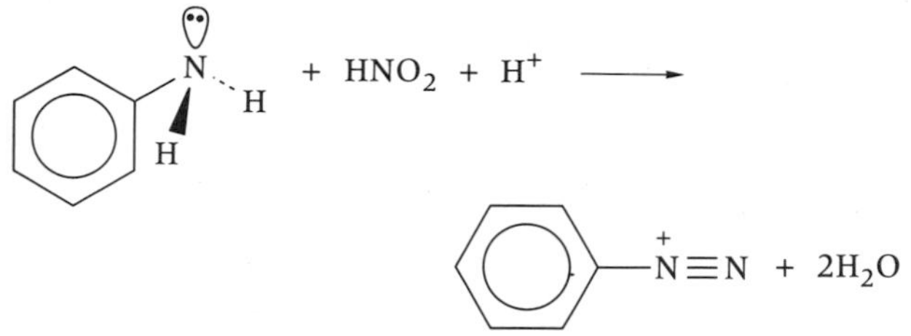

$$C_6H_5NH_2 + HNO_2 + H^+ \rightarrow C_6H_5N_2^+ + 2H_2O$$
a diazonium
ion

which is followed by

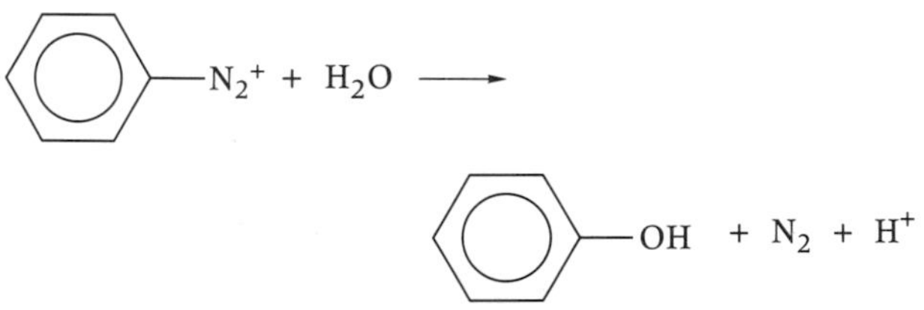

$$C_6H_5N_2^+ + H_2O \xrightarrow{T>5\,°C} C_6H_5OH + N_2 + H^+$$

Actually diazonium ions can decompose in other ways; but the change into an alcohol is often the main reaction.

Secondary amines perform a different reaction. They turn into *nitrosoamines*, but no nitrogen is given off. For example,

$$(C_2H_5)_2NH + HNO_2 \rightarrow (C_2H_5)_2N-N=O + H_2O$$
diethylamine $\qquad$ *N*-nitrosodiethylamine

Nitrosoamines are oily liquids, often with a distinct colour, and are dangerous carcinogens.

Tertiary amines do not react with nitrous acid.

122.11 What will be the effect of reacting (phenylmethyl)amine, $C_6H_5CH_2NH_2$, with nitrous acid at 10 °C?

122.7 Substitution reactions of diazonium ions

Here we will assume that by a judicious use of ice we manage to keep the temperature of the nitrous acid and amine mixture below 5 °C. This will ensure that the diazonium ions remain intact for many minutes. In practice it is the diazonium ions made from amine groups directly attached to a benzene ring that perform useful reactions, so we shall concentrate on them. Aliphatic diazonium ions, e.g. $C_2H_5N_2^+$, decompose almost as soon as they are made.

(a) Substitution reactions

If you think back to the equation that we wrote for phenylamine being converted into phenol, you should be able to persuade yourself that we could call this a substitution reaction. An OH group has been substituted for the N_2^+ group. A wide variety of other substitutions can be achieved just as easily. For example, in the *Sandmeyer reaction* we add cold potassium iodide solution to the benzenediazonium ion solution obtained from phenylamine and nitrous acid. An iodine atom replaces the N_2^+ group:

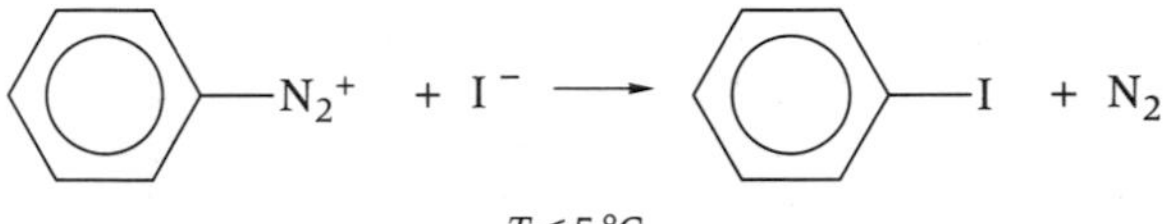

$$C_6H_5N_2^+ + I^- \xrightarrow{T<5\,°C} C_6H_5I + N_2$$
benzenediazonium $\qquad$ iodobenzene
ion

This way of introducing an iodine atom on to a benzene ring was invented by Traugott Sandmeyer. Between 1884 and 1890 he also discovered methods for making chloro- and bromobenzenes. Here are other Sandmeyer reactions.

To make *chlorobenzene* the recipe is to mix copper(I) chloride and concentrated hydrochloric acid with the benzenediazonium ion solution:

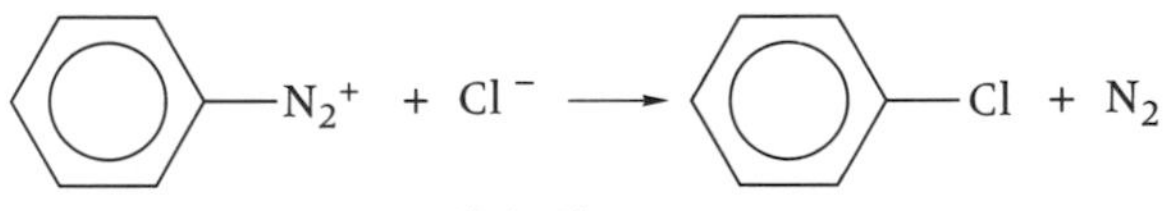

$$C_6H_5N_2^+ + Cl^- \xrightarrow{Cu_2Cl_2/HCl} C_6H_5Cl + N_2$$
chlorobenzene

To make bromobenzene, copper(I) bromide and hydrobromic acid are used.

Chlorobenzene can be made by a similar reaction, invented by Ludwig Gatterman. In the *Gatterman reaction*, copper powder is added to a solution of the benzenediazonium ion in hydrochloric acid. Chlorobenzene is made directly.

Nitriles can be made by substituting a cyanide, CN, group for the N_2^+ ion. The benzenediazonium ion solution is reacted with copper(I) cyanide:

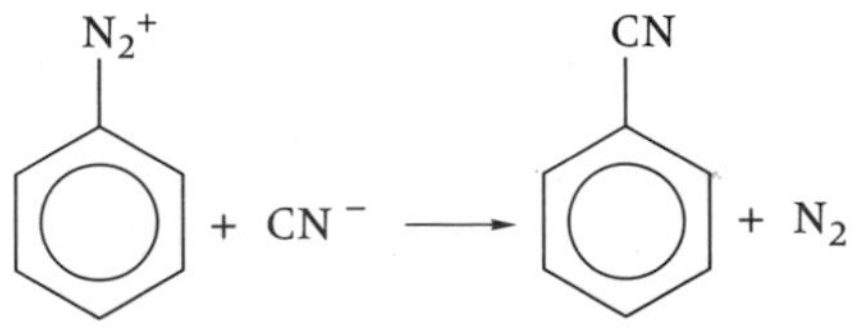

$$C_6H_5N_2^+ + CN^- \rightarrow C_6H_5CN + N_2$$
benzonitrile

One of the reasons why this reaction is important is that it provides a way of introducing an acid group on to a benzene ring. The acid is made in the usual way by acid or alkaline hydrolysis of the nitrile.

122.12 Suggest a method of making benzoic acid: (i) starting with phenylamine, (ii) starting with benzene.

122.8 Coupling reactions of benzenediazonium ions

Aromatic diazonium ions like the benzenediazonium ion are poor electrophiles. They will attack a benzene ring that has an activating group on the ring (see Unit 113). Phenol is one of the simplest examples. To carry out the reaction you would dissolve phenol in sodium hydroxide solution, cool it, and then add the benzenediazonium ion solution a little at a time. The reaction is immediate and you would see a strong yellow colour appear. This is caused by a precipitate of (4-hydroxyphenyl)azobenzene:

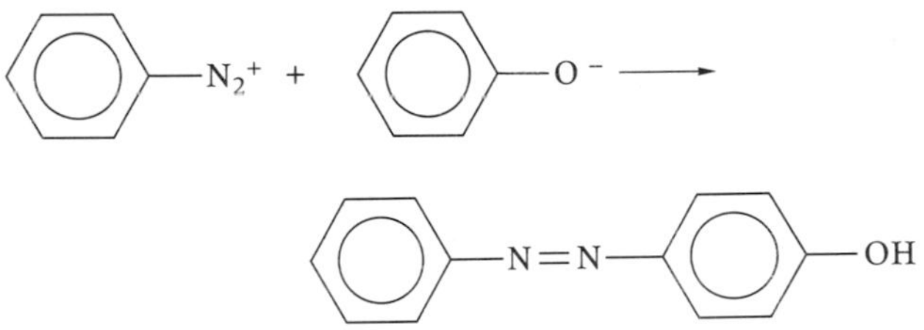

$$C_6H_5N_2^+ + C_6H_5O^- \rightarrow C_6H_5N_2C_6H_5OH$$
phenate ion   (4-hydroxyphenyl)azobenzene

A compound that has *azobenzene* as part of its name is based on the structure:

$$C_6H_5-N{=}N-C_6H_5$$
phenylazobenzene

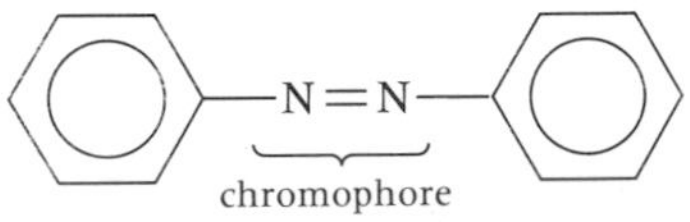

Electrons belonging to the two nitrogen atoms can interact with the π clouds on the two benzene rings. This causes delocalisation of the electrons and leads to a change in the energy levels available to them. It so happens that the energy gap between some of the levels is low enough that visible light can cause electrons to transfer between them. The double bonded nitrogen atoms are largely responsible for giving the azobenzenes their colours. A group like —N=N— is called a *chromophore*. Phenylazobenzene itself is a vivid orange-red.

A similar reaction takes place with naphthalen-2-ol, a naphthalene molecule with an OH group on it:

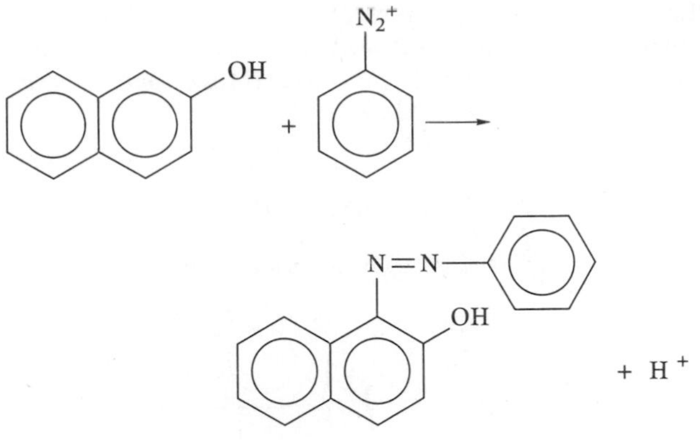

The product is a red solid.

When a benzenediazonium ion reacts with phenylamine, a somewhat different reaction takes place. Instead of attacking the ring, the ion bonds to the nitrogen atom of phenylamine and a yellow solid is made:

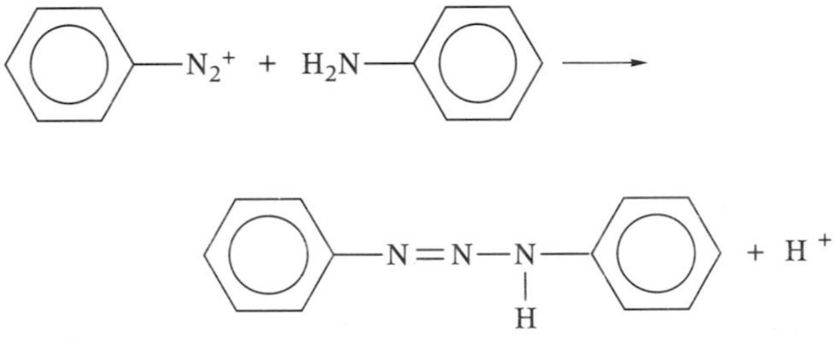

$$C_6H_5N_2^+ + C_6H_5NH_2 \rightarrow C_6H_5-N{=}N-NH-C_6H_5 + H^+$$

There are two main uses for the highly coloured substances made when benzenediazonium ions couple with aromatic molecules. We can use them as dyes or, sometimes, as indicators in acid–base titrations.

(a) Dyes

The dyeing industry has a long and, in some respects, unpleasant history. In Britain, much of the wealth of the country during the nineteenth century was built on the trade in cotton and woollen goods. The skills developed in the manufacture of dyes, and the methods used to dye clothes and other articles, played a large part in the success of the cotton and wool industries.

Unfortunately, the workers in the dye factories were exposed to contact with the dyes, many of which are now known to be strongly carcinogenic. A large number of workers died from the effects of handling the dyes. Owing to the dangers involved, you should *not* attempt to carry out dye-making reactions.

Some of the most effective dyes are based on the molecule called anthraquinone (Figure 122.4). The colours of the dyes depend on the groups attached to the parent molecule. Sometimes they can have extremely complicated structures.

The manufacture and use of dyes pose some interesting problems for chemists. In the first place, we have to discover how to build molecules providing the colour we want. However, not only must the colour be right, it must be resistant to fading, e.g. by being left in sunlight

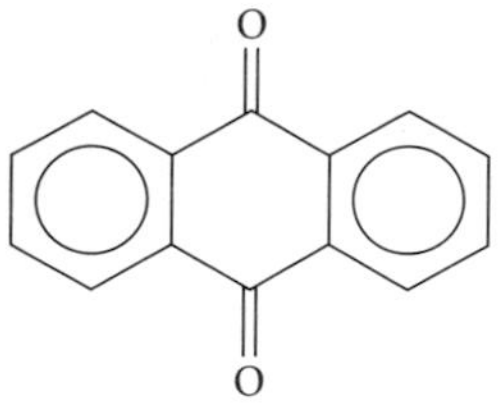

A single molecule of anthraquinone

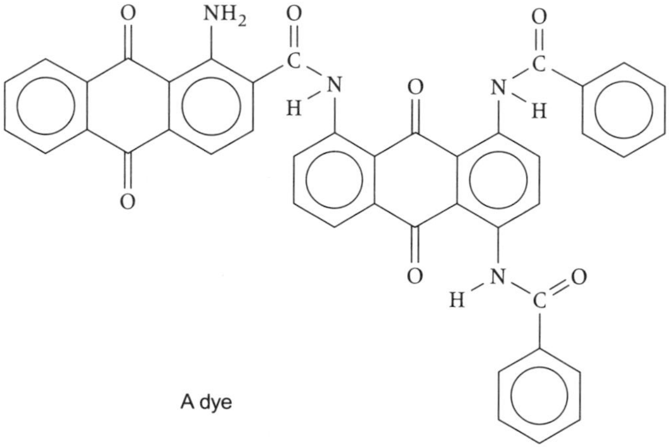

A dye

Figure 122.4 *An example of a dye based on anthraquinone*

or by frequent washing. Even if we solve these problems, there is still the problem of getting the dye to stick permanently to the fabric. This is no easy task because there is such a wide range of fabrics available now, ranging from cotton or wool through to artificial fibres like Nylon.

Essentially there are two ways of encouraging a dye to be permanent. Either you trap dye molecules between the strands of the fibres (the *diffusion* method), or you rely on attraction between ionic groups on the fibres and the dye molecules (the *affinity* method).

Some dyes are used to colour foodstuffs (Table 122.3).

(b) *Indicators*

A good acid–base indicator will change its colour as the concentration of hydrogen ions in a solution changes. Often an indicator will have one or more atoms that can gain or lose protons. For example, methyl orange is an azo dye that is yellow in alkali and red in acid. The arrangement of the electrons is upset by the arrival or departure of a proton on one of the nitrogen atoms, so the energy levels change and the colour changes:

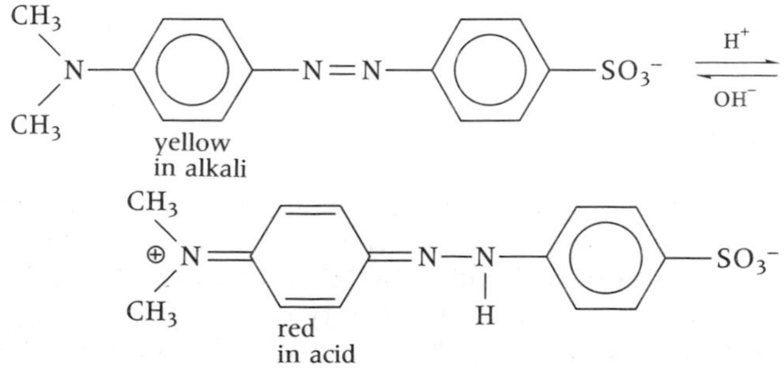

Table 122.3. Dyes commonly used in foods*

Tartrazine (orange-yellow)

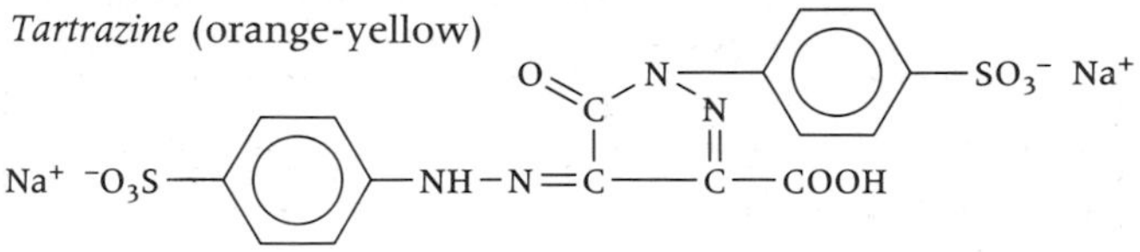

Uses: fizzy drinks, custard powder, chewing gum, jellies, ice lollies
E number 102

Sunset yellow FCF

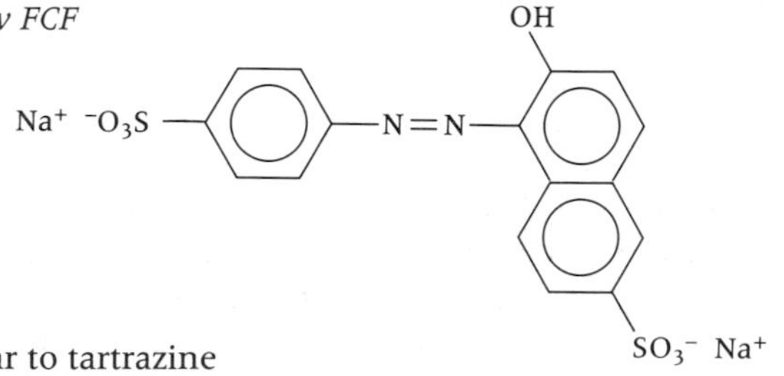

Uses: similar to tartrazine
E number 110

Ponceau (red)

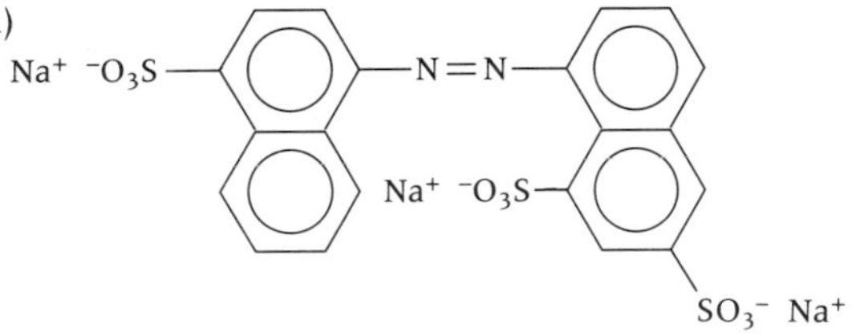

Uses: tinned fruit (e.g. cherries, strawberries), cake mixes
E number 124

Acid brilliant green

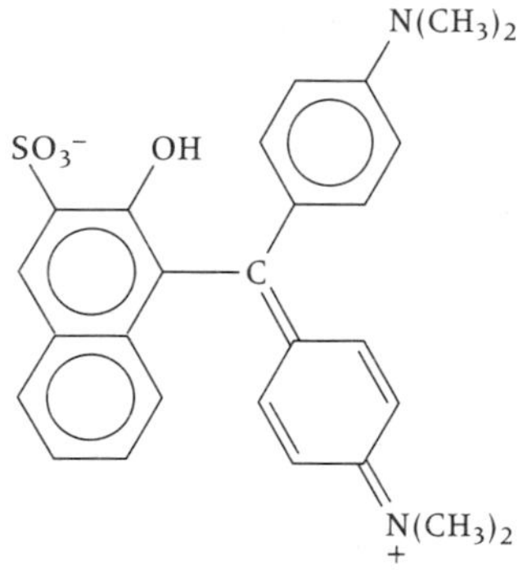

Uses: tinned peas
E number 142

*The use of azo dyes in foods has been blamed for hyperactivity in young children, and other medical problems. Absolute proof that they have harmful effects is hard to establish. This may be because they have no such effects. Alternatively only some individuals may be sensitive to them. The E number is an agreed code given to food additives by a number of countries. The additives and their codes should be shown on the packet or tin

122.13 Look back to Unit 113 to remind yourself whether an amine group on a benzene ring directs *ortho/para* or *meta*. Now see if you can predict the structure of the orange dye made when benzene-diazonium ions react with benzene-1,3-diamine.

122.9   Some reactions of phenyl-amine and other aromatic amines

Phenylamine is the simplest aromatic amine. We have seen that it can be converted into a benzenediazonium ion, and then into a wide range of different compounds. However, there are one or two reactions that you should know about that we have not discussed yet.

(a)   Halogenation

The first is a reaction that shows that the amine group activates a benzene ring.

With bromine, phenylamine reacts very quickly to give a white precipitate of 2,4,6-tribromophenylamine:

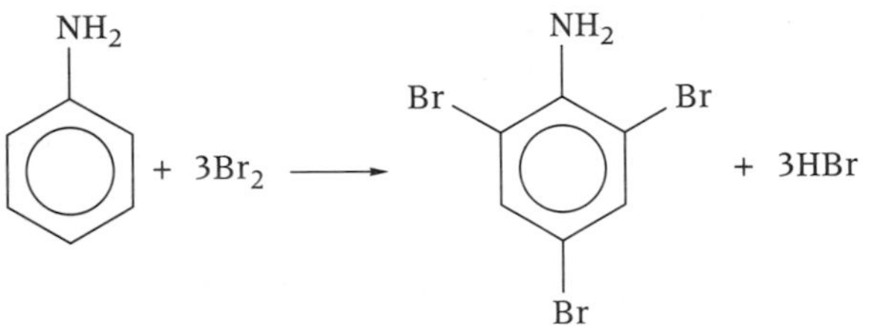

$$C_6H_5NH_2 + 3Br_2 \rightarrow C_6H_2Br_3NH_2 + 3HBr$$

This reaction is similar to that between phenol and bromine. The problem with the amine group is that in some ways it is too good at activating the ring. For example, if we wanted to make 4-bromophenylamine from phenylamine we have a problem. We cannot stop the bromine atoms appearing at the *ortho* (2) positions as well. There is a standard way of getting round the problem. It is to make a change to the amine group. The usual method is to acetylate it with ethanoyl chloride or ethanoic anhydride, e.g.

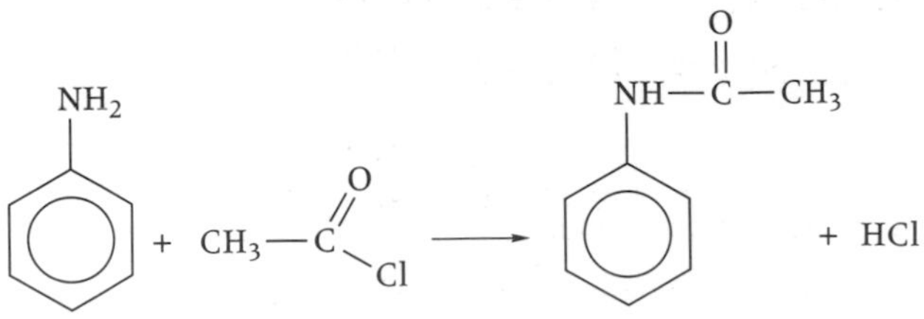

$$C_6H_5NH_2 + CH_3COCl \rightarrow C_6H_5NHCOCH_3 + HCl$$

When bromine is added (usually in a solution with ethanoic acid), the *para* position only is attacked:

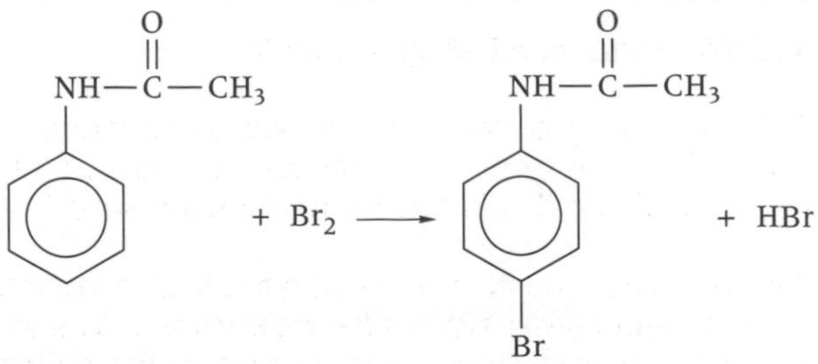

$$C_6H_5NHCOCH_3 + Br_2 \rightarrow BrC_6H_4NHCOCH_3 + HBr$$

Finally the acetylated group is converted back to an amine by hydrolysis:

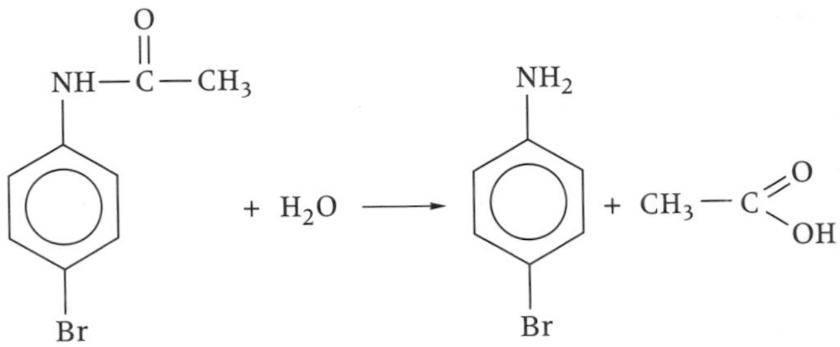

$$BrC_6H_4NHCOCH_3 + H_2O \rightarrow BrC_6H_4NH_2 + CH_3COOH$$

(b)   Nitration

Acetylation of the amino group is also used in the nitration of phenylamine. Nitrating mixture will convert phenylamine into a mixture of nitro products. These include a lot of 3-nitrophenylamine, which is made because the NH_2 group is protonated by the acid; the resulting NH_3^+ group directs *meta* (3 position). Also, the phenylamine is oxidised to a nasty collection of oily and tarry products. In short, the reaction is best avoided. If we acetylate the amine group first, the ring is no longer so reactive and it is less prone to oxidation. Acetylation *protects* phenylamine from oxidation. A mixture of 2-nitrophenylamine and 4-nitrophenylamine is produced after hydrolysis:

$$C_6H_5NHCOCH_3 \xrightarrow{\text{nitration}} NO_2C_6H_4NHCOCH_3$$

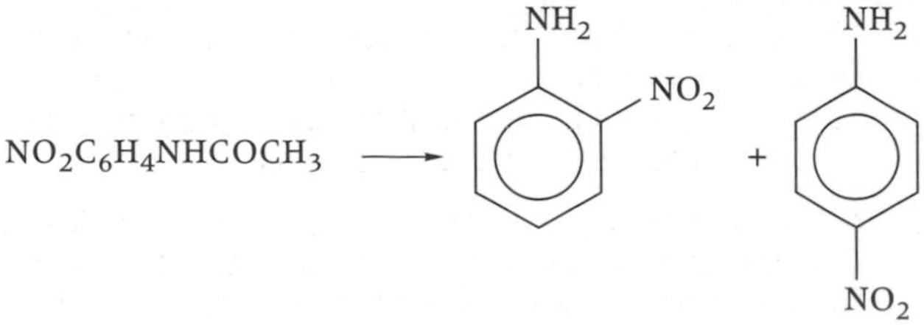

$$NO_2C_6H_4NHCOCH_3 \xrightarrow{\text{hydrolysis}}$$

$$NO_2C_6H_4NH_2 \quad + \quad NO_2C_6H_4NH_2$$
2-nitrophenylamine   4-nitrophenylamine

122.14 What is nitrating mixture?

122.15 If 25 g of nitrobenzene were converted into 15 g of phenylamine by reduction with tin and hydrochloric acid, what is the percentage yield?

122.16 This question is about a technique invented by Hofmann called *exhaustive methylation*. The idea is that the methylation of amines follows the pattern:

$$\text{primary} \xrightarrow{1} \text{secondary} \xrightarrow{2} \text{tertiary} \xrightarrow{3} \text{quaternary}$$
$$\text{amine} \qquad \text{amine} \qquad \text{amine} \qquad \text{amine}$$

In each of the three stages, 1 mol of the amine would react with 1 mol of chloromethane. For example, 1 mol of ethylamine reacted with chloromethane and completely converted to its quaternary salt would use up 3 mol of chloromethane:

$$CH_3CH_2NH_2 + 3CH_3Cl \rightarrow$$
$$CH_3CH_2N(CH_3)_3{}^+Cl^- + 2HCl$$

Now see if you can answer these questions.

(i) One mole of an amine of formula C_4H_9N was found to react with 2 mol of chloromethane. Was the amine primary, secondary or tertiary?

(ii) The amine did not react with bromine. What does this tell you about the bonding between the carbon atoms?

(iii) Try to discover a structure for the amine that fits the information.

Answers

122.1 (i) The Hofmann degradation decreases the number of carbon atoms in a molecule.

(ii) Benzamide, $C_6H_5CONH_2$.

122.2 React benzene with nitrating mixture.

122.3 First, pass chlorine through methylbenzene in the presence of ultraviolet light until sufficient (chloromethyl)benzene is made:

$$C_6H_5CH_3 + Cl_2 \rightarrow C_6H_5CH_2Cl + HCl$$

After separation, react this with an alcoholic solution of ammonia under pressure:

$$C_6H_5CH_2Cl + NH_3 \rightarrow C_6H_5CH_2NH_2 + HCl$$

122.4 A nitrogen atom is sufficiently electronegative to draw electron density away from a hydrogen atom. This leaves the hydrogen atom with a slight positive charge, which is able to hydrogen bond with a lone pair on the oxygen atom in a water molecule.

122.5 (i) A white ionic solid, $CH_3CH_2NH_3{}^+Cl^-$, would be left. We can call it ethylamine hydrochloride. You may find that amines are stored in the laboratory in the form of salts.

(ii) The amine is released when it is heated with an alkali. In this case ethylamine would be given off. This is equivalent to the reaction of ammonium salts giving off ammonia with alkali.

(iii) $C_2H_5NH_3{}^+ + OH^- \rightarrow C_2H_5NH_2 + H_2O$

122.6 The blue solution should become a much deeper blue. Amines can act as ligands. They give a deep blue colour similar to that found when ammonia is added to a solution of copper(II) ions.

122.7 According to the Lewis theory, acids are electron pair acceptors and bases are electron pair donors. With the lone pair on the nitrogen atom, amines are electron pair donors. That is, they are Lewis bases. If we wanted, we could classify many nucleophiles as Lewis bases.

122.8 The quaternary salt is ionic and when it dissolves the component ions are free to move through the water. The chloride and silver ions will give a white precipitate of silver chloride.

122.9 In the reaction with ethanoic anhydride, ethanoic acid is one of the products:

$$(CH_3CO)_2O + C_6H_5NH_2 \rightarrow$$
ethanoic   phenylamine
anhydride

$$C_6H_5NHCOCH_3 + CH_3COOH$$
N-phenyl-   ethanoic
ethanamide   acid

122.10 The ester phenyl benzoate is produced:

The reaction is:

$$C_6H_5COCl + C_6H_5OH \rightarrow C_6H_5COOC_6H_5 + HCl$$

122.11 Phenylmethanol is made, with nitrogen given off:

$$C_6H_5CH_2NH_2 + HNO_2 \rightarrow C_6H_5CH_2OH + N_2 + H_2O$$

122.12 (i) The route is as follows:

$$C_6H_5NH_2 \xrightarrow{\text{diazotise}} C_6H_5N_2{}^+ \xrightarrow{\text{KCN}}$$

$$C_6H_5CN \xrightarrow{\text{hydrolysis}} C_6H_5COOH$$

(ii) There are several methods. We could convert benzene into phenylamine and then proceed as in (i):

$$C_6H_6 \xrightarrow{\text{nitration}} C_6H_5NO_2 \xrightarrow{\text{Sn/conc. HCl}} C_6H_5NH_2$$

Alternatively, benzene could be converted into methylbenzene by a Friedel–Crafts reaction (chloromethane plus aluminium trichloride), followed by oxidation of the methyl side chain with alkaline potassium manganate(VII).

122.13 Amine groups activate a benzene ring and they direct *ortho* and *para*. With the two amine groups in the 1,3 positions, the best position for the benzene-diazonium ion to attack is *para* to one of them and *ortho* to the other. The orientation is also influenced by steric hindrance. The dye has the structure:

122.14 A mixture of concentrated sulphuric and nitric acids.

122.15 One mole of $C_6H_5NO_2$, 123 g, should give 1 mol of $C_6H_5NH_2$, 93 g. Thus 25 g of nitrobenzene should give

$93\,g \times 25\,g/123\,g = 18.9\,g$. The percentage yield is $15\,g/18.9\,g \times 100\% = 79\%$.

122.16 (i) It is a secondary amine.

(ii) There are no double or triple bonds.

(iii) The structure is a five-membered ring called pyrrolidine

Actually we have only covered one part of Hofmann's method. The quaternary amine produced at the end of the experiment is heated. This breaks it apart, and the products can be analysed. The nature of the products gives extra information about the structure of the original amine.

UNIT 122 SUMMARY

- Amines:
    (i) Have one or more NH_2 groups.
    (ii) Primary amines, RNH_2; secondary $RR'NH$; tertiary, $RR'R''N$.
    (iii) Are basic.
    (iv) Owing to the lone pair of electrons on the nitrogen, they can make salts, e.g. $RNH_3^+Cl^-$.
- Preparation:
    (i) Heat a halogenoalkane under pressure with ammonia in alcohol;

    e.g. $NH_3 + C_2H_5I \rightarrow C_2H_5NH_2 + HI$

    (ii) Reduction of nitriles;

    e.g. $C_2H_5CN + H_2 \xrightarrow{Ni,\ 140\,°C} C_2H_5CH_2NH_2$

    (iii) Hofmann degradation of amides: reflux an amide with a mixture of bromine and sodium hydroxide solution;

    e.g. $C_2H_5CONH_2 + OBr^- + 2OH^- \rightarrow$
    $\qquad\qquad C_2H_5NH_2 + CO_3^{2-} + H_2O + Br^-$

    This reaction decreases the number of carbon atoms by one.

    (iv) Reduction of a nitro group; especially for making phenylamine;

    $C_6H_5NO_2 + 3H_2 \xrightarrow{Sn,\ conc.HCl} C_6H_5NH_2 + 2H_2O$

    The product appears as a salt, which must be destroyed with alkali; separated by steam distillation.

Reactions

- As bases:
    (i) pH > 7 in water;

    e.g. $CH_3NH_2(aq) + H_2O(l) \rightleftharpoons$
    $\qquad\qquad CH_3NH_3^+(aq) + OH^-(aq)$

    Primary amines most basic, tertiary amines least basic.
    (ii) Form salts (dissolve in solutions of inorganic acids);

    e.g. $C_2H_5NH_2 + H^+ \rightarrow C_2H_5NH_3^+$

- With halogenoalkanes:
    Halogen atom displaced;

    e.g. $CH_3NH_2 + CH_3Cl \rightarrow (CH_3)_2NH + HCl$

    A quaternary amine can be made;

    $(CH_3)_3N + CH_3Cl \rightarrow (CH_3)_4N^+Cl^-$

- Acetylation and benzoylation:
    Amines react very easily with an acid chloride or benzoyl chloride;

    $C_2H_5NH_2 + CH_3COCl \rightarrow C_2H_5NHOCCH_3 + HCl$
    $CH_3NH_2 + C_6H_5COCl \rightarrow CH_3NHOCC_6H_5 + HCl$

- Nitrous acid:
    Non-aromatic amines decompose to alcohols, giving off nitrogen gas.

Reactions of phenylamine

- Nitrous acid:
  (i) Below 5 °C, the benzenediazonium ion is made from phenylamine and sodium nitrite in hydrochloric acid;

  $$C_6H_5NH_2 + HNO_2 + H^+ \rightarrow C_6H_5N_2^+ + 2H_2O$$

  This can take part in reactions: to make C_6H_5I, C_6H_5CN; to make dyes (coupling reactions), e.g. with phenate ions

  $$C_6H_5N_2^+ + C_6H_5O^- \rightarrow C_6H_5N_2C_6H_5O^- + H^+$$

  (ii) Above 5 °C, diazonium ions decompose; nitrogen bubbles off;

  $$C_6H_5NH_2 + HNO_2 \rightarrow C_6H_5OH + N_2 + H_2O$$

- Bromination:
  The amine group activates the ring. Immediate white precipitate;

  $$C_6H_5NH_2 + 3Br_2 \rightarrow C_6H_2Br_3NH_2 \qquad + 3HBr$$
  2,4,6-tribromophenylamine

- Acetylation:
  Is used to protect the amine group. For example, to make 4-bromophenylamine:

  $$C_6H_5NH_2 + CH_3COCl \rightarrow C_6H_5NHCOCH_3 + HCl$$
  $$C_6H_5NHCOCH_3 + Br_2 \rightarrow BrC_6H_4NHCOCH_3 + HBr$$
  $$BrC_6H_4NHCOCH_3 + H_2O \rightarrow BrC_6H_4NH_2 + CH_3COOH$$

- Nitration:
  Similar scheme: acetylate, nitrate, hydrolyse.

123

Amino acids and proteins

123.1 **What is an amino acid?**

An amino acid has *two* functional groups in its molecule. One is an amine group, NH$_2$, the other a carboxylic acid, COOH. You might realise that a molecule with an acidic and a basic group in the same molecule may have some unusual properties; and you would be right. To begin with, even the simplest amino acids are *solids*. (Up to now the early members of all the homologous series we have looked at have been gases or liquids.) Of rather more importance is that we find amino acids, or their derivatives, in biologically important systems such as proteins. Indeed, amino acids are the *fundamental building blocks of living things*.

If we are to understand life and its evolution from a scientific point of view, we have to understand how amino acids are made, and how they react. Another odd thing about the amino acids in living things is that (approximately) only 20 amino acids occur naturally (Table 123.1), although we can make many more in the laboratory. In addition to a number of chemical properties that they have in common, these 20 have a remarkable structural similarity. For example, they all have the amine group on the carbon atom next door to the acid group. This carbon atom is the alpha (α) carbon atom. As a consequence we call the naturally occurring amino acids the *alpha amino acids*. Also, apart from glycine they are all optically active, and the arrangement of the groups around the asymmetric carbon atom follows the same pattern; but more of this later.

123.1 What would happen if you were to mix glycine with a solution of nitrous acid at room temperature? Write an equation for the reaction.

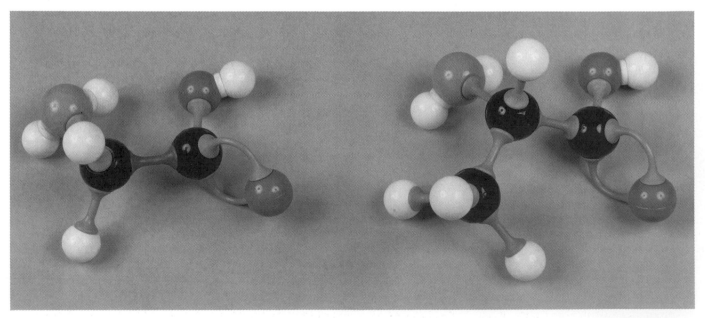

These models of glycine (left) and alanine (right) show the configurations of the naturally occurring amino acids (CORN rule).

Table 123.1. The 20 most important naturally occurring amino acids*†

Name	Code	Structure	Name	Code	Structure
Alanine	ala	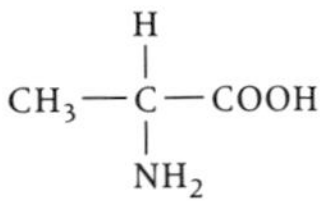 $CH_3-\overset{\overset{H}{\mid}}{\underset{\underset{NH_2}{\mid}}{C}}-COOH$	Leucine	leu	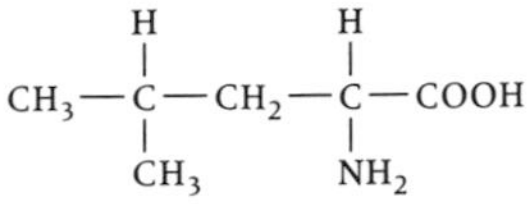 $CH_3-\overset{\overset{H}{\mid}}{\underset{\underset{CH_3}{\mid}}{C}}-CH_2-\overset{\overset{H}{\mid}}{\underset{\underset{NH_2}{\mid}}{C}}-COOH$
Arginine	arg	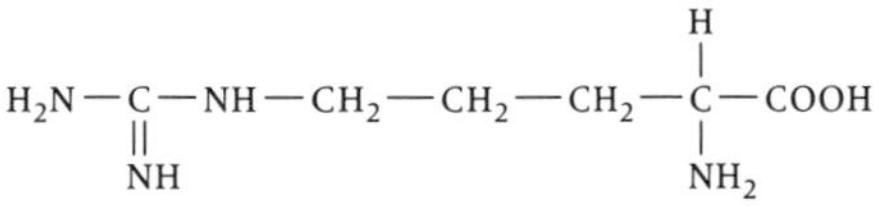 $H_2N-\overset{}{\underset{\underset{NH}{\parallel}}{C}}-NH-CH_2-CH_2-CH_2-\overset{\overset{H}{\mid}}{\underset{\underset{NH_2}{\mid}}{C}}-COOH$	Lysine	lys	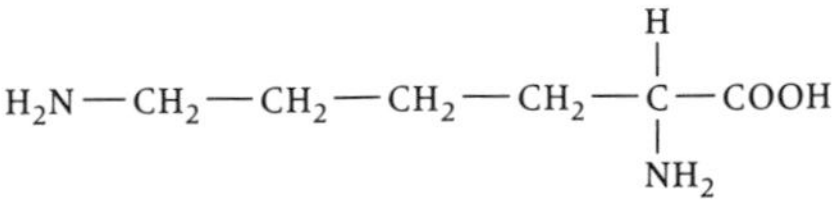 $H_2N-CH_2-CH_2-CH_2-CH_2-\overset{\overset{H}{\mid}}{\underset{\underset{NH_2}{\mid}}{C}}-COOH$
Asparagine	asn	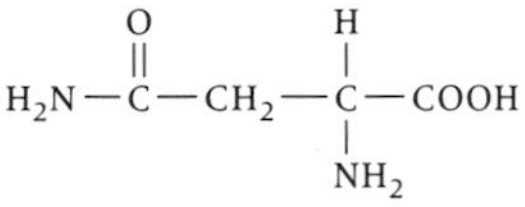 $H_2N-\overset{\overset{O}{\parallel}}{C}-CH_2-\overset{\overset{H}{\mid}}{\underset{\underset{NH_2}{\mid}}{C}}-COOH$	Methionine	met	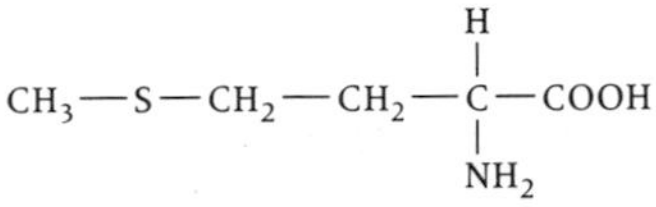 $CH_3-S-CH_2-CH_2-\overset{\overset{H}{\mid}}{\underset{\underset{NH_2}{\mid}}{C}}-COOH$
Aspartic acid	asp	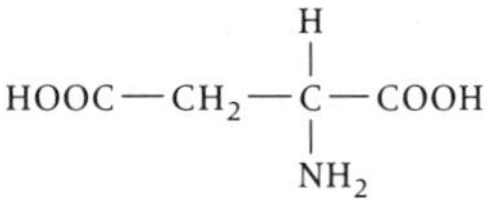 $HOOC-CH_2-\overset{\overset{H}{\mid}}{\underset{\underset{NH_2}{\mid}}{C}}-COOH$	Phenyl alanine	phe	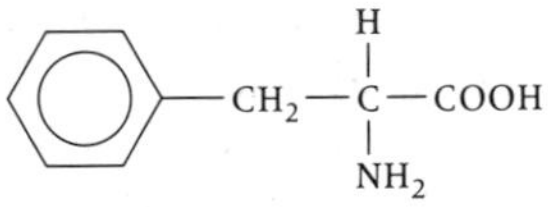 $\langle\text{benzene}\rangle-CH_2-\overset{\overset{H}{\mid}}{\underset{\underset{NH_2}{\mid}}{C}}-COOH$
Cysteine	cys	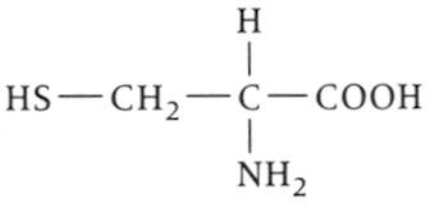 $HS-CH_2-\overset{\overset{H}{\mid}}{\underset{\underset{NH_2}{\mid}}{C}}-COOH$	Proline	pro	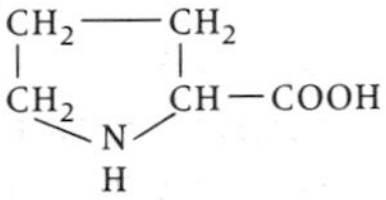 ring: CH_2-CH_2 / CH_2 $CH-COOH$ / N(H)
Glutamic acid	glu	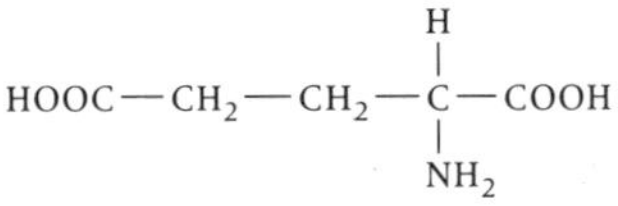 $HOOC-CH_2-CH_2-\overset{\overset{H}{\mid}}{\underset{\underset{NH_2}{\mid}}{C}}-COOH$	Serine	ser	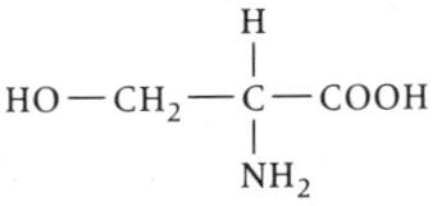 $HO-CH_2-\overset{\overset{H}{\mid}}{\underset{\underset{NH_2}{\mid}}{C}}-COOH$
Glutamine	gln	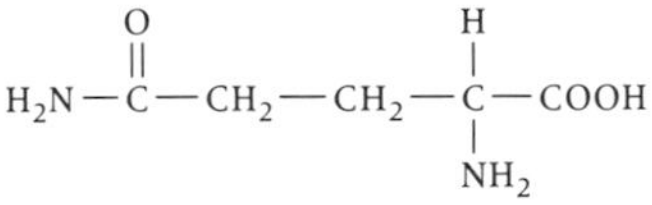 $H_2N-\overset{\overset{O}{\parallel}}{C}-CH_2-CH_2-\overset{\overset{H}{\mid}}{\underset{\underset{NH_2}{\mid}}{C}}-COOH$	Threonine	thr	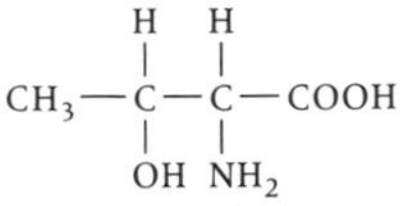 $CH_3-\overset{\overset{H}{\mid}}{\underset{\underset{OH}{\mid}}{C}}-\overset{\overset{H}{\mid}}{\underset{\underset{NH_2}{\mid}}{C}}-COOH$
Glycine	gly	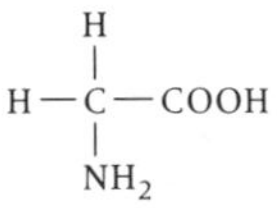 $H-\overset{\overset{H}{\mid}}{\underset{\underset{NH_2}{\mid}}{C}}-COOH$	Tryptophan	try	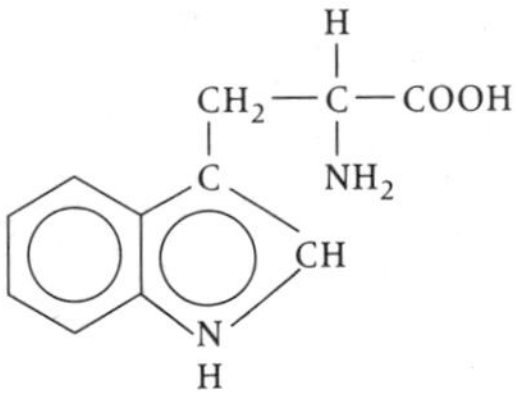 indole ring$-CH_2-\overset{\overset{H}{\mid}}{\underset{\underset{NH_2}{\mid}}{C}}-COOH$
Histidine	his	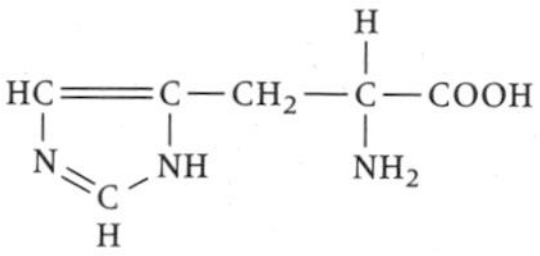 $HC=\overset{}{\underset{}{C}}-CH_2-\overset{\overset{H}{\mid}}{\underset{\underset{NH_2}{\mid}}{C}}-COOH$ (imidazole ring: $N=\underset{\underset{H}{}}{C}-NH$)	Tyrosine	tyr	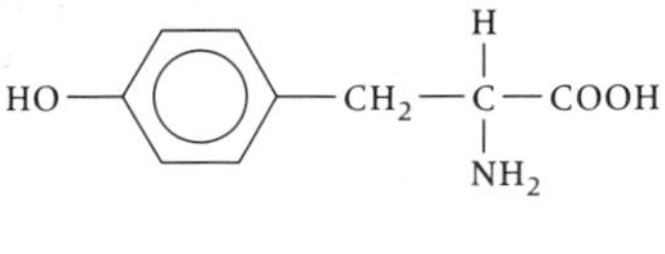 $HO-\langle\text{benzene}\rangle-CH_2-\overset{\overset{H}{\mid}}{\underset{\underset{NH_2}{\mid}}{C}}-COOH$
Isoleucine	ile	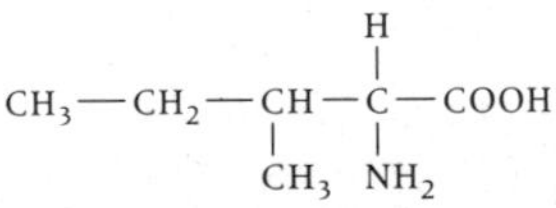 $CH_3-CH_2-\overset{\overset{H}{}}{\underset{\underset{CH_3}{\mid}}{CH}}-\overset{\overset{H}{\mid}}{\underset{\underset{NH_2}{\mid}}{C}}-COOH$	Valine	val	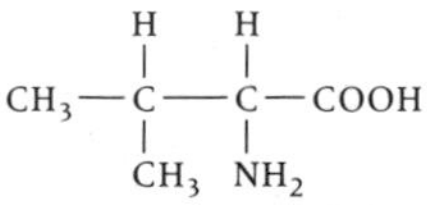 $CH_3-\overset{\overset{H}{\mid}}{\underset{\underset{CH_3}{\mid}}{C}}-\overset{\overset{H}{\mid}}{\underset{\underset{NH_2}{\mid}}{C}}-COOH$

*The table shows the non-ionic structures of the amino acids
†The amino acids arg, his, ile, leu, lys, met, phe, thr, try and val are all essential requirements in the diet of humans. Without them we die

123.2 Methods of making amino acids

The classical method of making an amino acid is to start with a carboxylic acid. In section 119.4 we found that the hydrogen atoms on the carbon atom next to the acid group, the α carbon, can be replaced by halogen atoms. For example, to make glycine we start with ethanoic acid, and convert it into chloroethanoic acid:

$$CH_3COOH + Cl_2 \rightarrow CH_2ClCOOH + HCl$$

This reaction is carried out by passing chlorine through the hot acid in sunlight, or with red phosphorus as a catalyst. It is possible to replace a hydrogen atom by bromine by a similar method. After separation, the chloroethanoic acid is treated with concentrated ammonia solution:

$$CH_2ClCOOH + NH_3 \rightarrow CH_2NH_2COOH + HCl$$
$$\text{glycine}$$

As you would expect, the more complicated amino acids require more sophisticated preparations, but it is unlikely that you will need to know anything of them.

123.2 Outline a method of making alanine.

123.3 Amino acids exist as dipolar ions

The formula of glycine, NH_2CH_2COOH, gives no indication of why it should be a white crystalline solid. However, in the crystal, glycine is really present as a *dipolar ion*, i.e. it has a positive and a negative charge on the molecule at the same time. Its structure is:

Similarly the other amino acids exist as dipolar ions. For example,

These dipolar structures survive when the amino acids dissolve in water, and they are responsible for some unusual properties. The dipolar ions are also known as *zwitterions*.

123.4 Amino acids can be both acidic and basic

Like ordinary amines, amino acids will take up protons from acids; but it is *not* the NH_2 group in the acid that is responsible for the reaction. If you look at the dipolar structure of glycine, you can see that the nitrogen is already protonated. When an amino acid meets a strong acid like hydrochloric acid, it is the COO^- group that accepts a proton:

$$NH_3{}^+CH_2COO^- + H_3O^+ \rightleftharpoons NH_3{}^+CH_2COOH + H_2O$$
$$\text{protonated}$$
$$\text{glycine}$$

In this reaction, the COO^- group is the basic group (proton acceptor).

Likewise, if we add a strong alkali to an amino acid solution, the $NH_3{}^+$ group is the acid (proton donor):

$$NH_3{}^+CH_2COO^- + OH^- \rightleftharpoons NH_2CH_2COO^- + H_2O$$

If you dissolve an amino acid in water, you will find that its pH is not 7 (neutral). The reason is that water can act as both an acid and a base. Therefore, it can react with a dipolar ion in two ways. If we use glycine as our example, either

$$NH_3{}^+CH_2COO^- + H_2O \rightleftharpoons NH_3{}^+CH_2COOH + OH^-$$

or

$$NH_3{}^+CH_2COO^- + H_2O \rightleftharpoons NH_2CH_2COO^- + H_3O^+$$

In the first case the COO^- group is acting as a stronger base than water. In the second reaction, the $NH_3{}^+$ group is acting as a stronger acid than water. In essence, a competition is set up when you dissolve an amino acid in water. If the first reaction wins over the second, the solution will give an alkaline pH; if the second reaction wins over the first, the solution will have an acidic pH. A solution of glycine has a pH very slightly less than 7. This tells us that the acidic nature of the $NH_3{}^+$ group wins over the basic nature of the COO^- group.

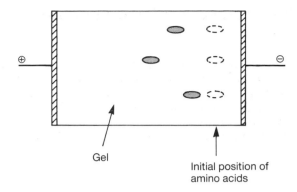

Figure 123.1 *The basis of an electrophoresis experiment. Amino acids, or substances built from them, move in an electric field. Whether they move towards the positive or negative electrode depends on the pH of the medium*

123.5 Electrophoresis and isoelectric points

We know from our work on electrolysis that positive ions in a solution will move towards the negative electrode (cathode), and negative ions towards the positive electrode (anode). In electrolysis we are interested in the changes that take place when the ions arrive at the electrodes. In an electrophoresis experiment we also have an anode and cathode dipping into a conducting medium, but the purpose of the experiment is different. We do not want the ions to be discharged at the electrodes; rather we want to study their *movement* towards the electrodes. A typical arrangement is to place a small sample of an amino acid on a strip of gel with an electrode at each end (Figure 123.1). If a voltage is applied between the electrodes (typically around 40 V) and the amino acid has an overall positive or negative charge, then it will travel towards one of the electrodes. After a suitable period of time, which will depend on the applied voltage and the nature of the amino acid, the experiment is stopped. The presence of the amino acid can be shown by dipping the gel into, or spraying it with, a dye.

Because amino acids move at different speeds through the gel, electrophoresis experiments can be used to separate mixtures of amino acids or similar compounds.

Actually it is possible to prevent an amino acid moving in an electrophoresis experiment. For example, the equilibrium

$$NH_3^+CH_2COO^- + H_2O \rightleftharpoons NH_2CH_2COO^- + H_3O^+$$

lies a little in favour of the right-hand side. If we add acid, the equilibrium will shift to the left and the glycine will exist (almost) entirely as dipolar ions. Because they have a positive charge at one end, and a negative charge at the other, they are effectively electrically neutral. Therefore they will not move towards either of the electrodes in an electrophoresis experiment.

The pH at which an amino acid exists completely as dipolar ions is the *isoelectric point*. Glycine has an isoelectric point a little over pH = 6.

123.6 The geometry of amino acids

First, let us look at the structure of alanine. If you have read Unit 110 on optical activity you should recognise that alanine has an asymmetric carbon atom. This means that the molecule can exist in two mirror image forms.

However, only one of the optical isomers is found in naturally occurring alanine. If you build a model of the naturally occurring isomer, look at it with the hydrogen atom on the asymmetric carbon atom pointing towards you. You should find that you see the arrangement

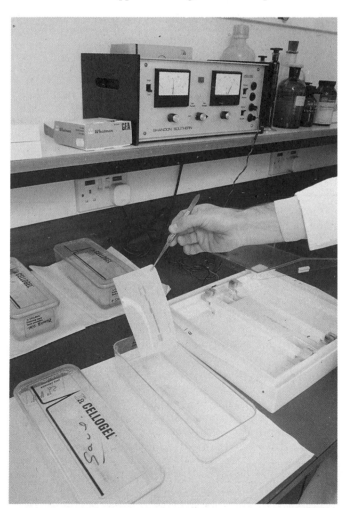

A typical laboratory setting for an electrophoresis experiment. In this case the sample is haemoglobin.

Amino acids and proteins 853

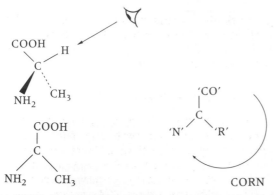

Figure 123.2 *The CORN rule. The view seen when looking down the C—H bond of alanine illustrates the CORN rule for naturally occurring amino acids*

illustrated in Figure 123.2. If we follow the groups around in a clockwise direction, with the acid group at the top, the order is COOH, CH_3, NH_2. We can generalise this pattern to cover any of the other naturally occurring amino acids. The way they differ from alanine is that instead of a CH_3 group on the asymmetric carbon, they have some other hydrocarbon group. We shall call this group R. In this way we can say that the arrangement around the asymmetric carbon is COOH, R, NH_2. This is often written in a shorthand way as CORN, thus giving us the *CORN rule*. All naturally occurring amino acids follow the CORN rule.

This 'handedness' of the amino acids is one of the reasons why specific patterns crop up in structures built from them, e.g. the α-helix in fibrous proteins and in DNA.

123.4 After separating the alanine made in the method of question 123.2 and putting a solution of it in a polarimeter, a student found that the amino acid was *not* optically active. Why was this?

123.7 Amino acids use a peptide link to join together

Amino acids have the ability to join together. The simplest case is where two glycine molecules join. We can imagine the reaction taking place by the loss of a water molecule between the two molecules:

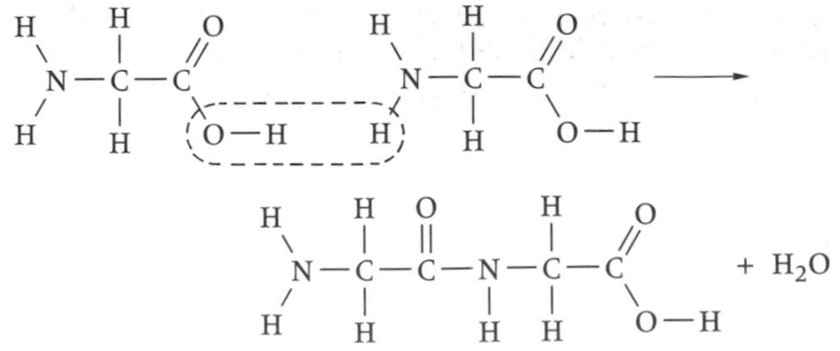

$$NH_2CH_2COOH + NH_2CH_2COOH \rightarrow$$
$$NH_2CH_2CONHCH_2COOH + H_2O$$
glycylglycine

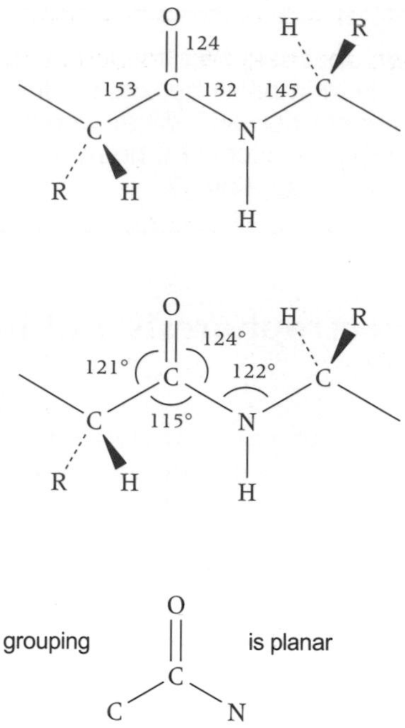

The product is glycylglycine. It is an example of a *dipeptide*. A peptide is held together by peptide links (Figure 123.3). You can find the peptide link in the middle of glycylglycine. It is an interesting collection of atoms because X-ray diffraction studies have shown its geometry to be rather special.

The first thing to notice is that the bond length between the carbon of the C=O group and the nitrogen atom of the NH group is 132 pm. This compares with a carbon–nitrogen bond length of 147 pm in an amine like methylamine. The shorter bond length tells us that the bond in the peptide link is *stronger* than normal. In turn this leads us to suspect that it might have some *double bond character*.

Now, if you look at the bond angles around the C=O group, you will find that they add up to 360°. This can only happen if the peptide link is *flat* (planar).

Once you know that a peptide is built from amino acids, you can begin to spot the bits of the amino acids that are left over in the peptide. These 'left over' bits are the *amino acid residues*. Glycylglycine has two residues. This is why it is called a *dipeptide* (*not* because it has two peptide links, which clearly it has not). Often it is convenient to write down the structure of a polypeptide by listing the amino acid residues by their shorthand names. Thus we can write glycylglycine as *gly.gly*. Similarly, a peptide with three amino acid residues will be called a tripeptide. If we persist in joining amino acids together we can make very long chains of residues. We then have a *polypeptide*. Actually, biochemical systems in Nature are far better at making polypeptides than we are in a laboratory.

Figure 123.3 *The bond lengths, in pm, and bond angles in the peptide link. The angles around the carbonyl group add up to 360°. This means that the bonds all lie in one plane: the grouping is flat. There is delocalisation of electrons over the carbon, oxygen and nitrogen atoms*

123.5 Here are three short peptide chains:

(a)

$$H_2N-CH_2-\overset{\overset{\displaystyle O}{\|}}{C}-\underset{\underset{\displaystyle H}{|}}{N}-\underset{\underset{\displaystyle CH_3}{|}}{\overset{\overset{\displaystyle H}{|}}{C}}-COOH$$

(b)

$$H_2N-CH_2-\overset{\overset{\displaystyle O}{\|}}{C}-\underset{\underset{\displaystyle H}{|}}{N}-CH_2-\overset{\overset{\displaystyle O}{\|}}{C}-\underset{\underset{\displaystyle H\ H-\overset{|}{C}-H}{|}}{N}-\overset{\overset{\displaystyle H}{|}}{C}-COOH$$

$$CH_3$$

(c)

$$H_2N-\underset{\underset{\displaystyle CH_3}{|}}{\overset{\overset{\displaystyle H}{|}}{C}}-\overset{\overset{\displaystyle O}{\|}}{C}-\underset{\underset{\displaystyle H}{|}}{N}-\underset{\underset{\displaystyle H-\overset{|}{C}-CH_3}{|}}{\overset{\overset{\displaystyle H}{|}}{C}}-\overset{\overset{\displaystyle O}{\|}}{C}-\underset{\underset{\displaystyle H}{|}}{N}-CH_2-COOH$$

$$CH_3$$

(i) How many amino acid residues do they contain?

(ii) Which of them are dipeptides, and which tripeptides?

(iii) Which amino acids have been used to build the chains?

123.8 Proteins

If the molecular mass of a polypeptide is very high, we prefer to call the molecule a *protein*. As a rule of thumb we can say that:

> **A protein is a polypeptide with a relative molecular mass of at least 10 000.**

However, this is only a guide. For example, the hormone insulin has a relative molecular mass of around 6000, but it is certainly best regarded as a protein. At the other end of the scale, some proteins are so large that they have relative molecular masses of well over 1 000 000. Proteins are of overwhelming importance in living systems. They are found in large amounts in cells and, in the form of enzymes, they are extremely efficient catalysts for hundreds of reactions that keep living

things working. To understand how proteins behave, we shall look at them in four different ways by concentrating on their *primary*, *secondary*, *tertiary* and *quaternary structures*.

(a) Primary structure

When we think about the primary structure of a protein, we are concerned with the way amino acid residues join together in chains. There are two aspects to this. First, the geometry of the peptide link is one of the key aspects of primary structure. Secondly, the *sequence* in which the amino acid residues appear in a chain is important. Insulin is made of 51 amino acid residues joined together. The order in which they occur is shown in Figure 123.4.

Presently we shall see how the sequence of amino acid residues can be discovered.

(b) Secondary structure

The orderly arrangement of parts (or all) of protein chains constitutes the secondary structure of a protein. Proteins do not simply lie about in long, straight, chains. Their X-ray diffraction patterns show that parts, or all, of the chains have orderly arrangements of one of two types. Sometimes the chains make a pleated and sometimes a spiral structure. These are the two secondary structures (Figures 123.5 and 123.6).

Both structures are caused by hydrogen bonds between portions of the amino acid residues. The —NH group of the peptide link on one residue can hydrogen bond to the carbonyl group on another residue. However, for the hydrogen bond to be successful, the two groups must be aligned in a special way. The best arrangement is the one which gives a large number of hydrogen bonds among all the residues. Especially, in fibrous proteins this arrangement causes the chains to twist and spiral into the geometry of the α-helix (alpha-helix), the structure of which is shown in Figure 123.6.

(c) Tertiary structure

The three-dimensional shape taken up by a protein is its tertiary structure (Figure 123.7). Proteins that spiral over their entire length are the *fibrous* proteins. In the *globular* proteins the chains spiral over short lengths, and then fold into a different pattern. This gives globular proteins a more rounded look. Although hydrogen bonding between parts of the chain is extremely important in globular proteins, their struc-

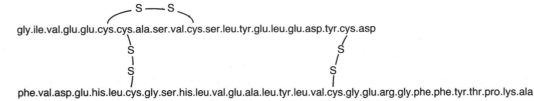

Figure 123.4 The order of amino acid residues in insulin

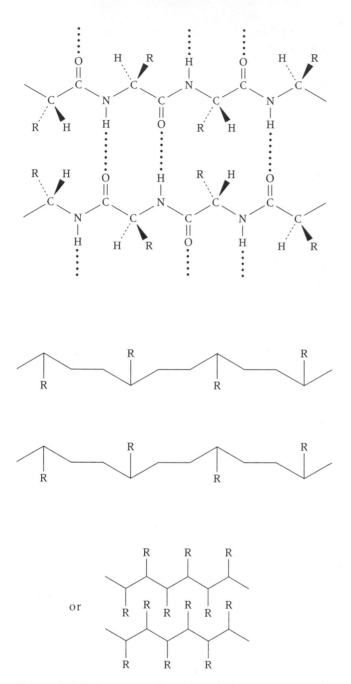

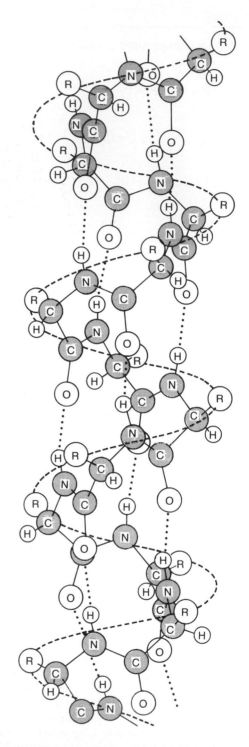

Figure 123.5 *In some proteins the chains are arranged as pleated sheets, held together by hydrogen bonds*

Figure 123.6 *Structure of the α-helix found in proteins such as α-keratin. Note the hydrogen bonds (shown by dotted lines) between different parts of the chain. These are responsible for the shape of the helix. (Taken from: Morrison, R. T. and Boyd, R. N. (1966).* Organic Chemistry, *Allyn and Bacon, Boston, figure 37.4, p. 1123)*

tures are also determined by other interactions. Especially, if the protein is to exist in an aqueous solution (e.g. in the cytoplasm of cells or in blood), then the outside of the protein should be attractive to water molecules (Figure 123.8). That is, the outside of the protein should have groups sticking out of it that can hydrogen bond with water molecules (i.e. hydrophilic groups). Likewise, parts of the chain that are not particularly attractive to water molecules (hydrophobic groups) will be wrapped up and hidden inside the globule. In addition there is also the possibility of attractions and repulsions between ionic groups along the chain. Yet another com-

plication is that some amino acid residues in different parts of a chain can be linked by bridging atoms. A common example is the disulphide bridge between two cysteine residues (see Figure 123.4).

(a)

Fibrous protein

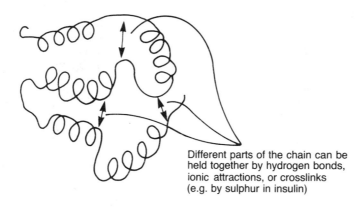

Different parts of the chain can be held together by hydrogen bonds, ionic attractions, or crosslinks (e.g. by sulphur in insulin)

Globular protein

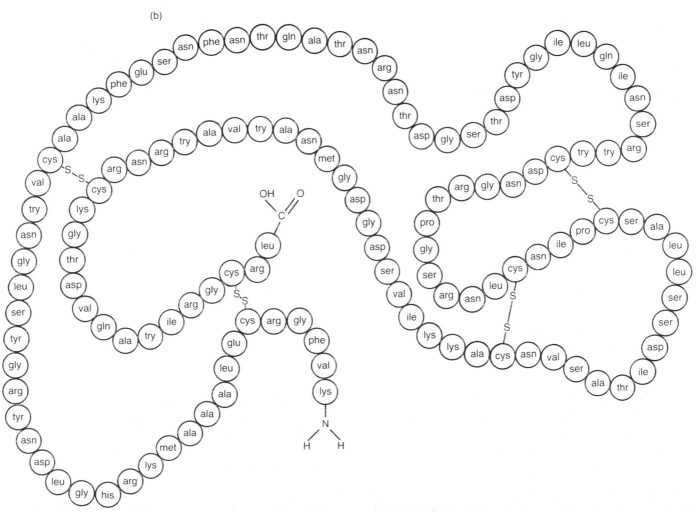

Figure 123.7 *(a) Fibrous proteins and globular proteins have different tertiary structures. (b) This illustrates the complexity of even a rather small enzyme, lysozyme. (Taken from: Boer, A. S. et al. (1971). Central Concepts of Biology, Macmillan, New York, figure 3.15, p. 38)*

Figure 123.8 *The shape of one of the haemoglobin protein chains (a β chain). This is an example of the tertiary structure of a protein. The dot (●) marks the position of the haem group. (Adapted from: Zubay, G. (1988). Biochemistry, Macmillan, New York, figure 2.12)*

(d) *Quaternary structure*

Many biologically active molecules consist of two or more protein chains packed together. For example, haemoglobin is built from four chains, shown in Figure 123.9 as two pairs (α_1 and α_2, β_1 and β_2). The way the four chains fit together is described as the quaternary structure of haemoglobin.

The quaternary structure of a protein depends on the possibilities for hydrogen bonding, and effective van der Waals bonds between groups on the individual chains.

If protein solutions are heated, or treated with concentrated acids or alkalis, the interactions between the various parts of the peptide chain are changed. We say that the protein has been *denatured*. A denatured protein will not only look different, it will lose its chemical activity. This is especially noticeable when enzymes lose their catalytic activity after heating. For example, compare the action of raw and cooked liver on hydrogen peroxide!

Not all proteins are made of peptides alone. For example, the portion of haemoglobin that takes up oxygen and transports it around our bodies has no peptide links. This non-peptide part of the protein is called

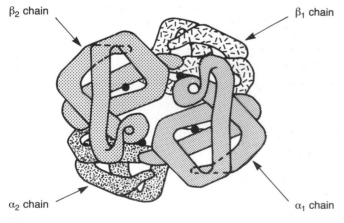

β₂ chain   β₁ chain

α₂ chain   α₁ chain

Figure 123.9 *The quaternary structure of haemoglobin consists of four individual chains clinging together. The dots (●) mark the positions of haem groups. (Adapted from: Zubay, op. cit.)*

the *prosthetic group*. Often, as in haemoglobin, the prosthetic group contains a metal ion (see Figure 105.10).

123.6 Are hydrophilic or are hydrophobic groups likely to be found on the outside of a globular protein?

123.7 In a globular protein the hydrocarbon side chains on the amino acid residues are often to be found arranged inwards. What type of bonding will attract them together.

123.9 How to discover the structure of a protein

If you were to attempt to discover the structure of a protein, you would be undertaking one of the most difficult tasks that befalls a chemist or biochemist. In theory there are two stages. First, find out the primary structure by analysing the chain. This can be done by a method known as N-terminal analysis. Secondly, discover the secondary and tertiary structures. X-ray diffraction plays a crucial role in this.

Owing to the complexity of proteins, both stages are exceedingly difficult and time consuming. We shall have to be content with a simplified description of them.

(a) *N-terminal analysis*

The method used by Frederick Sanger to discover the primary structure of insulin has become a standard procedure. It had been known for many years that a protein could be broken into its component amino acids by acid or alkaline hydrolysis. Hydrolysis breaks peptide links, so the individual amino acids can be set free. Although direct hydrolysis can tell us which amino acids are present, it does not give us their order in the chain. Sanger found a way round this problem by labelling the amino acid residue on the end of the chain with 2,4-dinitrofluorobenzene (DNFB). This molecule has the ability to bond strongly with the NH_2 group on the very last amino acid residue on a peptide chain. For example,

$$O_2N - \bigcirc - F$$
$$NO_2$$
DNFB

$$O_2N - \bigcirc - \overset{H}{\underset{NO_2}{N}} - CH_2 - COOH$$
DNFB + glycine

Once the DNFB reacts with the protein, the mixture is hydrolysed by refluxing with moderately concentrated hydrochloric acid. This breaks peptide links along the chain. We can represent the result of the hydrolysis like this:

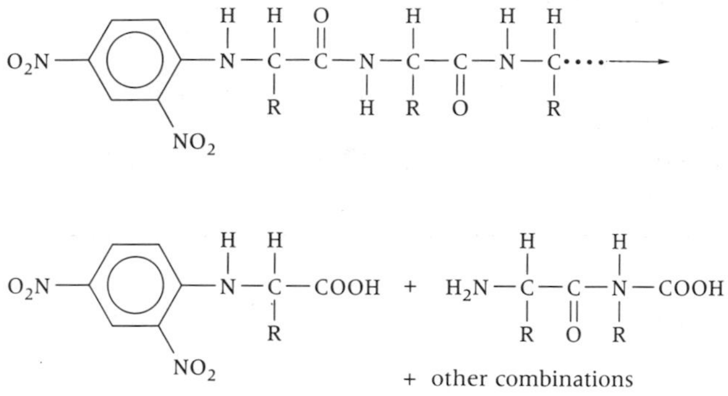

Peptide chains of different lengths are released, together with free amino acids. However, the important thing is that a short length of the chain stays bonded to the DNFB. These short chains are called 2,4-dinitrophenyl (DNP) peptides. By selecting the best solvent, the DNP peptides can be extracted from the bulk of the solution. They are then separated from one another by column chromatography. Once the fractions are separated, the next stage is to hydrolyse them with 6 mol dm^{-3} hydrochloric acid. This frees the individual amino acids along the DNP peptide chain. Finally the amino acids are identified by paper chromatography. By changing the conditions used in the hydrolysis and using different solvents in the chromatography, it is also possible to identify the terminal amino acid bonded to the DNFB.

(Incidentally, you might like to know the scale on which Sanger worked. For example, in the hydrolysis of the DNP peptides he used milligram quantities of the peptide with 5 *drops* of 6 mol dm^{-3} hydrochloric acid in a 50 cm^3 round bottomed flask.)

We can use some of Sanger's results to see how he began the task of unravelling the structure of insulin. Following a hydrolysis of a set of DNP peptides, he found that they all contained DNP-glycine. When the products of the hydrolysis were separated by column chromatography, he found four bands. The first band was DNP-glycine; the others were short peptide chains. Sanger showed that the composition of the first three fractions was:

Fraction	Composition
1	DNP-glycine
2	Isoleucine
3	Isoleucine and valine

These results show that the order of amino acids in the peptide chain originally bonded to the DNFB must be in the order gly–ile–val. Analysis of further fractions showed that the chain also had glutamic acid residues following the valine.

N-terminal analysis is not the only weapon that can be used in analysing proteins. As we know, DNFB reacts with the amine group on one end of a peptide chain; but it will not react with the carboxylic acid group on the other end of the chain. However, the enzyme carboxy-peptidase has the remarkable ability to chop off the amino acid having the acid group. Once free from the chain, this amino acid can be identified. The use of carboxypeptidase is called C-terminal analysis.

The determination of the order of amino acids can now be done semi-automatically, using chemicals and techniques that were unavailable to Sanger. For example, DNFB has largely been replaced by the chemical whose common name is dansyl chloride:

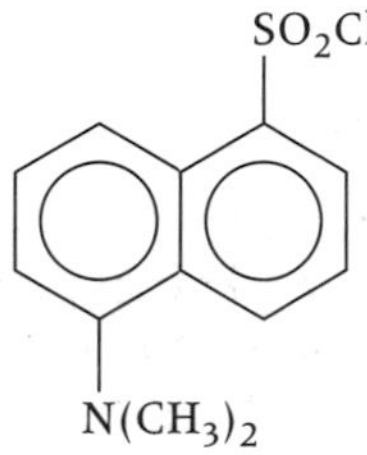

Like DNFB, this molecule binds to the terminal amino acid, but the products, of general formula

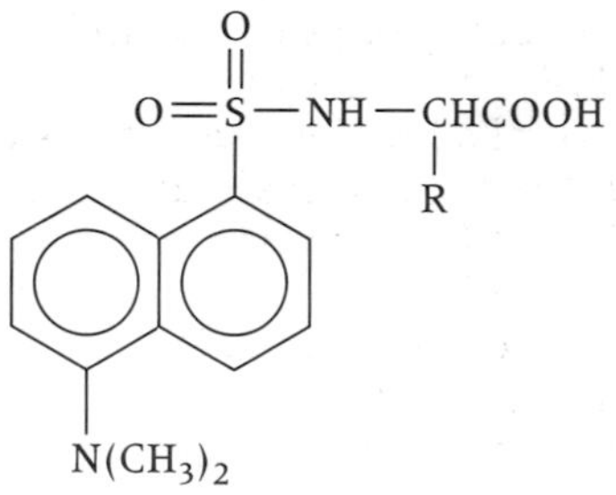

are fluorescent and can be readily identified by chromatography.

123.8 A peptide chain was subjected to N-terminal analysis using DNFB. In addition to DNP-glycine, the following combinations of amino acid residues were isolated: gly.val, val.leu, ala.phe.gly, leu.ala.phe. What is the order of amino acid residues in the peptide?

123.10 The X-ray patterns of proteins

Proteins that have a spiral, or helical, structure show many common features in their X-ray diffraction patterns. You can see examples in the photo. Notice that the patterns are symmetrical. This corresponds with the symmetrical nature of the helix. By analysing the pattern it is possible to calculate the key dimensions of the helix, for example the repeat length of the spiral.

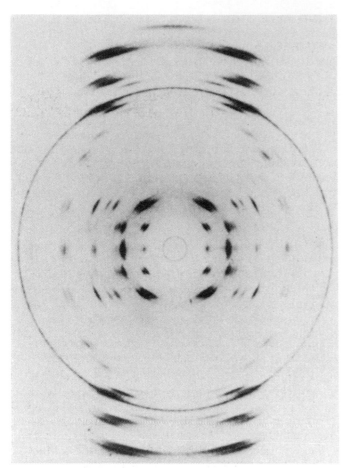

A X-ray diffraction pattern from a fibre of DNA. The pattern of spots is characteristic of helical structures. (The continuous circle is not due to DNA, but to a salt impurity.) Taken from Figure 13.2 of J. P. Glusker and K. N. Trueblood, Crystal Structure Analysis, *Oxford University Press, Oxford, 1985.*

The diffraction patterns of globular proteins are much more complicated. Even with modern computers, there is a huge amount of work that has to be done to tease out the structure of the protein from the myriad number of dots in the X-ray pattern.

123.11 How enzymes act as catalysts

An enzyme is a protein that acts as a highly efficient catalyst (Table 123.2). An enzyme normally has one particular site at which reactions take place. This is the *active site*. The tertiary structure of an enzyme is so complicated that molecules must have a very particular shape if they are to fit into the active site. Those molecules which do fit the active site are called the *substrate* molecules. You can see this idea in Figure 81.5. We have made no attempt to show any particular enzyme or substrate. The diagram showed the special relationship between the geometry of the active site and the substrate. You should now understand why enzymes are said to be *specific* catalysts: they only catalyse one reaction.

There are several ways of ruining an enzyme. The

Table 123.2. Some enzymes and their action*

Enzyme	Function
Amylase	Converts starch into sugars
Catalase	Converts hydrogen peroxide into oxygen and water
Invertase	Converts sucrose into glucose and fructose
Lysozyme	Attacks the polysaccharide cell walls of bacteria
Papain	Tenderises meat
Urease	Converts urea into carbon dioxide and ammonia

*Enzymes have been increasingly used in industrial processes, especially in the food industries. For example, you may be able to buy bottles of papain to sprinkle on meat before you cook it; invertase is widely used in the making of sweets and confectionery. Manufacturers of soap powders have, with some success, adapted certain enzymes so that they can withstand the temperatures used in washing machines. These enzymes have the ability to destroy the sorts of molecules found in natural dirts and stains, e.g. blood stains

easy way is to heat it. This permanently denatures the enzyme. If you do not want to be so drastic in your action, the addition of acid or alkali will also change the catalytic activity of an enzyme. This is because of the ability of hydrogen and hydroxide ions to add or remove protons from various parts of a peptide chain and thereby upset the interactions between the active site and the substrate.

You should not be surprised to learn that most enzymes work best over a very limited range of pH. Typically, a graph of enzyme activity against pH has the shape shown in Figure 123.10.

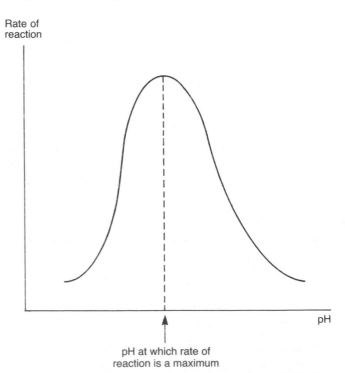

Figure 123.10 *Enzyme reactions have rates that are sensitive to pH. At too low or too high a pH, the enzyme loses its activity*

123.9 This is a question about the kinetics of enzyme reactions. Urease catalyses the breakdown of urea. When the concentration of urea is low, the reaction is first order in urea; but when the concentration of urea is high, the reaction is zeroth order in urea. Why is this?

Answers

123.1 Nitrogen gas would be given off:

$$NH_2CH_2COOH + HNO_2 \rightarrow HOCH_2COOH + N_2 + H_2O$$

123.2 Alanine has the structure shown in Table 123.1. A reasonable molecule with which to start the preparation would be propanoic acid. If this is treated with chlorine, then one of the α hydrogen atoms can be displaced. Finally the product is reacted with a concentrated solution of ammonia. The reactions are:

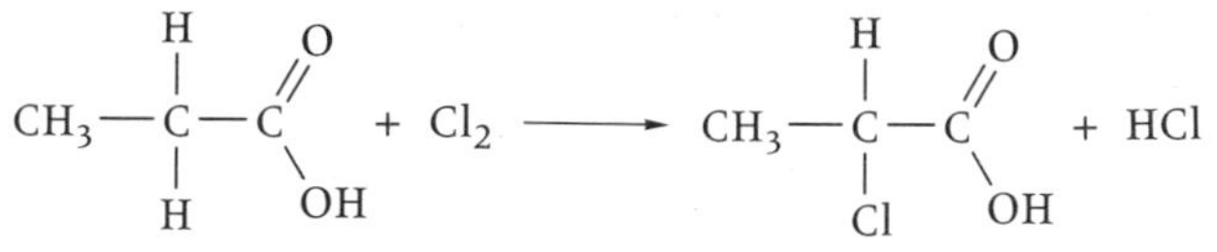

$$CH_3CH_2COOH + Cl_2 \rightarrow CH_3CHClCOOH + HCl$$

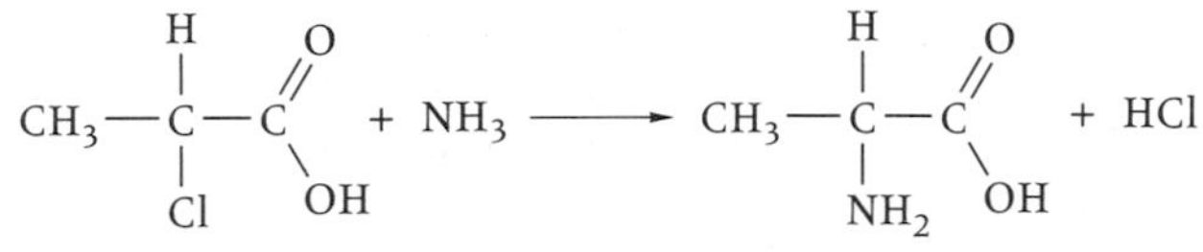

$$CH_3CHClCOOH + NH_3 \rightarrow \underset{\text{alanine}}{CH_3CHNH_2COOH} + HCl$$

123.3 We saw in section 123.4 that an amino acid can behave as an acid or a base. However, you would have to ask the student to say which part of the molecule was acting as the acid. The important thing is that the student should know that because the acid will exist as a dipolar ion, it is the NH_3^+ group that acts as the acid:

$$(CH_3)_2CHCHNH_3^+COO^- + OH^- \rightarrow (CH_3)_2CHCHNH_2COO^- + H_2O$$

123.4 The solution of alanine must have contained equal proportions of the two optical isomers of alanine. That is, it is a racemic mixture. The reason is that when the chlorine reacts with propanoic acid, both optical isomers of 2-chloropropanoic acid can be made. These two isomers will react to give the two mirror image forms of alanine.

123.5 (i), (ii) Chain (a) contains two residues, so it is a dipeptide; chains (b) and (c) contain three residues, so they are tripeptides.

(iii) Chain (a) is built from glycine and alanine (gly.ala); chain (b) from two glycines and valine (gly.gly.val); chain (c) from alanine, valine and glycine (ala.val.gly).

123.6 Hydrophilic. It is these groups that will enable the protein to be solvated by water molecules.

123.7 Van der Waals forces, the hydrophobic interactions, help to keep hydrocarbon fragments together.

123.8 The order of the residues is as follows: gly.val.leu.ala.phe.gly. (The first gly is the amine end of the chain; the last gly is the carboxylic acid end.) You might have given val.leu.ala.phe.gly.val as an alternative. The reason why this one is not right is that it would not give DNP-glycine. The fact that this is isolated tells us that glycine must be a terminal amino acid.

123.9 First, you have to realise that the reaction takes place at the active site on the urease molecules. When the concentration of urea is low, we can assume that at any time there are a good number of active sites unoccupied. Thus if we increase the concentration of urea slightly, more of the active sites will be used and more urea molecules are destroyed. This is the reason for the first-order kinetics. However, if the concentration of urea increases beyond a certain point, on average the active sites will all be full. If we then add more urea, the reaction rate will not increase: the new molecules cannot get to the active sites. When this happens we find that the rate does not change even though the concentration of urea changes, i.e. we have a zeroth-order reaction.

UNIT 123 SUMMARY

- Amino acids:
  - (i) Have two functional groups: $-NH_2$ and $-COOH$.
  - (ii) Show both basic and acidic nature.
  - (iii) Exist as bipolar ions, e.g. $NH_3^+CH_2COO^-$; hence their existence as solids.
  - (iv) Are the basic units of proteins.
  - (v) Only α-amino acids occur naturally (obey the CORN rule).
  - (vi) Are separated by electrophoresis.
- Preparation:
  From chlorocarboxylic acids;

  e.g. $CH_3COOH + Cl_2 \rightarrow CH_2ClCOOH + HCl$
  $CH_2ClCOOH + NH_3 \rightarrow CH_2NH_2COOH + HCl$
- Reactions:
  Show both acidity and basicity.
  Acidic nature is due to the NH_3^+ group.
  Basic nature is due to the COO^- group.

Summary – cont.
- Peptides:

  Consist of chains of amino acids. The peptide link is planar

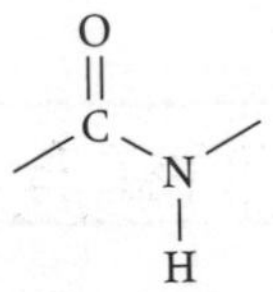

- Proteins:
  - (i) Are built from amino acids.
  - (ii) Can have very high molar masses.
  - (iii) Are essential components of biochemical systems.
  - (iv) Have primary, secondary, tertiary and quaternary structures.

    Primary: the order of amino acids in the protein chain.

    Secondary: how the chain of amino acid residues arrange themselves, often into an α-helix.

    Tertiary: how the chain folds to give the three-dimensional structure.

    Quaternary: how one or more chains fold to give the final structure of the protein.

- Structure:
  - (i) Order of amino acids is discovered by N-terminal analysis, e.g. by using DNFB.
  - (ii) Three-dimensional structure discovered by X-ray diffraction.

- Enzymes:
  - (i) Are specific catalysts.
  - (ii) Have active sites into which (usually) only one type of molecule (the substrate) will fit.

124

Deoxyribonucleic acid (DNA)

124.1 The structure of DNA

Deoxyribonucleic acid (DNA) is the substance that controls the development of all living things. Especially, DNA is responsible for handing on genetic information from one generation of a species to another. That is, DNA lies at the heart of the study of *genetics*.

The molar mass of DNA taken from different types of cell varies widely, but it is always very high, sometimes of the order $10^9\, \text{g mol}^{-1}$. This suggests that DNA is a polymeric material. The polymer is built from chains of *nucleotides*. One nucleotide consists of a phosphate group, a sugar molecule and a basic group containing nitrogen (Figure 124.1).

There are four different basic molecules found in DNA: *adenine, guanine, cytosine* and *thymine* (Figure 124.2). They all contain nitrogen atoms in a ring structure. Analysis of DNA shows that always the amounts of adenine and thymine are equal. Similarly, guanine and cytosine are found in equal amounts.

The sugar molecule in DNA is *deoxyribose*, a five-membered ring with three OH groups. *Ribonucleic acid* (RNA) is similar to DNA except that it is a polymer of nucleotides made from the sugar ribose rather than deoxyribose, and it has a slightly different set of bases: *adenine, guanine, cytosine* and *uracil* (Figure 124.2).

During the 1950s there were many groups of chemists and biochemists trying to determine how the nucleotides fitted together in DNA. Maurice Wilkins and, especially, Rosalind Franklin at King's College, London, were attempting to analyse the X-ray diffraction pattern of DNA (see photo on p. 860). This was no easy task, especially given that the calculations had to be done without the aid of computers. Rosalind Franklin was close to discovering the structure, but before she published her results, Francis Crick and James Watson (who worked in Cambridge) discovered the solution to the puzzle. They did not achieve this remarkable feat by analysing the diffraction pattern in detail. Rather, they realised that the pattern could be produced by a double helix; that is, two separate strands winding round each other. Linus Pauling had also understood that this was possible, but he had not given a satisfactory explanation of how the strands fitted together. Where Crick and Watson made their vital breakthrough was in building scale models of the nucleotides and joining them to represent short lengths of two strands of DNA. By attempting to fit the model together, they discovered how the strands could wind round in such a symmetrical fashion. The key to the pattern was that:

> **The two strands were held together by hydrogen bonds between pairs of bases.**

They found that the shapes of adenine and thymine made a beautiful match, allowing two hydrogen bonds between the molecules (Figure 124.3). Similarly, guanine and cytosine fitted together, but this time with three hydrogen bonds at work.

Their model fitted together precisely if the base pairs attached to the phosphate groups pointed towards the middle of the two strands. Each strand is an α-helix, which you can see in outline in Figure 124.4.

124.2 How DNA works

The magic of DNA is that the pattern of bases along the strands controls how amino acids join together to make proteins. (You may remember that the majority of bio-

(a)

··· —[Nucleotide]— [Nucleotide]— [Nucleotide]— ···

(b)   The pattern is

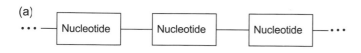

Figure 124.1 *A first look at DNA. (a) At the simplest level we can think of DNA as a polymer made of nucleotide units. (b) Each nucleotide is made from a phosphate + sugar + base*

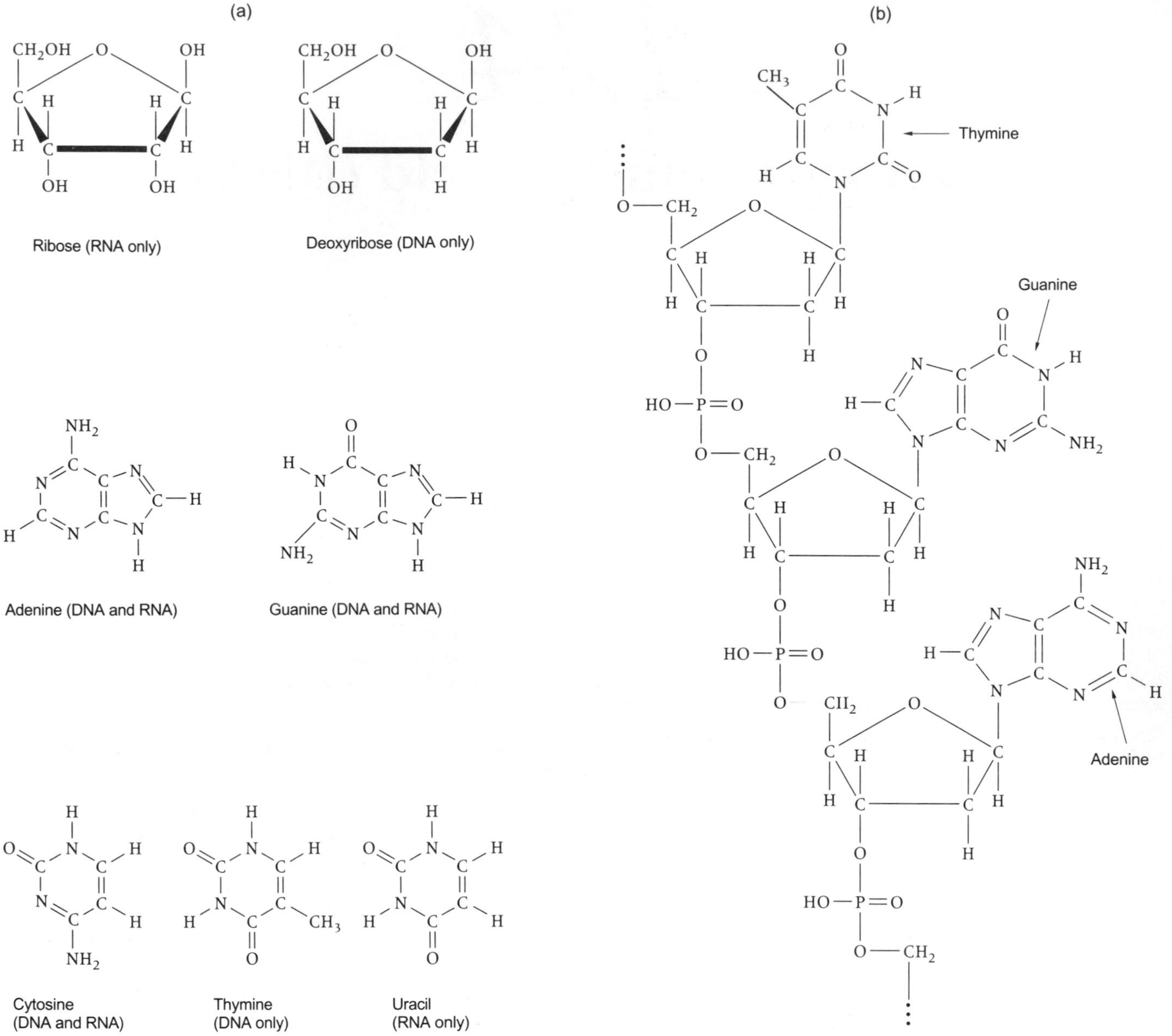

Figure 124.2 *(a) The two sugars and five bases found in DNA or RNA. (b) A fragment of a DNA chain*

logically important materials are proteins.) The pattern of bases along a strand of DNA determines the number, sequence and type of amino acid in the protein. Here is an outline of how this happens (see also Figure 124.5).

(a) Step 1: manufacture of messenger RNA

The two strands of DNA contained in the chromosomes of a cell unzip along part of their length. Every pattern of three bases along a strand acts as a code for an amino acid. Each pattern of three is called a *codon*. There are 64 codons (Table 124.1), but only 20 amino acids, so several codons code for the same amino acid. In addition, there are codons that control the place where the coding starts and stops.

The chain of amino acids that combine along the

length of the DNA strand makes a second chain of nucleotides. This new chain is *messenger RNA* (mRNA for short). mRNA has a different order of bases along its length than the original DNA. However, the important thing is that the structure of the mRNA, including the order of bases along its strands, is determined by the DNA.

(b) Step 2: RNA passes on information to make proteins

mRNA can pass into the cytoplasm of a cell. Here it attaches itself to small structures called ribosomes. Once joined to the ribosomes, we have *ribosomal RNA*. Yet another type of RNA in cells, called *transfer RNA* (tRNA), binds to ribosomal RNA. (tRNA is much smaller

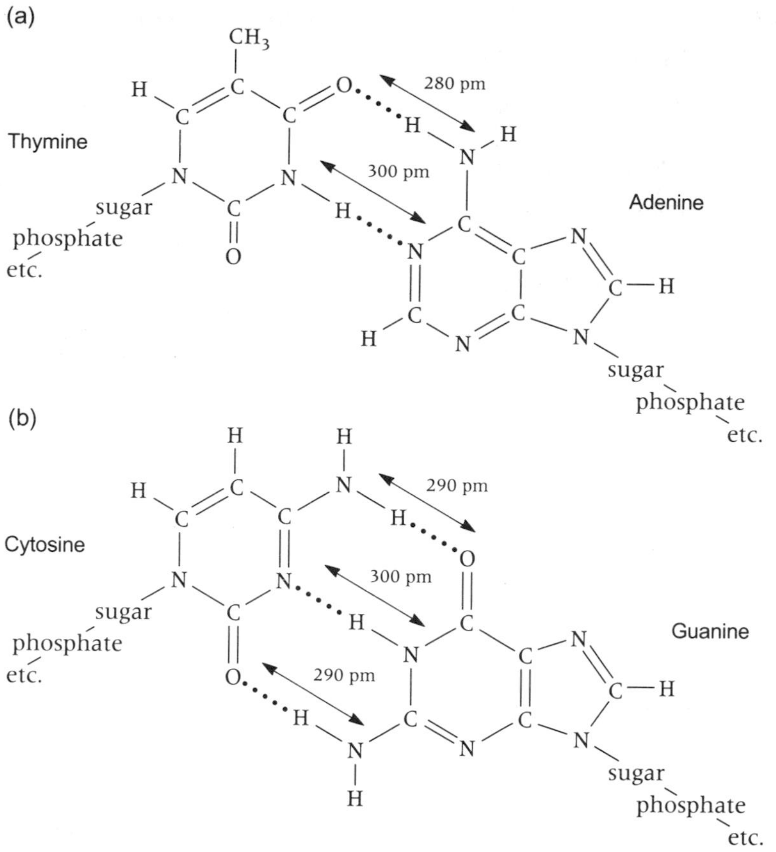

(a)

Thymine

sugar
phosphate
etc.

280 pm

300 pm

Adenine

sugar
phosphate
etc.

(b)

Cytosine

sugar
phosphate
etc.

290 pm

300 pm

290 pm

Guanine

sugar
phosphate
etc.

Figure 124.3 *How the bases hydrogen bond in DNA: (a) thymine H bonded to adenine; (b) cytosine H bonded to guanine*

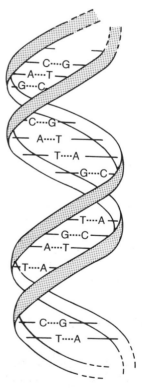

Figure 124.4 *A representation of how the two strands of DNA wind about one another. A, C, G and T are the four bases (see text). Dotted lines represent hydrogen bonds*

than either DNA or RNA.) Each molecule of tRNA carries with it a single amino acid. Now, the bases along the length of ribosomal RNA act as a code for different types of tRNA. This means that along the length of ribosomal RNA gather the various tRNA molecules, each with its own particular amino acid on the end.

(c) Step 3: amino acids on transfer RNA form peptide links

The amino acids are so close together that they join by making peptide links. As this happens, the tRNA minus its amino acid leaves the surface of the ribosomal RNA. The length of the peptide chain increases until one of the codons on the ribosomal RNA does not code for a tRNA with an amino acid. If, as is usually the case, the chain is very long, it wraps round to give a fibrous or globular structure.

DNA has the ability to replicate itself. It does this by unzipping and the two strands make copies of themselves. Over many generations of a species, random changes occur to DNA. Such changes are called *mutations*. Mutations can happen by chance (randomly), but others can be induced by physical or chemical factors, e.g. some types of radioactivity.

A mutation occurs when, after DNA unzips, the order of the bases in the new strands becomes slightly mixed up. Sometimes a mutation may be of little importance.

Deoxyribonucleic acid (DNA) 865

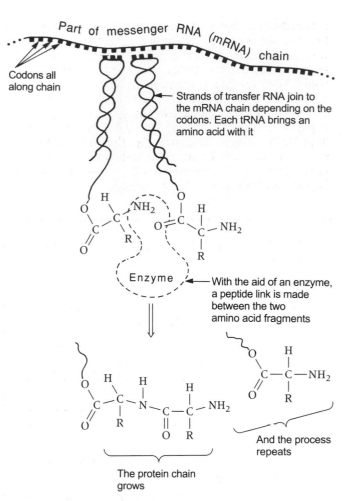

Part of messenger RNA (mRNA) chain

Codons all along chain

Strands of transfer RNA join to the mRNA chain depending on the codons. Each tRNA brings an amino acid with it

Enzyme

With the aid of an enzyme, a peptide link is made between the two amino acid fragments

And the process repeats

The protein chain grows

Figure 124.5 *How a protein chain is created from mRNA*

Table 124.1. The 64 codons in mRNA controlling protein synthesis*†

ACU	thr	AAU	asn	AUU	ile	AGU	ser
ACC	thr	AAC	asn	AUC	ile	AGC	ser
ACA	thr	AAA	lys	AUA	ile	AGA	arg
ACG	thr	AAG	lys	AUG	met	AGG	arg
CUU	leu	CCU	pro	CGU	arg	CAU	his
CUC	leu	CCC	pro	CGC	arg	CAC	his
CUA	leu	CCA	pro	CGA	arg	CAA	gln
CUG	leu	CCG	pro	CGG	arg	CAG	gln
GUU	val	GGU	gly	GCU	ala	GAU	asp
GUC	val	GGC	gly	GCC	ala	GAC	asp
GUA	val	GGA	gly	GCA	ala	GAA	glu
GUG	val	GGG	gly	GCG	ala	GAG	glu
UCU	ser	UUU	phe	UAU	tyr	UGU	cys
UCC	ser	UUC	phe	UAC	tyr	UGC	cys
UCA	ser	UUA	leu	UGG	tyr		
UCG	ser	UUG	leu				
UAA							
UAG	} start/stop information						
UGA							

*U = uracil; G = guanine; A = adenine; C = cytosine
†The first column of each pair gives you the pattern of three bases in each codon. The second column gives the amino acid corresponding to the codon. The three codons at the bottom do not code for an amino acid. Instead, they help to control the process, marking where the meaningful codons start and stop

For example, suppose that part of the DNA normally produces the codon UAU in RNA. Now imagine that the order of bases in DNA is upset so that instead of UAU we get UAC in the RNA. The change may have no effect on the ability of the RNA to produce the same protein as before. This is because both UAU and UAC code for the same amino acid, tyrosine. Mutations like this are neither good nor bad; they are neutral. It is likely that the great majority of changes in DNA are neutral.

However, if the change was from UAU to UAA, then things are very different. This is because UAA is one of the stop codons. If the mutation happened in a process controlling the development of a human embryo, then it could result in the child having a physical or mental disability; or it might result in a spontaneous abortion.

On the other hand, some changes in the order of bases might lead to changes in an organism that are beneficial; for example, allowing an enzyme to withstand a wider range of temperatures or pH before it is denatured.

124.3 Genetic engineering

DNA is remarkably efficient at making extremely complicated molecules. This fact has made DNA a popular substance to study in industry. Let us imagine that there is a market for a particular enzyme that we know is made by DNA in humans. The idea is first to discover the base sequences in the DNA strands that code for the molecule that we wish to make. Once this is done, the appropriate piece of DNA is removed from the strand, and combined with the DNA of another organism. If all goes well this organism will start to make the enzyme for us. This process is the basis of genetic engineering. Like an ordinary engineer, a genetic engineer attempts to design and build a molecule by choosing the right tools for the job. The tools are rather different though; in the case of the genetic engineer they are pieces of DNA. Over the last 20 or so years many sophisticated techniques have been developed that make each of these stages feasible. An important example is the manufacture of insulin. People who suffer from diabetes are unable to make their own insulin, and as a result their blood sugar level is not controlled properly. If it is left untreated, diabetes leads to blindness, liver failure and ultimately premature death. The disease can be treated reasonably effectively if diabetics regularly inject themselves with insulin.

Originally, insulin for diabetics was isolated from animals, a very costly and time consuming process. Now it is made by introducing the genetic information contained in parts of DNA into the bacterium *Escherichia coli*. The *E. coli* makes insulin in sufficient quantities to produce the (roughly) 2 tonnes of insulin needed by the population of diabetics each year.

However, you should not imagine that making chemicals by genetic engineering is easy. Even though the correct piece of DNA is extracted and introduced into another organism, it may still not behave as required. For example, the conditions in a laboratory bottle may be very different to those in the cells that contained the parent DNA. The first samples of insulin prepared by genetic engineering were hopeless for use by diabetics. The protein strand in the insulin wound itself in a different way to that of normal insulin. The result was an insoluble solid.

Even so, the potential rewards for harnessing the power of DNA are immense and many companies are successfully making chemicals using genetic engineering. Not all members of the public are convinced that this is a good thing. Some people think, perhaps with good reason, that the ability to control how DNA works could have terrible consequences. For example, there could be devastating effects if a new virus were developed, perhaps by accident, in genetic engineering experiments, and it were released into the general population. Owing to its novelty, the human immune system might not be able to deal with the invading virus. However, the spread of AIDS shows that this can happen without genetic engineering.

124.1 What is the significance of the analysis of DNA which shows that adenine and thymine appear in equal amounts in DNA (as do guanine and cytosine)?

124.2 RNA contains the base uracil instead of thymine. Draw a diagram to show how uracil hydrogen bonds to adenine in RNA.

124.3 DNA is said to be at the heart of the genetic code. What is the basis of the code?

124.4 The sequence of bases in a section of ribosomal RNA was

CCUAGUUAUUCCGUGCAU

(i) What would be the order of amino acids in the protein that this set of bases code for?

(ii) What, if anything, would be the result of the following highlighted changes in the sequence of bases: (a) CCUAGCUAUUCCGUGCAU; (b) CCUAGUUAGUCCGUGCAU?

Answers

124.1 This matches with Crick and Watson's idea that these two bases always appear hydrogen bonded together on different strands in DNA.

124.2

124.3 The genetic code is the order of the codons (groups of three bases) on the strands of DNA.

124.4 (i) CCUAGUUAUUCCGUGCAU codes for ala.ser.tyr.ser.val.his.
(ii) (a) CCUAGCUAUUCCGUGCAU codes for ala.ser.tyr.ser.val.his as well because both AGU and AGC code for ser. (b) The first two codons give ala.ser, but the third (UAG) is a start/stop instruction. This mutation is potentially disastrous. It will stop the production of the normal protein.

UNIT 124 SUMMARY

- DNA:
  - (i) Is deoxyribonucleic acid.
  - (ii) Controls the development of all living things.
  - (iii) Has a very high molar mass.
  - (iv) Is built from chains of nucleotides (combination of phosphate, sugar and base groups).
  - (v) Has two α-helix strands held together by hydrogen bonds between base pairs.
  - (vi) Only four bases are present: adenine, guanine, cytosine and thymine.

- Operation of DNA:
  - (i) Every pattern of three bases (codon) along a strand of DNA acts as a code for an amino acid.
  - (ii) The process of making polypeptide chains is: production of messenger RNA; RNA passes on information to make proteins; amino acids on transfer RNA form peptide links; thus the chain grows.
- Genetic code:
  Is the order of the codons on the strands of DNA.

125

Carbohydrates

125.1 What are carbohydrates?

In this unit we shall investigate three types of chemical that, in addition to proteins and DNA, play a very important role in living things. Carbohydrates are substances that contain three elements only: carbon, hydrogen and oxygen. The simplest ones are the sugars, which have the ability to link together to make polymers such as starch and cellulose.

The simplest carbohydrates are the *monosaccharides*, which all have the general formula $C_nH_{2n}O_n$. You will be familiar with some of them because we know them as sugars; for example when $n=6$ we have glucose, $C_6H_{12}O_6$. They all have OH groups in their molecules, together with either an aldehyde

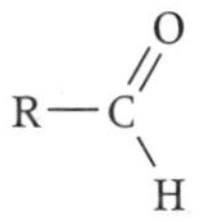

$$R-C\overset{\displaystyle O}{\underset{\displaystyle H}{<}}$$

Table 125.1. The naming of the common types of monosaccharide

Number of carbon atoms	Known as	Molecule contains	Known as
3	A triose	Aldehyde group	An aldose
5	A pentose	Ketone group	A ketose
6	A hexose		

Examples*		
Triose	Pentose	Hexose

Triose	Pentose		Hexose	
CHO | H—C—OH | CH₂OH	CHO | H—C—OH | H—C—OH | H—C—OH | CH₂OH	CHO | HO—C—H | H—C—OH | H—C—OH | CH₂OH	CHO | H—C—OH | HO—C—H | H—C—OH | H—C—OH | CH₂OH	CH₂OH | C=O | HO—C—H | H—C—OH | H—C—OH | CH₂OH
Glyceraldehyde $C_3H_6O_3$	Ribose $C_5H_{10}O_5$	Arabinose $C_5H_{10}O_5$	Glucose $C_6H_{12}O_6$	Fructose $C_6H_{12}O_6$

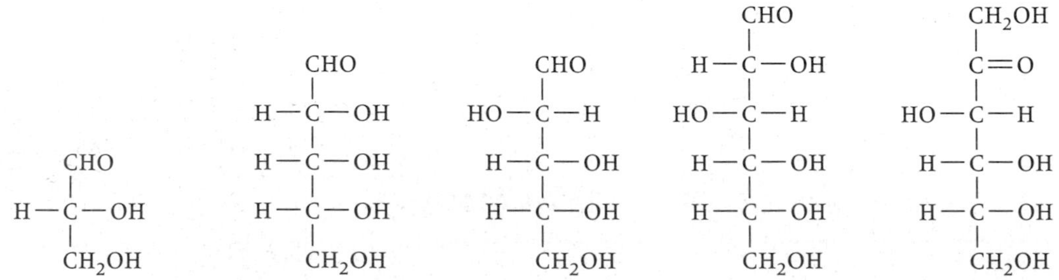

*The diagrams of the molecules have been drawn using the convention we met in Unit 110. You should imagine the vertical lines bending away from you behind the plane of the page; the horizontal lines represent bonds bending towards you out of the plane of the page. If we were to be precise we should add a D in front of their names, e.g. D-glucose. The D tells us about the configuration of the molecules, i.e. how the different groups are arranged. Throughout this unit all the sugars have the D configuration. This method of indicating configurations uses the labels D and L. It is based on comparing molecules with the two possible configurations of glyceraldehyde. It is an older method than the *R, S* scheme of Unit 110

or ketone

$$R - C \underset{R'}{\overset{O}{\parallel}}$$

group. As with many varieties of organic molecules, isomers are possible. In Table 125.1 you will find some of the more common sugar molecules listed. Each has its own particular name, but collectively they fit into groups according to the number of carbon atoms they possess. The naming system is also explained in the table.

125.2 The structures of carbohydrates

In Table 125.1 we have shown the sugars as straight-chain molecules. This is a useful way of showing the basic structure of the molecules, but in practice the pentoses and hexoses exist in different shapes. In most circumstances they occur naturally as *rings*. Figure 125.1 shows you diagrams of ribose and glucose.

125.3 Polysaccharides

For humans the most obvious property of sugars is that they taste sweet. The 'sugar' you might have at home to sweeten drinks and foodstuffs is sucrose (Figure 125.2). This is a sugar made from a molecule of glucose and a molecule of fructose. It is an example of a *disaccharide* (the prefix *di-* meaning two).

However, sugar molecules can join together to make polymer chains of great length. These are the *polysaccharides*, such as starch, cellulose (Figure 125.3) and glycogen. Animals, including the human variety, store carbohydrates in their bodies in the form of glycogen. It is a complicated polymer made from glucose monomer

A simple model of glucose which shows the puckered ring of five carbon and one oxygen atoms.

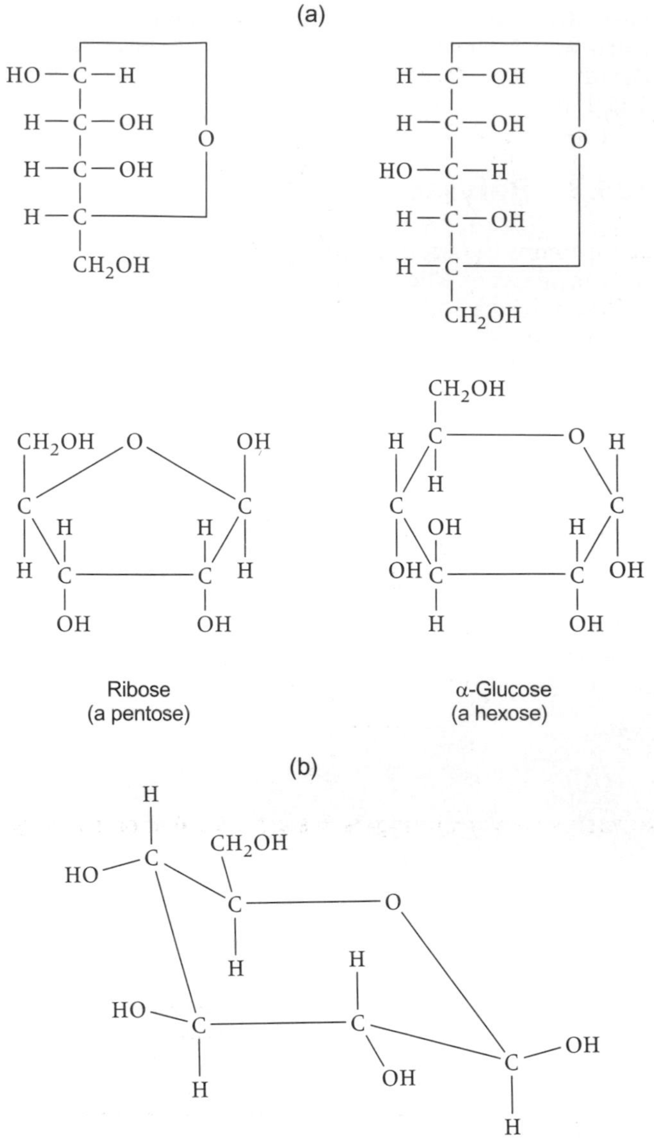

Figure 125.1 Diagrams showing the structures of two sugars. (a) Two ways of showing the structures of the sugars ribose and α-glucose. (b) This gives a more realistic impression of the ring in α-glucose

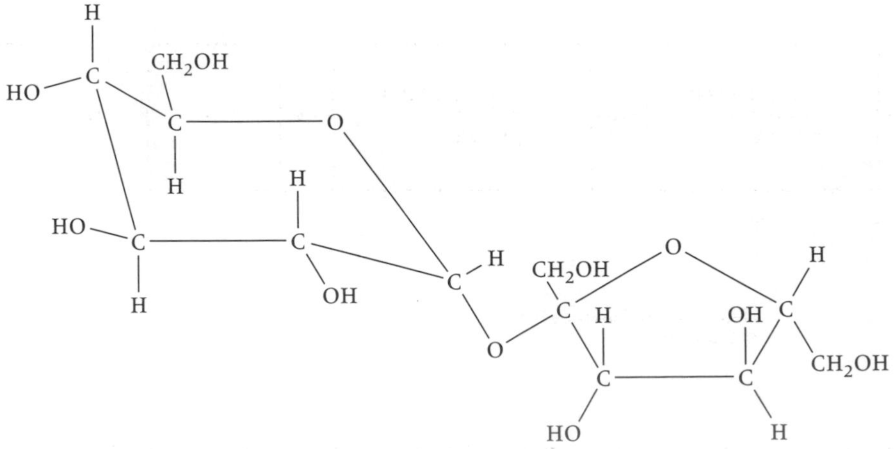

Figure 125.2 The structure of sucrose – a disaccharide

units. The virtue of glycogen is that, when energy is needed quickly in the body, it releases its glucose relatively easily.

Carbon dioxide, water vapour and sunlight are converted into glucose by chlorophyll in plants. Glucose can be used by plants to build starch and cellulose. Cellulose is the key ingredient of the fibres that are the basic building material of plants. Humans make use of starch molecules as a source of carbohydrate, but the metabolism of humans cannot convert cellulose into anything of nutritional value. However, cows and other ruminants are able to decompose cellulose. Hence cows rather than humans are to be found grazing on grass day after day.

Cellulose has uses apart from feeding cows (Table 125.2). It was one of the first materials to be converted into an artificial fibre, Rayon. You will find an outline of the process in Figure 125.4.

Table 125.2. Some uses of cellulose products

Cellulose	Gives a sense of smoothness to the taste of ice cream and whipped cream toppings
	As a harmless binding agent in medicines
	In products where water absorbency is important, e.g. disposable nappies
	Manufacture of artificial fibres such as Rayon
	Manufacture of thin, protective sheeting, especially Cellophane
Cellulose acetate	Manufacture of photographic film
Cellulose nitrate	Used in the explosives industry as gun cotton

125.4   Three properties of sugars

(a)   Hydrolysis

If you eat potatoes, crisps or bread, the enzyme salivary amylase in your saliva rapidly breaks down the starch into smaller units, especially glucose. It is the monomer

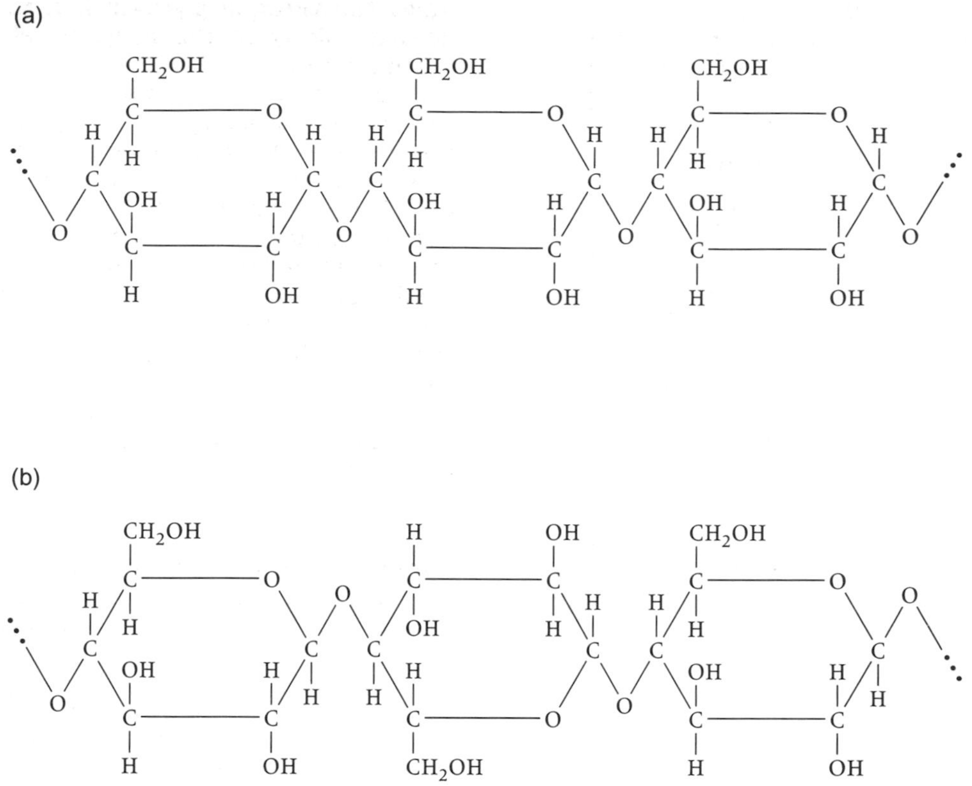

(a)

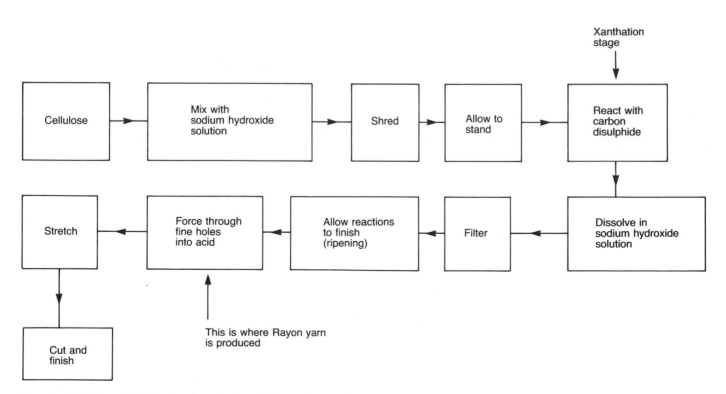

(b)

Figure 125.3 *The structures of starch and cellulose. (a) A typical fragment of a chain of sugars in starch. (b) A fragment of the chain of sugars in cellulose*

| Cellulose | → | Mix with sodium hydroxide solution | → | Shred | → | Allow to stand | → | React with carbon disulphide |

Xanthation stage →

| Stretch | ← | Force through fine holes into acid | ← | Allow reactions to finish (ripening) | ← | Filter | ← | Dissolve in sodium hydroxide solution |

This is where Rayon yarn is produced

Cut and finish

Figure 125.4 *Flowchart for the manufacture of Rayon from cellulose. The reaction with carbon disulphide, CS_2, called xanthation, has the effect of introducing sulphur atoms into the carbohydrate chains*

glucose molecules that can be absorbed and used to provide energy. In the laboratory we can break starch apart by heating it with water, acid or alkali. If the hydrolysis is completely successful, the final solution will only contain glucose molecules. If the hydrolysis is only partially successful, then we will have a mixture of fragments of the original starch molecules.

(b) *Reducing sugars*

Many sugars are reducing agents. Like the aldehydes that we discussed in Unit 118, a warm solution of glucose will reduce Fehling's solution. The tell-tale sign of the reaction is that the original blue colour of Fehling's solution becomes green, and finally a yellow-orange precipitate of copper(I) oxide is produced.

Glucose is an example of a *reducing sugar*, as is fructose; however, sucrose is a non-reducing sugar. Neither starch nor cellulose are easily reduced.

The disaccharide lactose is a sugar found in milk. If you have smelled sour milk, you will know that it has a particularly revolting odour. The odour is due to 2-hydroxypropanoic acid (lactic acid) and butanoic acid, which are produced when lactose is oxidised by bacteria.

(c) *Mutarotation*

The mutarotation of sugars is the spontaneous change in optical rotation of a solution of a sugar such as glucose. You will find details of mutarotation in section 110.8.

125.1 Look at the two diagrams below. They are called α-glucose and β-glucose. What is the difference between the two types of molecule?

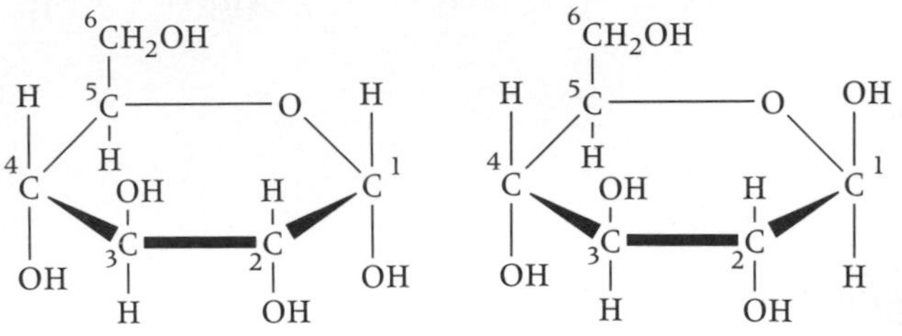

α-glucose   β-glucose

125.2 A student mixed a solution of starch with some Fehling's solution and warmed the two together. The student reported that there was no reaction, and felt pleased because she knew that books said that starch should show no reaction. However, she was not so pleased when having warmed the solution a little more, and leaving for a few minutes, she saw the blue solution turning green. What had happened?

Answers

125.1 The difference is the arrangement of the hydrogen atom and OH group on the carbon atom labelled 1. In α-glucose the OH group is below the ring, in β-glucose it is above the ring. These two isomers are called *anomers*. (They are also diastereoisomers; see section 110.4).

125.2 Starch undergoes hydrolysis, so after a while part of the starch is converted into smaller sugar units. If any glucose is released in the hydrolysis (which it will be after warming for many minutes), then this will reduce the Fehling's solution. Hence the colour change is not caused directly by the starch, but as the result of its hydrolysis.

UNIT 125 SUMMARY

- Carbohydrates:
  - (i) Contain carbon, hydrogen and oxygen only.
  - (ii) Can exist as sugars, or polymers such as starch and cellulose.
  - (iii) The monosaccharides have the general formula $C_nH_{2n}O_n$.
  - (iv) Polysaccharides consist of chains of sugar molecules.

Reactions

- Hydrolysis:
  Starch can be converted into individual sugars by heating with water or acid.
- Reducing sugars:
  Give an orange precipitate with Fehling's solution. Glucose is a reducing sugar, sucrose is not.

126

Vitamins, hormones, steroids and pharmaceuticals

126.1 Vitamins

Vitamins are one of the types of chemical that are essential ingredients of our diet. Without them we would develop a variety of *deficiency diseases*. The disease scurvy results from a lack of vitamin C in the diet. Scurvy was responsible for the extremely high death rate among sailors on long sea voyages during the seventeenth and eighteenth centuries. For example, a three-year voyage lead by Lord Anson in 1740 began with 961 men and finished with 335, a death rate of 65%. Even higher death rates were recorded. The symptoms of scurvy were described in 1725 like this:

> [the sufferers] have spots first red, then growing Livid and Blackish, infesting the Limbs and several Parts with an unusual lassitude or weariness, And who have dark red, itching, and corrupted Gums with a looseness of Teeth, that can't bear the least rub without bleeding . . . shifting Pains frequent about the Limbs and Gums: joined with a very unequal Pulse . . .

Although scurvy is often associated with sailors, the disease was widespread among populations on land as well. The very poor were prone to scurvy because they could not afford to buy food of any reasonable quality. However, the rich also developed scurvy because of eating a diet almost devoid of fruit and vegetables. For those who could afford it, scurvy could be cured by eating fruit or by drinking fresh lemon juice, both sources of vitamin C. In Ireland the poor traditionally ate a diet of potatoes, which happen to have a good deal of vitamin C in them. When the potato crop failed in the years 1845–1849, tens of thousands of people died, partly as a direct result of scurvy, and partly owing to scurvy making them susceptible to other fatal diseases like typhus.

Humans cannot make their own vitamin C, so it has to be obtained by eating the right types of food. This is true of other vitamins as well. Some vitamins can be synthesised naturally by our bodies. Vitamin D is made by a reaction in skin when it is in sunlight.

Table 126.1 gives you information about vitamins.

126.2 Hormones and steroids

Hormones act as chemical messengers controlling the activity and development of living things. In humans they are released in small amounts into the blood stream from a number of important sites, especially the pituitary gland. This gland takes up a small part of the brain (it weighs less than 1 g) but it is a vital source of over 20 hormones. The protein insulin, which we discussed in Unit 123, is a hormone that is secreted from the pancreas. It controls the concentration of glucose in blood ('blood sugar level'). In plants as well as in humans hormones control growth rate. A set of hormones called *oestrogens* and *androgens* control the sexual development of females and males respectively. Some examples of hormones are shown in Table 126.2.

From Figure 126.1 you can see that some hormones are peptides (built from amino acids); others are steroids. Steroids are compounds based on the combination of three cyclohexane rings and one five-membered ring, shown in Figure 126.2.

The advances made in our understanding of the structures and functions of hormones have had many repercussions. Among the most important has been the development of the female contraceptive pill. There are many different types of contraceptive pill, but most of them do not use the same hormones that are released naturally in females. Rather, slightly different molecules are used, which mimic the behaviour of the natural hormones. The aim is the same, to prevent conception taking place. One way of doing this is to use a substance that suppresses the monthly release of eggs during the menstrual cycle. Another method is to use a drug that blocks the action of one or more of the natural hormones, such as progesterone, which is vital for the maintenance of pregnancy. There have been many studies of the possible side effects that could result from the use of contraceptive pills. The evidence is not completely clear. Certain groups of women appear to be more susceptible to diseases, e.g. those who smoke heavily are likely to have a higher incidence of heart disease in later life. However, other groups show little or no increased risk.

Steroids are another important class of natural prod-

Table 126.1. Important vitamins

Vitamin	Source	Deficiency disease
A (retinol)	Fruit, vegetables	Lack of the vitamin leads to blindness and impairs growth

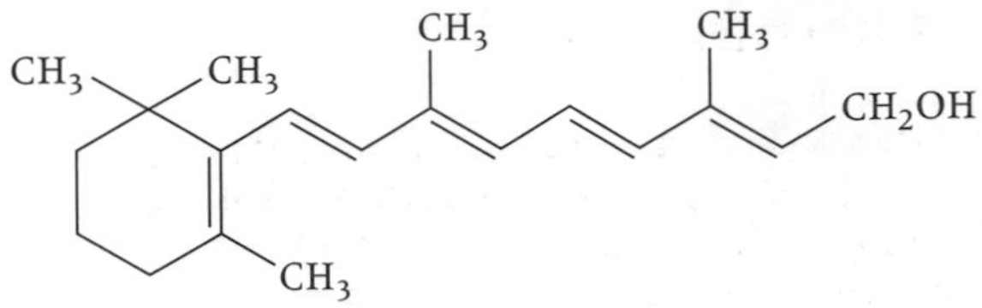

B₁ (thiamin)	Cereals	Beriberi: nervous system stops working properly

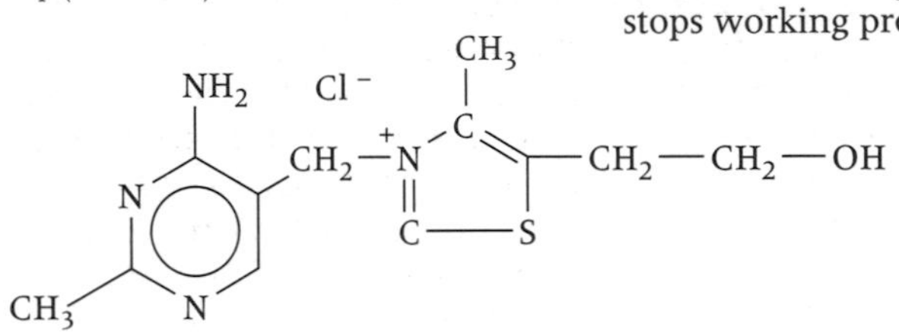		
C (ascorbic acid)	Fruit, vegetables	Scurvy: bleeding gums, pain in limbs, blindness

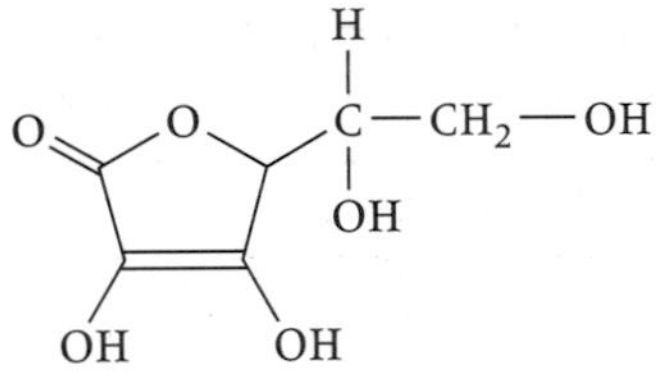		
D (cholecalci-ferol)	Action of sun-light on skin	Rickets: faulty bone growth, leading to deformed limbs

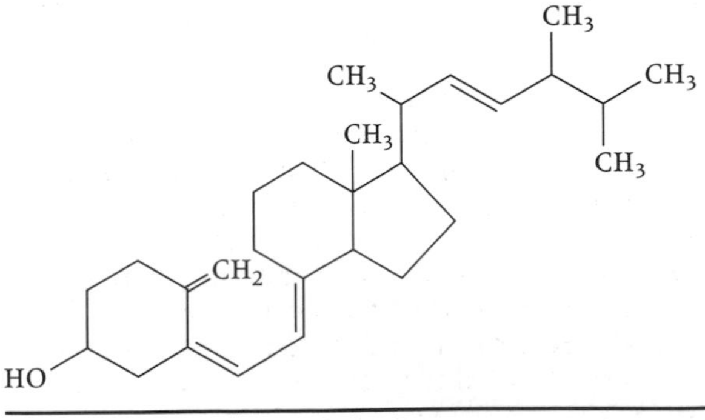

Table 126.2. Examples of hormones

Hormone	What it does
Adrenaline	The 'flight or fight hormone'; produced in response to abnormal degree of stress or excitement, especially in response to danger
Gibberellic acid	Encourages the growth of plant shoots
Insulin	Controls blood sugar level
Oestrogen	Prepares the uterus of a female for possible pregnancy
Oxytocin	Induces labour in pregnant women
Progesterone	Suppresses production of eggs from follicles in the female
Somatotrophin	Known as growth hormone, controls growth of humans
Testosterone	One of the androgen hormones secreted from the testes in males; helps to govern qualities linked to maleness, e.g. in sexual organs, body hair, muscle bulk

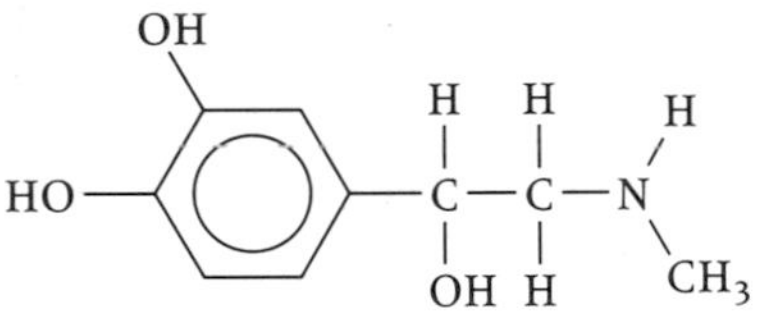

Adrenaline

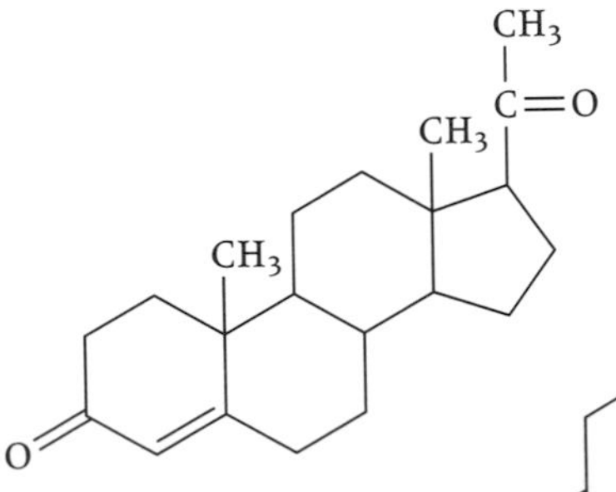

Progesterone

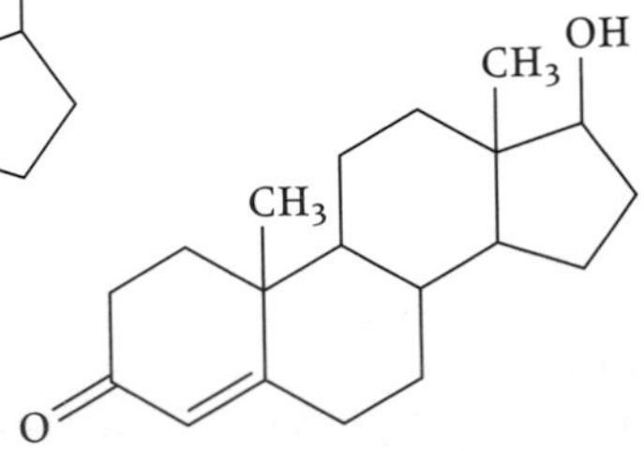

Testosterone

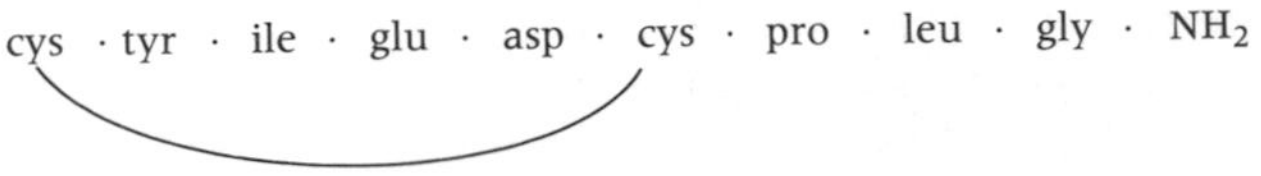

cys · tyr · ile · glu · asp · cys · pro · leu · gly · NH₂

Oxytocin: a hormone made from amino acids linked by peptide bonds

Figure 126.1 *Examples of hormones*

ucts. They all play some part in biochemical systems. Indeed, as we have seen, some of them are sex hormones.

In recent years the use of anabolic steroids among athletes has become a matter of great concern. Many of the androgens (male steroids) are involved in growth processes such as muscle formation. By taking them an athlete can become fitter and stronger more rapidly

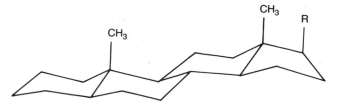

Figure 126.2 *The basic structure of a steroid. There are carbon atoms where the lines meet. The hydrogen atoms are not shown. The group R changes with the nature of the steroid. (Adapted from: Morrison, R. T. and Boyd, R. N. (1966). Organic Chemistry, 2nd edn, Allyn and Bacon, Boston, p. 521)*

than someone who does not take the steroid. However, long-term use of steroids can have very dangerous side effects (apart from more obvious effects like becoming extremely hairy).

Two other steroids that have gained a reputation for themselves are cholesterol and cortisone. (Strictly, cholesterol and other similar molecules that have an OH group should be called sterols.) These are shown in Figure 126.3. Cholesterol has been proposed as the chemical responsible for the hardening (blocking) of arteries, and therefore as a cause of heart attacks. It is also linked to the formation of gall stones. The evidence suggests that it is best to avoid eating large amounts of foods that are high in cholesterol, e.g. whites of eggs and animal fats. However, you will not be able to avoid cholesterol all together because it is made in the liver. It may be that some people are more liable to suffer from

Cholesterol

Cortisone

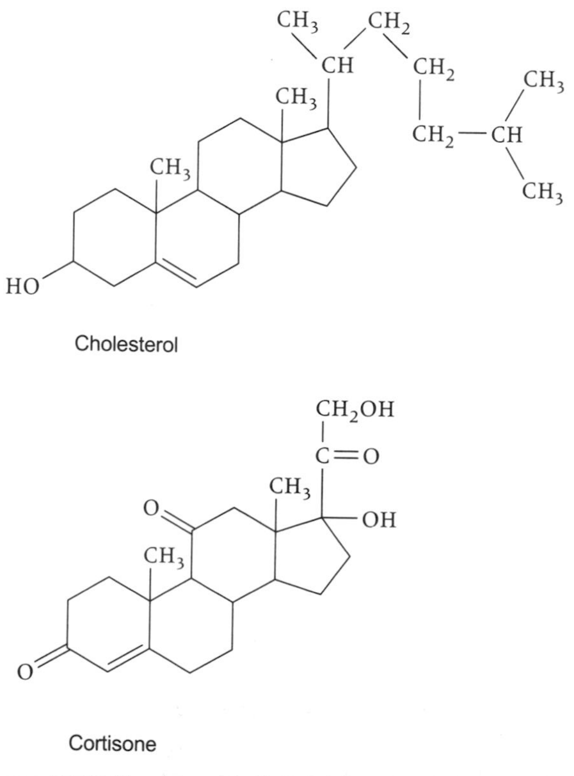

Figure 126.3 *Two steroids that are important in medicine*

heart disease because of the cholesterol produced in their own bodies rather than because they eat the wrong things. Cortisone is used by tens of thousands of people as treatment for arthritis, a very painful disease of the bones and joints.

126.3 Pharmaceuticals

Pharmaceuticals are chemicals that are prepared and sold with the intention of treating disease. You might be happy to call such chemicals by the shorter name: drugs. However, we tend to think of a drug as a substance that is not used to treat illness but, like heroin, is taken illegally by people simply for the short-term pleasurable effects it can give. Some pharmaceuticals are well known, particularly aspirin and paracetamol. Others are less common, but of extreme importance. Examples are listed in Table 126.3. Recently there has been increasing evidence not only that aspirin will suppress headaches and minor pains, but that its regular use helps to protect the user against heart disease.

The pharmaceutical industry is one of the most important parts of the chemical industry. The market for pharmaceuticals has increased until it is worth billions of pounds each year. Millions of people benefit from the treatment they receive for illnesses that even 10 years ago would have been impossible to treat. We often take it for granted that an operation in hospital, or a visit to the dentist, can be (almost) painless. In fact the advances made in the design of pharmaceuticals have been so successful that we tend to think that almost every complaint will be able to be treated. Nature is not so easily overcome! For example, when it was first widely used against bacterial infection during the Second World War, penicillin was almost 100% effective. Now a wide range of bacteria have adapted so that they are resistant to it. As a consequence, there is always a need for the pharmaceutical industry to adapt and design new, more powerful, anti-bacterial products.

Many pharmaceuticals act directly on the brain. Valium does this. It is widely prescribed for people who are over-anxious or depressed. Valium is not in itself addictive, but its long-term use makes people dependent on it. Some people have criticised doctors for prescribing valium, because it does little to change the reason why the patient is over-anxious. However, doctors may say that they are not able to treat the main cause, which might be due to emotional strains, e.g. of divorce or living in an over-crowded flat.

The *opiates* are drugs that cause addiction. Heroin and cocaine (Figure 126.4) are among the most famous (or infamous) examples. The chemistry of the brain adapts to the presence of the opiate in such a way that the person taking the drug feels an uncontrollable need to take more of it. Owing to their ability to induce an artificial feeling of well-being, the person taking heroin or cocaine quickly becomes addicted to them. Once this happens, the craving for the drug often causes the person to:

Table 126.3. Some important pharmaceuticals

Name	Structure	Treatment of
Aspirin	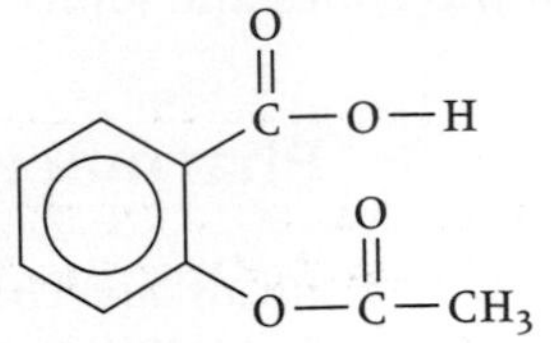	Mild pain and heart disease
Paracetamol	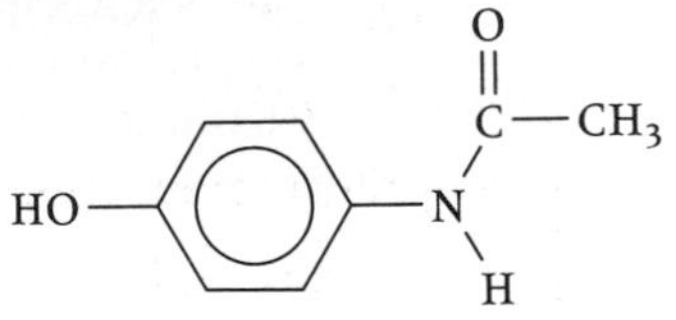	Mild pain
Timolol	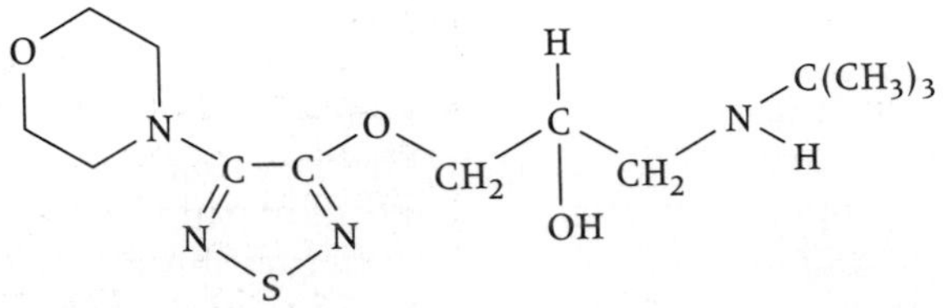	Glaucoma, an eye disease causing blindness
Morphine (for structure, see Figure 126.4)		Analgesic – masking of severe pain
Penicillin	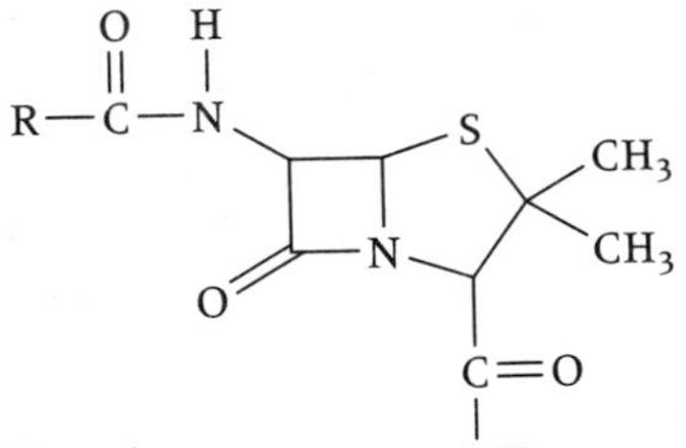 R varies	Bacterial infections
Nitroglycerin	CH_2ONO_2 $\mid$ $CHONO_2$ $\mid$ CH_2ONO_2	Angina, a common heart condition
Valium	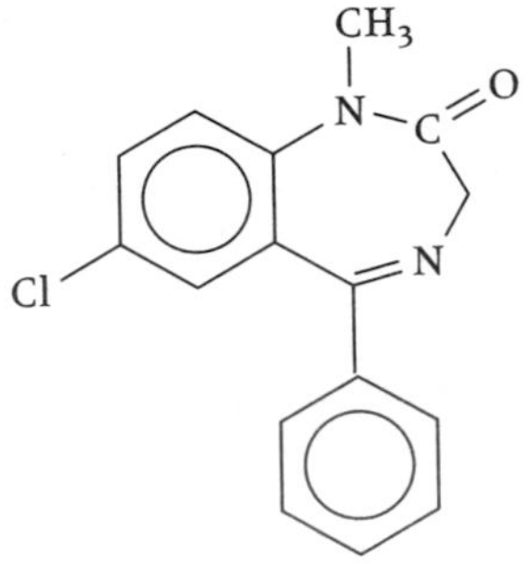	Feelings of anxiety or depression

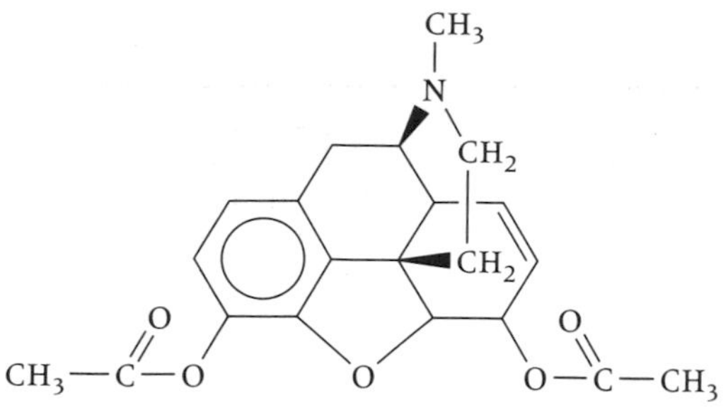

Heroin

(Morphine has OH groups in place of the CH₃COO groups)

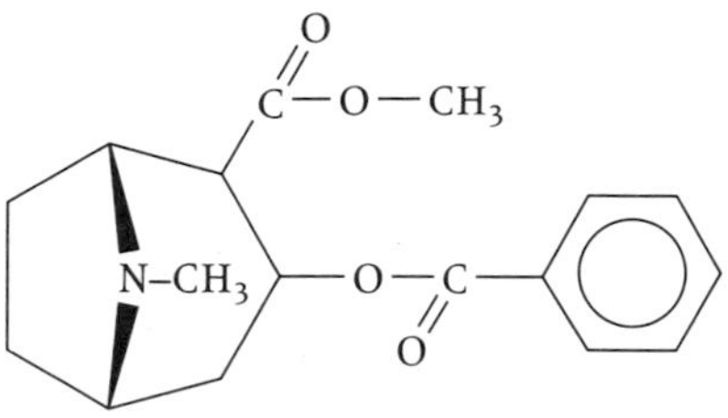

Cocaine

Figure 126.4 *Two most dangerous opiates*

(i) neglect their nutritional needs;
(ii) neglect ordinary rules of hygiene;
(iii) enter into a life of crime or prostitution in order to obtain the cash necessary to buy the drug.

As a result a drug addict often becomes thin and susceptible to illness. Especially, those injecting drugs are highly likely to become infected by AIDS. Unfortunately, women heroin addicts often become pregnant and their babies are born already showing the signs of addiction. For society at large, addiction to heroin and cocaine poses two linked problems: one of law and order, another of public health. The wise person avoids any contact with illegal drugs, or those using them.

126.1   Look at the structure of testosterone in Figure 126.1 and predict the result of reacting this hormone with:

(i) bromine;

(ii) sodium;

(iii) phosphorus pentachloride;

(iv) 2,4-dinitrophenylhydrazine.

126.2   Here is an account of the preparation of a well known analgesic.

Approximately 10 g of 2-hydroxybenzoic acid should be carefully added to a mixture of 10 cm³ of ethanoic anhydride and 10 cm³ ethanoic acid. (The latter mixture is used as an acetylating reagent.) The mixture should be refluxed for about 30 minutes. After cooling, the mixture should be poured into cold water and the white solid filtered off and recrystallised.

(i) Draw a diagram of 2-hydroxybenzoic acid.

(ii) Draw a diagram of ethanoic anhydride.

(iii) Draw a diagram of the product.

(iv) What is the name of the product?

(v) What is meant by the word 'analgesic'?

(vi) Predict how the product would react when refluxed with sodium hydroxide solution.

Answers

126.1   The key to answering this question is to realise that the chemistry of carbon compounds is mainly the chemistry of the functional groups. We can see three of them here: a double bond, as in an alkene; an OH group like that in an alcohol; and a >C=O group like that in a ketone. If you know the chemistry of these three groups, you can predict the outcome of the reactions.

(i) Bromine atoms will add on to each end of the double bond. (The bromine will be decolourised.)

(ii) Hydrogen will be given off when sodium reacts with an OH group.

(iii) Fumes of HCl will be given off. (This is the usual test for an OH group.)

(iv) A hydrazone will be made when the 2,4-dinitrophenylhydrazine reacts with the carbonyl group. (A standard reaction of the carbonyl group.)

126.2 (i)

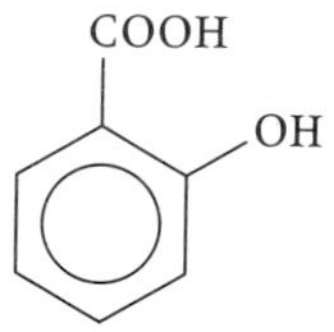

(ii)

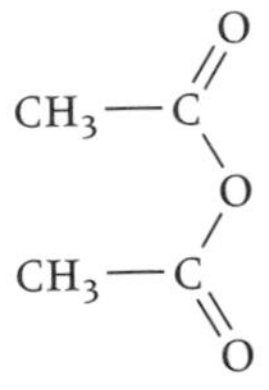

Answers – contd.

(iii)

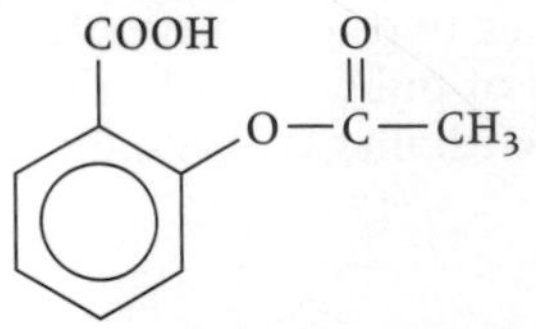

(iv) Its common name is aspirin.

(v) An analgesic relieves pain.

(vi) Aspirin is an ester, so we would expect it to undergo hydrolysis in the same way as do other esters (see Unit 120). The reaction is:

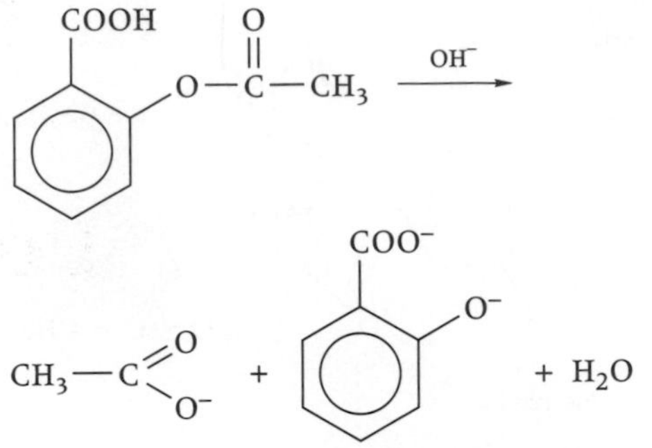

$$CH_3COOC_6H_4COOH \xrightarrow{OH^-} CH_3COO^- + {}^-OC_6H_4COO^- + H_2O$$

UNIT 126 SUMMARY

- Vitamins:
  Are essential for preventing deficiency diseases. See Table 126.1.
- Hormones and steroids:
  Act as chemical messengers, controlling the activity and development of living things. See Table 126.2.
- Pharmaceuticals:
  Are used to treat disease and to control biochemical processes. See Table 126.3.

127

Polymers

127.1 Monomers and polymers

The Swedish chemist Jacob Berzelius invented the word *polymeric* in 1832. He understood that some chemicals could have the same molecular formula as each other, and therefore the same empirical formula. These molecules were recognised to be isomers. However, chemists also knew that some molecules had the same empirical formulae, but different molecular formulae. An example that we are familiar with is ethyne, C_2H_2, and benzene, C_6H_6. They have the same empirical formula, CH, but different molecular formulae. At the time when Berzelius was working, chemists' knowledge of the structures of molecules was markedly different to ours. In particular, it was a great puzzle why molecules with the same empirical formula could have different chemical properties. Berzelius thought that such molecules belonged to a special class of isomers. He said:

> To describe this equality of composition coupled with a difference of properties I suggest that such substances be called polymeric.

In the years following this definition, chemists began to refer to benzene as a polymer of ethyne. Most chemists would not now think of benzene as a polymer. Our modern notion of a polymer includes the notion that it must have a *high relative molecular mass*. A typical example is poly(ethene), better known as polythene. It is a giant molecule made when thousands of ethene molecules join together to make a long chain. Ethene molecules are the *monomers*, which join to make the *polymer*:

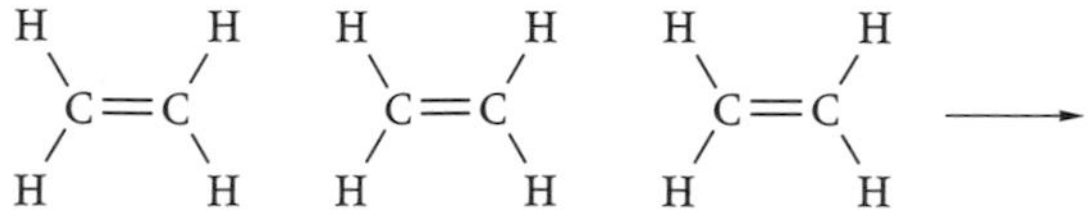

some separate ethene molecules,
i.e. monomers

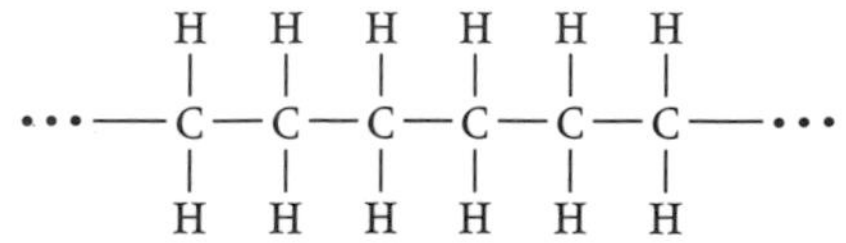

part of a polythene chain,
i.e. polymer

We can represent a polythene chain by the formula $-(C_2H_4)_n$. Here the subscript n means a very large number (e.g. 10 000) of C_2H_4 units joined together. You should be able to see that the empirical formula of polythene is CH_2, just as it is for ethene. It is often claimed that the first polymerisation reaction was performed in 1839 by a German pharmacist, E. Simon. He heated an extract of a resin from a tree with sodium carbonate solution and found a rubbery substance. Unbeknown to Simon, the resin extract contained styrene (systematic name phenylethene), and the rubbery substance was polystyrene (poly(phenylethene)):

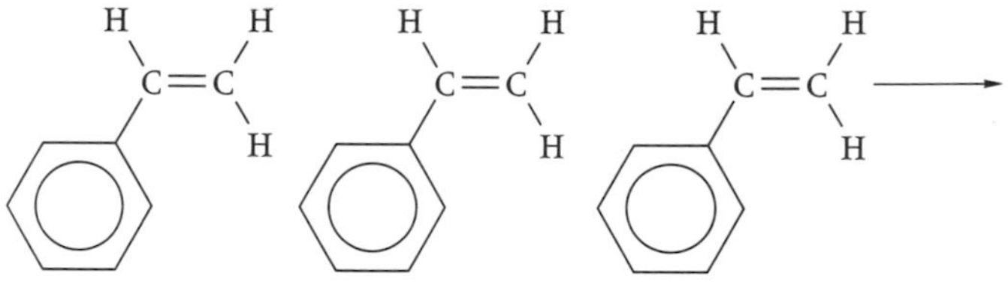

styrene monomers

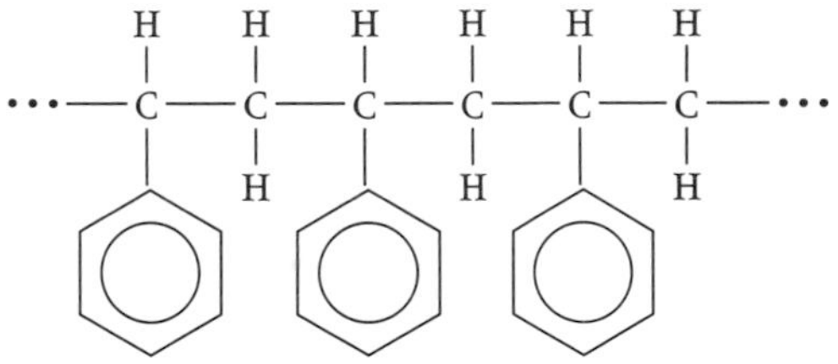

polystyrene

However, Richard Watson (Bishop of Llandaff) has a prior claim to that of Simon. Between 1782 and 1800, Watson reported in a series of 'Chemical Essays' that:

> The most transparent oil of turpentine, resembling naphtha, may be changed into an oil resembling petroleum by mixing it with a small portion of the acid of vitriol; with a larger proportion of the acid, the mixture becomes black and tenacious, like Barbadoes tar; and the proportions of the ingredients may be so adjusted, that the mixture will acquire a solid consistence, like asphaltum.

The 'tenacious tar' was a polymer.

Both models show a portion of a polyethene chain. In most situations, the model on the left is an idealised version. The chain is much more likely to twist as shown on the right.

127.1 Benzene can be called a trimer of ethyne. Can you remember examples of (i) an inorganic dimer, (ii) a trimer of ethanal? Briefly describe their structures.

127.2 Polymers can be made by addition reactions

You may have noticed that ethene and styrene molecules have a double bond. This feature of their structure is shared by many, but not all, molecules that will polymerise. An example of a polymerisation in which the monomers have no double bonds is the change from rhombic sulphur to plastic sulphur (see section 100.2):

S_8 rings

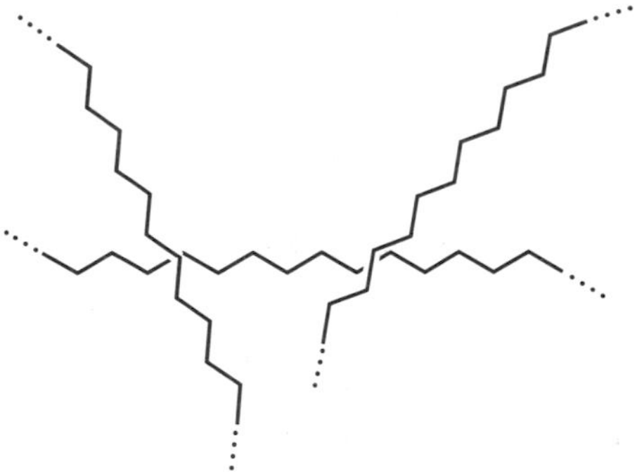

plastic sulphur

However, in the following pages we shall only deal with monomers that have at least one double bond.

Assuming that we have a suitable monomer, experience shows that there are two processes that will convert it into a polymer:

(**i**) an addition reaction, or
(**ii**) a condensation reaction.

In both cases, a catalyst may be used to help the reaction take place, and fairly high pressures and temperatures may be necessary. In this section we shall concentrate on addition reactions.

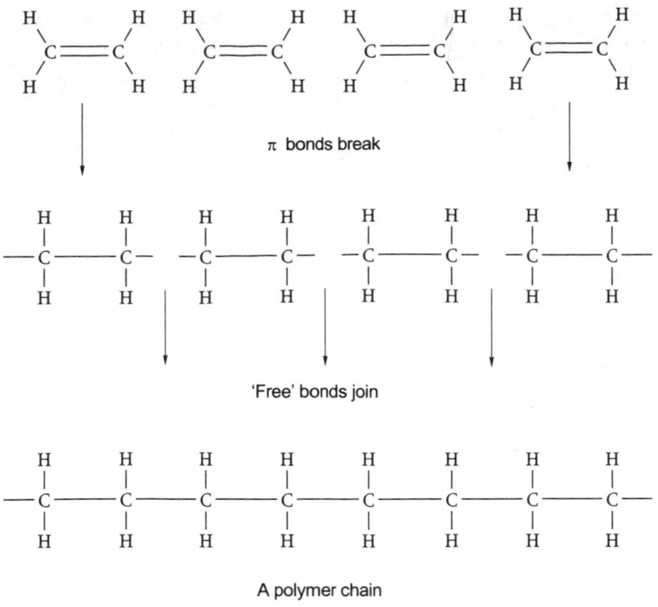

Figure 127.1 *Making poly(ethene) from ethene*

A typical example of an addition polymerisation is the conversion of ethene into poly(ethene) (often called polythene). If we adopt a simple view of the reaction, we can think of the π bond in each ethene molecule breaking to give two 'free bonds' at each end of the molecule. These bonds can be used to join neighbouring molecules, which therefore add together. This is the situation represented in Figure 127.1. However, this *is* a naive view of how the reaction actually takes place. To understand why things cannot be this simple, we should look at experimental evidence that comes from studies of the kinetics of polymerisation reactions. A crucial piece of information is that reactions like the polymerisation of ethene are catalysed by organic peroxides, which are a source of free radicals. (You will find details of free radicals in Unit 81.)

(a) Free radical addition

The three main steps in polymerisation are *initiation*, *propagation* and *termination*. For example, if we use benzoyl peroxide as an initiator of the polymerisation of fluoroethene, some of the reactions that occur are as follows:

(i) Initiation

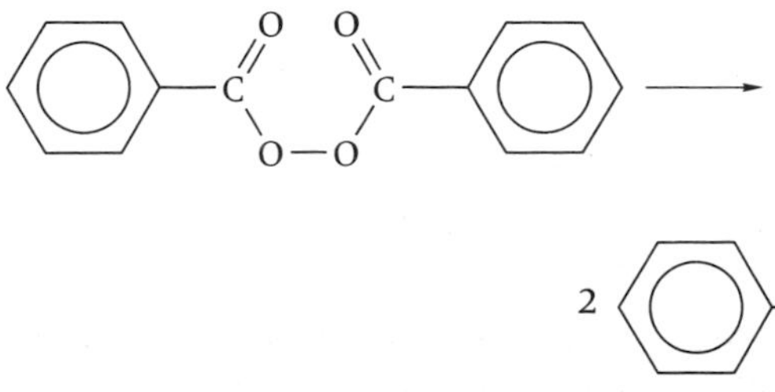

$$C_6H_5COO\!-\!OOCC_6H_5 \rightarrow 2C_6H_5COO\cdot$$

(ii) Propagation

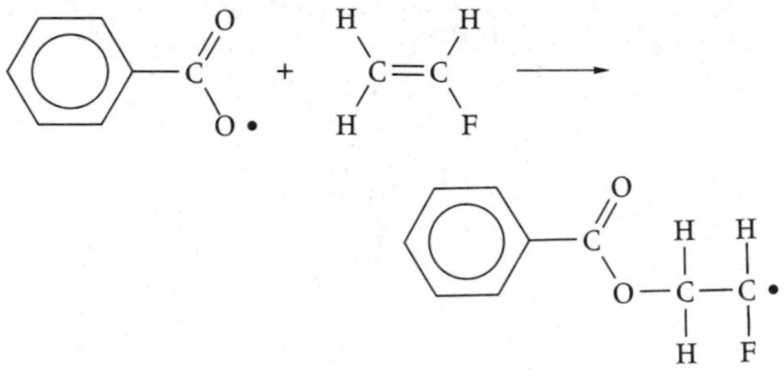

$$C_6H_5COO\cdot + CH_2\!=\!CHF \rightarrow C_6H_5COO\!-\!CH_2\!-\!CHF\cdot$$

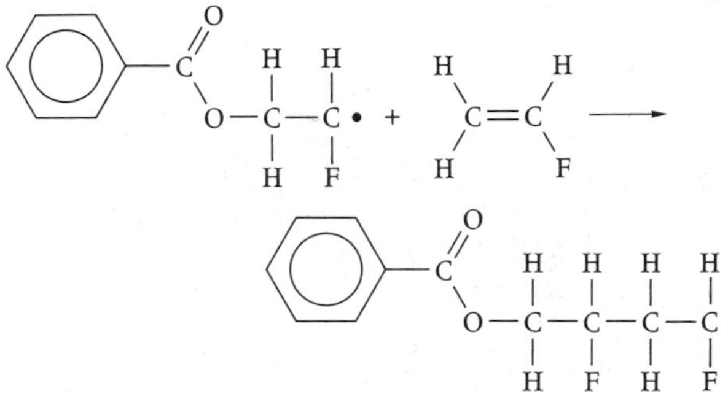

$$C_6H_5COO\!-\!CH_2\!-\!CHF\cdot + CH_2\!=\!CHF \rightarrow$$
$$C_6H_5COO\!-\!CH_2\!-\!CHF\!-\!CH_2\!-\!CHF\cdot$$

After many such reactions,

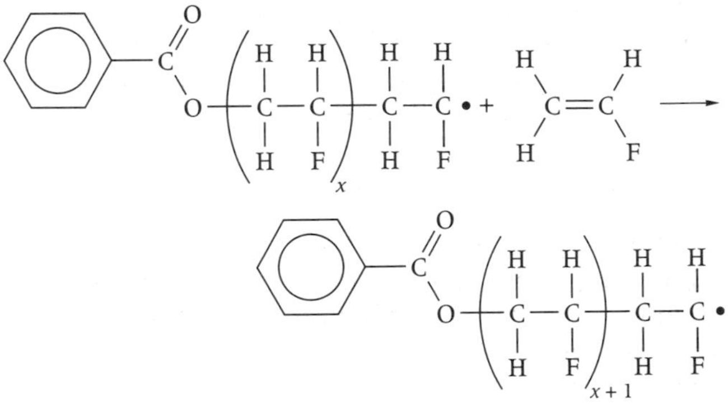

$$C_6H_5COO\!-\!(CH_2\!-\!CHF)_x\!-\!CH_2\!=\!CHF\cdot + CH_2\!=\!CHF \rightarrow$$
$$C_6H_5COO\!-\!(CH_2\!-\!CHF)_{x+1}\!-\!CH_2\!-\!CHF\cdot$$

(iii) Termination

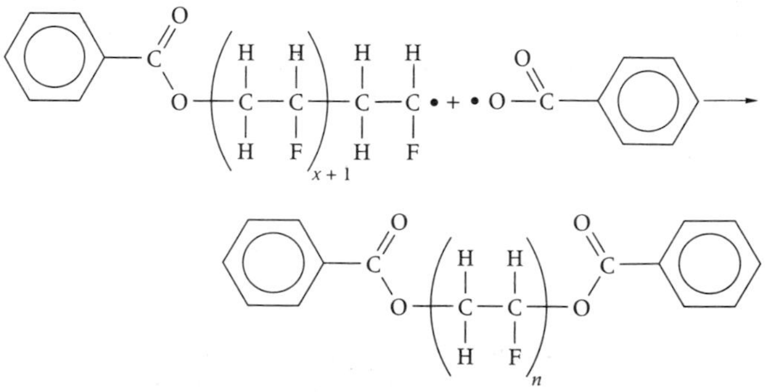

$$C_6H_5COO\!-\!(CH_2\!-\!CHF)_{x+1}\!-\!CH_2\!-\!CHF\cdot + \cdot OOCC_6H_5 \rightarrow$$
$$C_6H_5COO\!-\!(CH_2\!-\!CHF)_n\!-\!OOCC_6H_5$$

This is only one way in which termination can take place. Sometimes a chain can end by taking a hydrogen atom from another molecule. If this happens the polymer molecule might look like this:

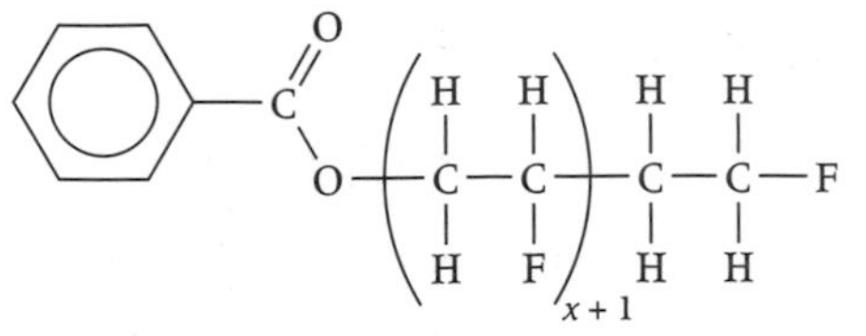

$$C_6H_5COO\text{---}(CH_2\text{---}CHF)_{\overline{x+1}}\, CH_2\text{---}CH_2F$$

We have not shown all the possible reactions that take place. One of the most important is the reaction of the growing polymer chain with solvent molecules. These reactions generally stop the chain growing, but the solvent molecule is itself turned into a free radical, which can induce further reactions. Sometimes this may be with an unreacted monomer, in which case another chain starts to grow. However, it can happen that part of a polymer chain is attacked along its length. This leads to the formation of side chains (Figure 127.2). Depending on the type of polymer needed, side chains can either be desirable or a nuisance.

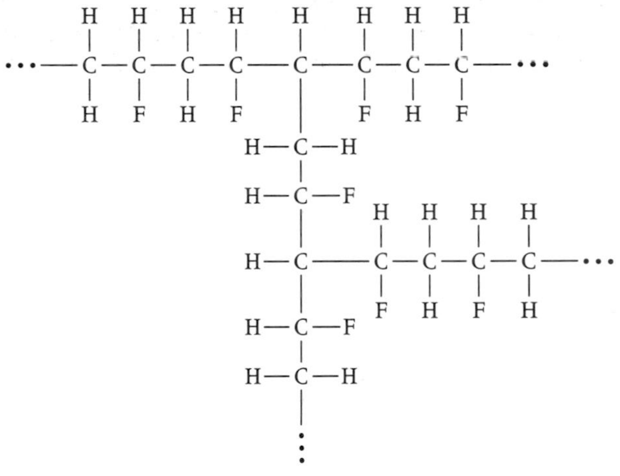

Figure 127.2 *Side chains can grow on a polymer*

Some monomers do not need the presence of a free radical initiator like benzoyl peroxide. Sometimes heat alone, or light, will cause the monomer itself to produce free radicals. Styrene, $C_6H_5CH\text{=}CH_2$

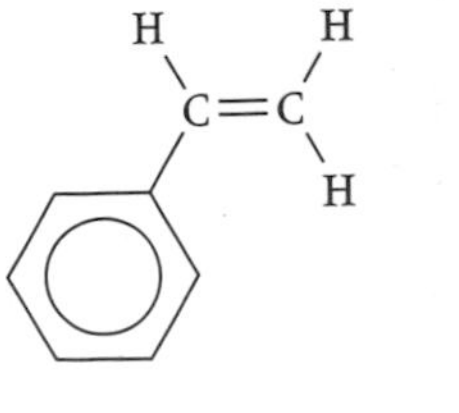

is well known for polymerising of its own accord if it is left in a clear glass bottle. For this reason monomers like styrene are kept in a solution with a second substance, which acts as an *inhibitor*. An inhibitor will react with

free radicals more efficiently than will the monomer; hence it prevents polymerisation. Monomers will usually react slowly with oxygen in air to make peroxides, which can then induce polymerisation. For this reason *antioxidants* are also added. The idea is that the oxygen will react with the antioxidant instead of with the monomer.

(b) *Ionic methods of addition*

Not all polymerisations take place because of free radicals. In many cases a polymer chain grows because of reactions between ions. Just as we have anions (negatively charged) and cations (positively charged) in general chemistry, so we have anionic and cationic polymerisation reactions. The key to both types is to find a substance that provokes a monomer into becoming an ion. We have seen reactions in which this sort of thing happens. For example, in Friedel–Crafts reactions (e.g. panel 113.1) where aluminium trichloride acts as a catalyst in the production of cations such as CH_3CO^+ with ethanoyl chloride, CH_3COCl. You may not be surprised to learn that aluminium trichloride can be used as an initiator of ionic polymerisations. So too can boron trifluoride, Grignard reagents and some acids.

As well as finding the appropriate initiator, the monomer itself must be capable of turning into an ion. This is best achieved by monomers that have groups which are fairly good at donating or withdrawing electrons.

Here is one example of *cationic polymerisation* that takes place when boron trifluoride is mixed with 2-methylpropene and a little water. A polymer known as butyl rubber is made. The first step in the reaction is for the boron trifluoride and water to react and release a hydrogen ion:

$$BF_3 + H_2O \rightarrow BF_3OH^- + H^+$$

The hydrogen ion can then react with the alkene to give a carbocation:

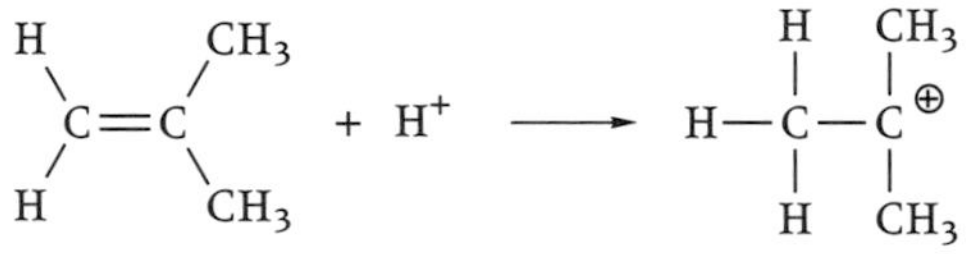

$$CH_2\text{=}C(CH_3)_2 + H^+ \rightarrow CH_3\text{---}C(CH_3)_2{}^+$$

It is this carbocation which will be attacked by the electrons in the π cloud of a neighbouring alkene molecule. If you look back to panel 112.2 you will find that we discovered just this type of thing happening in the addition of bromine to alkenes. The reaction ensures the production of a longer chain cation; hence the growth of the polymer chain:

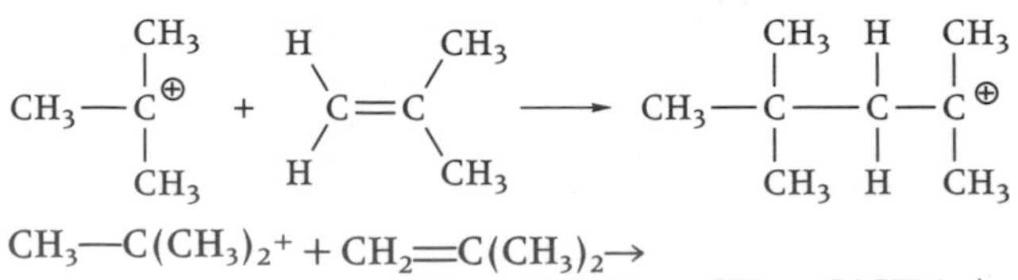

$$CH_3\text{---}C(CH_3)_2{}^+ + CH_2\text{=}C(CH_3)_2 \rightarrow$$
$$CH_3\text{---}C(CH_3)_2\text{---}CH_2\text{---}C(CH_3)_2{}^+$$

Eventually the chain will stop growing, either because the supply of monomer runs out, or because of a chain terminating reaction. For example the cation may combine with a negative ion, or revert to an alkene by the loss of a proton in a reaction like the reverse of the second reaction above.

In *anionic polymerisation*, the monomer unit carries a negative charge. Often the monomer is an alkene with a group attached that can stabilise the negative charge, i.e. the group helps to spread the charge. Examples are $CH=CHX$ with $X=$ —CN, —COOR, —COR, —$CH=CH_2$. The initiator must be capable of generating the negative charge on the monomer. Compounds made from a hydrocarbon and a highly electropositive metal may be used. A good example is $(CH_3)_3CLi$ (sometimes known as butyl-lithium), which produces the 2-methylpropane carbanion,

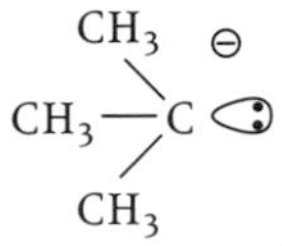

A typical reaction scheme using $(CH_3)_3CLi$ is, first,

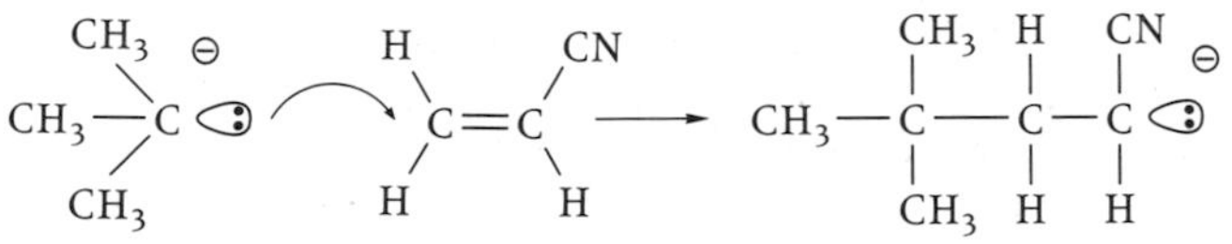

and then

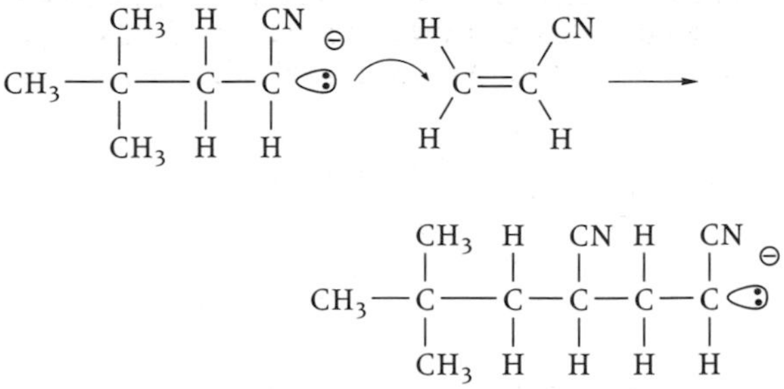

and then further reactions like this which lengthen the chain. Eventually a polymer of formula $(CH_3)_3C\!\!-\!\!(CH_2CHCN)_n$ results.

(c) Coordination addition

This is our third, and last, type of addition polymerisation. It is also one of the most important methods of polymerisation that has yet been discovered. The inventor of the method was K. Ziegler, and his research workers. However, Ziegler's name is linked with that of G. Natta, who extended the research and developed it into a reliable method for making particular types of polymer. Both men received the Nobel Prize for their work. The key to the method of making polymers by the Ziegler–Natta method is the use of a catalyst (Figure 127.3) consisting of a mixture of a transition metal

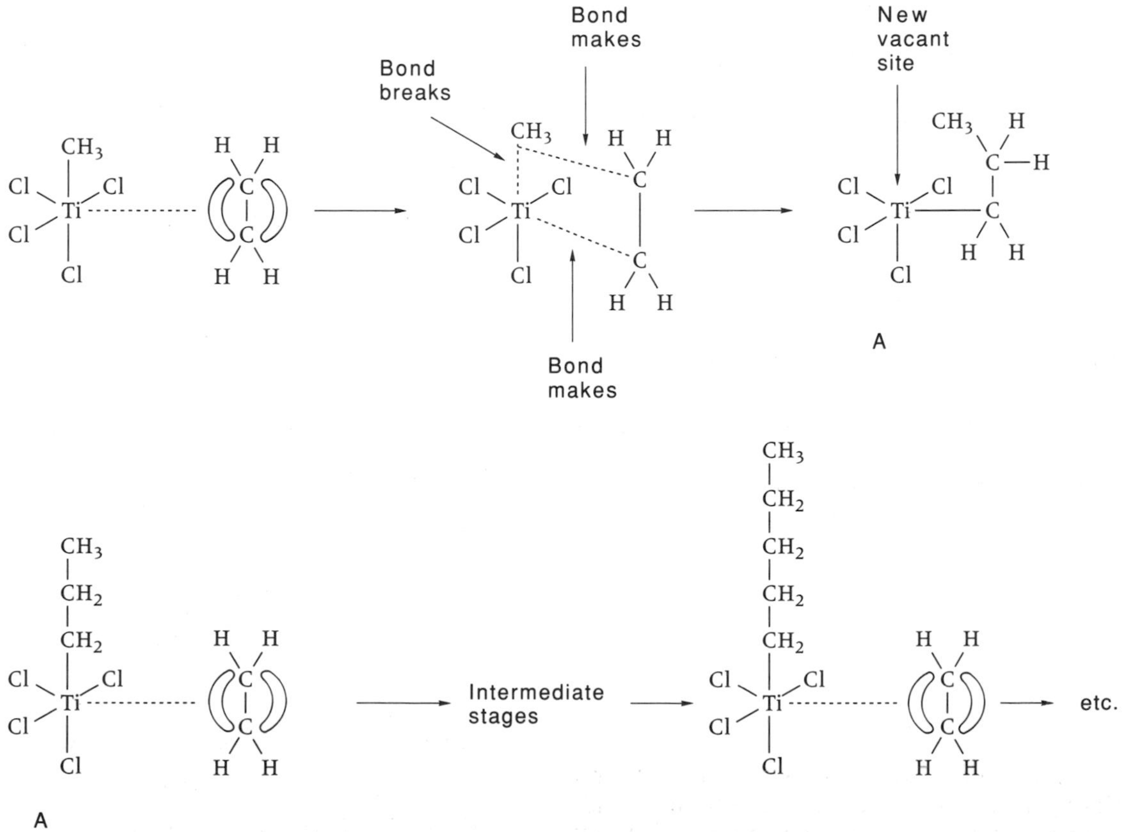

Figure 127.3 *Ziegler–Natta catalysts are used for making polymers*

salt, especially titanium(IV) chloride, $TiCl_4$, and an organometallic compound such as triethylaluminium, $Al(C_2H_5)_3$. The importance of the reaction is that it produces polymer chains in which the groups along the chain are arranged in very orderly ways.

To understand this point, think about the possible outcomes of polymerising propene. The carbon backbone of the polymer chain will have hydrogen atoms and methyl groups sticking out. However, there are three arrangements of the hydrogen atoms and methyl groups along the backbone. The first, and most likely, arrangement is a random one. Sometimes the methyl groups stick out on one side of the chain, sometimes on the other. This is the *atactic* arrangement:

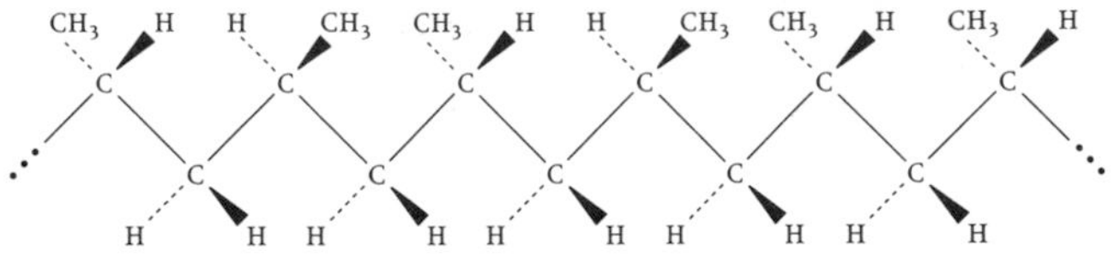

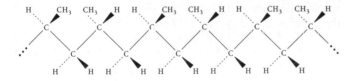

Secondly, there can be a more orderly arrangement where there is a regular pattern of the methyl groups on one side, then on the other. This is the *syndiotactic* arrangement:

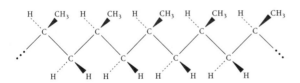

Thirdly, we have the extremely symmetrical arrangement where the methyl groups are always on the same side of the backbone. This is the *isotactic* arrangement:

The different types of polymer chain have different physical properties, which determine the uses to which we can put them. We shall take this up again in the next unit. For the present, let us assume that isotactic polymers have particularly desirable properties. This being so they are more valuable (both in terms of their possible uses, and in terms of money) than the other two kinds. The problem is that in the hurly burly of a free radical reaction it is impossible to control the way a monomer molecule joins to a growing polymer radical. All other things being equal, the atactic geometry is the most likely outcome. The beauty of Ziegler–Natta reactions is that they can reliably be used to prepare isotactic polymers, and even syndiotactic polymers. Similarly the branching of chains that can occur with other methods of polymerisation is avoided. The mechanism of the reaction is rather involved, but a key part of the process is that the alkenes make use of their π clouds of

electrons to bond to a vacant coordination site on the transition metal ion. Once it is held there, the π bond breaks and a σ bond is made with another alkene at a neighbouring coordination site. Providing that new alkene molecules are available, this process can continue almost indefinitely. The regular nature of the geometry of the polymer is a result of the constant way in which the alkenes bond to the transition metal ion.

127.2 If you see styrene being polymerised in your laboratory, you will find that the sample of styrene has to be washed with alkali and water before it is used. Why?

127.3 Draw a diagram showing a portion of a polystyrene polymer.

127.4 When butyl rubber is made, what is the rule that tells us with which carbon atom the hydrogen will combine?

127.3 Condensation polymers

(a) *Polyamides*

One of the most famous condensation polymers invented is Nylon. The American W. H. Carothers first made a Nylon polymer in February 1935. It can be made by mixing hexanedioic acid with hexane-1,6-diamine and heating the solid product. We can see the essential chemistry of the reaction if we write the reaction like this:

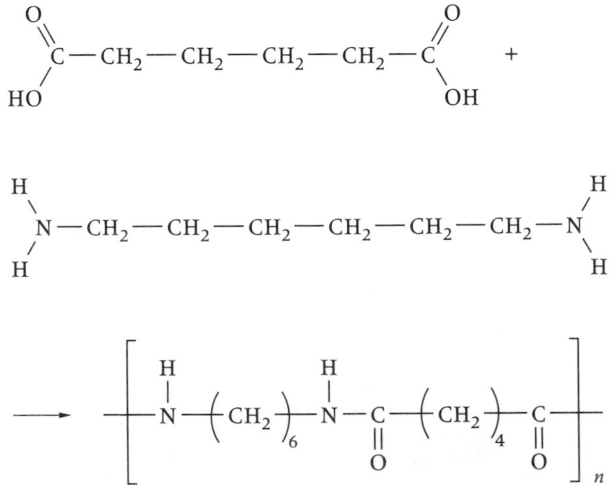

$$HOOC(CH_2)_4COOH + H_2N(CH_2)_6NH_2 \rightarrow$$
$$\text{---}NH(CH_2)_6NHCO(CH_2)_4CO\text{---}_n$$

For each amine and acid group that come together, a molecule of water is released (Figure 127.4). This is why

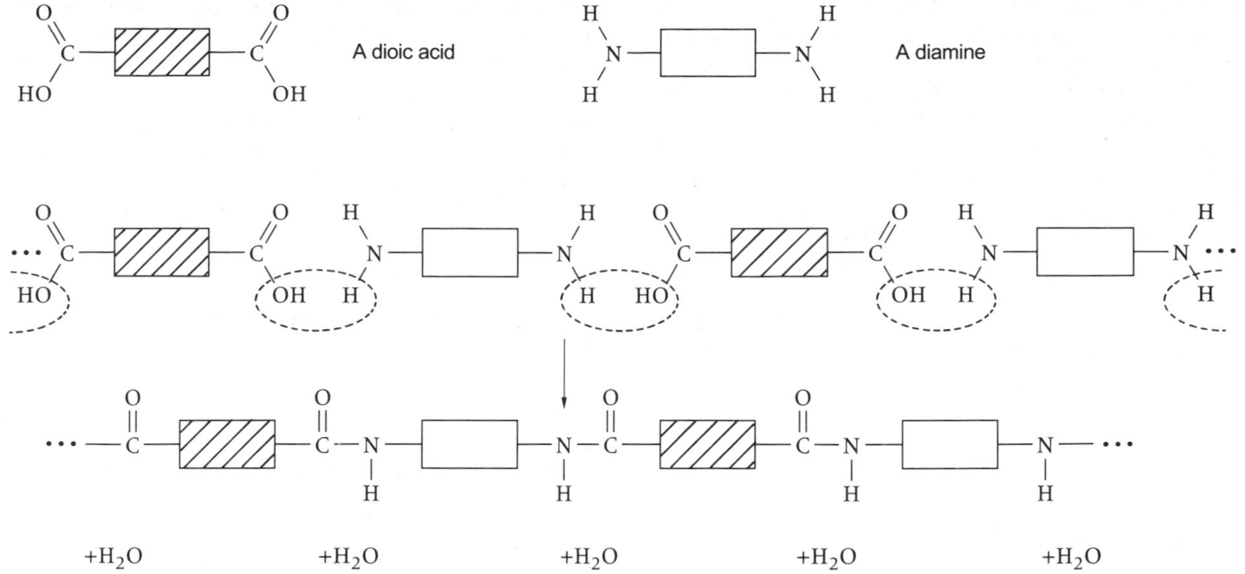

Figure 127.4 A block diagram illustrating how Nylon is made from a dioic acid and a diamine. The rectangles represent chains of CH_2 groups

the polymerisation is called a condensation reaction. This particular variety of Nylon is called Nylon-6,6. (The first of the pair of numbers gives the number of carbon atoms in the amine molecule, and the second the number in the acid.) It can be easily made in the laboratory by a rather different way outlined in panel 127.1 and Figure 127.5. Nylon-6,6 is an example of a *copolymer*. A copolymer is a polymer made from two different monomers.

Another type of Nylon, Nylon-6, is manufactured from a cyclic compound, caprolactam. On heating, the ring opens. This allows neighbouring molecules to link and produce a polymer chain:

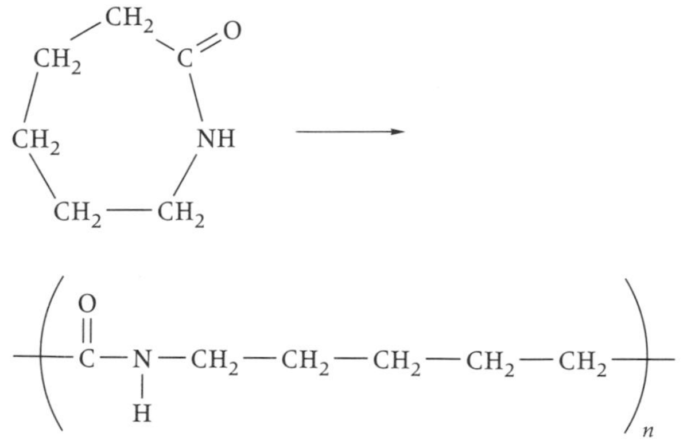

Nylon-6

Nylon polymers are examples of *polyamides* for the simple reason that they contain CONH groups.

Owing to the ease with which Nylon can be drawn into fibres, it has been widely used in manufacturing clothing. However, the fibres also have great strength, a quality that makes Nylon ropes and lines very useful in such different activities as rock climbing, air–sea rescue and fishing.

Panel 127.1

The Nylon rope trick
Warning: the chemicals used in this reaction are poisonous.

A simple method of making Nylon is to react hexanedioyl dichloride (adipyl chloride) with hexane-1,6-diamine. Hexanedioyl dichloride is an acid chloride, and in section 122.5 we found that acid chlorides readily react with amines. If the two reagents are mixed together directly, an exothermic reaction takes place, and a spongy mass of Nylon is made. It is better to make a solution of the dichloride in an organic solvent, and a solution of the amine in water made slightly alkaline. If you pour the aqueous solution carefully on to the organic solution, the polymer is made only where the two solutions meet. By gripping the Nylon layer with forceps you can draw out a thread, which can be wound round a test tube (Figure 127.5). If you wish, the thread can be cleaned by washing with ethanol and then drying in an oven.

(b) Polyesters

An ester is made when an alcohol reacts with an organic acid. If we wish to make a polyester, then we choose an alcohol with an OH group on both ends of the molecule. We react it with an acid that has an acid group at each end of its molecule. An important example is the reaction between ethane-1,2-diol (glycol) and benzene-1,4-dicarboxylic acid (terephthalic acid):

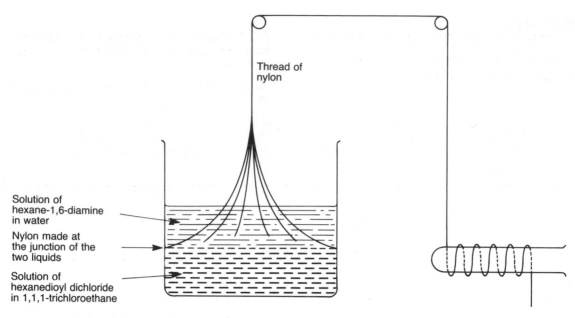

Thread of nylon

Solution of hexane-1,6-diamine in water

Nylon made at the junction of the two liquids

Solution of hexanedioyl dichloride in 1,1,1-trichloroethane

Figure 127.5 *The Nylon rope trick*

Polyesters have a large number of uses, particularly as fibres in clothing but increasingly as thermosetting plastics. We shall learn more about these plastics in the next unit.

$$OHCH_2CH_2OH + HOOCC_6H_4COOH \rightarrow$$
$$-(OCH_2CH_2OOCC_6H_4CO)_n$$

The product is known in industry as poly(ethylene terephthalate), and to shoppers under the trade names of Terylene and Dacron, as well as some others. In industry the methyl ester of benzene-1,4-dicarboxylic acid is used to prepare the polyester rather than the pure acid.

127.5 Suggest a formula for the ionic solid that is made in the first stage of Nylon-6,6 production. (Hint: the positive ion has a charge of +2, and the negative ion a charge of −2.)

127.6 Nylon-6,10 can be made by reacting hexane-1,6-diamine with decanedioyl dichloride, ClOC—(CH_2)_8—COCl. Draw a portion of the polymer chain.

127.7 Which of the following are copolymers: (i) Nylon-6; (ii) Nylon-6,10; (iii) poly(ethylene terephthalate)?

127.8 Nylons are closely related to some naturally occurring compounds. What are they?

Answers

127.1 (i) An inorganic dimer, Al_2Cl_6, is formed by aluminium trichloride (see section 94.4).

(ii) Ethanal, CH_3CHO, can turn into the cyclic trimer paraldehyde of formula $(CH_3CHO)_3$ (see section 118.12).

127.2 The washing removes the inhibitor.

127.3

127.4 This is Markovnikoff's rule (see section 112.4).

127.5 The salt has the formula
$[H_3N(CH_2)_6NH_3]^{2+}[OOC(CH_2)_4COO]^{2-}$.

127.6

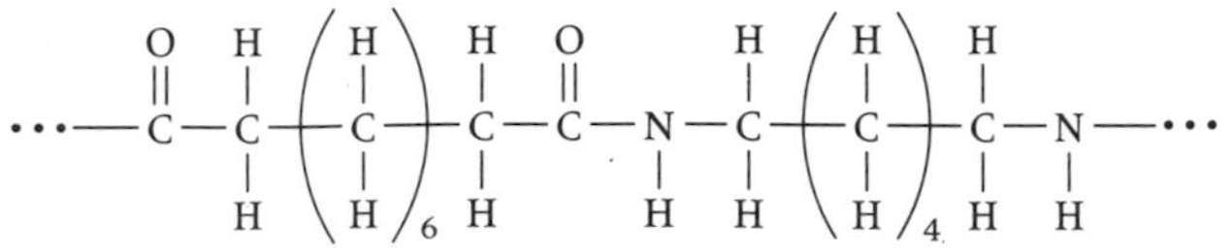

127.7 Nylon-6 is not a copolymer because it is made from only one type of monomer; the other two are copolymers.

127.8 Polypeptides, or proteins. They all have peptide links.

UNIT 127 SUMMARY

- Polymers:
  (i) Consist of chains of monomer units.
  (ii) Have very high molar masses.
- Preparation:
  (i) Can be by addition, or by condensation.
  (ii) Are often initiated by a catalyst.
  (iii) In free radical addition, polymerisation may be induced by peroxides.
  (iv) Ionic methods of addition are: anionic, in which carbocations induce polymerisation; cationic, in which carbanions induce polymerisation.
  (v) Coordination addition makes use of Ziegler–Natta catalysts (a mixture of a transition metal salt, especially titanium(IV) chloride, $TiCl_4$, and an organometallic compound such as triethylaluminium, $Al(C_2H_5)_3$).

- Polymer chains:
  Arrangements of groups along the carbon backbone of a chain can be:
  (i) Atactic, i.e. a random pattern.
  (ii) Syndiotactic, i.e. regular alternating pattern of groups.
  (iii) Isotactic, symmetrical arrangement where the attached groups are always on the same side of the backbone.
- Condensation polymers:
  Examples are
  (i) Polyamides, such as Nylon;

  e.g. Nylon-6,6
  $$+NH(CH_2)_6NHCO(CH_2)_4CO+_n$$
  (ii) Polyesters such as poly(ethylene terephthalate) (or Terylene);
  $$+OCH_2CH_2OOCC_6H_4CO+_n$$

128

Polymers and industry

128.1 Thermoplastic and thermosetting polymers

When they are hot, *thermoplastic* polymers can be moulded. When they cool down, they keep their new shape. If they are heated again, the whole process can be repeated, with the polymer being moulded into another shape.

Thermosetting polymers can also be moulded when they are heated. However, they set permanently to their new shape. Once they cool down, another bout of heating will have no effect, except to destroy them if the temperature gets too high.

To understand how these two different properties come about, we need to know more about the chemistry of polymers. First, the polymers that we met in the previous unit were all *linear* polymers. This does not mean that the polymer chains are literally all in straight lines. To begin with, the carbon backbone of the polymer takes up a zig-zag pattern. This is a result of the need to keep the tetrahedral bond angles about the

Unexpanded polystyrene can be used to make imitation 'glasses', whereas expanded polystyrene is more commonly used for packaging.

carbon atoms. Also, left to themselves, polymer chains will often coil up in more or less random spirals and other shapes. To explain why this is so we need to seek the help of thermodynamics. We shall simply say that the more random arrangement corresponds to a higher entropy than if the chains are all neatly lined up (Figure 128.1). However, if we were to analyse the system properly we should also take account of the balance between the demands of entropy and the enthalpy changes due to intermolecular forces.

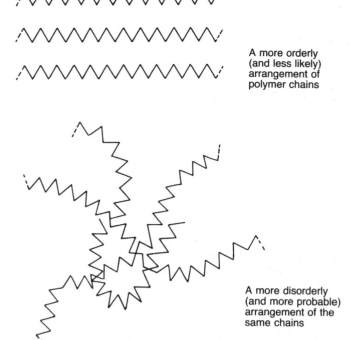

A more orderly (and less likely) arrangement of polymer chains

A more disorderly (and more probable) arrangement of the same chains

Figure 128.1 *Left to themselves, it is more likely that polymer chains will arrange themselves in a disorderly pattern*

Some would-be linear polymers turn out as *branched* polymers. Here, side chains grow off the main chain. One step further than branching brings us to *crosslinked* polymers. Two different chains are connected by short lengths of another chain, which in some cases may be only two or three atoms long. These are shown in Figure 128.2.

You should realise that a crosslinked polymer is likely to be more rigid than a linear or branched polymer. In industry, thermosetting polymers are designed to produce crosslinks when they are hot. This is done either by making the polymer with side chains that will react to make the crosslinks, or by adding a second substance that will produce the crosslinks with the original polymer chains.

The amount of the crosslinking that takes place will determine how rigid the final structure becomes. A nice example of this is the vulcanisation of rubber. Natural rubber is a soft, sticky substance, which gradually hardens to a useless mass, unless it is treated with sulphur, or some other more sophisticated chemical. Rubber is built from isoprene (systematic name methylbuta-1,3-diene) monomers:

a crosslinked portion of rubber

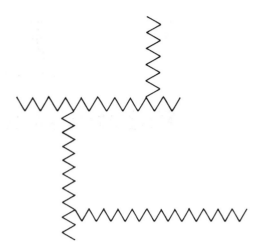

A branched polymer

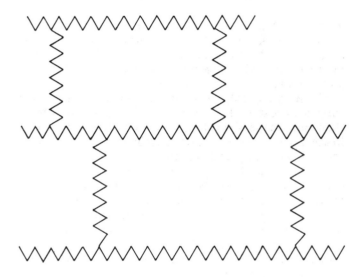

A crosslinked polymer

Figure 128.2 *Crosslinked polymers are more rigid than branched or linear polymers*

It was Dunlop (albeit by accident) who discovered that rubber heated with sulphur maintained its spring for very much longer periods of time. It also retained the shape into which it was moulded. Hence the discovery of rubber tyres, and eventually the use of rubber in such things as water hoses, shoes and tennis balls. Now, other chemicals are used to produce crosslinking in rubber, and inhibitors are added to prevent the rubber hardening with age. Many different polymers have been developed to replace natural rubber in a variety of applications. You will find some listed in Table 128.1 later in this unit.

128.2 Amorphous and crystalline polymers

It is extremely difficult to obtain single crystals of polymers. Usually the best we can do is to produce a

Rubber is manufactured from latex tapped from rubber trees. This rubber plantation is in Sri Lanka.

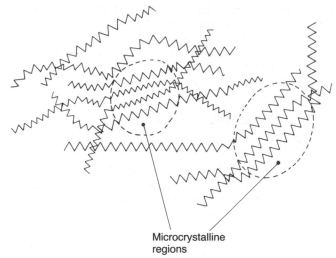

Figure 128.3 Typically, a polymer crystal will have microcrystalline regions surrounded by amorphous regions

(a) Amorphous polymers

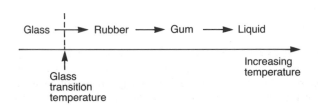

(b) Crystalline polymers

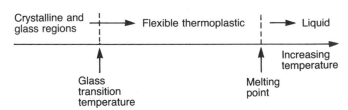

Figure 128.4 How amorphous and crystalline polymers behave as temperature changes. Notice that amorphous polymers have no clear melting points. (Adapted from: Allcock, H. R. and Lampe, F. W. (1981). Contemporary Polymer Chemistry, Prentice-Hall, Englewood Cliffs, NJ, figure 1.4, p. 11)

polymer in which the chains are packed in symmetrical arrangements in a number of small *microcrystalline* regions (Figure 128.3). Outside these regions the chains pack in more random fashions. If a solid polymer has its chains arranged randomly throughout the solid, it is called an *amorphous* polymer.

Crystalline and amorphous polymers behave differently when they are heated. At low temperatures (how low depends on the polymer) both exist as glassy materials. In this state the solid tends to shatter if it is hit. This is because the chains cannot move at all easily, so they cannot absorb the energy of the blow. Instead, the energy breaks the bonds between the chains, and sometimes the bonds in the chains. If the polymer is heated, it eventually softens and becomes more flexible. This happens at the glass transition temperature. After the transition temperature, crystalline and amorphous polymers behave differently. This is outlined in Figure 128.4, where you will see that only crystalline polymers have a well defined melting point. Heavily crosslinked polymers are a law unto themselves. As we have already said, heating has little effect on them until, at a high enough temperature, they are destroyed.

Softening is a sign that the polymer chains have gained enough energy to move slightly; for example, they may flex and twist. Now the energy of a blow on the polymer can be dispersed by the chains moving more violently and passing on the energy along, and across, chains. Hence the polymer will not shatter.

A polymer that has the ability to flex and twist its chains is an *elastomer*. Elastomers are widely used because they are able to withstand shocks and abrasions and will adopt their original shape after being distorted.

The properties of some polymers that are not good elastomers can be improved by mixing them with another material – a plasticiser. For example PVC (polyvinyl chloride or poly(chloroethene)) is a widely used polymer, but it is not a good elastomer until aromatic esters are mixed with it.

128.3 Mechanical properties of polymers

By controlling the length of a polymer chain and the extent of branching (or crosslinking), polymer scientists can control the degree of crystallinity of a polymer, and its elastomeric properties. The length and type of polymer chain will also affect the strength of the polymer; for example, how it responds to stress and strain. It is common practice for sections of a polymer to be given a stress–strain test. A force is applied to a sample; this is the stress. As the stress increases, the sample stretches; the amount of stretch is a measure of the strain. Different types of polymer give different stress–strain graphs. Typical examples are shown in Figure 128.5.

Apart from a very brittle polymer, there are two key stages in the experiment. At some stage the stress becomes so great that the polymer begins to yield; that is, it begins to stretch very easily for only a tiny increase in stress. This stage is called necking. The next important point is when the stress becomes so large that the sample breaks (the break point).

During the necking stage the polymer chains are drawn out from their coiled state and they tend to line up in more orderly arrangements. This increases the crystallinity of the sample, and with increasing crystallinity often comes high tensile strength. However, it is not so much the crystalline regions themselves that contribute to the strength, but the non-crystalline regions connecting them. High tensile strength is a quality required in fibres that are used in spinning or as ropes or cables. It is for this reason that polymer fibres are stretched during the final stages of their manufacture.

The mechanical properties of polymers can also be adapted by mixing different types of polymer, using plasticisers, or adding a *filler*. A filler can make a polymer more bulky, but as in the case of carbon added to rubber, it can also provide elements of crosslinking.

128.4 Manufacturing techniques

In this section we shall look at ways of making polymers. We can split the methods into four types depending on whether the polymer is to be used as thin sheets, for moulding, as fibres, or as coatings. We shall not be concerned with the details of the methods, only with the principles. The diagrams of the processes should show you the main points.

(a) Thin sheets

The key to many methods is for the polymer to be held in solution with a volatile solvent. The solution is spread over a roller and the solvent quickly evaporated to leave a thin layer of polymer. A more exotic method is to use a molten polymer, force it through a circular channel and blow compressed air into the middle

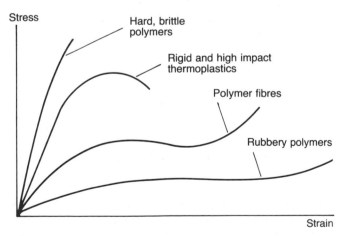

Figure 128.5 Four curves illustrating how polymers of different types behave. The lines end where the sample breaks. Temperature has a marked effect on the behaviour of a polymer. At low temperatures, the behaviour is more like the curve for hard, brittle polymers. At high temperatures, it is more like the rubbery polymer curve. (Diagram adapted from: Heaton, C. A. (ed.) (1986). The Chemical Industry, Blackie, Glasgow, figure 1.10, p. 36)

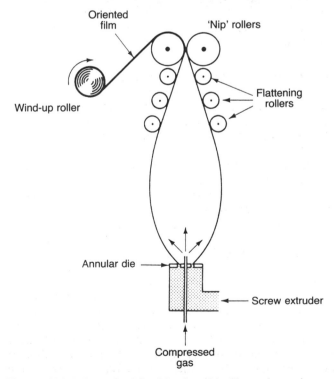

Figure 128.6 A method for blowing thin films of a polymer. (Taken from: Allcock and Lampe, op. cit., figure 20.9, p. 513)

(Figure 128.6). This stretches the polymer into a continuous film, which is rolled together and wound onto a drum.

(b) Moulding

There are several methods of moulding. Injection, blow and vacuum moulding all use a molten polymer and a pre-prepared mould (Figure 128.7). The mould has to be very carefully manufactured if a good quality product is to result, but once it is made the moulding process can be fully automated. This gives a speedy and reliable method of making a wide range of articles. Each method relies on the use of a thermoplastic or thermosetting polymer.

A different method of moulding is to use an expansion method. Here the polymer is left to expand into the shape of the mould. The expansion can be produced by a chemical reaction in the polymer mix, or by bubbling a gas into the polymer. In either method the final product consists of a matrix of polymer material and gas bubbles. Important examples of materials made by this method are polyurethane foam and expanded polystyrene. Polyurethane foams are made by mixing isocyanates, which have the structure OCN—(carbon chain)—NCO, and diols, which have OH groups at both ends of the molecule. If water is present, the isocyanate reacts with it, giving off carbon dioxide. This is the gas that produces the foaming. Polyurethane foams have been widely used as fillings for furniture, but they have got a bad name for themselves. Unfortunately they release highly poisonous fumes in a fire. Different types of foam are now being developed, as are effective fire retarding chemicals, which can be used to treat furnishings.

Expanded polystyrene is made using the second approach. Gas is bubbled through the polymer. Pentane has been used for this purpose, but increasingly CFCs (chlorofluorocarbons) have been used for making polymer foams. This use represents one of the major sources of CFCs in the atmosphere, and contributes to the depletion of the ozone layer.

(c) Fibres

One common method of making fibres is to pass the polymer material through a network of fine holes. The emerging strands of polymer can be wound together to give a fibre consisting of multiple strands. Figure 128.8 illustrates the method.

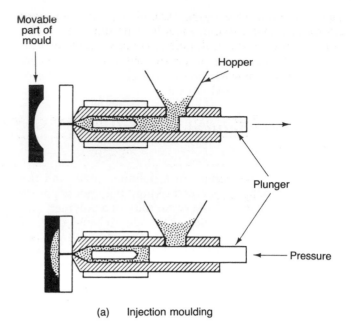

(a)   Injection moulding

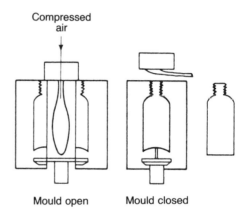

(b) Blow moulding

Figure 128.7 Two methods of moulding. (Taken from: Allcock and Lampe, op. cit., figures 20.21 and 20.22, p. 524)

(d) Coatings

One fairly simple method of coating is to heat the object to be coated, and then to dip it into a bath of polymer beads. With the right type of bead, the object will gain a layer of polymer. Sometimes the layer of polymer will

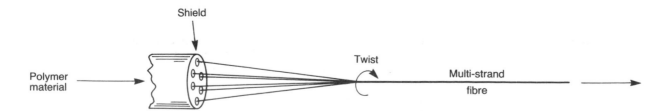

Figure 128.8 One method of making polymer fibres

have to be heated more strongly, or treated with another chemical, to make it adhere permanently to the object. On an industrial scale it is convenient to force the polymer through a narrow slit onto the object, which travels underneath it.

128.5 Uses of polymers

There are so many uses of polymers that we can only look at a brief selection of them. You will find details in Table 128.1.

Table 128.1. Some polymers and their uses

Polymer	Repeat unit	Properties and uses
Polyamides		
Nylon-6,6	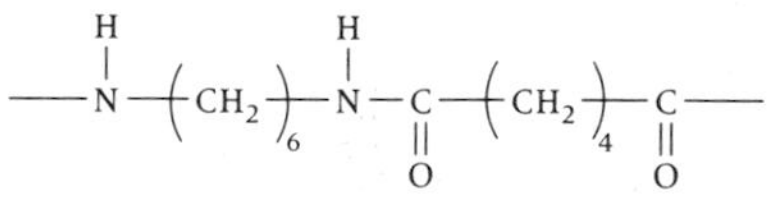	Textile fibre, threads, ropes Moulded gears and electrical insulation
Nylon-6,10	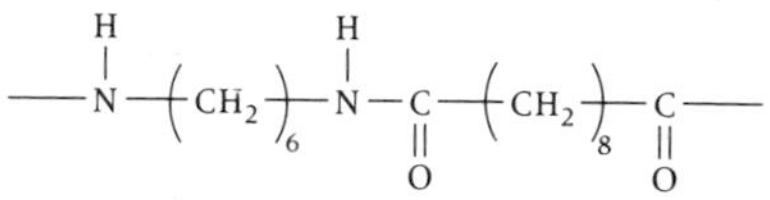	Sports equipment Bristles for brushes
Nomex	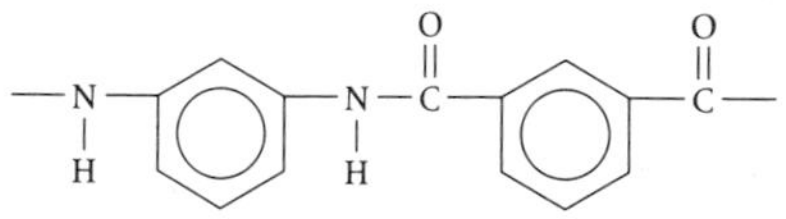	Heat resistant polymer in space suits Also for parachute cords
Polyesters and polycarbonates		
Terylene, Dacron	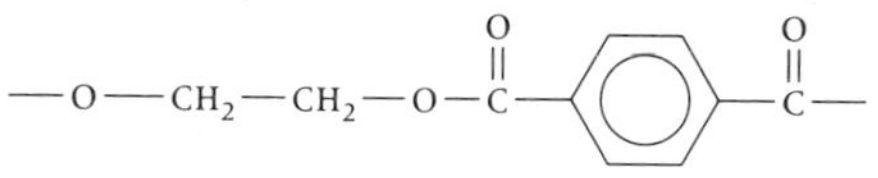	Textile fibre Basis for magnetic tape and photographic film
Lexan	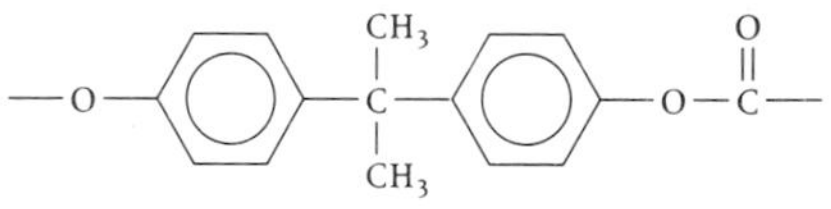	Tough and transparent Bullet proof windows, safety glass Food containers Car components
Polyethers		
Polyglycol 166	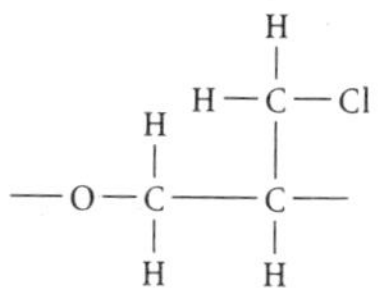	Making urethanes and speciality elastomers, e.g. for oil and fuel hoses, oil-well equipment
Delrin, Celcon	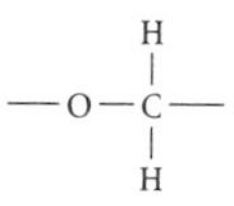	Tough plastic for gears, pipes, pens
Phenol-based		
Bakelite	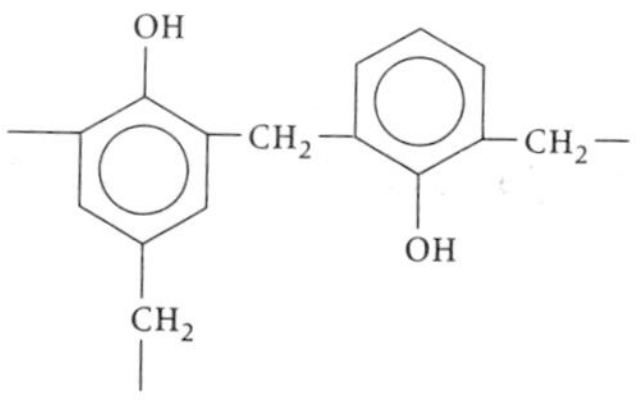	Hard thermosetting polymer Telephones, buttons, electrical insulators

Table 128.1 – cont.

Polymer	Repeat unit	Properties and uses
Poly(melamine-formaldehyde)	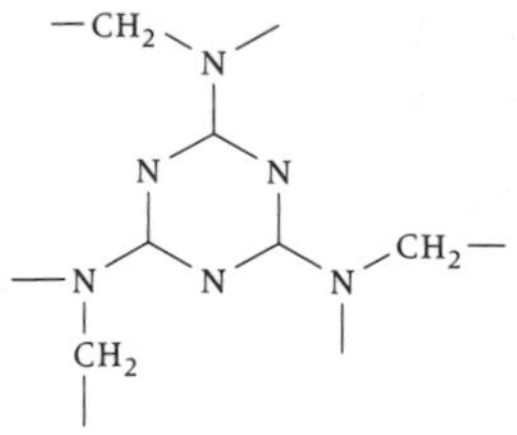	Laminated surfaces, e.g. table tops, cupboards

Polyurethanes

Polymer	Repeat unit	Properties and uses
Polyurethane	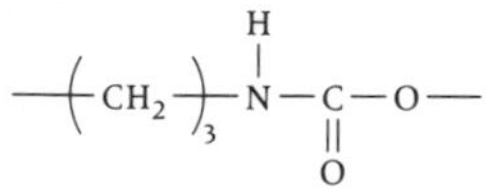 $-(CH_2)_3-N(H)-C(=O)-O-$	Foam rubber, synthetic leather
Lycra	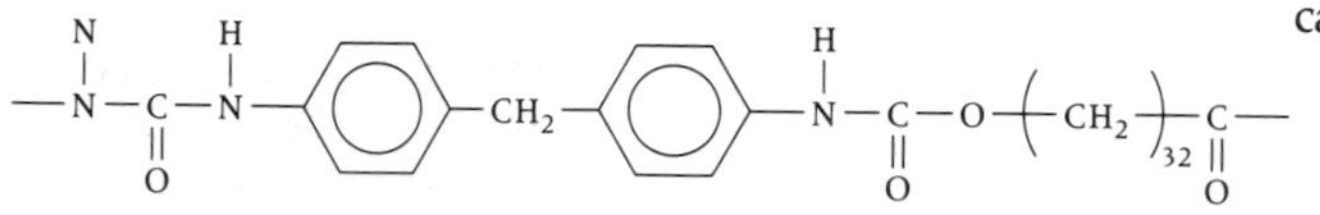	Expanded foam rubber, carpet underlays

Alkenes

Polymer	Repeat unit	Properties and uses
ABS* polymers contain these four types of repeat unit	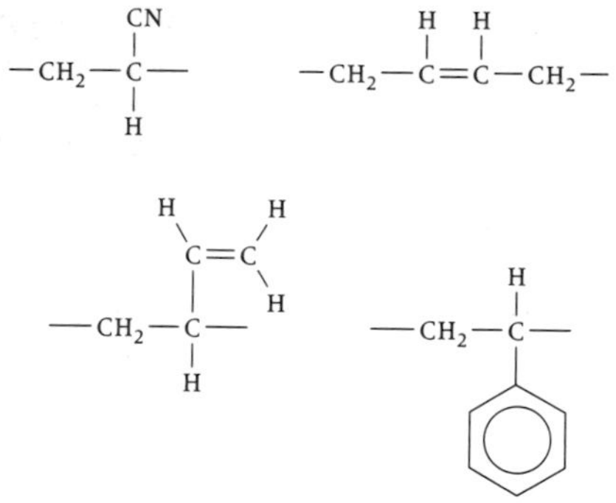 $-CH_2-C(CN)(H)-$; $-CH_2-C=C-CH_2-$; $-CH_2-C(H)-$ (with CH=CH$_2$) ; $-CH_2-C(H)-$ (with phenyl)	Tough structural plastic or rubber. Telephones, pipes, many moulded articles
Polybutadiene (butadiene rubber)	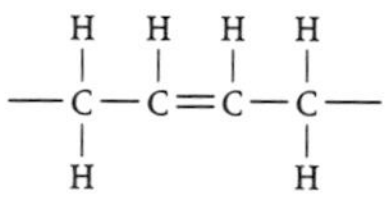 $-C(H)(H)-C(H)=C(H)-C(H)(H)-$	Alternative to natural rubber. Footwear, tyres, toys
Neoprene	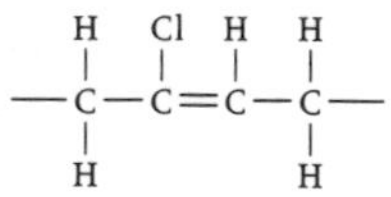 $-C(H)(H)-C(Cl)=C(H)-C(H)(H)-$	Adhesives, golf ball covers, liquid seals
Polythene† (polyethene)	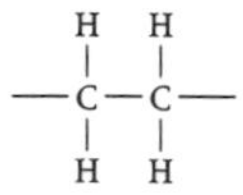 $-C(H)(H)-C(H)(H)-$	Tough plastic. Fibres, thin films, extrusion moulded objects, toys, bottles
Butyl rubber	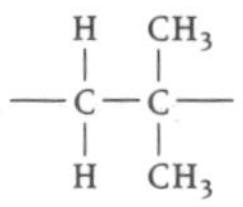 $-C(H)(H)-C(CH_3)(CH_3)-$	Tyre inner tubes, raincoats, seals

Table 128.1 – cont.

Polymer	Repeat unit	Properties and uses
Natural rubber (poly(*cis*-1,4-isoprene))	$-CH_2-C(CH_3)=CH-CH_2-$	After vulcanisation, used in car, and other, tyres
PTFE (poly(tetra-fluoroethene), Teflon)	$-CF_2-CF_2-$	Highly water repellent Non-stick cooking ware Industrial uses where very low friction needed
Polystyrene	$-CH_2-CH(C_6H_5)-$	Transparent, glass-like Wide variety of moulded and expanded objects Packing and insulation
Perspex (poly(methyl methacrylate))	$-CH_2-C(CN)(C(=O)OCH_3)-$	Transparent, glass-like Windows, fibre optics, illuminated signs
PVC (polyvinyl chloride)	$-CH_2-CHCl-$	Hard inflexible polymer With plasticiser, used in tubing, thin films, car seat covers, floor tiles
Alkynes		
Polyethyne (polyacetylene)	$-CH=CH-$ (delocalised electrons)	With iodine, an electrically conducting polymer
Inorganic		
Silicone rubber	$-O-Si(CH_3)_2-$	Seals, hoses, waterproofing, 'silicone grease'
Carbon fibres	Carbon layers with layers parallel to axis of fibre	Very high strength fibres, e.g. in aeroplane and boat building
Polythiazyl	$-S=N-$	An electrically conducting polymer Semiconductor at very low temperatures

*ABS = acrylonitrile–butadiene–styrene
†There are two main types of polyethene: low density polyethene (LDPE) has considerable branching; high density polyethene (HDPE) has no branching

128.6 Polymers and light

A reaction that is influenced by light is a *photochemical* reaction. Polymerisations can sometimes be started by photochemical reactions in which incoming light, especially ultraviolet light, produces free radicals. The *quantum yield* of a photochemical reaction is defined by

$$\text{quantum yield} = \frac{\text{number of molecules taking part in the reaction}}{\text{number of photons used}}$$

In polymer chemistry we can write this as

$$\text{quantum yield} = \frac{\text{number of monomer units consumed}}{\text{number of photons used}}$$

From the initiation and propagation steps in free radical polymerisation, you can see that one photon causing the dissociation of a peroxide molecule could result in a very long chain. A quantum yield of 1000 is not uncommon.

Initiation by light is extremely useful, but light can also bring about the destruction of polymers, i.e. photodecomposition. This is a problem with paints that may discolour when left in bright sunlight. In order to stop photodecomposition, *stabilisers* are added to polymers. One type of stabiliser makes use of chemicals that absorb ultraviolet radiation efficiently. This prevents the high energy photons from breaking the bonds of the polymer.

There are times when photodecomposition is highly desirable. Many items made from polymers are meant to be disposable, for example rubbish bin liners and other plastic bags. When they were first introduced, these plastics were not biodegradable; but now they can be made with chemicals added that are designed to decompose the polymer chains once they are in contact with light or air for a long period of time. Some of these chemicals work by releasing free radicals, which attack the polymer chains.

Polymers that are decomposed by light are used in the semiconductor industry, and by people who build their own electronics circuits, although the details of the methods are different. One method of making a circuit board uses an insulating board coated with a layer of copper and then a layer of polymer. A mask is made, which has black lines where the circuit is going to be, and is clear where electrical connections are not wanted. The mask is placed over the polymer and ultraviolet light is shone on the surface for some minutes. Where the light hits the polymer a photochemical reaction is set up, which degrades its structure. By washing the board in a chemical, often an alkali, the ruined polymer can be washed off, leaving unaffected polymer over the copper. The whole board is immersed in iron(III) chloride solution, which etches away the exposed copper. The next step is to remove the polymer from the remaining copper by washing with a solvent. The copper lines show the circuit into which the electronic components can be soldered.

> **128.1** The glass transition temperatures of three polymers are: Perspex, 105 °C; polystyrene, 100 °C; Nylon-6,6, 45 °C. What is likely to happen to a sample of each of them if it were hit by a hammer at (i) 0 °C, (ii) 20 °C, (iii) 80 °C, (iv) 120 °C?

The construction of printed circuit boards for computers depends on polymer chemistry as well as electronics.

128.2 This is an extract from a catalogue of scientific equipment:

Plastic Tubing: Polythene. Natural grade, medium wall, tough and flexible. Can be manipulated by softening in hot water.

(i) Does this mean that polythene is a thermosetting polymer?

(ii) What does it suggest about the glass transition temperature of polythene?

128.3 In an experiment on the free radical polymerisation of ethene, a lamp projected 10^{18} photons into the reaction vessel. At the end of the experiment, 0.28 g of monomer had been used up.

(i) How many moles of monomer molecules were used in the reaction?

(ii) How many molecules were used up?

(iii) What was the quantum yield?

Answers

128.1 The key is that, below the glass transition temperature, the samples would be brittle and shatter; above the transition temperature, they should deform, but not break. Therefore they would all shatter at 0 °C and 20 °C; only Nylon-6,6 would survive the blow at 80 °C. At 120 °C none of them should break.

128.2 (i) No. It is a thermoplastic because its shape can be changed as the temperature changes.

(ii) As it is flexible, the transition temperature must be below room temperature. In fact, it can be as low as -170 °C for some types of polythene.

128.3 (i) One mole of ethene, C_2H_4, has a mass of 28 g, so 0.01 mol were used.

(ii) The number of molecules is $0.01 \, \text{mol} \times 6.02 \times 10^{23} \, \text{mol}^{-1} = 6.02 \times 10^{21}$.

(iii) Quantum yield $= \dfrac{6.02 \times 10^{21}}{10^{18}} = 6020$

UNIT 128 SUMMARY

- Thermoplastic and thermosetting polymers:
  (i) Thermoplastic polymers can be moulded, heated and then remoulded.
  (ii) Thermosetting polymers can be moulded when heated, but once cold are permanently set.
- Crosslinking:
  Bonds made between separate polymer chains increase the rigidity of the polymer.
- Amorphous polymers:
  (i) Have random arrangements of chains.
  (ii) Melt over a wide range of temperatures.
- Crystalline polymers:
  (i) Have more orderly arrangements of chains.
  (ii) Have well defined melting points.
- Mechanical properties:
  (i) Can be changed by controlling the lengths of the chains and the amount of crosslinking.
  (ii) Increased crystallinity increases the tensile strength (useful in fibres).
  (iii) Fillers can increase the bulk of a polymer, and the amount of crosslinking.

- Elastomers:
  (i) The chains can flex and twist.
  (ii) Can withstand shocks and abrasions.
  (iii) Return to their original shape after distortion.
- Manufacturing techniques:
  Polymers are produced as
  (i) Thin sheets, e.g. by forcing air into an envelope of molten polymer.
  (ii) As moulded articles, using injection, blow and vacuum techniques.
  (iii) Foams, by passing gas into the polymerising mixture.
  (iv) Coatings for other objects, e.g. by hot dipping.
- Uses:
  See Table 128.1.
- Polymers and light:
  Polymerisation can be induced by light, or polymers can be made sensitive to light so that they decompose when left in light.

129

Fats, oils, soaps and detergents

129.1 What are fats and oils?

Fats are one class of material called *lipids*. Lipids are biologically active compounds that are insoluble in water. (We have discussed some lipids in previous units, e.g. vitamins and hormones.) Animals have always been used as food by humans, partly for meat, which is a source of protein, and fat, which is a source of energy. Animal fats have also been used as a source of other useful materials: for example, oils and waxes for lamps and candles. If animal fat is boiled with alkali, or reacted with superheated steam, it produces two major products:

(i) *Glycerol.* The systematic name for glycerol is propane-1,2,3-triol (Figure 129.1). It is an alcohol with three OH groups in its molecule. Glycerol is also called glycerine.
(ii) *Carboxylic acids.* These were first obtained from fats, and hence for a long time they were called *fatty acids*. The acids are different to those we discussed in Unit 119 in several respects. The most important is that many have a carbon chain up to 20 carbon atoms long, and some are even longer. You will find a list of them in Table 129.1, and the structures of some in Figure 129.1. Most are saturated (no carbon–carbon double bonds), but some are unsaturated (have one or more double bonds).

Given that fats can be broken into a mixture of an alcohol and acids, it follows that fats must be esters. However, they are more often called *glycerides*. Three examples are shown in Figure 129.2. The interesting thing about glycerides in general is that some are solids and some liquids. We call the liquid glycerides *oils*. Oils contain large proportions of triglycerides built from glycerol and unsaturated acids like oleic acid.

There is medical evidence to show that a high proportion of saturated fats in a person's diet brings an increased risk of heart disease. The unsaturated oils appear not to have the same dangerous effects, and the sales of margarines and butter substitutes that are 'high in polyunsaturated fats' have increased enormously in recent years.

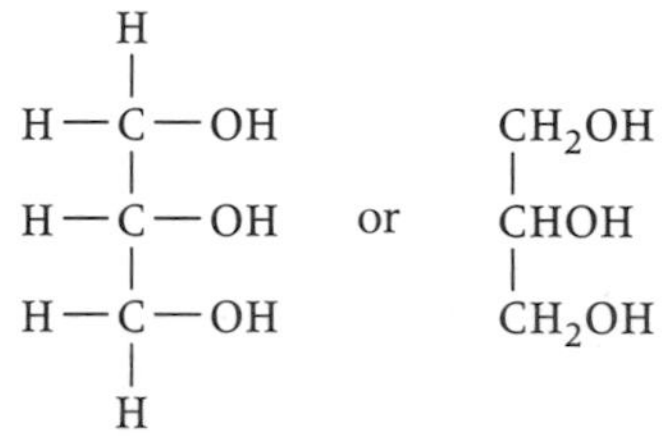

Glycerol
(propane-1,2,3-triol)

$CH_3 \left(CH_2 \right)_{16} COOH$
Stearic acid (saturated acid), $C_{17}H_{35}COOH$

$CH_3 \left(CH_2 \right)_5 CH = CH \left(CH_2 \right)_7 COOH$
Palmitoleic acid, $C_{15}H_{29}COOH$

$CH_3 \left(CH_2 \right)_7 CH = CH \left(CH_2 \right)_7 COOH$
Oleic acid, $C_{17}H_{33}COOH$

$CH_3 \left(CH_2 \right)_3 \left(CH_2CH = CH \right)_2 \left(CH_2 \right)_7 COOH$
Linoleic acid, $C_{17}H_{31}COOH$

Figure 129.1 *The structures of glycerol and of one saturated and three unsaturated acids from which fats are made*

Some naturally occurring esters are similar to fats but are not built on glycerol. These are known as *waxes*. They have been used for polishing furniture or softening leather, and making candles.

Table 129.1. Acids obtained from fats

Acid*	Formula
Saturated	
Butyric (butanoic)	C_3H_7COOH
Caproic	$C_5H_{11}COOH$
Lauric	$C_{11}H_{23}COOH$
Palmitic	$C_{15}H_{31}COOH$
Stearic	$C_{17}H_{35}COOH$
Arachidic	$C_{19}H_{39}COOH$
Unsaturated	
Oleic	$C_{17}H_{33}COOH$
Linoleic	$C_{17}H_{31}COOH$

*All these acids have systematic
names, but it is much easier to use their traditional ones

Many manufacturers market margarines which are high in polyunsaturated fats.

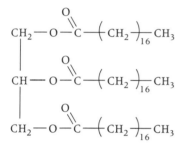

Tristearin

Tripalmitin

Triolein

Figure 129.2 Three glycerides. Tristearin (or stearin for short) together with tripalmitin are the main ingredients of (solid) fats. Triolein is a major component of oils, e.g. olive oil, palm oil, rapeseed oil

Bees are remarkable creatures, not least for their ability to build hexagonal closed-packed lattices of wax like the one shown here.

129.2 How are soaps and detergents made?

Strictly a detergent is a cleaning agent that will remove grease and grime from surfaces. However, it is more common for liquid cleaning agents to be called detergents, and solids to be called soaps. Soaps and detergents are used on an enormous scale, both in the home and in industry. Soap has a very long history, detergents somewhat shorter.

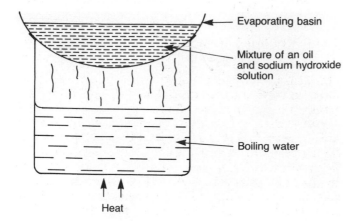

Figure 129.3 *A simple way of making soap*

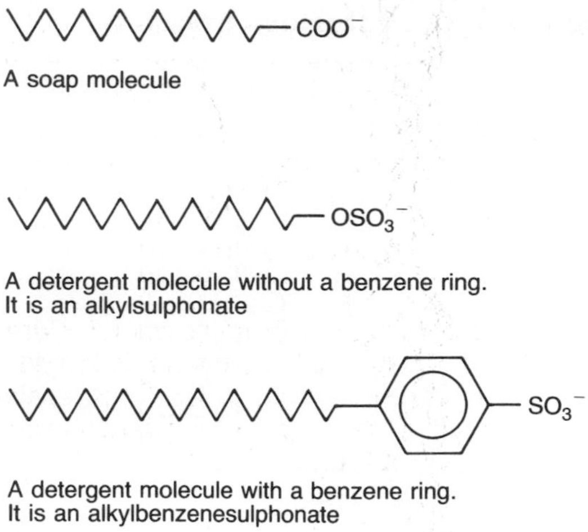

A soap molecule

A detergent molecule without a benzene ring. It is an alkylsulphonate

A detergent molecule with a benzene ring. It is an alkylbenzenesulphonate

Figure 129.4 *Soap and detergent molecules have a hydrocarbon tail and an ionic head. Soaps have a —COO⁻ group as the ionic head, which gives a precipitate with hard water. Sulphonate groups do not give precipitates*

You can make soap easily in the laboratory. All you have to do is to heat a fat or oil with sodium hydroxide solution. A suitable apparatus is shown in Figure 129.3. After a few minutes, and continuous stirring, the oil and water layers merge and thicken. If all goes well, the surface will begin to cake and on cooling you will obtain a yellow solid. This is soap, although it is likely to be impure. For example, it may contain unreacted alkali.

Essentially the same process has been used in industry to make soap, although much greater care has to be taken with the proportions of the ingredients that are mixed, the amount of heat supplied, and the removal of impurities. Also, many soaps have perfumes added, together with chemicals that improve their texture.

The reaction that takes place in soap making is called saponification. The alkali breaks the glyceride molecules into glycerol and anions of the fatty acids. The glycerol is helpful in that it gives soap a pleasant feel. The solid soap is a salt made between the anions and sodium ions from the sodium hydroxide:

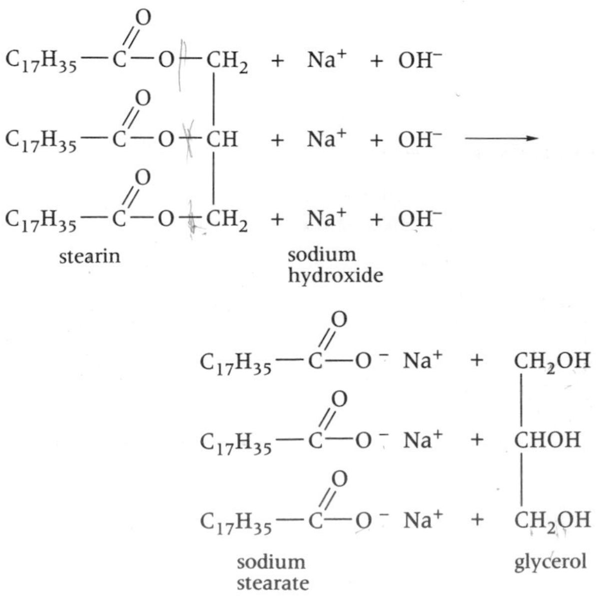

Sodium stearate is a soap. Other glycerides can be saponified in a similar manner. If sodium hydroxide is used, the resulting soap is fairly hard when it has been dried. If potassium hydroxide is used, the soap is softer.

A liquid detergent is made by a totally different method. Instead of having a carboxylic acid group on the end of a hydrocarbon chain, a sulphonate group is usually present. In some types the sulphonate is attached to a benzene ring on the end of the chain. Figure 129.4 shows you the two varieties.

It is extremely difficult to make the long hydrocarbon chains except through polymerisation reactions. This is why detergents were not widely available until the 1950s after polymer science had become quite sophisticated.

The reason why detergents are so useful is that they do not give precipitates with metal ions such as Na^+, K^+, Ca^{2+} or Mg^{2+}. The lack of precipitates with Ca^{2+} and Mg^{2+} is especially important. These ions are responsible for the hardness of water. Ordinary soap gives a precipitate with hard water; this is 'scum'. Detergents do not give a scum even in the hardest of water areas.

In the past detergents have gained a bad reputation for causing pollution of rivers and waterways. The early polymer chains used for detergent manufacture suffered a great deal of branching (Figure 129.5). The hydrocarbon side chains did not interfere with the

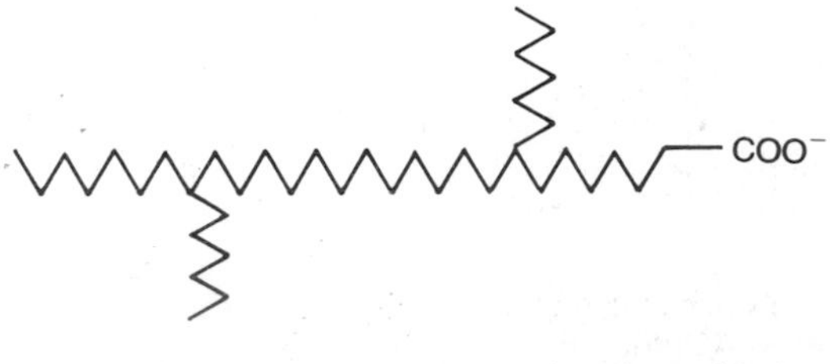

$$C_{17}H_{35}-\overset{O}{\overset{\|}{C}}-O-CH_2 + Na^+ + OH^-$$
$$C_{17}H_{35}-\overset{O}{\overset{\|}{C}}-O-CH + Na^+ + OH^- \longrightarrow$$
$$C_{17}H_{35}-\overset{O}{\overset{\|}{C}}-O-CH_2 + Na^+ + OH^-$$

stearin   sodium hydroxide

$$C_{17}H_{35}-\overset{O}{\overset{\|}{C}}-O^-Na^+ + CH_2OH$$
$$C_{17}H_{35}-\overset{O}{\overset{\|}{C}}-O^-Na^+ + CHOH$$
$$C_{17}H_{35}-\overset{O}{\overset{\|}{C}}-O^-Na^+ + CH_2OH$$

sodium stearate   glycerol

Figure 129.5 *Soap or detergent molecules with branched hydrocarbon tails can be the cause of a great deal of pollution*

cleaning power of the detergent, but they did prevent bacteria from attacking and breaking the chains. This meant that detergent molecules degraded very slowly. Now the amount of branching can be kept to a minimum. Unbranched chains are much more appetising to bacteria, so the detergents are more easily biodegraded.

129.1 A process used in soap manufacture, and one that you can try in the laboratory, is to add salt to the reaction mixture shortly before it is allowed to cool. This is *salting out*. Explain why adding salt helps the soap to precipitate.

129.2 Before the days of washing-up liquids, pans could be cleaned of fat by boiling them with a solution of washing soda (sodium carbonate). Explain why this worked. (Hint: look at section 94.1.)

129.3 Some people believe that the more scum they see in their bath, the dirtier they must have been. Are they right?

129.3 How do soaps and detergents clean?

We shall explain the cleaning action of both soaps and detergents by using a simplified diagram of each type of molecule. We shall show them like this:

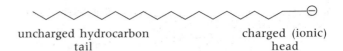

uncharged hydrocarbon tail   charged (ionic) head

This highlights the key features of the molecules: they have an uncharged hydrocarbon tail, and a charged (ionic) head. (We have shown the head carrying a negative charge; in practice, positively charged heads are also possible.) If you have read Unit 123 you should be familiar with the idea that the hydrocarbon tail is likely to be hydrophobic (water hating), and the head hydrophilic (water loving). Water molecules cannot solvate hydrocarbons efficiently, but they can solvate ions. On the other hand, grease and dirt is mainly organic in nature, so the tails will be able to mix happily with it. There is nothing to be gained by the ionic heads entering a ball of grease when they can be solvated by water molecules instead.

If you drop a little washing-up liquid on water, the detergent spreads out across the surface and some mixes into the body of the water. By adding more detergent there comes a point at which the molecules gather together into clumps called *micelles* (Figure 129.6). The tails stick inwards into the roughly spherical balls and the heads stick outwards into the water (where they can be solvated). Now let us suppose that a plate with a layer of grease on it is put into the solution. Tails of some detergent molecules will attach themselves to the grease. If the water is agitated slightly, the grease tends to lift off and fragment. This gives other detergent molecules the opportunity to connect to the grease particles. Shortly, the solution contains small

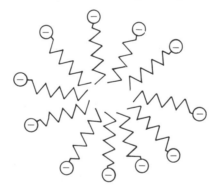

Detergent molecules gather together in micelles in water

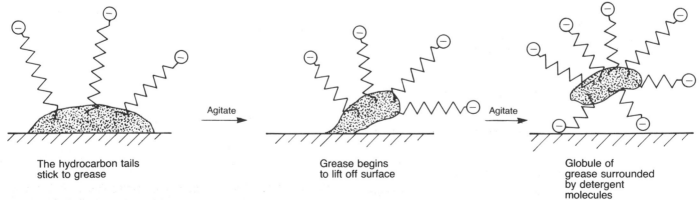

The hydrocarbon tails stick to grease   Agitate →   Grease begins to lift off surface   Agitate →   Globule of grease surrounded by detergent molecules

Figure 129.6 How a detergent cleans

globules of grease surrounded by detergent molecules. The attraction of the water molecules for the ionic heads holds the globules in solution. The electric charge on the heads means that two globules approaching one another are repelled and the layer of grease cannot re-form. The result is a clean plate and a solution containing the grease, which can be thrown away.

Liquid detergents are one class of a wide range of compounds called *surfactants*. As you might guess, surfactants change the properties of surfaces. For example, water will not wet the surface of grease or oil; they remain as two separate layers. Detergent molecules allow the two to mix by changing the behaviour of the surfaces. (Technically, we say that detergents reduce the interfacial surface tension.)

Other surfactants are used in stabilising foams, such as those used in fire fighting and in the extraction of metals by foam flotation. They can also be used to destabilise foams; for example, those that might be produced in industrial boilers. A large number of food products, cosmetics and paints have surfactants added to stabilise gels and emulsions.

129.4 Which type of bonding allows grease and the tails of soaps or detergents to mix?

129.5 Micelles and the globules of grease surrounded by detergent molecules are often of a colloidal size. What is the range of size for a colloid?

Answers

129.1 This is an example of the common ion effect. Soap is partially soluble in water. If excess sodium ions are added, then the equilibrium between dissolved ions and solid is shifted towards the solid. Hence we see more soap produced.

129.2 Sodium carbonate undergoes hydrolysis to give an alkaline solution. When the alkali is boiled with the fat, the glycerides are broken apart. This destroys some of the fat, but soap is also made. This also helps to clean the pan.

129.3 No. The amount of scum is mainly a reflection of how hard the bath water is.

129.4 Van der Waals bonding.

129.5 Between 1 and 1000 nm in diameter.

UNIT 129 SUMMARY

- Fats and oils:
  (i) Fats are a type of lipid.
  (ii) Fats are esters called glycerides. They are esters of propane-1,2,3-triol (glycerol) and long-chain carboxylic acids containing up to 20 carbon atoms.
  (iii) Unsaturated fats have double bonds between some of the carbon atoms in the carboxylic acid chains.
- Soaps and detergents:
  (i) A soap can be made by heating a fat or oil with sodium hydroxide solution (saponification).
  (ii) A soap has a COO^- group on the end of a hydrocarbon chain. This group gives a precipitate (scum) with Ca^{2+} ions in hard water.
  (iii) A liquid detergent may have an SO_3^- group attached to a benzene ring on the end of the chain. This does not give a scum with hard water.
- Cleaning action:
  (i) A soap or detergent molecule has a hydrophobic hydrocarbon tail and a hydrophilic ionic head.
  (ii) In water the tails gather together, with the ionic heads pointing outwards. Roughly spherical groups (micelles) are made.
  (iii) The tail is attracted to grease; the head to water.
  (iv) Grease goes into the bulk of water as small particles surrounded by the chains (tails in the grease, heads in water).
- Surfactants:
  (i) Surfactants change the surface tension of water.
  (ii) They are used to stabilise foams, and in the floth flotation extraction of metals.

130

Organic problems

130.1 What types of problem are there?

A simple-minded response to this question is 'hard and easy'. However, from the point of view of the person who sets the problems, they are meant to lead you to discovering one or more of the following:

(i) the formula of a compound;
(ii) its structure, i.e. the arrangement of the atoms;
(iii) the functional groups it contains;
(iv) how these groups react.

We shall now work through examples of the type that you might meet in a test or examination. However, before we begin you should know that you cannot succeed in finding answers to problems unless you are familiar with the majority of work that we have covered in the units on organic chemistry and spectroscopy.

130.2 Predicting structures from percentage compositions

We shall consider two examples of typical questions.

Example 1

A compound has the composition C 92.3%, H 7.7%, a relative molecular mass of 78 and burns with a very sooty flame. What might it be?

Before we work out its molecular formula, let us look at the percentage composition. Clearly this compound contains a great deal of carbon compared with hydrogen. This alone should make us think that it might be an aromatic compound, i.e. one containing a benzene ring. The smoky flame is consistent with this notion: it is a characteristic of unsaturated hydrocarbons, especially if they have a benzene ring.

Applying the method we developed in Unit 37, we have:

	Carbon	Hydrogen
100 g of compound contains/g	92.3	7.7
Number of moles present/mol	$\dfrac{92.3}{12}$ =7.7	$\dfrac{7.7}{1}$ =7.7
Ratio of moles	1	to	1

Empirical formula CH.
Relative molecular mass of one unit of CH is

$$M_r(CH) = 12 + 1 = 13$$

Therefore there are $78/13 = 6$ units in the molecular formula. The formula is C_6H_6. The compound is benzene.

Example 2

A substance has the percentage composition C 52.2%, H 13.0%, O 34.8%. It has a relative molecular mass of 46. It does not give white fumes of hydrogen chloride with phosphorus pentachloride. What is the compound?

Here we have a substance that contains oxygen as well as carbon and hydrogen. There are five homologous series to which such a substance could belong: alcohols, aldehydes, ketones, acids, or ethers. (Actually there are other organic compounds that contain oxygen, e.g. carbohydrates, but it is unlikely that you would be asked about them.) Even before we attempt to discover its structure, we can say that it cannot contain an OH group. This is because such compounds *do* give hydrogen chloride with phosphorus pentachloride. Therefore we are looking for an aldehyde, ketone, or ether.

Example 2 – cont.

	Carbon	Hydrogen	Oxygen
100 g of compound contains/g	52.2	13.0	34.8
Number of moles present/mol	$\frac{52.2}{12}$ $=4.35$	$\frac{13.0}{1}$ $=13.0$	$\frac{34.8}{16}$ $=2.18$
Ratio of moles	$\frac{4.35}{2.18}$ $=2$	$\frac{13.0}{2.18}$ $=6$	$\frac{2.18}{2.18}$ $=1$

Empirical formula C_2H_6O.
Relative molecular mass of one unit of C_2H_6O is

$$M_r(C_2H_6O) = 2 \times 12 + 6 \times 1 + 16 = 46$$

Therefore in this case the molecular formula is the same as the empirical formula, i.e. C_2H_6O.

Now we must try to fit the atoms together to make an aldehyde, ketone or ether. The place to start is with the functional group. That is, we write down:

aldehyde   ketone   ether

Left over H_5 H_6 H_6

We can discount the ketone because it requires three carbon atoms, and the molecular formula says only two are available. This time we try to fit the remaining atoms around the carbon atoms. It is best to try this for yourself by using models, but you should be able to see that it is impossible to fit all the atoms together for the aldehyde:

aldehyde   ether

Left over H_2 none

The compound is methoxymethane (dimethyl ether).

130.1 Another hydrocarbon also has the percentage composition C 92.3%, H 7.7%. It has a relative molecular mass of 26 and it too burns with a smoky flame. What is it?

130.2 Repeat example 2, but this time you are told that the compound does give hydrogen chloride with phosphorus pentachloride.

130.3 Using information from spectra

(a) Mass spectra

One of the most direct pieces of information about a compound can be obtained from a mass spectrum. The key thing here is to look for the peak at the highest mass to charge ratio. This will give you the relative molecular mass of the parent ion, i.e. the relative molecular mass

---

Example 3

The idealised mass spectrum of a hydrocarbon is shown in Figure 130.1. It does not react with bromine water. What is its formula?

The parent ion is at a mass of 114, so this is the relative molecular mass of the compound. From its lack of reaction with bromine, we know the compound is saturated, i.e. it is probably an alkane. (We shall assume that it is not a cyclic hydrocarbon such as cyclobutane or cyclohexane.) Hence its general formula must be C_nH_{2n+2}. We can now begin to make a guess at its actual formula, or do a systematic calculation. The first method is to say that as the relative molecular mass is less than 120, there must be fewer than 10 carbon atoms in the molecule, so let us guess at six. This would give us a relative molecular mass of $6 \times 12 + 14 \times 1 = 86$, which is too low. Another guess is to try eight carbon atoms: $8 \times 12 + 18 = 114$. This gives us the formula C_8H_{18}.

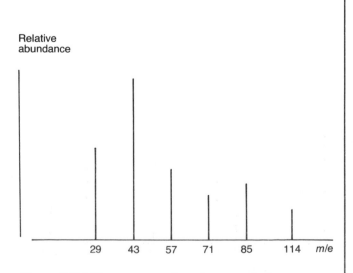

Figure 130.1 The mass spectrum for example 3

Example 3 – cont.
The systematic method is to say that n carbon atoms have a mass of $12n$ units, and $2n + 2$ hydrogen atoms give $2n + 2$ units. Therefore we have

$$12n + 2n + 2 = 114$$

so

$$14n = 112$$
$$n = 8$$

You should be careful about saying which alkane it is. In fact, the spectrum belongs to octane, but unless you knew a great deal about mass spectrometry you could not be sure that it was not an isomer, e.g. 3,4-dimethylhexane. The other peaks (the fragmentation pattern) belong to parts of chains that break up in their progress through the spectrometer. For example, the peak at mass 29 corresponds to $C_2H_5^+$, and that at 43 to $C_3H_7^+$.

of the substance. The spectrum may also provide a guide to the structure of the molecule.

(b) Infrared (vibrational) spectra

We found in Unit 27 that particular groups vibrate at fairly well defined frequencies. This means that, if we look at an infrared spectrum, it should tell us if these groups are present or not. You should take special notice of the bands in the following regions:

(i)   1600 to 1800 cm^{-1}. This is where the vibration of a carbonyl group appears. A strong peak in this region is characteristic of aldehydes, ketones, acids and amides.

(ii)   2720 cm^{-1}. Aldehydes show a strong band here owing to the vibration of the hydrogen atom bonded to the carbon atom of the carbonyl group.

(iii)   2900 and 1400 cm^{-1}. Carbon–hydrogen vibrations will also appear in spectra in this region.

(iv)   3200 to 3700 cm^{-1}. The vibrations of OH groups in alcohols appear here. If the group is involved in hydrogen bonding, then the vibration band is broad; if not, it is narrow and shifts to around 3600 cm^{-1}.

130.3 The mass spectrum of a liquid is shown in Figure 130.2. The liquid is known to be an aromatic ketone with percentage composition C 80.0%, H 6.7%, O 13.3%.

(i) Use the mass spectrum to determine the relative molar mass of the compound.

(ii) Combine this information with the percentage composition to discover the molecular formula.

(iii) Suggest a structure for the molecule. What is its name?

(iv) What fragment gives the peak at mass 105?

130.4 Identify the types of compound giving the spectra shown in Figure 130.3.

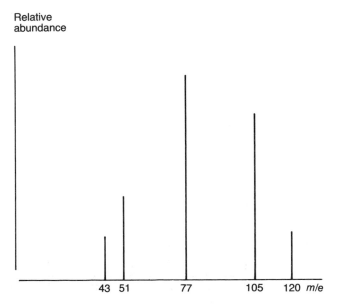

Relative abundance

43 51   77   105   120 m/e

***Figure 130.2** The mass spectrum for question 130.3*

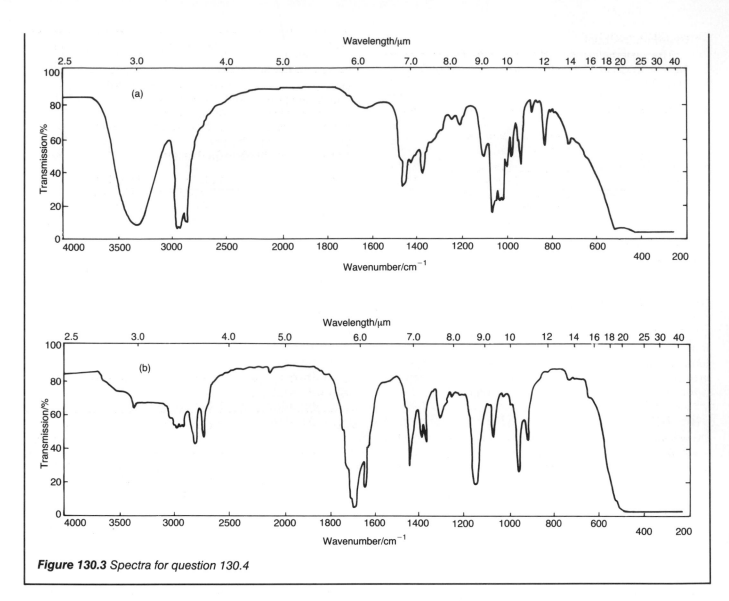

Figure 130.3 *Spectra for question 130.4*

130.4 **The results of chemical reactions**

There are a great many organic reactions that you might need to know, but some are more important than others. Tables 130.1 and 130.2 summarise them. You will have to look up the section reference to get information about reaction conditions and other details. The best course of action is to make sure you learn these reactions. You may find this tiresome, but it pays dividends in the long run. The best way to learn them is to commit two or three to memory each day, or every other day. In this way you will avoid last minute panic before exams. However, you will need to test yourself from time to time to make your brain establish the information in long-term memory.

One of the key principles that makes the study of functional groups fruitful is that, in the main, the reactions of one type of group do not interfere with the reactions of the others. We can make use of this idea to provide answers to questions that would otherwise be impossible to tackle.

Table 130.1. Important organic reactions

Test*	What it tests for	See section(s)
Alcoholic potassium cyanide	Converts halogenoalkane to nitrile; increases the number of carbon atoms by one	115.3
Alcoholic potassium hydroxide	Converts halogenoalkane to alkene	115.4
Alkaline potassium manganate(VII)	Alkene changed to diol; side chains on benzene oxidised to acid (—COOH)	112.3, 113.4
Ammonia given off with alkali	Amide, $RCONH_2$, or ammonium salt present	120.4
Bromine or bromine water decolourised	Unsaturation – especially alkenes, alkynes, phenol and phenylamine	112.4, 117.4, 122.9

Table 130.1 – cont.

Test*	What it tests for	See section(s)
Bromine in alkali	Hofmann degradation; amide converted to amine with one less carbon atom	120.4
Chlorine gas and light	Free radical reaction with alkanes; chlorination of an acid at the α-carbon; especially useful for making amino acids	113.3, 123.2
Chlorine gas and red phosphorus or iodine	Chlorination of an acid at the α-carbon; especially useful for making amino acids	123.2
2,4-Dinitrophenyl-hydrazine gives orange precipitates (hydrazones)	Aldehyde or ketone present	118.6
Ethanoyl chloride, CH_3COCl, or benzoyl chloride, C_6H_5COCl, give fumes of HCl	OH group in an alcohol reacts to give an ester; amines are acetylated or benzoylated	120.6
Hydrogenation, e.g. H_2/Ni	Unsaturation – especially alkenes, alkynes and arenes; converts nitriles to amines	112.4, 113.5, 120.7
Hydrogen halide, e.g. HCl	Addition to alkenes obeys the Markovnikoff rule	112.4
Hydrolysis (reaction with water, acid or alkali)	Converts nitriles to acids	120.7
Iodoform reaction; warm with iodine in alkali gives yellow precipitate, CHI_3	Presence of 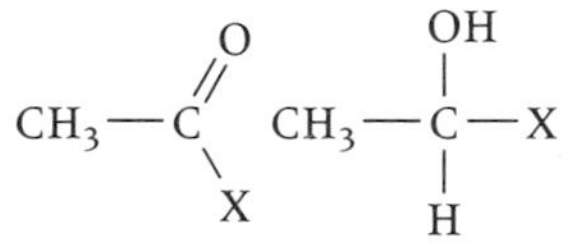 groups	118.9
Lithium tetrahydrido-aluminate(III), $LiAlH_4$	Reduces acids and acid chlorides to alcohols; esters, aldehydes and ketones to alcohols; nitriles to amines	119.3, 120.2, 120.6, 118.7, 122.2
Ozonolysis	Breaks double (and triple) bonds and produces aldehydes and/or ketones	112.5
Phosphorus penta-chloride, PCl_5, or sulphur dichloride oxide, $SOCl_2$, give fumes of HCl	Test for OH groups in alcohols or acids; OH group replaced by Cl	116.4, 119.3
Phosphorus(V) oxide, P_4O_{10}	Converts amides to nitriles	120.4

Test*	What it tests for	See section(s)
Sodium dichromate(VI)/H⁺ changes from orange to green	Oxidation, especially of primary alcohol to aldehyde or acid, secondary alcohol to ketone	116.3
Sodium nitrite in acid	Amines give diazonium ions	122.6
T<5 °C	Only aromatic diazonium ions are stable and make diazonium compounds	
T>5 °C	Change into alcohol and N_2 gas	

*The tests are given in (approximately) alphabetical order of the chief reagent

Table 130.2. Special reactions of aromatic compounds

Reaction	Comment	See section(s)
Substance gives a white precipitate with bromine	Phenol or phenylamine likely; OH and NH_2 activate a benzene ring	117.4, 122.9
A white precipitate appears on acidification	Precipitate likely to be benzoic acid	119.3
Concentrated nitric and sulphuric acids (nitrating mixture)	Introduces nitro group, —NO_2, into benzene ring	113.6
Halogen plus halogen carrier, e.g. $AlCl_3$, iron filings	Introduces halogen into benzene ring	113.6
Ethanoyl chloride, CH_3COCl, or benzoyl chloride, C_6H_5COCl, with $AlCl_3$*	Benzene ring is acetylated or benzoylated	113.6
Chloroalkane with $AlCl_3$*	Alkylation of a benzene ring; introduces a hydrocarbon side chain	113.6
Reduction with tin and hydrochloric acid	Nitro group reduced to amine	122.2

*These are examples of Friedel–Crafts reactions

Example 4

Predict how the molecule in Figure 130.4 would react with (i) bromine, (ii) lithium tetrahydrido-aluminate(III), (iii) phosphorus pentachloride.

We know (or we can find the information from Tables 130.1 and 130.2 and elsewhere) that (i) bromine will seek out and add to double bonds (but not benzene rings), (ii) lithium tetrahydrido-aluminate(III) will reduce a ketone to an alcohol, and (iii) phosphorus pentachloride will give off fumes of

hydrogen chloride with an alcohol and the OH is replaced by Cl; further, the

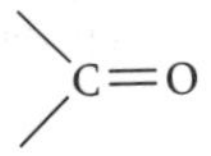

$$\begin{array}{c}\diagdown\\ \diagup\end{array}C{=}O$$

group is replaced by

$$\begin{array}{c}\diagdown\\ \diagup\end{array}\begin{array}{c}Cl\\ C\\ Cl\end{array}$$

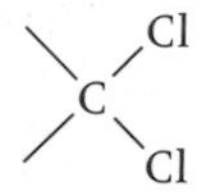

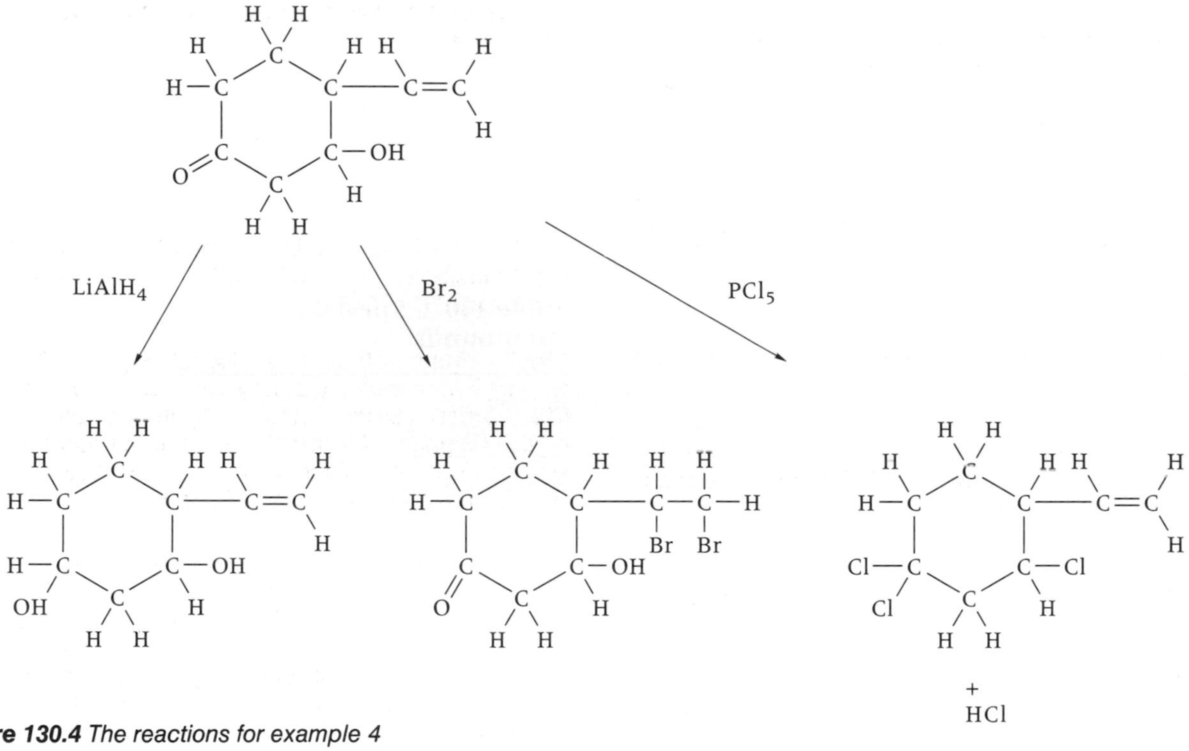

Figure 130.4 *The reactions for example 4*

130.5 A substance has molecular formula C_3H_5N. It shows no reaction with bromine water.

(i) There are two homologous series to which compounds containing nitrogen but no oxygen might belong. What are they?

(ii) Can you work out which series this molecule belongs to? What is its structure?

130.6 The same substance as in question 130.5 undergoes the following reactions (which are shown in shorthand):

$$C_3H_5N \xrightarrow{H^+} C_3H_6O_2 \xrightarrow{Cl_2/P} C_3H_5ClO_2 \xrightarrow{NH_3} C_3H_7NO_2$$

(i) Explain the first reaction.

(ii) Draw the structure of $C_3H_6O_2$. What is its name?

(iii) How does this substance react with chlorine and red phosphorus?

(iv) Draw the structure of $C_3H_5ClO_2$.

(v) Give the name and structure of $C_3H_7NO_2$. What type of substance is it?

(vi) Build a model of $C_3H_7NO_2$. There are two versions of the molecule. Explain why this is so.

130.7 Here is an outline reaction scheme:

$$C_6H_5O_2N \xrightarrow{Sn/HCl} C_6H_7N \xrightarrow[T>5\,°C]{NaNO_2,\ HCl} C_6H_6O$$
miscible in alkali, not in acid

$$C_6H_5O_2N \xrightarrow{Sn/HCl} C_6H_7N \xrightarrow[\substack{(i)\ NaNO_2,\ HCl \\ T<5\,°C \\ (ii)\ phenol}]{} C_{12}H_{11}N_2O$$
brightly coloured dye

(i) What can you deduce about the nature of $C_6H_5O_2N$, given the large proportion on carbon to hydrogen?

(ii) If you rearrange the molecular formula slightly, does this give you a recognisable molecule? What is its name?

(iii) What is the name and structure of C_6H_7N?

(iv) What is the name and structure of C_6H_6O? What else would you see during the reaction?

(v) Explain the observation about the miscibility in acid and alkali.

(vi) What is the special name given to the reaction producing the coloured dye? Draw the structure of the product. (You may need to look back at section 122.8.)

Answers

130.1 With the same percentage composition as in example 1, the empirical formula is also CH. This time there are $26/13 = 2$ units in the molecular formula. The substance is ethyne, C_2H_2.

130.2 The extra information tells us that the substance is an alcohol or an acid. If you try to write down structures, you will find that the only one that fits is ethanol, CH_3CH_2OH.

130.3 (i) The parent ion gives the relative molar mass as 120.

(ii)

	Carbon	Hydrogen	Oxygen
100 g of compound contains/g	80.0	6.7	13.3
Number of moles present /mol	$\dfrac{80.0}{12}$ $=6.7$	$\dfrac{6.7}{1}$ $=6.7$	$\dfrac{13.3}{16}$ $=0.83$
Ratio of moles	$\dfrac{6.7}{0.83}$ $=8$	$\dfrac{6.7}{0.83}$ $=8$	$\dfrac{0.83}{0.83}$ $=1$

Empirical formula C_8H_8O.
Molar mass of one unit of C_8H_8O is

$M(C_8H_8O)$
$= 8 \times 12\,g\,mol^{-1} + 8 \times 1\,g\,mol^{-1} + 1 \times 16\,g\,mol^{-1}$
$= 120\,g\,mol^{-1}$

so C_8H_8O is also the molecular formula of the compound.

(iii) We know that it is a ketone, so we start by writing down the functional group: $>C=O$. This takes out CO from the formula, leaving us with C_7H_8. We also know that it is aromatic, so it has a benzene ring present. Be careful here; there must be a group attached to the ring, which means that one of the six hydrogen atoms has been lost. Therefore we try to fit a C_6H_5 group (not C_6H_6) to the carbonyl group. If we do this we are left with CH_3 as the remaining atoms. This is just right, as it

represents a methyl group. The final structure is that of phenylethanone:

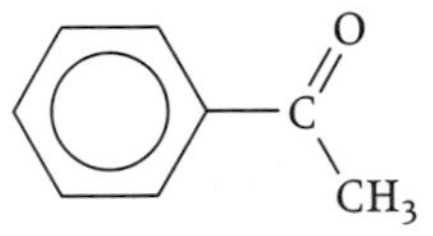

(iv) 105 is 15 mass units less than the parent mass. A $CH_3{}^+$ ion corresponds to this difference in mass, so the fragment is $C_6H_5CO^+$.

130.4 Figure 130.3a is an alcohol; Figure 130.3b is a carbonyl compound (actually an aldehyde). Note the characteristic broad band of hydrogen bonded OH groups around 3300 cm^{-1}, and the carbonyl stretch around 1700 cm^{-1}. There is also a C—H stretch characteristic of an aldehyde near 2720 cm^{-1}.

130.5 (i) Amines, RNH_2, and nitriles, RCN.
(ii) If you try drawing structures on paper you will find that an amine is only possible if there are carbon–carbon double bonds in the molecule. This is because an amine has an NH_2 group. This leaves C_3H_3, but then there are not enough hydrogen atoms to saturate the carbon atoms. The bromine water test tells us that these bonds are absent. The molecule is a nitrile, propanenitrile:

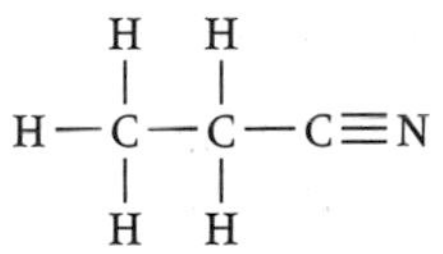

130.6 (i) Nitriles are hydrolysed to carboxylic acids.
(ii) Propanoic acid:

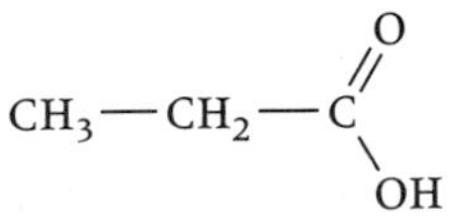

(iii), (iv) The α hydrogen atoms are lost one by one. The product is α-chloropropanoic acid:

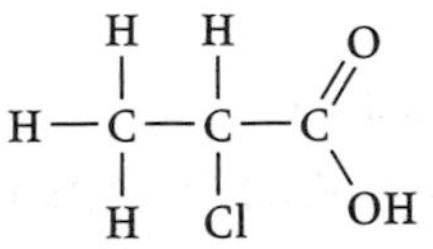

Answers – cont.

(v) This is the way in which amino acids are prepared. The product is alanine:

$$H-\overset{\underset{\displaystyle |}{H}}{\underset{\underset{\displaystyle |}{H}}{C}}-\overset{\underset{\displaystyle |}{H}}{\underset{\underset{\displaystyle |}{NH_2}}{C}}-\overset{\displaystyle O}{\underset{\displaystyle OH}{C}}$$

(vi) Alanine has an asymmetric carbon atom, i.e. it is chiral (optically active). There are two mirror image forms (enantiomers) of the molecule (see Unit 110).

130.7 (i) It is aromatic, i.e. contains a benzene ring.

(ii) $C_6H_5NO_2$, nitrobenzene.

(iii) This reaction converts nitrobenzene into phenylamine, $C_6H_5NH_2$.

(iv) This diazonium reaction converts an amine into a diazonium ion, in this case $C_6H_5N_2^+$. Above 5 °C (approx.) it reacts with water to give an alcohol, in this case phenol, C_6H_5OH. Phenol is a weak acid:

$$C_6H_5OH \rightleftharpoons C_6H_5O^- + H^+$$
partially   soluble
miscible

(v) In alkali, hydrogen ions are removed and the equilibrium is driven to the right. We see the phenol dissolving. In acid, the equilibrium shifts to the left, and a layer of phenol will appear.

(vi) A coupling reaction. The product is

UNIT 130 SUMMARY

- This unit is its own summary.

APPENDICES

B

Table of ionisation energies

Number of electrons removed

Element	1	2	3	4	5	6	7	8	9	10	11	12	13	14	15	16	17	18	19	20
Hydrogen	1312																			
Helium	2372	5250																		
Lithium	520	7298	11815																	
Beryllium	899	1757	14849	21006																
Boron	801	2427	3660	25026	32827															
Carbon	1086	2353	4620	6223	37830	47277														
Nitrogen	1402	2856	4578	7475	9445	53266	64360													
Oxygen	1314	3388	5300	7469	10989	13326	71334	84078												
Fluorine	1681	3471	6050	8408	11023	15164	17868	92038	106434											
Neon	2081	3952	6122	9370	12178	15238	19999	23069	115379	131431										
Sodium	513	4562	6912	9544	13353	16610	20115	25490	28934	141362	159074									
Magnesium	738	1451	7733	10540	13630	17995	21704	25656	31643	35462	169991	189367								
Aluminium	578	1817	2745	11577	14831	18378	23295	27459	31861	38457	42654	201270	222314							
Silicon	786	1577	3232	4356	16091	19785	23786	29252	33877	38733	45934	50511	235204	257920						
Phosphorus	1012	1903	2912	4957	6274	22233	25397	29854	35867	40965	45983	54072	59036	271798	296192					
Sulphur	1000	2251	3361	4564	7013	8496	27106	31670	36578	43138	48705	54481	62874	68230	311058	337126				
Chlorine	1251	2297	3822	5158	6542	9459	11018	33604	38600	43961	51067	57117	63362	72340	78096	352990	380756			
Argon	1521	2666	3931	5771	7238	8781	11995	13842	40760	46186	52002	59652	66199	72918	82472	88575	397602	427062		
Potassium	419	3051	4411	5877	7976	9649	11343	14942	16964	48575	54431	60699	68894	75948	83150	93399	99768	444897	476060	
Calcium	590	1145	4912	6474	8144	10496	12321	14207	18192	20385	57048	63333	70052	78792	86367	93978	104881	111635	494886	527759

All values are in kJ mol^{-1}. Data adapted from *Handbook of Chemistry and Physics*, CRC Press, Boca Raton, Florida, 1989

C

Table of atomic masses

In order of atomic number

Atomic number	Element	Atomic mass /g mol^{-1}
1	Hydrogen	1.0
2	Helium	4.0
3	Lithium	6.9
4	Beryllium	9.0
5	Boron	10.8
6	Carbon	12.0
7	Nitrogen	14.0
8	Oxygen	16.0
9	Fluorine	19.0
10	Neon	20.2
11	Sodium	23.0
12	Magnesium	24.3
13	Aluminium	27.0
14	Silicon	28.1
15	Phosphorus	31.0
16	Sulphur	32.1
17	Chlorine	35.5
18	Argon	39.9
19	Potassium	39.1
20	Calcium	40.1
21	Scandium	45.0
22	Titanium	47.9
23	Vanadium	50.9
24	Chromium	52.0
25	Manganese	54.9
26	Iron	55.9
27	Cobalt	58.9
28	Nickel	58.7
29	Copper	63.5
30	Zinc	65.4
31	Gallium	69.7
32	Germanium	72.6
33	Arsenic	74.9
34	Selenium	79.0
35	Bromine	79.9

In alphabetical order

Atomic number	Element	Atomic mass /g mol^{-1}
89	Actinium	227.0
13	Aluminium	27.0
51	Antimony	121.8
18	Argon	39.9
33	Arsenic	74.9
85	Astatine	210.0
56	Barium	137.3
4	Beryllium	9.0
83	Bismuth	209.0
5	Boron	10.8
35	Bromine	79.9
48	Cadmium	112.4
55	Caesium	132.9
20	Calcium	40.1
6	Carbon	12.0
58	Cerium	140.1
17	Chlorine	35.5
24	Chromium	52.0
27	Cobalt	58.9
29	Copper	63.5
9	Fluorine	19.0
87	Francium	223.0
31	Gallium	69.7
32	Germanium	72.6
79	Gold	197.0
72	Hafnium	178.5
2	Helium	4.0
1	Hydrogen	1.0
49	Indium	114.8
53	Iodine	126.9
77	Iridium	192.2
26	Iron	55.9
36	Krypton	83.8

In order of atomic number			In alphabetical order		
Atomic number	Element	Atomic mass /g mol^{-1}	Atomic number	Element	Atomic mass /g mol^{-1}
36	Krypton	83.8	57	Lanthanum	138.9
37	Rubidium	85.5	82	Lead	207.2
38	Strontium	87.6	3	Lithium	6.9
39	Yttrium	88.9			
40	Zirconium	91.2	12	Magnesium	24.3
			25	Manganese	54.9
41	Niobium	92.9	80	Mercury	200.6
42	Molybdenum	95.9	42	Molybdenum	95.9
43	Technetium	99.0			
44	Ruthenium	101.1	10	Neon	20.2
45	Rhodium	102.9	93	Neptunium	239.1
			28	Nickel	58.7
46	Palladium	106.4	41	Niobium	92.9
47	Silver	107.9	7	Nitrogen	14.0
48	Cadmium	112.4			
49	Indium	114.8	76	Osmium	190.2
50	Tin	118.7	8	Oxygen	16.0
51	Antimony	121.8	46	Palladium	106.4
52	Tellurium	127.6	15	Phosphorus	31.0
53	Iodine	126.9	78	Platinum	195.1
54	Xenon	131.3	94	Plutonium	239.1
55	Caesium	132.9	84	Polonium	210.0
			19	Potassium	39.1
56	Barium	137.3	91	Protactinium	231.0
57	Lanthanum	138.9			
58	Cerium	140.1	88	Radium	226.0
			86	Radon	222.0
			75	Rhenium	186.2
72	Hafnium	178.5	45	Rhodium	102.9
73	Tantalum	181.0	37	Rubidium	85.5
74	Tungsten	183.9	44	Ruthenium	101.1
75	Rhenium	186.2			
			21	Scandium	45.0
76	Osmium	190.2	34	Selenium	79.0
77	Iridium	192.2	14	Silicon	28.1
78	Platinum	195.1	47	Silver	107.9
79	Gold	197.0	11	Sodium	23.0
80	Mercury	200.6	38	Strontium	87.6
			16	Sulphur	32.1
81	Thallium	204.4			
82	Lead	207.2	73	Tantalum	181.0
83	Bismuth	209.0	43	Technetium	99.0
84	Polonium	210.0	52	Tellurium	127.6
85	Astatine	210.0	81	Thallium	204.4
			90	Thorium	232.0
86	Radon	222.0	50	Tin	118.7
87	Francium	223.0	22	Titanium	47.9
88	Radium	226.0	74	Tungsten	183.9
89	Actinium	227.0			
90	Thorium	232.0	92	Uranium	238.1
			23	Vanadium	50.9
91	Protactinium	231.0	54	Xenon	131.3
92	Uranium	238.1			
93	Neptunium	239.1	39	Yttrium	88.9
94	Plutonium	239.1	30	Zinc	65.4
			40	Zirconium	91.2

With some exceptions, the lanthanides (atomic numbers between 58 and 71), actinides (atomic numbers between 90 and 103) and elements following the actinides have been omitted

D

Values of some universal constants

Quantity	Symbol	Value and units
Avogadro constant	L	6.022×10^{23} mol^{-1}
Bohr radius	a_0	5.292×10^{-11} m
Boltzmann constant	k	1.381×10^{-23} J K^{-1}
Electron charge	$-e$	1.602×10^{-19} C
Electron mass	m_e	9.109×10^{-31} kg
Permittivity of vacuum	ε_0	8.854×10^{-12} C^2 N^{-1} m^{-2}
Planck constant	h	6.626×10^{-34} J s
Proton mass	m_p	1.673×10^{-27} kg
Speed of light in vacuum	c	2.998×10^8 m s^{-1}

E

Organic analysis

E.1. Two types of analysis

All chemicals can be analysed qualitatively or quantitatively. In qualitative analysis we seek to discover which elements, or groups of atoms, the chemical contains; in quantitative analysis we attempt to find out how much of each element is present. In this appendix you will find a brief summary of qualitative analysis as it has traditionally been applied to organic chemicals. Few practical details are provided. You should consult a specialist book if you need details of the methods. Modern techniques, such as infrared spectroscopy, nuclear magnetic resonance spectroscopy and mass spectrometry, are not described here. Please turn to the units on spectrometry for information about them.

Warning

On no account attempt any of the experiments outlined below without the guidance of your teacher or lecturer. Some of the reactions can be dangerous if not conducted with great care.

E.2 Qualitative analysis

The simplest tests are to discover:

(i)   if the compound is organic, i.e. if it contains at least the elements carbon and hydrogen;
(ii) if it contains nitrogen;
(iii) if it contains halogens;
(iv) if it contains sulphur.

It is not possible to make a simple test to discover the presence of oxygen.

(a) Test for carbon and hydrogen

A dry sample is mixed with dry copper(II) oxide and heated. Most organic compounds have some reducing power, and will reduce the copper(II) oxide to copper. When this happens, hydrogen atoms in the compound combine with part of the oxygen to make water, and carbon atoms are converted to carbon dioxide.

The presence of water can be confirmed using anhydrous copper(II) sulphate (which turns from white to blue), and lime water tests for carbon dioxide (the solution turns 'milky').

(b) Test for nitrogen, halogens and sulphur

There are two common methods for detecting nitrogen (as well as halogens and sulphur). One method is known as Middleton's test; the other is Lassaigne's test.

Middleton's test

The compound is heated with a mixture of anhydrous sodium carbonate and powdered zinc. The reaction should be done in a small, heat resistant, glass tube (an ignition tube). When the tube is red hot, it is plunged into cold water, causing the glass to break and allowing soluble compounds to dissolve in the water. *Note the warning given above!*

This reaction, known as Middleton's test, converts nitrogen in the compound into cyanide ions. It also releases halogens as free halide ions; sulphur is converted to zinc sulphide. The cyanide and halide ions dissolve in the water, while any zinc sulphide is left in the solid residue. The solution is filtered and the filtrate split into at least two portions.

To test for the presence of *nitrogen*, one portion is reacted with a little sodium hydroxide solution followed by iron(II) sulphate solution. Free cyanide ions bond to the iron(II) ions, forming a solution of hexacyanoferrate(II). On adding a little iron(III) chloride solution, followed by concentrated hydrochloric acid, a blue-green precipitate of 'Prussian blue', $Fe_4[Fe(CN)_6]_3$, confirms the presence of cyanide ions, and therefore nitrogen in the original sample. Actually, this test can be difficult to perform successfully: the amount of Prussian blue is often small and difficult to see. It can be masked by other coloured iron compounds, although the hydrochloric acid helps to destroy, for example, iron(II) and iron(III) hydroxides.

The test for *halogens* is as follows. To a second portion of the filtrate, add dilute nitric acid followed by silver nitrate solution. This is the normal test for halide ions: chlorides give a white precipitate, bromides a cream precipitate, and iodides a yellow precipitate. If it is uncertain whether the precipitate is a bromide or iodide, a third portion of the original filtrate can be mixed with chlorine water followed by a little 1,1,1-trichloroethane. The tell-tale purple colour of iodine in the organic layer confirms iodine.

The test for *sulphur* is done on the residue, and relies on the fact that sulphides will give off hydrogen sulphide when warmed with dilute hydrochloric acid. On warming with the acid, there is the likelihood of smelling the foul fumes of hydrogen sulphide. However, a simple chemical test is to place a piece of filter paper moistened by lead(II) nitrate solution in the vapour. If hydrogen sulphide is present, the paper will turn black owing to the formation of lead(II) sulphide.

Lassaigne's test

Here the organic compound is heated with a pellet of sodium, again in an ignition tube. The method is similar to Middleton's test in that the sodium converts nitrogen to cyanide, halogens to halides and sulphur to sulphide. The tube is broken under water and the solution filtered. This time cyanide ions, halide ions and sulphide ions all go into solution. (Sodium sulphide is far more soluble in water than zinc sulphide.) The tests for cyanide and halide ions are performed as in Middleton's method. The test for sulphide ions is different. A few drops of a solution of sodium nitroprusside, $Na_2Fe(NO)(CN)_5$, gives a violet colour if sulphide ions are present.

These tests are summarised in Figure E.1.

(a) Test for an organic substance (carbon and hydrogen)

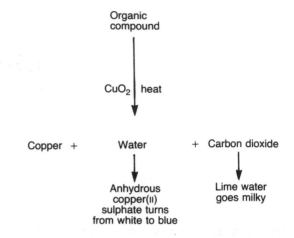

(b) Test for nitrogen, halogens and sulphur

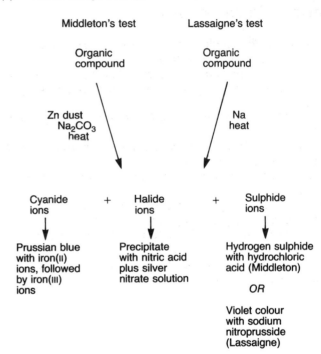

Figure E.1 Qualitative organic analysis – a summary of the tests

Bibliography

The following is a short list of books in which you can find further information.

Inorganic chemistry

Cotton, F. and Wilkinson, G. (1987). *Basic Inorganic Chemistry*, Wiley, New York
This book provides a more leisurely account of inorganic chemistry than its more heavy-weight parent:
— (1980). *Advanced Inorganic Chemistry*, Wiley, New York
Orgel, L. E. (1966). *An Introduction to Transition Metal Chemistry*, Methuen, London
A fine account of the main features of the d-block metals.
Vogel, A. I. (1973). *A Text Book of Macro and Semimicro Qualitative Analysis*, Longman, London
Vogel's book is full of practical and theoretical information on inorganic reactions. This is the place to look if, for example, you want to find out the test for a particular cation or anion.

Organic chemistry

Morrison, R. T. and Boyd, R. N. (1988). *Organic Chemistry*, Macmillan, New York
A book designed for degree courses in chemistry, but which gives excellent explanations of the organic chemistry covered in A-level courses.
Sykes, P. (1986). *A Guidebook to Mechanism in Organic Chemistry*, Longman, London
A classic book now in its sixth edition. A well written account of many reaction mechanisms.

Biochemistry

Zubay, G. (1988). *Biochemistry*, Macmillan, New York
This is a huge book, and considers deep matters befitting an undergraduate course in biochemistry. However, it is worth consulting, both for its careful explanations and for the beauty of the diagrams.

Examination questions

Inorganic chemistry

C1

(a) Part of the Periodic Table is given below:

I	II	III	IV	V	VI	VII
Li	Be	B	C	N	O	F
Na	Mg	Al	Si	P	S	Cl
K	Ca			As		
Rb	Sr			Sb		
Cs	Ba			Bi		

Of the elements in this table, write the name of the element which is most:

(i) electropositive
(ii) non-metallic

and the name of the element which forms:

(iii) a tetrachloride which is not hydrolysed by water
(iv) the most basic hydride having a pyramidal molecular shape
(v) two allotropes of low melting point
(vi) the least soluble s-block sulphate
(vii) an amphoteric oxide, X_2O_5
(viii) a series of double sulphates with Group I elements.

(b) (i) With an explanation, arrange the following bonds in order of increased bond polarity:

N—H, F—H, B—H, C—H

(ii) Describe the nature and direction of polarity of the bonds in trichloromethane and tetrachloromethane and explain why only one of these structures has a dipole moment.

(iii) Show how trichloromethane and propanone can form strong inter-molecular bonds explaining your answer in terms of bond polarity.

(c) (i) Draw bond diagrams for the N_2O_4 molecule.

(ii) Explain the principle of electron delocalisation using dinitrogen tetroxide as an example. Comment on the nitrogen to oxygen bond lengths and the geometrical shape of the molecule.

SU 1986 (2)

C2

Give an account of the chlorides of the elements sodium, magnesium, aluminium, silicon, phosphorus and sulphur.

You should consider their composition and bonding, methods of preparation and reactions with water, but your answer should not be confined to these points.

Describe how, given a pure sample, you would determine the formula of an oxide OR chloride of any ONE of these elements.

ULSEB Winter 1986 (3)

C3

This question concerns the hydrides of a range of elements.

(a) In each of the following cases, give the name of a hydride and write an equation for the reaction described.

(i) A hydride which hydrolyzes rapidly and extensively in cold water.
(ii) A hydride which is spontaneously flammable in air.
(iii) A hydride which, on electrolysis in the molten state, gives hydrogen at the anode.
(iv) A hydride which, in water, gives a dibasic acid.
(v) A hydride which, in water, gives a weak alkali.
(vi) A hydride which can act as a ligand with d-block metal ions.

(b) Give the names of TWO hydrides which combine with each other by addition and write an equation for the reaction.

(c) Suggest an explanation for the ability of carbon to form long chain hydrocarbons.

(d) The enthalpy of combustion of methane is highly exothermic. To what do you attribute the stability of methane in air?

(e) Suggest a reason for the large difference in boiling points of methane and water.

ULSEB Winter 1985 (2) (slightly adapted)

C4
(a) Describe how the following properties vary on descending groups IA and IIA of the periodic table and state how these properties differ between the two groups:

   (i) first and second ionisation energies;
   (ii) cationic radii;
   (iii) rates of reaction with cold water.

(b) For the same s-block elements, discuss:

   (i) the action of heat on their nitrates;
   (ii) the crystal structure of their chlorides, XY type only.

(c) Outline the *principles* involved in the manufacture of sodium hydroxide by the electrolysis of sodium chloride, explaining the reactions in terms of selective discharge of ions.

SU 1985 (2)

C5
(a) (i) State the conditions under which magnesium and calcium will react with water, and write balanced equations for the reactions.
   (ii) Explain any differences between the two reactions in terms of the atomic properties of the two metals.

(b) Compare the chemistries of magnesium and calcium with reference to the following:

   (i) the solubilities of their sulphates in water;
   (ii) the thermal stabilities of their carbonates;
   (iii) the reaction of their oxides with water.

(c) A mineral, which can be represented by the formula $Mg_xBa_y(CO_3)_z$, was analysed as described below.

From the results, calculate the formula of the mineral.

A sample of the mineral was dissolved in excess hydrochloric acid and the solution made up to $100 \, cm^3$ with water. During the process $48 \, cm^3$ of carbon dioxide, measured at $25 \, °C$ and 1 atmosphere pressure, were evolved.

A $25.0 \, cm^3$ portion of the resulting solution required $25.0 \, cm^3$ of EDTA solution of concentration $0.02 \, mol \, dm^{-3}$ to reach an end-point. A further $25.0 \, cm^3$ portion gave a precipitate of barium sulphate of mass $0.058 \, g$ on treatment with excess dilute sulphuric acid. You may assume that Group 2 metal ions form 1:1 complexes with EDTA.

Molar volume of any gas at $25 \, °C$ and 1 atmosphere pressure = $24 \, dm^3$).

AEB 1990 (1)

C6
This question concerns the elements boron and aluminium.

(a) Give the electronic configurations of boron and aluminium.

	1s	2s	2p			3s	3p		
B									
Al									

(b) Explain why

   (i) boron does not form simple compounds containing the B^{3+} ion.
   (ii) aluminium chloride is predominantly covalent whereas aluminium fluoride is predominantly ionic.

(c) (i) How do the hydroxides of boron and aluminium differ in their acid–base character?
   (ii) Briefly explain this difference in behaviour.

(d) Give and explain the structures of the chlorides of boron and aluminium at temperatures just above their vaporisation points.

ULSEB Winter 1986 (2)

C7
(a) Describe how you would prepare a specimen of **either** hydrated aluminium potassium sulphate (potash alum) **or** hydrated chromium(III) potassium sulphate (chrome alum) in the laboratory. Give essential experimental details and explain the chemistry involved.

(b) The standard electrode potential for Na^+/Na is $-2.71 \, V$, for $Al^{3+}/Al \, -1.66 \, V$, and for $Cu^{2+}/Cu \, +0.34 \, V$.

   (i) Explain what is meant by the term *standard electrode potential*.
   (ii) Explain how standard electrode potentials may be used to predict the reactions of the above three metals with dilute sulphuric acid.
   (iii) Explain why the prediction may be incorrect for aluminium.

(c) State the species present when aluminium sulphate is dissolved in water. Explain what happens when sodium carbonate solution is added to this solution.

UODLE 1987 (2)

C8
This question concerns the elements carbon, silicon, germanium, tin and lead in Group 4 of the Periodic Table.

(a) Copy the following table and give the electronic configuration of Ge and Sn^{2+}. The atomic numbers of germanium and tin are 32 and 50 respectively.

	1s	2s	2p	3s	3p	3d	4s	4p	4d	4f	5s	5p
Ge												
Sn^{2+}												

How do you account for the fact that tin forms the 2+ ion but germanium does not?

(b) The melting point of carbon is almost 4000 °C whereas that of tin is 232 °C. How do you account for this difference?

(c) Explain why PbO_2 liberates chlorine from concentrated hydrochloric acid but SnO_2 does not.

(d) Tetrachloromethane and tetrachlorosilane are both thermodynamically (energetically) unstable with respect to reaction with water, but only tetrachloromethane is kinetically stable.

  (i) Explain what is meant by thermodynamic stability and kinetic stability.
  (ii) How do you account for this difference in kinetic stability?

ULSEB 1988 (2) (slightly adapted)

C9

(a) Describe the appearance of lead, silicon and the allotropes of carbon and make a comparison, with explanations, of their electrical conductivities.

(b) Explain what is meant by *doping* and by *p*- and *n-type semiconductors*.

(c) Compare the bonding in each of the four following compounds and their reactions with water:

CCl_4; $SiCl_4$; and **either** $PbCl_2$ and $PbCl_4$ or $SnCl_2$ and $SnCl_4$.

(d) Give equations to illustrate

  (i) the reaction of $Pb^{2+}(aq)$ with $NaOH(aq)$,
  (ii) the reaction of $Pb^{2+}(aq)$ with $I^-(aq)$,
  (iii) one reducing reaction of $Sn^{2+}(aq)$.

WJEC 1986 (2)

C10

(a) Two isotopes of tin may be represented as $^{118}_{50}Sn$ and $^{120}_{50}Sn$.

  (i) State the number of protons, neutrons, and electrons in each of these two isotopes.
  (ii) Discuss, briefly, the meaning of the term isotope.
  (iii) Write down the electronic configuration of tin and assign the element to a periodic group.
  (iv) Suggest the type of bonding in tin(IV) chloride.

(b) Draw a fully labelled diagram of the apparatus which could be used to prepare a sample of tin(IV) chloride.

(c) Tin(II) chloride and mercury(II) chloride both form clear solutions in hydrochloric acid. Mercury(I) chloride, Hg_2Cl_2, is a white solid, sparingly soluble in this acid. Hg(I) disproportionates on warming. If mercury precipitates from solution it always appears black. The relative orders of magnitude of the standard electrode potentials are:

$$\begin{array}{cc} A & B \\ 2Hg^{2+} / Hg_2Cl_2 & > \quad Sn^{4+} / Sn^{2+} \\ \text{(most positive)} & \text{(most negative)} \end{array}$$

  (i) Show why the relative magnitudes of the stan-

dard electrode potentials A and B indicate that $Sn^{2+}(aq)$ will reduce $Hg^{2+}(aq)$.

  (ii) Write the *ionic* equation for the disproportionation of the Hg_2^{2+} ion.
  (iii) Predict what would be observed if tin(II) chloride solution is *slowly* added to mercury(II) chloride solution until the former is in excess.

SU 1985 (1)

C11

(a) Nitrogen and phosphorus both form an oxide with the empirical formula X_2O_5.

  (i) With the aid of diagrams describe the arrangement of atoms in each of these oxides.
  (ii) Give the electronic structure of N_2O_5.
  (iii) Discuss the electronic structure of the phosphorus atoms in the oxide of phosphorus; and
  (iv) name the compounds formed when each of the two oxides react with water.

(b) Describe how you would prepare a sample of phosphorus pentachloride from phosphorus trichloride.

(c) What would be observed if (i) phosphorus trichloride and (ii) bismuth(III) chloride were each added to separate samples of water, shaken until there was no further change, and then concentrated hydrochloric acid was added to the resulting mixture?

AEB 1987 (1)

C12

(a) Describe the manufacture of nitric acid **from ammonia**, with special reference to the physico-chemical principles and economic factors involved.

(b) How and under what conditions would you expect nitric acid to react with (i) sulphur and (ii) iron(II) carbonate?

(c) How and under what conditions would you expect ammonia to react with (i) copper(II) oxide and (ii) chlorine?

UODLE 1987 (2)

C13

(a) Draw a fully labelled diagram of the apparatus you would use to prepare and collect a pure sample of sulphur(VI) oxide (melting point 17 °C) in the laboratory using sulphur(IV) oxide (sulphur dioxide) as a starting material.

(b) The conversion of sulphur(IV) oxide to sulphur(VI) oxide in part (a) above is exothermic. Discuss the optimum industrial conditions of temperature and pressure for achieving an economic yield.

(c) Possible bond diagrams for sulphuric and nitric acids are:

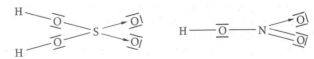

(i) What kind of bond is represented by the arrows?

(ii) Re-write the bond diagrams placing positive and negative symbols on the atoms which are charged.

(iii) Draw a bond diagram of the acid $H_2S_2O_7$ showing how this structure contains the same bonding features as the two examples in (c).

(d) When concentrated sulphuric and nitric acids are mixed, the reactions A and B below occur:

A $\quad H_2SO_4 + HNO_3 \rightarrow HSO_4^- + \underline{\quad}^+$

B $\quad\quad\quad\quad \underline{\quad}^+ \rightarrow \underline{\quad}^+ + H_2O$

(i) Re-write and complete the equations A and B.

(ii) What type of reaction is A?

(iii) What type of reaction is B?

(iv) Why is the final nitrogen(v) cation important in organic chemistry?

SU 1986 (2)

C14

In the stratosphere 30 km above the Earth's surface, ozone is being made continuously by the following reaction:

$O + O_2 \rightarrow O_3$; rate $= k[O][O_2]$.

The free oxygen atoms arise from the splitting of oxygen molecules by ultraviolet light from the Sun:

$O_2 \rightarrow 2O$.

(a) Calculate the rate at which ozone forms, given the following values:

$k = 3.9 \times 10^{-5} \, dm^3 \, mol^{-1} \, s^{-1}$,
$[O] = 3 \times 10^{-14} \, mol \, dm^{-3}$,
$[O_2] = 1.3 \times 10^{-4} \, mol \, dm^{-3}$.

(b) The temperature of the stratosphere is $-50\,°C$.

(i) State a typical value of air temperature at sea level.

(ii) Calculate the molar concentration of oxygen molecules (in $mol \, dm^{-3}$).
[One mole of gas occupies 24 dm^3 under the conditions at sea level.]

(iii) By using the collision theory of reaction kinetics and your answers to (b) (i) and (b) (ii), discuss qualitatively how the rate of ozone formation at sea level would compare with that in the stratosphere.

In fact, practically no ozone is formed in the lower atmosphere. Suggest a reason for this.

(c) It is assumed that the concentration of ozone in the stratosphere has remained roughly constant for many thousands of years but there is now some evidence that chlorofluorocarbons (CFCs—used as refrigerants and aerosol propellants) are causing the ozone concentration to decrease. It is not thought that their presence affects the rate of ozone formation, however.

What does this tell you about the role of CFCs in the other reactions involving ozone that must be occurring in the stratosphere?

UCLES 1990 (3)

C15

This question concerns the elements of Group 7: fluorine, chlorine, bromine and iodine.

(a) The word halogen means 'salt maker'. Explain why this is a suitable name for the elements of the Group.

(b) When sodium chloride is treated with concentrated sulphuric acid, a colourless gas, X, which fumes in moist air, is formed. When sodium iodide is treated in the same way a coloured vapour, Y, is produced.

(i) Identify X and Y.

(ii) Explain the difference in behaviour between the two sodium salts.

(iii) If 90% phosphoric(v) acid is used instead of sulphuric acid, a colourless gas is produced in each reaction. Explain why phosphoric(v) acid behaves differently from sulphuric acid.

(c) A number of oxoanions of chlorine are known; examples include ClO^-, ClO_3^- and ClO_4^-.

(i) ClO^- is formed when chlorine reacts with aqueous alkali. Write an ionic equation for this reaction.
Write an ionic equation for the reaction that $ClO^-(aq)$ undergoes when heated.

(ii) When $KClO_3$ is heated, it disproportionates into KCl and $KClO_4$. Write an equation for this reaction.

(iii) What is the oxidation state of chlorine in each of the following?
$KClO_3$; $KClO_4$; KCl.

(iv) Show how your answer to (c) (iii) is consistent with the ratio $KClO_4/KCl$ in your equation in (c)(ii).

ULSEB 1987 (2)

C16

When chlorine is bubbled through a concentrated aqueous solution of ammonium chloride, a yellow oily liquid, nitrogen trichloride, NCl_3, is formed, together with a solution of hydrochloric acid. Nitrogen trichloride is hydrolysed by aqueous sodium hydroxide, producing ammonia gas and a solution of sodium chlorate(I).

(a) Write balanced equations for the formation and hydrolysis of nitrogen trichloride.

(b) Draw the shape of the nitrogen trichloride molecule.

(c) Apart from peaks associated with solitary nitrogen atoms (at $m/e = 14$) and chlorine atoms (at $m/e = 35$ and $m/e = 37$), the mass spectrum of nitrogen trichloride contains 9 peaks arranged in 3 groups, ranging from $m/e = 49$ to $m/e = 125$. Predict the m/e values of all 9 peaks, and suggest a formula for the species responsible for each one.

UCLES 1990 (3)

C17

(a) Explain the term *metallic bonding* and show how it accounts for the electrical conductivity of metals.

(b) How does the electronic structure of a d-block (transition) element differ from that of a main group element? Illustrate your answer with reference to iron and calcium.

(c) 'The d-block elements show variable oxidation numbers in their compounds.' Discuss this statement, using iron to illustrate your answer.

When aqueous iron(II) sulphate is boiled with an excess of aqueous potassium cyanide a yellow solution is obtained. When chlorine is passed into the yellow solution it turns red and this red solution gives a dark blue colour on addition of aqueous potassium iodide in the presence of a few drops of starch solution. Neither the yellow nor the red solution gives a precipitate with aqueous sodium hydroxide. Explain these observations as far as you can.

UCLES 1986 (1)

C18

(a) (i) Indicate the important ores of iron and give the name and formula of the iron compound in each ore.
  (ii) Describe **briefly** the process by which iron ore is converted to pig (cast) iron in the blast furnace. Give the names of the materials used, indicating their roles, the types of reactions involved and the conditions used. Give balanced equations for the routes by which the products are formed.

(b) The oxides CO_2 and SiO_2 are both involved in blast furnace chemistry.

  (i) Compare the structures and physical state of these two oxides at ordinary temperatures and pressures.
  (ii) Write down the equilibria involving **each** of these oxides with quicklime, CaO, in the blast furnace, and state whether their equilibrium constants will be greater or less than one (1.0) in the direction written under blast furnace conditions.

(c) (i) Iron is a metallic conductor and has a body-centred cubic structure similar to sodium. Sketch this structure and describe the nature of the bonding in metallic iron.
  (ii) State **one** use of metallic iron as a catalyst and explain why transition metals are often good catalysts for chemical reactions.

(d) Iron has four stable isotopes whose relative abundances are tabulated below.

Isotope	^{54}Fe	^{56}Fe	^{57}Fe	^{58}Fe
Relative abundance/%	5.8	91.7	2.2	0.3

  (i) Draw a sketch of the mass spectrum that would be given by iron vapour, labelling and numbering the axes of the grid that you draw.
  (ii) Calculate the number of,

1. neutrons present in ^{54}Fe,
2. protons present in ^{57}Fe, and
3. electrons present in $^{56}Fe^+$.

WJEC 1990 (2)

C19

Hydrated chromium(III) chloride, $CrCl_3.6H_2O$, exists as three structural isomers. Their formulae can be represented as

  I $[Cr(H_2O)_6]^{3+}$ $3Cl^-$
  II $[Cr(H_2O)_5Cl]^{2+}2Cl^-.H_2O$
  III $[Cr(H_2O)_4Cl_2]^+$ $Cl^-.2H_2O$

(a) State and explain the type of bonding which occurs in I between chromium and water.

(b) Reaction with aqueous silver nitrate can be used to distinguish between the three compounds shown.

  (i) Give the name and appearance of the insoluble compound formed when aqueous silver nitrate reacts with each of the compounds shown.
  (ii) Explain how this reaction can be used to distinguish between the three chromium compounds, stating how the result would differ quantitatively in each case.

(c) (i) Which one of the compounds, I, II and III can exist as geometrical isomers?
  (ii) Draw diagrams to show the three-dimensional shapes of the complex ions which are geometrical isomers, and label them *cis* and *trans* as appropriate.

(d) A chromium(III) salt, **Y**, contains 19.96% Cr, 39.16% NH_3 and 40.88% Cl by mass.

  (i) Calculate the empirical formula of **Y**.
  (ii) What aqueous reagent would need to be added to hydrated chromium(III) chloride in order to prepare **Y**?
  (iii) What condition would be necessary to ensure that all the ligands in **Y** are identical?

AEB 1987 (2)

C20

(a) (i) Explain what is meant by amphoteric character for an oxide or a hydroxide.
  (ii) Indicate which of the following oxides show amphoteric character.

  Na_2O, BaO_2, Al_2O_3, P_4O_{10}, SnO, SiO_2.

  (iii) State **two** chemical differences between the oxides of metals and of non-metals.

(b) (i) State which **one** of the oxides listed in (a)(ii) above will, on treatment with dilute sulphuric acid, liberate hydrogen peroxide.
  (ii) Explain why this particular oxide is especially suitable for the preparation in this way of solutions of hydrogen peroxide.
  (iii) Give a balanced chemical equation for the reaction in (b)(i) above.

(iv) 1.6×10^{-3} mol of the oxide in (b)(i)–(iii) above was treated with excess dilute sulphuric acid, and excess aqueous potassium iodide was added to the aqueous hydrogen peroxide thus produced.
1. Give a balanced chemical equation for the reaction which takes place between hydrogen peroxide and iodide ion in acid solution.
2. Calculate the volume of aqueous sodium thiosulphate, $Na_2S_2O_3$, of concentration 0.100 mol dm^{-3} which would be required to react completely with the product of the reaction in (b)(iv)(1).

(c) Consider the following list of hydrides:

LiH, CH_4, NH_3, HF, NaH, SiH_4, PH_3, HCl.

(i) Indicate which of these hydrides show dominantly ionic bonding in the anhydrous state.
(ii) State which (if any) of the above hydrides are decomposed by interaction with pure water. Give an equation or equations for the reaction(s).
(iii) For those of the listed hydrides not decomposed by water, state which will give solutions which show
(1) acidic properties,
(2) basic properties.
(iv) State, and explain, whether or not it is possible by simple chemical means to distinguish between the hydrides of metals and non-metals.

WJEC 1989 (1)

C21
Suggest explanations for each of the following.

(a) Xenon forms fluorides, but helium does not.

(b) Lithium is the only alkali metal to react readily with nitrogen.

(c) Of the chlorides of the Group 2 elements only beryllium chloride is covalent.

(d) When aqueous sodium carbonate is added to a solution of aluminium ions, a precipitate of aluminium hydroxide is formed.

(e) Silicon tetrachloride is readily hydrolysed by water, but tetrachloromethane is not, in spite of the fact that energy would be liberated in each reaction.

(f) The acids $HClO$, $HClO_2$, $HClO_3$ and $HClO_4$ have different strengths in aqueous solution.

ULSEB Winter 1986 (3)

C22
Copper(I) chloride can be prepared as follows:
'1 g of copper filings is dissolved in dilute nitric acid forming a blue solution A and a colourless gas B. On evaporation of A to dryness, a blue solid C is formed. On further heating, C gives a black residue D and a brown gas E. The heating is continued until C is completely decomposed. D is then dissolved in concentrated hydrochloric acid giving a green solution F. A piece of copper foil, mass 2 g, is added to F and the mixture boiled for five minutes. After cooling, the copper foil is removed and the remaining colourless solution G poured into air-free cold water. The white precipitate which forms is copper(I) chloride and this is filtered, washed, and dried in a desiccator.'

(a) (i) Re-draft the information above in the form of a simple flow diagram and name the compounds A to E, labelling each compound.
(ii) How could you be sure that C had completely decomposed on heating?
(iii) Why is air-free water used to precipitate the product?
(iv) Name a drying agent which could be used in the desiccator.

(b) (i) Give the formula of the copper(II) anion in the green solution F and the formula of the copper(I) anion in the colourless solution G.
(ii) Write an ionic equation representing the reaction of the piece of copper foil with the solution F.
(iii) What would be the mass of the copper foil at the end? Explain this using the equation in (b)(ii).

(c) (i) Briefly discuss the bonding in copper metal and mention two physical properties of copper which can be explained by the nature of the bonding.
(ii) Name the species $Cu(H_2O)_4^{2+}$ and suggest two possible geometrical arrangements of the ligands round the central atom.

SU 1986 (1)

C23
(a) (i) Give the name of the ion $[Ni(H_2O)_6]^{2+}$ and draw a diagram to show its shape.
(ii) What is the nature of the metal–ligand bond?

(b) Ammonia is described as a 'monodentate' ligand and EDTA as a 'hexadentate' ligand. What do you understand by these terms?

(c) A simple colorimetric method may be used to determine the formulae of the complex ions formed by nickel and ammonia molecules, and by nickel and EDTA ions. (The formula of EDTA ions may be regarded as $EDTA^{4-}$.)

The following graphs indicate the absorption of light by the solutions as equimolar solutions of the reagents are mixed.

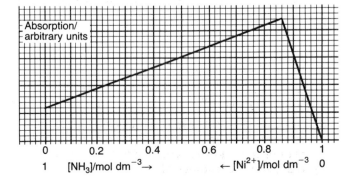

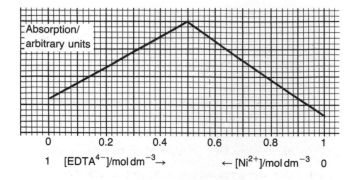

(i) Suggest a formula for the complex ion formed between nickel and ammonia.

(ii) Suggest a formula for the complex ion formed between EDTA and nickel.

(d) Write equations for the formation of the complex ions between
   (i) hydrated nickel ions and ammonia molecules,
   (ii) hydrated nickel ions and $EDTA^{4-}$ ions.

(e) Write an expression for the stability constant of the nickel–ammonia complex ion, stating the units.

ULSEB Winter 1985 (2)

Organic chemistry

D1

(a) What is meant by *isomerism* in organic chemistry? Use compounds of the following formulae to illustrate *three* types of isomerism.

   C_3H_6O $C_3H_6O_3$ C_4H_8 C_4H_{10}

(b) State and explain the requirement for a compound to exist as a pair of optical isomers, paying particular attention to the geometry of the carbon atoms.

(c) When *cis*-but-2-endioic acid is heated it readily forms a cyclic anhydride whereas the *trans* isomer only forms the same anhydride when heated to a higher temperature and for a longer time.

   (i) Draw the structure of the anhydride;
   (ii) suggest why the *trans* isomer is reluctant to form the anhydride.

(d) Explain why one isomer of the compound shown below is *not* optically active.

   $HO_2C.CHOH.CHOH.CO_2H$

SU 1986 (1)

D2

The members of the following pairs of isomeric compounds have different melting points or boiling points.

Indicate which member has the higher value and suggest reasons for the difference.

(a) 2-nitrophenol and 4-nitrophenol

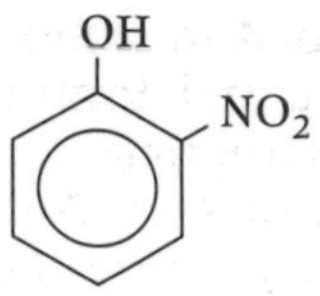

(b) (Z)-butenedioic acid and (E)-butenedioic acid

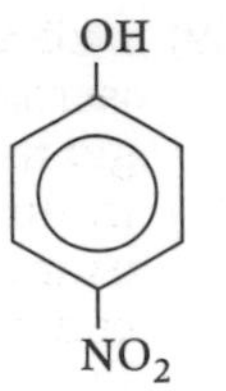

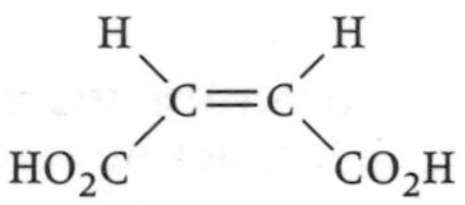

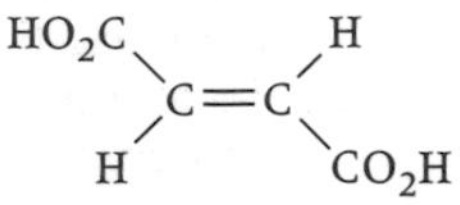

(c) 2-methylpropan-2-ol ($H_3C-C(CH_3)(CH_3)-OH$) and $CH_3CH_2CH_2CH_2OH$

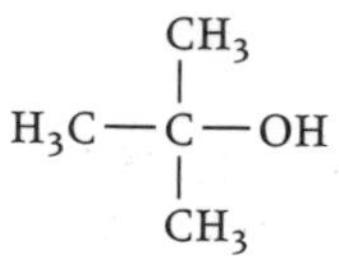

(d) CH_3CH_2OH and CH_3OCH_3

For each of (b), (c) and (d), discuss **one** other difference in **either** physical **or** chemical behaviour.

ULSEB 1985 (3)

D3

This question is largely concerned with tests on a test-tube scale, to distinguish between organic compounds.

As well as stating **what** reactions occur, it is essential to indicate what **experimental observations** should arise from them.

(a) Consider the compounds C_6H_5OH and $C_6H_5CH_2OH$; in the former the functional group is directly attached to the aromatic ring whereas in the latter the phenyl group acts merely as a substituent in an aliphatic hydroxy compound.

   (i) State **two** chemical reactions which could be used to distinguish between the compounds.
   (ii) State **one** further characteristic difference between these two compounds.

(b) Consider the compounds C_6H_5Cl and $C_6H_5CH_2Cl$; again, in the former, the functional group is directly attached to the aromatic ring whereas in the latter the phenyl group is simply a substituent in an aliphatic chloro compound.
   State **one** chemical reaction to distinguish between the compounds.

(c) The following four compounds have the same molecular formula, C_4H_8O.
   A $CH_3COCH_2CH_3$;   **B** $CH_3CH_2CH_2CHO$;
   C $CH_3CH=CHCH_2OH$;   **D** $CH_2=CHCH(OH)CH_3$

   (i) State **one** chemical reaction to distinguish between **A** and **B**.
   (ii) Write down the structural formula of another aldehyde structurally isomeric with **B**.
   (iii) Write down the structural formula of another primary alcohol structurally isomeric with **C**.

(iv) Indicate which (if any) of **A** to **D** show geometric isomerism, giving structural formulae for the geometric isomers if appropriate.

(v) Indicate which (if any) of **A** to **D** show optical isomerism, labelling any chiral (asymmetric) centres in a structural formula with an asterisk (*).

(vi) State **one** chemical reaction to distinguish between **C** and **D**.

(vii) State **one** chemical reaction which would be undergone by both **C** and **D** but **not** by either **A** or **B**.

WJEC 1990 (1) (slightly adapted)

D4

This question is about lactic acid, $CH_3CH(OH)CO_2H$, which occurs naturally, especially in sour milk.

(a) What is the systematic name for lactic acid?

(b) Draw this molecule in a three dimensional representation.

(c) What type of isomerism may be shown by this compound?

(d) (i) The compound can be synthesised from ethanol by the following route:

$$\begin{array}{ccc} & \mathbf{A} & \mathbf{B} \\ CH_3CH_2OH & \rightarrow CH_3CHO \rightarrow & CH_3CH(OH)CN \\ & & \mathbf{C} \\ & & \rightarrow CH_3CH(OH)CO_2H \end{array}$$

Give the reagents and conditions for the stages **A**, **B** and **C**.

(ii) Explain how the lactic acid synthesised in this way might differ from lactic acid isolated from sour milk.

(e) A reaction of biological importance involving lactic acid is its formation from pyruvic acid,

$$CH_3 - \underset{\underset{O}{\|}}{C} - CO_2H$$

This reaction occurs in muscle tissue during exercise, and is catalysed by the enzyme lactic acid dehydrogenase.

(i) What *type* of chemical reaction is this?

(ii) Suggest a reagent for carrying out this reaction in the laboratory.

ULSEB 1990 (1)

D5

Answer the following questions on the structural isomers of butanol by giving the appropriate letter(s).

$$\begin{array}{cc} CH_3CH_2CH_2CH_2OH & CH_3CH(OH)CH_2CH_3 \\ \mathbf{A} & \mathbf{B} \\ (CH_3)_3COH & (CH_3)_2CHCH_2OH \\ \mathbf{C} & \mathbf{D} \end{array}$$

(a) Which of the alcohols is/are (i) primary, (ii) secondary, (iii) tertiary?

(b) Which alcohol(s) contain(s) a chiral centre?

(c) Which alcohol(s) would react with acidified potassium dichromate(VI) to form (i) a ketone, (ii) a carboxylic acid, containing the **same number** of carbon atoms?

(d) (i) Describe how the Lucas test is performed.

(ii) State what result you would expect for the Lucas test with the alcohols you chose in (a)(i) and (a)(ii).

(e) Alcohols may be dehydrated to form alkenes.

(i) State the appropriate reaction conditions.

(ii) Give the letter(s) representing the alcohol(s) which would be dehydrated to form each of the following alkenes: but-1-ene, but-2ene, 2-methylpropene.

AEB 1986 (2) (slightly adapted)

D6

(a) Explain what is meant by *nucleophilic addition reaction* and *electrophilic addition reaction*. Illustrate your answer by reference to the mechanism of one reaction of the $\overset{}{>}C{=}O$ group and one of the $\overset{}{>}C{=}C\overset{}{<}$ group.

(b) State the reagents and conditions which are suitable for the hydrogenation of

(i) the $\overset{}{>}C{=}O$ group and

(ii) the $\overset{}{>}C{=}C\overset{}{<}$ group.

(c) Choose **two** other addition reactions of the $\overset{}{>}C{=}O$ group and **two** of the $\overset{}{>}C{=}C\overset{}{<}$ group, and give in each case the reagents and the formula of the product obtained by reaction with a named carbonyl compound or alkene.

(d) Describe a simple test-tube reaction to show the presence of the $\overset{}{>}C{=}C\overset{}{<}$ group. You should name the reagent(s) and describe the observations which you would expect to make.

(e) Describe a simple test-tube reaction to distinguish between an aldehyde and ketone. You should name the reagent(s) and describe what you would expect to observe in each case, explaining the reason for the difference in behaviour.

AEB 1987 (1)

D7

(a) (i) Give the reagents and conditions required to convert benzene into chlorobenzene.

(ii) Write an equation for the conversion of benzene into chlorobenzene.

(b) The structures of two compounds containing chlorine attached to a benzene ring are given below.

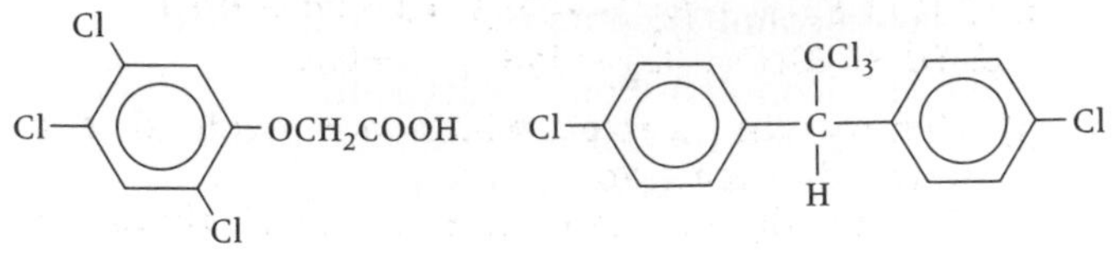

2, 4, 5–T   DDT

Although still used in the UK 2,4,5-T is banned in some countries because it may contain the toxic impurity dioxin.

The large-scale use of DDT is now banned in the UK.

(i) What is 2,4,5-T used for in the UK?
(ii) Describe what can happen if dioxin is accidently released into the air.
(iii) What was DDT once used for?
(iv) Why has DDT now been banned?

AEB 1989 (AS)

D8

This question is about benzoic acid.

(a) Give the reagent(s) and reaction conditions for each of the following methods of preparation of benzoic acid.

(i) $C_6H_5CH_3 \rightarrow C_6H_5CO_2H$
(ii) $C_6H_5CONH_2 \rightarrow C_6H_5CO_2H$
(iii) $C_6H_5CO_2C_2H_5 \rightarrow C_6H_5CO_2H$

Name the functional group of the reactant in (ii) and (iii).

(b) Benzoic acid dissolves readily in sodium hydroxide solution, but this solution turns milky when excess hydrochloric acid is added. Explain.

(c) The organic reactant in (a)(iii) was isotopically labelled at one of its oxygen atoms with oxygen–18. The resulting benzoic acid after reaction contained only normal oxygen–16. Write out the structure of the labelled reactant, showing the position of the label clearly. Write a mechanism for the reaction consistent with the benzoic acid being unlabelled.

ULSEB 1988 (2) (slightly adapted)

D9

This question is about organic compounds containing nitrogen.

(a) When phenylamine is treated with aqueous bromine a white precipitate is obtained.

(i) Write a balanced equation for this reaction, and give the structural formula of the organic product.
(ii) Classify this reaction mechanistically.
(iii) Give the formula of one other mono-substituted benzene compound that reacts in a similar way to phenylamine.

(b) Phenylamine can be prepared from nitrobenzene by treatment with tin/concentrated hydrochloric acid, addition of excess sodium hydroxide, followed by steam distillation.

(i) What is the role of the tin/concentrated hydrochloric acid?

(ii) What is the role of the sodium hydroxide?
(iii) Suggest an explanation why steam distillation is a sensible step before the final purification of the phenylamine.
(iv) Explain briefly how the resulting steam distillate (of water and phenylamine) would be treated to give pure phenylamine.
(v) Explain why freshly prepared phenylamine, which is colourless, darkens on standing.

(c) An orange-red dye can be synthesised by the following reaction sequence:

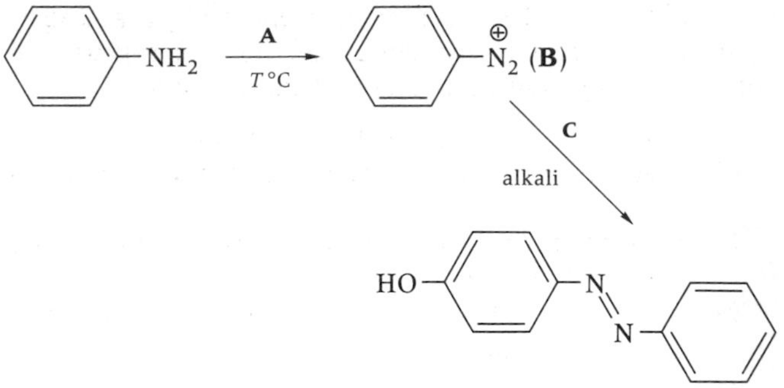

(i) Identify A and C, and the appropriate temperature ($T°C$) at which the first reaction should be carried out.
(ii) The ion, B, undergoes a number of useful reactions. Give an equation for any one such reaction.

ULSEB 1987 (2)

D10

(a) Compare the base strengths of phenylamine, ammonia and 1-aminobutane, giving reasons for the differences.

(b) (i) State the products of the reaction of 1-aminobutane with nitrous acid.
   (ii) Give the experimental conditions necessary for diazotisation of phenylamine and write an equation to show the formation of the diazonium ion.

(c) Phenylamine reacts with an acyl chloride, A, forming an organic derivative, B, of relative molecular mass 197.

(i) Identify A and B by giving their molecular and structural formulae. Explain your reasoning.
(ii) State the class of compound to which B belongs.
(iii) Suggest a suitable method of purification for compound B.
(iv) If 5.0 g phenylamine gave 8.0 g B calculate the percentage yield of B.

AEB 1987 (1)

D11

Aspirin and paracetamol are substances which are taken to relieve pain: they are said to be analgesics. Although their action in the body is complex, their structures are relatively simple:

aspirin   paracetamol

(a) *Name* the functional groups in each compound (excluding the benzene ring itself).

(b) Draw the *full* structures of the organic products formed when each compound is separately treated with hot aqueous sodium hydroxide.

(c) Aspirin (m.p. 135 °C) may be prepared by reacting 2-hydroxybenzoic acid (salicylic acid) with a mixture of ethanoic anhydride and glacial ethanoic acid.

   (i) Write an equation for the reaction;
   (ii) give an alternative reagent to ethanoic anhydride;
   (iii) outline a method for the purification of the aspirin produced;
   (iv) describe a method to determine the purity of a sample of aspirin.

(d) A container of aspirin tablets which has been exposed previously to damp air will have the pungent smell of vinegar when opened again. Explain this phenomenon and write an equation for the appropriate reaction.

SU 1986 (2)

D12
A sample of ethanal can be prepared in the laboratory as follows:
'To 12 cm³ water in a flask, 4 cm³ of concentrated sulphuric acid is slowly added with mixing and the apparatus is set up for distillation. 8 cm³ of ethanol is added to a solution containing 10 g sodium dichromate in 10 cm³ of water. This mixture is then poured into a dropping funnel which is part of the apparatus. The acid is then heated and the mixture containing the ethanol is slowly added so that the acid remains at its boiling point and an aqueous solution of the product distils over. The distillate is re-distilled and the fraction boiling between 20° and 23 °C is collected.'

(a) Draw carefully a fully labelled diagram of the apparatus required for the initial stage of this preparation.

(b) (i) Explain why the above experimental set-up must be used rather than *reflux* followed by distillation.
   (ii) Give reasons for the use of heat in the initial stage.

(c) If the density of ethanol is 0.79 g cm⁻³, calculate the maximum theoretical yield of ethanal in grams, stating any assumptions you make. (H = 1; C = 12; O = 16)

(d) Ethanol may be produced industrially from ethene by the Wacker process. The essential reactions are summarised below:

I. $C_2H_4 + PdCl_2 + H_2O \rightarrow CH_3CHO + Pd + 2HCl$
II. $Pd + 2HCl + \frac{1}{2}O_2 \rightarrow PdCl_2 + H_2O$

  (i) Identify the species which are oxidised and reduced in reaction I.
  (ii) How would the purity of a sample of ethanal be checked?

(e) Devise a scheme for making ethyl ethanoate using ethanal as the only organic starting material. State reagents and conditions.

SU 1985 (2)

D13
The preparation of a pure sample of benzoic acid by the reaction:

$C_6H_5COOC_2H_5(l) + NaOH(aq) \rightarrow$
$\qquad\qquad C_6H_5COO^-(aq) + C_2H_5OH(aq)$

involves the following stages.

Stage I
Place 5 cm³ of ethyl benzoate in a 100 cm³ round bottomed flask and add 40 cm³ of aqueous sodium hydroxide of concentration 2 mol dm⁻³. Reflux gently until the last signs of the ester layer have disappeared.

Stage II
Arrange the apparatus for distillation and collect about 10 cm³ of distillate.

Stage III
Cool the residue left in the distillation flask and add slowly, with stirring, 20 cm³ of concentrated hydrochloric acid. Cool the flask and filter off the precipitated crystals of benzoic acid. Wash the crystals with a little distilled water.

Stage IV
Recrystallise the benzoic acid from distilled water. Filter, wash, dry and weigh the crystals.

Stage V
Measure the melting point of your sample of benzoic acid. Mix a small sample of your crystals with an equal volume of a pure sample of benzoic acid. Determine the melting point of the mixture.

Stage VI
To determine the purity of your benzoic acid crystals, weigh accurately into a conical flask 0.30 g of the crystals. Pipette 50 cm³ of aqueous sodium hydroxide of concentration 0.100 mol dm⁻³ into the flask and titrate the mixture with aqueous hydrochloric acid of concentration 0.100 mol dm⁻³ using phenolphthalein indicator. Repeat. For each determination calculate the percentage of benzoic acid in your crystals.

Answer the following questions based upon this method.

(a) Why is a reflux condenser necessary when heating the ethyl benzoate with the sodium hydroxide solution?

(b) Assuming that the density of ethyl benzoate is 1.01 g cm⁻³ and that of pure ethanol is 0.79 g cm⁻³, calculate the theoretical volume of ethanol that should have been formed. In the light of your calculation suggest a reason for the instruction to collect $10 \, cm^3$ of distillate.

(c) The hydrolysis of ethyl benzoate in acidic solution reaches equilibrium. Explain why the conversion of ethyl benzoate goes to completion in alkaline solution.

(d) Explain in terms of structure and bonding why sodium benzoate is soluble in cold water whilst benzoic acid is relatively insoluble.

(e) The prepared sample of crystals was found to melt over the range 120.5–121.5 °C and the melting point range of the sample of the crystals mixed with pure benzoic acid was 120–121 °C. Explain carefully the information that these data give.

(f) Explain briefly why benzoic acid can be purified by recrystallisation from hot water. Suggest a reason why methanol would not be suitable.

(g) In the determination of the purity, 0.300 g of a sample of the crystals were used and $25.50 \, cm^3$ of the hydrochloric acid were required.

Calculate the percentage purity of the sample.

AEB 1987 (1)

D14

2-aminopropanoic acid is an amino acid.
It has the following formula.

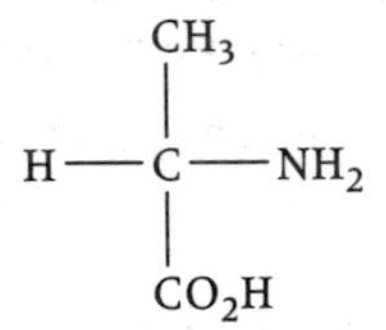

$$
\begin{array}{c}
CH_3 \\
| \\
H - C - NH_2 \\
| \\
CO_2H
\end{array}
$$

Amino acids are soluble in water since they exist as zwitterions and also form hydrogen bonds with water molecules.

(a) (i) Draw the zwitterionic form of 2-aminopropanoic acid.
   (ii) Draw diagrams to show how the carboxylic acid and the amino groups of 2-aminopropanoic acid form hydrogen bonds with water.

(b) The scheme below shows a suggested reaction scheme for the synthesis of 2-aminopropanoic acid from 2-hydroxypropanoic acid.

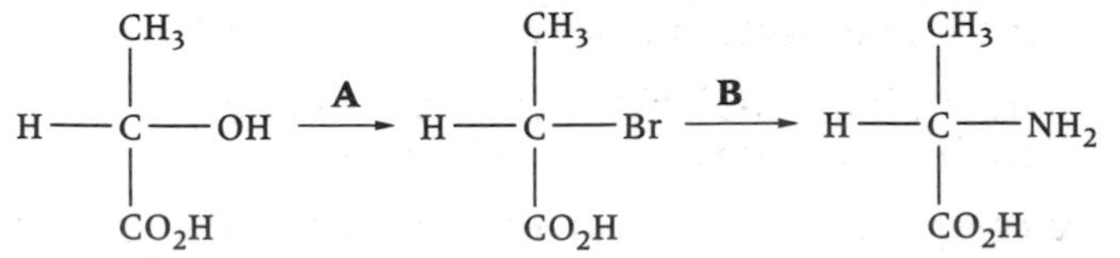

$$
\begin{array}{ccccc}
CH_3 & & CH_3 & & CH_3 \\
| & & | & & | \\
H-C-OH & \xrightarrow{A} & H-C-Br & \xrightarrow{B} & H-C-NH_2 \\
| & & | & & | \\
CO_2H & & CO_2H & & CO_2H
\end{array}
$$

Suggest the reagents and conditions needed to carry out the two steps A and B.

UCLES 1990 (2) (slightly adapted)

D15

(a) Explain the meanings of the following terms as applied to proteins:

   (i) the peptide bond,
   (ii) an amino acid residue,
  (iii) primary structure,
  (iv) the disulphide link.

(b) The technique of electrophoresis involves applying a potential difference across a starch gel. If the amino acid glycine is dissolved in the gel and a potential applied, it is observed that

   (i) in strongly acidic solution the amino acid moves towards the cathode,
   (ii) in strongly alkaline solution the amino acid moves towards the anode,
  (iii) at pH = 6 the glycine will not move in either direction.

Explain these observations.

(c) The determination of the protein content of food is an important step in judging the quality of the food. In such an experiment 25 g of a food sample was decomposed by heating with hot concentrated sulphuric acid to convert all the nitrogen from proteins into ammonium sulphate. The resulting solution was treated with sodium hydroxide solution which liberated sufficient ammonia gas to neutralise $10 \, cm^3$ of 0.1 M hydrochloric acid.

   (i) Calculate the percentage of nitrogen in the food.
   (ii) What other factors would you consider to be important in deciding on the quality of food protein?

UCLES 1989 (AS2)

D16

When monochromatic, plane-polarised light is passed through an aqueous solution of a sugar the angle of rotation of the light is dependent upon the structure of the molecule. A standard rotation can be measured which differs from molecule to molecule. Chiral enantiomers rotate the light in equal but opposite directions (indicated by + and −). The specific rotations of three sugars (measured at the same concentration and under the same conditions) are as shown below:

Glucose

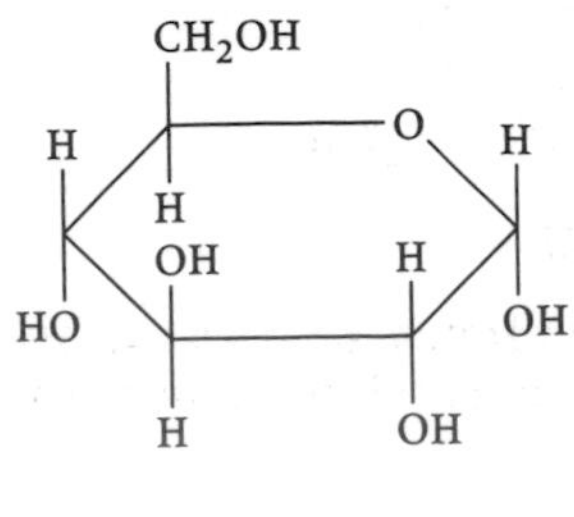

$$
\begin{array}{c}
CH_2OH \\
\end{array}
$$

specific rotation = +53°

Maltose

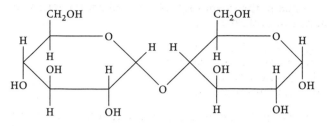

specific rotation = +137°

Sucrose

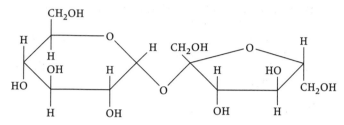

specific rotation = +67°

The following solutions were prepared under the same conditions, left for some time and their angles of rotation of polarised light measured:

Solution	Angle of rotation of light
25 cm³ glucose (initial reading)	+53°
25 cm³ glucose + 25 cm³ water (final reading)	+26.6°
25 cm³ maltose (initial reading)	+137°
25 cm³ maltose + 25 cm³ water containing the enzyme maltase (final reading)	+53°
25 cm³ sucrose (initial reading)	+67°
25 cm³ sucrose + 25 cm³ water containing the enzyme sucrase (final reading)	−20°

(a) What do you understand by the term *enantiomer*?

(b) Draw a diagram to represent the straight-chain form of glucose.

(c) (i) Why did the angle of rotation for glucose change from +53 °C to +26.5 °C when diluted with an equal quantity of water?

(ii) What do you think happened to the maltose when treated with the enzyme maltase?

(iii) Calculate the specific rotation of fructose.

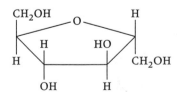

(d) Why is it that some carbohydrates, such as sucrose, are useful foods but others, such as cellulose, are not?

UCLES 1989 (AS2)

D17

(a) Draw as many conclusions as you can about the identity of each of the compounds A to D, using the information given below and giving your reasoning in each case.

(i) A ($C_6H_{12}O$) is a chiral molecule produced by the oxidation of an alcohol by boiling with acidified potassium dichromate(VI) solution for a considerable time.

(ii) B ($C_{10}H_{20}O_2$), when boiled with dilute sulphuric acid for some time produces two compounds, $C_2H_4O_2$ and $C_8H_{18}O$, the first one of which smells of vinegar whereas the second has little noticeable odour.

(iii) C (C_4H_8) can be produced in two different isomeric forms, both of which react with hydrogen to give butane.

(iv) D ($C_5H_{12}O$) is very resistant to oxidation but will react with phosphorus(v) chloride to give $C_5H_{11}Cl$.

(b) (i) One of the compounds A to D smells of oranges and is used in the preparation of perfumes and flavourings for food. Which one?

(ii) One of the compounds A to D could be used as the starting material for the manufacture of a plastic. Which one?

UCLES 1990 (AS1)

D18

(a) Explain what you understand by the terms *chirality* and *functional group* and give an example in each case.

(b) Outline, with suggested reagents and suitable conditions for each step, how you would carry out the conversion.

$$CH_3CH_2OH \rightarrow CH_3COOCH_2CH_3$$

(c) Consider the compound A whose structural formula is likely to be unfamiliar to you but which is similar to caprolactam from which nylon 6 is made.

From your knowledge of organic chemistry suggest the most likely organic product of the reaction of A with

(i) lithium tetrahydridoaluminate ($LiAlH_4$).

(ii) boiling aqueous sodium hydroxide.

By comparison with the production of nylon 6, by the opening of a caprolactam ring, it might be thought possible to produce a polymer from compound A.

(iii) Sketch the likely polymer unit that would be formed in such a reaction.

Examination questions 931

(d) Three alcohols, all $C_4H_{10}O$, were separately heated with acidified potassium dichromate(VI) solution. Draw the structural formulae of the organic products in each case.

UCLES 1989 (AS2)

D19

(a) Draw the full structural formula of E, 4-phenylbut-1-ene. Indicate on your diagram:

(i) one carbon–carbon bond which has length 154 nm;

(ii) one carbon–carbon bond which has length 139 nm;

(iii) one carbon–carbon bond which has length 133 nm.

(b) Using your knowledge of simple alkenes and arenes, predict equations and give the structural formulae of the products obtained when E reacts with

(i) hydrogen bromide;

(ii) bromine water;

(iii) hydrogen with a palladium or platinum catalyst at room temperature and pressure;

(iv) hydrogen with a nickel catalyst at high temperature and pressure.

(c) Illustrate the meaning of the term *optical isomerism*, using one of the products of the reactions in part (b) above.

UODLE 1987 (3)

D20

Warfarin is used to destroy rodents. The full structural formula of warfarin is shown below.

(a) Warfarin gives a positive tri-iodomethane (iodoform) test.

(i) What are the reagents and conditions used for the tri-iodomethane test?

(ii) Circle on a copy of the formula which part of the warfarin molecule gives the tri-iodomethane.

(b) Indicate with an arrow on your copy of the formula the part of the warfarin molecule which reacts with 2,4-dinitrophenylhydrazine.

(c) State and explain whether warfarin would give a positive reaction with $[Ag(NH_3)_2]^+$ or an alkaline Cu^{2+} complex (Fehling's solution).

(d) Warfarin reacts with aqueous sodium hydroxide.

(i) Name the group in warfarin which is attacked by the hydroxide ion.

(ii) Draw the structure of the product formed by this reaction.

(e) Warfarin reacts with phosphorus pentachloride.

(i) What gas will be produced?

(ii) Without drawing the whole warfarin molecule, give the formula of the new organic group produced.

UCLES 1989 (2)

D21

Answer EITHER A or B.

EITHER

A.

(a) Describe simple chemical tests that would enable you to distinguish between the compounds in each of the pairs below. You should describe the test and indicte the result on both compounds. When one of the compounds does not react, say so.

(i) Pentan-2-one, $CH_3COCH_2CH_2CH_3$, and pentan-3-one, $CH_3CH_2COCH_2CH_3$

(ii) Butan-1-ol, $CH_3CH_2CH_2CH_2OH$, and butan-2-ol, $CH_3CH_2CH(OH)CH_3$

(iii) Ethylamine, $C_2H_5NH_2$, and propanamide, $C_2H_5CONH_2$.

(b) Using no *organic* compounds other than the one listed, give syntheses, indicating reagents and essential reaction conditions, for the preparation of

(i) propanoic acid, $CH_3CH_2CO_2H$, from ethanol, CH_3CH_2OH.

(ii) phenylmethyl benzoate, $CO_2CH_2C_6H_5$ from methylbenzene, CH_3.

OR

B.

State Raoult's Law as it applies to mixtures of methanol (b.pt. 64 °C) and ethanol (b.pt. 78 °C) which behave ideally, and explain the reasons for this ideal behaviour.

Give a fully labelled diagram showing the relationship between boiling temperature and composition for mixtures of methanol and ethanol.

Give full practical details for the fractional distillation in the laboratory of a mixture of methanol and ethanol in which the mole fraction of methanol is 0.2 and, by reference to your temperature-composition diagram, explain the principles of the process.

At a particular temperature, the vapour pressures of pure methanol and pure ethanol are 81 mm Hg and 45 mm Hg, respectively. Calculate the partial pressure of each component above a mixture of 64 g of methanol and 46 g of ethanol at this temperature.

Mixtures of benzene (b.pt. 80 °C) and ethanol show a positive deviation from Raoult's Law. Give a fully labelled temperature–composition diagram for such mixtures and state and explain what happens when benzene is added to ethanol.

(Relative atomic masses: H = 1, C = 12, O = 16.)

ULSEB 1990 (2)

D22

A test tube contains a liquid, **X**, which is known to be one of the substances ethanal, methanal, or propanone.

(*a*) Describe chemical tests, excluding the one in part (*b*) below, which you would carry out on **X** to establish its identity. Your answer must include the chemical names of the reagents, balanced equations, and the observations in each case.

(*b*) Write an equation for the reaction between propanone and hydroxylamine, NH_2OH, drawing the *full* structural formula of the organic product. State the type of reaction which takes place.

(*c*) Ethanal will react according to the following scheme:

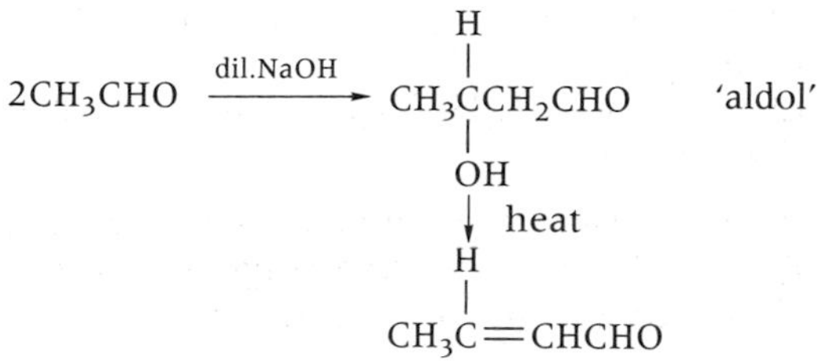

$$2CH_3CHO \xrightarrow{\text{dil.NaOH}} \underset{\substack{| \\ OH}}{\overset{\substack{H \\ |}}{CH_3CCH_2CHO}} \quad \text{'aldol'}$$

$$\downarrow \text{heat}$$

$$\underset{}{\overset{\substack{H \\ |}}{CH_3C}} = CHCHO$$

(i) What type of reaction occurs when 'aldol' is heated?
(ii) Write the structural formulae of the geometrical isomers of $CH_3CH = CHCHO$ and name them.
(iii) Methanal does *not* react with aqueous sodium hydroxide to give a product equivalent to 'aldol'. Write an equation for the reaction which does take place and explain the reaction type.

SU 1985 (1)

D23

Compound **C** is readily hydrolysed by moist air giving ethanoic acid and a gas **F** which gives dense white fumes with concentrated aqueous ammonia. Compound **C** reacts vigorously with an amine **D**, which is a dense oily liquid, forming a white solid **E** and also evolving gas **F**. If **D** is treated at between 5 and 10 °C with aqueous sodium nitrite and excess hydrochloric acid, and the mixture produced is poured into hot water, then phenol is formed.

(*a*) Give equations for the:

(i) hydrolysis of **C**, (ii) reaction of **C** with **D**, (iii) reaction of **F** with ammonia.

(*b*) Name **C** and **D**.

(*c*) (i) Give the structural formula of **E**.

(ii) Ring the grouping in the structural formula of **E** which is also present in a naturally occurring polymer.
(iii) Name this type of natural polymer

(*d*) If another amine **G** were treated with **C**, the solid product could be used to assist in identification of **G**.

(i) State how the solid product could be purified.
(ii) Explain how you could use this product to assist in identification of **G**.
(iii) Suggest a test whereby the identity of **G** could be confirmed.

(*e*) (i) Give the formula of the organic compound formed initially by reacting the amine **D** with aqueous sodium nitrite and excess hydrochloric acid.
(ii) Write an equation for a reaction of this compound (other than its conversion to phenol).

AEB Winter 1986 (2)

D24

Name and write structures for the compounds **A** to **H** in the reaction scheme below and give the reagents and conditions necessary to bring about each of the changes.

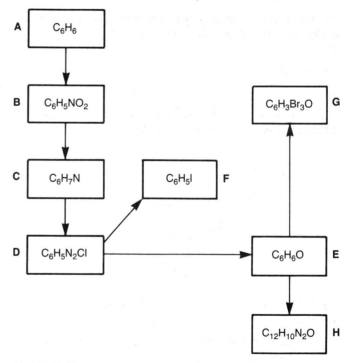

AEB 1988 (1)

D25

(*a*) An alcohol has the molecular formula C_4H_7OH. Oxidation of this alcohol gives a ketone. Ozonolysis of this alcohol gives methanal as one of the products.

Giving your reasoning in full, deduce the structural formula for this alcohol and predict whether it will be optically active.

(*b*) Outline the stages by which ethanal may be converted into 2-oxopropanoic acid (pyruvic acid) CH_3COCO_2H. Full practical details are not required, although you should specify reagents and any important conditions.

(c) A gaseous hydrocarbon **X** can exist as geometric isomers. $30\ cm^3$ of gaseous **X** were exploded with $200\ cm^3$ of oxygen. The residual gases were cooled to $25\ °C$ and found to occupy $140\ cm^3$. After treatment with aqueous sodium hydroxide, the final volume of gas remaining was $20\ cm^3$. Assuming that all volumes were measured at $25\ °C$ and the same pressure, deduce the molecular and structural formulae of hydrocarbon **X**.

OCSEB 1990 (3,4,5)

D26

A compound **P** contains C, 35.0%; H, 6.60%; Br, 58.4%, by mass. Treatment of **P** with warm aqueous sodium hydroxide gives **Q**, which on oxidation gives **R**. **R** readily forms a 2,4-dinitrophenylhydrazone, and both **Q** and **R** on treatment with iodine and alkali form triiodomethane. When **P** is treated with alcoholic sodium hydroxide it forms three isomeric alkenes **V**, **W** and **X**, each of which, on treatment with HBr, is converted back to **P**.

Identify the compounds **P**, **Q**, **R**, **V**, **W** and **X**, explaining your reasoning. Discuss the isomerism shown by the alkenes.

Suggest a synthesis for a compound of formula $C_5H_{10}O_2$ from **P**.

ULSEB 1986 (3)

D27

A hydrocarbon **F** reacts with chlorine under suitable conditions to give **G**. **G** contains 14.29% carbon, 1.19% hydrogen and 84.52% chlorine.

Careful hydrolysis of **G** with aqueous sodium hydroxide gives **H**. **H** reacts with ammoniacal silver nitrate to give silver and ammonium ethanedioate.

Addition of hydrogen chloride to **F** in the presence of aqueous mercury(II) chloride gives a well-known monomer, **J**. Further reaction of **J** with hydrogen chloride gives **K**, which has the empirical formula CH_2Cl. Hydrolysis of **K** with aqueous sodium hydroxide gives **L**, which reacts with ammoniacal silver nitrate to give silver and a salt, **M**.

Deduce the structural formulae of compounds **F**, **G**, **H**, **J**, **K**, **L** and **M**, and explain your reasoning.

UODLE 1987 (3)

D28

(a) An organic compound contains, by mass, 55.80% carbon, 7.04% hydrogen, and 37.16% oxygen. On treatment with sodium hydrogencarbonate solution it liberates carbon dioxide. It also undergoes an addition reaction with bromine in a 1:1 molar ratio whereby $1.00\ g$ of the compound reacts with $1.856\ g$ of bromine. From the above information

(i) calculate the empirical formula of the compound,

(ii) calculate the relative molecular mass of the compound,

(iii) write down molecular formulae for all of the isomeric structures which are consistent with these data,

(iv) indicate which (if any) of them will show geometric isomerism.

$$[A_r(H) = 1.01;\ A_r(C) = 12.01;$$
$$A_r(O) = 16.00;\ A_r(Br) = 79.90.]$$

(b) On hydrogenation **one** of the isomers (compound **A**) in (a)(iii) above is converted into a product **B**, but all of the other isomers are converted on hydrogenation into another compound, **C** which is *not* the same as compound **B**. From this information identify and write down structures for compounds **A**, **B** and **C**.

(c) Considering now **all** of the structural isomers in (a)(iii) above,

(i) write down structural formulae for each of the compounds which may be formed by their addition reactions with bromine,

(ii) write down structural formulae for **all** of the possible compounds which might be formed by their addition reactions with hydrogen bromide,

(iii) indicate the presence of any chiral centres in the products formed in (c)(i) and (ii) above by appending asterisks to the appropriate carbon atoms,

(iv) *briefly* describe how the racemate corresponding to **any one** of the above structures possessing chiral centres might be resolved into its optically active isomers.

WJEC 1989 (2)

D29

A is phenylamine (aniline) $C_6H_5NH_2$, a typical aromatic amine with the $—NH_2$ group directly attached to the ring. **B** is (phenylmethyl) amine, $C_6H_5CH_2NH_2$, which may be considered to have the properties of an aliphatic amine with the phenyl group as a substituent.

(a) Calculate the percentage by mass of C, H and N in both **A** and **B**.

$$[A_r(C) = 12.01,\ A_r(H) = 1.01,\ A_r(N) = 14.01.]$$

Comment on the usefulness of an elemental analysis giving C = 78.10%, H = 8.00% and N = 13.9% for distinguishing between **A** and **B**.

(b) (i) Outline the preparation of
1. **A** from nitrobenzene, $C_6H_5NO_2$
and
2. **B** from (phenylmethyl) bromide, $C_6H_5CH_2Br$

(ii) Give the conditions under which both **A** and **B** may be acylated, together with formulae for the acylating agent and for the products formed.

(c) (i) State the reagents and conditions required for the conversion of **A** to a diazonium salt. Give **one** example of an azo coupling reaction for the

diazonium salt and explain the significance of this reaction for the dyestuff industry.
(ii) How might the reaction in (c)(i) above be used to distinguish between **A** and **B**?

(d) The dissociation constant, K_b, for compound **B** is $10^{-5}\,mol\,dm^{-3}$. If an aqueous solution of **B** is titrated against dilute hydrochloric acid, state whether the pH at the end-point would be 7, less than 7, or greater than 7. State therefore which of the following indicators you would use to determine the end-point of the titration,

bromophenol blue (pH range 3.0–4.6),
bromothymol blue (pH range 6.2–7.6),
phenolphthalein   (pH range 8.0–10.0).

WJEC 1990 (2)

D30
(a) (i) An organic compound, **A**, containing the elements carbon, hydrogen, oxygen and chlorine gave a mass spectrum in which the most prominent peaks corresponded to relative molecular masses of 92 and 94 in a 3:1 ratio. Hydrolysis of **A** produced a compound **B**. After isolation and purification of **B** it was found that:
(1) it did not contain chlorine;
(2) it reacted with sodium hydrogencarbonate to produce carbon dioxide;
(3) on heating with dry soda lime a gas, **C**, was produced which burned with a non-luminous flame.

Name the functional group in **B** which can be deduced from this information, giving a brief explanation.

(ii) When 0.265 g of compound **A** was hydrolysed, the chloride ion released required 52.09 cm³ of 0.055 mol dm⁻³ aqueous silver nitrate solution for complete reaction. Calculate the number of chlorine atoms present in a molecule of **A** and identify **A**.
$[A_r(H) = 1.0;\quad A_r(C) = 12.0;\quad A_r(O) = 16.0;$
$A_r(Cl) = 35.5;\ A_r(Ag) = 107.9.]$

WJEC 1988 (1)

D31
(a) (i) State which molecule(s) in the following list show(s) optical activity:

A   HOOCCH=CHCOOH
B   CFClBrH
C   CH₃CH(OH)COOH
D   CH₃CH(Br)CH₃

(ii) Describe briefly how a polarimeter works and how it is used to detect optical activity.

(b) (i) Write down the systematic name of the following molecule:

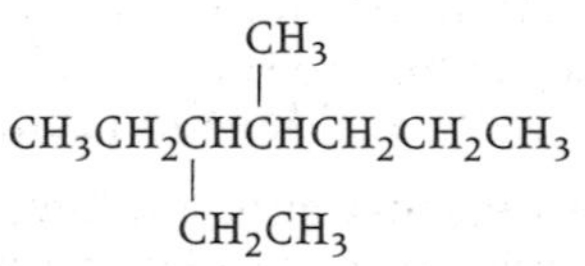

$$CH_3CH_2CHCHCH_2CH_2CH_3$$
with CH₃ above and CH₂CH₃ below

(ii) Draw a full structural formula for *cis*-but-2-ene.

(c) A hydrocarbon which is not a ring compound contains 85.7% carbon and 14.3% hydrogen by mass. The main peaks in its mass spectrum, in order of decreasing height, are as follows:

m/e 56;   29;   27;   15.

(i) Calculate the empirical formula of the compound.
$[A_r(H) = 1;\qquad A_r(C) = 12.]$
(ii) Calculate the molecular formula of the compound.
(iii) Deduce the structure of the compound, giving your reasons, and draw the structure below.

(d) Give **one** example of each of the following:

(i) a free radical;
(ii) an electrophile;
(iii) a nucleophile;
(iv) heterolytic bond fission;
(v) an elimination reaction.

(e) Distinguish between *addition polymerisation* and *condensation polymerisation*. Give **one** example of a single addition step for **each** type of polymerisation.

WJEC 1988 (1)

D32
Two organic compounds, **A** and **B**, are isomers with the composition by mass of carbon, 70.5%; hydrogen, 5.9%; oxygen, 23.6%. **A** is moderately soluble in water and **B** is a pleasant-smelling liquid. Their mass spectra are shown below.

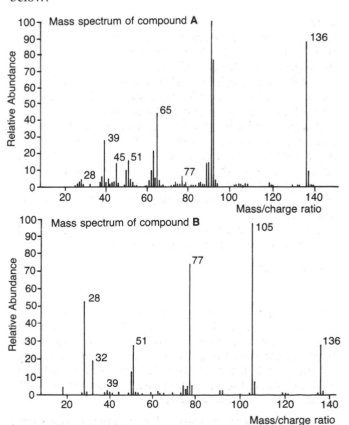

(a) (i) What is the empirical formula of **A** and **B**?
(Relative atomic masses: C = 12, O = 16, H = 1)
(ii) What is the molecular formula of **A** and **B**?
Justify your answer.

(b) Give the formulae of the molecular fragments corresponding to the following peaks:
Mass/charge ratio: 136; 105; 91; 77.

(c) What structural formulae would you predict for **A** and **B**?

(d) Describe **two** tests or chemical reactions in which the behaviour of **A** and **B** would differ.

N 1984 (3) (slightly adapted)

D33

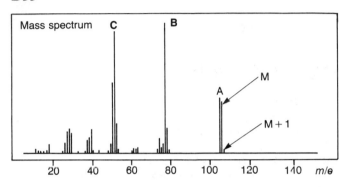

The mass spectrum above was obtained from a compound **S** which contains the elements carbon, hydrogen and oxygen only. The relative heights of the major peaks are:

M + 1	M	A	B	C
3	39	42	100	92

(a) Show by calculation how many carbon atoms are likely to be present in a molecule of **S**.

(b) What fragments have been lost in order to produce each of the following?

(i) A from M; (ii) B from A; (iii) C from B.

(c)

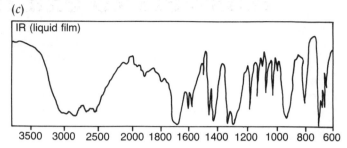

The infra-red spectrum above is of **T**, the product obtained when **S** is oxidised. What functional groups are responsible for the peaks labelled X and Y?

(d) Heating **T** with soda lime yields a liquid, **U**, the infra-red spectrum of which is shown below.

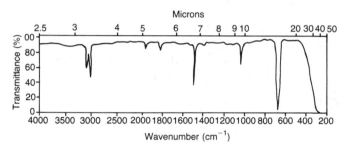

Deduce the identity of the three compounds **S**, **T** and **U**.

UCLES Winter 1987 (2) (slightly adapted)

Answers to examination questions

Answers are only given for those questions which have a numerical answer.

Inorganic chemistry

C5   (c) $MgBa(CO_3)_2$
C14 (a) $1.52 \times 10^{-22} \, mol \, dm^{-3} \, s^{-1}$;
         (b) (i) 20 °C; (ii) $9 \times 10^{-3} \, mol \, dm^{-3}$
C16 (c) 49, 51: $^{14}N^{35}Cl$, $^{14}N^{37}Cl$; 84, 86, 88: $^{14}N^{35}Cl_2$,
         $^{14}N^{35}Cl^{37}Cl$,$^{14}N^{37}Cl_2$; 119, 121, 123, 125:
         $^{14}N^{35}Cl_3$, $^{14}N^{35}Cl_2{}^{37}Cl$, $^{14}N^{35}Cl^{37}Cl_2$, $^{14}N^{37}Cl_3$
C18 (d) (ii) 28, 26, 25
C19 (d) (i) $Cr(NH_3)_6Cl_3$
C20 (iv) $33 \, cm^3$
C23 (c) (i) $Ni(NH_3)_6{}^{2+}$; (ii) $[Ni(EDTA)]^{2-}$

Organic chemistry

D10 (c) (iv) 75.5%
D12 (c) 6.05 g
D13 (b) $1.96 \, cm^3$; (g) 59%
D15 (c) (i)0.112%
D21 B Methanol, $p = 54$ mm Hg;
         Ethanol $p = 15$ mm Hg
D25 C_4H_8
D27 Empirical formula $CHCl_2$
D28 (i) C_2H_3O; (ii) 86.1
D29 (a) Compound **A**: C 77.42%, H 7.53%,
         N 15.05%; compound **B**: C 78.50%,
         H 8.41%, N 13.08%
D30 (a) (ii) $M_r(A) = 92.5$; A contains 1 mol Cl
D31 (c) Empirical formula CH_2
D32 (a) (i) C_4H_4O, $C_8H_8O_2$
D33 (a) 7

Subject index

cyclopentadiene, 676
cyclopropane, 710
cytosine, 683–5

d block elements, 536
d orbital splitting, 677–9
d orbitals
    in phosphorus, 616–17
    in silicon, 596–7
    in sulphur, 636–7
    in transition metals, 672
dansyl chloride, 859
DDT, 806–7
deactivating groups, 762
decarboxylation, 733
deficiency diseases, 873–4
dehydration
    of alcohols, 788
    of amides, 827
    of ethanamide, 824
dehydrohalogenation, 738
delocalisation of electrons
    in anthracene, 753
    in benzene, 709, 751
    in naphthalene, 753
denaturing of proteins, 858
deoxyribonucleic acid, 863–7
    X-ray diffraction, 859–60
deoxyribose, 863–4
detergents, 898–902
deuterium, 551
dextrorotation, 721
diagonal relationships in the Periodic
    Table, 549–50
diamond, structure, 591
diasterioisomers, 725
diazonium ions, 842–5
diborane
    reduction of alkenes, 785–6
    structure, 555, 586
dichlorine heptaoxide, 548, 659, 662
diethyl ether, see ethoxyethane
dilead(II) lead (IV) oxide, 602–3
dimercury(I) chloride, 699
2,4-dinitrofluorobenzene, 858–9
dinitrogen oxide, 610
dinitrogen pentaoxide, 610
dinitrogen tetraoxide, 610
dinitrogen trioxide, 610
2,4-dinitrophenylhydrazine, reaction
    with aldehydes and ketones, 803–5
dipolar ions, 852–3
dipole moments, of substituted benzene,
    763
disaccharides, 869
disproportionation reactions of halogens,
    651, 660
disulphur dichloride, 548, 642
DNA, see deoxyribonucleic acid
DNFB, see 2,4-dinitrofluorobenzene
DNP peptides, 859
Dobereiner's triads, 533
dolomite, 593
double bond, in ethene, 737
drug addiction, 876–7
dry ice, 592
dyes, 844–5

EDTA
    as a ligand, 673
    in water analysis, 564–5
elastomers, 890–1
electron deficient molecules, 555
electronegativity, periodicity of, 543
electrophiles, in organic chemistry, 717
electrophilic substitution, 762–7, 755–7
electrophilic substitution, summary, 756
electrophoresis, 853
elimination reactions of halogenoalkanes,
    774
enantiomers, 723, 726
enol group, test for, 795
enzyme activity, pH dependence, 860
enzymes,
    active sites, 860
    as catalysts, 860
Epsom salts, 578
esters, 825–6
    cleavage, 833
    hydrolysis, 826
    preparation from alcohols, 788
    reaction with ammonia, 827
    reduction, 826
ethanal
    manufacture, 799–80
    preparation, 800
ethanal trimer, 808
ethanamide
    dehydration, 824
    hydrolysis, 824
    preparation, 823–4
    reaction with nitrous acid, 823–4
ethane, 731–6
ethane-1,2-diamine, as a ligand, 673
ethane,1,2-diol, 740
ethanedioate ion, as a ligand, 673
ethanedioic acid, dehydration, 640
ethanoate ion, bonding, 811
ethanoic anhydride, preparation, 821–2
ethanol
    dehydration, 788
    esterification, 788
    oxidation, 786
    preparation, 783–6
    reactions of OH group, 787–8
ethanoyl l chloride
    preparation, 818, 821
    reaction with alcohols, 788
ethene
    addition reactions, 740–4
    oxidation, 739–40
    polymerisation, 744–5
    preparation, 738
    also see alkenes
ethers, 831–5
    cleavage, 833
    reaction with phosphorus
        pentachloride, 833
ethoxide ions, 772, 833
ethoxyethane, preparation, 832–3
ethyl benzoate
    preparation, 826
    reduction, 826
ethyl ethanoate, 788
    hydrolysis, 826

preparation, 825–6
reduction, 826
ethylamine
    preparation, 836–8
    reactions, 839–43
ethylenediaminetetraacetic acid, see
    EDTA
ethyne
    addition reactions, 747
    oxidation, 747
    polymerisation, 746–7
expanded polystyrene, 892
external compensation, of optical
    isomers, 725

fats, 741, 898–902
fatty acids, 898–9
Fehling's test, 806, 872
fermentation, 783–4
ferrocene, 676
fibrous proteins, 855–7
fire extinguishers, 585, 592
flash point, of ethers, 831
fluorapatite, 617
fluorides, 658–9
fluorine
    bond strength, 650
    extraction, 647
food colourings, 845
formalin, 797
fractional distillation of air, 627
Frasch process, 637–9
free radical reactions
    attack on arenes, 754
    of alkanes, 731–3
    polymerisation of alkenes, 744–5
free rotation about bonds, 716
Friedel–Crafts reactions
    acetylation, 820–2
    alkylation, 757
    preparation of ketones, 801
fructose, 868
fuel cell, 552
functional groups, introduction, 705–7

gallium arsenic phosphide, 583
Gattermann reaction, 843
genetic engineering, 866–7
glass, 601–2
glass transition temperature, 890
globular proteins, 855–7
glucose, 868
glyceraldehyde, 868
glycerides, 898
glycerol, making soap, 898
glycine
    preparation, 852
    also see amino acids
glycogen, 870–1
gold, 539
Gouy balance, 680
graphite, 591
greenhouse effect, 589–60, 630
Grignard reagents
    in preparing alcohols, 785
    in preparing alkanes, 734

in preparing carboxylic acids, 813
    preparation of, 733–4
Group 0, *see* noble gases
Group I, 569–74
Group IB, 539
Group II, 575–81
Group IIB, 697–701
Group III, 582–8
Group IV, 596–605
Group V, 616–26
Group VI, 636–46
Group VII, 647–53
guanine, 863–5
gun powder, 572
gypsum, 578–9

haber process, 606
haem group, 677
haemoglobin, 629, 858
halide ions, tests for, 651–2
halogenoalkanes, 768–74
    amine formation, 773
    dipole moments, 771
    elimination reaction, 774
    ether formation, 772–3
    hydrolysis, 772
    nitrile formation, 773
    nucleophilic attack, 768–73
    preparation, 768
    substitution reactions summary, 772
    Wurtz reaction, 773
halogenoarenes
    examples, 775
    preparation, 774–6
    reactions, 776
halogens, 647–53
    and the Periodic Table, 537
    displacement reactions, 650
    oxoacids, 661–2
    reactions with alkali, 651, 660
    reactions with hydrocarbons, 651
    reactions with metals, 650–1
    reactions with water, 651
    reactivities, 649–50
    redox reactions, 661
hard water, 564
heavy water, 552, 560
heroin, 877
heterolysis, 717
hexacyanoferrate(II) ions, 694
hexacyanoferrate(III) ions, 694
hexane-1,6-diamine, 884–5
hexanedioic acid, 884–5
high spin complexes, 680
Hofmann degradation, 823, 825, 838
homologous series
    ascending, 827
    descending, 823
    examples, 711–13
    introduction, 705
homolysis, 717
hormones, 873–5
hydration of ions, 559
hydrazine, 608–9
hydrazine, reaction with aldehydes and
        ketones, 803–5
hydrazones, 803–5

hydrides
    boiling points, 555
    of elements in the Periodic Table, 546
    of Group I, 573
    of Group IV, 599
    of metals, 554
    of non-metals, 554–5
hydroboration, preparation of alcohols,
        785–6
hydrocarbons
    aromatic, 751–61
    non-aromatic, 731–50
hydrofluoric acid, etching of glass, 658
hydroformylation, 746
hydrogen
    extraction of, 551–2
    preparation, 553
    properties, 551–4
hydrogen bonds
    in bases in DNA, 865
    in hydrogen fluoride, 654–6
    in proteins, 856
hydrogen bromide, preparation, 654–6
hydrogen chloride, preparation, 654
hydrogen cyanide, 665
hydrogen fluoride, 547, 654
hydrogen halides
    bond energies, 654–5
    properties, 654–64
hydrogen iodide, preparation, 654–5
hydrogen peroxide
    oxidation and reduction 633–4
    preparation, 634
    reaction with manganese(IV) oxide,
        693
    structure, 633
hydrogen sulphide, 641
hydrolysis
    of carbon tetrachloride, 600
    of nitriles, 827
    of silicon tetrachloride, 600

indicators, 845
inert gases, *see* noble gases
inert pair effect, 538–9
inhibitors, of polymerisation, 882
initiators, of polymerisation, 744
insulin
    amino acid sequence, 855
    discovery of structure, 858–9
inter-halogen compounds, 657
internal compensation, in optical isomers,
        725
interstitial hydrides, 554
iodides, 654–9
iodine
    manufacture, 648
    preparation, 649
    use of 5d orbitals, 657
iodine heptafluoride, 657
iodine pentafluoride, 657
iodine titrations, 643
iodine trichloride, 657
iodobenzene, preparation, 774–6
iodoethane, preparation, 768, 770–1
iodoform reaction, 806
ion exchange, 564–5

ionisation energies
    periodicity, 541
    of transition metals, 671–2
    table of values, 913
iron, 670–2, 694
iron alum, 586
iron carbonyl, 592
iron(II), test for, 694
iron(II) ammonium sulphate, 694
iron(II) chloride, 694
iron(II) sulphate, 694
iron(III), test for, 665, 694
iron(III) chloride, preparation, 694
isoelectric point, 853
isomerism
    chain, 613–16
    cis and trans in alkenes, 716, 737–8
    cis and trans in complex ions, 686
    functional group, 717
    geometrical, 686
    hydrate, 685–6
    in complex ions, 685–7
    in organic chemistry, 713–17
    ionisation, 685
    optical, *see* optical activity
isoprene, 745–6, 799, 889

keto–enol tautomerism, 806
ketones, 797–811
    cyanide addition, 802–3
    hydrogensulphite, 802–3
    preparation, 800–1
    reaction with phosphorus
        pentachloride, 807
    reduction, 785, 805

lactic acid, 803
laevorotation, 721
lanthanides, 535
laughing gas, 610
lead complexes, 600
lead(II) oxide, reactions, 602
lead(II) salts, solubilities, 600
lead(II) sulphide, 604
lead(IV) compounds, oxidising ability,
        600
lead(IV) oxide, reactions, 602
leaving groups, 771, 826
ligand field theory, 677–8
ligands, 673
light emitting diodes, 583
limestone, 593
lime water test for carbon dioxide, 592
lipids, 898
lithium
    comparison with magnesium, 550
    electrode potential and ionisation
        energy, 569–70
    reaction with water, 570
lithium tetrahydridoaluminate(III)
    compared with sodium
        tetrahydridoborate(III), 587
    preparation 555–6
    reduction of aldehydes, 805
    reduction of carboxylic acids, 814
    reduction of esters, 826
    reduction of nitriles, 827

sulphonation, 794
test for, 795
phenoxide ion, 764
phenylamine
conversion into phenol, 791
halogenation, 846
manufacture, 836
nitration, 846
preparation, 776, 838–9
reaction with amines, 842–3
with bromine, 846
phenylethanone, preparation, 801
phenylhydrazine, reaction with
aldehydes and ketones, 803–5
phenylhydrazones, 803–5
phosphates, 622–3
phosphine, 618
phosphoric(V) acid, 548, 621
preparation of hydrogen halides, 654
titrations, 622
phosphorus
extraction, 617–18
oxoacids, 621–2
use of 3d orbitals, 617
phosphorus pentachloride, 619–20
reactions with ethers, 833
test for OH groups, 620
phosphorus sulphides, 623–4
phosphorus trichloride, 548, 619
phosphorus trichloride oxide, 619
phosphorus triiodide, in hydrogen halides
preparation, 655
phosphorus(III) oxide, 621
phosphorus(V) oxide, 548, 621
dehydration of amides, 824
photochemical reactions, 896
photochemical smog, 611
photosynthesis, 629–30
pi bonds, in alkenes, 737
pi bonds, in benzene, 751
pi complex, 757
pK_b, of amines, 839
plane polarised light, 721–2
plaster of Paris, 579
plastic sulphur, 637
platinum
as a catalyst, 681
complexes, 686
polarimeters, 722
polyesters, 885–6
polyhydric alcohols, 781
polymer manufacture
blow moulding, 892
coatings, 892–3
fibres, 892
injection moulding, 892
moulding, 892
thin sheets, 891
polymerisation
addition reactions, 880–1
anionic polymerisation, 883
cationic polymerisation, 882–3
coordination addition, 883–4
free radical addition, 881–2
ionic addition, 882
of aldehydes, 807–8
of ethene, 744–5, 881

of ethyne, 747
polymers, 879–88
action of light, 896
amorphous, 889–90
atactic, 884
branching, 889
crosslinking, 889
crystalline, 889–90
in industry, 889–97
isotactic, 884
mechanical properties, 891
syndiotactic, 884
table of types and uses, 893–5
thermoplastic, 888
thermosetting, 888
use of fillers, 891
use of stabilisers, 896
poly(methanal), 807
polypeptides, 854–5
polysaccharides, 869–70
polystyrene, 745
polythene, 745, 881
polyurethane foams, 892
ponceau, 845
potash alum, 586
potassium, reaction with water, 570
potassium carbonate, 570–1
potassium dichromate(VI), as an
oxidising agent, 692–3
potassium hydrogencarbonate, 570–1
potassium hydroxide, 570
potassium manganate(VII)
as an oxidising agent, 693
preparation, 693
potassium nitrate, 572
potassium permanganate, see potassium
manganate(VII)
potassium peroxide, 570
potassium superoxide, 570
potassium thiocyanate, tests for iron(III),
694
primary alcohols, 783
primary amines, 836
progesterone, 874
propanone, manufacture, 800
prosthetic group, in proteins, 858
protecting groups, 846
protein structures
primary, 855
quaternary, 858
secondary, 855
tertiary, 855–6
proteins, 850–62
Prussian blue, 694
pseudohalides, 665–6
pseudohalogens, 665–6
PTFE, 745
putrescine, 838
PVC, 745

quantum yield, 896
quartz, 597
quaternary amines, 841

racemic mixtures, 725
rayon, 871
red lead, 602

redox potentials of complex ions, 689
redox reactions of halogens and their
oxoanions, 661
reducing sugars, 872
reduction
of acid chlorides, 820–1
of nitriles, 836
of nitro groups, 838
of silicon tetrachloride, 597
also see lithium
tetrahydridoaluminate(III)
resolution of optical isomers, 726
respiration, 629
rhombic sulphur, 637
ribonucleic acid, 863
ribose, 863–4, 868, 870
ribosomal RNA, 864
rock salt, crystal structure, 571
Rosenmund reaction, 801
rubber, 746, 889

s block metals, 536
compared with B metals, 539
salt hydrolysis, 560
Sandmeyer, reaction, 776, 843
saponification, 899–900
saturated fats, 898–9
scandium, 670–2
Schotten–Baumann reaction, 842
secondary alcohols, 783
secondary amines, 836
shielding, 538
side chains
in polymers, 882
oxidation of, 754
siderite, 593
sigma complex, 757
silane, 599
silica, 548, 597–601
silica gel, 601
silicates, 597–8
silicon
preparation, 597
reaction with steam, 601
use of 3d orbitals, 596–7
silicon tetrachloride, 548, 597, 600
silicones, 603
silver, as a catalyst, 797
silver mirror test, 805–6
silver nitrate test, 651–2
soaps, 898–902
soaps and detergents, cleaning action,
901–2
sodium tetrahydridoborate(III)
as a reducing agent, 556
comparison with lithium
tetrahydridoaluminate(III), 587
sodium, reaction with water, 570
sodium ammonium tartrate, optical
isomers, 723
sodium carbonate, 570–1
sodium chloride, 548, 571
sodium dichromate(VI), as an oxidising
agent, 692, 786
sodium hydrogencarbonate, 570–1
sodium hydroxide, 548, 570
sodium nitrate, 572

Index of names

Baeyer, Adolf von, 710
Balard, Antoine, 648
Bartlett, N., 668
Baumann, Eugen, 842
Bel, J. A. le, 723
Berzelius, Jacob, 879

Carothers, W. H., 884
Crafts, James, 752, 753, 820
Crick, Francis, 863

Davy, Humphry, 570, 647
Döbereiner, J. W., 533
Dorn, Friedrich, 668

Faraday, Michael, 759
Franklin, Rosalind, 863
Friedel, Charles, 752, 753, 820

Gatterman, Ludwig, 843
Grignard, Victor, 733

Hoff, J. A. van't, 723
Hofmann, A. W. von, 823, 847

Kekulé, Friedrich, 709

Markovnikoff, Vladimir, 742
Mendeléeff, Dimitri, 534

Mossan, Henri, 647

Natta, Giulio, 883
Newlands, J. A. R., 533

Pasteur, Louis, 723
Pauling, Linus, 863
Prout, William, 533

Ramsay, Sir William, 668
Rayleigh, Lord, 668

Sandmeyer, Traugott, 843
Sanger, Frederick, 858
Scheele, C. W., 647
Schotten, Carl, 842
Seaborg, G. T., 535
Simon, E., 879

Travers, M. W., 668

Watson, James, 863
Watson, Richard (Bishop of Llandaff), 879
Werner, Alfred, 685
Wilkins, Maurice, 863
Williamson, A. W., 833
Wurtz, A., 733

Ziegler, Karl, 883